British Isles

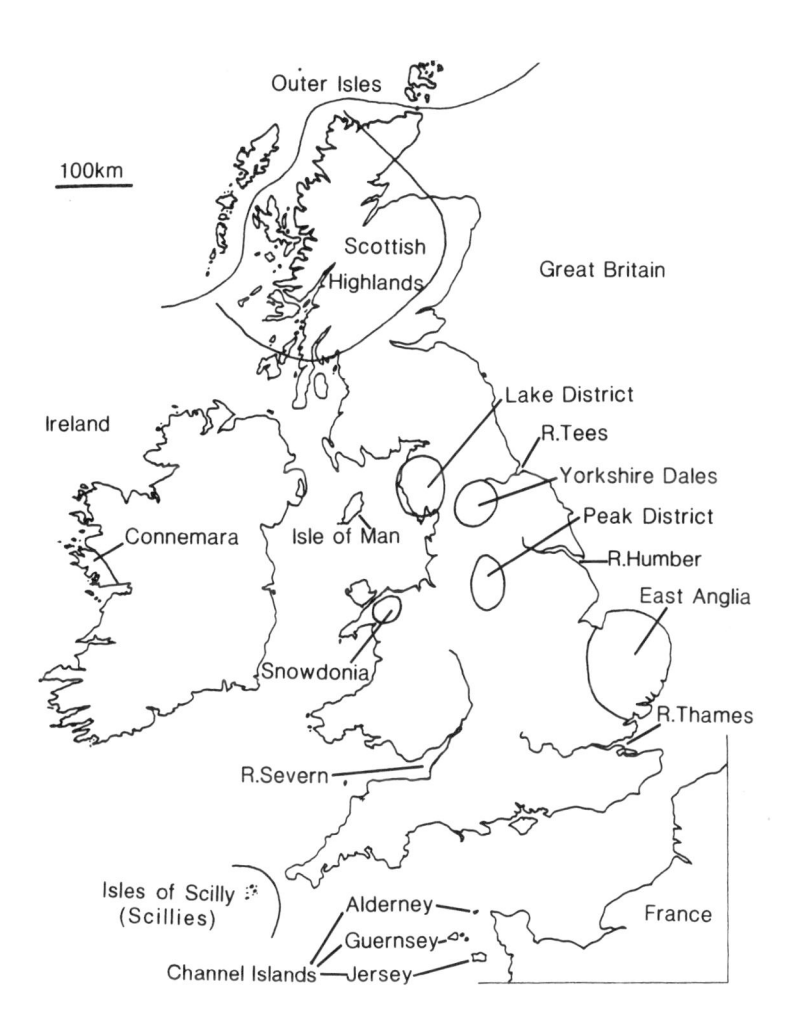

Field Flora of the British Isles

Field Flora of the British Isles provides a portable, comprehensive guide for those wishing to identify plants growing in the wild in the British Isles. Based on the author's now standard work, *New Flora of the British Isles (2nd edition, 1997)*, this abridged version offers the same complete coverage and user-friendly features in a compact format that is particularly suitable for use in the field. The reduction in the length of the book has been achieved by omitting the separate descriptions and many of the illustrations, while leaving the determination keys intact. Additional information on growth habit, habitat and distribution is provided as each taxon is identified. Where appropriate, information has been updated, but no taxonomic changes or extra taxa have been incorporated. Twenty-seven pages of essential illustrations are included to aid the user, together with a comprehensive glossary and a concise index.

CLIVE STACE is Professor of Plant Taxonomy at Leicester University, England, where his research work involves the taxonomy, evolution and breeding behaviour of diverse groups of flowering plants ranging from European grasses to tropical American Combretaceae. He has a special interest in hybridization, and edited *Hybridization and the Flora of the British Isles (1975)*. He is author of the textbook *Plant Taxonomy and Biosystematics (1980, 1989)* and of *New Flora of the British Isles (1991, 1997)*, which became the standard Flora of the British Isles soon after its first appearance.

FIELD FLORA OF THE BRITISH ISLES

CLIVE STACE

CAMBRIDGE UNIVERSITY PRESS

PUBLISHED BY THE PRESS SYNDICATE OF THE UNIVERSITY OF CAMBRIDGE
The Pitt Building, Trumpington Street, Cambridge, United Kingdom

CAMBRIDGE UNIVERSITY PRESS
The Edinburgh Building, Cambridge CB2 2RU, UK
40 West 20th Street, New York, NY 10011-4211, USA
477 Williamstown Road, Port Melbourne, VIC 3207, Australia
Ruiz de Alarcón 13, 28014 Madrid, Spain
Dock House, The Waterfront, Cape Town 8001, South Africa

http://www.cambridge.org

© Cambridge University Press 1999

This book is in copyright. Subject to statutory exception
and to the provisions of relevant collective licensing agreements,
no reproduction of any part may take place without
the written permission of Cambridge University Press.

First published 1999
Reprinted 2005

Printed in the United Kingdom at the University Press, Cambridge

Typeset by the author

A catalogue record for this book is available from the British Library

ISBN 0 521 65315 0

CONTENTS

PREFACE AND ACKNOWLEDGEMENTS	page vii
INTRODUCTION	ix
Taxonomic Scope	ix
Geographical Scope	x
Classification and Nomenclature	x
Identification Keys	x
Illustrations	xii
Conservation	xii
HOW TO USE THIS BOOK **(Black-edged)**	xv
SIGNS AND ABBREVIATIONS	xvi
THE DIAGNOSTIC KEYS	1
PTERIDOPHYTES	1
LYCOPODIOPSIDA (Clubmosses)	5
EQUISETOPSIDA (Horsetails)	8
PTEROPSIDA (Ferns)	11
PINOPSIDA (Gymnosperms) **(Black-edged)**	25
MAGNOLIOPSIDA (Angiosperms) **(Black-edged)**	35
MAGNOLIIDAE (Dicotyledons)	59
LILIIDAE (Monocotyledons)	517
GLOSSARY **(Black-edged)**	693
INDEX	711

PREFACE AND ACKNOWLEDGEMENTS

The idea behind this *Field Flora* is to present the essential data contained in my *New Flora of the British Isles* (1991; 2nd ed. 1997) in a more compact and portable form suitable for taking into the field or on holiday.

These 'essential data' are considered to include the means of identification of all the species and subspecies contained in the *New Flora* (i.e. the keys), together with a brief statement of the habitats and geographical areas in which they occur. The greatest reduction in space has been achieved by omitting the separate descriptions of the species and subspecies, except for an indication of the form of the plant (which has been added to the keys), and by greatly abbreviating the family descriptions. The other main space-saver has been omission of most of the figures. Some of the text has been updated (e.g. correction of Latin names or distributions in the light of recent discoveries), but no taxonomic changes or extra taxa have been introduced.

The format adopted for this *Field Flora* is by no means novel, for it resembles Floras in current use in several Continental countries, but it contrasts strongly with the compact Flora most used in this country (A.R. Clapham, T.G. Tutin & E.F. Warburg, *Excursion Flora of the British Isles*: ed. 1, 1959; ed. 2, 1968; ed. 3, 1981), which gives greater emphasis to common species and omits many of the less common aliens. The *Field Flora*, like the *New Flora*, is designed to enable field-botanists to identify all the plants that are found in the wild in the British Isles, hence the choice of format.

In certain genera it is difficult to use some of the key characters in the field, often due to their small size or to combinations not occurring at the same time of year. In these genera the keys do not switch to easier 'field characters', such as the colour of the leaves or the jizz of the plant, because the latter are often unreliable (leading to a correct determination in most rather than all cases). A greater reliance on 'field characters' can be made after long experience, but such features are usually difficult to describe in text and frequently serve to mislead or give false confidence to the less experienced.

All those botanists listed as giving me valuable assistance in writing the *New Flora* have of course been equally important in preparing this *Field Flora*, although their names are not repeated here. However, I must express my special indebtedness to Douglas Kent and Arthur Chater, who have proved again to be ever-reliable and unstinting sources of help and advice.

As in the case of my *New Flora*, I would be very glad to hear comments and criticisms from readers.

Ullesthorpe, Leicestershire CLIVE A. STACE
September 1998

INTRODUCTION

The following paragraphs are intended to explain the contents and arrangement of the *Field Flora*.

TAXONOMIC SCOPE

All species and subspecies of vascular plants (pteridophytes, gymnosperms and angiosperms) are included, as is traditional in British Floras. These are placed in classes (Lycopodiopsida, etc.), subclasses (Magnoliidae, etc.), families (Lauraceae, etc.) and genera. Below the family level, subfamilies or tribes (both in the case of Asteraceae and Poaceae) are defined only for those families with 20 or more genera.

Apomictic microspecies are covered in full in most genera, but not for the three notorious genera *Rubus*, *Taraxacum* and *Hieracium*, for which specialist accounts already exist or are in preparation by experts. In these genera a separation into relatively easily recognized groups of microspecies (here called sections) is provided instead.

The coverage of alien taxa has been as thorough and consistent as possible. To merit inclusion, an alien must be either naturalized (i.e. permanent and competing with other vegetation, or self-perpetuating) or, if a casual, frequently recurrent so that it can be found in most years. All this applies as much to garden-escapes or throwouts as to unintentionally introduced plants. Rarity, or the requirement of a highly specialized habitat, have not been taken into consideration (any more than is the case with natives). Cultivated species have been included if they are field-crops or forestry-crops or, in the case of trees only, ornamentals planted on a large scale. Exclusively garden plants, however abundant, whether crops or ornamentals, have not been covered, but most of the commoner taxa are included anyway because of their occurrence as escapes or throwouts. Also excluded are non-tree ornamentals planted en masse on new roadsides or in parks, etc. The aim of this set of criteria is to include **all taxa that the plant-hunter might reasonably be able to find 'in the wild' in any one year.** Any such plant, whether native, accidentally introduced or planted, affects wild habitats and is part of the ecosystem, and botanists and others might be expected to need or want to identify it.

Hybrids are given a more abbreviated treatment than species and subspecies. After each generic entry the existence of hybrids is indicated, but only the more common or interesting ones are named. All intergeneric hybrids are also mentioned. Only those hybrids that have attained distributions no longer tied to those of their parents (i.e. those that occur at least sometimes in the absence of both parents) are covered in the keys, where their parents are listed.

GEOGRAPHICAL SCOPE
This Flora deals with the British Isles, comprising Great Britain (England, Scotland and Wales), Ireland (Northern Ireland and Eire), the Isle of Man, and the Channel Islands (Bailiwicks of Jersey and Guernsey). The Bailiwick of Guernsey includes Guernsey, Alderney, Sark, Herm and various lesser islands. The above are always referred to in their correct, strict senses, rather than loosely, except that a distinction between Great Britain and the Isle of Man is made only where necessitated by particular patterns of distribution. The United Kingdom (Great Britain and Northern Ireland) is not referred to in the text.

The smallest geographical unit utilized is usually the vice-county. There are 111 of these in Great Britain and 40 in Ireland, with the Isle of Man and Channel Islands representing two others. All 153 are mapped and listed on the end-papers. The Isles of Scilly (Scillies) are part of W Cornwall, but they have a distinctive flora and are therefore frequently referred to separately.

CLASSIFICATION AND NOMENCLATURE
I have attempted to use the most up-to-date and accurate classification and nomenclature, but in the interests of stability I have adopted names that represent departures from normal usage in the British literature only where the evidence is unequivocal or overwhelming. The only synonyms given are names that are used in the 2 editions of *New Flora of the British Isles* but are no longer considered correct. A fuller list of synonyms can be found in those works and in D.H. Kent's *List of vascular plants of the British Isles* (1992).

The family sequence and circumscription followed is that of A. Cronquist, *An integrated system of classification of flowering plants* (1981), for the angiosperms, and L.N. Derrick, A.C. Jermy & A.M. Paul, in *Sommerfeltia* vol. 6 (1987), for the pteridophytes. Generic sequences are usually those of acknowledged authorities.

The abbreviations of authors used are those in the internationally accepted standard by R.K. Brummitt & C.E. Powell, *Authors of plant names* (1992). In the *Field Flora* the authority citations have been abbreviated as sanctioned in the *International code of botanical nomenclature* by omitting any authors before 'ex', e.g. *Clematis montana* Buch.-Ham. ex DC. becomes *C. montana* DC.

English names are given for all the species except those in certain complex genera. In all but a few cases I have used the names adopted in J.G. Dony, S.L. Jury & F.H. Perring, *English names of wild flowers*, ed. 2 (1986). About 1200 species that I have included are not listed in that work; English names for these have been gathered from many sources.

IDENTIFICATION KEYS
The method of identification provided is the diagnostic keys to families, genera, and species and subspecies. Except for the multi-access main key used in *Cotoneaster*, these are all dichotomous keys. In order to save space no line-gap is left between couplets, but alternate couplets are slightly indented to effect visual separation. Despite this appearance all the dichotomous keys are of the bracketed version, which I consider to be

INTRODUCTION

generally superior to (i.e. easier to use than) the indented type followed by some Floras.

In constructing the keys I have attempted to avoid as many as possible of the pitfalls that I have personally encountered over the years. Keys are a vital part of any Flora, yet are one of the most difficult aspects to master and they provide a frequent barrier for the beginner. Long keys are particularly daunting, so I have subdivided keys wherever necessary by providing a general key to a series (A, B, etc) of supplementary keys. Hence few keys contain more than 20 couplets and very few more than 30. In many cases 'difficult' characters are allowed for by providing two or more routes in a key. For example the (superior) ovaries of *Rosa* are liable to be wrongly scored as inferior, the (five) leaflets of *Lotus* are often mistaken for three plus two stipules, and the (white) petals of *Berteroa* often fade to yellow when dried. In these and in many other cases both alternatives are allowed for. A consequence of this is that the 'information' given in a number of keys to families and genera is sometimes taxonomically inaccurate. These keys are provided solely for the purposes of identification, and should not be used uncritically to compile descriptions of taxa. The keys to species and subspecies, however, should be free of any such misleading data. It is important to note that no account of variation outside the British Isles is taken; indeed, it is specifically excluded, and the reader must beware of using the descriptions and keys as definitions of the taxa on a world-wide or even European scale.

Certain conventions in terminology will become apparent after usage, especially if the *Glossary* is consulted. For example, 'above' and 'below' are used only to imply the upper and lower parts of a plant; upper and lower surfaces of an organ are referred to as 'upperside' and 'lowerside', or sometimes more specifically as adaxial and abaxial sides. The term 'leaf', unless otherwise stated, refers to the leaf-blade, excluding the petiole; this fact is especially important in the case of leaf length/breadth ratios.

I have assumed that the reader is familiar with the use of dichotomous and multi-access keys, but I provide here some hints that I have found very valuable in the past. *Before* starting on the keys it is important to examine in detail the structure of the flowers, making sure that the number, shape and arrangement of the various parts are fully ascertained. If the flowers are not all bisexual then the distribution of the sexes must be understood. The structure of the gynoecium usually presents the greatest problems; sectioning with a razor-blade vertically and transversely is often required. If fresh material is being collected, observations on underground parts, woodiness of stems and range of leaf-shape should be recorded. If possible, flowers (and fruits) of varying ages should be gathered. Mistakes are often made in distinguishing between a compound leaf (no buds in axils of leaflets) and a group of simple leaves (with buds, often very rudimentary, in axils). The use of insufficient magnification is a frequent cause of misidentification; a **lens with at least x10 magnification** is essential.

Many measurements given, especially those describing plant heights or lengths, should be prefixed 'normally', 'usually' or 'mostly'. It is often misleading to give ranges including the absolute extremes that have been encountered (e.g. a grass species 2-153cm high); usually the normal range is much more useful. More exceptional measurements are often given in brackets, e.g. 3-6(9)cm, but even these do not always represent the recorded extremes. In the case of trees, however, the maximum heights known in the British Isles are given. Measurements given without qualification are lengths; those separated by a multiplication sign (e.g. 3-6 x 1-2cm) are lengths and widths respectively.

In order to save space, few data are given beyond the features needed to identify the taxa. However, at each place in the keys leading to the identification of a species or subspecies, the English and Latin names (with authority) are presented, together with an indication of the form of the plant (e.g. tree to 25m; erect annual to 25cm) and the habitat and geographical distribution of the taxon.

In general, flowers are needed for identification by means of the keys, but there are some exceptions such as near the start of the General Key and in Keys A and B of the *Key to Families of Magnoliopsida*. Apart from non-flowering material, it is usually not possible to key out a range of abnormalities such as extreme horticultural variants (e.g. *flore pleno* or otherwise with more floral parts than usual, extremely dissected leaves or petals, unusual colour variants), abnormally tall or dwarfed plants, monstrosities such as many-headed *Plantago* and leafy-stemmed *Taraxacum*, plants with petaloid or leafy bracts, gall-induced variations, and various odd mutants (e.g. *Fraxinus* leaves with 1 leaflet). In the wild such plants usually occur with normal ones.

Finally, four tips. Firstly, before using a key to genera read carefully the family diagnosis and any notes that follow it, and before using a key to species read carefully the genus description and any subsequent notes. These descriptions and notes always contain useful data and sometimes vital ones, since special terms and conventions (e.g. 'spikelet length' in *Festuca* is not actually the total spikelet length) are often defined. Secondly, read the *whole* of *both* alternatives of each couplet before attempting to choose between them. Thirdly, if there is genuine doubt about which alternative to choose, follow both, as one will usually soon show itself to be unsuitable. Fourthly, if a nonsensical answer is obtained, check back to ensure that the frequent error of choosing the correct alternative but following the wrong subsequent route has not been committed.

ILLUSTRATIONS
In order to save space few illustrations are provided. The page on which each occurs is indicated in bold within a shaded strip in the right-hand margin of the text. The illustrations selected are those considered essential aids to identification. The photographs were all prepared by me; attribution of the authorship of the line drawings is made in the appropriate caption.

CONSERVATION
By far the greatest threat to our wild flora is the destruction of habitats, still continuing at a most alarming rate in the name of everything from 'agricultural policy' and 'economic development' to 'leisure activity'. When

populations of plants (or animals) are decimated they become highly vulnerable to secondary pressures, of which collecting is one. There can be little objection to the accumulation of a reference collection of plants, providing uncommon species are excluded and populations of even common ones are not significantly reduced. Indeed, a collection of accurately determined plants is the best way of learning them and of enabling identification of extra species encountered later. Often, however, only a small part of the plant (e.g. a basal leaf or a single flower) is needed for diagnostic purposes, and rarely are underground parts essential. Usually, even where they are, they can be adequately substituted by notes made in the field. It should be noted that it is actually illegal in Great Britain to uproot any wild plant, even common weeds, without the land-owner's permission, and there are more specific regulations governing Nature Reserves and rare species.

Since it is only botanists who have a good knowledge of our wild flora, it is vital that they consider themselves under a special obligation to protect it by example and by persuasion. No rare plant should be collected, damaged or disturbed (e.g. by trampling, or by 'arranging' the immediate surroundings during photography). Even species not generally rare are frequently rare in some areas; they should be respected there as much as species that are rare throughout the British Isles.

HOW TO USE THIS BOOK

It is strongly recommended that, before consulting the information in this book, the contents page, the introductory chapter and this page (including the reverse of it) be read carefully.

The Index at the end of the book should be used in order to look up a family or genus; both Latin and English names are indexed.

In order to identify a plant it is necessary first to decide whether it is a pteridophyte, gymnosperm or angiosperm. Many Floras purport to do this by means of keys, but in reality the questions posed (e.g. plant reproducing by spores; ovules enclosed in a carpel) amount to the same as the decision called for here. In practice it is best to become familiar with the range of form and structure found in the relatively few pteridophytes and gymnosperms, all other vascular plants being angiosperms. In the case of pteridophytes, the few that do not have divided fern-like leaves can easily be learnt, and in the case of gymnosperms, all have simple narrow leaves (except *Araucaria*) and woody female cones (except *Taxus* and *Juniperus*). It is especially crucial to distinguish between superficially similar but unrelated plants that provide pitfalls for the unwary. Well-known examples are mosses and Lycopodiopsida; *Equisetum* and *Hippuris*; *Lemna* and *Azolla*; *Isoetes* and *Littorella*; *Pilularia* and *Juncus*; and *Alnus* and conifers. If flowers, spore-bearing sporangia or woody cones are evident, then the task is an easy one. If not, familiarity and experience will soon prevent one from falling into traps such as the above. According to the decision, follow the generic or family keys starting on pages 1, 25 and 35, the positions of which are marked by black-edged pages. These will lead to a family or genus, which will provide further keys as necessary, enabling one to arrive at the genus, species and subspecies. Before using the keys the appropriate part of the introductory chapter should be studied.

Illustrations are numbered according to the page on which they appear, not in a sequence from 1 onwards. References to illustrations are given in the right-hand margin adjacent to the relevant taxon by sole means of a bold number in the shaded strip.

A glossary is placed after the diagnostic keys (marked by a black-edged page).

Signs and abbreviations are listed on the next page.

Maps and a ruler are provided on the end covers.

SIGNS AND ABBREVIATIONS

BI	- British Isles
CI	- Channel Islands
Br	- Great Britain
En	- England
Ir	- Ireland
Sc	- Scotland
Wa	- Wales
N, E, S, W, NE, etc.	- points of compass
C, M, MW, etc.	- central, Mid-, Mid-West, etc.
Leics, W Kent, etc.	- (vice-counties) see end papers
Jan, Feb, Mar, etc.	- months of year
agg.	- aggregate
auct.	- of various authors but not the original one
c.	- about
excl.	- excluding
FIG, Fig	- Figure (number following is the page number)
incl.	- including
intrd	- introduced
natd	- naturalized
R.	- River
sp., spp.	- species (singular and plural)
ssp., sspp.	- subspecies (singular and plural)
surv	- survivor
var., vars	- variety, varieties
$\pm$	- more or less
$>, <$	- more than, less than
$\geq$	- over and including; at least; not less than
$\leq$	- up to and including; at most; not more than
0	- absent
x	- times (2x, etc. = twice, etc.); or indicating a hybrid
2n=	- sporophytic chromosome number

KEYS TO GENERA OF PTERIDOPHYTES
(LYCOPODIOPSIDA, EQUISETOPSIDA, PTEROPSIDA)

General key
1. Leaves scale-like, in whorls fused into sheath at each node; stems jointed   **4/1. EQUISETUM**
1. Leaves not in a fused whorl at each node; stems not jointed   2
  - 2. Plants free-floating on water, with 2-lobed leaves on short stem   **19/1. AZOLLA**
  - 2. Plant rooted in solid substratum   3
3. Leaves simple but lobed almost to midrib or just pinnate, when mature covered on lowerside by dense reddish-brown scales   **15/3. CETERACH**
3. Leaves variously divided, not covered on lowerside by reddish-brown scales   4
  - 4. Leaves simple, not lobed or lobed <1/2 way to midrib   *Key A*
  - 4. Leaves compound, or simple and lobed >1/2 way to midrib (rarely a few ± simple)   5
5. Sporangia borne on leaves or parts of leaves or special branches distinctly different from vegetative leaves   *Key B*
5. Sporangia borne on normal foliage leaves   6
  - 6. Sori on margins of leaves either in protruding indusia or at least partly covered by indusium-like folded-over leaf-margin   *Key C*
  - 6. Sori on underside of leaves, sometimes near margin but then not covered by folded-over leaf-margin   *Key D*

Key A - Leaves simple, not lobed or lobed <1/2 way to midrib
1. Stem a rhizome or stolon, or very short and leaves single or tufted from ground; leaves usually >1cm   2
1. Stem elongated and aerial; leaves <1cm   6
  - 2. Leaves filiform, ≤5mm wide   3
  - 2. Leaves linear to ovate-elliptic, >5mm wide   5
3. Plant rhizomatous; leaves borne singly (often close together) and rolled in flat spiral when young   **9/1. PILULARIA**
3. Plant with very short corm-like stem; leaves 1-2 or in a rosette, not rolled in flat spiral when young   4
  - 4. Leaves borne in rosette, with sporangia at base on upperside   **3/1. ISOETES**
  - 4. Leaves 1-2; sporangia borne on spike-like special branches   **5/1. OPHIOGLOSSUM**
5. Leaves cordate at base, with sporangia borne in linear sori on lowerside   **15/1. PHYLLITIS**

5 Leaves cuneate at base; sporangia borne on spike-like special
  branches **5/1. OPHIOGLOSSUM**
    6 Leaves distinctly serrate along most of margin (x10 lens), the
      youngest ones with minute ligule near base on upperside;
      heterosporous **2/1. SELAGINELLA**
    6 Leaves entire, serrate only at base, or obscurely serrate along
      margin, without ligule; homosporous 7
7 Stems all ascending to erect, dividing into equal branches;
  sporangium-bearing leaves not in differentiated cones **1/1. HUPERZIA**
7 Main stems procumbent, with shorter branches; sporangium-bearing
  leaves in differentiated cones 8
    8 Branches flattened, with leaves in 2 alternating, opposite pairs
      1/4. DIPHASIASTRUM
    8 Branches not flattened, with leaves borne in whorls, alternately
      or spirally 9
9 Sterile and sporangium-bearing leaves similar, without either hair-
  points or scarious margins **1/2. LYCOPODIELLA**
9 Either sterile leaves with hair-points or sporangium-bearing leaves
  with scarious, toothed margins **1/3. LYCOPODIUM**

Key B - Leaves compound, or simple but lobed >1/2 way to midrib;
  sporangia borne on leaves or branches that are different from
  foliage leaves

1 Leaves simple and deeply lobed or 1-pinnate, the lobes or leaflets not
  or scarcely lobed 2
1 Leaves ≥2-pinnate or 1-pinnate with deeply lobed leaflets 5
    2 Stalk from ground bearing 1 pinnate vegetative branch and 1
      sporangium-bearing branch **5/2. BOTRYCHIUM**
    2 Stalks from ground either a vegetative leaf or a sporangium-
      bearing leaf 3
3 Sorus-bearing pinnae with distinct flat, green central region, the
  sori clearly marginal **8/1. PTERIS**
3 Sorus-bearing pinnae without green flat region, or if with one then
  sori clearly on its lowerside 4
    4 Sterile leaves triangular-ovate in outline, <2x as long as wide
      16/2. ONOCLEA
    4 Sterile leaves oblanceolate to lanceolate in outline, >3x as long as
      wide **18/1. BLECHNUM**
5 Stalks from ground each bearing very different vegetative and fertile
  branches **6/1. OSMUNDA**
5 Stalks from ground either a vegetative leaf or a sporangium-bearing
  leaf 6
    6 Sterile leaves >2-pinnate, finely divided, ± parsley-like 7
    6 Sterile leaves regularly 2-pinnate, or 1-pinnate with deeply lobed
      pinnae 8
7 Perennial with densely scaly rhizome; sori near leaf margin which is
  folded over to cover it **7/1. CRYPTOGRAMMA**
7 Annual with very short sparsely scaly rhizome; sori on leaf lowerside,

	not covered	**7/2. ANOGRAMMA**
8	Lowest pinna on each side bearing another pinna near its base	**8/1. PTERIS**
8	Lowest pinna on each side ± like upper ones, not bearing another pinna	9
9	Leaves borne singly spaced out along rhizome; fertile leaves green on upperside	**14/1. THELYPTERIS**
9	Leaves borne in tufts from apices of branches of rhizome; fertile leaves brown at maturity	**16/1. MATTEUCCIA**

Key C - Leaves compound, or simple but lobed >1/2 way to midrib; sporangia borne on edge of normal vegetative leaves

1	Sori a continuous line round margins of pinnules	2
1	Sori few-many discrete groups of sporangia, sometimes close together	3
2	Leaves 1-2-pinnate, tufted, ≤75cm excl. petiole; rhizomes short, scaly	**8/1. PTERIS**
2	Leaves (2)3-pinnate, borne singly, ≤2(5)m excl. petiole; rhizomes long, pubescent	**13/1. PTERIDIUM**
3	Rhizome trunk-like, >20cm thick, covered with old leaf-bases; some leaves >1m incl. petioles	**12/1. DICKSONIA**
3	Rhizome horizontal, <1cm thick, not covered with leaf-bases; leaves <50cm incl. petioles	4
4	Ultimate leaf-segments >5mm wide; indusia formed from folded-under flap of pinnule	**7/3. ADIANTUM**
4	Ultimate leaf-segments <5mm wide; indusia formed from tubular or 2-valved protruding outgrowth from pinnule	5
5	Distal part of petiole winged; rhizomes pubescent; mature indusia with protruding bristle, tubular	**10/2. TRICHOMANES**
5	Petiole not winged; rhizomes glabrous; indusia without protruding bristle, of 2 valves	**10/1. HYMENOPHYLLUM**

Key D - Leaves compound, or simple but lobed >1/2 way to midrib; sporangia borne on lowerside of normal vegetative leaves

1	Leaves simple, or 1-pinnate with the pinnae entire to toothed <1/2 way to midrib	2
1	Leaves 1-pinnate with the pinnae divided >1/2 way to midrib, or 2- or more-pinnate	6
2	Sori narrowly elliptic to linear	**15/2. ASPLENIUM**
2	Sori circular to very broadly elliptic	3
3	Indusium 0	4
3	Indusium present	5
4	Leaves regularly pinnate or nearly so	**11/1. POLYPODIUM**
4	Leaves (on 1 plant) very variably and irregularly pinnately lobed	**11/2. PHYMATOSORUS**
5	Pinnae <1.5cm wide, with sori in one row either side of midrib	**17/1. POLYSTICHUM**
5	Pinnae >1.5cm wide, with sori distributed ± evenly all over lowerside	**17/2. CYRTOMIUM**

	6	Sori linear to oblong or C- to V-shaped, >1.5x as long as wide 7
	6	Sori orbicular to broadly elliptic-oblong, <1.5x as long as wide 8

7 Sori linear to oblong, with the margin next to midrib straight
15/2. ASPLENIUM

7 Sori oblong to C- or V-shaped, with the margin next to midrib curved or bent **16/3. ATHYRIUM**

    8 Leaves borne singly spaced out along rhizome   9
    8 Leaves borne in tufts from apices of branches of rhizome   12

9 Leaves 2-pinnate, or 1-pinnate with the pinnae deeply lobed   10
9 Leaves 3-pinnate, at least at base   11

    10 Pinnae all ± parallel, the longest ones several removed from the basal one; indusium present   **14/1. THELYPTERIS**
    10 Lowest pair of pinnae bent back away from plane of others, the longest one the basal or the next to basal; indusium 0
        14/2. PHEGOPTERIS

11 Indusium 0   **16/4. GYMNOCARPIUM**
11 Indusium present, persistent   **16/5. CYSTOPTERIS**

    12 Indusium consisting of ring of hairs or narrow scales arching over sorus when young; petiole with joint c.1/3 way from base
        16/6. WOODSIA
    12 Indusium 0, vestigial or well developed and membranous; petiole not jointed   13

13 Indusium a flap-like hood; leaves slender, with few or 0 scales on petiole   **16/5. CYSTOPTERIS**
13 Indusium 0, vestigial or reniform or peltate; leaves often large and with many scales on petiole   14

    14 Pinnules with teeth contracted into very fine acuminate apices; indusium peltate   **17/1. POLYSTICHUM**
    14 Pinnules untoothed or with rounded to acute teeth; indusium 0, vestigial or reniform   15

15 Sori in rows on pinnules distinctly nearer margin than midrib; fresh leaves with faint lemon scent when crushed   **14/3. OREOPTERIS**
15 Sori either rather scattered on pinnules or in rows no nearer margin than midrib; fresh leaves without lemon scent   16

    16 Indusium reniform, very obvious; widespread   **17/3. DRYOPTERIS**
    16 Indusium 0 or vestigial; mountains of Sc   **16/3. ATHYRIUM**

LYCOPODIOPSIDA — CLUBMOSSES AND QUILLWORTS

Herbaceous plants with simple or sparingly branched stems and simple leaves with 1 vein. Sporangia homosporous or heterosporous, borne singly in leaf axils or on upperside of leaf near its base, the sporangium-bearing leaves often aggregated into cones. Gametophyte of homosporous species free-living, subterranean, mycorrhizal and saprophytic; gametophytes of heterosporous species much reduced and retained within spore, which lies on the ground.

1. LYCOPODIACEAE - *Clubmoss family*

Stems elongated, not, little or considerably branched, bearing roots and leaves without ligules. Moss-like plants whose leaves have true midribs and stomata.

1. HUPERZIA Bernh. - *Fir Clubmoss*
Stems all ascending to erect, dividing into equal, non-flattened branches; leaves 4-8mm, spirally arranged, often with bud-like outgrowths in their axils (these effect vegetative propagation); sporangium-bearing leaves not differentiated into cones, similar to sterile leaves.

1 Stems to 30cm. Native; heaths, moors and mountains
(*Fir Clubmoss* - **H. selago** (L.) Schrank & Mart.)   2
  2 Stems to 30cm, 6-12mm thick; leaves linear-lanceolate, ± patent, green when healthy. Common in NW Br, very scattered elsewhere   **H. selago ssp. selago**
  2 Stems to 10cm, 5-6mm thick; leaves ovate-lanceolate to narrowly ovate, appressed to stem, yellowish-green even when healthy. Rare in parts of Sc   **H. selago ssp. arctica** (Tolm.) Á. & D. Löve

2. LYCOPODIELLA Holub - *Marsh Clubmoss*
Stems procumbent, with non-flattened branches, giving rise to erect, fertile lateral stems; leaves 4-6mm, spirally arranged; sporangium-bearing leaves weakly differentiated into apical cones.

1 Procumbent stems to c.20cm; erect stems to 8(10)cm. Native; wet heaths, often on bare peaty soil; formerly frequent, now very local, extinct in C & E En   *Marsh Clubmoss* - **L. inundata** (L.) Holub

3. LYCOPODIUM L. - *Clubmosses*

Stems procumbent, with non-flattened branches, with erect sterile and fertile lateral stems; leaves spirally arranged or in whorls; sporangium-bearing leaves well differentiated into apical cones.

1 Leaves 3-5mm, erect to erecto-patent, with long white apical point, minutely toothed; cones (1-3) borne at apex of distinct peduncle 1.5-20cm. Procumbent stems to 1m or more. Native; heaths, moors, mountains; formerly frequent, now local and absent from much of lowlands *Stag's-horn Clubmoss* - **L. clavatum** L.
1 Leaves 4-10mm, patent to erecto-patent, acute, ± entire; cones (1) borne at apex of leafy stems. Procumbent stems to 60cm. Native; moors and mountains, often among *Calluna*; local in C & N Sc, Westmorland *Interrupted Clubmoss* - **L. annotinum** L.

4. DIPHASIASTRUM Holub - *Alpine Clubmosses*

Stems procumbent, often ± subterranean, with flattened erect branches arising in fan-like groups; leaves in alternating opposite pairs; dorsal leaves 2-4mm; sporangium-bearing leaves well differentiated into apical cones.

1 Ventral leaves petiolate, c.0.5mm wide, with >1mm free from stem; lateral leaves fused to stem for c.1/2 their length; cones at apices of normal leafy shoots. Procumbent stems to 50(100)cm. Native; moors and mountains among grass and *Calluna*; locally common in N & W *Alpine Clubmoss* - **D. alpinum** (L.) Holub
1 Ventral leaves sessile, c.1mm wide, with <1mm free from stem; lateral leaves fused to stem for c.2/3 their length; cones at apices of sparsely-leafed peduncles. Native; heaths and lowland moors; very rare in Sc and N En *Issler's Clubmoss* - **D. complanatum** (L.) Holub

2. SELAGINELLACEAE - *Lesser Clubmoss family*

Stems elongated, little or considerably branched, bearing roots on end of special leaf-less branches or on small corm-like swelling at base of stem; leaves serrate, with microscopic outgrowths (ligules) on upperside near base.

1. SELAGINELLA P. Beauv. - *Lesser Clubmosses*

1 Stems decumbent to procumbent, to 15cm, the branches not flattened, bearing erect fertile branches to 6(10)cm with terminal rather ill-defined cones; leaves all of 1 sort. Native; damp places on mountains; locally common in N & C BI
Lesser Clubmoss - **S. selaginoides** (L.) P. Beauv.
1 Stems procumbent, to 1m, the branches dorsiventrally flattened, bearing well-defined cones increasing in length with age and apparently not terminal; leaves of 2 distinct sorts. Intrd-natd;

shrubberies and damp shady places; scattered in S & W
Krauss's Clubmoss - **S. kraussiana** (Kunze) A. Braun

3. ISOETACEAE - *Quillwort family*

Stems short and corm-like, bearing roots at base and a rosette of long, erect, ± subulate leaves with minute ligule on upperside near base. Similar only to certain angiosperms (notably *Lobelia, Littorella, Subularia, Juncus*); in absence of sporangia the leaves with 4 air-cavities seen in transverse section (only 1 in *I. histrix*) and the peculiar corm-like 2-3-lobed stem are diagnostic.

1. ISOETES L. - *Quillworts*

1 Plant only seasonally submerged, with leaves Oct-Jun; leaves 1-4(10)cm x c.1mm; leaf-bases dark, shiny, horny, persistent. Native; sandy or peaty hollows where water lies in winter; extremely local in W Cornwall and CI   *Land Quillwort* - **I. histrix** Bory
1 Plant submerged for all or most of year, with leaves Jan-Dec; leaf-bases not dark, shiny and horny, not persistent   2
   2 Megaspores 530-700 microns across, with blunt, anastomosing tubercles on outer face; leaves stiff, remaining apart when plant removed from water; leaves 8-25(40)cm x 2-5mm. Native; in upland lakes; locally frequent in N & W   *Quillwort* - **I. lacustris** L.
   2 Megaspores 440-550 microns across, with acute spines on all faces; leaves flaccid, falling together when plant removed from water; leaves 4-15cm x 2-3mm. Native; in similar places to *I. lacustris*, rarer in N, commoner in SW
Spring Quillwort - **I. echinospora** Durieu

I. lacustris x *I. echinospora* = **I. x hickeyi** W.C. Taylor & Luebke is very rare.

EQUISETOPSIDA — HORSETAILS

Herbaceous rhizomatous perennials; aerial stems elongated, jointed, simple or bearing whorls of branches at nodes; leaves simple, with 1 vein, borne in whorls and fused into sheath round stem. Sporangia homosporous, borne in clusters under peltate specialized branches which are packed into well defined terminal cones. Gametophyte free living, green and photosynthetic.

4. EQUISETACEAE - *Horsetail family*

The jointed ridged aerial stems without branches or with whorls of branches at each node, and with whorls of leaves forming a fused sheath at each node, are unmistakable. The cones are terminal either on the normal green vegetative stems or on special unbranched brownish or whitish stems produced earlier than the vegetative ones.

1. EQUISETUM L. - *Horsetails*

1 Stems brown or whitish, simple, with cone at apex   2
1 Stems green, simple or branched, with or without cone at apex   5
   2 Leaf-sheaths with (15)20-30(40) teeth; cones (2)4-8cm. Vegetative stems to 1.5(2)m; fertile stems to 25(30)cm. Native; damp shady places; frequent, mostly in S   *Great Horsetail* - **E. telmateia** Ehrh.
   2 Leaf-sheaths with 3-20 teeth; cones 1-4cm   3
3 Leaf-sheaths with teeth united into 3-6 obtuse lobes at least at some nodes. Stems to 50(80)cm. Native; damp woods, hedgerows and stream-banks in lowlands, open moorland in uplands; frequent, mostly in N & W   *Wood Horsetail* - **E. sylvaticum** L.
3 Leaf-sheaths with (3)6-20 separate (sometimes slightly adherent) acute teeth   4
   4 Base of stem and leaf-sheaths usually tinged green; green branches very soon produced. Stems to 50cm. Native; banks of rivers and streams and flushed grassy areas, mostly upland; local in N BI   *Shady Horsetail* - **E. pratense** Ehrh.
   4 Green colour absent from stem and leaf-sheaths; branches normally not produced. Vegetative stems to 80cm; fertile stems to 20(30)cm. Native; grassy places, damp places and rough and cultivated ground; abundant   *Field Horsetail* - **E. arvense** L.
5 Leaf-sheaths normally with conspicuous black bands near top and bottom; teeth falling off before or as soon as shoots fully expanded   6
5 Leaf-sheaths with 0-1 conspicuous black bands, or black ± all over;

1. EQUISETUM

teeth present on fully expanded shoots 7
- 6 Leaf-sheaths c. as long as wide, without teeth from very early on; stems perennial. Stems to 1m. Native; in ditches and on river-banks; scattered and decreasing **Rough Horsetail - E. hyemale** L.
- 6 Leaf-sheaths distinctly longer than wide, with teeth until shoots fully expanded; stems to 60cm, wholly or largely dying down in winter. Native; dunes and banks by sea; Co Wexford and Co Wicklow (*E. hyemale* x *E. ramosissimum*)
  Moore's Horsetail - E. x moorei Newman
- 7 Stems perennial with previous year's cones persisting; cones obtuse to apiculate at apex 8
- 7 Stems annual; previous year's cones not persisting; cones rounded at apex 10
  - 8 Teeth of leaf-sheaths with broad scarious margins each much wider than black centre, obtuse at maturity; stem-ridges 4-10. Stems to 20(80)cm. Native; dune slacks, river-banks, lake-shores, wet stony mountain sites; scattered in N & W BI, very rare in C & S En **Variegated Horsetail - E. variegatum** F. Weber & D. Mohr
  - 8 Teeth of leaf-sheaths at least near tip with narrow scarious margins no wider than black centre, tapering to fine point; stem-ridges 8-20 9
- 9 Stems usually ± well branched; spores fertile; central hollow of stem mostly ≥1/2 as wide as stem. Stems to 1m. Probably intrd-natd; in long grass; S Lincs and N Somerset
  Branched Horsetail - E. ramosissimum Desf.
- 9 Stems not or sparsely branched; spores sterile; central hollow of stem mostly <1/2 as wide as stem. Stems to 1m. Native; sandy damp places; very scattered in Ir and Sc, Cheshire (*E. hyemale* x *E. variegatum*)
  Mackay's Horsetail - E. x trachyodon A. Braun
  - 10 Stem internodes white, often ≥1cm wide, with 18-40 ridges; stems with whorls of branches ± to top (see couplet 2) **E. telmateia**
  - 10 Stem internodes green, mostly <1cm wide, with 4-30 ridges, if >18 then at least top part of stem without whorls of branches 11
- 11 Branches regularly branched again; teeth of leaf-sheaths united into 3-6 lobes (fewer than stem-ridges) (see couplet 3) **E. sylvaticum**
- 11 Branches 0, or present but not or sparsely and irregularly branched again; teeth of leaf-sheaths not fused, as many as stem-ridges 12
  - 12 Stem-internodes with central hollow >3/4 as wide as stem, with 10-30 ridges (usually >20 in stems >8mm wide). Stems to 1.5m. Native; ponds, ditches, marshes, in or by water; common
    Water Horsetail - E. fluviatile L.
  - 12 Stem-internodes with central hollow <3/4 (usually c.1/2 or less) as wide as stem, with 4-20 ridges (stems rarely >8mm wide) 13
- 13 Stem with peripheral hollows c. same size as central hollow; stem-internodes with 4-9(12) ridges. Stems to 60cm. Native; wet or marshy ground; very common **Marsh Horsetail - E. palustre** L.
- 13 Stem with peripheral hollows <1/2 size of central hollow; stem-internodes with 6-20 ridges 14

14 Internodes of branches mostly 3-angled, the lowest shorter than adjacent leaf-sheath on main stem (see couplet 4) **E. pratense**

14 Internodes of branches mostly 4(or more)-angled, the lowest as long as to longer than adjacent leaf-sheath on main stem 15

15 Cones always produced on green stems; stem-internodes with central hollow c.1/2 as wide as stem. Stems to 1m. Native; wet places; scattered (*E. fluviatile* x *E. arvense*) **Shore Horsetail - E. x litorale** Rupr.

15 Cones only exceptionally produced on green stems; stem-internodes with central hollow <1/2 as wide as stem (see couplet 4) **E. arvense**

6 other hybrids occur; all are rare and sterile.

PTEROPSIDA — FERNS

Herbaceous plants with simple or variously branched, frequently subterranean stems (or rhizomes) and usually much divided (rarely simple) leaves with several to many veins, the veins usually much branched and often anastomosing. Sporangia homosporous or rarely heterosporous, often grouped in specialized regions (*sori*) on lowerside of leaves or on specialized leaves or parts of leaves, never in cones, often covered by special flaps of tissue (*indusia*). Gametophyte of homosporous species free living, either non-green, subterranean and mycorrhizal, or green and photosynthetic; gametophytes of heterosporous species much reduced and retained within spore, which lies on or under water.

5. OPHIOGLOSSACEAE - *Adder's-tongue family*

Rhizome short or corm-like, without scales; leaves borne singly, with erect stem-like petiole and sterile blade often plus 1 fertile blade; sterile blade simple and entire or 1-pinnate, not spirally coiled when young; fertile blade a simple spike or a panicle.

1. OPHIOGLOSSUM L. - *Adder's-tongues*
Sterile blade simple, entire; fertile blade a simple spike of sunken sporangia.

1 Sterile blade rarely >2cm, linear to narrowly elliptic, the vein-islets without free vein endings; spores ripe Jan-Mar. Leaves to 2cm. Native; very short turf by sea; local in Guernsey and Scillies
Least Adder's-tongue - **O. lusitanicum** L.
1 Sterile blade rarely <2cm, oblong-elliptic to broadly so, the vein-islets with minute free vein-endings within them; spores ripe Apr-Aug 2
  2 Sterile blade mostly 3-3.5cm; sporangia 6-14 either side of spike. Leaves to 10cm. Native; barish or grassy places near sea; very scattered round coasts *Small Adder's-tongue* - **O. azoricum** C. Presl
  2 Sterile blade mostly 4-15cm; sporangia 10-40 either side of spike. Leaves to 30(45)cm. Native; grassland, dune-slacks, open woods, mostly in lowlands; frequent *Adder's-tongue* - **O. vulgatum** L.

2. BOTRYCHIUM Sw. - *Moonwort*
Sterile blade pinnate; fertile blade a panicle of axes with sessile but not sunken sporangia.

1 Leaves to 30cm. Native; dry grassland, mostly in uplands; throughout

Br and Ir, especially N & W Br *Moonwort* - **B. lunaria** (L.) Sw.

6. OSMUNDACEAE - *Royal Fern family*

Rhizomes rather short, thick, not scaly, suberect to ascending; leaves in tufts at apices of rhizome-branches, spirally coiled when young, with stout ± erect petiole and large 2-pinnate blade, the lower parts sterile, the upper parts often fertile.

1. OSMUNDA L. - *Royal Fern*

1 Leaves (incl. petiole) (0.3)0.6-2(4)m. Native; fens, bogs, wet woods and heaths on peaty soil; throughout most of BI but absent from large areas of E Br *Royal Fern* - **O. regalis** L.

7. ADIANTACEAE - *Maidenhair Fern family*

Not recognisable as a family on superficial characters, but each of the 3 genera is very distinctive.

1. CRYPTOGRAMMA R. Br. - *Parsley Fern*
Rhizome short; leaves tufted at apices of rhizome-branches, of 2 sorts (fertile and sterile), 2-4-pinnate; sori oblong, becoming contiguous when mature, protected by continuous reflexed leaf-margin.

1 Leaves (incl. petiole) to 30cm. Native; rocky places on acid soils on mountains; local in N & W *Parsley Fern* - **C. crispa** (L.) Hook.

2. ANOGRAMMA Link - *Jersey Fern*
Plant annual; rhizome very short; leaves tufted, of 2 sorts (fertile and sterile), 1-3-pinnate; sori linear along veins, without indusium; leaf-margin flat.

1 Leaves (incl. petiole) to 10cm. Native; damp shady hedgebanks; local in CI *Jersey Fern* - **A. leptophylla** (L.) Link

3. ADIANTUM L. - *Maidenhair Fern*
Rhizome rather short; leaves produced singly but close together, all of 1 sort, 1-3-pinnate; sori suborbicular at tips of pinnules, covered by discrete reflexed flaps of pinnule.

1 Leaves (incl. petiole) to 25(45)cm. Native; limestone cliffs, grykes and rock crevices near sea; locally frequent in W
Maidenhair Fern - **A. capillus-veneris** L.

8. PTERIDACEAE - *Ribbon Fern family*

Rhizomes short, scaly; the pinnate (or 2-pinnate) leaves with long narrow pinnae, the lowest of which on each side give rise to an extra pair, are diagnostic.

1. PTERIS L. - *Ribbon Ferns*

1 Leaves (incl. petiole) to 75cm. Intrd-natd; on walls, old buildings and rock-faces; scattered in S & C En   ***Ribbon Fern* - P. cretica** L.

9. MARSILEACEAE - *Pillwort family*

Rhizomes long, thin, pubescent, not scaly; *Pilularia* could be mistaken only for an angiosperm, but the spirally coiled filiform young leaves and globose spore-bearing leaves distinguish it.

1. PILULARIA L. - *Pillwort*

1 Leaves 3-8(15)cm x c.0.5mm. Native; on silty mud by lakes, ponds and reservoirs; scattered and decreasing   ***Pillwort* - P. globulifera** L.

10. HYMENOPHYLLACEAE - *Filmy-fern family*

Rhizomes thin, glabrous or pubescent, not scaly; the very thin leaves with protruding marginal sori in tubular or valve-like indusia are unique.

1. HYMENOPHYLLUM Sm. - *Filmy-ferns*
Rhizomes filiform, glabrous; petiole not winged; leaves 1-2-pinnate; indusium of 2 valves, without protruding bristle.

1 Veins ending just short of leaf-margin (x10 lens), with cells mostly c.1.5x as long as wide; indusium valves conspicuously dentate. Leaves (incl. petiole) 2-5(10)cm. Native; shaded, damp rock-faces and tree-trunks; local in W Br and Ir, E Sussex
      ***Tunbridge Filmy-fern* - H. tunbrigense** (L.) Sm.
1 Veins mostly reaching leaf-margin (x10 lens), with cells mostly ≥2x as long as wide; indusium valves entire. Leaves (incl. petiole) 3-10(20)cm. Native; similar places and distribution to *H. tunbrigense*, but not E Sussex   ***Wilson's Filmy-fern* - H. wilsonii** Hook.

2. TRICHOMANES L. - *Killarney Fern*
Rhizomes thin, pubescent; petiole winged at least distally; leaves 2-3-pinnate; indusium tubular, with protruding bristle when mature.

1 Veins ending just short of leaf-margin (x10 lens). Leaves (incl. petiole)

7-45cm. Native; very sheltered, damp rock-faces, often near waterfalls or at cave entrances; extremely local in W, formerly commoner, still surviving as gametophytes in parts of En and Wa where sporophytes no longer develop   *Killarney Fern* - **T. speciosum** Willd.

11. POLYPODIACEAE - *Polypody family*

Rhizomes extended, scaly; the only ferns with pinnate or deeply pinnately lobed leaves with ± parallel-sided pinnae/lobes and orbicular to elliptic sori without indusia.

1. POLYPODIUM L. - *Polypodies*
Leaves pinnate to very deeply pinnately lobed right to base of blade; sori in row on either side of midribs of pinnae, not sunken.

Microscopic examination is necessary for certain identification. The key identifies only plants with full, turgid spores; plants with spores empty and shrivelled are hybrids. Ten sporangia per plant should be measured to obtain mean figures.

1 Leaf-blades mostly ≤2x as long as wide; pinnae usually narrowly acute, often markedly serrate; sporangia mixed with hairs (paraphyses) which are ≥0.5mm. Leaves (incl. petiole) to 40cm. Native; mostly on base-rich rocks; scattered in W BI, rare in E
Southern Polypody - **P. cambricum** L.
1 Leaf-blades >2x (up to c.6x) as long as wide; pinnae rounded to acute, usually subentire; sporangia without paraphyses   2
  2 Sori usually orbicular; mature leaves parallel-sided in proximal 1/3-1/2; annulus at yellow sporangium stage dark orange-brown, with mean of 10-14 thick-walled cells. Leaves (incl. petiole) to 25cm. Native; on rocks, walls, tree-trunks and banks, often on acid soils; frequent to common   *Polypody* - **P. vulgare** L.
  2 Sori usually broadly elliptic; mature leaves scarcely parallel-sided; annulus at yellow sporangium stage pale buff to golden-brown, with mean of 7-9 thick-walled cells. Leaves (incl. petiole) to 40cm. Native; similar places and distribution to *P. vulgare* but more calcicole   *Intermediate Polypody* - **P. interjectum** Shivas

Hybrids occur in all 3 combinations and are not rare where 2 spp. occur together; all are sterile.

2. PHYMATOSORUS Pic. Serm. (*Phymatodes* C. Presl) - *Kangaroo Fern*
Leaves pinnately lobed to varying extent on 1 plant, but base of blade gradually cuneate and not lobed; sori in row on either side of main midrib and of midribs of main lobes, sunken.

1 Leaves (incl. petiole) to 60cm. Intrd-natd; shady walls and damp

places in woods; S Kerry, Guernsey and Scillies (*Phymatodes diversifolia* (Willd.) Pic. Serm.)   **Kangaroo Fern - P. diversifolius** (Willd.) Pic. Serm.

12. DICKSONIACEAE - *Tree-fern family*

Rhizomes very thick, suberect, trunk-like, with dense hairs but no scales; the only natd tree-ferns.

1. DICKSONIA L'Hér. - *Australian Tree-fern*

1   Rhizome eventually to 15m but often <1m; leaves up to 200 x 60cm. Intrd-natd; woods and shady places in SW En and SW Ir
Australian Tree-fern - **D. antarctica** Labill.

13. DENNSTAEDTIACEAE - *Bracken family*

Rhizomes very extensive, deeply subterranean, pubescent but without scales; leaves borne singly, without scales, (2-)3-pinnate, with characteristic scent when crushed, with strong, long petiole and rhachis, resembling an aerial stem and leaves.

1. PTERIDIUM Scop. - *Bracken*

1   Leaves (incl. petiole) stiffly erect, to 3(5)m. Native
(*Bracken* - **P. aquilinum** (L.) Kuhn)   2
   2   Leaves often >1m; pinnae unfolding sequentially acropetally in spring; pinna midribs horizontal to downward-curved at least when young; petioles of emerging leaves with abundant white hairs and some reddish hairs, all gradually lost during maturity. Woods, heaths and moors, usually on acid dry soils; abundant
P. aquilinum ssp. aquilinum
   2   Leaves relatively small and wiry, ≤1m, often with only 1 main pair of pinnae each ± as large as rest of leaf, the 3 portions unfolding ± simultaneously in spring; pinna midribs upturned and ± straight; petioles of emerging leaves with 0 or few white hairs but abundant reddish hairs, all soon lost. Presumed relic of native *Pinus* woodland in open woodland and on moors; very scattered in C & N Sc (*P. pinetorum* C.N. Page & R.R. Mill)
P. aquilinum ssp. **latiusculum** (Desv.) C.N. Page

14. THELYPTERIDACEAE - *Marsh Fern family*

Recognizable by the submarginal ± round sori with 0 or flimsy and quickly withering indusia on somewhat *Dryopteris*-like leaves.

1. THELYPTERIS Schmidel - *Marsh Fern*
Rhizomes long and slender, bearing leaves singly at intervals or sometimes in sparse tufts; leaves ± of 2 sorts, straight, the fertile usually longer and with narrower pinnules recurved at margins; lowest pair of pinnae usually ≥1/2 as long as longest pair and parallel with it; petiole c. as long as blade or longer.

1   Leaves (incl. petiole) to 80cm (sterile) or 1.5m (fertile). Native; marshes and fens, usually shaded; very scattered, decreasing, mostly in E Anglia   *Marsh Fern* - **T. palustris** Schott

2. PHEGOPTERIS (C. Presl) Fée - *Beech Fern*
Rhizomes long and slender, bearing leaves singly at intervals; leaves of 1 sort, with blade bent sharply away from plane of petiole; lowest pair of pinnae ± as long as longest pair, reflexed; petiole as long as blade or longer.

1   Leaves (incl. petiole) to 50cm. Native; damp woods, shady rocky places and banks, on acid soils; frequent in W & N, absent from most of C, S & E En   *Beech Fern* - **P. connectilis** (Michx.) Watt

3. OREOPTERIS Holub - *Lemon-scented Fern*
Rhizomes short and stout, bearing leaves in tuft at apex; leaves of 1 sort, straight; lowest pair of pinnae extremely short, much <1/2 as long as longest; petiole much shorter than blade.

1   Leaves (incl. petiole) to 1.2m. Native; damp shady places and woods on acid soil; frequent, especially in N & W
               Lemon-scented Fern - **O. limbosperma** (All.) Holub

15. ASPLENIACEAE - *Spleenwort family*

Rhizomes short or very short, with scales; distinguished by the elongated (often very long) sori either with a straight-sided indusium or with indusium 0 but whole leaf lowerside covered with scales; most spp. are small and/or have relatively little-divided leaf-blades.

1. PHYLLITIS Hill - *Hart's-tongue*
Leaves simple, entire, with few or 0 scales on lowerside; sori linear, long, strictly parallel with each other and with lateral veins, each apparent sorus actually 2 lying adjacent, the pair with a linear indusium on each side with facing openings.

1   Leaves (incl. petiole) to 60cm but often much less. Native; shady moist rocky places, banks, walls and woods; common
               Hart's-tongue - **P. scolopendrium** (L.) Newman

For hybrids see under *Asplenium*.

2. ASPLENIUM L. - *Spleenworts*

Leaves 1-3-pinnate or sparsely and irregularly divided into linear segments, with few or 0 scales on lowerside; sori oblong to linear, usually parallel with lateral veins of pinnae or pinnules, single and each with 1 indusium.

1 Leaves irregularly and sparsely divided into linear segments, to 8(15)cm incl. petiole. Native; in rock-crevices in acid areas, mostly upland; very local in W & N Br, W Galway
Forked Spleenwort - **A. septentrionale** (L.) Hoffm.
1 Leaves 1-3-pinnate; pinnules or pinnae not linear 2
  2 Leaves 1-pinnate 3
  2 Leaves 2-3-pinnate at least in part 7
3 Distal part of rhachis conspicuously green-winged; larger pinnae >12mm, ≥2x as long as wide. Leaves (incl. petiole) to 40cm (but often much less). Native; walls, cliffs and rock-crevices close to sea; frequent on coasts except E & S En   *Sea Spleenwort* - **A. marinum** L.
3 Rhachis not or scarcely winged; pinnae ≤12mm, ≤2x as long as wide 4
  4 Rhachis green, not winged or with extremely narrow green wings. Leaves (incl. petiole) to 15(20)cm. Native; base-rich upland rock-crevices; rather local in W & N Br (*A. trichomanes-ramosum* L.)
Green Spleenwort - **A. viride** Huds.
  4 Rhachis blackish, with very narrow scarious wings. Leaves (incl. petiole) to 20(40)cm. Native; on rocks and walls
(*Maidenhair Spleenwort* - **A. trichomanes** L.) 5
5 Pinnae oblong-triangular, usually attached to rhachis in centre of base, usually conspicuously serrate, minutely pubescent on lowerside. Calcicole; very local in W Br
A. trichomanes ssp. pachyrachis (H. Christ) Lovis & Reichst.
5 Pinnae suborbicular to oblong, often attached to petiole at proximal corner, usually subentire, ± glabrous 6
  6 Pinnae suborbicular to rhombic, often asymmetric due to uneven basal bulges, up to 8mm, flat to concave on upperside; sori ≤2mm, ≤6(9) per pinna; rhizome-scales ≤3.5mm; spores (23)29-36(42) microns. Calcifuge; local in N & W Br, Co. Down
A. trichomanes ssp. trichomanes
  6 Pinnae oblong, usually ± symmetric, up to 12mm, convex to flat on upperside; sori ≤3mm, ≤9(12) per pinna; rhizome-scales ≤5mm; spores (27)34-43(50) microns. Calcicole, incl. mortar in walls in acidic areas; common   **A. trichomanes ssp. quadrivalens** D.E. Mey.
7 Longest pinnae at c. middle of blade, the basal pair slightly to considerably shorter; petioles usually much shorter than blade. Leaves (incl. petiole) to 30cm. Native; rocks, hedgebanks and walls; W & SW, common in CI and SW En
Lanceolate Spleenwort - **A. obovatum** Viv.
7 Longest pinnae clearly the basal ones; petioles nearly as long as to longer than blade 8
  8 Petioles green; blade 1-2-pinnate; indusia with fringed margins. Leaves (incl. petiole) to 8(15)cm. Native; rocks and walls with

        base-rich substratum; common   ***Wall-rue*** - **A. ruta-muraria** L.
- 8   Petioles reddish-brown to blackish; blade (2-)3-pinnate; indusia subentire   9
- 9   Blade and pinnae acute to very shortly acuminate; spores (32)39-45(52) microns across. Leaves (incl. petiole) to 50cm. Native; rocky places in woods, banks and open sites, walls; common

                                                    Black Spleenwort - **A. adiantum-nigrum** L.
- 9   Blade and pinnae very long-acuminate; spores (25)30-35(39) microns across. Leaves (incl. petiole) to 50cm. Native; dry banks and rock-faces mostly on limestone and near sea; very local in S Ir

                                                    Irish Spleenwort - **A. onopteris** L.

Hybrids are quite numerous but all are very rare and scarcely contribute to identification problems. *A. adiantum-nigrum*, *A. obovatum* and *A. trichomanes* also form very rare hybrids with *Phyllitis scolopendrium* (X *Asplenophyllitis* Alston). All the hybrids are sterile and ± obviously intermediate.

3. CETERACH Willd. - *Rustyback*
Leaves simple but lobed ± to base or pinnate, often former distally and latter proximally, densely covered on lowerside with eventually reddish-brown scales; sori linear, without indusium, becoming merged.

- 1   Leaves (incl. petiole) to 15(20)cm. Native; base-rich crevices and mortar cracks in walls; common in S & W, rare in N & E

                                                    Rustyback - **C. officinarum** Willd.

16. WOODSIACEAE - *Lady-fern family*

The family in our flora most defying definition (except by reference to vascular architecture); the 6 genera differ widely in superficial characters.

1. MATTEUCCIA Tod. - *Ostrich Fern*
Rhizomes short; leaves borne in apical tufts, of 2 sorts; sterile leaves 2-pinnate or nearly so; fertile leaves shorter, with longer petiole, 1-pinnate, not green; sori in 1-2 contiguous rows on each pinna, protected by tightly inrolled leaf-margin.

- 1   Sterile leaves (incl. petiole) up to 1.5m, superficially resembling *Oreopteris limbosperma*; fertile leaves up to 60cm. Intrd-natd; shady places; scattered in N   ***Ostrich Fern*** - **M. struthiopteris** (L.) Tod.

2. ONOCLEA L. - *Sensitive Fern*
Rhizomes long; leaves borne singly, of 2 sorts; sterile leaves 1-pinnate or nearly so with winged rhachis and entire to lobed pinnae; fertile leaves shorter, 2-pinnate or nearly so, with non-green pinnae; sori in compact groups protected by tightly inrolled leaf-margin.

1 Sterile leaves (incl. petiole) up to 1m; fertile leaves up to 80cm. Intrd-
natd; shady places; scattered in W   *Sensitive Fern* - **O. sensibilis** L.

3. ATHYRIUM Roth - *Lady-ferns*

Rhizomes short; leaves borne in apical tufts, of 1 sort, 2-pinnate with deeply lobed to crenate pinnules; sori on lowerside of pinnules, not protected by inrolled leaf-margin, with 0, inconspicuous or well-developed indusium.

1 Sori oblong to curved with well-developed J- or C-shaped indusium. Leaves (incl. petiole) to 1(1.5)m. Native; damp woods and hedgebanks, mountain screes, marshes; common
Lady-fern - **A. filix-femina** (L.) Roth
1 Sori orbicular, with 0 or vestigial indusium   2
  2 Leaves erect, oblanceolate to narrowly elliptic, up to 30(70)cm incl. petiole c.1/8-1/4 as long as blade; sori commoner near base of blade. Native; acid gullies, boulder-slopes and scree, rarely below 600m; local in C & N Sc   *Alpine Lady-fern* - **A. distentifolium** Opiz
  2 Leaves spreading, bent out at or just above base of blade, irregularly narrowly oblong-elliptic, up to 20(30)cm incl. petiole ≤1/8 as long as blade; sori commoner near apex of blade. Native; damp acid rocky places at 1040-1140m, very local in C Sc
Newman's Lady-fern - **A. flexile** (Newman) Druce

4. GYMNOCARPIUM Newman - *Oak Ferns*

Rhizomes long; leaves borne singly, of 1 sort, 2-3-pinnate, with petiole much longer than blade, with ± triangular blade; sori on underside of pinnules rather near margin, orbicular to elliptic, without indusium.

1 Leaves to 40cm (incl. blackish petiole); blade yellowish- to mid-green, non-glandular; basal pinnae each nearly as large as rest of blade. Native; damp woods and shady rocks, often in ± acid humus-rich soil; frequent in N & W Br, rare elsewhere
Oak Fern - **G. dryopteris** (L.) Newman
1 Leaves to 50cm (incl. greenish-brown petiole); blade dull green, with small glands that are also on rhachis and top of petiole; basal pinnae each c.1/2 as large as rest of blade. Native; open scree-slopes or rocky places on limestone, natd on walls; local in En and Wa, rare elsewhere   *Limestone Fern* - **G. robertianum** (Hoffm.) Newman

5. CYSTOPTERIS Bernh. - *Bladder-ferns*

Rhizomes long with leaves borne singly, or short with leaves in a terminal tuft; leaves of 1 sort, 2-3(4)-pinnate; sori orbicular, on lowerside of pinnules, not protected by inrolled leaf-margin, with flap-like indusium that becomes reflexed to expose sporangia.

1 Rhizomes elongated, bearing leaves singly; leaves (incl. petiole) to 45cm, triangular-ovate, the lowest pair of pinnae the longest. Native; shady wet basic rock-ledges, gullies and scree above 700m; local in

C Sc *Mountain Bladder-fern* - **C. montana** (Lam.) Desv.
1 Rhizome short, bearing terminal tuft of leaves; leaves narrowly
  oblong to lanceolate, the longest pinnae near the middle of leaf   2
  2 Spores rugose; adjacent pinnules and pinnae strongly overlapping.
    Leaves (incl. petiole) to 20(25)cm. Native; basic rocks in sea-caves
    and stream-gorges; very rare in CE Sc
                         Dickie's Bladder-fern - **C. dickieana** R. Sim
  2 Spores spinose; adjacent pinnules and pinnae scarcely overlapping.
    Leaves (incl. petiole) to 25(45)cm. Native; shady basic rocks and
    walls, incl. mortar in acid areas; common in N & W, very
    scattered in S & E   *Brittle Bladder-fern* - **C. fragilis** (L.) Bernh.

6. WOODSIA R.Br. - *Woodsias*

Rhizomes short, with leaves borne in a terminal tuft; leaves of 1 sort, 1-pinnate with deeply lobed pinnae or 2-pinnate on proximal pinnae; petiole with a visible joint c.1/3 way up (an eventual abscission point); sori orbicular, on lowerside of pinnules, not protected by inrolled leaf-margin, with an indusium consisting of basal ring of hairs or narrow scales.

1 Leaves (incl. petiole) to 10(15)cm, pubescent, with scales (c.2-3mm) on
  lowerside; pinnae 7-15 on each side, the longest ones oblong or ovate-
  oblong and c.1.5-2x as long as wide with 3-8 lobes on each side.
  Native; crevices in mostly neutral rocks, 360-720m; very local in
  Caerns, Cumberland, C & S Sc   *Oblong Woodsia* - **W. ilvensis** (L.) R.Br.
1 Leaves (incl. petiole) to 8(15)cm, rather sparsely pubescent, with
  scales (c.1-2mm) only on petiole and on rhachis of blade and pinnae;
  pinnae 5-10 on each side, the longest ones triangular-ovate and
  c.1-1.5x as long as wide, with 1-4 lobes on each side. Native; crevices
  in mostly basic rocks, 580-920m; very local in Caerns and C Sc
                              Alpine Woodsia - **W. alpina** (Bolton) Gray

17. DRYOPTERIDACEAE - *Buckler-fern family*

Rhizomes short, densely scaly; familiar ferns with usually 2-3-pinnate densely tufted leaves (1-pinnate in some spp.) with peltate or reniform indusia attached at 1 point only.

1. POLYSTICHUM Roth - *Shield-ferns*

Leaves 1-2-pinnate, of 1 sort; sori in a row down each side of pinna or pinnules or sometimes of main lobes of pinnules; indusium peltate, attached in centre.

1 Leaves 1-pinnate with shallowly toothed (not lobed, except for 1 basal
  lobe) pinnae, even when producing sori   2
1 Leaves 1-pinnate with deeply lobed pinnae to 2-(3)-pinnate   3
  2 Basal pinnae not or scarcely shorter than longest ones; leaves
    (incl. petiole) up to 120 x 15cm. Intrd-natd; shady laneside bank;

1. POLYSTICHUM

Surrey   *Western Sword-fern* - **P. munitum** (Kaulf.) C. Presl
2   Basal pinnae markedly shorter than longest ones; leaves (incl. petiole) up to 30(60) x 6cm. Native; basic rock-crevices, scree and ravines mostly above 600m; local in N & NW
Holly-fern - **P. lonchitis** (L.) Roth
3   Lowest pinnae nearly as long as longest pinnae; proximal pinnules on each pinna shortly stalked, with a blade ± right-angled at base. Leaves (incl. petiole) to 1.5m. Native; woods and hedgebanks in moist places; frequent in SW, rarer in NE
Soft Shield-fern - **P. setiferum** (Forssk.) Woyn.
3   Lowest pinnae c.1/2 as long as longest pinnae; proximal pinnules on each pinna sessile or pinnules not even differentiated, with a blade acute-angled at base. Leaves (incl. petiole) to 1m. Native; similar places to *P. setiferum* but more upland and northern
Hard Shield-fern - **P. aculeatum** (L.) Roth

Hybrids in all 3 combinations of native spp. occur rarely; they are sterile.

2. CYRTOMIUM C. Presl (*Phanerophlebia* C. Presl) - *House Holly-fern*
Leaves 1-pinnate, of 1 sort; sori scattered all over lowerside of pinnae; indusium peltate, attached in centre.

1   Leaves (incl. petiole) to 1.2m. Intrd-natd; on and by walls, among maritime rocks and in other shady places; scattered in W Br, rare elsewhere (*Phanerophlebia falcata* (L.f.) Copel.)
House Holly-fern - **C. falcatum** (L.f.) C. Presl

3. DRYOPTERIS Adans. - *Buckler-ferns*
Leaves 2-3(4)-pinnate, usually of 1 sort; sori in a row down each side of pinnules or sometimes of main lobes of pinnules; indusium reniform, attached at the notch.

1   Leaves 1-pinnate with deeply lobed pinnae, to 2-pinnate with pinnules lobed to c.1/2 way to midrib   2
1   Leaves 2-pinnate with pinnules lobed nearly to midrib, to 3(-4)-pinnate   7
　2   Leaves ± of 2 sorts, the fertile longer and more erect, lanceolate-oblong, parallel-sided for most of length; pinnae with <15 pinnules or lobes each side; pinnules with mucronate teeth. Leaves (incl. petiole) to 60(100)cm. Native; wet heaths, dune-slacks, marshes and fens; very local and decreasing in SE En, Renfrews
Crested Buckler-fern - **D. cristata** (L.) A. Gray
　2   Leaves of 1 sort, lanceolate-elliptic, scarcely parallel-sided; pinnae with >15 pinnules each side; pinnules with acute, obtuse or 0 teeth   3
3   Leaf dull green, ± mealy due to many minute glands on both surfaces; lowest pinna with proximal 3-4 pinnules on both sides ± same size. Leaves (incl. petiole) to 75cm. Native; in limestone

crevices, grykes and scree; very local in Wa and NW En
Rigid Buckler-fern - **D. submontana** (Fraser-Jenk. & Jermy) Fraser-Jenk.

3 Leaf clear green (of various shades), not or slightly glandular; lowest pinna with pinnules successively smaller from base distally, or just the 2 proximal ± same size 4

    4 Pinnules evenly lobed nearly 1/2 way to midrib; petiole c.1/3-1/2 as long as blade. Leaves (incl. petiole) to 75cm. Native; formerly in woods in W Ir, ?Dunbarton; extinct. Derived from *D. affinis* x *D. expansa* *Scaly Buckler-fern* - **D. remota** (Döll) Druce

    4 Pinnules entire to lobed distinctly <1/2 way to midrib; petiole usually ≤1/3 as long as blade; common 5

5 Pinnules parallel-sided for most of length, broadly rounded to ± truncate (but often toothed) at apex; pinnae with dark blotch where they join rhachis; petioles with dense golden scales. Leaves (incl. petiole) to 1.5m. Native; similar places to *D. filix-mas*; frequent to common. Derived from ≥3 diploid spp. Several sspp. have been recognised but are difficult to define
Scaly Male-fern - **D. affinis** (Lowe) Fraser-Jenk.

5 Pinnules distinctly tapering, rounded to ± obtuse at apex; pinnae without dark blotches at base; petioles with sparse to ± dense greyish- or pale-brown scales 6

    6 Rhizome not or little branched, hence leaf-tufts scattered; pinnules with erect or convergent teeth at apex; sori c.1.5mm across, along ± whole length of pinnules, with non-glandular indusia with edges not tucked under. Leaves (incl. petiole) to 1.2m. Native; woods, hedgebanks, ditches, mountains in open or shade; common *Male-fern* - **D. filix-mas** (L.) Schott

    6 Rhizome well branched, hence leaf-tufts in groups; pinnules with teeth spreading fan-wise at apex; sori c.1mm across, ± confined to proximal 1/2 of pinnules with ± glandular indusia with edge well tucked under before sporangial dehiscence. Leaves (incl. petiole) to 0.5(-1.2)m. Native; rocky places on mountains; local above 240m in N & W *Mountain Male-fern* - **D. oreades** Fomin

7 Leaf dull green, ± mealy due to many minute glands on both surfaces; lowest pinna with proximal 3-4 pinnules on both sides ± same size (see couplet 3) **D. submontana**

7 Leaf clear green (of various shades), not or slightly glandular on upperside; lowest pinna with pinnules successively smaller from base distally 8

    8 Petioles with dense golden scales at least in proximal 1/2; pinnules evenly lobed ≤5/6 way to midrib; extinct (see couplet 4) **D. remota**

    8 Petioles variously clothed but not with dense golden scales; leaves 2-3(4)-pinnate, if only 2-pinnate then at least lowest pinnules on lowest pinnae lobed ± to base; common 9

9 Pinnules distinctly concave on upperside, with numerous minute glands on lowerside and sometimes upperside. Leaves (incl. petiole) to 75cm. Native; moist shady woods, ravines and hedgebanks; local in Ir and W & S Br *Hay-scented Buckler-fern* - **D. aemula** (Aiton) Kuntze

9 Pinnules flat or convex on upperside, glands 0 or rare on lowerside
(except sometimes on indusia) and 0 on upperside 10

  10 Scales on petiole usually with very distinct dark centre
(sometimes uniformly pale in small upland plants); pinnules
usually convex on upperside; indusia glandular. Leaves (incl.
petiole) to 1(1.5)m. Native; woods, hedgebanks, ditches, heaths
and mountains; common

                               Broad Buckler-fern - **D. dilatata** (Hoffm.) A. Gray

  10 Scales on petiole uniformly pale or dark brown, or somewhat
darker in centre (the dark suffusing outwards, not in a distinct
zone); pinnules usually flat; indusia glandular or not (if glandular
then scales not pale) 11

11 Leaf-blades ovate-triangular, with all pinnae ± in 1 plane; petioles
with scales mid- to dark-brown or with darker centres; lowest pinnae
with lowest pinnule on basal side usually ≥1/2 as long as its pinna.
Leaves (incl. petiole) to 80(100)cm. Native; cool, often damp places in
woods and mountain crevices and scree; locally frequent in N & W
Br   *Northern Buckler-fern* - **D. expansa** (C. Presl) Fraser-Jenk. & Jermy

11 Leaf-blades narrowly triangular-lanceolate, with lower pinnae
twisted into ± horizontal plane; petioles with pale brown scales;
lowest pinnae with lowest pinnule on basal side <1/2 as long as its
pinna. Leaves (incl. petiole) to 80(100)cm. Native; damp or wet
woods, marshes, fens and wet heaths; frequent throughout most of
Br and Ir   *Narrow Buckler-fern* - **D. carthusiana** (Vill.) H.P. Fuchs

Hybridization is widespread but, except for *D. affinis* x *D. filix-mas* = **D. x complexa** Fraser-Jenk., rare. Most hybrids are sterile, but some produce good spores with an unreduced chromosome number that develop into new sporophytes without fertilization. Such taxa, e.g. *D. affinis* and *D. remota*, are best treated as spp.

18. BLECHNACEAE - *Hard-fern family*

Rhizomes short, scaly; the sori having openings towards the midrib, and the separate fertile and sterile leaves being coriaceous and 1-pinnate, are diagnostic.

1. BLECHNUM L. - *Hard-ferns*

1 Leaves (incl. petiole) to 50(75) x 4cm; pinnae entire, the sterile up to
2 x 0.5cm, attached to rhachis over whole width of their base. Native;
woods, heaths, moors, grassy and rocky slopes on acid soils;
common   *Hard-fern* - **B. spicant** (L.) Roth

1 Leaves (incl. petiole) to 100 x 20cm; pinnae subentire, the sterile up
to 10 x 2.5cm, greatly narrowed at base and attached to rhachis over
short width. Intrd-natd; shady places and by streams; rare in SW

& W *Chilean Hard-fern* - **B. cordatum** (Desv.) Hieron.

19. AZOLLACEAE - *Water Fern family*

Stems very slender, branching, floating on water, with hanging, simple roots, without scales; leaves on 2 opposite sides of stem, small, 2-lobed.

1. AZOLLA Lam. - *Water Fern*

1 Stems 1-5(10)cm; leaves up to 2.5 x 1.5mm. Intrd-natd; ponds, canals and dykes, often covering water surface like a *Lemna* (and often with it); frequent but sporadic in C & S *Water Fern* - **A. filiculoides** Lam.

PINOPSIDA — CONIFERS, GYMNOSPERMS

Trees or shrubs with simple, usually evergreen leaves. Male sporangia borne on sporophylls arranged in male cones. Female sporangia borne in naked ovules either borne singly and terminally or borne on cone-scales in female cones. Female gametophyte greatly reduced and retained in ovule. Fertilized ovule (seed) retained on sporophyte until ripe.

VEGETATIVE KEY TO GENERA OF PINOPSIDA

1	Some leaves at least 1cm wide	**23/1. ARAUCARIA**
1	All leaves <1cm wide	2
2	Leaves in opposite pairs or in threes at each node	3
2	Leaves 1 at each node, borne spirally but often apparently 2-ranked	6
3	At least some leaves in threes, not appressed to stem (some other cultivated Cupressaceae key out here)	**22/5. JUNIPERUS**
3	Leaves in opposite pairs, usually ± appressed to stem (some cultivated *Juniperus* spp. key out here)	4
4	Leaves obtuse; ultimate branchlets not flattened, spreading in 3 dimensions	**22/1. CUPRESSUS**
4	Leaves acute to acuminate; ultimate branchlets flattened, mainly or wholly spreading in 1 plane	5
5	Foliage with a resinous or oily scent when crushed	**22/2 & 3. X CUPRESSOCYPARIS & CHAMAECYPARIS**
5	Foliage with a sweet aromatic scent when crushed	**22/4. THUJA**
6	Leaves long and needle-like, borne in groups of 2-5 on very dwarf short-shoots	**20/7. PINUS**
6	Leaves all borne singly on long-shoots or most borne in clusters of >10 on short-shoots	7
7	Most leaves borne in dense clusters on short-shoots	8
7	All leaves borne singly on long-shoots	9
8	Leaves deciduous, dorsiventrally flattened	**20/5. LARIX**
8	Leaves evergreen, 3-5-angled in section	**20/6. CEDRUS**
9	Leaves tapered from broad base to narrowly acute apex, not dorsiventrally flattened	10
9	Leaves narrowed at base and apex, flattened or not	11
10	Trunk with thick spongy outer bark; leaves wider than thick	**21/2. SEQUOIADENDRON**
10	Trunk with thin stringy outer bark; leaves thicker than wide	**21/3. CRYPTOMERIA**

- 11 Trunk with thick spongy outer bark; leading shoots with reduced ± scale-like leaves **21/1. SEQUOIA**
- 11 Trunk with thin scaly outer bark; leading shoots with ± full-sized leaves 12
  - 12 Winter vegetative buds green; leaves and wood without resin-ducts **24/1. TAXUS**
  - 12 Winter vegetative buds brown; leaves and wood normally with resin-ducts 13
- 13 Leaves with distinct slender, short, green petiole ± appressed (but not fused) to twig; resin-duct 1 per leaf **20/3. TSUGA**
- 13 Leaves sessile, though often much narrowed at base and sometimes borne on brown, petiole-like peg; resin-ducts usually >1 per leaf, sometimes 0-1 14
  - 14 Leaves borne on distinct brown, petiole-like pegs remaining on twig when leaf falls **20/4. PICEA**
  - 14 Leaves sessile on twigs or on slightly raised cushions 15
- 15 Winter buds conical, shiny, sharply pointed **20/2. PSEUDOTSUGA**
- 15 Winter buds rounded at apex **20/1. ABIES**

KEY TO FAMILIES OF PINOPSIDA

- 1 Some leaves at least 1cm wide **23. ARAUCARIACEAE**
- 1 All leaves <1cm wide 2
  - 2 Leaves in opposite pairs or in 3s at each node **22. CUPRESSACEAE**
  - 2 Leaves 1 at each node and borne spirally (but often apparently 2-ranked), or tightly clustered in groups of 2-many on short-shoots 3
- 3 Vegetative buds with green bud-scales; usually dioecious; ovules solitary, surrounded by red succulent upgrowth at seed maturity **24. TAXACEAE**
- 3 Vegetative buds with brown bud-scales or proper bud-scales 0; monoecious; ovules borne in cones that become woody at seed maturity 4
  - 4 Vegetative buds with brown bud-scales; bracts and cone-scales distinct **20. PINACEAE**
  - 4 Vegetative buds without proper bud-scales; bracts and cone-scales completely fused, their distinction difficult **21. TAXODIACEAE**

20. PINACEAE - *Pine family*

Evergreen or deciduous resiniferous trees; distinguished from other conifers by the combination of spirally borne, very narrow leaves, brown, scaly buds, ovules borne in cones, and each cone-scale with a bract below it.

Some genera bear short lateral stems (short-shoots) of very limited length, in addition to the normal extension shoots (long-shoots). 'Twigs' refers to stems of the previous 1-2 years' growth.

20. PINACEAE

1 Leaves borne in groups of 2-5 on very dwarf short-shoots borne in axils of small scale-leaves **7. PINUS**
1 Leaves all borne singly on long-shoots, or most borne in dense clusters on short-shoots 2
   2 Most leaves borne in dense clusters on short-shoots 3
   2 All leaves borne singly on long-shoots 4
3 Leaves deciduous, flattened; female cones <5cm, eventually falling whole **5. LARIX**
3 Leaves evergreen, 3-5-angled in section; female cones >5cm, eventually disintegrating on tree **6. CEDRUS**
   4 Leaves with a distinct slender, short, green petiole appressed to twig; female cones <3cm **3. TSUGA**
   4 Leaves sessile though usually narrowed at base and sometimes borne on a brown petiole-like peg; female cones >3cm 5
5 Leaves borne on distinct brown, petiole-like pegs; female cones pendent, falling whole; bracts not protruding beyond cone-scales **4. PICEA**
5 Leaves sessile or borne on slightly raised cushions 6
   6 Winter buds conical, sharply pointed; female cones pendent, falling whole; bracts 3-lobed, protruding beyond cone-scales but not reflexed **2. PSEUDOTSUGA**
   6 Winter buds rounded to obtuse; female cones erect, disintegrating on tree; bracts not protruding beyond cone-scales, or protruding, 1-lobed and reflexed **1. ABIES**

1. ABIES Mill. - *Firs*

Evergreen; leaves borne only on long-shoots, single, sessile, falling to leave disc-like scars, with 2 resin-ducts, dorsiventrally flattened; female cones erect, the scales deciduous at maturity.

1 Leaves on upperside of twigs all strongly divergent, leaving centre parting exposing the twig axis 2
1 Leaves on upperside of twigs not divergent so as to expose the twig axis 3
   2 Twigs conspicuously pubescent (hairs >0.25mm); buds not resinous; female cones with exserted bracts. Tree to 48m. Intrd-natd; planted, often self-sown *European Silver-fir* - **A. alba** Mill.
   2 Twigs minutely pubescent (hairs <0.25mm); buds resinous; female cones with included bracts. Tree to 60m. Intrd-natd; planted, occasionally self-sown

Giant Fir - **A. grandis** (D. Don) Lindl.
3 Leaves grey-green on upperside, bent or curved, appressed to twig near base and then widely divergent; female cones mostly >15cm. Tree to 47m. Intrd-natd; planted, occasionally self-sown

Noble Fir - **A. procera** Rehder
3 Leaves dark to bright green on upperside, ± straight, those along midline not divergent from twig axis and overlying it; female cones

mostly <15cm. Tree to 44m. Intrd-natd; planted, rarely self-sown
Caucasian Fir - **P. nordmanniana** (Steven) Spach

2. PSEUDOTSUGA Carrière - *Douglas Fir*
Evergreen; vegetative buds sharply pointed; leaves borne only on long-shoots, sessile, single, falling to leave slightly raised cushions, with 2 resin-ducts, dorsiventrally flattened; female cones pendent, falling whole, with exserted, forward-pointing, 3-toothed bracts.

1 Tree to 65m. Intrd-natd; planted, occasionally self-sown
Douglas Fir - **P. menziesii** (Mirb.) Franco

3. TSUGA (Antoine) Carrière - *Hemlock-spruces*
Evergreen; vegetative buds subacute to rounded; leaves borne only on long-shoots, single, distinctly petiolate, falling to leave short brown pegs, with 1 resin-duct, dorsiventrally flattened; female cones pendent, falling whole, with included bracts.

1 Tree to 46m. Intrd-natd; planted, often self-sown
Western Hemlock-spruce - **T. heterophylla** (Raf.) Sarg.

4. PICEA A. Dietr. - *Spruces*
Evergreen; vegetative buds acute to rounded; leaves borne only on long-shoots, single, sessile, falling singly to leave distinct brown pegs, with 0-2 resin-ducts, dorsiventrally flattened or 4-angled in section; female cones pendent, falling whole, with minute bracts.

1 Leaves dorsiventrally flattened 2
1 Leaves about as thick as wide or thicker than wide, 4-angled in section 3
  2 Twigs glabrous; leaves mostly 15-25mm, sharply pointed at apex; female cones 6-10cm. Tree to 55m. Intrd-natd; much planted, often self-sown *Sitka Spruce* - **P. sitchensis** (Bong.) Carrière
  2 Twigs pubescent; leaves 8-18mm, obtuse to rounded but shortly mucronate; female cones 3-6cm. Tree to 31m. Intrd-surv; planted
Serbian Spruce - **P. omorika** (Pancic) Purk.
3 Leaves bright to dark green with faint stripes, with a resinous smell when crushed; female cones 10-20cm. Tree to 46m. Intrd-natd; planted, occasionally self-sown *Norway Spruce* - **P. abies** (L.) H. Karst.
3 Leaves bluish-green with rather conspicuous stripes, with an unpleasant smell when crushed; female cones ≤7.5cm 4
  4 Twigs glabrous; leaves 8-18mm. Tree to 31m. Intrd-surv; planted
White Spruce - **P. glauca** (Moench) Voss
  4 Twigs pubescent; leaves 15-25mm. Tree to 27m. Intrd-surv; planted *Engelmann Spruce* - **P. engelmannii** Engelm.

5. LARIX Mill. - *Larches*

Deciduous; leaves borne singly on leading long-shoots and in dense clusters on lateral short-shoots, sessile, falling singly, with 2 resin-ducts, dorsiventrally flattened; male cones borne singly, <2cm; female cones erect, eventually falling whole, with bracts exserted at anthesis but ± concealed at maturity.

1 Female cone-scales erect, not with recurved tips; leaves with inconspicuous greenish stripes on lowerside. Tree to 46m. Intrd-natd; much planted, commonly self-sown *European Larch* - **L. decidua** Mill.
1 Female cone-scales with patent or somewhat recurved tips; leaves with rather conspicuous greyish or whitish stripes on lowerside 2
    2 Female cones ovoid, usually c.1.25-1.5x as long as wide. Native; planted, often originating anew (*L. decidua* x *L. kaempferi*)
Hybrid Larch - **L. x marschlinsii** Coaz
    2 Female cones broadly ovoid to subglobose, usually 1-1.25x as long as wide. Tree to 37m. Intrd-natd; much planted, sometimes self-sown *Japanese Larch* - **L. kaempferi** (Lindl.) Carrière

6. CEDRUS Trew - *Cedars*

Evergreen; leaves borne singly on leading long-shoots and in dense clusters on lateral short-shoots, sessile, falling singly, with 2 resin-ducts, 3-5-angled in section; male cones borne singly, >2cm; female cones erect, disintegrating on tree, with minute bracts.

1 Tip of tree pendent; leaves on short-shoots mostly 3-3.5cm. Tree to 37m. Intrd-natd; planted, occasionally self-sown
Deodar - **C. deodara** (D. Don) G. Don
1 Tip of tree stiffly curved to 1 side or erect; leaves on short-shoots mostly 1.5-2.5cm 2
    2 Leaves mostly 1.5-2.5cm, abruptly tapered to c.0.2mm ± translucent tip; twigs usually glabrous; female cones mostly 7-10cm. Tree to 40m. Intrd-surv; planted
Cedar-of-Lebanon - **C. libani** A. Rich.
    2 Leaves mostly 1.5-2cm, rather gradually tapered to c.0.5mm translucent tip; twigs very shortly pubescent; female cones mostly 5-8cm. Tree to 39m. Intrd-natd; planted, rarely self-sown
Atlas Cedar - **C. atlantica** (Endl.) Carrière

7. PINUS L. - *Pines*

Evergreen; leaves borne in groups of 2, 3 or 5 on very dwarf short-shoots, sessile, with 2-many resin-ducts, semicircular to variously angled in section; male cones borne in clusters; female cones eventually falling whole, with minute bracts.

On very young plants long leaves are borne singly on long-shoots, but otherwise long-shoots bear only scale-leaves.

1 Leaves in pairs 2

1 Leaves in threes or fives 8
   2 Leaves mostly >10cm 3
   2 Leaves mostly <10cm 5
3 Bud-scales recurved at apex; female cones 8-22cm. Tree to 35m. Intrd-natd; planted, often self-sown   *Maritime Pine* - **P. pinaster** Aiton
3 Bud-scales appressed at apex; female cones 3-9cm. Tree to 42m. Intrd-natd; planted, often self-sown **(P. nigra** J.F. Arnold) 4
   4 Crown wide with long side branches; leaves dark green, 8-12cm, rather stiff   *Austrian Pine* - **P. nigra ssp. nigra**
   4 Crown columnar with short side branches; leaves bright green, 10-18cm, rather flexible   *Corsican Pine* - **P. nigra ssp. laricio** Maire
5 Upper part of trunk pale orange-red; cone-scales ± without prickle on outer face; leaves glaucous, twisted. Tree to 36m. Native; Highlands of C & N Sc, commonly natd elsewhere in BI
                                                       Scots Pine - **P. sylvestris** L.
5 Upper part of trunk grey-brown to red-brown; cone-scales usually with distinct prickle on outer face; leaves green, twisted or not 6
   6 Leaves distinctly twisted; cone-scales with rather long slender prickle. Tree to 25m. Intrd-natd; planted, rarely self-sown
                                            Lodgepole Pine - **P. contorta** Loudon
   6 Leaves scarcely twisted; cone-scales with very short stout prickle 7
7 Buds obtuse to subacute; leaves ≤8cm; female cones ≤5cm; resin-ducts just below leaf surface. Shrub to 4m. Intrd-surv; planted
                                        Dwarf Mountain-pine - **P. mugo** Turra
7 Buds acuminate; leaves often >8cm; female cones often >5cm; resin-ducts deep-seated in leaf   (see couplet 4) **P. nigra**
   8 Leaves in threes 9
   8 Leaves in fives 10
9 Leaves 10-15cm, bright green, rarely >1mm wide. Tree to 41m. Intrd-natd; planted in SW, rarely self-sown *Monterey Pine* - **P. radiata** D. Don
9 Leaves (10)15-25cm, dull green, ≤2mm wide. Tree to 40m. Intrd-surv; planted   *Western Yellow-pine* - **P. ponderosa** P. & C. Lawson
   10 Twigs with minute hairs at base of short-shoots; tip of female cone-scales scarcely thickened. Tree to 40m. Intrd-natd; planted, rarely self-sown   *Weymouth Pine* - **P. strobus** L.
   10 Twigs glabrous; tip of female cone-scales distinctly thickened 11
11 Leaves 7-12cm, directed towards tip of shoot; cones 8-15cm. Tree to 30m. Intrd-surv; planted   *Macedonian Pine* - **P. peuce** Griseb.
11 Leaves 8-20cm, pendent; cones mostly 15-25cm. Tree to 35m. Intrd-natd; planted, rarely self-sown   *Bhutan Pine* - **P. wallichiana** A. B. Jacks.

21. TAXODIACEAE - *Redwood family*

Evergreen or deciduous resiniferous trees; distinguished from Pinaceae in the indistinct bracts fused to the cone-scales; the usually >2 male and female sporangia; and the buds lacking proper bud-scales.

1 Leaves dorsiventrally flattened, not decurrent at base, spreading in
   1 plane on opposite sides of twigs **1. SEQUOIA**
1 Leaves tapering from decurrent base to apex, not flattened, disposed
   ± equally around stem 2
   2 Trunk with thick, spongy outer bark; leaves wider than thick;
      female cones >4cm, without recurved points at apex of scale
   2. SEQUOIADENDRON
   2 Trunk with thin, stringy outer bark; leaves thicker than wide;
      female cones <3cm, with recurved points at apex of scales
   3. CRYPTOMERIA

1. SEQUOIA Endl. - *Coastal Redwood*
Evergreen; trunk with thick, spongy, red-brown outer bark; leaves dorsiventrally flattened, not decurrent; male cones solitary at apex of twigs; female cones without recurved point at apex of scale, with 15-20 cone-scales.

1 Tree to 43m. Intrd-natd; planted, rarely self-sown (or suckering?)
   Coastal Redwood - **S. sempervirens** (D. Don) Endl.

2. SEQUOIADENDRON Buchholz - *Wellingtonia*
Evergreen; trunk with thick, spongy, red-brown outer bark; leaves tapering from decurrent base to pointed apex; male cones solitary at apex of twigs; female cones without recurved point at apex of scale, with 25-40 cone-scales.

1 Tree to 53m. Intrd-surv; planted
   Wellingtonia - **S. giganteum** (Lindl.) Buchholz

3. CRYPTOMERIA D. Don - *Japanese Red-cedar*
Evergreen; trunk with thin, stringy, red-brown outer bark; leaves tapering from decurrent base to pointed apex; male cones lateral, in groups behind apex of twigs; female cones with recurved points at apex of scale, with 20-30 cone-scales.

1 Shrub or tree to 37m. Intrd-natd; planted, rarely self-sown
   Japanese Red-cedar - **C. japonica** (L.f.) D. Don

22. CUPRESSACEAE - *Juniper family*

Evergreen resiniferous trees or shrubs; distinguished from other conifers by the opposite or whorled leaves and female cone-scales. Juvenile foliage is needle-like and patent, with leaves usually in whorls of 3 but sometimes opposite. Mature foliage is scale-like and appressed with opposite leaves.

1 Female cone berry-like at maturity, dispersed whole; leaves usually
   borne in threes **5. JUNIPERUS**
1 Female cone dry and woody at maturity, the scales opening to release
   the seeds; leaves usually opposite 2

2 Most female cones ≥2cm; leaves obtuse; ultimate branchlets not flattened, spreading in 3 dimensions **1. CUPRESSUS**
2 Female cones <2cm; leaves acute to acuminate; ultimate branchlets flattened, mainly or wholly spreading in 1 plane 3
3 Female cones elongated, with flattened scales; leaves with sweet aromatic smell when crushed **4. THUJA**
3 Female cones ± globose, with peltate scales; leaves with resinous or oily smell when crushed 4
    4 Female cones ≤12mm **3. CHAMAECYPARIS**
    4 Female cones mostly 15-20mm **2. X CUPRESSOCYPARIS**

1. CUPRESSUS L. - *Cypresses*
Twigs spreading in 3 dimensions, not flattened in 1 plane; juvenile foliage lost at very young stage; mature foliage with leaves opposite, scale-like, obtuse, appressed to twig; monoecious; female cones ripening in second year, ± globose, with 4-7 decussate pairs of peltate, woody cone-scales each with 8-20 narrowly winged seeds.

1 Tree to 37m. Intrd-natd; planted, self-sown in SW and CI
    Monterey Cypress - **C. macrocarpa** Gordon

2. X CUPRESSOCYPARIS Dallim. (*Cupressus* x *Chamaecyparis*) - *Leyland Cypress* (see *Chamaecyparis* key)
Variously intermediate between the parental genera; more similar to *Chamaecyparis* in vegetative characters but near half-way in cone characters; ultimate branchlets flattened, mainly spreading in 1 plane but some leaving it; female cones usually with 4 pairs of cone-scales and c.5-6 seeds per scale.

1 Tree to 34m. Intrd-surv; planted (*Cu. macrocarpa* x *Ch. nootkatensis*)
    Leyland Cypress - **X C. leylandii** (A. B. Jacks. & Dallim.) Dallim.

3. CHAMAECYPARIS Spach - *Cypresses*
Twigs spreading in 1 plane, flattened; juvenile foliage usually lost at very young stage; mature foliage with leaves opposite, ± scale-like, acute to acuminate, ± appressed to twig; monoecious; female cones usually ripening in first year, ± globose, with 3-6 decussate pairs of peltate, woody cone-scales each with 2-5 narrowly winged seeds.

Key to Chamaecyparis and X Cupressocyparis
1 Female cones with small prickle on each cone-scale; ultimate branchlets with white markings beneath 2
1 Female cones with prominent central conical spine on each cone-scale; ultimate branchlets without white markings beneath 3
    2 Leaves usually acuminate, with inconspicuous glands on back; female cones c.5-6mm. Tree to 26m. Intrd-natd; planted, occasionally self-sown
        Sawara Cypress - **Ch. pisifera** (Siebold & Zucc.) Siebold & Zucc.
    2 Leaves usually acute or narrowly acute, the dorsal row with a

conspicuous translucent gland on back; female cones c.7-9mm.
Tree to 41m. Intrd-natd; much planted, commonly self-sown
Lawson's Cypress - **Ch. lawsoniana** (A. Murray bis) Parl.
3 Ultimate branchlets pendent; female cones 8-12mm. Tree to 30m.
Intrd-surv; planted   *Nootka Cypress* - **Ch. nootkatensis** (Lamb.) Spach
3 Ultimate branchlets patent to erect; female cones mostly 15-20mm
(see above) **X Cu. leylandii**

4. THUJA L. - *Red-cedars*
Differs from *Chamaecyparis* in the elongated female cones with usually 10-12 flattened cone-scales each with a recurved apical spike and 2-3 seeds.

1 Tree to 42m. Intrd-natd; much planted, frequently self-sown
Western Red-cedar - **T. plicata** D. Don

5. JUNIPERUS L. - *Junipers*
Twigs spreading in 3 dimensions, not flattened in 1 plane; foliage all juvenile; leaves erect to patent, usually borne in whorls of 3; usually dioecious; female cones berry-like, with succulent ± fused cone-scales; seeds not winged, 1 per cone-scale, retained within cone at dispersal.

1 Tree or shrub to 7(16)m. Native; little planted (*Common Juniper* - **J. communis** L.)   2
   2 Leaves up to 15mm, erecto-patent, acute to obtuse with a mucronate tip. Procumbent matted shrub. W & NW Br and Ir, mostly in upland areas   **J. communis ssp. nana** (Hook.) Syme
   2 Leaves up to 20mm, ± patent, acuminate with a sharp point   3
3 Leaves 1-1.5mm wide, loosely and irregularly spaced, with a greyish-white stomatal band. Spreading shrub to erect tree. Very local in Br and Ir, on limestone and acid soils   **J. communis ssp. communis**
3 Leaves 1.3-2mm wide, closely and regularly spaced, with a pure white stomatal band. Low compact shrub. Maritime cliffs in SW Br
J. communis ssp. hemisphaerica (J. & C. Presl) Nyman

23. ARAUCARIACEAE - *Monkey-puzzle family*

Evergreen resiniferous trees; differs from other conifers in the broad many-veined leaves.

1. ARAUCARIA Juss. - *Monkey-puzzle*

1 Tree to 30m. Intrd-natd; commonly planted, very rarely self-sown
Monkey-puzzle - **A. araucana** (Molina) K. Koch

24. TAXACEAE - *Yew family*

Evergreen non-resiniferous trees or shrubs; distinguished from other conifers by the lack of resin and the single ovules with a red succulent aril after fertilization.

1. TAXUS L. - *Yew*

1 Large bush or spreading tree to 28m. Native; very local throughout Br and Ir, very rare in Sc, also very widely grown and self-sown
Yew - **T. baccata** L.

MAGNOLIOPSIDA — ANGIOSPERMS, FLOWERING PLANTS

Trees, shrubs, climbers, herbs or variously reduced plants of extremely varied growth-form. Male sporangia borne in specialized organs (stamens) grouped 1-many in male or bisexual flowers. Female sporangia borne in ovules enclosed 1-many together in carpels; carpels grouped 1-many in female or bisexual flowers. Female gametophyte greatly reduced and retained in ovule. Fertilized ovule (seed) retained within carpel on sporophyte until ripe, then dispersed separately or within carpel (fruit).

KEYS TO FAMILIES OF MAGNOLIOPSIDA
(Dicotyledons and Monocotyledons)

Before using these keys the section on Identification Keys in the Introduction should be read carefully. In particular, it is essential to work out fully the structure of the flowers *before* starting on the keys. The keys have been made as user-friendly as possible, but it must be admitted that pitfalls still abound. Only after considerable experience should the supplementary keys (A, B, etc.) be used directly; it is advisable to start always with the General key. In either case it is of course of paramount importance to arrive at the correct supplementary key; hence the General key should be used very carefully and a good working knowledge of it be built up.

The distinctions between 2 perianth whorls that are similar as opposed to different, and between sepaloid and petaloid perianth segments, are subjective. Wherever the answer is considered equivocal both alternatives are allowed for here. In some cases, however, the inner and outer whorls of perianth segments are quite different, but a lens is needed to discover this due to their small size (e.g. *Empetrum*, *Ruscus*) and careless observation will produce the wrong answer.

Because of the variation shown, many families are keyed out in several different positions in up to 5 different supplementary keys. In all cases where only 1 genus from a family containing >1 genus is referred to, that genus is stated.

General key
1   Plants consisting of floating or submerged ± undifferentiated pad-like fronds ≤10(15)mm, sometimes with narrow stalk-like part at 1 end, with or without roots dangling in water (rarely stranded temporarily on mud) (beware *Azolla*)   **152. LEMNACEAE**

1   If plants free-floating then with clearly differentiated stems and leaves   2
- 2   Aquatics with some leaves or parts of leaves modified as small bladders to catch minute animals (*Utricularia*)   **132. LENTIBULARIACEAE**
- 2   Leaves never modified as small bladders   3

3   Aquatic or mud plants with at least some leaves in whorls of ≥3, the leaves linear or ± so or divided into linear segments   ***Key A***

3   If aquatic or on mud then leaves not whorled and/or not linear or with linear segments   4
- 4   Woody plant parasitic on aerial parts of trees, the roots buried in living host branches   **89. VISCACEAE**
- 4   If growing on aerial parts of trees then merely epiphytic, with roots not buried in living host branches   5

5   Trees with unbranched stem and terminal rosettes of huge pinnate or palmate leaves; or seedlings with leaves ribbed alternately on each surface (scarcely natd aliens not treated here)   **(ARECACEAE)**

5   If trees then not with single terminal rosette of compound leaves; if seedlings then not with leaves ribbed alternately on each surface   6
- 6   Plant consisting of 1-few rosettes of many linear simple leaves usually ≥1m, either borne on ground or at tips of woody branches   **164. AGAVACEAE**
- 6   If leaves all in 1 or few rosettes then much <1m and often not linear   7

7   Plant consisting of dense hemispherical mass often ≥1m across, with narrow pineapple-like leaves with strongly spiny margins   **160. BROMELIACEAE**

7   If plant in a dense hemispherical spiny mass then without pineapple-like leaves   8
- 8   All inflorescences entirely replaced by vegetative propagules (bulbils or plantlets)   ***Key B***
- 8   At least some inflorescences bearing flowers (or fruits)   9

9   Perianth 0, or of 1 whorl, or of ≥2 whorls or a spiral of similar segments   10

9   Perianth of 2 (rarely more) distinct whorls or rarely a spiral, the inner and outer differing markedly in shape/size/colour   12
- 10   Perianth ± corolla-like, usually white or distinctly (often brightly) coloured   ***Key C***
- 10   Perianth ± calyx-like or bract-like (greenish to brownish, or scarious or much reduced) or 0   11

11   Trees or shrubs, sometimes very short or procumbent but then with woody stems producing growth in subsequent years   ***Key D***

11   Herbs, with non-woody (or only basally woody) stems dying to ground level after 1(-few) year(s)   ***Key E***
- 12   Flowers all male or all female, the non-functional parts 0 or extremely vestigial   ***Key F***
- 12   At least some flowers bisexual or both male and female flowers present   13

13   Petals fused at base for varying distances to apex   14

13	Petals free, or rarely ± fused just at basal point, or rarely fused above base but free at base	15
	14 Ovary superior (hypogynous or perigynous)	**Key G**
	14 Ovary inferior (epigynous) or partly so	**Key H**
15	Ovary inferior (epigynous) or partly so	**Key I**
15	Ovary superior (hypogynous or perigynous)	16
	16 Carpels and styles free, or carpels fused just at base	**Key J**
	16 Carpels and/or styles fused wholly or for greater part, or carpel 1	**Key K**

Key A - Aquatic or mud plants with whorled leaves, the leaves either linear or divided into linear segments (beware *Equisetum*, with whorled lateral branches)

1	Leaves simple, sometimes prominently toothed	2
1	Leaves divided (forked or pinnate)	6
	2 Tip of stems erect and emergent at flowering; leaves in whorls of 4-12	3
	2 Stems always submerged (unless stranded by drought); leaves opposite or in whorls of 3-6(8)	4
3	Leaves entire, 6-12 per whorl	**123. HIPPURIDACEAE**
3	Leaves with few to many pricklets on margins, 4-8 per whorl (*Galium*)	**134. RUBIACEAE**
	4 Stipules free from leaf-base	**149. ZANNICHELLIACEAE**
	4 Leaves without stipules, often with sheathing leaf-base or with 2 minute scales at base in axil	5
5	Leaves with distinctly widened base shortly sheathing stem	**148. NAJADACEAE**
5	Leaves wider near base than near apex but narrowed at extreme base, sometimes slightly clasping stem but not sheathing it	**142. HYDROCHARITACEAE**
	6 Leaves forked 1-4 times	**29. CERATOPHYLLACEAE**
	6 Leaves 1-2-pinnate	7
7	Leaves with flat segments; flowers conspicuous, >15mm across (*Hottonia*)	**71. PRIMULACEAE**
7	Leaves with filiform segments; flowers inconspicuous, <6mm across (*Myriophyllum*)	**81. HALORAGACEAE**

Key B - Plants with all inflorescences entirely replaced by vegetative propagules

1	Leaves reniform, palmately lobed (*Saxifraga*)	**76. SAXIFRAGACEAE**
1	Leaves linear to lanceolate, entire	2
	2 Inflorescences proliferating by producing small plantlets in place of flowers	3
	2 Inflorescences replaced by axillary solitary or terminal clusters of small solid structures ('bulbils')	4
3	Stems with a central pith; leaves unifacial, flattened-cylindrical (*Juncus*)	**155. JUNCACEAE**
3	Stems hollow; leaves bifacial, flat or inrolled or infolded	**157. POACEAE**

     4   Bulbils sessile, forming ± globose compact terminal head
         (*Allium*) or ± globose mass at ground level (*Gagea*)   **162. LILIACEAE**
     4   Bulbils on slender stalks, solitary in leaf-axils (*Lysimachia*)
                                                             71. PRIMULACEAE

Key C - Perianth of 1 whorl, or of ≥2 whorls or a spiral of ± similar
         segments, ± corolla-like (usually white or distinctly coloured)
         (beware taxa with distinct inner and outer perianth whorls but 1 or
         other soon falling, e.g. Papaveraceae)
1   Flowers bisexual or monoecious   2
1   Flowers dioecious   32
     2   Ovary inferior (flowers epigynous)   3
     2   Ovary superior (flowers hypogynous to perigynous)   14
3   Stamens ≥6   4
3   Stamens ≤5   5
     4   Leaves ovate or broadly so, cordate at base; ovary 6-celled;
         styles/stigmas 6   **26. ARISTOLOCHIACEAE**
     4   Leaves narrow, gradually tapered to base; ovary 3-celled;
         styles/stigmas 1 or 3   **162. LILIACEAE**
5   Stamens 1-3   6
5   Stamens 4-5 (some may drop very early)   8
     6   Perianth with 5 segments; ovule 1 (or 0) per cell
                                                        137. VALERIANACEAE
     6   Perianth with 6 segments; ovules numerous per cell   7
7   Style obvious, with 3 obvious branches or 3 separate stigmas;
     stamens 3   **163. IRIDACEAE**
7   Styles 0; stamens 1 or 2   **166. ORCHIDACEAE**
     8   Leaves simple, in whorls of ≥4   **134. RUBIACEAE**
     8   Leaves not in whorls of ≥4   9
9   Petals free, inserted on top of ovary   10
9   Petals fused into tube at least proximally   12
     10 Stamens 4; carpels and styles 1; herbaceous perennial
         (*Sanguisorba*)   **77. ROSACEAE**
     10 Stamens 5; carpels 2-5; styles 2 or 5, or if 1 then woody evergreen 11
11   Fruit a 2-celled dry schizocarp; styles 2; mostly herbaceous, if woody
     then leaves simple, entire and with a clear midrib   **111. APIACEAE**
11   Fruit a berry; styles 1 or 5; mostly woody   **110. ARALIACEAE**
     12 Flowers borne in open raceme-like cymes   **88. SANTALACEAE**
     12 Flowers borne in dense capitula   13
13   Stamens 4, free, ± exserted; ovary and fruit surrounded by epicalyx
                                                        138. DIPSACACEAE
13   Stamens 5, their anthers fused into tube round style; ovary and fruit
     not surrounded by epicalyx   **139. ASTERACEAE**
     14 Tree or shrub or woody climber   15
     14 Herb   19
15   Tepals 10; stamens 5   **98. STAPHYLEACEAE**
15   Tepals 3-8; stamens 6-many or flowers female   16
     16 Climber or scrambler   17

16	Strongly self-supporting shrub	18
17	Leaves simple; stipules scarious, fused round stem; stamens 6-9	
		49. POLYGONACEAE
17	Leaves usually ternate or pinnate; stipules 0; stamens >10 (*Clematis*)	
		30. RANUNCULACEAE
	18 Tepals free, creamy- to greenish-yellow; flowers dioecious; leaves evergreen, sweetly scented when crushed **25. LAURACEAE**	
	18 Tepals fused into tube (or arising from tubular hypanthium), purple; flowers bisexual; leaves deciduous, not sweetly scented	
		84. THYMELAEACEAE
19	Carpels ≥2, free or ± so	20
19	Carpels 1, or >1 and fused	25
	20 Aquatic plant with finely dissected submerged and entire floating leaves **28. CABOMBACEAE**	
	20 Not an aquatic plant with dissected submerged and entire floating leaves	21
21	Carpels each with 1-2 ovules	22
21	Carpels each with several to many ovules	23
	22 Flowers in racemes; fruit succulent **43. PHYTOLACCACEAE**	
	22 Flowers solitary or in cymes; fruit not succulent	
		30. RANUNCULACEAE
23	Tepals 1(-2), white; inflorescence a forked spike just above water-surface **143. APONOGETONACEAE**	
23	Tepals ≥5, usually coloured; inflorescence not a forked spike	24
	24 Leaves simple, linear, without petiole; inflorescence an umbel	
		140. BUTOMACEAE
	24 Leaves compound and/or with well developed petiole; inflorescence not an umbel **30. RANUNCULACEAE**	
25	Stamens >10	26
25	Stamens 3-9	27
	26 Leaves compound **30. RANUNCULACEAE**	
	26 Leaves simple, entire **43. PHYTOLACCACEAE**	
27	Ovary 1-3-celled, each cell with 2-many ovules	28
27	Ovary 1-celled (or 3-celled with 2 cells ± aborted and empty), with 1 ovule	29
	28 Ovary 1-celled; tepals and stamens 5 (*Glaux*) **71. PRIMULACEAE**	
	28 Ovary 3-celled; tepals and stamens 4 or 6 **162. LILIACEAE**	
29	Leaves pinnate (*Sanguisorba*) **77. ROSACEAE**	
29	Leaves simple, usually entire	30
	30 Perianth >15mm across, with tube >2cm (scarcely natd aliens not treated here) **(NYCTAGINACEAE)**	
	30 Perianth <10mm across, with tube <1cm (often 0)	31
31	Emergent aquatic with blue flowers; tepals and stamens always 6	
		161. PONTEDERIACEAE
31	Flowers blue; tepals mostly 5; stamens (4)8(-9) **49. POLYGONACEAE**	
	32 Flowers in dense capitula closely surrounded by ≥1 row of bracts **139. ASTERACEAE**	
	32 Flowers not in dense capitula	33

33	Tepals 4 or 6	34
33	Tepals 5	35
	34 Tree or shrub; tepals 4; leaves elliptic	**25. LAURACEAE**
	34 Herb; tepals 6; leaves reduced to scales, replaced by cylindrical cladodes (*Asparagus*)	**162. LILIACEAE**
35	Leaves simple, entire or ± so	**49. POLYGONACEAE**
35	At least leaves on stem pinnate or deeply pinnately lobed	36
	36 Tepals free; basal leaves pinnate (*Trinia*)	**111. APIACEAE**
	36 Tepals fused into tube proximally; basal leaves simple and ± entire (*Valeriana*)	**137. VALERIANACEAE**

Key D - Perianth 0 or of 1 or more whorls or a spiral of ± similar segments, ± calyx-like or bract-like (usually greenish or brownish or scarious); plants with woody stems

1	Leaves pinnate or ternate	2
1	Leaves simple, often deeply lobed	7
	2 Stems spiny (*Aralia*)	**110. ARALIACEAE**
	2 Stems not spiny	3
3	Shrubs <2m; flowers in dense capitula	4
3	Trees >2m; flowers various	5
	4 Stems ± erect; flowers in clusters of ± bell-shaped capitula (*Artemisia*)	**139. ASTERACEAE**
	4 Woody stems procumbent; flowers in spherical capitula solitary at apex of erect herbaceous stem (*Acaena*)	**77. ROSACEAE**
5	Leaves alternate or spiral; fruit a nut, sometimes with wing ± surrounding it; monoecious	**39. JUGLANDACEAE**
5	Leaves opposite or ± so; fruit 1 or 2 achenes each with elongated wing on 1 side; often dioecious or ± so	6
	6 Stamens c.8; fruit of 2 achenes each with elongated wing	**101. ACERACEAE**
	6 Stamens 2; fruit of 1 achene with elongated wing	**127. OLEACEAE**
7	Leaves opposite or ± so	8
7	Leaves alternate or spiral	12
	8 Plants dioecious; male and female flowers in dense usually elongated catkins (*Salix*)	**63. SALICACEAE**
	8 Plants dioecious or not; flowers sometimes in pendent panicles (if so not dioecious) but not in dense catkins	9
9	Leaves entire	10
9	Leaves palmately lobed or serrate	11
	10 Most leaves >5cm; tepals fused into tube proximally (*Coprosma*)	**134. RUBIACEAE**
	10 Leaves all <3cm; tepals 0 or ± free (*Buxus*)	**92. BUXACEAE**
11	Leaves palmately lobed, the lobes entire to serrate; fruit of 2 winged achenes	**101. ACERACEAE**
11	Leaves not lobed, serrate; fruit a berry (*Rhamnus*)	**94. RHAMNACEAE**
	12 Stems scrambling or climbing	13
	12 Plant a self-supporting shrub or tree, but sometimes dwarf	14
13	Leaves evergreen, palmately veined and often palmately lobed;	

MAGNOLIOPSIDA 41

    fruit a black or rarely yellow berry (*Hedera*)   **110. ARALIACEAE**
13 Leaves deciduous, pinnately veined, not lobed; fruit and achenes surrounded by white succulent tepals (*Muehlenbeckia*)
    49. POLYGONACEAE
    14 Flowers borne on inside of hollow receptacles that become succulent in fruit (*Ficus*)   **37. MORACEAE**
    14 Flowers not borne on inside of hollow receptacles   15
15 At least male flowers (if dioecious then female flowers also) in pendent or rigid catkins or in pendent tassels or globular heads   16
15 Flowers not in catkins, if in tight groups then not pendent   22
    16 Male and female flowers in separate, spherical, pendent capitula; leaves with petiole hollow at base and forming cap over axillary bud   **34. PLATANACEAE**
    16 Flowers not in spherical pendent capitula; base of petiole not concealing axillary bud   17
17 Leaves densely mealy; male and female flowers in same catkin (*Atriplex*)   **45. CHENOPODIACEAE**
17 Leaves glabrous to pubescent, not mealy; male and female flowers in separate catkins or heads (or dioecious)   18
    18 Leaves dotted with translucent glands, with strong aromatic scent when crushed   **40. MYRICACEAE**
    18 Leaves not gland-dotted   19
19 Fresh stems and leaves with latex; fruits red to black, succulent (*Morus*)   **37. MORACEAE**
19 Latex absent; fruits not succulent   20
    20 Ovary 1-celled, with many ovules; fruit a capsule with many plumed seeds; or flowers all male   **63. SALICACEAE**
    20 Ovary 2-6-celled, each cell with 1-2 ovules; fruit a nut, sometimes borne in husk formed from enlarged scales   21
21 Ovary 3- or 6-celled, with 3-9 styles   **41. FAGACEAE**
21 Ovary 2-celled, with 2 styles   **42. BETULACEAE**
    22 Leaves palmately lobed   23
    22 Leaves not palmately lobed   24
23 Leaves peltate; fruit a capsule; ovary superior (*Ricinus*)
    93. EUPHORBIACEAE
23 Leaves not peltate; fruit a berry; ovary inferior (*Fatsia*)
    110. ARALIACEAE
    24 Leaves with dense ± sessile scales at least on lowerside, appearing mealy   **80. ELAEAGNACEAE**
    24 Leaves without scales, glabrous to pubescent   25
25 Leaves <20 x 2mm, succulent (*Suaeda*)   **45. CHENOPODIACEAE**
25 At least most leaves >20mm and >2mm wide, not succulent   26
    26 Flowers in compound umbels, epigynous (*Bupleurum*)
    111. APIACEAE
    26 Flowers not in compound umbels, hypogynous or perigynous   27
27 Stamens 8-12, or flowers all female with 1-celled ovary   28
27 Stamens 4-5, or flowers all female with 2-4-celled ovary   29
    28 Dioecious; tepals longer than perianth-tube and falling from

  it after flowering   **25. LAURACEAE**
 28 Flowers bisexual; tepals shorter than perianth-tube and not falling separately from it   **84. THYMELAEACEAE**
29 Flowers all bisexual; ovary 1-celled; fruit a winged achene; leaves deciduous   **35. ULMACEAE**
29 Flowers monoecious or bisexual to dioecious; ovary 2-4-celled or 0; fruit succulent or ± so; leaves evergreen   30
  30 Shrub to 5m; fruit black, without persistent styles (*Rhamnus*)
     94. RHAMNACEAE
  30 Dwarf shrub to 25cm; fruit whitish, with 2 persistent horn-like styles (*Pachysandra*)   **92. BUXACEAE**

Key E - Perianth 0 or of 1 or more whorls or a spiral of ± similar segments, ± calyx-like or bract-like (usually greenish or brownish or scarious); herbs
1 Flowers numerous in dense capitula closely surrounded by sepal-like bracts   **139. ASTERACEAE**
1 Flowers not in dense capitula, or if so then not closely surrounded by sepal-like bracts   2
  2 Leaves in whorls of ≥4 (*Rubia*)   **134. RUBIACEAE**
  2 Leaves not whorled or in whorls of 3   3
3 Leaves at least partly opposite or whorled; aquatic or marsh plants with floating, procumbent or very weakly ascending stems   4
3 Leaves all alternate or all basal, or if some or all opposite then plant not aquatic, or if so then stems self-supporting   12
  4 Leaves fused in opposite pairs, forming succulent sheath round stem   **45. CHENOPODIACEAE**
  4 Leaves not fused in succulent sheath round stem   5
5 Tepals 0; stamens 1-4 (or flowers female)   6
5 Tepals 4-6 on at least some flowers; stamens 4-12 (or flowers female)   8
  6 Flowers bisexual; stamens 4; fruits on stalks >10mm; only upper leaves opposite   **147. RUPPIACEAE**
  6 Monoecious; male flowers with 1-2 stamens; fruits on stalks ≤10mm; ± all leaves opposite or in 3s   7
7 Stigmas linear, 2 per ovary; ovary developing into 4 nutlets in 2 pairs
     124. CALLITRICHACEAE
7 Stigmas peltate, 1 per carpel; 1-4(more) carpels per flower developing into nutlet   **149. ZANNICHELLIACEAE**
  8 Tepals 5-6 (or 0 in female flowers)   9
  8 Tepals 4   10
9 Flowers monoecious, hypogynous; tepals 5 in male flowers, 0 in female flowers; female flowers and fruit with 2 prominent basal bracteoles   **45. CHENOPODIACEAE**
9 Flowers bisexual, perigynous; tepals 6; flowers without bracteoles
     83. LYTHRACEAE
  10 Flowers in terminal flat-topped cymes; stamens 8 (*Chrysosplenium*)   **76. SAXIFRAGACEAE**
  10 Flowers solitary or in spikes in leaf-axils; stamens 4   11

MAGNOLIOPSIDA

11 Flowers in long-stalked axillary spikes   **146. POTAMOGETONACEAE**
11 Flowers solitary and sessile in leaf-axils (*Ludwigia*)   **86. ONAGRACEAE**
    12 Flowers greatly reduced, arranged in units largely composed of leafy or membranous scaly bracts, with perianth 0 or represented by bristles or minute scales, aerial; leaves linear, grass-like, sheathing the stem proximally   13
    12 Flowers with obvious structure, mostly with perianth, if greatly reduced with 0 or obscure perianth then not arranged in units as above and often subaquatic or on water surface; leaves various   14
13 Flowers with bract above as well as below (if not then stems hollow); stems usually with hollow internodes, circular or rarely compressed or ± quadrangular in section; leaf-sheaths usually with free overlapping margins   **157. POACEAE**
13 Flowers never with bract above; stems usually with solid internodes, often ± triangular in section; leaf-sheaths usually cylindrical, with fused margins   **156. CYPERACEAE**
    14 Aquatic or marsh plants with linear leaves   15
    14 If leaves linear then plants not in water or marshes; if aquatic then leaves not linear   26
15 Leaves all basal; inflorescence a tight capitate mass on long leafless stem   **154. ERIOCAULACEAE**
15 If leaves all basal, then inflorescence not a single terminal tight capitate mass   16
    16 Flowers very small, many tightly packed in dense spherical or elongated conspicuous clusters   17
    16 Flowers not many together in dense clusters   19
17 Flowers bisexual; fresh leaves with strong spicy scent when crushed (*Acorus*)   **151. ARACEAE**
17 Flowers unisexual, the male and female in clearly separated parts of inflorescence; leaves without spicy scent   18
    18 Flowers in globose heads   **158. SPARGANIACEAE**
    18 Flowers in cylindrical spikes   **159. TYPHACEAE**
19 Leaves very thin, ribbon- or thread-like, mostly subaquatic   20
19 Leaves thicker, not ribbon- or thread-like   22
    20 Flowers bisexual, in stalked spikes   **146. POTAMOGETONACEAE**
    20 Flowers dioecious or monoecious, in stalked or sessile spathes   21
21 Flowers dioecious, in short- or long-stalked spathes; tepals 3; freshwater (*Vallisneria*)   **142. HYDROCHARITACEAE**
21 Flowers monoecious, in sessile spathes; tepals 0; marine   **150. ZOSTERACEAE**
    22 Tepals 4-5, or 0   23
    22 Tepals 6   24
23 Tepals 5 (or 0 in female flowers); leaves alternate   **45. CHENOPODIACEAE**
23 Tepals 4; leaves all basal (*Subularia*)   **64. BRASSICACEAE**
    24 Flowers in branched cymes, sometimes compact   **155. JUNCACEAE**
    24 Flowers in a simple terminal raceme   25
25 Leaves on stems 0 or few and near base, without pore at apex;

	flowers numerous, without bracts **145. JUNCAGINACEAE**
25	Leaves several on stems, with prominent pore at apex; flowers <12, with bracts **144. SCHEUCHZERIACEAE**
26	Flowers small, in dense spike on axis, the axis sometimes extended distally as sterile projection, with large spathe at base often partly or wholly obscuring flowers **151. ARACEAE**
26	If flowers in single dense spike then without large spathe at base 27
27	Stems entirely rhizomatous, producing large (usually >1m across) leaves and huge (usually >50cm) elongated dense panicles **82. GUNNERACEAE**
27	If stems entirely rhizomatous then leaves <10cm and flowers solitary or in whorls or umbels 28
28	Inflorescence consisting of units arranged in umbels, each unit consisting of several male flowers (each of 1 stamen) and 1 female flower (of 1 stalked ovary) all surrounded by 4-5 conspicuous glands; plants with copious white latex (*Euphorbia*) **93. EUPHORBIACEAE**
28	Inflorescence not consisting of units as above; plants without copious white latex 29
29	Leaves opposite; stems procumbent to weakly ascending 30
29	If leaves opposite then stems self-supporting and ± erect 32
30	Tepals 5, free **48. CARYOPHYLLACEAE**
30	Tepals 3-4, fused 31
31	Leaves ≤15mm; perianth 4-lobed, greenish (*Nertera*) **134. RUBIACEAE**
31	Leaves >20mm; perianth 3-lobed, brownish-purple (*Asarum*) **26. ARISTOLOCHIACEAE**
32	Plants aquatic or in wet bogs; leaves simple, entire or ± so **146. POTAMOGETONACEAE**
32	Plants usually on dry ground, if in marshes or bogs then leaves not simple and entire 33
33	Perianth of 6 lobes or segments, 3 in outer and 3 in inner whorl 34
33	Perianth lobes or segments 1, or 2-5, or 4-5 in each of 2 whorls, rarely 6 and then not in 2 whorls of 3 37
34	Leaves or leaf-like organs linear, without basal lobes 35
34	Leaves broader than linear, or if not then with basal lobes 36
35	Dioecious; leaves reduced to scales, their normal function replaced by clusters of 4-10(more) cladodes (*Asparagus*) **162. LILIACEAE**
35	Flowers bisexual; leaves reduced to scales or not, but not replaced by clusters of cladodes **155. JUNCACEAE**
36	Twining climber; ovary inferior; fruit a red berry **165. DIOSCOREACEAE**
36	Not climbing; ovary superior; fruit an achene (*Rumex*) **49. POLYGONACEAE**
37	Flowers epigynous, semi-epigynous, or perigynous with deeply concave hypanthium 38
37	Flowers hypogynous, or perigynous with flat to saucer-shaped hypanthium, or flowers all male (dioecious) 44
38	Ovary with 3-8 cells; stigmas 3-8 39

MAGNOLIOPSIDA

38 Ovary with 1-2 cells; styles/stigmas 1-2 40
39 Leaves simple, entire; inflorescence small, axillary (*Tetragonia*)
    44. AIZOACEAE
39 Leaves 1-2-pinnate, inflorescence large, conspicuous, terminal
   (*Aralia*)   **110. ARALIACEAE**
   40 Tepals 4, or 4 plus 4 epicalyx segments beneath; stamens 1-4, or ≥8 41
   40 Tepals 5; stamens 5 42
41 Stamens 1-4, or ≥10; fruit 1-many achenes   **77. ROSACEAE**
41 Stamens 8; fruit a capsule (*Chrysosplenium*)   **76. SAXIFRAGACEAE**
   42 Ovary 2-celled; fruit a 2-celled schizocarp   **111. APIACEAE**
   42 Ovary 1-celled; fruit a 1-celled achene 43
43 Leaves linear or ± so   **88. SANTALACEAE**
43 Leaves ovate to lanceolate or deltate (*Beta*) **45. CHENOPODIACEAE**
   44 Tepals 2; stamens 12-18, conspicuous (*Macleaya*)
    32. PAPAVERACEAE
   44 Tepals 1, or 3-5, or 5 with 5 epicalyx segments beneath (or
     sometimes 0 in female flowers), rarely 2 and then stamens also 2 45
45 Leaves opposite; flowers often dioecious 46
45 Leaves alternate; flowers rarely dioecious 50
   46 Leaves deeply palmately lobed to ± palmate **36. CANNABACEAE**
   46 Leaves simple, at most toothed 47
47 Leaves ≥1.5cm; stigmas branched or conspicuously papillose; at
   least male flowers in axillary spikes (catkins) 48
47 Leaves <1.5cm; stigmas ± smooth or minutely papillose; flowers
   not in spikes 49
   48 Leaves usually with strong stinging hairs; tepals 4; fruit an
    achene (*Urtica*)   **38. URTICACEAE**
   48 Leaves without stinging hairs; tepals 3; fruit a 2-celled capsule
    (*Mercurialis*)   **93. EUPHORBIACEAE**
49 Tepals 3; leaves with conspicuous stipules (*Koenigia*)
   49. POLYGONACEAE
49 Tepals 4-5; leaves without stipules   **48. CARYOPHYLLACEAE**
   50 Gynoecium composed of 1 carpel with 1 ovule, producing a
    1-seeded fruit, or all flowers male 51
   50 Gynoecium composed of 2-many (often fused) carpels, with
    many ovules in total, producing a many-seeded fruit or many
    1-seeeded fruits 55
51 Leaves palmately lobed to base or ± so (*Cannabis*) **36. CANNABACEAE**
51 Leaves simple, entire or pinnately lobed but not ± to base 52
   52 Leaves with well-developed stipules fused into short tube round
    stem   **49. POLYGONACEAE**
   52 Leaves without stipules 53
53 Tepals (2)3-5, scarious, with 3-5 often similar bracteoles just below;
   fruit often dehiscent   **46. AMARANTHACEAE**
53 Tepals 0-5, if >1 then herbaceous; flowers often without bracteoles;
   fruit always indehiscent 54
   54 Tepals 1 or 3-5 or 0; styles 2-3, each with smooth to papillate

	stigma	**45. CHENOPODIACEAE**
54	Tepals 4; style 1; stigma much branched	**38. URTICACEAE**
55	Ovary of 2 carpels completely fused; tepals 4	**64. BRASSICACEAE**
55	Ovary of 2-many carpels, free or <1/2 fused, if >1/2 fused then tepals 5	56
	56 Each carpel with several to many ovules, producing several to many seeds	57
	56 Each carpel with 1-2 ovules, producing 1 seed	58
57	Leaves palmate or ± so; flowers >1cm across (*Helleborus*)	**30. RANUNCULACEAE**
57	Leaves 2-ternate to 2-pinnate; flowers <1cm across (*Astilbe*)	**76. SAXIFRAGACEAE**
	58 Carpels fused, elongated distally into sterile column ending in 1 style with 5 stigmas	**106. GERANIACEAE**
	58 Carpels free, each with 1 stigma	59
59	Tepals 5, with 5 epicalyx segments just beneath (*Sibbaldia*)	**77. ROSACEAE**
59	Tepals 4-5, without epicalyx	**30. RANUNCULACEAE**

Key F - Perianth of 2 (rarely more) distinct whorls or rarely a spiral, the inner and outer differing markedly in shape/size/colour; flowers all male or all female

1	Tree or shrub, sometimes procumbent but with distinctly woody stems	2
1	Herb	11
	2 Leaves pinnate; tree with white flowers (*Fraxinus*)	**127. OLEACEAE**
	2 Leaves simple, or if pinnate then shrub with yellow flowers	3
3	Leaves all scale-like, functionally replaced by leaf-like stem outgrowths bearing flowers on their faces (*Ruscus*)	**162. LILIACEAE**
3	Leaves photosynthetic, not bearing flowers on their faces	4
	4 Stems procumbent; leaves *Erica*-like, with revolute margins	**66. EMPETRACEAE**
	4 Stems ascending, spreading or erect; leaves not *Erica*-like	5
5	Leaves stipulate, either pinnate or simple and entire	**77. ROSACEAE**
5	Leaves without stipules, or if stipulate then simple and serrate	6
	6 Leaves palmately lobed (*Ribes*)	**74. GROSSULARIACEAE**
	6 Leaves entire to serrate or shallowly spinose-pinnately lobed	7
7	Leaves gland-dotted, <2cm (scarcely natd aliens not treated here)	**(MYRSINACEAE)**
7	Leaves not gland-dotted, most or all >2cm	8
	8 Corolla white; at least some leaves usually with spine-tipped teeth	**91. AQUIFOLIACEAE**
	8 Corolla purple or greenish-yellow; leaves not spiny	9
9	Flowers ± hypogynous, with large nectar-secreting disc at base	**90. CELASTRACEAE**
9	Flowers clearly epigynous or perigynous, the inferior ovary or hypanthium distinct even in male flowers	10

MAGNOLIOPSIDA

10 Leaves closely serrate, usually stipulate (*Rhamnus*)
94. RHAMNACEAE
10 Leaves entire or remotely serrate, without stipules
87. CORNACEAE
11 Climbing plant bearing tendrils (*Bryonia*) **62. CUCURBITACEAE**
11 Plant not climbing, without tendrils 12
   12 Aquatic or marsh plants with all leaves in basal rosette(s) 13
   12 If aquatic or marsh plants then all leaves not in basal rosettes 14
13 Petals 3, free, conspicuous, white **142. HYDROCHARITACEAE**
13 Petals 4, fused, inconspicuous, ± scarious (female flowers present low down in leaf-rosette, easily missed) (*Littorella*)
125. PLANTAGINACEAE
   14 Sepals 3; petals 3 15
   14 Sepals and petals not both 3 16
15 Leaves reduced to scales, replaced by bunches of ± cylindrical cladodes (*Asparagus*) **162. LILIACEAE**
15 Leaves flat, with 2 basal lobes (*Rumex*) **49. POLYGONACEAE**
   16 Flowers in dense capitula closely surrounded by ≥1 row of bracts **139. ASTERACEAE**
   16 Flowers not in dense capitula though sometimes crowded 17
17 At least some leaves ternate to pinnate 18
17 All leaves simple (sometimes lobed) 21
   18 Basal leaves simple; stem-leaves 1-pinnate (*Valeriana*)
137. VALERIANACEAE
   18 All leaves compound, at least some 2-3-pinnate or 1-3-ternate 19
19 All leaves 1-ternate (*Fragaria*) **77. ROSACEAE**
19 At least some leaves 2-3-pinnate or 2-3-ternate 20
   20 Stamens ≤10; carpels usually 2; plant <1m (*Astilbe*)
76. SAXIFRAGACEAE
   20 Stamens >10; carpels usually 3; plant usually ≥1m (*Aruncus*)
77. ROSACEAE
21 Leaves petiolate, truncate to cordate at base 22
21 Leaves sessile or ± so, rounded to cuneate at base 23
   22 Leaves not lobed, entire; tepals 5, all ± white; plant >(25)100cm (*Fallopia*) **49. POLYGONACEAE**
   22 Leaves lobed, serrate; sepals 5, green; petals 5, white; plant <25cm (*Rubus*) **77. ROSACEAE**
23 Leaves alternate; petals 4; stamens 8 or ovaries 4 (*Sedum*)
75. CRASSULACEAE
23 Leaves opposite; petals 5; stamens usually 10 or ovary 1
48. CARYOPHYLLACEAE

Key G - Perianth of 2 (rarely more) distinct whorls or rarely a spiral, the inner and outer differing markedly in shape/size/colour; petals fused at base for varying distances to apex; ovary present in at least some flowers, superior
1 Stems twining, not green; leaves reduced to small scales, not green
116. CUSCUTACEAE

1	If stems twining then leaves expanded and green	2
	2 Plant wholly lacking green colour, yellow to brown, sometimes red- or purple-tinged **129. OROBANCHACEAE**	
	2 Plant with obvious green colouring	3
3	Leaves linear to subulate, all in basal rosette under water or on mud; flowers inconspicuous, solitary on pedicels or densely clustered on leafless scapes	4
3	Leaves not both linear to subulate and all in basal rosette	6
	4 Flowers bisexual, solitary; stamens included in corolla (*Limosella*) **128. SCROPHULARIACEAE**	
	4 Flowers in dense heads or spikes on scapes, or if solitary then unisexual and stamens exserted from corolla	5
5	Tepals 4, densely fringed at apex; flowers in capitate clusters, unisexual **154. ERIOCAULACEAE**	
5	Tepals >4, not fringed at apex; flowers either bisexual and in spikes, or unisexual and solitary **125. PLANTAGINACEAE**	
	6 Basal leaves peltate, succulent, glabrous; inflorescence a terminal raceme (*Umbilicus*) **75. CRASSULACEAE**	
	6 Leaves not peltate and succulent	7
7	Flowers pea-like, zygomorphic; petals 5, 1 upper, 2 lateral, and 2 lower fused to form keel; stamens 10 **79. FABACEAE**	
7	Flowers not pea-like with 5 petals and 10 stamens	8
	8 Sepals 5 (3 outer small, 2 inner large); petals 3; stamens 8 **97. POLYGALACEAE**	
	8 Not with the combination sepals 5, petals 3, stamens 8	9
9	Stamens >10	10
9	Stamens ≤10	12
	10 Carpels 5-many, usually ± fused; leaves stipulate **55. MALVACEAE**	
	10 Carpel 1; leaves without stipules	11
11	Tree or shrub; flowers numerous in dense clusters, each with perianth <1cm **78. MIMOSACEAE**	
11	Herb; flowers in elongated raceme or sometimes few, each with perianth >1cm (*Consolida*) **30. RANUNCULACEAE**	
	12 Stamens 2	13
	12 Stamens >2	16
13	Leaves all in a basal rosette; flowers solitary on erect pedicels (*Pinguicula*) **132. LENTIBULARIACEAE**	
13	Plant not with all leaves in a basal rosette and flowers solitary on erect pedicels	14
	14 Ovary 4-celled, each cell with 1 ovule **122. LAMIACEAE**	
	14 Ovary 2-celled, each cell with 2-many ovules	15
15	Perianth actinomorphic; fruit a berry or capsule with ≤4 seeds **127. OLEACEAE**	
15	Perianth slightly to strongly zygomorphic; fruit a capsule with usually >4 seeds **128. SCROPHULARIACEAE**	
	16 Ovary 4-celled with 1 ovule per cell; fruit a cluster of 1-seeded nutlets; plant bisexual	17
	16 Ovary not 4-celled with 1 ovule per cell, or if so then fruit a	

MAGNOLIOPSIDA

berry and plant dioecious ... 19
17 Leaves alternate; flowers usually in cymes spirally coiled when young; stems not square in section **120. BORAGINACEAE**
17 Leaves opposite; flowers not in spirally coiled cymes; stems usually ± square in section ... 18
   18 Ovary scarcely lobed at flowering, with terminal style and capitate stigma **121. VERBENACEAE**
   18 Ovary deeply lobed at flowering, with usually basal style and (1-)2 linear stigmas **122. LAMIACEAE**
19 Sepals 2; petals fused only at base **47. PORTULACACEAE**
19 Sepals usually >2, if 2 then petals fused for >1/2 of length ... 20
   20 Tree or shrub (sometimes very dwarf) ... 21
   20 Herb; if stems woody then climbing or trailing ... 31
21 Leaves opposite or whorled ... 22
21 Leaves alternate or spiral ... 26
   22 Stamens 5-10 ... 23
   22 Stamens 4 ... 24
23 Stamens 5; leaves mostly >1cm wide; flowers >15mm across **113. APOCYNACEAE**
23 Stamens 8 or 10, or if 5 then leaves all <5mm wide and flowers <10mm across **67. ERICACEAE**
   24 Flowers dull, brownish (*Plantago*) **125. PLANTAGINACEAE**
   24 Flowers white or brightly coloured ... 25
25 Flowers ≥2.5cm, in loose inflorescences **128. SCROPHULARIACEAE**
25 Flowers <2.5cm, in dense inflorescences **126. BUDDLEJACEAE**
   26 Dioecious, either stamens or ovary rudimentary ... 27
   26 Flowers bisexual, with functional stamens and ovary ... 28
27 Leaves gland-dotted, <2cm, serrate; fruit purple (scarcely natd aliens not treated here) **(MYRSINACEAE)**
27 Leaves not gland-dotted, most >2cm, entire or with spiny teeth; fruit red to yellow **91. AQUIFOLIACEAE**
   28 Stamens 4 **126. BUDDLEJACEAE**
   28 Stamens 5-10 ... 29
29 Plant a dense cushion <10cm with solitary flowers on erect stalks <5cm; stamens 5, with anthers opening by slits **70. DIAPENSIACEAE**
29 Plant rarely a dense cushion, if so then stamens 8 or 10 with anthers opening by apical pores ... 30
   30 Stamens 5; flowers actinomorphic **114. SOLANACEAE**
   30 Stamens 8 or 10, or if 5 then flowers slightly zygomorphic **67. ERICACEAE**
31 Inner perianth-segments (corolla-lobes) and stamens on same radius 32
31 All or most inner perianth-segments (corolla-lobes) and stamens on alternating radii ... 33
   32 Styles 5; stigmas linear **50. PLUMBAGINACEAE**
   32 Styles 1; stigma capitate **71. PRIMULACEAE**
33 Aquatic or bog plant with showy corollas densely fringed distally with hairs or narrow serrations **117. MENYANTHACEAE**
33 If aquatic or marsh plant then corollas not fringed or fringed at base

of lobes 34
- 34 Leaves all in basal rosette; corolla violet with yellow centre and yellow anthers **130. GESNERIACEAE**
- 34 If leaves all in basal rosette then corolla not violet with yellow centre 35
- 35 Flowers in compact spikes; perianth scarious and brownish; stamens long-exserted (*Plantago*) **125. PLANTAGINACEAE**
- 35 If flowers in compact spikes then not with both scarious perianth and long-exserted stamens 36
  - 36 Flowers in axils of spiny bracts; sepals 4, 2 large and 2 small; corolla with 3-lobed lower lip and 0 upper lip **131. ACANTHACEAE**
  - 36 If bracts spiny then sepals not 2 large and 2 small and corolla not with large lower and 0 upper lip 37
- 37 Leaves opposite 38
- 37 Leaves alternate 41
  - 38 Ovary 1-celled 39
  - 38 Ovary 2-celled, or ovaries 2 and each 1-celled 40
- 39 Ovule 1 (scarcely natd aliens not treated here) **(NYCTAGINACEAE)**
- 39 Ovules many **112. GENTIANACEAE**
  - 40 Ovaries 2, with common style expanded into ring below stigma **113. APOCYNACEAE**
  - 40 Ovary 1, with styles not expanded into ring below stigma **128. SCROPHULARIACEAE**
- 41 Stamens normally 6-8; inner or outer tepals with keel on abaxial side, otherwise similar to others; ovary 1-celled, with 1 ovule (*Fallopia*) **49. POLYGONACEAE**
- 41 Stamens 3-5; outer tepals sepaloid, inner petaloid; ovary 1-5-celled, with total of ≥4 ovules 42
  - 42 Ovary 3-celled; stigmas 3 on 1 style **118. POLEMONIACEAE**
  - 42 Ovary 1-5-celled, if 3-celled then stigmas 1-2 per style 43
- 43 Stamens 3 or 4, at least in most flowers **128. SCROPHULARIACEAE**
- 43 Stamens 5, at least in most flowers 44
  - 44 Styles 2, or style 1 but divided distally into 2 branches/long stigmas 45
  - 44 Style 1, with 1-2 short stigmas at apex 46
- 45 Ovary 2-lobed, each lobe with 1 style; corolla 1.5-2.5mm (*Dichondra*) **115. CONVOLVULACEAE**
- 45 Ovary not lobed, with 1 style divided distally; corolla 6-10mm **119. HYDROPHYLLACEAE**
  - 46 Ovary 1-2(3)-celled, with total of 4(-6) ovules; stems trailing or climbing **115. CONVOLVULACEAE**
  - 46 Ovary 2(-5)-celled, each cell with many ovules; stems usually not trailing or climbing 47
- 47 Flowers distinctly zygomorphic; corolla divided >1/2 way to base; anthers not forming close cone round style (*Verbascum*) **128. SCROPHULARIACEAE**
- 47 Flowers actinomorphic or slightly zygomorphic; corolla divided

MAGNOLIOPSIDA

<1/2 way to base or if >1/2 way to base then flowers actinomorphic
and anthers forming close cone round style **114. SOLANACEAE**

Key H - Perianth of 2 (rarely more) distinct whorls or rarely a spiral, the inner and outer differing markedly in shape/size/colour; petals fused at base for varying distances to apex; ovary present in at least some flowers, inferior or partly so

1 Stamens 8 to numerous (or 4-5 but appearing 8-10 because filaments split to base with each 1/2 bearing one 1/2-anther) 2
1 Stamens ≤5 (sometimes alternating with staminodes) 4
   2 Tree (*Eucalyptus*) **85. MYRTACEAE**
   2 Herb or shrub <2m 3
3 Flowers 5 in 1 terminal cluster, greenish; leaves ternate; herb
136. ADOXACEAE
3 Flowers 1-several, axillary, white to red; leaves simple; shrub (*Vaccinium*) **67. ERICACEAE**
   4 Flowers in dense heads surrounded by row(s) of sepal-like bracts 5
   4 Flowers not in dense heads, or if so then with only 2 bracts at base 8
5 Ovary 2-celled; sepals 4-5, fused proximally 6
5 Ovary 1-celled; calyx represented by a cup, or by often >5 narrow teeth, bristles, scales or hairs 7
   6 Leaves alternate; stamens borne on receptacle; ovules numerous
133. CAMPANULACEAE
   6 Leaves whorled; stamens borne on corolla-tube; ovules 1 per cell (*Sherardia*) **134. RUBIACEAE**
7 Anthers usually 5, fused into tube round style; ovary and fruit not enclosed in epicalyx **139. ASTERACEAE**
7 Anthers 4, not fused; ovary and fruit enclosed in tubular epicalyx
138. DIPSACACEAE
   8 Stamens 1-3 (vestigial if flowers female) 9
   8 Stamens 4-5 (0 or vestigial if flowers female) 11
9 Flowers numerous, ± closely packed in flat or domed inflorescences, <1cm across **137. VALERIANACEAE**
9 Flowers rather few in spikes or racemes, or solitary, >2cm across 10
   10 Stamens 3, with obvious filaments and anthers **163. IRIDACEAE**
   10 Stamens 2, with sessile anthers (*Cypripedium*) **166. ORCHIDACEAE**
11 Leaves opposite or in whorls 12
11 Leaves alternate or spiral 13
   12 Leaves ≥4 per node, or 2 per node and each stipulate
134. RUBIACEAE
   12 Leaves 2 per node, without stipules **135. CAPRIFOLIACEAE**
13 Flowers unisexual; stamens 3 (0 in female flowers), 2 with 2 pollen-sacs and 1 with 1 **62. CUCURBITACEAE**
13 Flowers bisexual; stamens 5, all with 4 pollen-sacs 14
   14 Stamens opposite corolla-lobes; stigma 1 (*Samolus*)
71. PRIMULACEAE
   14 Stamens alternating with corolla-lobes; stigmas 2-5
133. CAMPANULACEAE

Key I - Perianth of 2 (rarely more) distinct whorls or rarely a spiral, the inner and outer differing markedly in shape/size/colour; petals free, or rarely fused just at basal point, or fused near apex but free at base; ovary present in at least some flowers, inferior or partly so

1	Petals >8	2
1	Petals ≤6 (unless *flore pleno*)	3
	2 Aquatic plant with floating flowers and leaves (*Nymphaea*) **27. NYMPHAEACEAE**	
	2 Terrestrial plant; leaves very succulent **44. AIZOACEAE**	
3	Sepals and petals each 3	4
3	Sepals and petals each 2 or 4-6	7
	4 Flowers zygomorphic; stamens 1-2 **166. ORCHIDACEAE**	
	4 Flowers actinomorphic; stamens 3-12 (or 0 in female flowers)	5
5	Outer whorl of tepals sepaloid; stamens 9-12 (or 0 in female flowers); plant aquatic **142. HYDROCHARITACEAE**	
5	Both whorls of tepals petaloid; stamens 3-6; plant terrestrial	6
	6 Stamens 6 (*Galanthus*) **162. LILIACEAE**	
	6 Stamens 3 (*Libertia*) **163. IRIDACEAE**	
7	Tree or shrub	8
7	Herb, rarely woody at base	15
	8 Stamens 4-5 (or 0 in female flowers)	9
	8 Stamens 8-many	11
9	Flowers in true umbels, arising at 1 point, sometimes the umbels further aggregated **110. ARALIACEAE**	
9	Flowers solitary or in racemes or panicles, sometimes in corymbose umbel-like clusters but not arising at 1 point	10
	10 Origin of sepals separated from ovary wall by saucer- to cup-shaped hypanthium **74. GROSSULARIACEAE**	
	10 Flowers without hypanthium (sepals arising direct from ovary wall) **87. CORNACEAE**	
11	Leaves alternate	12
11	Leaves opposite or whorled	13
	12 Fruit a woody capsule; style 1 (*Leptospermum*) **85. MYRTACEAE**	
	12 Fruit surrounded by succulent or pithy hypanthium; styles (1)2-many **77. ROSACEAE**	
13	Fruit a capsule; styles 2-4, free or united proximally **73. HYDRANGEACEAE**	
13	Fruit a berry; style 1	14
	14 Hypanthium extending 5-16mm beyond ovary apex; flowers pendent on long pedicels (*Fuchsia*) **86. ONAGRACEAE**	
	14 Hypanthium not or scarcely extending beyond ovary apex; flowers not pendent nor on long pedicels (*Luma*) **85. MYRTACEAE**	
15	Leaves simple, all or most >1m across; inflorescence an elongated very dense panicle >50cm **82. GUNNERACEAE**	
15	If leaves >50cm then compound; if inflorescence >30cm then not dense and elongated	16
	16 Flowers in umbels	17
	16 Flowers not in umbels, but sometimes corymbose	19

MAGNOLIOPSIDA

17 Styles 2; fruit a dry 2-celled schizocarp **111. APIACEAE**
17 Styles 1 or 5; fruit a succulent drupe or berry 18
   18 Plant with a single terminal umbel; fruit red; leaves simple (*Cornus*) **87. CORNACEAE**
   18 Plant with umbels in large panicles; fruit dark purple to black; leaves 1-2-pinnate (*Aralia*) **110. ARALIACEAE**
19 Sepals 5 20
19 Sepals 2 or 4 21
   20 Petals bright yellow; leaves pinnate **77. ROSACEAE**
   20 Petals not bright yellow, or if so then leaves simple **76. SAXIFRAGACEAE**
21 Sepals 2; petals yellow, <1cm; fruit a capsule opening transversely (*Portulaca*) **47. PORTULACACEAE**
21 If sepals 2 then petals white; if petals yellow then ≥1cm; if fruit a capsule then not opening transversely 22
   22 Fruit an achene without bristles, with persistent sepals; leaves subsessile (*Haloragis*) **81. HALORAGACEAE**
   22 Fruit a many-seeded capsule, or if a 1-2-celled achene then with many hooked bristles and without persistent sepals, and leaves long-petiolate **86. ONAGRACEAE**

Key J - Perianth of 2 (rarely more) distinct whorls or rarely a spiral, the inner and outer differing markedly in shape/size/colour; petals free or rarely fused just at basal point, or fused near apex but free at base; >1 ovary present in at least some flowers, superior, free or fused just at extreme base

1 Tree with flowers borne in pendent unisexual globose clusters **34. PLATANACEAE**
1 Herb or shrub; flowers not in pendent unisexual globose clusters 2
   2 Sepals and petals each 3 3
   2 Sepals and petals each >3 5
3 Carpels 3; stamens 3; flowers <3mm across (*Crassula*) **75. CRASSULACEAE**
3 Carpels ≥6; stamens ≥6 (or 0 in female flowers); flowers >5mm across 4
   4 Carpels with numerous ovules, forming follicles in fruit; sepals purple-tinged green; leaves linear **140. BUTOMACEAE**
   4 Carpels with 1 ovule forming achenes in fruit, or if with >1 ovules forming follicles in fruit then sepals green and leaves ovate with cordate base **141. ALISMATACEAE**
5 Leaves opposite and/or very succulent **75. CRASSULACEAE**
5 Leaves all basal, alternate or spiral, not or only slightly succulent 6
   6 Shrub **77. ROSACEAE**
   6 Herb, sometimes with ± woody surface rhizome 7
7 Flowers perigynous; origin of sepals (often of stamens and petals too) separated from base of ovary by obvious flat or saucer- to bowl-shaped hypanthium 8
7 Flowers hypogynous or ± so, the stamens, petals and sepals free and arising directly at base of carpels 9

8	Carpels 2, each with many ovules; fruit a pair of follicles; leaves simple, unlobed; petals bright pink (*Bergenia*)	**76. SAXIFRAGACEAE**
8	Carpels 3-many, or if 2 then each with 1 ovule; fruit 1-many achenes; plant not with both simple unlobed leaves and pink petals	**77. ROSACEAE**
9	Carpels with 1 ovule; fruit an achene	**30. RANUNCULACEAE**
9	Carpels with few to many ovules; fruit a follicle	10
10	Flowers ≥7cm across; seeds black (fertile) and red (sterile) usually mixed in each follicle	**51. PAEONIACEAE**
10	Flowers <7cm across; seeds not red and black mixed together	11
11	Carpels 5-many	**30. RANUNCULACEAE**
11	Carpels 2-4	12
12	Stamens 5-10 (or 0 in female flowers); carpels rarely >2	**76. SAXIFRAGACEAE**
12	Stamens >10 (or 0 in female flowers); carpels usually >2	13
13	Leaves palmate or deeply palmately lobed	**30. RANUNCULACEAE**
13	Leaves 2-3-pinnate (*Aruncus*)	**77. ROSACEAE**

Key K - Perianth of 2 (rarely more) distinct whorls or rarely a spiral, the inner and outer differing markedly in shape/size/colour; petals free, or rarely fused just at basal point, or fused near apex but free at base; ovary present in at least some flowers, superior, of 1 carpel or of >1 wholly or mostly fused carpels

1	Plant yellowish-brown, lacking green pigment in all parts	**69. MONOTROPACEAE**
1	Plant with at least leaves or stems at least partly green	2
2	Leaves modified to form tubular 'pitchers' with small blade ('hood') at entrance, all in basal rosette	**56. SARRACENIACEAE**
2	Leaves not modified as 'pitchers'	3
3	Leaves covered on upperside with very sticky glandular hairs, reddish, all basal	**57. DROSERACEAE**
3	Leaves not all basal and covered with very sticky glandular hairs	4
4	Flowers zygomorphic	5
4	Flowers actinomorphic	19
5	At least 1 sepal or petal with conspicuous basal spur or pouch	6
5	Flowers without basal spur(s)	10
6	Spur(s) formed from petal(s); ovary 1-celled	7
6	Spur(s) formed from sepal(s); ovary 3-5-celled	8
7	Leaves pinnate or ternate, without stipules; sepals 2; petals 4	**33. FUMARIACEAE**
7	Leaves simple, stipulate (the stipules sometimes deeply divided or pinnate); sepals 5; petals 5	**59. VIOLACEAE**
8	Sepals 3; petals 5 but apparently 3 due to fusion of 2 pairs of laterals	**109. BALSAMINACEAE**
8	Sepals 5; petals 5	9
9	Upper sepal with spur fused to pedicel; ovary 5-celled, with distal sterile beak (*Pelargonium*)	**106. GERANIACEAE**

MAGNOLIOPSIDA

9 Upper 1-3 sepals with free spurs; ovary 3-celled, without distal beak
108. TROPAEOLACEAE
   10 Stamens 8 or 10, all or all but 1 with filaments fused into tube   11
   10 Stamens 3-many, free   12
11 Flowers with sepal uppermost (on top-line); stamens 8; anthers opening by pores   **97. POLYGALACEAE**
11 Flowers with petal uppermost (on top-line); stamens 10; anthers opening by slits   **79. FABACEAE**
   12 Tree, or shrub >1m high; ovary 3-celled   13
   12 Herb, or shrub <1m high; ovary 1-2-celled   14
13 Leaves palmate, opposite; flowers white to red
100. HIPPOCASTANACEAE
13 Leaves 1-2-pinnate, alternate; flowers yellow   **99. SAPINDACEAE**
   14 Stamens 3; petals 3, filiform, brown (*Tolmiea*)
76. SAXIFRAGACEAE
   14 Stamens 5-many; petals >3, not filiform, white or pink   15
15 Ovary 2-celled ; petals 4, the lower 2 and upper 2 forming 2 different pairs   16
15 Ovary 1-celled; petals 4-6(8), if only 4 then upper one, lateral two and lower one of 3 different forms   17
   16 Leaves palmate; capsule >2cm (scarcely natd aliens not treated here)   **(CAPPARACEAE)**
   16 Leaves simple to pinnate; capsule <2cm   **64. BRASSICACEAE**
17 Petals 4-6(8), at least some deeply lobed ; stamens 7-many (*Reseda*)
65. RESEDACEAE
17 Petals 5, or apparently 4 due to fusion of lower 2, entire to very shallowly lobed or toothed   18
   18 Leaves paripinnate (scarcely natd aliens not treated here)
(CAESALPINIACEAE)
   18 Leaves ternate (*Thermopsis*)   **79. FABACEAE**
19 Stamens >12, >2x as many as petals   20
19 Stamens 1-12, ≤2x as many as petals   28
   20 Petals ≥9   21
   20 Petals ≤6   22
21 Aquatic plant with leaves and flowers at or just above water surface (*Nuphar*)   **27. NYMPHAEACEAE**
21 Terrestrial plant with succulent leaves   **44. AIZOACEAE**
   22 Stamens with filaments fused into tube round styles
55. MALVACEAE
   22 Stamens free or ± united into bundles, but not forming tube round styles   23
23 Ovary with 1-2(5) ovules per cell   24
23 Ovary with several to many ovules per cell   25
   24 Ovary 5-celled; inflorescence stalks fused to narrowly oblong papery bract   **54. TILIACEAE**
   24 Ovary 1-celled; inflorescence stalks not fused to long papery bract   **77. ROSACEAE**
25 Leaves simple, entire   26

25	Leaves simple and toothed or lobed, or compound	27
	26 Leaves with translucent and/or coloured sessile glands; styles 3 or 5	**53. CLUSIACEAE**
	26 Leaves without sessile glands, style 0 or 1	**58. CISTACEAE**
27	Sepals 2(-3), sepaloid, usually falling early	**32. PAPAVERACEAE**
27	Sepals (3)4-5, petaloid, not falling early	**30. RANUNCULACEAE**
	28 Perianth in 3-6 whorls each of 2-4 segments	**31. BERBERIDACEAE**
	28 Perianth in 2 whorls each of ≥2 segments	29
29	Stems herbaceous, sometimes woody just at base	30
29	Stems wholly or mostly woody	54
	30 Sepals 2; petals 5	31
	30 Sepals ≥3; petals as many as sepals or fewer	32
31	Stems twining or sprawling, >1m; underground tubers present; capsule indehiscent (scarcely natd aliens not treated here)	**(BASELLACEAE)**
31	Stems decumbent to erect, <50cm; underground tubers 0; capsule dehiscent	**47. PORTULACACEAE**
	32 Sepals 3; petals 2-3	33
	32 Sepals >3; petals >3	35
33	Ovary 1-celled, with 1 ovule; sepals and petals both sepaloid or petaloid (*Rumex*)	**49. POLYGONACEAE**
33	Ovary 3-celled, each cell with 2-many ovules; petals petaloid; sepals sepaloid	34
	34 Leaves opposite, stipulate, <1cm; flowers <5mm across	**52. ELATINACEAE**
	34 Leaves alternate, with sheathing base but without stipules, >1cm; flowers ≥10mm across	**153. COMMELINACEAE**
35	Ovary 1-celled, at least apically, with 1-many ovules	36
35	Ovary 2-10-celled throughout with 1-many ovules per cell	42
	36 Stamens 5, alternating with 5 conspicuous deeply divided staminodes (*Parnassia*)	**76. SAXIFRAGACEAE**
	36 Deeply divided staminodes 0	37
37	Ovary with 1 ovule	38
37	Ovary with few to many ovules	40
	38 Leaves without stipules; flowers showy, usually pink or blue, >4mm across	**50. PLUMBAGINACEAE**
	38 Leaves stipulate; flowers inconspicuous, green, yellowish-green or white	39
39	Leaves simple, entire	**48. CARYOPHYLLACEAE**
39	Leaves compound or conspicuously lobed	**77. ROSACEAE**
	40 Style 1, ≥1mm, with 1 or 5 stigmas at apex	**68. PYROLACEAE**
	40 Styles ± 0 or 2-5, free, if 1 then either <1mm or divided into 3 distally	41
41	Ovary with ovules on parietal placentas; styles fused proximally; leaves without stipules	**61. FRANKENIACEAE**
41	Ovary with ovules on free-central placenta; styles free, or 0, or rarely fused proximally and then leaves stipulate	**48. CARYOPHYLLACEAE**
	42 Flowers perigynous, with tubular or cup- or bowl-shaped	

MAGNOLIOPSIDA

  hypanthium bearing sepals and petals at apex   43
 42 Flowers hypogynous, with sepals and petals borne at base of ovary   45
43 Hypanthium becoming hard and protective at fruiting; carpels 2, each with 1 ovule   **77. ROSACEAE**
43 Hypanthium not becoming hard and protective; carpels 2(-4), each with many ovules   44
 44 Sepals and petals each usually 6, with epicalyx segments outside; leaves ± all on stem, at least the lower opposite or in whorls of 3   **83. LYTHRACEAE**
 44 Sepals and petals each 5, without epicalyx segments; leaves all or nearly all basal (*Rodgersia*)   **76. SAXIFRAGACEAE**
45 Stem-leaves all in single whorl of 3-8, with 1 flower above (*Paris*)   **162. LILIACEAE**
45 Stem-leaves (if present) not all in 1 whorl   46
 46 Petals and sepals each 4, at least in most flowers   47
 46 Petals and sepals each 5 in all flowers   50
47 Stamens 4 or 6   48
47 Stamens 8   49
 48 Stamens 4; capsule dehiscing by 8 valves (*Radiola*)   **96. LINACEAE**
 48 Stamens (4-)6; capsule dehiscing by 2 valves or breaking transversely or indehiscent   **64. BRASSICACEAE**
49 Leaves simple, entire, stipulate   **52. ELATINACEAE**
49 Leaves 2-3-pinnately lobed, without stipules (*Ruta*)   **104. RUTACEAE**
 50 Fruit a 5-celled schizocarp with 1 seed per cell, elongated distally into sterile column   51
 50 Fruit a 5- to many-seeded capsule, without a sterile column   52
51 Leaves stipulate; petals white to red, blue or purple   **106. GERANIACEAE**
51 Leaves without stipules; petals yellow   **107. LIMNANTHACEAE**
 52 Ovary 2-celled; styles 0 or 2 (*Saxifraga*)   **76. SAXIFRAGACEAE**
 52 Ovary 5-celled; styles 5   53
53 Capsule opening by 5 valves; leaves ternate or palmate   **105. OXALIDACEAE**
53 Capsule opening by 10 valves; leaves simple, entire   **96. LINACEAE**
 54 Sepals 3; petals 2-3   55
 54 Sepals and petals each >3   57
55 Woody climber (*Fallopia*)   **49. POLYGONACEAE**
55 Procumbent to erect shrub   56
 56 Leaves <3mm wide; flowers in small axillary clusters   **66. EMPETRACEAE**
 56 Leaves replaced by leaf-like flat cladodes >3mm wide, bearing 1-2 flowers on their surface (*Ruscus*)   **162. LILIACEAE**
57 Stamens 2x as many as petals   58
57 Stamens <2x as many as petals at least on most flowers   62
 58 Woody stems ± procumbent; leaves >6cm, simple, serrate (*Bergenia*)   **76. SAXIFRAGACEAE**
 58 Woody stems erect or ± so; leaves usually compound or lobed   59

59 Leaves simple, serrate; ovules many in each cell (scarcely natd aliens not treated here) **(CLETHRACEAE)**
59 Leaves compound, or simple and entire, or lobed with the lobes entire; ovules 1-2 per cell 60
  60 Ovary 2-celled; leaves opposite **101. ACERACEAE**
  60 Ovary 4-6-celled; leaves alternate or opposite 61
61 Fruit a 4(-5)-celled capsule, not winged; leaves simple, or ternate, or pinnate or pinnately lobed with segments <1cm wide **104. RUTACEAE**
61 Fruit a group of 1-6 winged achenes; leaves pinnate with leaflets mostly >1cm wide **103. SIMAROUBACEAE**
  62 Leaves opposite 63
  62 Leaves alternate 67
63 Stems procumbent; ovary 1-celled **61. FRANKENIACEAE**
63 Stems erect; ovary >1-celled 64
  64 Leaves simple and not lobed 65
  64 Leaves pinnate 66
65 Fruit a dehiscent capsule with orange seeds; leaves without stipules (*Euonymus*) **90. CELASTRACEAE**
65 Fruit a black non-dehiscent berry; leaves with herbaceous stipules at least when young **94. RHAMNACEAE**
  66 Sepals 5, free; stamens 5; ovules many in each cell; fruit an inflated capsule **98. STAPHYLEACEAE**
  66 Sepals 4, fused into tube proximally; stamens 2; ovules 2 in each cell; fruit a winged achene (*Fraxinus*) **127. OLEACEAE**
67 Petals and sepals each 4; stamens 6; ovules many in each cell **64. BRASSICACEAE**
67 Petals and sepals each 5, or if 4 then stamens 4 and ovules 1-2 per cell 68
  68 Leaves <5mm, ± scale-like **60. TAMARICACEAE**
  68 Leaves >5mm, not scale-like 69
69 Leaves with strongly revolute margins, densely rusty-tomentose on lowerside; stamens mostly 6-8 (*Ledum*) **67. ERICACEAE**
69 Leaves with margins not or scarcely revolute, not rusty-tomentose (sometimes white-tomentose) on lowerside; stamens 4-5 70
  70 Leaves palmate or palmately lobed **95. VITACEAE**
  70 Leaves pinnate, pinnately lobed or unlobed 71
71 Petals purple; ovary 1-celled with many ovules **72. PITTOSPORACEAE**
71 Petals not purple; ovary 1-5-celled, each cell with 1-2 ovules 72
  72 Ovary 1-celled with 1 ovule **102. ANACARDIACEAE**
  72 Ovary 2-5-celled, each cell with 1-2 ovules 73
73 Leaves evergreen, coriaceous and shiny, at least some with strong marginal spines; petals white **91. AQUIFOLIACEAE**
73 Leaves deciduous, not coriaceous or shiny, without spines on margin; petals yellowish-green 74
  74 Fruit a black non-dehiscent berry; leaves with herbaceous stipules at least when young **94. RHAMNACEAE**
  74 Fruit a dehiscent capsule with red seeds; leaves with spinose stipules at least when young (*Celastrus*) **90. CELASTRACEAE**

MAGNOLIIDAE — DICOTYLEDONS

Often trees or shrubs; commonly with secondary thickening from a permanent vascular cambium; vascular bundles usually in a ring in the stem; primary root commonly persisting; leaves usually with pinnate or palmate major venation and reticulate minor venation; flower parts mostly in fours or fives; pollen grains mostly radially symmetrical, commonly with 3 pores and/or furrows; cotyledons normally 2; endosperm typically nuclear or cellular. Numerous exceptions to all the above occur.

25. LAURACEAE - *Bay family*

Evergreen trees or shrubs; easily recognized by the laurel-like aromatic leaves and dioecious flowers with 4-lobed perianth.

1. LAURUS L. - *Bay*

1 Shrub or tree exceptionally to 18m. Intrd-natd; probably bird-sown in scrub and on cliffs near sea in S **Bay - L. nobilis** L.

26. ARISTOLOCHIACEAE - *Birthwort family*

Perennial herbs; 2 distinctive genera easily recognized by their cordate leaves and weird flowers with 1-whorled tubular perianth.

1 Aerial stem ≤10cm, densely pubescent; petiole much longer than leaf-blade; flower solitary, terminal; perianth 3-lobed **1. ASARUM**
1 Aerial stem >10cm, ± glabrous; petiole shorter than leaf-blade or ± absent; flowers 1-8, axillary; perianth 1-lobed **2. ARISTOLOCHIA**

1. ASARUM L. - *Asarabacca*
Rhizomatous herb with aerial stems bearing scales, usually 2 apparently opposite foliage leaves, and 1 terminal flower; flowers actinomorphic; perianth bell-shaped, 3-lobed; stamens 12, in 2 whorls, arising from base of stylar column.

1 Flowers 12-15mm, close to soil on short pedicel. Intrd-natd; woods; now rare and very scattered over Br *Asarabacca* - **A. europaeum** L.

2. ARISTOLOCHIA L. - *Birthworts*

Aerial stems not or little branched, bearing numerous alternate leaves and 1-8 flowers in each leaf-axil; flowers zygomorphic; perianth with rounded swollen base, narrowly tubular upper part and 1-lobed apex; stamens 6, in 1 whorl, arising from side of stylar column.

1 Rhizomatous; petioles longer than pedicels; flowers 2-3.5cm x <5mm, (1)2-8 in each leaf-axil. Stems to 1m. Intrd-natd; rough ground, very scattered over En, Wa and S Sc **Birthwort - A. clematitis** L.
1 With an underground tuber; petioles much shorter than pedicel; flowers 2.5-5cm x <5mm, 1 in each leaf-axil. Stems to 60cm. Intrd-natd; chalky slopes; Surrey **Smearwort - A. rotunda** L.

27. NYMPHAEACEAE - *Water-lily family*

Aquatic perennial herbs with stout rhizomes, with large long-petioled leaves and flowers on or near water surface.

1 Leaf-veins forming a reticulum near leaf-margin; petals white, outermost much longer than the usually 4 sepals **1. NYMPHAEA**
1 Leaf-veins forking near leaf-margin, not re-joining; petals yellow, shorter than the usually 5 sepals **2. NUPHAR**

1. NYMPHAEA L. - *White Water-lilies*

Mature leaves rarely submerged, if so similar to floating leaves; leaf-veins forming a reticulum near leaf-margin; sepals (3-)4(5), green to reddish-brown externally; petals white, inserted at a range of levels on side of ovary, the outer longer than sepals; stamens 46-125, inserted on side of ovary above petals; ovary subglobose; stigmatic rays 9-25, projecting as curved, horn-like processes; fruit ripening under water.

1 Petals 12-33, the outer 2-8.5cm. Native; in lakes, ponds, dykes and slow-flowing rivers (***White Water-lily* - N. alba** L.)
   2 Leaves 9-30cm; flowers 9-20cm across, opening wide; stamens borne almost to top of ovary and leaving scars to top of fruit; stigmatic rays usually >14; fruit usually obovoid. Throughout Br and Ir **N. alba ssp. alba**
   2 Leaves 9-13cm; flowers 5-12cm across, usually never opening wide; stamens not borne on upper part of ovary; stigmatic rays usually <16; fruit usually subglobose. W Ir and N & W Sc
N. alba ssp. occidentalis (Ostenf.) Hyl.

2. NUPHAR Sm. - *Yellow Water-lilies*

Mature leaves usually floating and submerged, the latter thinner and with undulate margins; leaf-veins forking and not rejoining near leaf-margin; sepals 5(-7), yellowish-green; petals 9-26, yellow, shorter than sepals; petals and stamens inserted at base of ovary; stamens 37-200; ovary bottle-shaped;

2. NUPHAR

stigmatic rays 7-24, not projecting or projecting as flat, obtuse bulges; fruit ripening above water.

1 Leaves to 40 x 30cm, erect, held above water surface; petioles ± terete. Intrd-natd; scattered from Surrey to Easterness
Spatter-dock - **N. advena** (Aiton) W.T. Aiton
1 Leaves horizontal, floating on water surface; petioles dorsiventrally compressed or triangular in section with rounded angles 2
  2 Leaves to 40 x 30cm, with 23-28 lateral veins on each side; stigmatic disc 10-15mm across, circular or slightly crenate at margin, with 9-24 rays; flowers 3-6cm across. Native; in lakes, ponds, dykes, rivers; frequent *Yellow Water-lily* - **N. lutea** (L.) Sm.
  2 Leaves with 11-22 lateral veins on each side; stigmatic disc 6-11mm across, crenate to lobed at margin, with 7-14 rays; flowers 1.5-4cm across 3
3 Leaves to 18 x 14cm, with 15-22 lateral veins on each side; stigmatic disc 7.5-11mm across, crenate at margin, with 9-14 rays; stamens 60-100; pollen <25% fertile. Native; scattered in N En and Sc, Merioneth (*N. lutea* x *N. pumila*)
Hybrid Water-lily - **N. x spenneriana** Gaudin
3 Leaves to 17 x 12.5cm, with 11-18 lateral veins on each side; stigmatic disc 6-8.5mm across, distinctly lobed at margin, with 7-12 rays; stamens 37-65; pollen >90% fertile. Native; in ponds and lochs; local in Sc, Salop *Least Water-lily* - **N. pumila** (Timm) DC.

28. CABOMBACEAE - *Water-shield family*

Aquatic perennial herbs; the opposite, palmately dissected, submerged leaves, peltate floating leaves and flowers with 6 white tepals are unique.

1. CABOMBA Aubl. - *Water-shields*

1 Stems to 2m; submerged leaves 3-5cm. Intrd-natd; canals; N Hants
Carolina Water-shield - **C. caroliniana** A. Gray

29. CERATOPHYLLACEAE - *Hornwort family*

Submerged rootless aquatic perennial herbs; distinguished from other subaquatic plants by the whorled, bifid leaves.

1. CERATOPHYLLUM L. - *Hornworts*

1 Leaves dark green, rigid, forked 1-2x; fruit 4-5mm, smooth or slightly warty, apiculate to long-spined at apex, often also with 2 long spines or tubercles at base. Stems to 1m. Native; in ponds, ditches and slow

rivers; scattered, rare in N   ***Rigid Hornwort* - C. demersum** L.
1 Leaves mid green, not rigid, forked 3-4x; fruit 4-5mm, conspicuously warty, apiculate to obtuse at apex, without basal spines. Stems to 1m. Native; in ponds and ditches, mostly near sea; very local in En, Wa and CI, Co Down   ***Soft Hornwort* - C. submersum** L.

30. RANUNCULACEAE - *Buttercup family*

Herbaceous annuals or perennials, sometimes woody climbers; very variable in floral morphology, but most genera have spirally arranged leaves without stipules and produce a head of achenes or follicles from each flower, which is always hypogynous and often with the sepals more conspicuous than the petals.

1 Woody climber; leaves opposite; perianth of 4 segments in 1 whorl **12. CLEMATIS**
1 Herbaceous, not climbing; leaves alternate or whorled; perianth rarely of 4 segments, often in 2 whorls   2
   2 Ovary with many ovules; fruit a follicle, capsule or berry   3
   2 Ovary with 1 ovule; fruit an achene   11
3 Flowers with 1 carpel   4
3 Flowers with (2)3 or more carpels   5
   4 Flowers actinomorphic, whitish; sepals not spurred; fruit a berry **8. ACTAEA**
   4 Flowers zygomorphic, usually blue, sometimes white or pink; the upper sepal with a conspicuous spur; fruit a follicle **7. CONSOLIDA**
5 At least 1 of the petals or sepals conspicuously hooded or spurred   6
5 Petals and sepals not spurred or hooded   7
   6 Flowers zygomorphic, the upper sepal hooded, only the 2 upper petals spurred **6. ACONITUM**
   6 Flowers actinomorphic, the sepals not hooded, each of the 5 petals spurred **16. AQUILEGIA**
7 Flowers white to reddish, bluish or greenish, sometimes tinged purple   8
7 Flowers yellow   9
   8 Annuals; sepals bluish; follicles fused up to apex; leaves divided into fine linear entire segments **5. NIGELLA**
   8 Perennials; sepals green, white, violet or purple; follicles free or fused at base only; leaves divided into wider, toothed segments **3. HELLEBORUS**
9 Stem-leaves (bracts) 3, in a whorl just below flower **4. ERANTHIS**
9 Stem-leaves 1-many, not whorled and usually not just below flower   10
   10 Leaves deeply lobed; perianth of 2 whorls, the inner consisting of small nectaries **2. TROLLIUS**
   10 Leaves simple, finely toothed; perianth of 1 whorl **1. CALTHA**
11 Perianth of 1 whorl   12

30. RANUNCULACEAE

11	Perianth apparently of 2 whorls	14
	12 Stamens longer than perianth; flowering stems with alternate leaves and many flowers	**17. THALICTRUM**
	12 Stamens shorter than perianth; flowering stems with 1-few flowers and whorled leaves or bracts	13
13	Styles scarcely elongating in fruit; whole plant glabrous to shortly pubescent	**9. ANEMONE**
13	Styles greatly elongating in fruit and becoming feathery; whole plant with long hairs	**11. PULSATILLA**
	14 Outer whorl of apparent perianth of 3 segments; inner whorl of perianth segments blue or white	**10. HEPATICA**
	14 Outer whorl of apparent perianth usually of 5 segments, if of 3 then inner whorl yellow	15
15	Flowers solitary on leafless stem; leaves all linear, in a basal rosette; sepals spurred	**15. MYOSURUS**
15	Flowers on usually branched, leafy stems; lowest leaves not linear; sepals not spurred	16
	16 Petals red with a blackish basal blotch	**14. ADONIS**
	16 Petals yellow or white, without dark basal blotch	**13. RANUNCULUS**

1. CALTHA L. - *Marsh-marigold*
Herbaceous perennials; leaves spirally arranged, simple, without stipules; flowers solitary or in few-flowered cymes, without a whorl of bracts below, actinomorphic; perianth of 1 whorl of 5-8(10) petaloid yellow sepals; stamens numerous; carpels 5-15, free, spirally arranged; fruit a follicle.

1 Stems to 60cm, erect to procumbent. Native; marshes, ditches and by ponds and streams; common   *Marsh-marigold* - **C. palustris** L.

2. TROLLIUS L. - *Globeflower*
Herbaceous perennials; leaves borne spirally, palmate or deeply palmately lobed, without stipules; flowers solitary or in few-flowered cymes, without a whorl of bracts below, actinomorphic; perianth of 2 whorls; sepals 5-15, petaloid, yellow; petals 5-15, in the form of narrow, small nectaries; stamens numerous; carpels numerous, free, spirally arranged; fruit a follicle.

1 Stems to 70cm. Native; damp places in grassland or woods, often upland; local in N and Wa   *Globeflower* - **T. europaeus** L.

3. HELLEBORUS L. - *Hellebores*
Herbaceous perennials; leaves spirally arranged or all basal, ternate to palmate with long, toothed leaflets, the lateral ones joined at base, without stipules; flowers in few- to many-flowered cymes, or solitary, without whorl of bracts below, actinomorphic, appearing very early in season; perianth of 2 whorls; sepals 5, usually not brightly coloured; petals 5-12, in the form of small tubular nectaries; stamens numerous; carpels 2-5, usually slightly fused at base; fruit a follicle.

1 Not rhizomatous; stems lasting from 1 spring to next, many-flowered; bracts ± entire; leaves all on stem 2
1 Rhizomatous; stems lasting from winter to late spring, 2-4-flowered; bracts deeply divided; leaves all basal 3
   2 Leaves palmate, the leaflets with short acute to obtuse teeth; flowers bell-shaped; fresh plant stinking when crushed. Stems to 80cm. Probably native; woods and scrub on calcareous soils; very local in S Br, natd further N *Stinking Hellebore* - **H. foetidus** L.
   2 Leaves ternate, the leaflets with subulate ± sharp teeth; flowers cup-shaped; fresh plant not stinking when crushed. Stems to 1.2m. Intrd-natd; marginal habitats; very scattered in En, Jersey *Corsican Hellebore* - **H. argutifolius** Viv.
3 Follicles fused for c.1/4 their length, sessile; flowers mostly 3-5cm across, pale green. Stems to 40cm. Native; woods and scrub on calcareous soils; very local in S & C Br, natd further N *Green Hellebore* - **H. viridis** L.
3 Follicles free to base, shortly stalked; flowers mostly 5-7cm across, yellowish-green to purplish. Stems to 60cm. Intrd-natd; woods and parks; very scattered in Br *Lenten-rose* - **H. orientalis** Lam.

4. ERANTHIS Salisb. - *Winter Aconite*

Herbaceous perennials, with underground tubers; leaves all basal except bracts, palmate or deeply palmately lobed, without stipules; flowers solitary, with whorl of 3 leaf-like bracts just below, actinomorphic; perianth of 2 whorls; sepals usually 6, petaloid, yellow; petals usually 6, in the form of small tubular nectaries; stamens numerous; carpels usually 6, free; fruit a follicle.

1 Aerial stems to 15cm. Intrd-natd; woods, parks and roadsides; scattered in Br *Winter Aconite* - **E. hyemalis** (L.) Salisb.

5. NIGELLA L. - *Love-in-a-mist*

Annuals; leaves spirally arranged, pinnate, without stipules; flowers solitary or in few-flowered cymes, with whorl of bracts below, actinomorphic; perianth of 2 whorls; sepals 5, petaloid, blue; petals 5, in the form of clawed nectaries; stamens numerous; carpels usually 5, fused up to apex; fruit a capsule.

1 Stems to 50cm. Intrd-natd; often persisting on waste ground and rubbish tips; En and Wa *Love-in-a-mist* - **N. damascena** L.

6. ACONITUM L. - *Monk's-hoods*

Herbaceous perennials; leaves spirally arranged, palmate or deeply palmately lobed, without stipules; flowers in terminal racemes, each in axil of small bract, zygomorphic; perianth of 2 whorls; sepals 5, petaloid, the upper one forming an elongated hood; petals 2-10, in the form of nectaries, the 2 upper large and enclosed in the sepal-hood, the others very small or absent; stamens numerous; carpels 3(-5), ± fused at base; fruit a follicle.

6. ACONITUM

1 Flowers yellow or cream; upper sepal >2x as high as wide; leaves not divided to base. Stems to 1m. Intrd-natd; occasional by streams and in woods; C & N Br (*A. vulparia* Rchb.)
Wolf's-bane - A. lycoctonum L.
1 Flowers blue or blue and white; upper sepal <2x as high as wide; at least lower leaves divided to base   2
    2 Pedicels pubescent to densely pubescent; flowers usually blue; upper sepal ± as high as wide, gradually tapered into forward-projecting spur. Stems to 1.5m. Native; shady places by streams; probably only SW Br, but natd over rest of Br in similar places or rough ground   **Monk's-hood - A. napellus** L.
    2 Pedicels glabrous to sparsely pubescent; flowers blue and white or blue; upper sepal distinctly higher than wide, abruptly narrowed into forward-projecting spur. Stems to 1.5m. Intrd-natd; damp shady places; frequent in Sc, rare in Wa and En (*A. napellus* x *A. variegatum* L.) **Hybrid Monk's-hood - A. x cammarum** L.

7. CONSOLIDA (DC.) Gray - *Larkspurs*

Annuals; leaves spirally arranged, palmate with finely divided segments, without stipules; flowers in terminal racemes, each in axil of bract and with bracteoles on pedicel, zygomorphic; perianth of 2 whorls, blue, pink or white; sepals 5, petaloid, the upper one long-spurred; petals 4, elaborately shaped, the upper two fused and with a nectariferous spur enclosed in the sepal-spur; stamens numerous; carpel 1; fruit a follicle.

1 Stems to 1m. Intrd-natd; common garden escape in BI, often persistent   **Larkspur - C. ajacis** (L.) Schur

8. ACTAEA L. - *Baneberry*

Herbaceous perennials; leaves borne spirally, ternate or pinnate, without stipules; flowers in terminal racemes, each in axil of a small bract, actinomorphic; perianth usually of 2 whorls; sepals (3)4(-5), petaloid, white; petals 4-6 or 0, petaloid, without nectar; stamens numerous; carpel 1; fruit a berry.

1 Stems to 60cm. Native, limestone pavement and sparse woodland on limestone; local in N En, intrd elsewhere   **Baneberry - A. spicata** L.

9. ANEMONE L. - *Anemones*

Herbaceous perennials; leaves all basal, palmate, palmately lobed or ternate, without stipules; flowers solitary or few, with a whorl of 3 leaf-like bracts some way below, or in terminal few-flowered cymes, actinomorphic; perianth of 1 whorl of 5-20 petaloid sepals; stamens numerous; carpels numerous, free, spirally arranged; fruit an achene, the style remaining shorter than fertile portion.

1 Stem >30cm, usually branched with several flowers; sepals densely silky-pubescent on lowerside; autumn flowering. Intrd-natd; old

garden sites or escape or throwout; rare throughout BI
(*A. hupehensis* (Lemoine) Lemoine x *A. vitifolia* DC.)

Japanese Anemone - **A. x hybrida** Paxton

1 Stem <30cm, simple, with 1(-3) flowers; sepals glabrous to sparsely pubescent on lowerside; spring flowering 2

  2 Sepals yellow, mostly 5. Stems to 30cm. Intrd-natd; garden throwout or escape in shady places; very scattered in En and Sc

Yellow Anemone - **A. ranunculoides** L.

  2 Sepals white to pink or blue, mostly 6 or more 3

3 Sepals (5)6-7(9), mostly white or pinkish. Stems 5-30cm. Native; woodland, hedgerows, open grassland in wetter districts; common

Wood Anemone - **A. nemorosa** L.

3 Sepals (8)10-15(20), mostly blue 4

  4 Rhizomes elongated; sepals and basal leaves sparsely pubescent on lowerside; head of ripe achenes erect. Stems to 30cm. Intrd-natd; woodland, hedgerows and rough ground; scattered in Br

Blue Anemone - **A. apennina** L.

  4 Rhizomes short, tuberous; sepals and basal leaves glabrous on lowerside; head of ripe achenes pendent. Stems to 30cm. Intrd-natd; woodland, hedgerows and rough ground; very scattered in En, Man   *Balkan Anemone* - **A. blanda** Schott & Kotschy

10. HEPATICA - *Liverleaf*

Herbaceous perennials; leaves all basal, distinctively 3-lobed, without stipules; flowers solitary, actinomorphic; perianth apparently of 2 whorls, but inner whorl of 6-10 blue or white segments is calyx and outer whorl of 3 sepal-like segments is bracts; stamens numerous; carpels numerous, free, spirally arranged; fruit an achene, the style remaining shorter than fertile portion.

1 Petioles and pedicels to 15cm. Intrd-natd; shady places: E Gloucs and SE Yorks   *Liverleaf* - **H. nobilis** Schreb.

11. PULSATILLA Mill. - *Pasqueflower*

Herbaceous perennials; leaves all basal, 2-pinnate, without stipules; flowers solitary, with a whorl of 3 almost leaf-like bracts just below, actinomorphic; perianth of 1 whorl of 6 petaloid sepals, deep violet-purple; stamens numerous, the outer sterile and nectariferous; carpels numerous, free, spirally arranged; fruit an achene, the style becoming greatly elongated and feathery.

1 Stems to 30cm. Native; dry calcareous grassland; very local in C & E En   *Pasqueflower* - **P. vulgaris** Mill.

12. CLEMATIS L. - *Traveller's-joys*

Woody climbers; leaves opposite, pinnate or ternate or rarely simple, without stipules, the petioles and rhachis twining round supports; flowers in axillary cymes, often with 2 small opposite bracteoles below,

12. CLEMATIS

actinomorphic; perianth of 1 whorl of usually 4 petaloid sepals; stamens numerous; carpels numerous, free, spirally arranged; fruit an achene, the style becoming greatly elongated and feathery or not so.

1 Leaves ternate, the primary divisions serrate. Deciduous climber to 6m. Intrd-natd; rarely escaping or persisting over hedges and walls; very scattered in Br N to Dunbarton, Man
Himalayan Clematis - **C. montana** DC.
1 Leaves pinnate, the primary divisions divided further or not 2
   2 Flowers yellow. Deciduous climber to 4m. Intrd-natd; Man
Orange-peel Clematis - **C. tangutica** (Maxim.) Korsh.
   2 Flowers white to blue or purple 3
3 Flowers blue to purple; styles glabrous, not elongating in fruit. Deciduous climber to 4m. Intrd-natd; hedges; very scattered in C & S En   *Purple Clematis* - **C. viticella** L.
3 Flowers white or cream; styles pubescent, elongating in fruit 4
   4 Leaves 1-pinnate; sepals pubescent on both surfaces. Deciduous climber to 30m. Native; hedgerows, scrub and woodland on base-rich soils; locally abundant in En and Wa N to SW Yorks, natd elsewhere   *Traveller's-joy* - **C. vitalba** L.
   4 Leaves 2-pinnate; sepals pubescent only on lowerside. Deciduous climber to 6m. Intrd-natd; cliffs and dunes near sea; S En, Caerns   *Virgin's-bower* - **C. flammula** L.

13. RANUNCULUS L. - *Buttercups*

Herbaceous perennials or annuals, some aquatic; leaves spirally arranged, simple to much divided, with or without stipules; flowers usually in cymes, sometimes solitary and leaf-opposed, without a whorl of bracts below, actinomorphic; perianth of 2 whorls; sepals 3 or 5, sepaloid; petals 5 or fewer by reduction (to 0) or 7-12, white or yellow, petaloid, usually with a small nectar-secreting pit (nectar-pit) on inner face; stamens usually numerous, sometimes 5-10; carpels numerous, free, spirally arranged; fruit an achene, the style remaining shorter than fertile portion.

General key
1 Sepals 3; petals 7-12; many roots modified as whitish, swollen tubers with rounded apices. Stems to 25cm. Damp meadows, lawns, woods, hedgebanks and by water; common (*Lesser Celandine* - **R. ficaria** L.) 2
1 Sepals 5; petals usually 5, sometimes <5 or many; root tubers rarely present, if so then with finely tapering apices 5
   2 Tubers absent from leaf-axils after flowering period 3
   2 Cream-coloured tubers present in leaf-axils after flowering period 4
3 Petals 10-20 x 4-9mm. Native; throughout BI   **R. ficaria ssp. ficaria**
3 Petals 18-25 x 9-15mm. Intrd-natd; garden escape, rare in S En
R. ficaria ssp. chrysocephalus P.D. Sell
   4 Well-developed achenes 0-6 per head; petals mostly <15mm; pollen-grains mostly empty. Native; almost throughout BI
R. ficaria ssp. bulbilifer Lambinon

    4   All or ± all achenes well developed; petals mostly >15mm; pollen-
        grains mostly full. Intrd-natd; garden escape scattered in S
                            R. ficaria ssp. ficariiformis (F.W. Schultz) Rouy & Foucaud
5   Petals yellow *Key A*
5   Petals white, often yellow at base 6
    6   Plant erect, terrestrial; largest leaves basal; achenes not
        transversely ridged, often *flore pleno*. Stems to 60cm. Intrd-natd;
        damp places and by streams; scattered in N
                            ***Aconite-leaved Buttercup* - R. aconitifolius** L.
    6   Plant rarely erect, often aquatic; basal leaves 0; achenes
        transversely ridged, never *flore pleno* *Key B*

Key A - Sepals 5; petals yellow, usually 5
1   Leaves entire or toothed at margin, unlobed 2
1   At least some leaves divided at least 1/4 way to base 8
    2   Flowers 2-5cm across; achenes c.2.5mm. Stems to 120cm. Native;
        marshes and pondsides; scattered throughout BI but mostly
        only natd ***Greater Spearwort* - R. lingua** L.
    2   Flowers <2(2.5)cm across; achenes 1-2(2.3)mm 3
3   Lower leaves cordate at base 4
3   All leaves cuneate at base 5
    4   Stems erect; achenes tuberculate. Stems to 40cm. Native; marshy
        ground; Gloucs
                            ***Adder's-tongue Spearwort* - R. ophioglossifolius** Vill.
    4   Stems procumbent to decumbent; achenes smooth. Native;
        wet places. N Sc and Co Clare
                            R. flammula ssp. minimus (A. Benn.) Padmore
5   Stems erect to ± procumbent, usually rooting only at lower nodes;
    beak c.1/8-1/10 as long as rest of achene; widest petals usually >4mm
    across. Stems to 50cm. Native; wet places
    (***Lesser Spearwort* - R. flammula** L.) 6
5   Stems procumbent, rooting at all or most nodes; beak c.1/3-1/6 as
    long as rest of achene; widest petals usually <4mm across 7
    6   Lowest leaves persistent, with well developed leaf-blade.
        Common **R. flammula ssp. flammula**
    6   Lowest leaves falling early, with leaf-blade 0 or much reduced.
        NW Sc and NW Ir
                            R. flammula ssp. scoticus (E.S. Marshall) A.R. Clapham
7   Beak c.1/5 -1/6 as long as rest of achene; basal leaves usually >1.5mm
    wide. Native; barish lake shores; local in N Br (*R. flammula* x
    R. reptans) ***Loch Leven Spearwort* - R. x levenensis** Gornall
7   Beak c.1/3-1/4 as long as rest of achene; basal leaves usually <1.2mm
    wide. Native; bare lake shores; extinct except for occasional non-
    persistent rearrivals; Sc ***Creeping Spearwort* - R. reptans** L.
    8   Sepals strongly reflexed at anthesis 9
    8   Sepals not reflexed at anthesis 14
9   Plant perennial with swollen stem-base. Stems to 40cm. Native; dry
    grassland and dunes; common ***Bulbous Buttercup* - R. bulbosus** L.

13. RANUNCULUS

9 Plant annual, without swollen stem-base   10
- 10 Achenes c.1mm, on elongated receptacle, ± smooth on sides. Stems to 60cm. Native; marshy fields, ditches, ponds and streamsides; frequent   *Celery-leaved Buttercup* - **R. sceleratus** L.
- 10 Achenes >2mm, on ± spherical receptacle, usually with tubercles or spines on sides   11

11 Achenes 5-8mm including beak of 2-3mm; longest spines on sides of achene at least 1mm. Stems to 40cm. Intrd-natd; weed of cultivated ground; SW En   *Rough-fruited Buttercup* - **R. muricatus** L.

11 Achenes ≤5mm including beak of <1.5mm; longest tubercles on sides of achene <1mm or tubercles 0   12
- 12 Flowers <8mm across; receptacle glabrous. Stems to 40cm. Native; open ground, especially near coast; frequent in S, very local elsewhere   *Small-flowered Buttercup* - **R. parviflorus** L.
- 12 Flowers >9mm across; receptacle pubescent   13

13 Achenes with tubercles (if present) confined to edge of faces, close to border; beak 0.2-0.5(0.6)mm. Stems to 40cm. Native; grassland and cultivated land mostly near coast; frequent in SE, very local elsewhere   *Hairy Buttercup* - **R. sardous** Crantz

13 Achenes with tubercles (if present) covering faces; beak (0.5)0.6-1mm. Stems to 40cm. Intrd-natd; cultivated ground in extreme SW En   *St Martin's Buttercup* - **R. marginatus** d'Urv.
- 14 Plant with procumbent stems rooting at nodes. Stems to 60cm. Native; wet grassland, woods, streamsides, marshes and duneslacks, and as a weed; abundant   *Creeping Buttercup* - **R. repens** L.
- 14 Stems usually not procumbent, not rooting at nodes   15

15 Plant annual; receptacle pubescent; achenes with conspicuous spines. Stems to 60cm. Native; cultivated ground, cornfields; rare in En   *Corn Buttercup* - **R. arvensis** L.

15 Plant perennial; receptacle glabrous; achenes glabrous or pubescent but without spines   16
- 16 Some roots swollen into spindle-shaped tubers; achenes borne in an elongated head. Stems to 40cm. Native; grassy places damp in winter; Jersey   *Jersey Buttercup* - **R. paludosus** Poir.
- 16 All roots thin; achenes borne in a ± spherical head   17

17 Basal leaves with a reniform outline, glabrous to sparsely appressed-pubescent; achenes pubescent. Stems to 40cm. Native; woods and hedgebanks; frequent in S & C Br, very scattered elsewhere   *Goldilocks Buttercup* - **R. auricomus** L.

17 Basal leaves with a polygonal or polygonal-rounded outline, conspicuously pubescent; achenes glabrous. Stems to 1m. Native; grassland, especially damp and calcareous; abundant
                                          Meadow Buttercup - **R. acris** L.

Key B - Sepals 5; petals white, 5; basal leaves 0 (subgenus *Batrachium*)
1 Laminar (floating or aerial) leaves the only ones present   2
1 Capillary (normally submerged) leaves present   6
- 2 Receptacle pubescent; leaves divided usually >1/2 way into

30. RANUNCULACEAE

   3(-5) main lobes   3
 2 Receptacle glabrous; leaves divided <1/2 way into 3-5(7) main lobes   5
3 Petals >5.5mm, contiguous at anthesis; achenes narrowly winged and borne on ovoid receptacle when completely mature. Heterophyllous, or with capillary leaves only, or rarely with laminar leaves only. Native; ditches and ponds near sea, often brackish; scattered   **Brackish Water-crowfoot - R. baudotii** Godr.
3 Petals <6mm, not contiguous at anthesis; achenes not winged, borne on ± spherical receptacle   4
 4 Leaves usually 3-lobed, rarely 5-lobed; petals ≤4.5mm, c.1.5x as long as sepals; pedicels strongly reflexed when fruit ripe. Heterophyllous or leaves all laminar. Native; wet mud, ditches and ponds; S & W En and Wa

       Three-lobed Crowfoot - R. tripartitus DC.
 4 Leaves often 5-lobed; petals ≤6mm, c.2x as long as sepals; pedicels remaining erect when fruit ripe. Heterophyllous or leaves all laminar. In and by pools in New Forest (S Hants) (*R. omiophyllus* x *R. tripartitus*)

       New Forest Crowfoot - R. x novae-forestae S.D. Webster
5 Leaf-lobes broadest at their base; leaf-sinuses very open, obtuse to subacute; petals <4.5mm, little longer than sepals. Leaves all laminar. Native; on mud and in shallow water; frequent

       Ivy-leaved Crowfoot - R. hederaceus L.
5 Leaf-lobes broadest above their base; leaf-sinuses narrowly acute; petals >4.5mm, 2-3x as long as sepals. Leaves all laminar. Native; on mud and in shallow ponds and streams; S & W Br, SW Sc, S Ir and Sark (CI)   **Round-leaved Crowfoot - R. omiophyllus** Ten.
 6 Plant with both laminar and capillary leaves   7
 6 Plant with only capillary leaves   13
7 Petals ≤6mm, not contiguous at anthesis   8
7 Petals 5-20mm, contiguous at anthesis   9
 8 Laminar leaves usually 3-lobed, rarely 5-lobed; petals ≤4.5mm, c.1.5x as long as sepals; pedicels strongly reflexed when fruit ripe   (see couplet 4) **R. tripartitus**
 8 Laminar leaves often 5-lobed; petals ≤6mm, c.2x as long as sepals; pedicels remaining erect when fruit ripe

       (see couplet 4) **R. x novae-forestae**
9 Achenes glabrous when immature, narrowly winged and borne on ovoid receptacle when completely mature   (see couplet 3) **R. baudotii**
9 Achenes usually pubescent when immature, not winged at maturity, borne on ± spherical receptacle   10
 10 Pedicel in fruit rarely >50mm, shorter than petiole of opposed laminar leaf; petals <10mm, with circular nectar-pit. Heterophyllous or with capillary leaves only. Native; ponds, ditches, canals and slow rivers; frequent

       Common Water-crowfoot - R. aquatilis L.
 10 Pedicel in fruit usually >50mm, usually longer than petiole of

13. RANUNCULUS

        opposed laminar leaf; petals usually >10mm, with pear-shaped nectar-pit 11
11 Leaves intermediate between laminar and capillary usually common, mostly as capillary leaves with some slightly flattened segments; mature achenes not formed. Heterophyllous. Native; slow rivers; scattered in Br, Co Antrim (*R. peltatus* x *R. fluitans*)
        ***Kelso Water-crowfoot* - R. x kelchoensis** S.D. Webster
11 Leaves intermediate between laminar and capillary rarely present; fertile achenes regularly produced 12
   12 Capillary leaves rigid or flaccid, shorter than adjacent stem internode, with markedly divergent segments. Heterophyllous or with capillary leaves only. Native; ponds, ditches, canals and slow rivers; frequent   ***Pond Water-crowfoot* - R. peltatus** Schrank
   12 Capillary leaves flaccid, usually longer than adjacent stem internode, with ± parallel segments. Heterophyllous. Native; rivers, usually swift-flowing; frequent in Ir, Wa and W En   ***Stream Water-crowfoot* - R. penicillatus** (Dumort.) Bab. ssp. **penicillatus**
13 Leaves rigid, circular in outline, the segments all lying in 1 plane. Leaves all capillary. Native; ponds, canals and slow-flowing rivers, mainly base-rich; sparsely scattered over C & S BI
        ***Fan-leaved Water-crowfoot* - R. circinatus** Sibth.
13 Leaves with segments not lying in 1 plane 14
   14 Achenes narrowly winged and borne on ovoid receptacle when completely mature   (see couplet 3) **R. baudotii**
   14 Achenes not winged at maturity, borne on ± spherical receptacle 15
15 Petals usually <6mm, with crescent-shaped nectar-pit, not contiguous at anthesis. Leaves all capillary. Native; ponds, ditches, canals and slow rivers; scattered but often common
        ***Thread-leaved Water-crowfoot* - R. trichophyllus** Chaix
15 Petals >5mm, with circular or pear-shaped (rarely crescent-shaped) nectar-pit, contiguous at anthesis 16
   16 Receptacle pilose to glabrous; mature leaves longer than adjacent internode, rarely >4x forked, with ± parallel segments. Leaves all capillary. Native; mostly larger rivers of moderate flow-rate; scattered in En, Wa and S Sc, Co Antrim
        ***River Water-crowfoot* - R. fluitans** Lam.
   16 Receptacle densely pubescent; mature leaves shorter to longer than adjacent internode, usually some >4x forked, with divergent to ± parallel segments 17
17 Pedicel in fruit rarely >50mm; petals <10mm, with circular nectar-pit   (see couplet 10) **R. aquatilis**
17 Pedicel in fruit usually >50mm; petals usually >10mm, with pear-shaped (rarely crescent-shaped) nectar-pit 18
   18 Mature achenes not formed; pedicels not elongating. Leaves all capillary. (*R. fluitans* x *R. trichophyllus* & *R. fluitans* x *R. aquatilis*) Native; rivers, usually swift-flowing; very scattered in En, Wa and S Sc   ***Wirtgen's Water-crowfoot* - R. x bachii** Wirtg.
   18 Fertile achenes regularly produced from each flower; pedicels

   elongating in fruit 19
19 Ultimate segments of well-developed leaves <100
             (see couplet 12) **R. peltatus**
19 Ultimate segments of well-developed leaves >100, often >200.
  Leaves all capillary. Native; rivers, usually swift-flowing; En, Wa,
  N Ir and S Sc (see couplet 12)
     R. penicillatus ssp. **pseudofluitans** (Syme) S.D. Webster

Apart from those covered above, c.10 hybrids in subgenus *Batrachium* (Key B) occur where 2 spp. grow or once grew together. All are largely sterile.

14. ADONIS L. - *Pheasant's-eye*

Annuals; leaves spirally arranged, 3-pinnate, without stipules; flowers ± solitary, without a whorl of bracts below, actinomorphic; perianth of 2 whorls; sepals 5, sepaloid; petals 5-8, red with dark base, petaloid, not nectariferous; stamens numerous; carpels numerous, free, spirally arranged; fruit an achene, the style remaining shorter than fertile portion.

1 Stems to 40cm. Intrd-natd; formerly a natd weed of cultivated
  ground in S, now mostly a rare casual *Pheasant's-eye* - **A. annua** L.

15. MYOSURUS L. - *Mousetail*

Annuals; leaves all basal, simple, without stipules; flowers solitary, without a whorl of bracts below, actinomorphic; perianth of 2 whorls; sepals 5(-7), ± sepaloid, each with a small basal spur; petals 5(-7), greenish, in the form of tubular nectaries; stamens 5-10; carpels numerous, free, spirally arranged; fruit an achene, the style remaining shorter than fertile portion.

1 Leaves 1-8cm; scapes to 10cm. Probably native; damp arable
  ground; rare in En *Mousetail* - **M. minimus** L.

16. AQUILEGIA L. - *Columbines*

Herbaceous perennials; leaves spirally arranged, 2-ternate, without stipules; flowers in cymes, without a whorl of bracts below, actinomorphic; perianth of 2 whorls; sepals 5, petaloid; petals 5, petaloid, blue, sometimes pink or white, each with a long, backwards-directed nectariferous spur; stamens numerous, the inner c.10 flat and sterile; carpels 5(-10), free; fruit a follicle.

1 Stems to 1m, branched; leaves pubescent on lowerside; petal-spur
  15-22mm, strongly hooked at end; follicles 15-20mm. Native;
  woods, fens and damp calcareous grassland and scree; local in BI,
  also a frequent escape *Columbine* - **A. vulgaris** L.
1 Stems to 25cm, often simple; leaves glabrous to sparsely pubescent
  on lowerside; petal-spur 10-16mm, gently curved; follicles 12-17mm.
  Intrd-natd; planted on rock-ledges at c.900m, Angus, known since
  1895 *Pyrenean Columbine* - **A. pyrenaica** DC.

17. THALICTRUM L. - *Meadow-rues*

Herbaceous perennials; leaves spirally arranged, pinnate to ternate, stipulate; flowers in racemes of compound inflorescences without a whorl of bracts below, actinomorphic; perianth of 1 whorl of 4 small but ± petaloid sepals; stamens numerous, more conspicuous than sepals; carpels 2-15, free; fruit an achene, the style remaining shorter than fertile portion.

1 Inflorescence a simple raceme; plant rarely >15cm. Stems rarely >15cm. Native; grassy and rocky places on mountains; local in N & NW   *Alpine Meadow-rue* - **T. alpinum** L.
1 Inflorescence branched with >1 flower per branch; plant usually >15cm   2
   2 Filaments thickened, wider than anthers, white to lilac or pink. Stems to 1m. Intrd-natd; garden throwout in grassy places; scattered in Br   *French Meadow-rue* - **T. aquilegiifolium** L.
   2 Filaments thin, narrower than anthers, yellowish   3
3 Sepals pink to lilac, c. as long as stamens. Stems to 1m. Intrd-natd; in grassy area in Cambs   *Chinese Meadow-rue* - **T. delavayi** Franch.
3 Sepals yellow, much shorter than stamens   4
   4 Inflorescence diffuse; stamens ± pendent; leaflets not or little longer than wide; achenes with 8-10 ribs. Stems to 1.2m. Native; in varied, usually calcareous habitats; scattered, but absent from C & SE En except as a garden throwout
    Lesser Meadow-rue - **T. minus** L.
   4 Inflorescence ± dense; stamens held stiffly erect to patent; leaflets much longer than wide; achenes with 6 ribs. Stems to 1.5m. Native; fens, streamsides and wet meadows; very scattered, mostly in E En   *Common Meadow-rue* - **T. flavum** L.

31. BERBERIDACEAE - *Barberry family*

Shrubs with yellow wood; distinguishable by the shrubby often spiny habit, yellow perianth of several whorls, 1-celled ovary and succulent berry.

1 Stems with spines; leaves simple   **1. BERBERIS**
1 Stems without spines; leaves pinnate   **2. MAHONIA**

1. BERBERIS L. - *Barberries*

Shrubs; stems with (1)3(-7)-partite spines bearing short-shoots in their axils; leaves deciduous or evergreen, simple; flowers in axillary racemes, fascicles or panicles, or solitary; perianth of 4-5 whorls each of 3 segments, various shades of yellow; stamens 6; fruit a few-seeded, red to purple-black, often bloomed berry.

1 Leaves deciduous to semi-evergreen, entire to serrate, the apex and teeth without spines or with weak spines much <1mm; fruit red   2
1 Leaves evergreen, entire to spinose-toothed, but apex on most or all

leaves with sharp spine ≥1mm; fruit bluish-black   5
- 2 Leaves 2.5-6cm, with >10 teeth on each side; flowers in pendent racemes. Deciduous shrub to 3m. Probably intrd-natd; hedges and rough ground, scattered over BI   **Barberry - B. vulgaris** L.
- 2 Leaves 1-3cm, entire or with <10 teeth on each side; flowers in 1-many-flowered fascicles or panicles   3

3 Leaves with several teeth on each side; flowers numerous in dense panicles. Deciduous shrub to 2m. Intrd-natd; on chalk, on walls, in hedges; scattered in Br   **Clustered Barberry - B. aggregata** C.K. Schneid.

3 Leaves usually entire; flowers in loose fascicles of 1-6   4
- 4 Twigs minutely pubescent; spines mostly 3-partite; leaves oblanceolate. Semi-evergreen shrub to 1m. Intrd-natd; sometimes bird-sown in a range of habitats in C & S

                 ***Mrs Wilson's Barberry* - B. wilsoniae** Hemsl.
- 4 Twigs glabrous; spines mostly simple; leaves obovate. Deciduous shrub to 2m. Intrd-natd; occasionally bird-sown in En and Wa

                 ***Thunberg's Barberry* - B. thunbergii** DC.

5 Flowers in fascicles without or with very short common peduncle   6
5 Flowers in racemes with common peduncle   9
- 6 Leaves >3cm, with >3 spinose teeth on each margin   7
- 6 Leaves <3cm, with <3 spinose teeth on each margin   8

7 Leaves linear to linear-lanceolate; twigs ± terete; fruit with ± sessile stigma. Evergreen shrub to 1.5m. Intrd-natd; on walls, in hedges and on riverbanks; scattered in C & S Br

                 ***Gagnepain's Barberry* - B. gagnepainii** C.K. Schneid.

7 Leaves narrowly elliptic; twigs with raised ridges; fruit with distinct but very short style. Evergreen shrub to 3m. Intrd-natd; walls and roadsides; scattered in Br   ***Chinese Barberry* - B. julianae** C.K. Schneid.
- 8 Leaves elliptic to obovate, with flat margins. Evergreen shrub to 2m. Intrd-natd; on commonland in W Norfolk, occasionally elsewhere in Br   **Box-leaved Barberry - B. buxifolia** Lam.
- 8 Leaves narrowly elliptic, with revolute margins. Evergreen shrub to 3m. Intrd-natd; occasionally self-sown in hedges and waysides; scattered (*B. darwinii* x *B. empetrifolia* Lam.)

                 ***Hedge Barberry* - B. x stenophylla** Lindl.

9 Spines <1cm, 3-7-partite; leaves 1-3cm; flowers orange. Evergreen shrub to 3m. Intrd-natd; occasionally bird-sown, scattered in BI

                 ***Darwin's Barberry* - B. darwinii** Hook.

9 Spines 1-3cm, 1-3-partite; leaves 2.5-8cm; flowers yellow. Semi-evergreen shrub to 3m. Intrd-natd; hedges in S Somerset, sometimes bird-sown elsewhere in S En   ***Great Barberry* - B. glaucocarpa** Stapf

2. MAHONIA Nutt. - *Oregon-grapes*

Shrubs; stems without spines; leaves evergreen, pinnate; flowers in axillary fasciculate racemes, yellow, structure as in *Berberis*; fruit a blue-black, bloomed berry with few seeds.

1 Somewhat stoloniferous shrub with ascending stems to 1.5m; leaflets

5-9, 3-8cm, c.2x as long as wide, glossy on upperside, not papillose on lowerside, with c.5-15 spinulose teeth on each side. Intrd-natd; scrub, woodland and hedges throughout Br

Oregon-grape - **M. aquifolium** (Pursh) Nutt.

1 Strongly stoloniferous shrub to 50cm, with decumbent stems; leaflets 5-7, <2x as long as wide, dull on upperside, minutely papillose on lowerside, with c.8-22 spinulose teeth on each side. Intrd-natd; woodland; Wilts, Cambs and Man (*M. aquifolium* x *M. repens* (Lindl.) G. Don) *Newmarket Oregon-grape* - **M. x decumbens** Stace

32. PAPAVERACEAE - *Poppy family*

Herbaceous annuals or perennials usually with white or yellow latex; the 2 sepals, 4(-6) showy petals, and distinctive latex usually distinguish this family, but *Macleaya* lacks petals.

1 Petals 0; flowers numerous in large panicles **7. MACLEAYA**
1 Petals conspicuous; flowers solitary or in <10-flowered inflorescences 2
   2 Sap watery; sepals fused, shed as a hood as flower opens; receptacle raised above base of ovary (flowers perigynous) **6. ESCHSCHOLZIA**
   2 Sap a white to orange latex; sepals often adherent but not fused; flowers hypogynous 3
3 Capsule >10x as long as wide; stigma with 2 lobes 4
3 Capsule <6x as long as wide; stigma with >3 lobes or rays 5
   4 Flowers >3cm across, 1-2 per leaf-axil; capsule >10cm, 2-celled **4. GLAUCIUM**
   4 Flowers <3cm across, mostly in umbels of >3; capsule <6cm, 1-celled **5. CHELIDONIUM**
5 Lobes and teeth of leaves tipped with long weak spine **3. ARGEMONE**
5 Leaves not spiny 6
   6 Petals white to red or mauve; style absent; stigma a 4-20-rayed disk **1. PAPAVER**
   6 Petals yellow; style present; stigma 4-6-lobed **2. MECONOPSIS**

1. PAPAVER L. - *Poppies*

Annuals or perennials with white latex unless otherwise stated; leaves glabrous to pubescent; flowers solitary; petals red or mauve to white; capsule 1-celled, with 4-20 incomplete placentae projecting inwards, opening by pores just below the persistent stigma; stigma sessile, a 4-20-rayed flat disk.

1 Tufted perennials with basal rosette(s) of leaves at flowering 2
1 Annuals usually without basal leaves at flowering 3
   2 Petals <45mm; capsule narrowly obovoid to clavate, widest immediately below stigma. Stems to 60cm. Intrd-natd; walls,

roadsides and rough ground; scattered throughout Br
Atlas Poppy - **P. atlanticum** (Ball) Coss.

2 Petals >45mm; capsule obovoid, widest c.3/4 way from base to apex. Stems to 1m. Intrd-natd; various habitats on well-drained soils in En and Sc (*P. orientale* auct.)
Oriental Poppy - **P. pseudoorientale** (Fedde) Medw.

3 Stem-leaves strongly glaucous, clasping stem at base, toothed or lobed but not to near midrid. Stems to 50(100)cm. (***Opium Poppy - P. somniferum*** L.)   4

3 Stem-leaves green to slightly glaucous, not clasping stem, mostly divided as far as midrib   5

   4 Glabrous or very sparsely bristly; leaves rather shallowly lobed; stigma-rays never overlapping. Intrd-natd; common casual over most of BI, often persistent   **P. somniferum ssp. somniferum**

   4 Stems, leaves and sepals conspicuously bristly; leaves rather deeply lobed; capsule relatively narrow, with overlapping stigma-rays. Intrd-casual; rare on tips etc. in Br
   P. somniferum ssp. setigerum (DC.) Arcang.

5 Capsule with long stiff hairs; filaments dilated distally   6
5 Capsule glabrous; filaments not dilated   7

   6 Capsule <1.5cm, c. as long as wide. Stems to 50cm. Probably native; arable fields and waste places; scattered in C & S Br, mainly in E & S En on calcareous soils
Rough Poppy - **P. hybridum** L.

   6 Capsule >1.5cm, at least some >2x as long as wide. Stems to 45cm. Probably native; arable fields and waste places; scattered in C & S Br, mainly in C & S En on light soils
Prickly Poppy - **P. argemone** L.

7 Capsule <2x as long as wide. Stems to 60(80)cm. Native; arable ground, roadsides and waste places; throughout BI, common in C & S Br   *Common Poppy* - **P. rhoeas** L.

7 Capsule >2x as long as wide. Stems to 60cm. Probably native; similar places to *P. rhoeas* (***Long-headed Poppy - P. dubium*** L.)   8

   8 Latex white or cream, becoming brown to black when dry; upper leaves with ultimate lobes often >1.5mm wide; anthers brownish- to bluish-black. Most of BI   **P. dubium ssp. dubium**

   8 Latex yellow or quickly turning yellow on exposure to air, becoming reddish when dry; upper leaves with ultimate lobes rarely >1.5mm wide; anthers often yellow. Sparsely scattered in C & S, mostly S En   **P. dubium ssp. lecoqii** (Lamotte) Syme

P. rhoeas x *P. dubium* = ***P. x hungaricum*** Borbás occurs very rarely.

2. MECONOPSIS Vig. - *Welsh Poppy*

Perennials with yellow latex; leaves nearly glabrous; flowers solitary; petals yellow; capsule 1-celled, with 4-6 incomplete placentae projecting inwards, opening by elongated pores at apex; style short but distinct; stigma ± capitate, 4-6-lobed.

1 Stems to 60cm. Native; shady places among rocks or under trees; Wa, SW En and scattered in Ir, also natd in these areas and elsewhere *Welsh Poppy* - **M. cambrica** (L.) Vig.

3. ARGEMONE L. - *Mexican Poppy*
Annuals with yellow latex; leaves with a weak spine at tip of lobes and teeth; flowers solitary, with usually 2 leaf-like bracts just below; petals 4-6, yellow; capsule 1-celled, with 4-6 incomplete placentae projecting inwards, opening by 4-6 elongated pores at apex; style short but distinct; stigma ± capitate, 4-6-lobed.

1 Stems to 90cm, usually much less. Intrd-casual; wool- or grain-alien and garden escape in waste places; very scattered in S & C En
Mexican Poppy - **A. mexicana** L.

4. GLAUCIUM Mill. - *Horned-poppies*
Biennials to perennials with yellow latex; leaves glaucous, the lower pubescent; flowers solitary; petals yellow; capsule 2-celled, opening from above ± along its length by 2 valves and leaving the seeds embedded in septum; style ± 0; stigma ± capitate, 2-lobed.

1 Stems to 90cm. Native; on maritime shingle and less often other substrata; common on coasts of En, Wa and CI, local in Ir and S Sc
Yellow Horned-poppy - **G. flavum** Crantz

5. CHELIDONIUM L. - *Greater Celandine*
Perennials with orange latex; leaves nearly glabrous; flowers in umbels of (2)3-6; petals yellow; capsule 1-celled, opening from below along whole length by 2 valves; style short but distinct; stigma ± capitate, 2-lobed.

1 Stems to 90cm. Possibly native; hedgerows, walls and other marginal habitats; throughout most of BI *Greater Celandine* - **C. majus** L.

6. ESCHSCHOLZIA Cham. - *Californian Poppy*
Annuals to perennials with watery sap; leaves glabrous, glaucous; flowers solitary; petals yellow to orange; capsule 1-celled, opening from below along its whole length by 2 valves; style very short; stigma deeply 4-6-lobed.

1 Stems to 60cm. Intrd-natd; frequent casual on tips and roadsides, natd and perennating on dunes, walls and cliff-tops in Guernsey, rarely elsewhere *Californian Poppy* - **E. californica** Cham.

7. MACLEAYA R. Br. - *Plume-poppies*
Rhizomatous perennials with orange latex; leaves glabrous or sparsely pubescent, white-glaucous on lowerside; flowers small, in crowded, large, terminal panicles; petals 0; capsule 1-celled, opening from above by 2 valves; style very short; stigma deeply 2-lobed.

1 Stems to 2.5m. Intrd-natd; garden escape and throwout, long persisting; En, Ir and S Sc (*M. cordata* (Willd.) R. Br. x *M. microcarpa* (Maxim.) Fedde)   *Hybrid Plume-poppy* - **M. x kewensis** Turrill

33. FUMARIACEAE - *Fumitory family*

Herbaceous annuals or perennials with watery sap; the distinctive zygomorphic flowers with 2 sepals, 4 petals and 2 tripartite stamens are unique.

1 Flowers with 2 spurred petals (the upper and lower)   **1. DICENTRA**
1 Flowers with 1 spurred petal (the upper)   2
  2 Fruit a 1-seeded achene   **5. FUMARIA**
  2 Fruit a dehiscent capsule, usually with >1 seed   3
3 Annual; leaves with tendrils; flowers <8mm   **4. CERATOCAPNOS**
3 Perennial; leaves without tendrils; flowers >8mm   4
  4 Stems branched, arising from ± cylindrical stock; flowers cream to yellow   **3. PSEUDOFUMARIA**
  4 Stems simple, arising from ± globose tuber; flowers white to purple   **2. CORYDALIS**

1. DICENTRA Bernh. - *Bleeding-hearts*
Perennials, with short branched rhizome; aerial stems with terminal inflorescence; leaves all or mostly basal; flowers pink; upper and lower petals spurred at base; fruit a capsule; seeds several, with an aril.

1 Flowering stems to 30cm. Intrd-natd; shady places, especially by streams; mainly in W & N Br   *Bleeding-heart* - **D. formosa** (Haw.) Walp.

2. CORYDALIS DC. - *Corydalises*
Perennials; stems usually simple; inflorescence 1, terminal; leaves present on aerial stem; flowers white to purple or yellow; upper petal spurred at base; fruit a capsule; seeds several, with aril.

1 Flowers yellow; leaves 2-4-pinnate, fern-like; subterranean tuber 0. Stems to 25cm. Intrd-natd; on walls; Surrey and S Hants   *Fern-leaved Corydalis* - **C. cheilanthifolia** Hemsl.
1 Flowers white to purple; leaves 2-3-ternate, not fern-like; stems arising from subterranean tuber   2
  2 Bracts narrowly lobed; stem with large scale below lowest leaf (often low down). Stems to 20cm. Intrd-natd; woods and hedges; scattered over En and Wa   *Bird-in-a-bush* - **C. solida** (L.) Clairv.
  2 Bracts ± entire; stem without large scale below leaves. Stems to 20cm. Intrd-natd; habitat and distribution as for *C. solida* but much less common   *Hollowroot* - **C. cava** (L.) Schweigg. & Körte

3. PSEUDOFUMARIA Medik. - *Corydalises*
Perennials, arising from ± cylindrical stock; stems branched; inflorescences several, terminal and leaf-opposed; flowers cream to yellow; upper petal spurred at base; fruit a capsule; seeds several, with aril.

1 Petioles ridged but not winged; flowers 12-18mm including basal spur 2-4mm, yellow; seeds shiny. Stems to 30cm. Intrd-natd; walls and stony places over most of BI *Yellow Corydalis* - **P. lutea** (L.) Borkh.
1 Petioles very narrowly winged; flowers 10-18mm including basal spur 1-2mm, cream with yellow tip; seeds matt. Stems to 30cm. Intrd-natd; walls and stony places; S Br *Pale Corydalis* - **P. alba** (Mill.) Lidén

4. CERATOCAPNOS Durieu - *Climbing Corydalis*
Annuals; stems branched, climbing by means of leaf-tendrils; inflorescences several, leaf-opposed; flowers pale cream; upper petal spurred at base; fruit a capsule; seeds (1)2-3, with aril.

1 Stems to 75cm, scrambling; flowers 4-6mm including basal spur to 1mm. Native; woods and other shady places often on rocks; scattered over most of BI *Climbing Corydalis* - **C. claviculata** (L.) Lidén

5. FUMARIA L. - *Fumitories*
Annuals to 1m but often much less; stems much-branched, scrambling, thin; leaves all on stem, 2-4-pinnate; inflorescences leaf-opposed racemes; flowers white to purple; upper petal spurred at base; upper and lateral petals darker coloured at tip; fruit a 1-seeded achene; seeds without aril.

All spp. are similar in appearance and are distinguished by inflorescence, flower and fruit characters. The upper petal has a dorsal ridge and lateral margins which may be bent upwards to ± conceal the ridge or spread laterally to reveal it. The lower petal appears parallel-sided to strongly spathulate in top or bottom view according to the relative expansion of the margins distally. Flower colours given ignore the often very dark petal tips. Flower length is measured from end of basal spur to tip of longest petal on fresh material; dried material has smaller floral parts (especially sepals). Fruit shape and measurements refer to those seen in widest profile on mature, dried fruits. It is essential to base determinations on well-grown, non-shaded material in early or mid flowering season. Late or shade-grown plants may be very atypical, with short (often cleistogamous) paler petals, longer narrower sepals, relatively long bracts and less or more recurved fruiting pedicels.

1 Flowers ≥9mm; sepals (2)3-6.5mm; lower petal usually ± parallel-sided, rarely subspathulate 2
1 Flowers ≤9mm; sepals <4mm; lower petal distinctly spathulate 10
    2 Lower petal with broad margins; fruit 2.75-3 x 2.75-3mm; flowers 12-14mm. Native; arable land and waste places; Cornwall and Scillies *Western Ramping-fumitory* - **F. occidentalis** Pugsley
    2 Lower petal with narrow margins; fruit ≤2.75 x 2.5mm; flowers

	usually mostly <12mm	3
3	Fruiting pedicels rigidly recurved to patent; sepals 4-6(6.5)mm	4
3	Fruiting pedicels erecto-patent, often not rigid; sepals (2)3-5mm	6

  4 Petals purple; upper petal with erect margins ± concealing tip of dorsal ridge. Flowers 10-13mm. Native; arable and waste ground and hedges; rare and very sparsely scattered over BI, mainly W *Purple Ramping-fumitory* - **F. purpurea** Pugsley

  4 Petals usually creamy-white, sometimes pink to red; upper petal with spreading margins not concealing dorsal ridge. Flowers 10-13(14)mm. Native; arable and waste ground and hedges (*White Ramping-fumitory* - **F. capreolata** L.) 5

5 Bracts usually shorter than fruiting pedicels; fruit c.2 x 2.mm, with rounded to very obtuse apex. CI **F. capreolata ssp. capreolata**

5 Bracts ± equalling fruiting pedicels; fruit c.2.5 x 2.5mm, with truncate apex. Scattered over most of Br and Ir, mostly in S & W

        F. capreolata ssp. babingtonii (Pugsley) P.D. Sell

	6 Lower petal subspathulate	7
	6 Lower petal ± parallel-sided	8
7	Flowers 9-10mm (see couplet 9) **F. muralis**	
7	Flowers 10-11mm. Very rare in S (*F. muralis* x *F. officinalis*)	

        Painter's Fumitory - **F. x painteri**

  8 Sepals 2-3 x 1-2mm, with sharp forward-pointing teeth all round margin; fruit distinctly rugose when dry. Flowers (8)9-11(12)mm. Native; arable and waste ground and hedgebanks; scattered in BI, mainly W *Tall Ramping-fumitory* - **F. bastardii** Boreau

  8 Sepals 3-5 x 1.5-3mm, with outward-pointing teeth in basal part of margin only or subentire; fruit smooth to finely rugose when dry 9

9 Flowers (8)9-11(12)mm; sepals usually dentate near base; racemes c. as long as peduncles. Native; arable and waste ground and hedgebanks; scattered over most of BI, the commonest *Fumaria* in W and CI *Common Ramping-fumitory* - **F. muralis** W.D.J. Koch

9 Flowers (10)11-13mm; sepals subentire to denticulate near base; racemes distinctly longer than peduncles. Native; cultivated ground; very rare in S En and Guernsey

        Martin's Ramping-fumitory - **F. reuteri** Boiss.

	10 Flowers ≥6mm; sepals ≥1.5mm	11
	10 Flowers 5-6mm; sepals 0.5-1.5mm	15

11 Bracts at least as long as pedicels, often longer. Flowers 6-7mm. Native; arable land, frequent on well-drained soils in SE En, rare elsewhere *Dense-flowered Fumitory* - **F. densiflora** DC.

11 Bracts shorter than pedicels 12

  12 Fruits truncate to retuse at apex, distinctly wider than long. Flowers (6)7-8(9)mm. Native; cultivated and waste ground; all over BI, the commonest *Fumaria* in Sc and E & C En (*Common Fumitory* - **F. officinalis** L.) 13

  12 Fruits rounded to subacute at apex, narrower than to as wide as long 14

13 Well-formed racemes >20-flowered; sepals 2-3.5 x 1-1.5mm.

Frequent **F. officinalis** ssp. **officinalis**
13 Racemes <20-flowered; sepals 1.5-2 x 0.75-1mm. Perhaps commoner than ssp. *officinalis* on light soils in E En, rare elsewhere
F. officinalis ssp. **wirtgenii** (W.D.J. Koch) Arcang.
14 Sepals 2-3(3.5) x 1-2mm, with sharp forward-pointing teeth all round margin; fruit distinctly rugose when dry
(see couplet 8) **F. bastardii**
14 Sepals (2.7)3-5 x 1.5-3mm, with outward-pointing teeth in basal part of margin only; fruit smooth to finely rugose when dry
(see couplet 9) **F. muralis**
15 Corolla white to pale pink; bracts at least as long as fruiting pedicels. Native; arable land usually on chalk; E, SE & SC En
Fine-leaved Fumitory - **F. parviflora** Lam.
15 Corolla pink; bracts shorter than fruiting pedicels. Native; arable land usually on chalk; SE & SC En, casual N to Angus
Few-flowered Fumitory - **F. vaillantii** Loisel.

3 other hybrids have been recorded very rarely.

34. PLATANACEAE - *Plane family*

An unmistakable deciduous tree, with palmately lobed alternate leaves, and monoecious flowers and fruits in stalked spherical clusters.

1. PLATANUS L. - *Planes*

1 Tree to 44m. Intrd-natd; abundantly planted as street and park tree, especially in S En, and often producing seedlings (?*P. occidentalis* L. x *P. orientalis* L.) *London Plane* - **P. x hispanica** Münchh.

35. ULMACEAE - *Elm family*

Unmistakable deciduous trees with asymmetric leaf-bases, flowers in small axillary clusters produced before leaves, and achenes with 2 wide wings extending beyond both base and apex.

1. ULMUS L. - *Elms*

An extremely difficult genus, having been interpreted in widely different ways. A complex hybrid origin of many taxa is probable, but parentages are often obscure. This account treats the 4 most distinctive native taxa as spp. Dutch Elm Disease has killed most trees in much of S and C Br, including the Midlands where the greatest variation occurs. Over large areas the elm population exists solely or largely as hedgerow suckers, and is now largely unidentifiable.

Only leaves from the middle of short-shoots in high summer should be used for identification; leaves from long-shoots, suckers, epicormic shoots or Lammas shoots must be avoided. Leaf lengths are measured from the base of the longer side of the leaf-blade to its apex. Tree outlines refer to mature, solitary specimens. All taxa except *U. glabra* and some *U. x vegeta* produce abundant suckers.

1 Pedicels ≥3x as long as flowers and most ≥1x as long as fruits; leaves glabrous or very softly pubescent. Tree to 21m. Intrd-natd; vigorously suckering from old trees in wood in Cards
European White-elm - **U. laevis** Pall.
1 Pedicels shorter than flowers and very much shorter than fruits; leaves glabrous or harshly pubescent 2
   2 Rust-coloured hairs abundant on buds; leaves >7cm, very rough on upperside, with >12 pairs of lateral veins; petiole <3mm, most of it overlapped by base of long-side of leaf-blade. Tree to 37m. Native; woods and hedgerows, especially on limestone; widely spread but much commoner in N & W *Wych Elm* - **U. glabra** Huds.
   2 Rust-coloured hairs 0 or present on buds; leaves usually <7cm, if >7cm smooth on upperside, with c.5-18 pairs of lateral veins; petiole usually >(3)5mm, not or partly overlapped by base of leaf-blade 3
3 Leaves usually >7cm, with length x width >28(cm); rust-coloured hairs often present on buds 4
3 Leaves usually <7cm, with length x width <28(cm); rust-coloured hairs 0 on buds (except in *U. glabra* hybrids) 5
   4 Leaves almost 2x as long as wide, acuminate at apex, with 12-18 pairs of lateral veins; tree outline broadly obovate to orbicular, with long branches from low down. Tree to 32m. Native; hedgerows and copses; C En and E Anglia, and widely planted (*U. glabra* x *U. minor*) *Huntingdon Elm* - **U. x vegeta** (Loudon) Ley
   4 Leaves distinctly <2x as long as wide, acute to shortly acuminate at apex, with 10-14 pairs of lateral veins; tree outline obovate, narrow below but with spreading branches above. Tree to 32m. Native; hedgerows; C & S En and CI, and widely planted (?*U. glabra* x *U. minor*, or *U. glabra* x *U. minor* x *U. plotii*)
Dutch Elm - **U. x hollandica** Mill.
5 Leaf width/length ratio >0.75; leaves usually rough on upperside; tree outline obovate to oblong, with strong branches at all levels. Tree to 33m. Probably native; hedgerows; C & S Br, intrd in Sc, Ir and CI *English Elm* - **U. procera** Salisb.
5 Leaf width/length ratio <0.75; leaves usually smooth on upperside, if rough then tree outline very narrow with no strong branches 6
   6 Tree outline narrow but irregular, with leading shoot arching or pendent; strong branches 0, all partly pendent; short-shoots mostly continuing growth as long-shoots; leaves usually rough on upperside. Tree to 20m. Native; hedgerows; C En, possibly E Wa *Plot's Elm* - **U. plotii** Druce
   6 Tree outline various but never as last, with erect leading shoot; some strong branches usually present, only the lower or 0

pendent; short-shoots rarely continuing growth; leaves smooth on upperside. Tree to 31m. Native; hedgerows and copses (**U. minor** Mill.) 7
7 Outline ± spreading, with major branches <1/2 way up tree at least some of which become horizontal or pendent; leaves usually strongly asymmetric at base. Kent to S Wa and C En

Small-leaved Elm **- U. minor ssp. minor**
7 Outline very narrow, or broad at the top, always with all main branches ascending, with no or very few major branches in lower 1/2 of tree; leaves usually weakly asymmetric at base 8
   8 Trunk persisting to tree apex; branches numerous, slightly ascending. Guernsey, probably intrd into rest of CI and Br, much planted   *Jersey Elm* **- U. minor ssp. sarniensis** (C.K. Schneid.) Stace
   8 Trunk ending short of tree apex; branches few, the lowest steeply ascending. Cornwall to S Hants

Cornish Elm **- U. minor ssp. angustifolia** (Weston) Stace

Many trees occur that are intermediate between 2 or more spp. in appearance and are presumed to be hybrids.

36. CANNABACEAE - *Hop family*

Perennial climbing or annual usually dioecious herbs with palmately lobed to ± palmate leaves; small inconspicuous flowers with small to ± absent perianth; and fruit an achene subtended by persistent bract. Differs from Urticaceae in the 5 (not 4) tepals and stamens and 2 (not 1) styles.

1 Erect annual; at least upper leaves alternate   **1. CANNABIS**
1 Rampant perennial climber; leaves all opposite   **2. HUMULUS**

1. CANNABIS L. - *Hemp*

Upper leaves alternate, lower sometimes opposite; inflorescences axillary clusters towards stem apex, the male looser, the female ripening to form irregular groups of achenes and bracts.

1 Stems to 2.5m. Intrd-casual; birdseed-alien frequent on tips, and in parks and farms where birds are fed; much of BI   *Hemp* **- C. sativa** L.

2. HUMULUS L. - *Hop*

Leaves all opposite; male inflorescences loose spreading axillary panicles; female inflorescences small dense capitate clusters ripening to cone-like fruiting heads, in rather loose panicles.

1 Scrambling to 8m. Native; hedgerows, scrub and fen-carr; S Br and CI, and widely natd almost throughout BI   *Hop* **- H. lupulus** L.

37. MORACEAE - *Mulberry family*

Trees or shrubs with latex; leaves simple, deciduous, often palmately lobed; flowers small and inconspicuous, monoecious, crowded into dense heads or hollow receptacles; perianth of 1 whorl of 4-5 free segments; male flowers with 4-5 stamens; female flowers with 1-celled ovary with 1 ovule; fruiting head a mass of drupes surrounded by succulent perianth or succulent receptacle.

1 Fruiting head raspberry-like in appearance; stipule-scar on 1 side of stem; latex watery **1. MORUS**
1 Fruiting head ± pear-shaped; stipule-scar completely encircling stem; latex milky **2. FICUS**

1. MORUS L. - *Mulberries*
Leaves mostly unlobed; stipules separate, their scars not encircling stem; fruiting head raspberry-like.

1 Tree to 14m. Intrd-natd; waste ground and walls, bird-sown from cultivated trees; natd by Thames in Middlesex, occasional seedlings elsewhere *Black Mulberry* - **M. nigra** L.

2. FICUS L. - *Fig*
Latex milky; leaves mosty deeply palmately lobed; stipules fused, their scars encircling stem; fruiting head pear-shaped.

1 Spreading shrub or tree to 10m. Intrd-natd; waste ground and walls, especially by rivers; natd in S Br, S Ir and CI, rarely as far N as C Sc, all or mostly from imported fruit *Fig* - **F. carica** L.

38. URTICACEAE - *Nettle family*

The 3 genera appear very different vegetatively, but are characterized by their inconspicuous, unisexual flowers with 4 perianth segments, 4 stamens, 1-celled superior ovary with 1 ovule, 1 style and densely branched stigma.

1 Leaves opposite, usually toothed and with stinging hairs, stipulate at least when young; stems erect **1. URTICA**
1 Leaves alternate, entire, without stinging hairs, without stipules; stems procumbent to decumbent 2
   2 Stems decumbent but not rooting at nodes; leaves mostly >10mm; flowers crowded **2. PARIETARIA**
   2 Stems procumbent and rooting at nodes; leaves rarely >6mm; flowers solitary **3. SOLEIROLIA**

1. URTICA L. - *Nettles*
Annual or perennial with erect stems; leaves opposite, stipulate, toothed

and normally with stinging hairs; flowers in dense axillary usually elongate inflorescences, monoecious or dioecious; perianth of free segments, 2 inner longer than 2 outer and enclosing fruit.

1 Rhizomatous and/or stoloniferous usually dioecious perennial to 1.5m; terminal leaf-tooth longer than adjacent laterals. Native; in many habitats, especially woodland, fens, cultivated ground and where animals defecate; abundant *Common Nettle* - **U. dioica** L.
1 Monoecious annual to 60cm; terminal leaf-tooth about as long as adjacent laterals. Probably native; cultivated and waste ground; frequent *Small Nettle* - **U. urens** L

2. PARIETARIA L. - *Pellitories-of-the-wall*

Perennial with decumbent stems; leaves alternate, entire, softly pubescent, without stipules; flowers in dense axillary short inflorescences, mostly unisexual, usually some bisexual; perianth of equal segments fused at least at base and enclosing fruit.

1 Stems to 40(110)cm, usually procumbent to ascending; bracts fused at base; perianth of female flowers 2-2.3mm in fruit; perianth of bisexual flowers becoming tubular and ≥3mm in fruit; achenes 1-1.2mm. Native; on walls, rocks, cliffs and steep hedgebanks; frequent *Pellitory-of-the-wall* - **P. judaica** L.
1 Stems to 70(160)cm, usually erect; bracts free; perianth of female flowers 2.7-3mm in fruit; perianth of bisexual flowers remaining bell-shaped and ≤3mm in fruit; achenes 1.5-1.8mm. Intrd-natd; in old neglected woods and hedgerows; rare in SE En
Eastern Pellitory-of-the-wall - **P. officinalis** L.

3. SOLEIROLIA Gaudich. - *Mind-your-own-business*

Perennial with procumbent stems rooting at nodes; leaves alternate, entire, sparsely pubescent, without stipules; flowers solitary in leaf-axils, monoecious; perianth of equal segments fused at least at base and enclosing fruit.

1 Stems to 20cm, very slender. Intrd-natd; on damp shady walls and banks; frequent in S En and CI, very scattered elsewhere
Mind-your-own-business - **S. soleirolii** (Req.) Dandy

39. JUGLANDACEAE - *Walnut family*

Trees with pinnate, deciduous leaves; flowers monoecious, epigynous, the male with 3-many stamens and in pendent catkins, the female with 1-celled ovary with 1 ovule and in pendent catkins or 1-few in stiff clusters; perianth small and inconspicuous; fruit a drupe or winged nut.

1 Leaflets entire, aromatic when crushed; fruits 1-few in rigid clusters,

    with green husk and hard-shelled 'nut' inside   **1. JUGLANS**
1 Leaflets serrate, not aromatic; fruits several-many in pendent clusters,
   with broad suborbicular wing around nut   **2. PTEROCARYA**

1. JUGLANS L. - *Walnut*

Leaves with 3-9 entire, aromatic, ovate leaflets; female flowers 1-few in stiff clusters; fruits drupes, with green outer husk and hard-shelled edible 'nut' inside.

1 Tree to 24m, not suckering at base. Intrd-natd; often surviving,
   self-sown only in warmer parts; most of Br   ***Walnut* - J. regia** L.

2. PTEROCARYA Kunth - *Wingnuts*

Leaves with (7)15-27(41) serrate, non-aromatic, lanceolate leaflets; female flowers in pendent catkins; fruits nuts with broad suborbicular wings derived from bracteoles.

1 Tree to 35m, usually suckering. Intrd-natd; field borders and
   embankments, occasionally spreading in the wild
              ***Caucasian Wingnut* - P. fraxinifolia** (Poir.) Spach

40. MYRICACEAE - *Bog-myrtle family*

Aromatic shrubs; flowers normally dioecious, sometimes monoecious, in stiff catkins; perianth 0; male flowers with usually 4 stamens; female flowers with 1-celled ovary with 1 ovule; fruit a drupe or narrowly winged nut.

1. MYRICA L. - *Bog-myrtles*

1 Shrub to 1.5(2)m; current season's twigs sparsely pubescent; catkins
   on twigs of previous season that do not continue growth; fruit a
   2-winged nut. Native; wet moorland and heathland, bogs and
   fens; throughout most of Br and Ir, mostly NW   ***Bog-myrtle* - M. gale** L.
1 Shrub to 2m; current season's twigs densely pubescent; catkins on
   twigs of previous season that continue growth and leaf production;
   fruit a globose drupe. Intrd-natd; wet heathland in N & S Hants
   (*M. cerifera* auct.)   ***Bayberry* - M. pensylvanica** Duhamel

41. FAGACEAE - *Beech family*

Catkin-bearing monoecious trees distinguished from other such families by the 3- or 6-celled ovary with 3-9 styles, and by the 1-3(6) nuts surrounded at the base by a *cupule* formed from fused scales.

1 Male flowers in elongated catkins; nuts terete or with rounded
   corners; winter buds obtuse to rounded at apex   2

1 Male flowers 1-many in short heads; nuts sharply 3-angled; winter
  buds acute at apex   3
  2 Male flowers in stiff catkins; cupule strongly spiny, completely
    enclosing 1-3(6) nuts during development   **3. CASTANEA**
  2 Male flowers in pendent catkins; cupule not spiny, enclosing
    only lower part of 1 nut   **4. QUERCUS**
3 Male flowers 1-3 in stiff clusters; nuts usually 3 per cupule
                                                      2. NOTHOFAGUS
3 Male flowers numerous in pendent tassels; nuts usually 2 per cupule
                                                           1. FAGUS

1. FAGUS L. - *Beech*
Leaves deciduous; male flowers numerous in pendent heads, with 8-16 stamens; female flowers usually 2 together in erect, pedunculate clusters; nuts sharply 3-angled, 1-2 per cupule.

1 Tree to 46m. Native; well drained soils on chalk and soft limestone
  and sometimes acid sandstone; SE Wa and S En, also very widely
  planted and natd over all BI   *Beech* - **F. sylvatica** L.

2. NOTHOFAGUS Blume - *Southern Beeches*
Leaves deciduous, often slightly asymmetric at base; male flowers 1-3 in stiff clusters, with 8-40 stamens; female flowers 1-3 in ± sessile clusters; nuts sharply 3-angled, usually 3 per cupule.

1 Winter buds c.4mm; leaves coarsely and doubly serrate, with
  7-11 pairs of lateral veins. Tree to 30m. Intrd-natd; widely planted
  for forestry in W, often self-sown   *Roblé* - **N. obliqua** (Mirb.) Blume
1 Winter buds c.10mm; leaves finely serrate to subentire or crenate,
  with 14-24 pairs of lateral veins. Tree to 26m. Intrd-natd; planted
  and self-sown as for *N. obliqua* (*N. nervosa* (Phil.) Krasser)
                            Rauli - **N. alpina** (Poepp. & Endl.) Oerst.

The above 2 spp. hybridise to produce saplings where grown together.

3. CASTANEA Mill. - *Sweet Chestnut*
Leaves deciduous; flowers in long rather stiff insect-pollinated catkins, mostly male but female at base, the male with 10-20 stamens, the female usually in groups of 3; nuts with 2-4 rounded angles, 1-3(more) per cupule.

1 Tree to 35m. Intrd-natd; planted throughout BI, setting seed and
  natd ± only in S En and CI   *Sweet Chestnut* - **C. sativa** Mill.

4. QUERCUS L. - *Oaks*
Leaves deciduous or evergreen; male flowers in pendent catkins, with 4-12 stamens; female flowers 1-few in stiff pedunculate or sessile clusters; nuts (acorns) terete, 1 per cupule.

1 Fruit cupule with erecto-patent to reflexed scales; at least terminal
   buds surrounded by persistent stipules   2
1 Fruit cupule with short appressed scales; buds not surrounded by
   persistent stipules   3
   2 Tree without green leaves in winter; leaf-lobes obtuse to acute
      or shortly apiculate. Tree to 40m. Intrd-natd; often natd on acid
      sands in S Br, scattered N to C Sc   ***Turkey Oak* - Q. cerris** L.
   2 Tree usually with some green leaves in winter; leaf-lobes
      mucronate to aristate. Tree to 35m. Intrd-natd; self-sown (and
      then widely segregating) in SE En (*Q. x pseudosuber* Santi;
      Q. cerris x *Q. suber* L.)   ***Lucombe Oak* - Q. x crenata** Lam.
3 Leaves evergreen, coriaceous, grey-tomentose on lowerside even
   at maturity. Tree to 25m. Intrd-natd; self-sown in S & C En, Wa, S Ir
   and CI   ***Evergreen Oak* - Q. ilex** L.
3 Leaves deciduous, not coriaceous, glabrous to sparsely or patchily
   pubescent on lowerside at maturity   4
   4 Leaves with acuminate to aristate leaf-lobes; nut with shell
      tomentose inside. Tree to 34m. Intrd-natd; much planted,
      especially on shallow sandy soils, often self-sowing; En, Wa
      and CI   ***Red Oak* - Q. rubra** L.
   4 Leaves with obtuse to rounded leaf-lobes; nut with shell
      glabrous inside   5
5 Petiole <1cm; leaf-base cordate with distinct auricles; leaf glabrous
   or with simple hairs on lowerside; peduncle 2-9cm, glabrous. Tree
   to 40m. Native; almost throughout BI, on a wide range of soils but
   especially deep rich ones in S & E   ***Pedunculate Oak* - Q. robur** L.
5 Petiole >1cm; leaf-base cuneate to cordate but without auricles; leaf
   with simple hairs along midrib and also with some stellate hairs on
   lowerside; peduncle 0-2(4)cm, with clustered hairs. Tree to 42m.
   Native; almost throughout Br and Ir, epecially on shallow, sandy,
   acid soils in N & W   ***Sessile Oak* - Q. petraea** (Matt.) Liebl.

Q. petraea x *Q. robur* = ***Q. x rosacea*** Bechst. is fertile and occurs throughout BI
in areas where 1 or both parents occur, sometimes commoner than either.

42. BETULACEAE - *Birch family*

Distinguished from other monoecious catkin-bearing trees or shrubs by the
simple leaves and 2-celled ovary with 2 styles.

1 Fruits winged, in compact cone-like structure formed from dried-
   out bracts; male flowers 3 per bract   2
1 Fruits not winged but each with enlarged lobed or laciniate bracts
   at base; male flowers 1 per bract   3
   2 Bracts of fruiting cones falling from axis with the fruits,
      distinctly 3-lobed; male catkins opening with the leaves; stamens
      2, the lobes of each well separated   **1. BETULA**

2 Bracts of fruiting cones persistent, falling with the whole cone, obscurely 5-lobed; male catkins opening before the leaves; stamens 4, the lobes of each slightly separated **2. ALNUS**
3 Winter buds rounded at apex; fruits 1-several in short clusters, each surrounded by cupule of laciniate bracts **4. CORYLUS**
3 Winter buds acute; fruits several in pendent catkins; each with large 3-lobed bract **3. CARPINUS**

1. BETULA L. - *Birches*

Trees or shrubs; male flowers 3 per bract, with 2 bracteoles per group, with minute perianth and 2 stamens; female flowers in stiff erect catkin, 3 per bract, with 2 bracteoles per group, without perianth; fruits winged, in compact cone-like structure which disintegrates from axis at maturity to release fruits and strongly 3-lobed bracts.

1 Leaves <2cm, ± orbicular, obtuse to truncate at apex; petiole <5mm; male catkins not exposed in winter, erect at least until anthesis, <1cm. Shrub to 1m. Native; upland moors and bogs on peat; rare in N En, local in C & N mainland Sc *Dwarf Birch* - **B. nana** L.
1 Leaves >2cm, ± ovate, subacute to acuminate at apex; petiole >5mm; male catkins exposed in winter, pendent, >2cm 2
  2 Leaves acuminate at apex, distinctly doubly serrate with prominent primary teeth, truncate to broadly cuneate at base, those on mature shoots glabrous. Tree to 30m. Native; on light, mostly acid soils, usually in drier places than *B. pubescens*; common, especially in S *Silver Birch* - **B. pendula** Roth
  2 Leaves acute to subacute at apex, rather evenly or irregularly serrate without obvious primary teeth, rounded to cuneate at base, glabrous to pubescent. Tree to 24m. Native; in similar places to *B. pendula* but favouring wetter and more peaty soils; common, especially in N (*Downy Birch* - **B. pubescens** Ehrh.) 3
3 Usually a tree; young twigs and petioles usually pubescent, sometimes glabrous; leaves mostly >3cm; each wing of fruit wider than body. Throughout BI but ± confined to lowlands in N

B. pubescens ssp. pubescens
3 Usually shrubby; young twigs and petioles pubescent or ± glabrous; leaves mostly <3cm; each wing of fruit not or little wider than body. Upland areas of N Br **B. pubescens ssp. tortuosa** (Ledeb.) Nyman

B. pendula x *B. pubescens* = **B. x aurata** Borkh. occurs in many areas, often in absence of 1 or both parents. It varies in characters between the 2 parents, in chromosome number, and in fertility (fully fertile to highly sterile). Its frequency is hard to determine due to uncertainty of parental limits. *B. pubescens* x *B. nana* = **B. x intermedia** Gaudin also occurs within the area of *B. nana*.

2. ALNUS Mill. - *Alders*

Trees; male flowers 3 per bract, with 4 bracteoles per group, with minute perianth and 4 stamens; female flowers in small stalked groups, 2 per bract, with 4 bracteoles per group, without perianth; fruits narrowly winged, in compact cone-like structure from which only fruits are released at maturity; fruits bracts persistent, obscurely 5-lobed.

1 Leaves rounded to cordate at base, regularly crenate-dentate; female flower-groups 1-3 on common stalk. Tree to 28m. Intrd-natd; frequently planted, especially in S, rarely self-sown but natd in N Somerset *Italian Alder* - **A. cordata** (Loisel.) Duby
1 Leaves truncate to cuneate at base, irregularly (often doubly) serrate; female flower-groups 3-8 on common stalk 2
 2 Leaves broadly obtuse to retuse at apex, sticky when young, not or scarcely paler on lowerside, with 4-8 pairs of lateral veins. Tree to 29m. Native; damp woods and by lakes and rivers; common *Alder* - **A. glutinosa** (L.) Gaertn.
 2 Leaves subacuminate to subacute at apex, not sticky when young, distinctly paler on lowerside, with 7-15 pairs of lateral veins 3
3 Edges of leaves flat; ends of last year's twigs shortly pubescent. Tree to 23m. Intrd-natd; planted especially on poor wet soils in N, often proliferating from suckers, occasionally self-sown
Grey Alder - **A. incana** (L.) Moench
3 Extreme edges of leaves narrowly but strongly rolled under; last year's twigs glabrous. Tree to 19m. Intrd-natd; planted and naturally regenerating in Argyll *Red Alder* - **A. rubra** Bong.

A. glutinosa x *A. incana* = **A. x hybrida** Rchb. occurs among plantings of the parents, sometimes having originated *in situ*.

3. CARPINUS L. - *Hornbeam*
Trees; male flowers 1 per bract, without bracteoles, without perianth, with c.10 stamens; female flowers in pendent catkin, 2 per bract, with 2 bracteoles per group, with minute irregularly toothed perianth; fruits not winged, in conspicuous pendent catkin, each nut subtended by much enlarged 3-lobed bract.

1 Tree to 32m. Native; common on clay soils in SE En, extending to Mons and Cambs; much planted elsewhere *Hornbeam* - **C. betulus** L.

4. CORYLUS L. - *Hazels*
Small trees or large shrubs; male flowers 1 per bract, with 2 bracteoles (fused to bract) per group, without perianth, with c.4 stamens; female flowers in small sessile, bud-like group, 2 per bract, with 2 bracteoles per group, with minute irregularly lobed perianth; fruits not winged, large and edible, each nut surrounded by much enlarged, laciniate, fused bracts.

1 Bracts around ripe nut forming ± cylindrical structure slightly narrowed beyond nut apex. Several-stemmed shrub or sometimes small tree to 10m. Intrd-natd; grown for its nuts, sometimes found

as a relic and rarely self-sown   *Filbert* - **C. maxima** Mill.
1 Bracts around ripe nut forming ± bell-shaped structure not reaching
   or not narrowed beyond nut apex   2
   2 Bracts slightly shorter to slightly longer than ripe nut, laciniate
      <1/2 way to base, with shaggy hairs on outside only near base.
      Several-stemmed shrub to 6(12)m. Native; hedgerows, scrub
      and woodland; common   *Hazel* - **C. avellana** L.
   2 Bracts c.2x as long as ripe nut, laciniate >1/2 way to base, with
      long shaggy hairs on outside nearly to apex. Tree to 26m with
      1 trunk. Intrd-surv; grown in plantation on dunes in S Lancs for
      many years, now as street tree in En   *Turkish Hazel* - **C. colurna** L.

C. avellana x *C. maxima* has been reported in the wild from Suffolk and many cultivated nut-trees are this.

43. PHYTOLACCACEAE - *Pokeweed family*

Tall herbaceous perennials with flowers in leaf-opposed racemes; perianth of 1 whorl of 5 free segments, usually petaloid; stamens 7-16; carpels 6-10 in 1 whorl; fruit succulent, blackish, berry-like, 6-10-lobed.

1. PHYTOLACCA L. - *Pokeweeds*

1 Carpels 6-9, free or united just at base; stamens 7-15, in 1-2 whorls.
   Stems to 1.5(2)m. Intrd-natd; bird-sown in disturbed ground;
   occasional in S   *Indian Pokeweed* - **P. acinosa** Roxb.
1 Carpels ± completely united (best observed in flower); stamens
   12-16, in 2 whorls. Distinction from *P. acinosa* uncertain. Intrd-natd;
   bird-sown in Jethou (CI)   *Chinese Pokeweed* - **P. polyandra** Batalin

44. AIZOACEAE - *Dewplant family*

Most spp. are easily recognized by their succulent leaves ('iceplants') and colourful, many-petalled daisy-like epigynous to perigynous flowers with usually many stamens and ovary with c.3-20 cells. Identification is much easier in the fresh state.

1 Leaves >15mm wide, strongly flattened, abruptly petiolate   2
1 Leaves <15mm wide, usually <2x as wide as thick, sessile or
   gradually petiolate   3
   2 Leaves alternate; petals 0   **9. TETRAGONIA**
   2 Leaves opposite; petals present   **1. APTENIA**
3 Fruit succulent; seeds embedded in mucilage; stigmas c.8-20
   8. CARPOBROTUS
3 Fruit woody, without copious mucilage; stigmas <8   4

4	Leaves c. as long as wide, triangular in section, sparsely but conspicuously toothed on all 3 angles	**4. OSCULARIA**
4	Leaves >2x as long as wide, triangular in section to terete, entire or minutely toothed	5
5	Leaves widest and thickest near apex, with 3 acute angles with a conspicuous (c.0.5mm wide), translucent, shallowly and irregularly toothed border on each	**7. EREPSIA**
5	Leaves not or scarcely wider or thicker near apex, terete or with 3 rounded or rarely acute angles each with a very narrow, ± entire, translucent border	6
6	Main stems long-procumbent, often rooting at nodes, mat-forming	7
6	Stems upright to ascending or rarely ± procumbent, stiff and strongly woody below, forming upright or spreading shrub	8
7	Leaves covered with rounded whitish papillae; young stems with white, patent or reflexed hairs	**6. DROSANTHEMUM**
7	Leaves not papillose; stems not hairy	**5. DISPHYMA**
8	All leaves <3cm long and/or <5mm thick; styles stout (c.0.4mm wide at mid length); stigmas tuft-like; flowers (20)30-50mm across	**3. LAMPRANTHUS**
8	Some leaves ≥3cm long and ≥5mm thick; styles filiform (c.0.1mm wide at mid length); stigmas tapered; flowers 18-30mm across	**2. RUSCHIA**

1. APTENIA N.E. Br. - *Heart-leaf Iceplant*

Perennial; leaves flattened, succulent, entire, petiolate; flowers axillary and terminal, solitary; sepals 4; petals numerous; stigmas 4; capsule woody, opening by 4 valves, with axile placentation.

1 Stems sprawling to 60cm; flowers (5)10-18mm across, purplish-red. Intrd-natd; escaping on walls and dry ground in Scillies, W Cornwall and CI   *Heart-leaf Iceplant* - **A. cordifolia** (L.f.) Schwantes

2. RUSCHIA Schwantes - *Shrubby Dewplant*

Dwarf shrubs, strongly woody below; leaves triangular in section with rounded angles to ± terete, entire, sessile, opposite pairs united into common sheath; flowers terminal, usually 3-4 together; sepals 5; petals numerous; stigmas 4-5; capsule woody, opening by 4-5 wingless valves, with parietal placentation.

1 Stems ± erect to ± procumbent, to 80cm; flowers 18-30mm across, purplish-red. Intrd-natd; escaping on sea-cliffs in Scillies
   Shrubby Dewplant - **R. caroli** (L. Bolus) Schwantes

3. LAMPRANTHUS N.E. Br. - *Dewplants*

Dwarf shrubs, strongly woody below; leaves triangular in section with acute angles to ± terete, entire or nearly so, sessile, opposite pairs shortly united; flowers terminal, 1-many together; sepals 5; petals numerous; stigmas 5; capsule woody, opening by 5 winged valves, with parietal placentation. A large critical S African genus.

1   Erect, bushy, to 30cm; leaves (6)10-15(20) x 1.5-5mm, falcate, mucronate at apex, conspicuously dotted; flowers 3.5-4.5cm across, pale pink. Intrd-natd; on walls, hedgebanks and cliffs; Pembs, Scillies   *Sickle-leaved Dewplant* - **L. falciformis** (Haw.) N.E. Br.
1   Erect, bushy, to 60cm; leaves (10)20-40mm, not or slightly falcate, acute or acuminate to obtuse or obscurely mucronate at apex, less conspicuously dotted; flowers (2)3-5cm across, pale pink. Intrd-natd; similar places to *L. falciformis*; W Cornwall, E & W Cork, CI
Rosy Dewplant - **L. roseus** (Willd.) Schwantes

4. OSCULARIA Schwantes - *Deltoid-leaved Dewplant*

Perennial; leaves sharply triangular in section, with conspicuous distant teeth on all 3 angles, opposite pairs shortly united; flowers terminal, 1-3 together; sepals 5; petals numerous; stigmas 5; capsule woody, opening by 5 narrowly winged valves, with parietal placentation.

1   Stems with spreading branches to 50cm; flowers 10-20mm across, pink. Intrd-natd; escaping or persisting on walls and banks; Scillies, Guernsey   *Deltoid-leaved Dewplant* - **O. deltoides** (L.) Schwantes

5. DISPHYMA N.E. Br. - *Purple Dewplant*

Perennial, with procumbent, succulent, somewhat woody main stems; leaves triangular in section with rounded angles to ± terete, entire, smooth, sessile, opposite pairs very shortly united; flowers terminal on short, ± erect, lateral branches, mostly solitary; sepals 5; petals numerous; stigmas 5; capsule rather spongy, opening by 5 winged valves, with parietal placentation.

1   Main stems to 1m, rooting at nodes; flowers 2.5-4(5)cm across, reddish-purple. Intrd-natd; walls, cliffs and sandy places; CI to W Suffolk and Anglesey   *Purple Dewplant* - **D. crassifolium** (L.) L. Bolus

6. DROSANTHEMUM Schwantes - *Pale Dewplant*

Perennial, with thin, procumbent, woody main stems; leaves ± terete, entire, covered in rounded papillae, sessile, opposite pairs not united; flowers terminal on short, ± erect, lateral branches, mostly solitary; sepals 5; petals numerous; stigmas mostly 5; capsule rather woody, opening by mostly 5 winged valves, with parietal placentation.

1   Main stems to 80cm, scarcely rooting at nodes; flowers 12-25mm across, pinkish-mauve. Intrd-natd; walls, rocks and cliffs; Scillies, W Cornwall, CI   *Pale Dewplant* - **D. floribundum** (Haw.) Schwantes

7. EREPSIA N.E. Br. - *Lesser Sea-fig*

Perennial; leaves sharply triangular in section, with translucent, shallowly and irregularly toothed edges (especially the abaxial), opposite pairs shortly united; flowers terminal, 1-3 together; sepals 5; petals numerous; stigmas 5-6; capsule woody to spongy, opening by 5-6 narrowly winged valves, with parietal placentation.

1 Stems erect to ascending, to 30cm; flowers 10-15mm across, reddish. Intrd-natd; spreading vegetatively and from seed; in quarry in Scillies *Lesser Sea-fig* - **E. heteropetala** (Haw.) Schwantes

8. CARPOBROTUS N.E. Br. - *Hottentot-figs*
Perennial; leaves triangular in section with acute angles, with translucent edges, entire or minutely toothed, opposite pairs shortly united; flowers terminal, solitary; sepals 5; petals numerous; stigmas c.8-20; fruit succulent, indehiscent, the seeds embedded in mucilage, with parietal placentation. A critical genus; the identity of many plants natd in BI is still far from clear.

1 Petals yellow, often becoming pinkish as they wither
(see couplet 3) **C. edulis**
1 Petals pink to purple from first 2
   2 Leaves thickest close to apex (scimitar-shaped), distinctly narrower than thick; ovary flat to depressed on top; receptacle abruptly narrowed into pedicel. Stems procumbent, to 2m; flowers mostly 7-10(12)cm across, pinkish-purple. Intrd-natd; on rocks, cliffs and sand near sea; rare in Devon, Cornwall and Scillies *Sally-my-handsome* - **C. acinaciformis** (L.) L. Bolus
   2 Leaves ± equally thick for most of length, about as wide as thick; ovary flat to raised on top; receptacle gradually tapered into pedicel 3
3 Flowers mostly 4.5-10cm across; petals paler pink or yellow at base; ripe fruit little or not longer than wide. Stems procumbent, to 3m; flowers mostly 4.5-10cm across; petals yellow to pinkish-purple. Intrd-natd; on rocks, cliffs and sand near sea; CI, Scillies to N Wa and E Suffolk, very local in W Lancs, Man and S & E Ir
Hottentot-fig - **C. edulis** (L.) N.E. Br.
3 Flowers mostly 3.5-6cm across; petals ± white at base; ripe fruit usually distinctly longer than wide. Stems procumbent, to 2m; flowers mostly 3.5-6cm across, pinkish-purple. Intrd-natd; rocks and cliffs by sea; CI, E Suffolk, Wigtowns
Angular Sea-fig - **C. glaucescens** (Haw.) Schwantes

9. TETRAGONIA L. - *New Zealand Spinach*
Annual; leaves alternate, flat, ± succulent, entire, abruptly narrowed to long petiole; flowers axillary, mostly solitary; sepals 4-5; petals 0; stigmas 3-8; fruit ± woody, indehiscent, ridged, with 1 seed in each of 3-8 cells.

1 Stems procumbent to ascending, to 1m; flowers very inconspicuous, yellow-green. Intrd-casual; rubbish-tips and waste ground; S En
New Zealand Spinach - **T. tetragonioides** (Pall.) Kuntze

45. CHENOPODIACEAE - *Goosefoot family*

Mainly dull-coloured weedy plants usually recognizable by the 1-whorled herbaceous perianth, 1-celled superior or semi-inferior ovary with 1 ovule, and 2-3 styles.

1 Leaves fused in opposite pairs, forming succulent sheath round stem, with 0 or very short free part 2
1 Leaves not fused to form succulent sheath round stem, usually alternate, with distinct free leaf-blade 3
   2 Annual, easily uprooted in entirety **8. SALICORNIA**
   2 Perennial, with procumbent rhizomes at or just below soil surface giving rise to aerial stems **7. SARCOCORNIA**
3 Leaves <5mm wide, thick and succulent, entire 4
3 Leaves not both <5mm wide and succulent, mostly neither, flat, often lobed or toothed at margin 5
   4 Leaves ending in a spine, plant bristly **10. SALSOLA**
   4 Leaves acute to obtuse but without a spine; plant glabrous **9. SUAEDA**
5 Flowers unisexual; fruits surrounded by 2 enlarged bracteoles 6
5 Flowers bisexual, or bisexual and female; fruits without bracteoles, usually surrounded by persistent tepals 7
   6 Stigmas 2; bracteoles almost free to ± completely fused, if fused >1/2 way then leaves mealy-white and cuneate at base **5. ATRIPLEX**
   6 Stigmas 4-5; bracteoles ± completely fused; leaves green, at least some ± truncate at base **4. SPINACIA**
7 Tepal 1; achene laterally compressed, narrowly winged **3. CORISPERMUM**
7 Tepals (2)4-5; achene not compressed, not winged 8
   8 Tepals at fruiting with very short transverse wing or tubercle abaxially **2. BASSIA**
   8 Tepals at fruiting without transverse tubercle or wing abaxially, but often with longitudinal keel 9
9 Ovary semi-inferior; receptacle becoming swollen at fruiting **6. BETA**
9 Ovary superior; receptacle not becoming swollen at fruiting **1. CHENOPODIUM**

1. CHENOPODIUM L. - *Goosefoots*

Annual or perennial herbs; leaves flattened, often mealy, entire, toothed or lobed; bracteoles 0; flowers bisexual or some female; tepals mostly 4-5, persistent and surrounding fruit, with or without abaxial longitudinal keel.

Vegetatively extremely plastic, especially in habit and leaf shape. Testa sculpturing is important and sometimes essential for identification. It can be examined under >x20 magnification after removal of the pericarp, which may be effected either by rubbing in the hand or sometimes only after boiling and dissection. The orientation of the seed is also important: either 'vertical', with long axis parallel to the length of the flower; or 'horizontal', with it at right angles to the length of the flower. *C. album* is extremely variable and may very closely resemble typical variants of several other spp. These often occur mixed with *C. album* and are often frosted before seeding.

1 Stems glandular-pubescent, at least towards apex; plant aromatic 2
1 Stems glabrous or mealy, not glandular-pubescent; plant not aromatic, sometimes stinking 6
   2 Tepals fused >1/2 way, net-veined on outside. Stems to 50cm. Intrd-casual; tips and fields as wool-alien; rather rare in S Br
Scented Goosefoot - **C. multifidum** L.
   2 Tepals not fused or fused <1/2 way, not net-veined on outside 3
3 Flower clusters in ± sessile axillary racemes or panicles; at least some seeds horizontal; larger leaves rarely <3cm. Stems to 1m. Intrd-natd; waste places and tips, mostly casual; occasional in S Br
Mexican-tea - **C. ambrosioides** L.
3 Flower clusters sessile and solitary in leaf axils; seeds vertical; larger leaves rarely >3cm 4
   4 Tepals rounded abaxially, not keeled, not meeting at margins and only partially concealing fruit. Stems to 50cm. Intrd-casual; tips and fields as wool-alien; occasional in S Br, Midlothian
Clammy Goosefoot - **C. pumilio** R. Br.
   4 Tepals prominently keeled abaxially, ± concealing the fruit 5
5 Tepals ± truncate and variably toothed at apex in side view; keel entire. Stems to 50cm. Intrd-casual; tips and fields as wool-alien; rather rare in S En *Keeled Goosefoot* - **C. carinatum** R.Br.
5 Tepals with long beak at apex; keel deeply laciniate along most of length. Stems to 50cm. Intrd-casual; tips and fields as wool-alien; rather rare in S En *Crested Goosefoot* - **C. cristatum** (F. Muell.) F. Muell.
   6 Stems woody at least below; inflorescence branchlets ending in bare weakly spinose points. Stems to 1m. Intrd-casual; fields and tips as wool-alien; rather rare in En
Nitre Goosefoot - **C. nitrariaceum** (F. Muell.) Benth.
   6 Stems herbaceous; branchlets not bare and spinose at tips 7
7 Flowers in a spike of globose sessile axillary heads >5mm across; perianth turning red and succulent at fruiting. Stems to 50cm. Intrd-casual; fields and waste places; sporadic in Br and Ir
Strawberry-blite - **C. capitatum** (L.) Ambrosi
7 Flowers in racemes or panicles of heads usually <5mm across; perianth not turning red and succulent at fruiting 8
   8 Rhizomatous perennial; stigmas 0.8-1.5mm. Stems to 50cm. Native; roadsides, pastures and by farm buildings; scattered and locally common **C. bonus-henricus** L. - *Good-King-Henry*
   8 Annual; stigmas <0.8mm 9
9 Fruiting perianths mostly longer than wide, with vertical or oblique seeds; inflorescence glabrous 10
9 Fruiting perianths wider than long, with horizontal seeds; inflorescence mealy or glabrous 12
   10 Leaves green on upperside, mealy-grey on lowerside. Stems to

50cm. Possibly native; waste places on rich soils, often near sea, casual elsewhere; very local in BI, mostly S & E En
Oak-leaved Goosefoot - **C. glaucum** L.
10 Leaves green (to reddish) on both surfaces   11
11 Tepals of lateral fruits in each cluster fused <1/2 way; leaves usually strongly toothed or lobed. Stems to 80cm; seeds brown. Native; cultivated and waste ground, often near sea, and on tips; frequent in En, local elsewhere   *Red Goosefoot* - **C. rubrum** L.
11 Tepals of lateral fruits fused to near apex; leaves usually entire to sparsely lobed or toothed. Stems to 30cm. Native; by dykes and in barish pastures near sea; local in SE En and CI
Saltmarsh Goosefoot - **C. chenopodioides** (L.) Aellen
 12 Leaves weakly cordate at junction with petiole, at least on some leaves. Stems to 1m. Probably intrd-natd; waste and arable ground; rare in Br, mainly S
Maple-leaved Goosefoot - **C. hybridum** L.
 12 Leaves cuneate at junction with petiole   13
13 Seeds distinctly acutely keeled at edges; tepals minutely denticulate. Stems to 1m; seeds black. Probably native near sea in S En and very scattered casual in waste places, especially in S En and CI   *Nettle-leaved Goosefoot* - **C. murale** L.
13 Seeds with subacute to rounded unkeeled edges; tepals entire   14
 14 Leaves entire or at most with 1 obscure tooth or lobe on each side 15
 14 At least lower leaves distinctly toothed and/or lobed   18
15 Leaves green (or reddish) on both surfaces, not mealy; stems square in section. Stems to 1m. Native; waste and cultivated ground; common in S, rare in N   *Many-seeded Goosefoot* - **C. polyspermum** L.
15 Leaves mealy-grey at least on lowerside; stems ± terete to ridged   16
 16 Leaves ovate-trullate, <2.5cm; tepals rounded abaxially; plant stinking like rotten fish. Stems to 40cm. Probably native; barish places near sea; rare in S En and CI, casual elsewhere
Stinking Goosefoot - **C. vulvaria** L.
 16 Leaves linear to triangular-ovate, rarely ovate-trullate, the largest usually >2.5cm; tepals keeled abaxially; plant not stinking   17
17 Leaves linear to linear-oblong, densely mealy-grey on lowerside, distinctly mucronate at apex, mostly with only 1(-2) pairs of lateral veins visible; petiole <1cm. Stems to 1m. Intrd-casual; tips and waste ground; rather rare in S Br
Slimleaf Goosefoot - **C. desiccatum** A. Nelson
17 Leaves usually oblong to ovate, trullate or triangular, less densely mealy-grey on lowerside, rarely mucronate at apex, mostly with >2 pairs of lateral veins visible; longest petioles usually >1cm   24
 18 Stems, leaves and flowers not or very sparsely mealy; leaves trullate to triangular. Stems to 1m; seeds black. Probably intrd-natd; waste and cultivated ground; rare in S, mostly casual
Upright Goosefoot - **C. urbicum** L.
 18 At least flowers and small branchlets conspicuously mealy; leaves various   19

19 Testa with conspicuous regular pits delimited by pronounced reticulum of ridges 20
19 Testa irregularly pitted to almost featureless, often with radial and/or tangential furrows, sometimes with regular reticulum of slightly raised ridges but with flat (not concave) areas within 22
  20 Lower leaves rarely distinctly 3-lobed; tepals with strong wing-like keel along whole length; testa with almost isodiametric honeycomb-like pitting. Stems to 1.5m. Intrd-casual; tips and waste ground; rare in En and S Wa
Pitseed Goosefoot - **C. berlandieri** Moq.
  20 Lower leaves usually distinctly 3-lobed; tepals weakly keeled or strongly so in distal part only; testa with radially-elongated pitting 21
21 Plant not stinking like rotten fish; lower leaves with elongated central lobe 2-3x as long as side lobes; seeds <1.4mm across. Stems to 1m. Native; waste and arable ground; Cl, S & E En and S Wa, casual elsewhere *Fig-leaved Goosefoot* - **C. ficifolium** Sm.
21 Plant stinking like rotten fish; lower leaves with short central lobe little longer than side lobes; seeds >1.4mm across. Stems to 1m. Intrd-casual; tips and waste ground from wool and birdseed; rare in En and Wa *Foetid Goosefoot* - **C. hircinum** Schrad.
  22 Plant to 2m; young shoots usually extensively coloured reddish-purple; larger leaves ≤14cm, ± always some >6cm, ovate-trullate to ovate-triangular. Stems to 2m. Intrd-casual; tips and waste places; scattered in En *Tree Spinach* - **C. giganteum** D. Don
  22 Plant usually less robust, if >1.5m leaves <6cm; plant green or variously red-tinged or -striped, but not with young shoots extensively reddened 23
23 Seeds (1.3)1.5-2mm in longest diameter, rarely <1.5mm in shortest diameter; perianth often blackish-green, rather sparsely mealy; leaves trullate. Stems to 1.5m. Intrd-casual; tips and waste ground; S En *Soyabean Goosefoot* - **C. bushianum** Aellen
23 Seeds ≤1.5mm in longest diameter, <1.5mm in shortest diameter; perianth usually green to grey, variously mealy; leaves various 24
  24 Leaves not or little longer than wide, ovate-trullate, not lobed or with 2 basal shallow lobes; tepals often fused to c.1/2 way; infloresence densely mealy. Stems to 1.5m. Intrd-natd; tips and waste ground; scattered in Br, mainly S
Grey Goosefoot - **C. opulifolium** W.D.J. Koch & Ziz
  24 Leaves distinctly longer than wide; tepals usually fused to <1/2 way; inflorescence variously mealy 25
25 Leaves narrowly oblong, with ± parallel sides, often obtuse, with shallow teeth but not lobed; seeds with ratio of longest to shortest diameters usually >1.15. Stems to 1.5m. Intrd-casual; tips and waste places; rather rare in Br, mainly S *Striped Goosefoot* - **C. strictum** Roth
25 Leaves various but not with parallel sides, usually acute; seeds with ratio of longest to shortest diameters usually <1.15 26
  26 Seeds with subacute edges as seen in narrowest profile. Stems

to 1.5m. Extremely variable. Native; waste and cultivated
ground; abundant **Fat-hen - C. album** L.

26 Seeds with obtuse to rounded edges as seen in narrowest profile 27

27 Plant often >1m, frequently tinged with red; seeds usually <1.25mm
in longest diameter but flowering very late, usually frosted before
seeding; tepals slightly keeled. Stems to 2m. Intrd-casual; tips and
waste places; scattered in Br *Probst's Goosefoot* - **C. probstii** Aellen

27 Plant <1m, not tinged with red (except rarely in leaf-axils); seeds
often >1.25mm in longest diameter, usually produced well before
frosts; tepals usually well keeled. Stems to 1m. Intrd-natd; tips,
cultivated ground and waste places); scattered in Br

Swedish Goosefoot - **C. suecicum** Murr

2. BASSIA All. - *Summer-cypress*

Annual herb; leaves flat, not mealy, entire, at least the lower pubescent;
bracteoles 0; flowers bisexual or some female; tepals 5, developing small
transverse wing or tubercle abaxially at fruiting.

1 Leaves linear to lanceolate; whole plant often becoming purplish-
red in autumn, to 1m. Intrd-natd; tips, waste places and by salted
roads; scattered in Bl, mainly En *Summer-cypress* - **B. scoparia** (L.) Voss

3. CORISPERMUM L. - *Bugseed*

Annual herb; leaves flat, not mealy, entire, glabrous or sparsely pubescent;
bracteoles 0; flowers bisexual; tepal 1; achene with narrow wing round
margin, laterally compressed.

1 Leaves linear to very narrowly elliptic or linear-lanceolate, 1-veined;
whole plant often becoming purplish-red in autumn, to 60cm. Intrd-
natd; casual or rarely persistent in sandy, mostly coastal places;
scattered in S Br *Bugseed* - **C. leptopterum** (Asch.) Iljin

4. SPINACIA L. - *Spinach*

Usually annual herb; leaves flat, not mealy, glabrous; bracteoles present
with female flowers, developing abaxial spine at fruiting; flowers unisexual,
male in dense spikes or panicles, female axillary; tepals 4-5 in male flowers,
0 in female flowers.

1 Leaves ovate to triangular, often with 1 basal acute lobe each side.
Stems to 1m. Intrd-casual; tips and waste places; rather rare in En,
over-recorded for *Beta vulgaris* *Spinach* - **S. oleracea** L.

5. ATRIPLEX L. - *Oraches*

Annual herbs or perennial shrubs; leaves flattened, often mealy, entire,
toothed or lobed; flowers mostly unisexual, the male with 5 tepals, the
female usually with 0 tepals but 2 bracteoles enlarging and partly
concealing fruit.

Vegetatively extremely plastic, especially in habit, mealiness and leaf-shape. Young plants often closely resemble *Chenopodium* spp., but note sex-distribution; in fruit the genus is very distinct. Lower leaves, which are often lost before fruit ripens, and ripe fruit and surrounding bracteoles are both important diagnostic characters, but populations usually contain a range of individuals of different ages.

1 Shrubs 2
1 Annual herbs 3
    2 Lower leaves opposite; bracteoles fused to >1/2 way. Shrub to 1m. Native; saline mud and sand, often by pools or dykes, rarely on sea-cliffs; coasts of Br N to S Sc, E Ir and CI
Sea-purslane - **A. portulacoides** L.
    2 All leaves alternate; bracteoles fused only at base. Stems to 2.5m. Intrd-natd; planted as wind-break by sea in S En and CI, spreading vegetatively only in CI   *Shrubby Orache* - **A. halimus** L.
3 Bracteoles with small apical lobe much exceeded by 2 adjacent laterals, fused ± to apex. Stems to 30cm. Native; drier, barish parts of salt-marshes; S Essex   *Pedunculate Sea-purslane* - **A. pedunculata** L.
3 Bracteoles without 3 such apical lobes, rarely fused to >1/2 way 4
    4 Bracteoles orbicular to broadly elliptic, entire, papery, present with only some female flowers. Stems to 2m. Intrd-casual; tips and waste places; occasional   *Garden Orache* - **A. hortensis** L.
    4 Bracteoles not orbicular to broadly elliptic, angled or toothed, herbaceous to ± woody, present with all female flowers 5
5 Bracteoles hardened (cartilaginous) in basal part at fruiting; ultimate venation of leaves thick, dark green against lighter background (fresh material) 6
5 Bracteoles not hardened at fruiting, remaining herbaceous or becoming spongy; ultimate venation of leaves very thin, not green 7
    6 Bracteoles 2-5mm, with 3-9 acute teeth in distal 1/2, not tuberculate abaxially. Stems to 60cm. Intrd-casual; rather frequent wool-alien   *Australian Orache* - **A. suberecta** Verd.
    6 Bracteoles 6-7mm, with irregular mostly obtuse teeth around middle, usually tuberculate abaxially. Stems to 30(50)cm. Native; lower parts of sandy beaches; most of BI
Frosted Orache - **A. laciniata** L.
7 Lower leaves linear to linear-lanceolate, entire to toothed but without distinct basal lobes; coastal except as casual. Stems to 1.5m. Native; saline, usually sandy places near sea, rarely inland; round most of BI   *Grass-leaved Orache* - **A. littoralis** L.
7 Lower leaves lanceolate to triangular or trullate, with distinct basal lobes; coastal or inland 8
    8 Bracteoles fused for >1/3 their length, often to c.1/2 way 9
    8 Bracteoles fused at base only, for <1/4 way 11
9 Lower leaves lanceolate to trullate, acutely cuneate at base, with forwardly directed basal lobes; bracteoles herbaceous at base; coastal and inland. Stems to 1m. Native; disturbed and waste ground of all types; common   *Common Orache* - **A. patula** L.

9 Lower leaves triangular to trullate, truncate to obtusely cuneate
at base, with laterally or forwardly directed basal lobes; bracteoles
thickened and spongy at base; coastal only 10
  10 Bracteoles sessile, 4-10mm, not foliaceous distally. Stems to
    90cm. Native; sandy or shingly beaches; all round BI
                      Babington's Orache - **A. glabriuscula** Edmondston
  10 Some bracteoles stalked, up to 20mm and foliaceous distally.
    Stems to 90cm. Native; exposed coastal beaches; Sc, N En and
    Man (*A. glabriuscula* x *A. longipes*)
                      Taschereau's Orache - **A. x taschereaui** Stace
11 Some bracteoles >10mm and foliaceous distally, with stalks ≥5mm   12
11 Bracteoles all <10mm, rarely foliaceous distally, with stalks ≤5mm   13
  12 Bracteoles to 25mm, strongly foliaceous distally, united only
    at base, with stalks to 25(30)mm. Stems to 90cm. Native; in
    taller saltmarsh vegetation; very local on coasts of Br
                      Long-stalked Orache - **A. longipes** Drejer
  12 Bracteoles to 20mm, sometimes foliaceous distally, the smaller
    ones often united nearly to 1/2 way, with stalks to 10mm
                      (see couplet 10) **A. x taschereaui**
13 Lower leaves trullate, cuneate at base; lower littoral zone of coasts
only. Stems to 10(15)cm. Native; margins of sea inlets; N & W Sc,
Cheviot   *Early Orache* - **A. praecox** Hülph.
13 Lower leaves triangular, ± truncate at base; coastal and inland   14
  14 Bracteoles 2-6(8)mm, sessile. Stems to 1m. Native; roadsides,
    waste places and cultivated ground, often in saline habitats;
    common   *Spear-leaved Orache* - **A. prostrata** DC.
  14 Bracteoles 3.5-9mm, always some with stalks to 1mm and
    often to 5mm. Stems to 80cm. Native; coastal estuarine and
    sometimes inland saline areas; scattered round Br (*A. prostrata*
    x *A. longipes*)   *Kattegat Orache* - **A. x gustafssoniana** Tascher.

4 other hybrids occur but are rare.

6. BETA L. - *Beets*

Annual to perennial herbs; roots often swollen; leaves flattened, not mealy, usually ± entire; bracteoles 0; flowers bisexual; tepals 5, persistent; ovary semi-inferior.

1 Tepals whitish-yellow, erect in fruit; stigmas 3. Perennials. Stems to
  1m. Intrd-natd; persisting or relict in waste places and on tips; S En,
  Co Dublin   *Caucasian Beet* - **B. trigyna** Waldst. & Kit.
1 Tepals green or purplish-red, incurved in fruit; stigmas usually 2.
  Stems to 1.5m (*Beet* - **B. vulgaris** L.)   2
  2 Usually sprawling perennials; lower leaves mostly <10cm; lower
    bracts mostly 10-35mm. Native; shores and waste ground by sea;
    most coasts of BI   *Sea Beet* - **B. vulgaris ssp. maritima** (L.) Arcang.
  2 Usually erect annuals or biennials; lower leaves mostly >10cm;
    lower bracts mostly 2-20mm; cultivated casual or relic   3

3 Grown for large foliage; roots not to moderately swollen
Foliage Beet - **B. vulgaris ssp. cicla** (L.) Arcang.
3 Grown for greatly swollen roots   *Root Beet* - **B. vulgaris ssp. vulgaris**

7. SARCOCORNIA A. J. Scott - *Perennial Glasswort*

Dwarf subshrubs; leaves fused in opposite pairs, forming succulent sheath round stem which appears composed of short segments; flower ± immersed in row of segments at ends of main stem ('terminal spike') and branches in 2 opposite groups in each segment, each group with 3 flowers in a ± straight transverse row, the centre one completely separating and c. as large as the laterals.

1 Aerial stems to 30cm, some fertile, some not, erect to decumbent, usually little-branched, becoming yellowish to reddish, arising from thin extensive rhizomes; terminal spike 10-40mm; fertile segments 3-4mm, 3-4.5mm wide at narrowest point; anthers c.1.5mm. Native; middle and upper parts of salt-marshes; scattered in S & C En and Wa, Co Wexford   *Perennial Glasswort* - **S. perennis** (Mill.) A. J. Scott

8. SALICORNIA L. - *Glassworts*

Like *Sarcocornia* but annuals and with 1-3 flowers in each group, the central 1 of each group of 3 extending much further apically than the laterals, so that the group is triangular in shape.

An extremely difficult genus, the problems arising mainly from great phenotypic plasticity and the inbreeding nature of the plants, which tend to form numerous distinctive local populations. At least 20-30 'sorts' can be distinguished in SE En; possibly only 3 spp. (*S. pusilla*; *S. europaea* agg.; *S. procumbens* agg.) should be recognized. Identification of segregates within the latter 2 aggregates should be attempted only on several fresh well-grown plants from unshaded populations developing ripe fruit (Sep-Oct).

1 Flowers 1 per group; fertile segments disarticulating when fruit ripe, <2 x 2mm. Stems to 25cm, becoming orangy- or purplish-pink; branches ± straight; terminal spike <10mm; lower fertile segments 1-1.5mm, 1-1.5mm wide at narrowest point. Native; drier parts of salt-marshes; S Br and S Ir   *One-flowered Glasswort* - **S. pusilla** Woods
1 Flowers mostly 3 per group; fertile segments not disarticulating, >2 x 2mm   2
   2 Anthers 0.2-0.5(0.6)mm; stamens 1(-2); central flower distinctly larger (c.2x) than 2 laterals; fertile segments with distinctly convex sides; seeds 1-1.7mm (**S. europaea** L. agg.)   3
   2 Anthers (0.5)0.6-0.9mm; stamens (1-)2; all 3 flowers about same size; fertile segments with straight or sometimes slightly convex or concave sides; seeds (1.3)1.5-2.3mm (**S. procumbens** Sm. agg.)   5
3 Apical edge of fertile segments with scarious border 0.1-0.2mm wide; plants deep shiny-green becoming reddish-purple. Stems to 40cm; branches ± straight; terminal spike (5)10-30(40)mm; lower fertile segments 1.9-3.5mm, 2-4mm wide at narrowest point. Native;

mostly middle and upper parts of salt-marshes; round coasts of BI
Purple Glasswort - **S. ramosissima** Woods
3 Apical edge of fertile segments with scarious border ≤0.1mm wide; plants lighter duller green becoming yellowish-green sometimes suffused
with pinkish-purple 4
  4 Plant glaucous, matt, becoming dull yellowish-green, not reddening or only slightly so around flowers; branches simple, the lowest usually <1/2 as long as main stem, curving upwards distally. Stems to 40cm; terminal spike 10-40(45)mm; lower fertile segments 2.5-4.5mm, 2.8-4(5)mm wide at narrowest point. Native; on bare mud in salt-pans and beside channels; S & E En *Glaucous Glasswort* - **S. obscura** P.W. Ball & Tutin
  4 Plant clear green, not matt, usually becoming yellowish-green suffused with pink or red; branches usually branched, lowest usually >1/2 as long as main stem, ± straight.. Stems to 35cm; terminal spike 10-50(60)mm; lower fertile segments 2.5-4mm, 3-4.5mm wide at narrowest point. Native; at all levels in salt-marshes; round coasts of BI *Common Glasswort* - **S. europaea** L.
5 Lower fertile segments ≤3(3.5)mm, ≤3.5(4)mm wide at narrowest point, plant becoming brownish-purple or -orange with diffuse red tinge. Stems to 25cm; branches few; terminal spike 12-40mm; lower fertile segments (1.8)2-3(3.5)mm, 1.8-3.5mm wide at narrowest point. Native; in middle and upper parts of salt-marshes; scattered in S Br and S Ir, Cheviot *Shiny Glasswort* - **S. nitens** P.W. Ball & Tutin
5 Lower fertile segments 3-6mm, 3-6mm wide at narrowest point; plant becoming pale green to yellowish, sometimes tinged purple 6
  6 Terminal spike ± cylindrical, of 6-15(22) fertile segments; plant becoming yellowish-green to bright yellow. Stems to 40cm; terminal spike (15)25-80(100)mm; lower fertile segments ± cylindrical, 3-5 x 3-6mm. Native; on lower parts of salt-marshes and along dykes and runnels; most coasts of Br and Ir
Yellow Glasswort - **S. fragilis** P.W. Ball & Tutin
  6 Terminal spike usually tapering, of 12-30 fertile segments; plant becoming dull green, dull yellow or yellowish-brown. Stem to 45cm; terminal spike (25)50-120(200)mm; lower fertile segments ± cylindrical, 3-6 x 3-6mm. Native; on lower parts of salt-marshes and along dykes and runnels; most coasts of Br and Ir
Long-spiked Glasswort - **S. dolichostachya** Moss

S. pusilla x *S. ramosissima* occurs rarely in En.

9. SUAEDA Scop. - *Sea-blites*

Annual herbs or perennial shrubs; leaves linear, ± flat on upperside, rounded on lowerside, succulent, 1-veined, entire, acute to obtuse; bracteoles 2-3, minute, scarious; flowers bisexual and female; tepals 5, persistent and partly concealing fruit.

1 Evergreen branching shrub to 1.2m; leaves obtuse, rounded at base; stigmas 3. Native; sand and shingle beaches and dry upper parts of salt-marshes; coasts of SE En *Shrubby Sea-blite* - **S. vera** J.F. Gmel.
1 Simple to much-branched annual herb to 30(75)cm; leaves acute to subacute, not contracted or rounded at base; stigmas usually 2. Native; middle and lower parts of salt-marshes, often with *Salicornia* spp.; round coasts of BI *Annual Sea-blite* - **S. maritima** (L.) Dumort.

10. SALSOLA L. - *Saltworts*

Annual somewhat woody herbs; leaves linear or linear-triangular, subterete to ± flattened, succulent, entire, spine-tipped; bracteoles 2, ± leaf-like; flowers bisexual; tepals 5, persistent and concealing fruit, usually developing a horizontal wing; seeds horizontal.

1 Stems erect to straggly, to 50(100)cm; leaves widest at base, tapered to spine-tipped apex (**S. kali** L.) 2
   2 Plant mostly straggly, to 50cm, usually hispid; leaves to 40 x 2mm; tepals stiff, with distinct midrib and apical spine, winged in fruit. Native; natural and disturbed maritime sandy places; round coasts of BI *Prickly Saltwort* - **S. kali ssp. kali**
   2 Plant mostly erect, to 1m, usually ± glabrous; leaves to 70 x c.1mm; tepals soft, with obscure midrib and usually no apical spine, often only ridged in fruit. Intrd-natd; casual in waste places, persistent in S Essex (ssp. *ruthenica* (Iljin) Soó)
Spineless Saltwort - **S. kali ssp. iberica** (Sennen & Pau) Rilke

46. AMARANTHACEAE - *Pigweed family*

Mainly dull-coloured weedy plants distinguished from Chenopodiaceae by the brownish, scarious perianth and often dehiscent fruit.

1. AMARANTHUS L. - *Pigweeds*

A difficult, wholly alien genus which should be collected late in the year or grown on to obtain fruit. It is important to distinguish between bracteoles and tepals (which can be similar), and male and female flowers. In key and descriptions 'tepals' refers to those of female flowers only. All spp. occur on rough and waste ground and on tips.

1 Flowering stems leafy to apex, the flowers borne in axillary clusters 2
1 Flowering stems leafless towards apex, the flowers borne in dense spike-like terminal panicles (often also in axillary clusters further back) 7
   2 Bracteoles c.2x as long as perianth, with ± spiny tip. Stems to 60cm. Intrd-natd; frequent casual occasionally persisting; Br, mainly S, rarely Ir *White Pigweed* - **A. albus** L.
   2 Bracteoles shorter than perianth, not with spiny tip 3
3 Tepals 4-5 4

1. AMARANTHUS

3 Tepals 3   5
    4 Tepals subequal, obovate to spathulate; fruit indehiscent. Stems to 70cm. Intrd-casual; infrequent; S Br

Indehiscent Pigweed - **A. standleyanus** Covas

    4 Tepals unequal, ovate to elliptic; fruit transversely dehiscent. Stems to 50cm. Intrd-casual; infrequent; Br, mainly S, rarely Ir

Prostrate Pigweed - **A. blitoides** S. Watson

5 Tepals shorter than fruit, with apical point <0.5mm and ± straight. Stems to 70cm. Intrd-casual; infrequent; S Br

Short-tepalled Pigweed - **A. graecizans** L.

5 Tepals longer than fruit, with apical point >0.5mm and usually bent outwards or hooked   6

    6 Tepals with fine, straight to outwardly curved (rarely >90°) apical point. Stems to 50cm. Intrd-casual; frequent; En and Sc

Thunberg's Pigweed - **A. thunbergii** Moq.

    6 Tepals with stiffly hooked (mostly 180-360°) apical point. Stems to 40cm. Intrd-casual; infrequent; En

Cape Pigweed - **A. capensis** Thell.

7 Plant wholly male. Stems to 1m. Intrd-casual; infrequent; S En

Dioecious Amaranth - **A. palmeri** S. Watson

7 Plant with female (and usually male) flowers   8
    8 Tepals (2-)3, much shorter than fruit; fruit indehiscent   9
    8 Tepals (3-)5, nearly as long as to longer than fruit; fruit dehiscent or not   10

9 Leaves obtuse; stems very shortly pubescent at least near apex; fruit inflated (seed much smaller than cavity). Stems to 60cm. Intrd-natd; infrequent casual in S & C Br, natd in CI

Perennial Pigweed - **A. deflexus** L.

9 Leaves usually retuse; stems ± glabrous; fruit not inflated. Stems to 60cm. Intrd-natd; infrequent casual in S Br, annually in Guernsey

Guernsey Pigweed - **A. blitum** L.

    10 Fruits indehiscent or irregularly dehiscent   11
    10 Fruits transversely dehiscent   12

11 Tepals tapered to very acute apex; bracteoles longer than perianth. Stems to 1m. Intrd-natd; casual in Br, sometimes natd in E Anglia, Scillies   *Indehiscent Amaranth* - **A. bouchonii** Thell.

11 Tepals rounded to retuse at apex, obovate to spathulate; bracteoles shorter than perianth   (see couplet 4) **A. standleyanus**

    12 Tepals all tapered to acute apex   13
    12 At least the inner tepals obovate to spathulate, rounded to retuse at apex though often with mucro   16

13 Terminal inflorescence often >30cm, heavy, usually brightly (green, red or yellow) coloured; seeds pale or dark brown; bracts not exceeding styles; longest bracteoles of female flowers 1-1.5x as long as perianth   14

13 Terminal inflorescence <30cm, not massive, usually green, rarely red; seeds dark brown; bracts usually exceeding styles; longest bracteoles of female flowers c.2x as long as perianth   15

14 Inflorescence stiff; bracts with stout midrib; styles thickened
at base. Stems to 1m. Intrd-casual; rather infrequent; S En
Prince's-feather - **A. hypochondriacus** L.
14 Inflorescence lax; bracts with slender midrib; styles not
thickened at base. Stems to 1m. Intrd-natd; fairly frequent
casual, very rarely natd; S En   *Purple Amaranth* - **A. cruentus** L.
15 Inflorescence stiff; bracts 5-5.5mm, with stout midrib; styles
thickened at base; tepals 3-5. Stems to 1m. Intrd-casual; infrequent
from soyabean waste; S En   *Powell's Amaranth* - **A. powellii** S. Watson
15 Inflorescence lax; bracts 3-5mm, with slender midrib; styles not
thickened at base; tepals 5. Stems to 1m. Intrd-natd; frequent casual,
very rarely natd; Br, CI, rare in Ir   *Green Amaranth* - **A. hybridus** L.
  16 Inflorescence long, pendent or trailing, often red; tepals obovate
  or broadly spathulate to elliptic, strongly overlapping at margins.
  Stems to 80cm. Intrd-casual; fairly frequent in Br and CI,
  mainly S   *Love-lies-bleeding* - **A. caudatus** L.
  16 Inflorescence erect or weakly pendent, rarely red; tepals
  narrowly oblong to spathulate, not or weakly overlapping at
  margins   17
17 Tepals with midrib ending below apex (though apex often
mucronate); stems at base of flowering region pubescent to densely
so. Stems to 1m. Intrd-natd; frequent casual, sometimes natd;
scattered in BI   *Common Amaranth* - **A. retroflexus** L.
17 Tepals with midrib extending beyond apex into mucro; stems at
base of flowering region glabrous to sparsely pubescent   18
  18 All flowers female; tepals dissimilar, the outer longer and
  tapering to acute apex   (see couplet 7) **A. palmeri**
  18 Male and female flowers mixed; tepals all similar. Stems to 1m.
  Intrd-casual; infrequent in Br, mainly S
  Mucronate Amaranth - **A. quitensis** Kunth

47. PORTULACACEAE - *Blinks family*

Herbaceous annuals to perennials; distinguished from Caryophyllaceae by
the 2 (not 5) sepals.

1 Petals yellow; stamens 6-c.12; capsule opening transversely; seeds >5
1. PORTULACA
1 Petals white to pink; stamens 3-5; capsule opening vertically;
seeds 1-3   2
  2 Stem-leaves 1 pair, opposite   **2. CLAYTONIA**
  2 Stem-leaves several pairs, opposite or alternate   **3. MONTIA**

1. PORTULACA L. - *Common Purslane*
Stems with several pairs of opposite to subopposite (or some alternate)
leaves with usually bristle-like stipules; flowers sessile, in groups of 1-3
with group of leaves just below; sepals and petals falling before fruiting;

petals yellow; stamens 6-c.12; ovary semi-inferior; fruit a many-seeded capsule, opening transversely.

1   Procumbent to erect, ± succulent annual to 50cm. Intrd-natd; weed of arable ground in CI and Scillies, rare casual in En
Common Purslane - **P. oleracea** L.

2. CLAYTONIA L. - *Purslanes*
Stems with 1 pair of opposite leaves; stipules 0; flowers stalked, several in terminal cyme; sepals persistent until after seed dispersal; petals 5, free, white or pink; stamens 5; ovary superior; fruit a 1-seeded capsule, opening vertically.

1   Stem-leaves fused to form cup-like structure at base of inflorescence; petals <5mm, white, entire or slightly notched. Annual to 30cm. Intrd-natd; weed of cultivated and waste ground; scattered throughout BI *Springbeauty* - **C. perfoliata** Willd.
1   Stem-leaves sessile but not fused; petals >5mm, pink or white, deeply notched. Annual to 40cm. Intrd-natd; barish damp shady places; scattered, mostly in N & W *Pink Purslane* - **C. sibirica** L.

3. MONTIA L. - *Blinks*
Stems with several alternate or opposite leaves; stipules 0; flowers stalked, in terminal or axillary groups of 1-3 or several in terminal cyme; sepals persistent until after seed dispersal; petals 5, white or pink, fused at base or free; stamens 3-5; ovary superior; fruit a usually 3-seeded capsule, opening vertically.

1   Stems unbranched; stem-leaves alternate; basal leaves long-petiolate; petals >6mm, pink; seeds minutely rough. Stoloniferous perennial to 20cm. Intrd-natd; wet banks and rocks by river; Lanarks
Small-leaved Blinks - **M. parvifolia** (Moç.) Greene
1   Stems branched; stem-leaves opposite; basal leaves 0; petals <2mm, white; very variable seed sculpturing. Annual to perennial; stems 1-20(50)cm. Native; many kinds of damp places, from streams to seasonally damp hollows (*Blinks* - **M. fontana** L.)   2
  2   Seeds smooth on faces and margin, very shiny. Locally common in NW BI **M. fontana ssp. fontana**
  2   Seeds tuberculate at least on margin, dull or somewhat shiny   3
3   Seeds with broad rounded tubercles on faces and margin, dull. Scattered over most of BI, rare in N, locally common in S & C Br
M. fontana ssp. chondrosperma (Fenzl) Walters
3   Seeds smooth near centre of faces, somewhat shiny   4
  4   Seeds with ≥3 rows of narrow long-pointed tubercles along margin. Scattered over most of BI, locally common in SW, rare in Sc **M. fontana ssp. amporitana** Sennen
  4   Seeds with usually only 1-4 rows of broad short-pointed tubercles along margin. Scattered over most of Br and Ir, but rare in

S & E En   M. fontana ssp. variabilis Walters

48. CARYOPHYLLACEAE - *Pink family*

Herbaceous annuals to perennials, sometimes woody at base; mostly recognisable by the opposite leaves without stipules, 4-5 sepals and petals, 8 or 10 stamens, and many-seeded capsule with free-central placentation, but exceptions to all these occur. The leaves are usually narrow and the petals white and often bifid.

General key
1	Stipules present, at least partly scarious	*Key A*
1	Stipules 0	2
	2 Sepals free or joined only at extreme base	*Key B*
	2 Sepals joined to form distinct calyx-tube	*Key C*

Key A - Stipules present, at least partly scarious
1	Leaves alternate	**12. CORRIGIOLA**
1	At least lower leaves opposite or whorled	2
	2 Petals conspicuous, about as wide as sepals or wider	3
	2 Petals inconspicuous and much narrower than sepals	4
3	Styles 5; capsule opening by 5 valves; petals white	**16. SPERGULA**
3	Styles 3; capsule opening by 3 valves; petals usually pink, sometimes white	**17. SPERGULARIA**
	4 Stigmas 3; fruit with >1 seed	**15. POLYCARPON**
	4 Stigmas 2; fruit with 1 seed	5
5	Sepals conspicuous, white, spongy; fruit a dehiscent capsule	**14. ILLECEBRUM**
5	Sepals inconspicuous, greenish to brownish, thin; fruit an indehiscent achene	**13. HERNIARIA**

Key B - Stipules 0; sepals free or joined only at extreme base
1	Flowers perigynous; styles 2; fruit an indehiscent achene	**11. SCLERANTHUS**
1	Flowers hypogynous; styles >2; fruit with >1 seed, usually dehiscent	2
	2 Capsule-teeth or -valves as many as styles (but sometimes slightly bifid), or flowers male	3
	2 Capsule-teeth or -valves 2x as many as styles	6
3	Leaves and capsules succulent; seeds ≥3mm; maritime	**3. HONCKENYA**
3	Plant not succulent; seeds <3mm	4
	4 Petals bifid >3/4 way to base	**8. MYOSOTON**
	4 Petals entire or slightly emarginate, or 0	5
5	Most flowers with 3(-4) styles and 5 sepals, and most capsules with 3(-4) teeth, or flowers male	**4. MINUARTIA**
5	Styles 4-5; capsule with 4-5 valves	**10. SAGINA**
	6 Petals 0	7

	6 Petals present	8
7	Styles 3	**5. STELLARIA**
7	Styles 5	**7. CERASTIUM**
	8 Petals irregularly toothed or jagged	**6. HOLOSTEUM**
	8 Petals entire to deeply bifid	9
9	Petals bifid ≥1/4 way to base	10
9	Petals entire to slightly emarginate	12
	10 Styles 4-5	**7. CERASTIUM**
	10 Styles 3	11
11	Petals divided ≤1/2 way; alpine	**7. CERASTIUM**
11	Petals divided ≥1/2 way	**5. STELLARIA**
	12 Styles 4	**9. MOENCHIA**
	12 At least most flowers with 3 styles	13
13	Seeds smooth, shiny, with small, lobed oil-body near hilum	**2. MOEHRINGIA**
13	Seeds tuberculate or papillose, without oil-body	**1. ARENARIA**

Key C - Stipules 0; sepals joined to form distinct calyx-tube

1	Styles 2 (rarely 3 on few flowers)	2
1	Styles 3-5 (rarely 2 on few flowers), or flowers male	6
	2 Calyx-tube scarious at joints between lobes	3
	2 Calyx-tube herbaceous all round	4
3	Calyx cylindrical or bell-shaped; seeds flattened, with convex and concave sides, the hilum on the latter; bracteoles placed close to base of calyx and forming epicalyx	**24. PETRORHAGIA**
3	Calyx bell-shaped; seeds reniform or comma-shaped, the hilum in the notch; bracteoles not close to base of calyx and hence not forming epicalyx	**25. GYPSOPHILA**
	4 Calyx-tube with 5 longitudinal wings	**23. VACCARIA**
	4 Calyx-tube without wings	5
5	Calyx-tube with 2-6 sepal-like bracteoles appressed around base; scales absent from base of petal-limb	**26. DIANTHUS**
5	Bracteoles absent; 2 small scales present on inner face of base of petal-limb	**22. SAPONARIA**
	6 Calyx-teeth longer than -tube, extending beyond petals	**18. AGROSTEMMA**
	6 Calyx-teeth shorter than -tube, falling short of petal tips	7
7	Fruit a black berry	**21. CUCUBALUS**
7	Fruit a capsule	8
	8 Styles (2)3(-5), or flowers male; capsule with 2x as many teeth as styles	**20. SILENE**
	8 Styles 5; capsule with 5 teeth	**19. LYCHNIS**

SUBFAMILY 1 - ALSINOIDEAE (genera 1-11). Leaves opposite, without stipules; sepals free or joined only at extreme base; petals mostly 4-5, small to medium; carpophore 0; fruit a capsule or achene.

1. ARENARIA L. - *Sandworts*
Annuals to perennials; sepals 5; petals 5, white, entire or ± so; stamens 10; styles 3(-5); fruit a capsule opening by 6(-10) teeth; seeds without oil-body.

1 Petals shorter than sepals. Stems to 30cm. Native; open ground on well-drained soils, especially sand and limestone (*Thyme-leaved Sandwort* - **A. serpyllifolia** L.) 2
1 Petals longer than sepals 3
   2 Relatively robust; sepals 3-4mm; petals 1.6-2.7mm; capsule flask-shaped, with 'neck', with brittle walls after seed dispersal, 3-3.5 x 1.5-2.5mm; seeds c.0.5-0.6.5mm. Throughout BI (ssp. *lloydii* (Jord.) Bonnier) **A. serpyllifolia ssp. serpyllifolia**
   2 Relatively slender; sepals 2.5-3.1(3.5)mm; petals 1.1-1.6mm; capsule conical, straight-sided, with flexible walls after seed dispersal, 2.5-3 x 1.2-1.5mm; seeds c.0.4-0.5mm. Throughout Bl
A. serpyllifolia ssp. leptoclados (Rchb.) Nyman
3 Leaves all petiolate, almost all borne on procumbent much-branched stems rooting at nodes. Intrd-natd; damp rocky places and on paths and walls; scattered over BI *Mossy Sandwort* - **A. balearica** L.
3 Upper leaves sessile, borne on ± erect stems; stems not rooting 4
   4 Petals >10mm; leaves 10-20mm; sepals pubescent over whole surface on lowerside. Stems to 20(30)cm. Intrd-natd; among *Pteridium*; Guernsey *Large-flowered Sandwort* - **A. montana** L.
   4 Petals <10mm; leaves 3-6mm; sepals glabrous or pubescent at base 5
5 Leaves not succulent, with distinct midrib, the margins pubescent for >1/3 from base; sepals with hairs on lower part of margin and back. Stems to 5cm. Native; limestone cliffs; Co Sligo
Fringed Sandwort - **A. ciliata** L.
5 Leaves slightly succulent, with obscure midrib, the margins glabrous or pubescent only in lower 1/3; sepals glabrous or with a few basal hairs. Stems to 6cm. Native (**A. norvegica** Gunnerus) 6
   6 Perennial with many non-flowering shoots; leaves obovate; flowers c.9-10mm across; styles 3-5. Base-rich scree and river shingle; extremely local in NW Sc and W Ir
Arctic Sandwort - **A. norvegica ssp. norvegica**
   6 Annual or biennial with few non-flowering shoots; leaves narrowly elliptic to narrowly ovate; flowers c.11-23mm across; styles 3. Bare limestone in MW Yorks
English Sandwort - **A. norvegica ssp. anglica** G. Halliday

2. MOEHRINGIA L. - *Three-nerved Sandwort*
Mostly annuals; sepals 5; petals 5, white, entire; stamens 10; styles 3(-4); fruit a capsule opening by 6(-8) teeth; seeds with distinctive oil-body.

1 Stems decumbent to erect, diffusely branched, to 40cm. Native; shady places in woods and hedgebanks; throughout most of BI
Three-nerved Sandwort - **M. trinervia** (L.) Clairv.

3. HONCKENYA Ehrh. - *Sea Sandwort*
Subdioecious ± succulent perennials; sepals 5; petals 5, greenish-white, entire; stamens 10 in male flowers; styles 3(-5) in female flowers; fruit a globose capsule opening by 3 valves.

1 Decumbent to erect flowering stems to 25cm arising from extensive stolons or rhizomes. Native; bare maritime sand and sandy shingle; common round coasts of BI   *Sea Sandwort* - **H. peploides** (L.) Ehrh.

4. MINUARTIA L. - *Sandworts*
Annuals or perennials; sepals usually 5; petals usually 5 or 0, white, entire; stamens 10 or fewer; styles 3(-5); fruit a capsule opening by 3(-5) teeth.

1 Slender annual without non-flowering shoots. Stems to 20cm. Native; dry bare stony places, walls, arable land; scattered in En, E Wa, CI, intrd in Ir   *Fine-leaved Sandwort* - **M. hybrida** (Vill.) Schischk.
1 Mat- or cushion-forming perennials with non-flowering shoots   2
    2 Petals usually 0 or minute, rarely ≥1/2 as long as sepals; nectaries 10, conspicuous at base of sepals. Stems forming dense cushions to 8cm. Native; rocky damp ledges and slopes, mostly on mountains; N & C Sc   *Cyphel* - **M. sedoides** (L.) Hiern
    2 Petals ≥1/2 as long as sepals; nectaries vestigial   3
3 Leaves indistinctly 1-veined; pedicels glabrous. Stems to 10cm. Native; calcareous flushes above 450m; Upper Teesdale, Durham
                       Teesdale Sandwort - **M. stricta** (Sw.) Hiern
3 Leaves distinctly 3-veined; pedicels glandular-hairy   4
    4 Petals shorter than sepals. Stems to 6cm. Native; base-rich rocks on mountains; very local in N & C Sc
                  Mountain Sandwort - **M. rubella** (Wahlenb.) Hiern
    4 Petals as long as or longer than sepals   5
5 Sepals 3-veined; usually laxly tufted. Stems to 15cm. Native; base-rich rocky places and sparse grassland, often on lead-mine spoil; local in N BI, SW En and C & S Wa   *Spring Sandwort* - **M. verna** (L.) Hiern
5 Sepals 5(-7)-veined; usually densely tufted. Stems to 5cm. Native; non-calcareous rocks above 500m; W Cork and S Kerry
              Recurved Sandwort - **M. recurva** (All.) Schinz & Thell.

5. STELLARIA L. - *Stitchworts*
Annuals or perennials; sepals 5; petals 5, sometimes fewer or 0, bifid, white; stamens 10 or fewer; styles 3; fruit a capsule opening by 6 teeth.

1 At least lower leaves petiolate; stems often terete to only weakly and smoothly ridged, sometimes square in section   2
1 All leaves sessile; stems strongly ridged to square in section   7
    2 Petals >1.5x as long as sepals; stems ± equally pubescent all round. Stems to 60cm. Native; damp woods and shady streamsides (*Wood Stitchwort* - **S. nemorum** L.)   3
    2 Petals <1.5x as long as sepals; stems glabrous or with 1(-2) lines

	of hairs down each internode	4
3	Bracts gradually decreasing in size at each higher node of inflorescence, those at 2nd node ≥1/3 as long as those at 1st; seeds with rounded hemispherical marginal papillae. Scattered in N & W Br	**S. nemorum ssp. nemorum**
3	Bracts abruptly shorter above 1st node of inflorescence, those at 2nd node ≤1/3 as long as those at 1st; seeds with cylindrical marginal papillae. Scattered in Wa	**S. nemorum ssp. montana** (Pierrat) Berher

4 Stems glabrous, square in section; sepals tapered-acute
(see couplet 8) **S. uliginosa**
4 Stems with 1(-2) lines of hairs down each internode, terete or
weakly ridged; sepals abruptly acute to obtuse   5
5 Sepals mostly 5-6.5mm; stamens ≥8; seeds mostly >1.3mm. Stems
to 80cm. Native; shady, usually damp places; scattered in Br, mainly
S, very rare in W Ir   *Greater Chickweed* - **S. neglecta** Weihe
5 Sepals mostly <5mm; stamens mostly <8; seeds mostly <1.3mm   6
6 Stamens 3-8; sepals mostly >3mm; seeds mostly >0.8mm; petals
usually present. Stems to 50cm. Native; ubiquitous weed of
cultivated and open ground *Common Chickweed* - **S. media** (L.) Vill.
6 Stamens 1-3; sepals mostly <3mm; seeds mostly <0.8mm; petals
usually absent. Stems to 40cm. Native; coastal dunes and shingle,
inland on bare sandy soil; scattered in BI, mainly S
Lesser Chickweed - **S. pallida** (Dumort.) Crép.
7 Bracts entirely herbaceous; petals bifid c.1/2 way to base. Stems to
60cm. Native; woods and shady hedgerows; common
Greater Stitchwort - **S. holostea** L.
7 Bracts entirely scarious or with wide scarious margins; petals bifid
much >1/2 way to base   8
8 Petals distinctly shorter than sepals; leaves mostly <15mm. Stems
to 40cm. Native; streamsides, ditches, wet tracks and ruts, often
on acid soils; throughout BI   *Bog Stitchwort* - **S. uliginosa** Murray
8 Petals as long as to longer than sepals; leaves mostly >15mm   9
9 Bracts and outer sepals pubescent at margins; flowers c.5-12mm
across. Stems to 80cm. Native; grassy, often dry places; throughout
BI   *Lesser Stitchwort* - **S. graminea** L.
9 Bracts and sepals glabrous; flowers c.12-18mm across. Stems to
60cm. Native; usually base-rich marshes and fens; scattered in Br
N to C Sc, C Ir (*S. palustris* Retz.)   *Marsh Stitchwort* - **S. glauca** With.

6. HOLOSTEUM L. - *Jagged Chickweed*

Annuals; sepals 5; petals 5, irregularly toothed, white; stamens usually 3-5; styles 3; fruit a capsule opening by 6 teeth.

1 Stems ± erect, little-branched, to 20cm, glandular-pubescent. Possibly
native; arable fields, banks and walls on well-drained soils; extinct
since 1930   *Jagged Chickweed* - **H. umbellatum** L.

7. CERASTIUM L. - *Mouse-ears*

Annuals or perennials; sepals 4-5; petals 4-5, sometimes 0, retuse or emarginate to bifid to 1/2 way, white; stamens 4, 5 or 10; styles 3-5(6); fruit a capsule opening by 2x as many teeth as styles.

1 Perennial, with usually ± procumbent non-flowering shoots as well as more erect flowering shoots   2
1 Annual, with all shoots eventually producing flowers   10
   2 Styles mostly 3; capsule-teeth mostly 6. Mat-forming perennial to 15cm. Native; moist rocky places on mountains; mainland N Sc   *Starwort Mouse-ear* - **C. cerastoides** (L.) Britton
   2 Styles 5; capsule-teeth 10   3
3 Petals ≤1.7(2)x as long as sepals. Tufted or matted perennial to 50cm. Native (*Common Mouse-ear* - **C. fontanum** Baumg.)   4
3 Petals >(1.7)2x as long as sepals   6
   4 Seeds 0.8-1mm, with tubercles c.0.1mm across at base; petals c.1.3-1.7x as long as sepals. Rocky mountain sites above 900m; very local in Angus   **C. fontanum ssp. scoticum** Jalas & P.D. Sell
   4 Seeds 0.4-0.9mm, with tubercles c.0.05mm across at base; petals c.0.8-1.2(2)x as long as sepals   5
5 Lower stem internodes with hairs all round; leaves pubescent on both sides. Grassland, and open waste and cultivated ground; very common   **C. fontanum ssp. vulgare** (Hartm.) Greuter & Burdet
5 Lower stem internodes glabrous or with 1-2 lines of hairs; leaves ± glabrous or very sparsely pubescent with most hairs on margin or lowerside midrib. Damp grassy and open ground; mostly in N
   C. fontanum ssp. holosteoides (Fr.) Salman, Ommering & de Voogd
   6 Leaves and upper parts of stems with predominantly densely matted or long (>1mm) shaggy hairs   7
   6 Leaves and stems with predominantly short (<1mm) ± straight hairs; long shaggy hairs less frequent or 0   8
7 Leaves linear to narrowly elliptic, densely tomentose with matted hairs; most stems with >4 flowers. Rampant mat-forming perennial to 40cm. Intrd-natd; dry places; widespread in BI
   Snow-in-summer - **C. tomentosum** L.
7 At least lower leaves elliptic to ovate or obovate, with many long shaggy hairs; stems with 1-4 flowers. Tufted or matted perennial to 20cm. Native; rock outcrops and ledges on mountains; mainland C & N Sc, very local S to N Wa   *Alpine Mouse-ear* - **C. alpinum** L.
   8 Leaves linear to narrowly oblong; flowering stems with short leafy shoots in some leaf axils. Loose mat-forming perennial to 30cm. Native; dry grassland; most of Br but rare in W, local in Ir   *Field Mouse-ear* - **C. arvense** L.
   8 Leaves narrowly to broadly elliptic; flowering stems without conspicuous axillary shoots   9
9 Leaves narrowly elliptic to elliptic, either sparsely pubescent or with both glandular and non-glandular hairs. Tufted or matted perennial to 15cm. Native; rock ledges and outcrops on mountains; N Wa, C &

N mainland Sc　　　　　　　　　　*Arctic Mouse-ear* - **C. arcticum** Lange
9　Leaves elliptic to broadly elliptic, densely glandular-pubescent.
　　Tufted or matted perennial to 15cm. Native; serpentine rocks on Unst,
　　Shetland
　　　　　Shetland Mouse-ear - **C. nigrescens** (H.C. Watson) H.C. Watson
　　10　Sepals with long eglandular hairs projecting beyond sepal apex　　11
　　10　Sepals without eglandular hairs projecting beyond sepal apex　　12
11　Inflorescence compact at fruiting; fruiting pedicels mostly shorter
　　than sepals; glandular hairs abundant on sepals. Stems to 45cm.
　　Native; open places in natural and artificial habitats; common
　　　　　　　　　　　　　　　　　　Sticky Mouse-ear - **C. glomeratum** Thuill.
11　Inflorescence lax at fruiting; fruiting pedicels mostly longer than
　　sepals; glandular hairs 0. Stems to 30cm. Possibly native; chalk
　　grassland and barish places on banks; very local in SE & C En
　　　　　　　　　　　　　　　　　　Grey Mouse-ear - **C. brachypetalum** Pers.
　　12　Bracts completely herbaceous; petals and stamens usually 4;
　　　　capsule-teeth usually 8. Stems to 30cm. Native; dry, open sandy
　　　　places; common round coasts, very local inland
　　　　　　　　　　　　　　　　　　　Sea Mouse-ear - **C. diffusum** Pers.
　　12　Bracts with scarious tips; petals and stamens usually 5; capsule-
　　　　teeth usually 10　　　　　　　　　　　　　　　　　　　　　　　13
13　Uppermost bracts scarious for at least apical 1/3; petals distinctly
　　shorter (c.2/3 as long) than sepals. Stems to 20cm. Native; dry open
　　places on sandy or calcareous soils, especially dunes; frequent
　　　　　　　　　　　　　　　　　　Little Mouse-ear - **C. semidecandrum** L.
13　Uppermost bracts scarious for at most apical 1/4; petals c. as long
　　as sepals. Stems to 12cm. Native; calcareous bare ground or open
　　grassland; very scattered in Wa and S & C En
　　　　　　　　　　　　　　　　　　　Dwarf Mouse-ear - **C. pumilum** Curtis

Several sterile hybrids occur, especially involving *C. arvense* and *C. fontanum*; most are rare but *C. arvense* x *C. tomentosum* is frequent.

8. MYOSOTON Moench - *Water Chickweed*
Perennials; sepals 5; petals 5, white, bifid almost to base; stamens 10; styles 5; fruit a capsule opening by 5 slightly bifid teeth.

1　Plant straggling, decumbent to ascending, to 1m. Native; marshes,
　　ditches and banks of water-courses; fairly common in En and Wa
　　　　　　　　　　　　　Water Chickweed - **M. aquaticum** (L.) Moench

9. MOENCHIA Ehrh. - *Upright Chickweed*
Annuals; sepals 4; petals 4, entire, white; stamens 4; styles 4; fruit a capsule opening by 8 teeth.

1　Erect to procumbent, glabrous, ± glaucous annual to 12cm. Native;
　　barish places on sandy or gravelly ground; local in En and Wa, CI
　　　　　　Upright Chickweed - **M. erecta** (L.) P. Gaertn., B. Mey. & Scherb.

10. SAGINA L. - *Pearlworts*
Annuals or perennials; leaves linear; sepals 4-5; petals 4-5, sometimes 0, entire, white or whitish; stamens 4, 5, 8 or 10; styles 4-5; fruit a capsule opening by 4-5 valves.

1 Annual, with all shoots producing flowers 2
1 Perennial, tufted or mat-forming, with short non-flowering shoots 4
   2 Leaves obtuse, sometimes minutely mucronate with point <0.1mm. Stems to 15cm. Native; damp sand-dunes, rocky places and cliff-ledges on barish soil; round coasts of BI, rarely inland by salted roads or on mountains in Sc *Sea Pearlwort* - **S. maritima** Don
   2 Leaves abruptly protracted into fine point >0.1mm. Stems to 15cm. Native (*Annual Pearlwort* - **S. apetala** Ard.) 3
3 Sepals erect to erecto-patent in fruit, at least the outer subacute; most seeds >1/3mm. Dry bare ground on heaths and paths; scattered
                                                                 S. apetala ssp. apetala
3 Sepals patent in fruit, subobtuse; most seeds <1/3mm. Paths, walls and bare cultivated ground; common **S. apetala ssp. erecta** F. Herm.
   4 Petals 0 or <1/2 as long as sepals 5
   4 Petals 3/4-2x as long as sepals 6
5 Plant a densely compact small cushion with rigidly recurved leaves hiding the stems; seed rarely produced. Stems to 2cm. Enigmatic; S Aberdeen in 1878; not seen wild since
                                            Boyd's Pearlwort - **S. boydii** F.B. White
5 Stems usually partly visible; leaves not or weakly recurved; seed abundantly produced (see couplet 9) **S. procumbens**
   6 Leaves at uppermost nodes usually >3x shorter than those at lower nodes; petals c.2x as long as sepals. Stems to 15cm. Native; damp, open sandy and peaty soil; locally frequent
                                          Knotted Pearlwort - **S. nodosa** (L.) Fenzl
   6 Leaves at uppermost nodes usually <2x shorter than those at lower nodes; petals <1.25x as long as sepals 7
7 Sepals with glandular hairs. Stems to 10cm. Native; dry open ground on sand or gravel; scattered in most of BI
                                       Heath Pearlwort - **S. subulata** (Sw.) C. Presl
7 Sepals glabrous 8
   8 Plant compactly tufted, with basal leaf-rosette only in first year. Stems to 3cm. Native; bare ground near mountain tops; C Sc
                                      Snow Pearlwort - **S. nivalis** (Lindblad) Fr.
   8 Plant a large tuft or mat-forming, with conspicuous basal leaf-rosette 9
9 Sepals 4(5); stamens 4(5). Stems to 20cm. Native; paths, lawns, ditchsides and short turf; very common
                                    Procumbent Pearlwort - **S. procumbens** L.
9 Sepals (4)5; stamens >5, usually 10 10
   10 Leaves with apical point usually >0.2mm; ripe capsule 2.5-3mm
                                              (see couplet 7) **S. subulata**
   10 Leaves with apical point 0-0.2mm; ripe capsules 3-4mm 11

11 Capsules mostly 3.5-4mm, ≥1.5x as long as sepals, fully fertile; stems not or little rooting at nodes. Stems to 10cm. Native; barish ground and rock ledges on mountains; C & N mainland Sc
Alpine Pearlwort - **S. saginoides** (L.) H. Karst.
11 Capsules mostly 3-3.5mm or not developing, ≤1.5x as long as sepals, usually with few seeds; stems usually rooting at nodes. Intermediate between its parents and partially fertile. Native; in range of *S. saginoides* (*S. saginoides* x *S. procumbens*)
Scottish Pearlwort - **S. x normaniana** Lagerh.

S. subulata x *S. procumbens* = **S. x micrantha** É. Martin has been found rarely in Wa and Sc.

11. SCLERANTHUS L. - *Knawels*
Annuals to perennials; leaves linear, fused at base in opposite pairs; flowers greenish, inconspicuous; sepals usually 5; petals 0; fertile stamens (2)5-10; styles 2; fruit an achene.

1 Biennial to perennial to 20cm, with sterile shoots at flowering time; sepals subacute to obtuse, with white border c.0.3-0.5mm wide. Native (*Perennial Knawel* - **S. perennis** L.) 2
1 Annual or biennial to 20cm, without sterile shoots at flowering time; sepals acute to subacute, with whitish border c.0.1mm wide. Native; dry open sandy ground (*Annual Knawel* - **S. annuus** L.) 3
   2 Stems ascending to erect; achene (incl. sepals) (3)3.5-4.5mm. Doloritic rocks in Rads **S. perennis ssp. perennis**
   2 Stems procumbent to ± ascending; achene (incl. sepals) 2-3(3.5)mm. Sandy heaths; Norfolk and Suffolk
S. perennis ssp. prostratus P.D. Sell
3 Achene (incl. sepals) 3.2-4.5(5.5)mm, with divergent sepals when ripe. Scattered over most of BI **S. annuus ssp. annuus**
3 Achene (incl. sepals) 2.2-3.0(3.8)mm, with parallel to convergent sepals when ripe. Frequent in Suffolk
S. annuus ssp. polycarpos (L.) Bonnier & Layens

SUBFAMILY 2 - PARONYCHIOIDEAE (genera 12-17). Leaves opposite, alternate or whorled, stipulate; sepals 5, free; petals 5, often very small; carpophore 0; fruit a capsule or achene.

12. CORRIGIOLA L. - *Strapwort*
Leaves alternate; flowers slightly perigynous; petals ± equalling sepals, white, sometimes red-tipped, entire; stamens 5; stigmas 3, sessile, capitate; fruit an achene.

1 Glabrous ± glaucous annual with 1-many decumbent stems to 25cm. Native; on sand and gravel by ponds; S Devon, infrequent casual elsewhere mainly by railways in En *Strapwort* - **C. litoralis** L.

13. HERNIARIA L. - *Ruptureworts*

Leaves opposite; petals shorter than sepals, filiform; stamens 5; stigmas 2, capitate, sessile or on very short styles or common style; fruit an achene.

1 Calyx and leaves with many hairs on surfaces. Annual to 30cm. Intrd-natd; casual on open waste ground, sometimes natd; En
Hairy Rupturewort - **H. hirsuta** L.
1 Calyx and leaves glabrous or with hairs at margins only   2
  2 Fruit acute to subacute, distinctly exceeding sepals; seeds c.0.5-0.6mm. Annual to perennial to 30cm. Native; dry sandy ground; E En   *Smooth Rupturewort* - **H. glabra** L.
  2 Fruit obtuse to subobtuse, scarcely or not exceeding sepals; seeds c.0.7-0.8mm. Native; maritime sandy and rocky places
(*Fringed Rupturewort* - **H. ciliolata** Melderis)   3
3 Stems hairy on 1 (usually upper) side; leaves ovate to obovate. W Cornwall, Guernsey and Alderney   **H. ciliolata ssp. ciliolata**
3 Stems hairy all round; leaves narrowly elliptic. Jersey
H. ciliolata ssp. subciliata (Bab.) Chaudhri

14. ILLECEBRUM L. - *Coral-necklace*

Leaves opposite; flowers hypogynous; petals much shorter than sepals, filiform; stamens 5; stigmas 2, sessile, capitate; fruit a 1-sided capsule opening by 5 valves adherent at apex.

1 Glabrous annual with procumbent to decumbent stems to 20cm, often reddish. Native; damp sandy open ground; rare in SC & SW En, rarely natd or casual elsewhere   *Coral-necklace* - **I. verticillatum** L.

15. POLYCARPON L. - *Four-leaved Allseed*

Leaves opposite or 4-whorled; flowers hypogynous; petals much shorter than sepals, often emarginate; stamens (1)3-5; stigmas 3, on separate styles; fruit a capsule opening by 3 valves.

1 Almost glabrous annual with erect to ascending much-branched stems to 25cm. Native; open sandy and waste ground near sea; CI, SW En, casual elsewhere   *Four-leaved Allseed* - **P. tetraphyllum** (L.) L.

16. SPERGULA L. - *Spurreys*

Leaves opposite but often appearing whorled due to axillary leaf clusters; flowers hypogynous; petals c. as long as sepals or slightly longer, white, entire; stamens 5-10; stigmas 5, on separate styles; fruit a capsule opening by 5 valves.

1 Leaves linear, furrowed on lowerside; seeds little flattened, without wing or with circum-equatorial wing <1/10 as wide as actual seed. Stems to 40(60)cm. Native; calcifuge usually on sandy cultivated ground; throughout BI   *Corn Spurrey* - **S. arvensis** L.
1 Leaves linear, not furrowed on lowerside; seeds strongly flattened,

with circum-equatorial wing >1/4 as wide as actual seed. Stems to 30cm. Intrd-natd; on sandy or peaty arable land; E Sussex and Kildare *Pearlwort Spurrey* - **S. morisonii** Boreau

17. SPERGULARIA (Pers.) J. & C. Presl - *Sea-spurreys*

Leaves opposite; flowers hypogynous; petals shorter to longer than sepals, (white to) pink, entire; stamens (0)5-10; stigmas 3, on separate styles; fruit a capsule opening by 3 valves.

1 At least some seeds with distinct circum-equatorial wing 2
1 All seeds unwinged 3
    2 Flowers (7)10-12(13)mm across; stamens (0-)10, if fewer than 10 the remainder represented as staminodes. Stems to 40cm. Native; sandy and muddy saline places; common on coasts, very rarely inland *Greater Sea-spurrey* - **S. media** (L.) C. Presl
    2 Flowers (4)5-8mm across; stamens (0)2-7(10); staminodes 0(-3). Native; sandy and muddy maritime places and inland saline areas; common *Lesser Sea-spurrey* - **S. marina** (L.) Griseb.
3 Flowers mostly >8mm across; sepals mostly >4mm; capsule mostly >5mm 4
3 Flowers mostly <8mm across; sepals mostly <4mm; capsule mostly <5mm 5
    4 Plant glabrous to glandular-pubescent only in inflorescence, not woody below (see couplet 2) **S. media**
    4 Plant glandular-pubescent over most of stem, woody below. Stems to 35cm. Native; walls and rocky maritime places; locally common on coasts of S & C BI *Rock Sea-spurrey* - **S. rupicola** Le Jol.
5 Seeds ≥0.6mm; stipules on young shoots fused for ≥1/3 their length
(see couplet 2) **S. marina**
5 Seeds <0.6mm; stipules on young shoots fused for <1/4 their length 6
    6 Upper bracts not much different from stem-leaves; most pedicels much longer than capsules. Stems to 25cm. Native; sandy or gravelly ground; common *Sand Spurrey* - **S. rubra** (L.) J. & C. Presl
    6 Upper bracts much shorter than stem-leaves or represented ± only by stipules; most pedicels shorter than capsules. Stems to 20cm. Possibly native; dry sandy and waste places by sea; frequent in CI, rare in Cornwall, rare casual elsewhere in S Br
Greek Sea-spurrey - **S. bocconei** (Scheele) Graebn.

S. rupicola x *S. marina* has been found rarely in En, Wa and Ir.

SUBFAMILY 3 - CARYOPHYLLOIDEAE (genera 18-25). Leaves opposite, without stipules; sepals fused to form calyx-tube and -lobes; petals 5, mostly medium to large, usually differentiated into a narrow proximal claw within and widened distal limb above calyx-tube; stamens 10; petals, stamens and ovary usually elevated above receptacle on sterile axis (*carpophore*); fruit a capsule or berry.

18. AGROSTEMMA L. - *Corncockle*
Bracteoles not forming epicalyx; petal limb ± rounded to retuse, without scales, purple or rarely white; carpophore 0; styles 5; fruit a capsule opening by 5 teeth.

1 Erect simple to slightly branched annual to 1m. Intrd-casual; cultivated and waste ground; formerly common in cornfields, now very rare casual in Br   *Corncockle* - **A. githago** L.

19. LYCHNIS L. - *Catchflies*
Bracteoles not forming epicalyx; petal-limb entire or 2-4-lobed, with 2 scales at base, scarlet, purple or rarely white; styles 5; fruit a capsule opening by 5 teeth.

1 Leaves and stems with dense white woolly hairs. Stems to 1m. Intrd-natd; frequent escape sometimes natd in waste places; En and Wa   *Rose Campion* - **L. coronaria** (L.) Murray
1 Leaves and stems glabrous to pubescent but not covered with white hairs   2
   2 Petal-limb divided much >1/2 way into 4 narrow lobes. Stems to 75cm. Native; marshy fields and other damp places; throughout BI   *Ragged-Robin* - **L. flos-cuculi** L.
   2 Petal-limb divided up to c.1/2 way into 2 lobes   3
3 Stem-leaves ovate, sparsely pubescent, with cordate base. Stems to 1m. Intrd-natd; rough and marginal ground; casual occasionally natd in S & C Br   *Maltese-Cross* - **L. chalcedonica** L.
3 Stem-leaves linear to elliptic-lanceolate or -oblanceolate, glabrous or pubescent only on margins, not cordate at base   4
   4 Petal-limb divided much <1/2 way, with conspicuous scales at base; carpophore >3mm at fruiting. Stems to 60cm. Native; cliffs and rocky places; extremely local in Wa and Sc, also rare escape   *Sticky Catchfly* - **L. viscaria** L.
   4 Petal-limb divided c.1/2 way, with minute knob-like scales at base; carpophore <2mm at fruiting. Stems to 20cm. Native; mountain serpentine or heavy-metalliferous rocks; extremely rare in Cumberland and Angus   *Alpine Catchfly* - **L. alpina** L.

20. SILENE L. - *Campions*
Bracteoles not forming epicalyx; petal-limb entire to 2-lobed, with or without scales at base, white to red or purple or yellow; styles 3 or 5; fruit a capsule opening by 2x as many teeth as styles.

1 Calyx 3-6mm; flowers arranged in whorl-like groups at each inflorescence node. Stems to 80cm. Native; dry grassy heathland; E Anglia, rare casual elsewhere   *Spanish Catchfly* - **S. otites** (L.) Wibel
1 Calyx 5-30mm, flowers not in whorl-like groups   2
   2 Plant with only male flowers   3
   2 Plant with at least some female or bisexual flowers   6

3	Calyx 20-veined, strongly inflated	11
3	Calyx 10-veined, not or scarcely inflated	4

    4 Dwarf cushion-plant with 1 flower per stem. Stems to 10cm.
Native; mountain rock-ledges and scree; N Wa, Lake District,
C & NW Sc, rare in N Ir   *Moss Campion* - **S. acaulis** (L.) Jacq.

    4 Stems branched, with >1 flower   5

5	Upper part of inflorescence sticky; calyx-teeth with broad scarious margins	15
5	Upper part of inflorescence not or scarcely sticky; calyx-teeth with 0 or narrow scarious margins	8
	6 Styles 5; capsule-teeth 10	7
	6 Styles 3; capsule-teeth 6	9
7	Flowers all bisexual; carpophore >5mm. Stems to 50cm. Intrd-casual; occasional escape on tips and in waste places; scattered in S Br *Rose-of-heaven* - **S. coeli-rosa** (L.) Godr.	
7	Flowers dioecious (sometimes smut-infected stamens present in female flowers); carpophore <2mm	8

    8 Corolla white; capsule-teeth erect. Stems to 1m. Native; banks,
roadsides, waste and cultivated ground, mostly on light soils in
the open; most of BI   *White Campion* - **S. latifolia** Poir.

    8 Corolla red or pink (rarely white); capsule-teeth revolute. Stems
to 1m. Native; woods and hedgerows, usually in shady places
or on open cliffs in lowlands but on rock-ledges and scree-slopes
in mountains; throughout BI   *Red Campion* - **S. dioica** (L.) Clairv.

9	Calyx 20-30-veined, strongly inflated at fruiting	10
9	Calyx 10-veined, usually not or scarcely inflated at fruiting	13

    10 Calyx 30-veined, cross-connecting veins not apparent; annual.
Stems to 35cm. Native; sandy places, mainly maritime dunes; CI,
E Anglia, intrd and scattered elsewhere *Sand Catchfly* - **S. conica** L.

    10 Calyx 20-veined, wih conspicuous cross-connecting veins;
perennial   11

11 Bracts largely herbaceous; capsule with patent to revolute teeth.
Stems to 30cm. Native; rocky and shingly coasts, cliffs, lake-shores
and streamsides in mountains; most coasts of BI, rather rare in
mountains   *Sea Campion* - **S. uniflora** Roth

11 Bracts scarious; capsule with erect to erecto-patent teeth. Stems to
80cm. Grassy places, open and rough ground
(*Bladder Campion* - **S. vulgaris** (Moench) Garcke)   12

    12 Rhizomes absent; lower stem-leaves elliptic to narrowly ovate;
petals white, capsule 6-9(11)mm; seeds 1-1.6mm. Native;
throughout most of BI   **S. vulgaris ssp. vulgaris**

    12 Rhizomes present; lower stem-leaves linear-lanceolate to
lanceolate; petals pinkish to greenish; capsule 11-13mm; seeds
1.6-2.1mm. Intrd-natd; Plymouth (S Devon)
    S. vulgaris ssp. macrocarpa Turrill

13	Perennial, with non-flowering shoots at flowering time	14
13	Annual, without non-flowering shoots	17

    14 Dwarf cushion-plant; flowers 1 per stem   (see couplet 4) **S. acaulis**

14 Stems branched, with >1 flower 15
15 Calyx 4-7mm; petal-limb shortly 4-lobed or -toothed. Stems to
    30cm. Intrd-natd; on rocks in uplands; M Perth, W Sutherland,
    MW Yorks   *Alpine Campion* - **S. quadrifida** (L.) L.
15 Calyx 9-24mm; petal-limb deeply bifid 16
    16 Carpophore about as long as capsule; petal-limb with small
        knob-like scales at base. Stems to 70cm. Intrd-natd; roadside
        banks and quarries; W Kent, rare casual elsewhere
                                    Italian Catchfly - **S. italica** (L.) Pers.
    16 Carpophore <1/2 as long as capsule; petal-limb with conspicuous
        acute scales at base. Stems to 80cm. Native; dry grassy or bare
        places; very locally common in Br N to N Wa and Peak District,
        E Sc, CI   *Nottingham Catchfly* - **S. nutans** L.
17 Stems, leaves and calyx glabrous. Stems to 40cm. Intrd-casual; tips
    and waste places; scattered in En and Jersey
                                    Sweet-William Catchfly - **S. armeria** L.
17 Stems, leaves and calyx pubescent 18
    18 Calyx >20mm. Stems to 50cm. Native; sandy arable soils; very
        scattered in En, Wa and E Sc, intrd in E Ir
                                    Night-flowering Catchfly - **S. noctiflora** L.
    18 Calyx <18mm 19
19 Calyx >12mm, with short sparse hairs. Stems to 45cm. Intrd-casual;
    occasional escape; scattered in S En   *Nodding Catchfly* - **S. pendula** L.
19 Calyx <12mm, with long ± patent hairs. Stems to 45cm. Native;
    waste places, cultivated land and open sandy ground; scattered in S
    Br and CI N to Yorks, intrd in Ir   *Small-flowered Catchfly* - **S. gallica** L.

S. vulgaris x *S. uniflora* is frequent where the parents meet, which is rather rarely, and is fertile. *S. latifolia* x *S. dioica* = ***S. x hampeana*** Meusel & K. Werner is highly fertile and common wherever the parents meet; it has pink petals.

21. CUCUBALUS L. - *Berry Catchfly*

Bracteoles not forming an epicalyx; petal-limb bifid, with 2 scales at base, white; styles 3; fruit a berry.

1 Shortly pubescent perennial with much-branched sprawling stems to
   1m. Intrd-natd; rough grassy places; rare casual, natd in W Norfolk
                                    Berry Catchfly - **C. baccifer** L.

22. SAPONARIA L. - *Soapworts*

Bracteoles not forming an epicalyx; petal-limb ± entire, with 2 scales at base, pink (to white); styles 2; fruit a capsule opening by 4 teeth.

1 Stems erect to ascending, glabrous; flowers in rather compact
   corymbose cymes, c.2.5cm across; calyx glabrous, 15-20mm. Intrd-
   natd; waste ground, roadsides and grassy places, often *flore pleno*,
   often well natd; throughout most of BI   *Soapwort* - **S. officinalis** L.

1 Stems procumbent to decumbent, pubescent; flowers in rather lax
dichasia, c.1cm across; calyx pubescent, 7-12mm. Intrd-natd; walls
and stony banks, natd in a few places; En and Wa
Rock Soapwort - **S. ocymoides** L.

23. VACCARIA Wolf - *Cowherb*
Bracteoles not forming an epicalyx; petal-limb entire or shortly bifid,
without scales, pink; styles 2; fruit a capsule opening by 4 teeth.

1 Glabrous glaucous erect annual to 60cm. Intrd-natd; casual on tips
and waysides and in parks, occasionally natd; scattered in BI
Cowherb - **V. hispanica** (Mill.) Rauschert

24. PETRORHAGIA (DC.) Link - *Pinks*
Bracteoles close to base of calyx, forming epicalyx; calyx-tube scarious at
joints between lobes; petal-limb emarginate, without scales, pink; styles 2;
fruit a capsule opening by 4 teeth.

1 Mat-forming perennials; flowers not in compact heads; epicalyx of
(2-)4(5) whitish bracts; calyx <8mm, ± bell-shaped. Stems to 35cm.
Intrd-natd; waste places; rare casual in Br, natd near Tenby (Pembs)
Tunicflower - **P. saxifraga** (L.) Link
1 Erect annuals; flowers in compact heads of (1)3-11; epicalyx of
several pairs of brown bracts; calyx >8mm, cylindrical 2
  2 Seeds tuberculate on surface. Stems to 50cm. Native; dry grassy
places; very rare in SC En, S Wa and CI, rarely natd elsewhere
Childing Pink - **P. nanteuilii** (Burnat) P.W. Ball & Heywood
  2 Seeds reticulate on surface. Stems to 50cm. Probably native; dry
banks; Beds and W Norfolk
Proliferous Pink - **P. prolifera** (L.) P.W. Ball & Heywood

25. GYPSOPHILA L. - *Baby's-breath*
Bracteoles not close to base of calyx and forming epicalyx; calyx-tube
scarious at joints between lobes; petal-limb retuse or entire, without scales,
usually white, sometimes pink; styles 2; fruit a capsule opening by 4 teeth.

1 Erect perennial to 1m with deep, strong tap-root. Intrd-natd; rough
grassy places on sandy soils; E Kent, Man and M Ebudes, casual
elsewhere   *Baby's-breath* - **G. paniculata** L.

26. DIANTHUS L. - *Pinks*
Bracteoles close to base of calyx, forming epicalyx; calyx-tube not scarious at
joints between lobes; petal-limb variously divided, without scales, usually
pink to red; styles 2; fruit a capsule opening by 4 teeth.

1 Flowers mostly in compact cymose clusters of usually >3 surrounded
by involucre of leaf-like bracts; epicalyx nearly as long as to longer
than calyx 2

1 Flowers solitary or in lax cymes, sometimes 2-3 close together but not surrounded by involucre of leaf-like bracts; epicalyx <3/4 as long as calyx 3
   2 Annual to biennial, without sterile shoots at flowering time; inflorescence pubescent. Stems to 60cm. Native; dry grassy places; very scattered in S & C Br, intrd further N, W Cork, CI
Deptford Pink - **D. armeria** L.
   2 Perennial, with sterile shoots at flowering time; inflorescence glabrous. Stems to 50cm. Intrd-natd; escape or throwout on tips and waste places; Br, mainly S   *Sweet-William* - **D. barbatus** L.
3 Petal-limb divided >1/4 way to base into narrow lobes 4
3 Petal-limb divided <1/4 way to base into narrow or triangular teeth 5
   4 Leaves on sterile shoots mostly >2cm, narrowly acute; inner bracteoles of epicalyx broadly obovate, shortly apiculate at apex. Stems to 30cm. Intrd-natd; on old walls or banks; C & S Br, commoner than *D. caryophyllus*   *Pink* - **D. plumarius** L.
   4 Leaves on sterile shoots mostly <1.5cm, obtuse to subacute; inner bracteoles of epicalyx oblong to obovate, apiculate to cuspidate at apex. Stems to 25cm. Intrd-natd; grassy coastal dunes; Jersey
Jersey Pink - **D. gallicus** Pers.
5 Stems shortly pubescent below; flowers scentless; inner bracteoles of epicalyx cuspidate. Stems to 45cm. Native; dry grassland; scattered in Br N to Banffs   *Maiden Pink* - **D. deltoides** L.
5 Stems glabrous; flowers scented; inner bracteoles of epicalyx apiculate 6
   6 Flowers <30mm across; calyx <20mm. Stems rarely >20cm. Native; limestone rock crevices and cliff-ledges; Cheddar Gorge, N Somerset, rare escape elsewhere
Cheddar Pink - **D. gratianopolitanus** Vill.
   6 Flowers >30mm across; calyx >20mm. Stems to 60cm. Intrd-natd; on old walls; S & C Br   *Clove Pink* - **D. caryophyllus** L.

Many garden escapes are hybrids involving *D. caryophyllus*, *D. plumarius* and *D. gratianopolitanus*, and are very difficult to determine.

49. POLYGONACEAE - *Knotweed family*

Herbaceous annuals to perennials or woody climbers; usually easily recognized by the fused, often scarious stipules sheathing stem; the perianth of 1-2 usually ± similar whorls of 3-6 tepals (2-3 per whorl) persistent in fruit; and the 1-celled ovary with 1 basal ovule

1 Tepals 3; tiny annual with mostly subopposite leaves   **2. KOENIGIA**
1 Tepals mostly ≥4; annual to perennial with alternate leaves 2
   2 Tepals 4; leaves reniform   **9. OXYRIA**
   2 Tepals mostly 5-6; leaves very rarely reniform 3
3 Tepals 6, in 2 whorls of 3 4
3 Tepals mostly 5 5

4	Stamens 6; inner 3 tepals enlarging and enclosing achene; achene not winged; leaves pinnately veined	**8. RUMEX**
4	Stamens 9; tepals remaining small in fruit; achene 3-winged; leaves palmately veined	**7. RHEUM**
5	Stems woody, twining	6
5	Stems entirely or mostly herbaceous though often shrub-like, sometimes twining but then entirely herbaceous	7
	6 Leaves <2cm, rounded at base	**6. MUEHLENBECKIA**
	6 Leaves >2cm, truncate to cordate at base	**5. FALLOPIA**
7	Outer 3 tepals with longitudinal wing or strong keel	**5. FALLOPIA**
7	Tepals not winged, rarely weakly keeled	8
	8 Inflorescences ≤6-flowered, all axillary	**4. POLYGONUM**
	8 Inflorescences mostly >6-flowered, some or all terminal	9
9	Achene >2x as long as perianth, not winged; leaves sagittate; filaments flattened or winged	**3. FAGOPYRUM**
9	Achene usually <2x as long as perianth, if >2x as long then strongly winged; leaves very rarely sagittate; filaments not winged or flat	
		1. PERSICARIA

1. PERSICARIA Mill. - *Knotweeds*
Annuals to perennials, rhizomatous, stoloniferous or neither; inflorescence many-flowered, terminal and axillary, spike-like to subcapitate or paniculate; tepals mostly 5, not winged, petaloid, not enlarging in fruit; stamens 8 (or 4-7 by reduction); style 1 and divided into 2 or 3, or 3; stigmas small, capitate to clavate; achene lenticular (biconvex to biconcave), 3-angled or 3-winged.

1	Stigmas 3 in all flowers; stamens 8; perennials	2
1	Stigmas 2 in all or most flowers; stamens often <8 in at least some flowers; annuals (perennial in *P. amphibia*)	9
	2 Inflorescence unbranched; stamens exserted	3
	2 Inflorescence a branched panicle; stamens included	5
3	Lower part of inflorescence with bulbils instead of flowers; lower leaves cuneate at base. Stems to 30cm. Native; grassland and rock-ledges on mountains; Sc and N En, very rare in N Wa and SW Ir	
	Alpine Bistort - **P. vivipara** (L.) Ronse Decr.	
3	Bulbils 0; lower leaves truncate to cordate at base	4
	4 Petioles of basal and lower stem-leaves winged above; flowers usually pale pink; stems unbranched. Stems to 80(100)cm. Native; grassy places; throughout most of Br and Ir but intrd in Ir and much of S Br *Common Bistort* - **P. bistorta** (L.) Samp.	
	4 Petioles unwinged; flowers usually red; stems usually branched. Stems to 1m. Intrd-natd; grassy and rough places; C & W Ir, CI, scattered in Br *Red Bistort* - **P. amplexicaulis** (D. Don) Ronse Decr.	
5	Tepals <2.5mm at flowering	6
5	Tepals >2.5mm at flowering	7
	6 Leaves tomentose on lowerside, with densely matted, twisted hairs; achene with 3 wings; perianth shrivelling in fruit. Stems	

to 1.5m. Intrd-natd; rough ground; Wastwater (Cumberland)
Chinese Knotweed - **P. weyrichii** (Maxim.) Ronse Decr.
6 Leaves glabrous to densely pubescent on lowerside, with ± straight hairs; achene not winged; perianth becoming succulent in fruit and ≤3mm. Stems to 1.5m. Intrd-natd; rough ground; Argyll and W Kent   *Soft Knotweed* - **P. mollis** (D. Don) H. Gross
7 Tepals fused to c.1/4-1/2 way, usually pinkish. Stems to 1m. Intrd-natd; damp shady places; sparsely scattered over BI
Lesser Knotweed - **P. campanulata** (Hook. f.) Ronse Decr.
7 Tepals free ± to base, white   8
  8 Styles (+ stigmas) <0.5mm; lower leaves cuneate at base, <3.5cm wide. Stems to 1m. Intrd-natd; on river shingle; by R. Dee near Ballater (S Aberdeen)   *Alpine Knotweed* - **P. alpina** (All.) H. Gross
  8 Styles (+ stigmas) >0.5mm; lower leaves truncate to cordate at base, >3.5cm wide. Stems to 1.5m. Intrd-natd; grassy places and roadsides; scattered over Br and Ir
Himalayan Knotweed - **P. wallichii** Greuter & Burdet
9 Stems with recurved prickles. Stems to 1m. Intrd-natd; by streams; S Kerry   *American Tear-thumb* - **P. sagittata** (L.) Nakai
9 Stems without prickles, usually glabrous   10
  10 Flowers in ± globose heads; styles 3. Stems to 40cm. Intrd-natd; cultivated and waste ground, rare casual in Br, natd in SC & SW En   *Nepal Persicaria* - **P. nepalensis** (Meisn.) H. Gross
  10 Flowers in cylindrical or tapering inflorescences; styles 2(-3)   11
11 Perennial with strong rhizomes or stolons, often aquatic; stamens often (not always) exserted at anthesis. Stems to 60cm. Native; in water, wet places, river banks and on rough ground; throughout BI
Amphibious Bistort - **P. amphibia** (L.) Delarbre
11 Annual, often rooting at lower nodes; stamens always included   12
  12 Sessile or stalked glands present on perianth and/or peduncle   13
  12 Glands 0 on perianth and peduncle   16
13 Glands on peduncle with stalks much longer than heads. Stems to 1m. Intrd-casual; on tips and waste ground; S En
Pinkweed - **P. pensylvanica** (L.) M. Gómez
13 Glands on peduncle ± 0 or sessile or with stalks shorter than heads   14
  14 Inflorescence lax, the flowers mostly separated; leaves with sharp peppery taste (usually strong, sometimes faint) when fresh. Stems to 75cm. Native; damp places and shallow water; throughout BI   *Water-pepper* - **P. hydropiper** (L.) Delarbre
  14 Inflorescence dense, the flowers crowded; leaves without peppery taste   15
15 Achenes biconcave to planoconcave or rarely (<1%) 3-angled. Stems to 1m. Native; waste, cultivated and open, especially damp ground; throughout BI   *Pale Persicaria* - **P. lapathifolia** (L.) Delarbre
15 Achenes biconvex to planoconvex or (c.10-60%) 3-angled. Stems to 80cm. Native; waste, cultivated and open ground; throughout BI
Redshank - **P. maculosa** Gray
  16 Inflorescence dense, the flowers crowded; leaves often with

	dark blotch (see couplet 15) **P. maculosa**
16	Inflorescence lax, the flowers mostly separated; leaves never with dark blotch 17

17 Lower leaves usually <5x as long as wide, 12-30mm wide; achene 2.5-3.5mm. Stems to 75cm. Native; damp places and shallow water; rare and very scattered over En, Wa and NE Ir (*P. laxiflora* (Weihe) Opiz) *Tasteless Water-pepper* - **P. mitis** (Schrank) Assenov

17 Leaves usually >5x as long as wide, 2-15mm wide; achene 2-2.5mm. Stems to 40cm. Native; damp fields, ditches and pondsides; very scattered over most of Br and Ir

Small Water-pepper - **P. minor** (Huds.) Opiz

Several hybrids with reduced fertility have been reported between the weedy spp. in couplets 14-17, but all are apparently rare.

2. KOENIGIA L. - *Iceland-purslane*

Annuals; inflorescence of 1-few flowers, terminal and axillary; tepals 3, not winged, not enlarging in fruit; stamens 3; styles 2; stigmas capitate; achene with 3 rounded angles.

Superficially like *Montia* or *Peplis*, but quite different in all details.

1 Stems to 6cm, erect, branched. Native; damp stony ground >500m; Skye and Mull (N & M Ebudes) *Iceland-purslane* - **K. islandica** L.

3. FAGOPYRUM Mill. - *Buckwheats*

Annuals or herbaceous perennials; flowers in terminal and axillary panicles; tepals 5, petaloid, not winged or keeled, not enlarging in fruit; stamens 8; styles 3, long; stigmas capitate, small; achene with 3 acute angles, far exserted.

1 Annual to 60cm; perianth 2.5-4mm; achene 5-7mm. Intrd-natd; casual on tips and waste ground, sometimes persisting; most of BI

Buckwheat - **F. esculentum** Moench

1 Perennial to 1(2)m; perianth 2.5-3.5mm; achene 6-8mm. Intrd-natd; by road; Pembs *Tall Buckwheat* - **F. dibotrys** (D. Don) H. Hara

4. POLYGONUM L. - *Knotgrasses*

Annuals or perennials, with strong tap-root; leaves small, narrowed at base; inflorescences ≤6-flowered, axillary; tepals 5, ± petaloid, not or slightly keeled, not enlarging in fruit; stamens 8; stigmas 3, capitate, small, almost sessile; achene with 3 rounded angles.

For correct identification plants must possess ripe achenes but not be so old that all lower leaves are gone.

1	Uppermost bracts not exceeding flowers, usually partly scarious	2
1	All bracts exceeding flowers, leaf-like	3
	2 Achene c.3mm, shiny; tepals erect, green to apex in midline; stems often erect. Annual to 1m. Intrd-casual; rather frequent	

grain- and wool-alien on tips, waste places and shoddy-fields; scattered in Br *Red-knotgrass* - **P. patulum** M. Bieb.

2 Achene c. 2mm, ± dull; tepals divergent and wholly pink at apex; stems procumbent to ascending. Annual to 50cm. Intrd-casual; status in Br as for *P. patulum*
Lesser Red-knotgrass - **P. arenarium** Waldst. & Kit.

3 Stems strongly woody below; stipules at upper nodes at least as long as internodes, with 6-12 branched veins. Procumbent, glaucous perennial to 50cm. Native; sandy beaches; very rare and sporadic in CI, S En and Waterford *Sea Knotgrass* - **P. maritimum** L.

3 Stems not or slightly woody below; stipules at upper nodes much shorter than internodes, with wholly or mostly unbranched veins 4

    4 Achene shiny, slightly to much longer than perianth. Procumbent annual or perennial to 1m. Native; sandy beaches; scattered round coasts of BI *Ray's Knotgrass* - **P. oxyspermum** Ledeb.

    4 Achene dull, shorter than to slightly longer than perianth 5

5 Leaves of main and lateral stems similar in size; tepals fused for ≥1/3; achene 1.5-2.5mm, with 2 convex and 1 concave sides. Procumbent annual to 30(50)cm. Native; all sorts of open ground; common *Equal-leaved Knotgrass* - **P. arenastrum** Boreau

5 Leaves of lateral stems much smaller than those of main stem; tepals fused for ≤1/4; achene 2.5-4.5mm, with 3 concave sides 6

    6 Leaves <4mm wide; tepals narrowly oblong, gaping near apex to reveal achene. Usually ± erect annual to 30cm. Native; cornfields and other arable land; rare in S & SE En, extremely rare N to C Sc *Cornfield Knotgrass* - **P. rurivagum** Boreau

    6 Larger leaves >5mm wide; tepals oblong-obovate, overlapping almost to apex 7

7 Leaves narrowly ovate to narrowly elliptic, with petioles ± included within stipules; achene ≤3.5mm. Procumbent to scrambling annual to 2m. Native; all sorts of open ground; commonest sp. of genus except in N Sc *Knotgrass* - **P. aviculare** L.

7 Larger leaves obovate to narrowly so, with petioles well exserted from stipules; achene >3mm. Procumbent to scrambling annual to 1m. Native; all sorts of open ground; Shetland and Orkney (commonest sp. of genus there), scattered S to SW Sc
Northern Knotgrass - **P. boreale** (Lange) Small

5. **FALLOPIA** Adans. - *Knotweeds*

Annuals to robust herbaceous or woody perennials; inflorescences terminal and axillary, simple to paniculate; tepals 5, ± petaloid, the outer 3 keeled or winged and enlarging to protect fruit; stamens 8; styles 3; stigmas ± capitate or much divided; achene with 3 rounded angles.

1 Rhizomatous herbaceous perennial; stigmas finely divided, flowers functionally dioecious 2
1 Twining; annual or woody perennial; rhizomes 0; stigmas capitate; flowers all bisexual 4

2   Leaves rarely >12cm, glabrous on lowerside, truncate at base, cuspidate at apex. Stems to 2(3)m. Intrd-natd; waste places, tips and by roads, railways and rivers; frequent to common
Japanese Knotweed - **F. japonica** (Houtt.) Ronse Decr.
2   Leaves often >12cm, sparsely pubescent on lowerside, cordate to cordate-truncate at base, acute to ± acuminate at apex   3
3   Leaves often >20cm, distinctly cordate at base, with scattered long flexuous hairs on lowerside. Stems to 3(4)m. Intrd-natd; similar places to *F. japonica* but rarer; scattered in Br and Ir
Giant Knotweed - **F. sachalinensis** (Maxim.) Ronse Decr.
3   Leaves rarely >20cm, weakly cordate to subtruncate at base, with numerous short stout hairs on lowerside. Stems to 3(4)m. Native; with or sometimes without parents; scattered in BI
(*F. japonica* x *F. sachalinensis*)
Conolly's Knotweed - **F. x bohemica** (Chrtek & Chrtková) J.P. Bailey
4   Woody perennial; the larger inflorescences well branched. Stems to many m. Intrd-natd; persistent throwout or relic, rarely well natd, in scrubby places or hedges; scattered over BI
Russian-vine - **F. baldschuanica** (Regel) Holub
4   Annual; inflorescences with 1 main axis   5
5   Fruiting pedicels 1-3mm; achene 4-5mm, dull. Stems to 1(1.5)m. Native; waste and arable ground; common
Black-bindweed - **F. convolvulus** (L.) Á. Löve
5   Fruiting pedicels 3-8mm; achene 2.5-3mm, shiny. Stems to 2m. Native; hedges and thickets; rare and very local in S En
Copse-bindweed - **F. dumetorum** (L.) Holub

F. japonica x *F. baldschuanica* has been found in Middlesex.

6. MUEHLENBECKIA Meisn. - *Wireplant*
Woody sprawling or climbing perennials; inflorescences short axillary or terminal racemes, dioecious; tepals 5, fused for ≥1/4 from base, enlarging and becoming white and succulent in fruit, not keeled or winged; stamens 8, styles 3, stigmas much divided; achene with 3 rounded angles.

1   Stems to several m; leaves <2cm. Intrd-natd; cliffs, walls and rough ground and in hedges; CI, Scillies, extreme SW En, S Hants
Wireplant - **M. complexa** (A. Cunn.) Meisn.

7. RHEUM L. - *Rhubarbs*
Tall rhizomatous herbaceous perennials; leaves large, mostly basal, palmately veined; flowers in large terminal and axillary panicles; tepals 6, ± petaloid, not winged or keeled, not enlarging in fruit; stamens usually 9; anthers versatile; stigmas 3, subsessile, capitate, papillate; achene with 3 acute angles, with 3 broad membranous wings.

1   Leaves glabrous, very shallowly lobed with entire, obtuse to rounded lobes; flowering stems to 1.5m, glabrous; flowers cream.

Intrd-natd; commonly grown vegetable often persisting as relic or throwout; scattered throughout BI *Rhubarb* - **R. x hybridum** Murray

1 Leaves sparsely pubescent, distinctly lobed with >5 acute, dentate lobes; flowering stems to 2m, pubescent; flowers reddish. Intrd-natd; ornamental occasionally found as a relic or outcast in grassy places; very scattered in En *Ornamental Rhubarb* - **R. palmatum** L.

8. RUMEX L. - *Docks*

Usually herbaceous perennials (rarely annual or biennial), sometimes ± rhizomatous; inflorescences terminal and axillary racemes or panicles with whorled flowers; tepals 6, ± sepaloid, not keeled or winged, the inner usually enlarging in fruit and often with swollen tubercle on face; stamens 6; anthers basifixed; styles 3; stigmas deeply divided; achene with 3 acute angles.

'Tepals' refer to the inner 3 at fruiting.

General Key

1 Lower leaves sagittate to hastate, acid-tasting; flowers mostly unisexual 2
1 Leaves not sagittate or hastate, not acid-tasting; flowers bisexual 8
   2 All leaves with distinct petiole, most c. as wide as long. Stems to 50cm. Intrd-natd; on banks, old walls and rough ground; very scattered in Br *French Sorrel* - **R. scutatus** L. 131
   2 Upper leaves sessile, most distinctly longer than wide 3
3 Upper leaves clasping stem; basal lobes pointed ± basally (sagittate); tepals becoming much longer than achene. Stems to 60(120)cm (*Common Sorrel* - **R. acetosa** L.) 4 131
3 Upper leaves not clasping stem; basal lobes mostly pointed laterally or ± forward (hastate); tepals not or scarcely longer than achene. Stems to 30cm. Native; heathy open ground, short grassland and cultivated land, mostly on acid sandy soils; throughout BI (*Sheep's Sorrel* - **R. acetosella** L.) 7 131
   4 Plant papillose to shortly pubescent (hairs <0.3mm) on all vegetative parts. Stems to 30(50)cm. Native; coastal dunes and grassy or stony places; NW, W & S Ir, N Sc, SW En, SW Wa
R. acetosa ssp. **hibernicus** (Rech. f.) Akeroyd
   4 Plant usually ± glabrous, with papillae ± confined to basal margin of leaves 5
5 Inflorescence with well-branched branches. Stems to 120cm. Intrd-natd; grown as vegetable; escape in Herts and E Suffolk
R. acetosa ssp. **ambiguus** (Gren.) Á. Löve
5 Inflorescence with simple branches 6
   6 Leaves thick, succulent; basal leaves c.2x as long as wide; stem-leaves (1)2-4. Stems to 20(30)cm. Native; sea-cliffs; W Cornwall, Cards, Co Clare
R. acetosa ssp. **biformis** (Lange) Valdés Berm. & Castrov.
   6 Leaves thin, not succulent; basal leaves usually 2-4x as long as wide; stem-leaves often >4. Stems to 60(100)cm. Native; in wide

range of grassy places; very common **R. acetosa** ssp. **acetosa**
7 Tepals forming a loose cover round ripe achene (easily rubbed off
by rolling between finger and thumb) **R. acetosella** ssp. **acetosella**
7 Tepals tightly adherent to ripe achene (not able to be rubbed off)
R. acetosella ssp. **pyrenaicus** (Pourr.) Akeroyd
    8 All tepals lacking swollen tubercle *Key A*
    8 At least 1 tepal with distinct swollen tubercle on outer face 9
9 Tepals with 1-several teeth (each >0.5mm) on each side *Key B*
9 Tepals entire to crenate *Key C*

Key A - All tepals lacking swollen tubercle
1 Tepals with long hooked teeth. Stems to 60cm. Intrd-casual; wool-alien; scattered in En and Sc *Hooked Dock* - **R. brownii** Campd. 13
1 Tepals entire or nearly so 2
    2 Tepals distinctly longer than wide. Stems to 2m. Native;
seasonally flooded ground by Loch Lomond (Stirlings and
Dunbarton) *Scottish Dock* - **R. aquaticus** L. 13
    2 Tepals c. as long as wide 3
3 Plant rhizomatous; lower leaves <1.5x as long as wide. Stems to
70(100)cm. Intrd-natd; relic of old cultivation in grassy places by
roads, streams and old buildings; scattered in Br S to Staffs
Monk's-rhubarb - **R. pseudoalpinus** Höfft 13
3 Plant not rhizomatous; leaves mostly >2x as long as wide. Stems
to 1.2m. Native; damp open and grassy ground often by water; Br S
to S Lancs, Staffs *Northern Dock* - **R. longifolius** DC. 13

Key B - At least one tepal with distinct swollen tubercle; all tepals with 1-several teeth
1 Tepals mostly <4mm 2
1 Tepals mostly >4mm 3
    2 Tepals mostly <3mm, with teeth c. as long (>2mm); tubercle
acute distally; anthers 0.4-0.6mm. Stems to 40(100)cm. Native;
edges of ponds, ditches, gravel-pits and in marshy fields,
usually flooded at times; scattered in Br N to S Sc, very
scattered in Ir *Golden Dock* - **R. maritimus** L. 13
    2 Tepals mostly >3mm, with teeth much shorter (<2mm); tubercle
obtuse distally; anthers 0.9-1.3mm. Stems to 60(100)cm. Native;
similar places to *R. maritimus* but rarely with it; very local in S
& E En N to MW Yorks, Mons *Marsh Dock* - **R. palustris** Sm. 13

FIG 131 - Fruiting tepals of ***Rumex***. 1, *R. obovatus*. 2, *R. longifolius*.
3, *R. pseudoalpinus*. 4, *R. scutatus*. 5, *R. aquaticus*. 6, *R. confertus*.
7, *R. cristatus*. 8, *R. frutescens*. 9, *R. patientia*. 10, *R. pulcher*.
11, ***R. dentatus***. 12, *R. hydrolapathum*. 13, *R. crispus*. 14, *R. x pratensis*.
15, *R. obtusifolius* var. *obtusifolius*. 16, var. *microcarpus*.
17, *R. salicifolius*. 18, *R. brownii*. 19, *R. acetosella*. 20, *R. acetosa*.
21, *R. maritimus*. 22, *R. rupestris*. 23, *R. conglomeratus*.
24, *R. sanguineus*. 25, *R. palustris*. Drawings by Hilli Thompson.

1-16 8mm   5mm 17-25

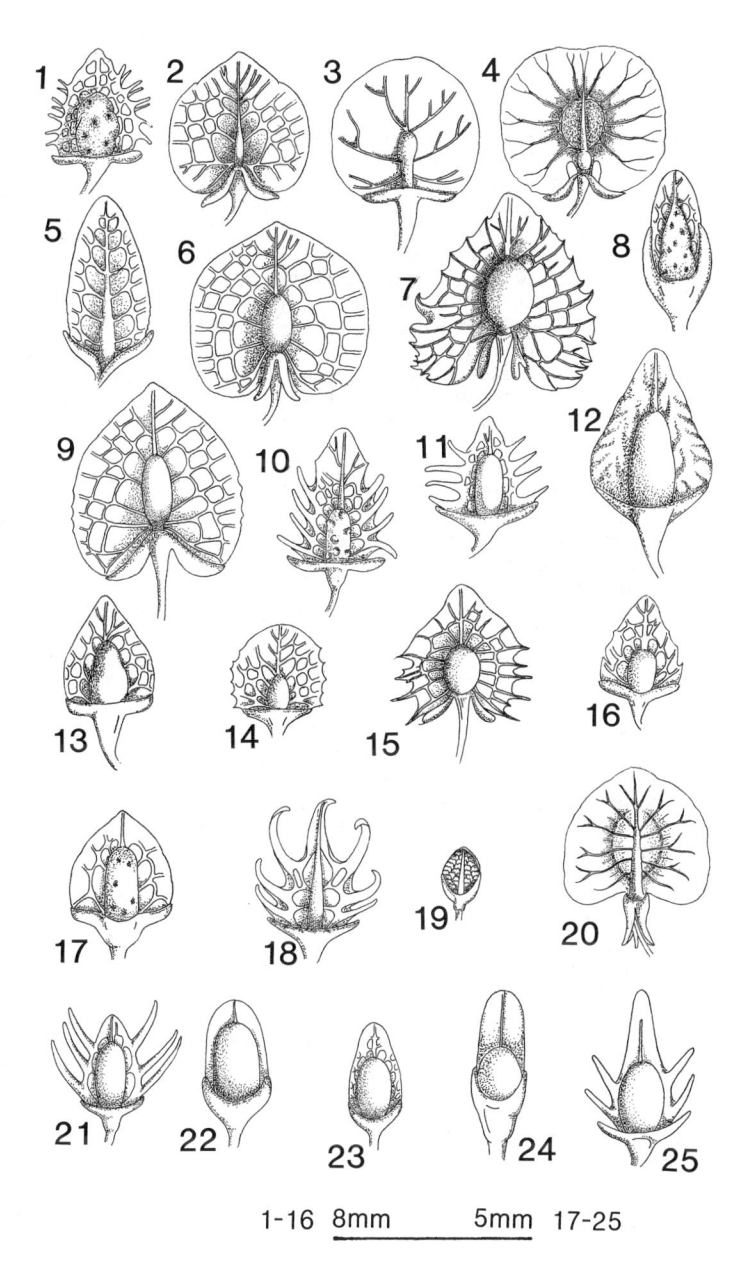

FIG 131 - see caption opposite

3	Tepals broadly ovate to suborbicular, >5mm wide (excl. teeth). Stems to 2m. Intrd-natd; on waste ground in SE En and S Wa, casual elsewhere in S En *Greek Dock* - **R. cristatus** DC.
3	Tepals ovate-triangular, <4mm wide (excl. teeth) 4
	4 Tubercles on tepals coarsely warty 5
	4 Tubercles on tepals ± smooth 6
5	Perennial; leaves usually constricted just below middle (violin-shaped), rounded to cordate at base; at least some branches arising at >60°. Stems to 40(50)cm. Native; dry grassy places; locally frequent in S & C BI, intrd in Ir *Fiddle Dock* - **R. pulcher** L.
5	Annual to biennial; leaves rarely violin-shaped, rounded to cuneate at base; branches arising at <60°. Stems to 40(70)cm. Intrd-casual; grain-alien near mills and warehouses; very scattered in Br *Obovate-leaved Dock* - **R. obovatus** Danser
	6 Annual with basal leaves to 12cm; usually all 3 tepals with well-developed tubercle. Stems to 70cm. Intrd-casual; wool-alien; very scattered in Br *Aegean Dock* - **R. dentatus** L.
	6 Perennial with basal leaves to 40cm; usually only 1 tepal with well-developed tubercle. Stems to 1(1.2)m. Native; grassland, by roads and rivers, waste and cultivated ground; abundant *Broad-leaved Dock* - **R. obtusifolius** L.

Key C - At least 1 tepal with distinct swollen tubercle; all tepals entire to crenate

1	Part of pedicel above joint shorter than tepals 2
1	Part of pedicel above joint c. as long as or longer than tepals 3
	2 Tepals entire, with ± smooth tubercles; branches few, arising at <45 degrees from main stem. Stems to 30cm. Intrd-natd; on coastal dunes; SW En and S Wa, casual elsewhere *Argentine Dock* - **R. frutescens** Thouars
	2 Tepals with some short teeth, with warty tubercles; branches numerous, arising at ≥45° from main stem (see Key B, couplet 5) **R. pulcher**
3	Tepals mostly <5mm 4
3	Tepals mostly >5mm 11
	4 Tepals <3mm 5
	4 Tepals mostly >3mm 6
5	All 3 tepals with well-developed oblong tubercle. Stems to 60(100)cm. Native; damp places, especially by ponds and rivers; thoughout BI *Clustered Dock* - **R. conglomeratus** Murray
5	1 tepal with well-developed ± globose tubercle, other 2 with 0 or rudimentary tubercle. Stems to 60(100)cm. Native; damp shady places, mostly in woods or hedgerows or by water; common *Wood Dock* - **R. sanguineus** L.
	6 Lower leaves ovate-oblong, strongly cordate at base; tepals often with some short teeth (see Key B, couplet 6) **R. obtusifolius**
	6 Lower leaves narrowly oblong, narrowly elliptic or lanceolate, cuneate to subcordate at base; tepals entire 7

7	Tepals not or scarcely wider than tubercles. Stems to 50(70)cm. Native; damp places on sand or rocks by sea; Anglesey, S Wa, SW En and CI **Shore Dock - R. rupestris** Le Gall	131
7	Tepals much wider than tubercles 8	
8	Leaves not undulate; stems with long branches flowering later than main stem, often procumbent at base, to 50(100)cm. Intrd-natd; waste land by docks, railways, canals and tips; very scattered in Br and Ir **Willow-leaved Dock - R. salicifolius** T. Lestib.	131
8	Lower leaves tightly undulate; stems with short ± erect branches, erect, to 1(1.2)m. Native (**Curled Dock - R. crispus** L.) 9	131
9	Achene 1.3-2.5mm; tubercles usually <2.5mm, unequal, often only 1 developed; maritime and inland. Waste, rough, cultivated and marshy ground; abundant **R. crispus ssp. crispus**	
9	Achene 2.5-3.5mm; tubercles ≤3.5mm, usually subequal; maritime 10	
10	Stems usually <1m; inflorescence dense in fruit; on coastal shingle, dunes and saltmarshes. Scattered on coasts of BI **R. crispus ssp. littoreus** (J. Hardy) Akeroyd	
10	Stems often >1m; inflorescence lax in fruit; on tidal estuarine mud. Locally common in S Ir and S Br **R. crispus ssp. uliginosus** (Le Gall) Akeroyd	
11	Tepals nearly as wide as long to wider, ± rounded at apex, only 1 with well-developed tubercle 12	
11	Tepals distinctly longer than wide, tapered to rounded to obtuse apex, usually all 3 with ± well-developed tubercle 14	
12	Basal leaves <1.5x as long as wide, deeply cordate at base; petiole longer than leaf. Stems to 1.2m. Intrd-natd; on roadside; E Kent **Russian Dock - R. confertus** Willd.	131
12	Basal leaves >2x as long as wide, cuneate to cordate at base; petiole shorter than leaf 13	
13	Lower leaves ± cordate at base; lateral veins arising at >60° from midrib (see Key B, couplet 3) **R. cristatus**	
13	Lower leaves cuneate to truncate at base; lateral veins arising at <60 degrees from midrib. Stems to 2m. Intrd-natd; waste places by docks and breweries; scattered in En **Patience Dock - R. patientia** L.	131
14	Tubercles >3mm; robust waterside plant to 2m with basal leaves >60cm. Native; by lakes, rivers, canals, ditches and marshes; scattered in BI N to Banffs **Water Dock - R. hydrolapathum** Huds.	131
14	Tubercles <3mm; basal leaves <50cm 15	
15	Basal leaves ovate-oblong, strongly cordate at base (see Key B, couplet 6) **R. obtusifolius**	
15	Basal leaves narrowly oblong, narrowly elliptic or lanceolate, cuneate to subcordate at base (see couplet 8) **R. crispus**	

Hybrids may be found where 2 or more spp. grow together, but mostly occur as single or few plants. Nearly 40 combinations have been recorded. All are to some degree sterile, most highly so, with undeveloped achenes. The 2 commonest and least sterile are *R. conglomeratus* x *R. sanguineus* = **R. x ruhmeri** Hausskn. and *R. crispus* x *R. obtusifolius* = **R. x pratensis** Mert. 131

W.D.J. Koch.

9. OXYRIA Hill - *Mountain Sorrel*
Herbaceous perennials; inflorescence a terminal panicle; tepals 4, sepaloid, not keeled or winged, the inner enlarging in fruit but without tubercles; stamens 6; anthers versatile; styles 2; stigmas deeply divided; achene biconvex.

1   Stems to 30cm, erect, little branched, with 0(-2) leaves; basal leaves long-stalked, reniform. Native; damp rocky places on mountains; N Wa, NW En, Sc, rare in Ir   *Mountain Sorrel* - **O. digyna** (L.) Hill

50. PLUMBAGINACEAE - *Thrift family*

Perennial, mostly coastal herbs with leaves all basal, simple, entire and without stipules. Distinguished by the actinomorphic 5-merous flowers; calyx fused below and scarious distally; corolla fused only at base and with 5 stamens borne at its base; and 5 styles.

1   Flowers in dense hemispherical heads with tubular sheath of fused scarious bracts beneath   **2. ARMERIA**
1   Flowers in branching cymes, the ultimate units of 1-5 flowers with 3 scale-like bracts   **1. LIMONIUM**

1. LIMONIUM Mill. - *Sea-lavenders*
Aerial stems branched, the ultimate branches consisting of small clusters (*spikelets*) of 1-5 flowers with 3 bracts (outer, middle, inner) below, the spikelets aggregated into *spikes* occupying ends of branches; flowers blue to purple or lilac; styles glabrous.

Spp. in *L. binervosum* agg. (couplets 8-15) are obligate apomicts and difficult to distinguish. Several plants from a population must be examined, as extreme individuals often cannot be identified; geographical location is an important aid. Several sspp. can be recognized in each of the 4 more widespread spp., as indicated, but these are not keyed out here.

1   Leaves distinctly pinnately veined   2
1   All obvious veins (1-9) arising separately from petiole-like base; vein-branches from midrib 0 or indistinct   5
   2   Upper part of stem and leaf-midribs pubescent. Stems to 80(100)cm. Intrd-natd; in rough ground mostly near sea; few sites in SE En   *Florist's Sea-lavender* - **L. latifolium** (Sm.) Kuntze
   2   Plant glabrous   3
3   Leaves rounded to emarginate at apex, dying off before autumn. Intrd-natd; on cliffs at Rottingdean and garden escape elsewhere in E Sussex and in Dorset   *Rottingdean Sea-lavender* - **L. hyblaeum** Brullo
3   Leaves acute to obtuse (and mucronate) at apex, dying in autumn or

1. LIMONIUM   135

    winter; salt-marshes   4
- 4 Spikes mostly 1-2cm, with >4 spikelets/cm; outer bract 1.7-3mm; anthers yellow. Stems to 40(60)cm. Native; muddy salt-marshes; locally common around coasts of CI and Br N to C Sc
Common Sea-lavender - **L. vulgare** Mill.
- 4 Longest spikes 2-5cm, with ≤3 spikelets/cm; outer bract 3-4mm; anthers reddish-brown. Stems to 40(60)cm. Native; similar distribution to *L. vulgare* in Br but rarer, frequent in Ir
Lax-flowered Sea-lavender - **L. humile** Mill.
- 5 Stems with numerous well-branched non-flowering lateral branches below; outer bract scarious except on midline. Stems to 30cm. Native; drier parts of salt-marshes; coasts of N (E & W) Norfolk and Lincs   *Matted Sea-lavender* - **L. bellidifolium** (Gouan) Dumort.
- 5 Stems with 0 or few or little-branched non-flowering lateral branches below; outer bract herbaceous for most of width   6
  - 6 Leaves obovate-spathulate, 11-25mm wide, with 5-7(9) obvious veins; inflorescence widest and ± flat at top; Channel Islands only   7
  - 6 Leaves linear-oblong to oblanceolate-spathulate, rarely obovate-spathulate, 5-15(25)mm wide, with 1-3(5) obvious veins; inflorescence widest below top and tapered above; widespread. Native (*Rock Sea-lavender* - **L. binervosum agg.**)   8
- 7 Outer bract (1.8)1.9-2.4(2.9)mm; calyx (3.8)4-4.6(5.5)mm. Stems to 30(45)cm. Native; on rocks by sea; Plemont Point (Jersey)
Broad-leaved Sea-lavender - **L. auriculae-ursifolium** (Pourr.) Druce
- 7 Outer bract (2.6)3-4(4.2)mm; calyx (4.1)4.8-5.5(6.3)mm. Stems to 20(25)cm. Native; on maritime rocks and dunes; Alderney and Jersey   *Alderney Sea-lavender* - **L. normannicum** Ingr.
- 8 Whole of stem strongly tuberculate (rough)   9
- 8 Stem not tuberculate (smooth) or tuberculate only above   10
- 9 Stems to 35cm, up to 1(1.5)mm thick; leaves up to 10mm wide; spikes 9-24mm, rather lax, with 3-7 spikelets in lowest cm; outer bract 1.9-2.7mm. On rocks and scree; S of Land's End (W Cornwall)
L. loganicum Ingr.
- 9 Stems to 36cm, 1.2-1.7mm thick; leaves up to 13.5mm wide; spikes 7-24(29)mm, dense, with 3-11 spikelets in lowest cm; outer bract 2.3-3.7mm. Cliffs and scree (often limestone) and salt-marshes; ssp. **recurvum** in Portland (Dorset); ssp. **portlandicum** Ingr. in Portland (Dorset) and N Kerry; ssp. **pseudotranswallianum** Ingr. in Co Clare; and ssp. **humile** (Girard) Ingr. in E & W Donegal, Cumberland and Wigtowns   **L. recurvum** C.E.Salmon
- 10 ± all stems branched from low down   11
- 10 Some stems unbranched in lower half   12
- 11 Stems very short and thin, to 7(13.5)cm, up to only 0.8mm thick; spikes (5)8-19mm, dense; outer bract 1.9-2.6mm; inner bract 3.5-4.9mm. Limestone cliff; Saddle Point (Pembs)   **L. parvum** Ingr.
- 11 Stems relatively tall, to 50(70)cm, mostly 3-3.5mm thick; spikes (5)10-25(40)mm, lax or dense; outer bract 2.3-3.6mm; inner bract 4.7-5.9mm. Cliffs and salt-marshes; ssp. **procerum** in E Ir, Wa, SW En

and Rottingdean (E Sussex); ssp. **devoniense** Ingr. near Torquay
(S Devon); and ssp. **cambrense** Ingr. near Pembroke (Pembs)

            L. procerum (C.E. Salmon) Ingr.

12 Mean petal width 1.2-1.5mm. Stems to 40cm, up to 1.2mm thick; spikes up to 19mm, dense, with often >10 spikelets in lowest cm. Limestone cliffs; Giltar Point (Pembs)

          L. transwallianum (Pugsley) Pugsley

12 Mean petal width >1.8mm	13
13 Mean length of outer bract ≥3mm	14
13 Mean length of outer bract <3mm	15

14 Mean length of inner bract <5mm. Stems relatively short, to 20(32)cm, rarely >2mm thick; leaves 3-13mm wide; spikes (5) 10-25(33)mm; outer bract concealing inner bract near end of spike. Rocks; St David's Head (Pembs) **L. paradoxum** Pugsley

14 Mean length of inner bract >5mm. Stems relatively tall, to 40cm, up to 2.8mm thick; leaves up to 22mm wide; spikes 10-35mm; outer bract not concealing inner bract near end of spike. Chalk cliffs and shingle; Dorset **L. dodartiforme** Ingr.

15 Spikes (8)13-25(45)mm, rather lax, with 2-5(6) spikelets in lowest cm. Stems relatively tall, to 50(70)cm. Cliffs (often chalk), rocky places and salt-marshes; ssp. **binervosum** in S E Kent and E Sussex; ssp. **cantianum** Ingr. in N E Kent; ssp. **anglicum** Ingr. from E Norfolk to N Lincs; ssp. **saxonicum** Ingr. in N Essex; ssp. **mutatum** Ingr. at Lannacombe (S Devon); and ssp. **sarniense** Ingr. in CI

         L. binervosum (G. E. Sm.) C.E. Salmon

15 Spikes (4)7-15(29)mm, dense to somewhat lax, with (4)6-8(10) spikelets in lowest cm. Stems relatively short, to 30cm. Cliffs, beaches and salt-marshes; ssp. **britannicum** in N (E & W) Cornwall; ssp. **coombense** Ingr. in S Devon and S E Cornwall; ssp. **transcanalis** Ingr. in N Devon and Pembs; and ssp. **celticum** Ingr. from Caerns to Westmorland **L. britannicum** Ingr.

L. vulgare x *L. humile* = **L. x neumanii** C.E. Salmon occurs in S & E En and NW Wa with both parents; it is partially fertile.

2. ARMERIA Willd. - *Thrifts*

Leaves numerous, very narrow; aerial stems unbranched, terminating in dense hemispherical inflorescence with tubular sheath of fused bracts beneath; flowers pink (rarely white); styles pubescent below.

1 Leaves linear-oblanceolate, some >3mm wide, 3-5(7)-veined, glabrous; flower-heads 20-30mm wide; calyx-teeth 1.5-2mm, incl. terminal point c.1/2 that. Stems to 60cm. Native; fixed dunes; W & S Jersey

       Jersey Thrift - **A. arenaria** (Pers.) Schult.

1 Leaves linear, <2mm wide, 1(-3)-veined, usually with hairs at least on margins; flower-heads 15-25mm wide; calyx teeth ≤1mm incl. very short mucro. Native (*Thrift* - **A. maritima** Willd.) 2

 2 Stems usually pubescent, to 30cm; bract-sheath ≤15mm;

outermost bracts (excl. sheath) shorter than inner. Salt-marshes, saline turf, rocks and cliffs by sea, and inland on mountain rocks and spoil-heaps; common on coasts of BI and on mountains in N Wa, N En and Sc **A. maritima ssp. maritima**

2 Stems glabrous, to 55cm; bract-sheath 12-25mm; outermost bracts (excl. sheath) usually at least as long as inner. Lowland inland rough pasture; rare in S Lincs
A. maritima ssp. elongata (Hoffm.) Bonnier

A. maritima x *A. arenaria* occurs in W Jersey where habitats of the parents meet. It is fertile and intermediate.

51. PAEONIACEAE - *Peony family*

Perennial herbs; some roots strongly tuberous; distinguished by the (bi)pinnate to (bi)ternate leaves, and the solitary or few large flowers with 5-8, usually red petals, very numerous stamens, 2-5, free carpels, and follicles with black fertile and red sterile seeds.

1. PAEONIA L. - *Peonies*

1 Basal leaves with 9(-20) elliptic to ovate segments; follicles (2)3-5, 2-5cm. Stems to 60cm; flowers 8-14cm across. Intrd-natd; on limestone on Steep Holm Island (N Somerset) and Flat Holm Island (Glam) *Peony* - **P. mascula** (L.) Mill.
1 Basal leaves with c.17-30 narrowly elliptic to lanceolate segments; follicles 2-3, 2-3.5cm. Intrd-natd; relic or throwout in marginal and rough ground; scattered in Br N to C Sc *Garden Peony* - **P. officinalis** L.

52. ELATINACEAE - *Waterwort family*

Small ± aquatic annuals with opposite, simple, entire leaves; superficially like Portulacaceae, but differs in many floral characters (e.g. sepals 3-4); distinguished from Caryophyllaceae by the completely compartmented ovary and herbaceous stipules (absent or scarious in Caryophyllaceae).

1. ELATINE L. - *Waterworts*

1 Pedicels at least as long as flowers; sepals and petals 3(-4); stamens 6(8); capsule with 3(-4) valves. Native; in ponds and on wet mud; very local and widely scattered in BI
Six-stamened Waterwort - **E. hexandra** (Lapierre) DC.
1 Pedicels 0 or very short; sepals and petals 4; stamens 8; capsule with 4 valves. Native; in ponds and small lakes; rare and very local in N & W Br and NE Ir *Eight-stamened Waterwort* - **E. hydropiper** L.

53. CLUSIACEAE - *St John's-wort family*

Perennial or rarely annual herbs or small shrubs; easily recognized by entire, opposite leaves without stipules, 5 yellow petals, and numerous stamens usually grouped in bundles.

1. HYPERICUM L. - *St John's-worts*

1 Stems and leaves conspicuously pubescent 2
1 Stems glabrous; leaves glabrous to pubescent 3
    2 Stems procumbent and rooting at nodes below, soft; glands on sepals reddish. Densely pubescent stoloniferous perennial with ascending to erect flowering stems to 40cm. Native; bogs, pondsides and streamsides on acid soil; local in W & S BI, rare or absent in C & E   *Marsh St John's-wort* - **H. elodes** L.
    2 Stems erect, rooting only near base, stiff; glands on sepals black. Pubescent perennial with stems to 1m. Native; open woodland, river-banks and damp grassland; frequent in most of Br
        Hairy St John's-wort - **H. hirsutum** L.
3 Glands on leaves, sepals and petals all translucent and inconspicuous, or 0   4
3 Some glands on leaves, sepals and/or petals black 11
    4 Delicate herb; leaves rarely >1.5 x 0.5cm. Erect glabrous annual or perennial to 20cm with basal buds. Probably intrd-natd; very rare in W and SW Ir   *Irish St John's-wort* - **H. canadense** L.
    4 Stems woody; leaves rarely <2 x 1cm 5
5 Styles 5; petals >(16)20mm; stamens c.1/2-3/4 as long as petals 6
5 Styles 3; petals <20mm; stamens c. as long as or longer than petals 8
    6 Rhizomatous; aerial stems to 60cm, not or little branched. Glabrous evergreen shrub. Intrd-natd; on hedgebanks and in shrubberies; scattered   *Rose-of-Sharon* - **H. calycinum** L.
    6 Not rhizomatous; aerial stems to 1.7m, much branched 7
7 Sepals acute to obtuse; stems persistently 4-ridged. Erect semi-evergreen glabrous shrub to 1.7m. Intrd-natd; in riverside woodland; W Cork   *Irish Tutsan* - **H. pseudohenryi** N. Robson
7 Sepals rounded at apex, sometimes apiculate; stems soon becoming terete. Erect semi-evergreen glabrous shrub to 1.5m. Intrd-natd; garden escape; scattered in Br
        Forrest's Tutsan - **H. forrestii** (Chitt.) N. Robson
    8 Petals >15mm; styles >3x as long as ovary at flowering; leaves strongly smelling of goats when bruised. Erect ± deciduous glabrous shrub to 1.5m. Intrd-natd; in shady places; scattered
        Stinking Tutsan - **H. hircinum** L.
    8 Petals <15mm; styles <3x as long as ovary at flowering; leaves often scented but without strong goat-like smell when bruised 9
9 Petals shorter than to c. as long as sepals; styles <5mm, c.1/2 as long as ovary; fruit succulent when ripe. Erect ± deciduous glabrous shrub to 0.8m. Native; damp woods and shady hedgebanks;

	locally frequent	*Tutsan* - **H. androsaemum** L.
9	Petals longer than sepals; styles >5mm, at least as long as ovary; fruit dry when ripe	10
	10 Petals <4.5mm wide; sepals <2.5mm wide, not overlapping laterally. Rhizomatous evergreen glabrous shrub to 1.5m. Intrd-natd; 3 shady places in N En	
	Turkish Tutsan - **H. xylosteifolium** (Spach) N. Robson	
	10 Petals >4.5mm wide; sepals >2mm wide, overlapping laterally. Erect ± deciduous glabrous shrub to 2m. Intrd-natd; shady places; sparsely scattered	*Tall Tutsan* - **H. x inodorum** Mill.
11	Stems with 4 ridges, sometimes 2 strong and 2 weak; ridges often winged	12
11	Stems with 0-2 ridges; ridges not winged	16
	12 Stems with 2 weak and 2 strong ridges. Partially fertile hybrid backcrossing to give a range of intermediates. Native; grassland; sparsely scattered in Br (*H. maculatum* x *H. perforatum*)	
	Des Etangs' St John's-wort - **H. x desetangsii** Lamotte	
	12 Stems square in section, with 4 strong ridges	13
13	Petals pale yellow, not red-tinged, <7.5mm, c. as long as sepals; stems broadly winged (0.25-0.5mm). Erect rhizomatous glabrous perennial to 60cm. Native; marshes, riverbanks and damp meadows; frequent	*Square-stalked St John's-wort* - **H. tetrapterum** Fr.
13	Petals bright yellow, sometimes red-tinged, >7.5mm, ≥2x as long as sepals; stems not or narrowly winged (≤0.25mm)	14
	14 Stems narrowly winged; sepals acute. Erect rhizomatous glabrous perennial to 60cm. Native; marshy fields and streamsides; local in SW En and W Wa	*Wavy St John's-wort* - **H. undulatum** Willd.
	14 Stems not winged; sepals obtuse. Erect rhizomatous glabrous perennial to 60cm. Native; grassy, usually moist places (*Imperforate St John's-wort* - **H. maculatum** Crantz)	15
15	Inflorescence-branches arising at c.30°; black glands on petals mainly as superficial dots; sepals (1.7)2-3mm wide, entire. Scattered in Sc, very rare in En	**H. maculatum ssp. maculatum**
15	Inflorescence-branches arising at c.50°; black glands on petals mainly as superficial lines; sepals 1.2-2mm wide, denticulate. Locally frequent	**H. maculatum ssp. obtusiusculum** (Tourlet) Hayek
	16 Leaves at least sparsely pubescent on lowerside. Erect perennial to 1m. Native; open woodland, hedgebanks and rocky slopes; local in En and Wa	*Pale St John's-wort* - **H. montanum** L.
	16 Leaves glabrous	17
17	Sepals unequal, 3 longer and wider than the other 2; petals <2x as long as sepals. Procumbent to ascending glabrous perennial to 20cm. Native; open woods, hedgebanks and dry heathland mostly on acid soils; frequent	*Trailing St John's-wort* - **H. humifusum** L.
17	Sepals ± equal; petals ≥2x as long as sepals	18
	18 Stems not ridged; leaves without black glands	19
	18 Stems with 2 ridges; leaves with black glands near margin	20
19	Leaves on main stem triangular-ovate, widest near base, ± sessile,	

± clasping stem. Erect glabrous perennial to 60cm. Native; dry open woodland, hedgebanks and heathland, usually on acid soils; locally common *Slender St John's-wort* - **H. pulchrum** L.
19 Leaves broadly elliptic to orbicular, widest near middle, distinctly petiolate, not clasping stem. Decumbent to erect glabrous perennial to 30cm. Intrd; natd on limestone quarry rock-ledge in NW Yorks
Round-leaved St John's-wort - **H. nummularium** L.
  20 Margins of sepals fringed with stalked black glands; leaves ± without translucent glands. Erect to ascending glabrous perennial to 40cm. Native; rocky acid slopes; rare in SW En, Wa and CI *Toadflax-leaved St John's-wort* - **H. linariifolium** Vahl
  20 Margins of sepals without stalked black glands; leaves with several to numerous translucent glands   21
21 Margins of sepals entire, acute at apex. Erect, glabrous, rhizomatous perennial to 80cm. Native; dry grassland, banks and open woodland; common   *Perforate St John's-wort* - **H. perforatum** L.
21 Margins of sepals denticulate towards apex, with apical apiculus
(see couplet 12) **H. x desetangsii**

H. humifusum x *H. linariifolium* occurs in CI.

54. TILIACEAE - *Lime family*

Deciduous trees; the fragrant flowers, 2-15 in cymes whose stalk is fused to large, narrowly oblong papery persistent bracteole later dispersed with fruit, and the numerous stamens ± coherent in 5 bundles are diagnostic.

1. TILIA L. - *Limes*

1 Leaves pubescent on lowerside; flowers 2-4(6) per cyme; fruit strongly ribbed. Tree to 34m. Native; woods and copses on base-rich soils; very local and scattered in En and Wa, more widespread intrd
Large-leaved Lime - **T. platyphyllos** Scop.
1 Leaves glabrous except for dense hair-tufts in vein axils on lowerside; flowers 4-15 per cyme; fruit not or slightly ribbed   2
  2 Cymes held obliquely erect above foliage; leaves mostly 3-6cm, with scarcely prominent tertiary veins on upperside. Tree to 38m. Native; woods on rich soils; locally common in En and Wa, intrd more widely   *Small-leaved Lime* - **T. cordata** Mill.
  2 Cymes pendent among foliage; leaves mostly 6-9cm, with prominent tertiary veins on upperside. Tree to 46m. Native; woods with both parents; rare from Herefs to NE Yorks, widely intrd (*T. x europaea* L.; *T. platyphyllos* x *T. cordata*)
Lime - **T. x europaea** L.

55. MALVACEAE - *Mallow family*

Annual to perennial herbs or sometimes shrubs; easily recognized by the flowers with 5 petals ± free but often fused at extreme base, with stamens united into a tube round the carpels, and usually with an epicalyx of 3-c.13 sepal-like segments below calyx

1	Epicalyx absent	2
1	Calyx-like epicalyx present below true calyx	5
	2 Flowers pink to purple; stigmas linear	**7. SIDALCEA**
	2 Flowers yellow or white; stigmas capitate	3
3	Leaves palmately veined; nutlets with several seeds	**8. ABUTILON**
3	Leaves pinnately veined; nutlets 1-seeded	4
	4 Petals white; calyx not curved over fruit; carpels winged	**2. HOHERIA**
	4 Petals yellow; calyx curved over fruit; carpels not winged	**1. SIDA**
5	Carpels 5; fruit a capsule	**9. HIBISCUS**
5	Carpels ≥6; fruit breaking into nutlets	6
	6 Epicalyx-segments 6-10	7
	6 Epicalyx-segments 3	8
7	Staminal-tube terete, pubescent	**5. ALTHAEA**
7	Staminal-tube 5-angled, glabrous	**6. ALCEA**
	8 Epicalyx-segments free to base	**3. MALVA**
	8 Epicalyx-segments fused below	**4. LAVATERA**

1. SIDA L. - *Queensland-hemps*
Annuals to perennials, woody or not; epicalyx 0; petals yellowish; carpels 5-14; stigmas capitate; fruit breaking into numerous 1-seeded nutlets.

1 Annual to perennial to 70cm, often woody at base; leaves widest near base; petioles ≤3cm; carpels 5. Intrd-casual; on tips; very local in C & S En *Prickly Mallow* - **S. spinosa** L.

1 Usually woody perennial to 2m, but smaller annual here; leaves widest near middle; petiole <1cm; carpels 8-10(14). Intrd-casual; on tips; very local in S En *Queensland-hemp* - **S. rhombifolia** L.

2. HOHERIA A. Cunn. - *New Zealand Mallow*
Woody shrubs; epicalyx 0; petals white; carpels 5(-6); stigmas capitate; fruit breaking into numerous 1-seeded nutlets.

1 Evergreen shrub potentially to 10m; leaves 7-14cm, ovate or elliptic. Intrd-natd; relic or self-sown in old gardens or on rough and marginal ground; Man *New Zealand Mallow* - **H. populnea** A. Cunn.

3. MALVA L. - *Mallows*
Annual to perennial herbs; epicalyx of 3 segments free to base; petals pink to purple (or white); carpels numerous; stigmas linear; fruit breaking into numerous 1-seeded nutlets.

Fully mature fruit is essential for correct determination.

1	Nutlets smooth or faintly reticulate with low rounded ridges	2
1	Nutlets strongly reticulate with sharp ridges	5
	2 Petals bright pink (or white), >16mm, ≥3x as long as sepals; upper leaves usually very deeply divided	3
	2 Petals pinkish- or whitish-mauve, <16mm, <3x as long as sepals; leaves shallowly lobed	4
3	Epicalyx-segments >3x as long as wide; calyx, epicalyx and pedicels with only simple hairs. Perennial to 80cm. Native; grassy banks and fields; throughout most of BI *Musk-mallow* - **M. moschata** L.	
3	Epicalyx-segments <3x as long as wide; calyx, epicalyx and pedicels with many stellate hairs. Perennial to 1.2m. Intrd-natd; grassy and waste places; SE En and Glam *Greater Musk-mallow* - **M. alcea** L.	
	4 Nutlets smooth, shortly pubescent; petals usually 2-3x as long as sepals. Annual to 80cm. Native; rough and waste ground, waysides; frequent in C & S Br and CI, very scattered elsewhere *Dwarf Mallow* - **M. neglecta** Wallr.	
	4 Nutlets obscurely ridged, glabrous or nearly so; petals usually <2x as long as sepals. Annual to 80cm. Intrd-natd; rough and waste ground, casual, sometimes persisting; scarce in C & S En *Chinese Mallow* - **M. verticillata** L.	
5	Perennial; petals (12)20-30mm, >2x as long as sepals, usually bright pinkish-purple. Perennial to 1(1.5)m. Native; waste and rough ground, by roads and railways; common in lowland En, Wa and CI, scattered elsewhere *Common Mallow* - **M. sylvestris** L.	
5	Annual or biennial; petals 4-12mm, <2x as long as sepals, usually pale pinkish- or whitish-mauve	6
	6 Epicalyx-segments narrowly ovate, ≤3x as long as wide; staminal-tube pubescent. Annual or biennial to 50cm. Intrd-natd; waste places, casual, sometimes persisting; mainly S En, very rare elsewhere *French Mallow* - **M. nicaeensis** All.	
	6 Epicalyx-segments linear-lanceolate, >3x as long as wide; staminal-tube glabrous or nearly so	7
7	Calyx with marginal hairs c.1mm; some pedicels >1cm at fruiting; angle of nutlets between dorsal and lateral surfaces sharp but not winged. Annual to 50cm. Intrd-natd; waste places, casual, sometimes persistent; very scattered *Small Mallow* - **M. pusilla** Sm.	
7	Calyx glabrous or with marginal hairs <0.5mm; pedicels all <1cm at fruiting; angle of nutlets between dorsal and lateral surfaces forming narrow wavy wing. Annual to 50cm. Intrd-natd; waste places, casual, sometimes persisting; scattered in Br *Least Mallow* - **M. parviflora** L.	

4. LAVATERA L. - *Tree-mallows*

Annual to perennial, woody or not; epicalyx of 3 segments fused below; petals pink to purple (or white); carpels numerous; stigmas linear; fruit breaking into numerous 1-seeded nutlets.

1 Central axis of fruit expanded above to form umbrella-like disc
  concealing nutlets. Annual to 1(1.2)m. Intrd-casual; escape or
  throwout, also grain- and wool-alien; scattered in S En
                                    Royal Mallow - **L. trimestris** L.
1 Central axis of fruit not expanded above; nutlets clearly visible from
  above unless obscured by calyx   2
  2 Epicalyx <3/4 as long as calyx. Annual or biennial to 1(2)m.
    Intrd-casual; rather infrequent wool-alien; very scattered in En
                                Australian Hollyhock - **L. plebeia** Sims
  2 Epicalyx 3/4 as long as to longer than calyx   3
3 Flowers 1 in each leaf-axil. Perennial (± shrub) with woody stems to
  2.5m. Intrd-natd; escape or throwout; scattered in S Br (*L. olbia* auct.)
                                Garden Tree-mallow - **L. thuringiaca** L.
3 Flowers 2-several in each leaf-axil   4
  4 Epicalyx expanding to longer than calyx in fruit; petals deep
    pinkish-purple with dark stripes. Biennial to 3m, woody below.
    Native; rocks, cliff-bottoms and waste ground near sea; coasts of
    BI N to C Sc, probably intrd in E Br   *Tree-mallow* - **L. arborea** L.
  4 Epicalyx slightly shorter than to as long as calyx in fruit; petals
    lilac. Annual or biennial to 1(1.5)m. Native; rough and waste
    ground by sea; frequent in Scillies and CI, rare casual elsewhere
                                Smaller Tree-mallow - **L. cretica** L.

5. ALTHAEA L. - *Marsh-mallows*

Annual to perennial herbs; epicalyx of 6-10 segments fused below; petals
pink to purple; staminal-tube terete, pubescent; carpels numerous; stigmas
linear; fruit breaking into numerous 1-seeded nutlets.

1 Perennial with erect stems to 1.5m; upper leaves shallowly 3-5-lobed;
  petals 15-20mm. Native; brackish ditches and grassland near sea;
  locally common on coasts of Br N to Lincs and S Wa
                                Marsh-mallow - **A. officinalis** L.
1 Annual with erect to decumbent stems to 60cm; upper leaves deeply
  3-5-lobed or palmate; petals 12-16mm. Probably intrd-natd; field and
  wood borders, N Somerset and W Kent, casual elsewhere
                                Rough Marsh-mallow - **A. hirsuta** L.

6. ALCEA L. - *Hollyhock*

Biennial to perennial herbs; epicalyx of 6-7 segments fused below; petals
various shades of red, yellow or white; staminal-tube 5-angled, glabrous;
carpels numerous; stigmas linear; fruit breaking into numerous 1-seeded
nutlets.

1 Stems erect, to 3m; leaves shallowly to rather deeply 3-9-lobed; petals
  25-50mm. Intrd-natd; escape or throwout on tips and waste ground;
  scattered over much of BI   *Hollyhock* - **A. rosea** L.

7. SIDALCEA Benth. - *Greek Mallows*
Perennial herbs; epicalyx 0; petals pinkish-purple or white; carpels numerous; stigmas linear; fruit breaking into numerous 1-seeded nutlets.

1 Stems to 1.5m, somewhat pubescent above; leaves pubescent on upperside; sepals 8-12mm; petals 10-25mm, pinkish-purple, rarely white. Intrd-natd; sometimes persisting as throwout or escape; very scattered over Br   *Greek Mallow* - **S. malviflora** (DC.) Benth.
1 Stems to 80cm, ± glabrous; leaves ± glabrous on upperside; sepals 4-6mm; petals 8-15mm, white to cream. Intrd-natd; rarely persisting as throwout or escape; Dunbarton   *Prairie Mallow* - **S. candida** A. Gray

8. ABUTILON Mill. - *Velvetleaf*
Annual herb; epicalyx 0; petals yellow; carpels numerous; stigmas capitate; fruit composed of 5 almost separate, several-seeded nutlets.

1 Erect annual to 1m; leaves suborbicular, acuminate, cordate, not lobed; petals 7-13mm; nutlets each with slender beak 2-3mm. Intrd-casual; waste places and tips, scattered over En and Wa
Velvetleaf - **A. theophrasti** Medik.

9. HIBISCUS L. - *Bladder Ketmia*
Annual herb; epicalyx of 10-13 segments free ± to base; petals pale yellow with violet patch at base; carpels 5, fused; stigmas capitate; fruit a dehiscent capsule with 5 many-seeded cells.

1 Erect to decumbent annual to 50cm; leaves mostly palmate with 3(-5) lobes; petals 15-25mm; calyx enlarging and inflated round fruit. Intrd-casual; waste places and tips; scattered over En, Wa and CI
Bladder Ketmia - **H. trionum** L.

56. SARRACENIACEAE - *Pitcherplant family*

Herbaceous perennials; leaves all in basal rosette, consisting of tubular insectivorous 'pitchers' with small leaf-blade ('hood') at entrance; flowers solitary, terminal, on leafless pedicels from ground level, with showy petals.

1. SARRACENIA L. - *Pitcherplant*

1 Pitchers ≤30cm, green, marbled with red, curved, decumbent; hood erect; pedicels ≤60cm; flowers c.5cm across, with purplish-red petals and sepals. Intrd-natd; wet peat-bogs; C Ir, a few less permanent sites in En and Sc   *Pitcherplant* - **S. purpurea** L.

57. DROSERACEAE - *Sundew family*

Herbaceous perennials, leaves all in basal rosette, reddish and covered with sticky hairs (insectivorous); flowers in simple cyme, on leafless peduncle from ground level; petals 5-8, 4-6mm, white, free.

1. DROSERA L. - *Sundews*

1 Leaf-blade ± orbicular, abruptly narrowed into pubescent petiole. Inflorescences to 10(25)cm, the peduncles straight and arising centrally. Native; wet acid peaty places; over most of BI, often common in N & W   *Round-leaved Sundew* - **D. rotundifolia** L.
1 Leaf-blade obovate to linear-oblong, gradually narrowed into ± glabrous petiole   2
    2 Peduncles straight, rising from near centre of leaf-rosette, often much longer than leaves. Inflorescences to 18(30)cm. Native; wet acid peaty places; locally common in Br and Ir, mostly in W
          Great Sundew - **D. anglica** Huds.
    2 Peduncles curved at base, arising from sides of leaf-rosette, often little longer than leaves. Inflorescences to 5(10)cm. Native; wet acid peaty places; locally common in Br and Ir, mostly in W
          Oblong-leaved Sundew - **D. intermedia** Hayne

D. rotundifolia hybridizes rarely with the other 2 spp.

58. CISTACEAE - *Rock-rose family*

Annuals or woody evergreen dwarf perennials; recognizable by the opposite leaves (at least below) with stellate hairs; 5 free sepals, the 2 outer smaller than the 3 inner; numerous stamens; and 1-celled ovary with many ovules on 3 placentae, 0 or 1 style, and stigma 1 (and then 3-lobed) or 3.

1 Annuals; style very short or 0   **1. TUBERARIA**
1 Woody perennials; style at least as long as ovary   **2. HELIANTHEMUM**

1. TUBERARIA (Dunal) Spach - *Spotted Rock-rose*
Annual with basal leaf-rosette; leaves with 3 obvious veins; stipules 0; petals yellow, usually with brownish-red blotch at base; style 0 or very short.

1 Stems ± procumbent to erect, to 30cm but often <10cm; Native; dry barish ground near sea; W & SW Ir, NW Wa, Jersey and Alderney
          Spotted Rock-rose - **T. guttata** (L.) Fourr.

2. HELIANTHEMUM Mill. - *Rock-roses*
Dwarf straggly or bushy woody perennials; leaves with 1 obvious vein; stipules present or 0; petals yellow or white; style at least as long as ovary.

1 Stipules 0; style strongly S-shaped, shorter than stamens; flowers <15(20)mm across. Stems to 25cm. Native; rocky limestone pastures (*Hoary Rock-rose* - **H. oelandicum** (L.) Dum. Cours.)   2
1 Stipules present; style slightly kinked near base, longer than stamens; flowers mostly >(15)20mm across   4
    2 Leaves glabrous to subglabrous on upperside; flowers 1-3(5) per cyme. NW Yorks (*H. canum* ssp. *levigatum* M. Proctor)
                       H. oelandicum ssp. levigatum (M. Proctor) D.H. Kent
    2 Leaves pubescent to sparsely so on upperside; flowers (2)3-6 per cyme   3
3 Leaves ± persistent on lower parts of non-flowering shoots, usually sparsely pubescent on upperside, mostly >10mm. W Ir (*H. canum* ssp. *piloselloides* (Lapeyr.) M. Proctor)
                       H. oelandicum ssp. piloselloides (Lapeyr.) Greuter & Burdet
3 Leaves not persistent on lower parts of non-flowering shoots, usually pubescent on upperside, often <10mm. Very local in N Wa, S Wa and NW En (*H. canum* (L.) Hornem.)
                       H. oelandicum ssp. incanum (Willk.) G. López
    4 Petals white; leaves grey-tomentose on upperside. Stems to 50cm. Native; dry limestone grassland; N Somerset and S Devon
                       White Rock-rose - **H. apenninum** (L.) Mill.
    4 Petals yellow; leaves green-pubescent on upperside. Stems to 50cm. Native; base-rich grassland; common in suitable places over most of Br, E Donegal
                       Common Rock-rose - **H. nummularium** (L.) Mill.

H. nummularium x *H. apenninum* = ***H. x sulphureum*** Schltdl. occurs with both parents; it has pale yellow petals and is usually highly fertile.

59. VIOLACEAE - *Violet family*

Annual or perennial herbs; immediately recognizable by the distinctive zygomorphic flowers, the lowest petal having a backward-directed spur into which are inserted spurs from the 2 lowest stamens.

1. VIOLA L. - *Violets*

1 Stipules ovate to linear-lanceolate, finely toothed; lateral 2 petals spreading horizontally; style not or gradually thickened distally, often hooked (subg. *Viola* - *Violets*)   2
1 Stipules ± leaf-like, at least some deeply lobed; lateral 2 petals directed upwards; style with ± globose apical swelling with hollow in one side (subg. *Melanium* (DC.) Hegi - *Pansies*)   12
    2 Style straight above, with oblique apex; leaves orbicular, obtuse to rounded at apex. Aerial stems 0. Native; bogs, fens, marshes, wet heaths and woods (*Marsh Violet* - **V. palustris** L.)   3
    2 Style hooked or with lateral beak at apex; leaves acute to obtuse

at apex 4
3 Leaves rounded at apex, with glabrous petioles; bracteoles often below middle of pedicel. Throughout most of Br and Ir
V. palustris ssp. palustris
3 Leaves obtuse to subacute, usually with pubescent petioles; bracteoles usually above or near middle of pedicel. Local in Ir, S & W Wa and W En **V. palustris ssp. juressi** (Wein) P. Fourn.
  4 Leaves (petioles and blades) and usually capsules pubescent 5
  4 Leaves and capsules glabrous 7
5 Sepals acute. Stems 0-5(10)cm. Native; short turf or barish places on limestone; Durham, Westmorland and MW Yorks
Teesdale Violet - **V. rupestris** F.W. Schmidt
5 Sepals obtuse to ± rounded 6
  6 Creeping stolons present; flowers sweet-scented; hairs on petioles mostly <0.3mm, reflexed or appressed. Stems only as stolons. Native; woodland, scrub and hedgerows, mostly on base-rich soils; most of BI, but often intrd *Sweet Violet* - **V. odorata** L.
  6 Stolons 0; flowers not scented; hairs on petioles mostly 0.3-1mm, patent. Stems 0. Native; calcareous pastures and open scrub; Br N to C Sc, rare in C Ir *Hairy Violet* - **V. hirta** L.
7 Plant with basal leaf-rosette; leaves ≤1.3x as long as wide 8
7 Basal leaf-rosette 0; leaves ≥1.4x as long as wide 9
  8 Sepal-appendages >1.5mm, corolla-spur paler than petals. Stems 0-20cm. Native; wide range of woods and grassland; common
Common Dog-violet - **V. riviniana** Rchb.
  8 Sepal-appendages <1.5mm; corolla-spur darker than petals. Stems 0-20cm. Native; woods and hedgebanks; common in En, scattered in Wa and Ir, rare in Sc
Early Dog-violet - **V. reichenbachiana** Boreau
9 Corolla-spur <2x as long as sepal-appendages; roots creeping underground, sending up stems at intervals. Stems to 25cm. Native; fens; Oxon, Cambs and Hunts, C & W Ir
Fen Violet - **V. persicifolia** Schreb.
9 Corolla-spur >2x as long as sepal-appendages; all stems arising from central tuft 10
  10 Petals cream to greyish-violet; leaves lanceolate to narrowly ovate, rounded to cuneate at base. Stems to 20cm. Native; dry heaths; local in S En, S & W Wa and C & S Ir *Pale Dog-violet* - **V. lactea** Sm.
  10 Petals clear blue; leaves ovate to narrowly ovate, truncate to cordate at base. Stems to 30(40)cm. Native
  (*Heath Dog-violet* - **V. canina** L.) 11
11 Stems procumbent to ascending; leaves <2x as long as wide, cordate; stipules of middle leaves <1/3 as long as petiole. Dry or wet heaths and dunes, fens; local **V. canina ssp. canina**
11 Stems erect; leaves c.2x as long as wide, truncate to subcordate; stipules of middle leaves ≤1/2 as long as petiole; flowers relatively large. Fens; Hunts and Cambs **V. canina ssp. montana** (L.) Hartm.
  12 Spur 10-15mm. Stems to 30cm. Intrd-natd; grassy places; frequent

	in Sc, rare elsewhere in Br	*Horned Pansy* - **V. cornuta** L.
12	Spur <7mm	13

13 Flowers usually >3.5cm vertically across, with strongly overlapping petals. Stems to 30cm. Intrd-natd; rough or cultivated land or on tips; scattered in lowland Br *Garden Pansy* - **V. x wittrockiana** Kappert

13 Flowers <2.5(3.5)cm vertically across, with not or slightly overlapping petals 14

    14 Plant perennial, with stems creeping underground 15

    14 Plant annual to perennial; creeping underground stems 0 16

15 Flowers >(1.5)2cm vertically across; terminal segment of stipules scarcely wider than others, entire. Stems to 20cm. Native; upland pastures and rocky places; upland areas of Wa, Sc and N En, very local in S Somerset and C & S Ir *Mountain Pansy* - **V. lutea** Huds.

15 Flowers often <2cm vertically across; terminal segment of stipules distinctly wider than others, often slightly crenate. Stems to 20cm. Native; frequent on maritime dunes in W Br and most of Ir, and inland by lakes in N Ir and on heaths of Norfolk and Suffolk

    Wild Pansy - **V. tricolor** L. **ssp. curtisii** (E. Forst.) Syme

    16 Corolla 4-8mm vertically across, concave. Stems to 10cm. Native; short turf on sandy soil by sea; Scillies and CI

    Dwarf Pansy - **V. kitaibeliana** Schult.

    16 Corolla ≥8mm vertically across, ± flat when fresh; widespread 17

17 Corolla 8-20mm vertically across, usually yellow or cream (rarely with blue), usually shorter than calyx; terminal segment of stipules usually strongly crenate, ± leaf-like; projection below stigma-hollow 0 or indistinct. Stems to 40cm. Native; weed of cultivated and waste ground; common *Field Pansy* - **V. arvensis** Murray

17 Corolla 10-25(35)mm vertically across, usually with some blue or violet, usually longer than calyx; terminal segment of stipules entire to obscurely crenate, scarcely leaf-like; projection below stigma-hollow distinct. Stems to 40cm. Native; waste, marginal and cultivated ground; common *Wild Pansy* - **V. tricolor** L. **ssp. tricolor**

Numerous hybrids have been recorded in both subgenera; they vary from sterile to partially fertile. Those involving *V. canina* or *V. riviniana* with other spp. of subg. *Viola* and with each other, and those involving *V. tricolor* with other spp. of subg. *Melanium*, are most frequent.

60. TAMARICACEAE - *Tamarisk family*

Deciduous shrubs or small trees; easily recognizable by the small scale-like leaves; and small flowers in long catkin-like racemes, with 5 white to pink petals and 5 stamens joined at base to lobed or stellate nectariferous disc.

1. TAMARIX L. - *Tamarisks*

1 Racemes 2-5cm x 4-5mm; petals 1.5-2mm. Shrub or small tree to 3m.

1. TAMARIX 149

Intrd-natd; sandy places by sea, persistent but very rarely self-sown; coasts of CI and Br N to N Wa and Suffolk, rarely elsewhere
Tamarisk - **T. gallica** L.
1 Racemes 3-7cm x 5-9mm; petals 2.5-3mm. Shrub or small tree to 3m. Intrd-surv; as *T. gallica*, but much rarer and never self-sown; S coast of En *African Tamarisk* - **T. africana** Poir.

61. FRANKENIACEAE - *Sea-heath family*

Evergreen procumbent perennial with stems woody at base; recognizable by the small and heather-like leaves; and small flowers with 5 pink petals with small scale at junction of claw and expanded limb and usually 6 stamens in 2 whorls of 3.

1. FRANKENIA L. - *Sea-heath*

1 Stems to 35cm; leaves 3-7mm. Native; on sandy or silty barish ground on drier parts of salt-marshes; coasts of CI and SE Br, sporadic elsewhere in En and Wa *Sea-heath* - **F. laevis** L.

62. CUCURBITACEAE - *White Bryony family*

Annuals or herbaceous perennials; stems scrambling or climbing; distinctive in the epigynous, unisexual (dioecious or monoecious) flowers with unusual stamens (apparently 3 (rarely 5), 2 with 2 pollen-sacs and 1 with 1); and succulent berry-like or cucumber-like fruit.

1 Dioecious perennials; fruit <1cm, red **1. BRYONIA**
1 Monoecious annuals or perennials; fruit >2cm, green to yellow 2
   2 Tendrils absent; fruit densely hispid when ripe **2. ECBALLIUM**
   2 Tendrils present; fruit glabrous to sparsely pubescent or hispid when ripe 3
3 Male flowers several together; central axis of anther extending beyond pollen-sacs **3. CUCUMIS**
3 Male flowers solitary; central axis of anther not extending beyond pollen-sacs 4
   4 Leaves pinnately lobed; corolla divided much >1/2 way
4. CITRULLUS
   4 Leaves palmately lobed; corolla divided <1/2 way **5. CUCURBITA**

1. BRYONIA L. - *White Bryony*

Dioecious perennials with tuberous roots; stems climbing, with simple tendrils; flowers several together; corolla divided >1/2 way, greenish-white; fruit a red globose, ± glabrous berry.

1 Stems to c.5m. Native; in scrub and hedgerows, mostly on well-

drained base-rich soils; local in En and Wa, intrd elsewhere

White Bryony - **B. dioica** Jacq.

2. ECBALLIUM A. Rich. - *Squirting Cucumber*

Monoecious perennials with tuberous roots; stems trailing; tendrils 0; male flowers several together; female flowers solitary; corolla divided <1/2 way, yellowish; fruit ellipsoid, green, densely hispid, explosive when ripe.

1 Stems to 50cm, hispid. Intrd-natd; rare casual in waste and cultivated land in CI and S Br, sometimes persisting in Jersey and extreme S En   *Squirting Cucumber* - **E. elaterium** (L.) A. Rich.

3. CUCUMIS L. - *Cucumbers*

Monoecious annuals with non-tuberous roots; stems climbing with simple tendrils; male flowers several together; female flowers solitary; corolla divided <1/2 way, yellow; fruit ellipsoid to cylindrical, very variable in surface texture, green to yellow, glabrous to pubescent.

1 Leaves not or shallowly palmately lobed; corolla-tube 8-28mm; fruit 3->30cm, usually ellipsoid. Stems to 1m. Intrd-casual; tips and sewerage works; mainly S En   *Melon* - **C. melo** L.
1 Leaves shallowly to rather deeply palmately lobed; corolla-tube 34-65mm; fruit 10->30cm, much longer than wide. Stems to 1(2)m. Intrd-casual; tips and sewerage works; rather rare in S En

Cucumber - **C. sativus** L.

4. CITRULLUS Schrad. - *Water Melon*

Monoecious annuals with non-tuberous roots; stems climbing with simple or branched tendrils; male and female flowers solitary; corolla divided >1/2 way, yellow; fruit globose to slightly elongated, green to yellow, smooth, ± glabrous.

1 Stems to 1m; leaves with 3 main lobes, the central one forming most of leaf. Intrd-casual; rare on tips, frequent at sewerage works; S Br

Water Melon - **C. lanatus** (Thunb.) Matsum. & Nakai

5. CUCURBITA L. - *Marrows*

Monoecious annuals with non-tuberous roots; stems climbing with usually branched tendrils; male and female flowers solitary; corolla divided <1/2 way, yellow; fruit globose to cylindrical, green to yellow, smooth to warty, finally glabrous.

1 Female pedicels thickened and fluted; fruit very variable, usually longer than wide. Stems to 2m. Intrd-casual; common on tips and at sewerage works; mostly S Br   *Marrow* - **C. pepo** L.
1 Female pedicels narrow and terete; fruit usually huge and depressed-globose. Stems to 2m. Intrd-casual; tips and sewerage works; rather rare in S En   *Pumpkin* - **C. maxima** Lam.

63. SALICACEAE - *Willow family*

Deciduous, dioecious, trees or shrubs; flowers much reduced, in racemose catkins, each in axil of single bract, each with either cup-like perianth or 1-2 small nectaries at base; seeds with long plume of hairs arising from base.

1 Flowers with 1-2 nectaries; cup-like perianth 0; bracts entire; winter buds with 1 outer scale **2. SALIX**
1 Flowers with cup-like perianth; nectaries 0; bracts toothed; winter buds with several outer scales **1. POPULUS**

1. POPULUS L. - *Poplars*

Trees; winter buds with several outer scales; flowers appearing before leaves, with toothed or deeply divided bracts, with cup-like perianth; nectaries 0; stamens 5-c.60.

The most important characters are the general habit of the tree and the shape of leaves on the dwarf lateral shoots (short-shoots) or extension shoots (long-shoots); leaves on suckers or epicormic shoots are often very different. The sex of the tree can also be significant. Apart from the need for mature leaves, most characters are best observed in late spring, when shoot pubescence, sex and tree shape are all obvious. Several taxa, notably *P. x canadensis*, are represented in BI by a number of cultivars; these are sometimes very distinctive, but they are not treated here.

1 Leaves of long-shoots densely tomentose on lowerside, sometimes becoming glabrous late in season 2
1 Leaves of long-shoots glabrous to slightly pubescent on lowerside 3
    2 Leaves of long-shoots distinctly palmately 3-5-lobed; bracts subentire to dentate; trees usually female. Tree to 24m. Intrd-natd; often well natd from dense sucker growth, especially on coastal dunes; scattered in BI *White Poplar* - **P. alba** L.
    2 Leaves of long-shoots not or scarcely palmately lobed, coarsely and bluntly dentate; bracts laciniate; trees mostly male. Tree to 46m. Possibly native; rarely natd, probably never arising anew in BI but often among native taxa in damp woods and by streams; scattered throughout BI (*P. alba* x *P. tremula*)
Grey Poplar - **P. x canescens** (Aiton) Sm.
3 Leaves of long- and short-shoots coarsely and bluntly sinuate-dentate; bracts with long silky hairs. Tree to 24m. Native; woods and hedges, often forming suckering thickets; throughout BI *Aspen* - **P. tremula** L.
3 Leaves shortly and regularly serrate; bracts glabrous or shortly pubescent 4
    4 Not balsam-scented; leaves only slightly paler on lowerside; petioles strongly laterally compressed (*Black-poplars*) 5
    4 Buds and young leaves distinctly balsam-scented; leaves much paler on lowerside; petioles nearly terete (*Balsam-*, and *Black- x Balsam-poplar* hybrids) 6
5 Leaves without sessile glands on leaf-blade near top of petiole,

broadly cuneate (to truncate) at base, not minutely pubescent at margin (often slightly pubescent on face); tree either fastigiate with erect branches or with spreading branches and usually large burrs on trunk. Tree to 36m. Ssp. **betulifolia** (Pursh) Dippel is native; in fields by streams and ponds, typically in river flood-plains; scattered throughout most of En and Wa, C Ir, planted elsewhere; several fastigiate cultivars are commonly grown   ***Black-poplar*** - **P. nigra** L.

5 Many leaves with 1-2 small sessile glands on leaf-blade near top of petiole, truncate or subcordate to broadly cuneate at base, minutely pubescent at margin when young; tree narrow or broad but not fastigiate and not with burrs on trunk. Trees of various habit, to >40m, those with spreading crowns differing from *P. nigra* in usually largely up-swept, not down-curved, lower branches (not twigs). Intrd-natd; many cultivars (each of 1 sex) grown for ornament, sometimes natd by pollination between cultivars; throughout BI (*P. nigra* x *P. deltoides*)   ***Hybrid Black-poplar*** - **P. x canadensis** Moench

   6 Leaves serrate, the tip of each serration reaching nearer leaf apex than base of the notch distal to it (*Black-* x *Balsam-poplar* hybrids)   7

   6 Leaves crenate, the tip of each crenation not reaching nearer leaf apex than the base of the depression distal to it (*Balsam-poplars*)   9

7 Leaves ovate-elliptic to -rhombic, rounded to cuneate at base. Narrow, subfastigiate tree to 33m. Intrd-natd; formerly much planted, still found surviving, rarely suckering; Br, mainly S (*P. nigra* 'Italica' x *P. laurifolia* Ledeb.)   ***Berlin Poplar*** - **P. x berolinensis** (K. Koch) Dippel

7 Leaves broadly to triangular-ovate, truncate to cordate at base   8

   8 Young branchlets conspicuously angled or ridged; young branchlets and petioles not or sparsely pubescent; suckers not produced; trees male or female. Spreading tree to 40m. Intrd-surv; formerly much planted, still found surviving; Br, mainly S (*P. deltoides* Marshall x *P. trichocarpa*)
                             Generous Poplar - **P. x generosa** A. Henry

   8 Young branchlets ± rounded; young branchlets and petioles conspicuously pubescent; mature trees producing abundant suckers; trees all female. Spreading tree to 30m. Intrd-natd; planted in damp woods and by rivers and ponds, often natd by suckering; Br, mainly S (*P. x candicans* auct.; *P. deltoides* x *P. balsamifera*)   ***Balm-of-Gilead*** - **P. x jackii** Sarg.

9 Tree with subfastigiate outline; lower branches arising at c.45°, with few or no pendent twigs; trees all female. Tree to >30m with very narrow outline, without suckers. Intrd-surv; now 1 of the most planted poplars in Br (*P. trichocarpa* x *P. balsamifera*)
                             Hybrid Balsam-poplar - **P. 'Balsam Spire'**

9 Tree rather narrowly pyramidal; some lower branches spreading and with pendent twigs; trees male or female   10

   10 Young branchlets conspicuously angled or ridged; leaves usually widest not much above petiole apex and abruptly rounded there, with mostly truncate to subcordate base; suckers scarce. Tree to 35m with narrow outline. Intrd-natd; frequently planted in Br,

natd by suckers in Lanarks
Western Balsam-poplar - **P. trichocarpa** Hook.
10 Young branchlets scarcely angled or ridged; leaves usually widest well above petiole apex and gradually rounded there, with mostly rounded base; suckers often abundant. Tree to 25m with narrow outline. Intrd-surv; planted but less common than *P. trichocarpa*   *Eastern Balsam-poplar* - **P. balsamifera** L.

Hybrids are very common, notably within and between the *Black-* and *Balsam-poplars*. Apart from those covered above, others occasionally arise naturally in Br by crossing between planted trees, and can become natd.

2. SALIX L. - *Willows*

Trees or dwarf to tall shrubs; winter buds with 1 outer scale; flowers appearing before, with or after leaves, with entire bracts, with 1-2 nectaries; cup-like perianth 0; stamens 1-5(12).

Identification is often made difficult by the extensive degree of hybridization (64 combinations known in BI, of which 10 are hybrids between 3 spp. and 1 is a hybrid between 4 spp.). The 14 hybrids treated fully here are not necessarily the most common, but those likely to be found without parents nearby. Most of these are crosses between the taller lowland species that have been planted for ornament and basket-work.

In many cases catkins and both young and mature leaves are desirable for identification; keys based on catkins alone are not satisfactory. Mature leaves should be collected in Jul-Sep, before senescence; abnormally vigorous shoots (e.g. suckers) should be avoided. In this account the so-called catkin-scales (scales associated with each flower) are termed bracts, which they truly are. 'Twigs' refer to those that have completed 1-2 years' growth, not current growth. Except locally, where plants have been clonally propagated, the spp. are generally represented by both sexes roughly equally, but many of the hybrids are much more commonly, or only, female. Bisexual catkins are not rare, especially in hybrids, including some of those noted as 'female only'.

General key

1 Shrubs <80(100)cm high (stems may be longer but not erect) *Key A*
1 Tree or shrub >1m high 2
  2 Shrub <1.5(2)m high *Key B*
  2 Tree or shrub >2m high 3
3 Leaves closely and finely serrate (>(3)5 serrations/cm at leaf midpoint); bracts yellowish (except *S. daphnoides* and *S. acutifolia*) *Key C*
3 Leaves entire, obscurely crenate-serrate, or coarsely and irregularly serrate (<4(7) serrations/cm at leaf midpoint); bracts dark or dark-tipped (except *S. x mollissima*, *S. elaeagnos* and *S. udensis*) 4
  4 Leaves <3x as long as wide, or mostly so *Key D*
  4 Leaves >3x as long as wide, or mostly so *Key E*

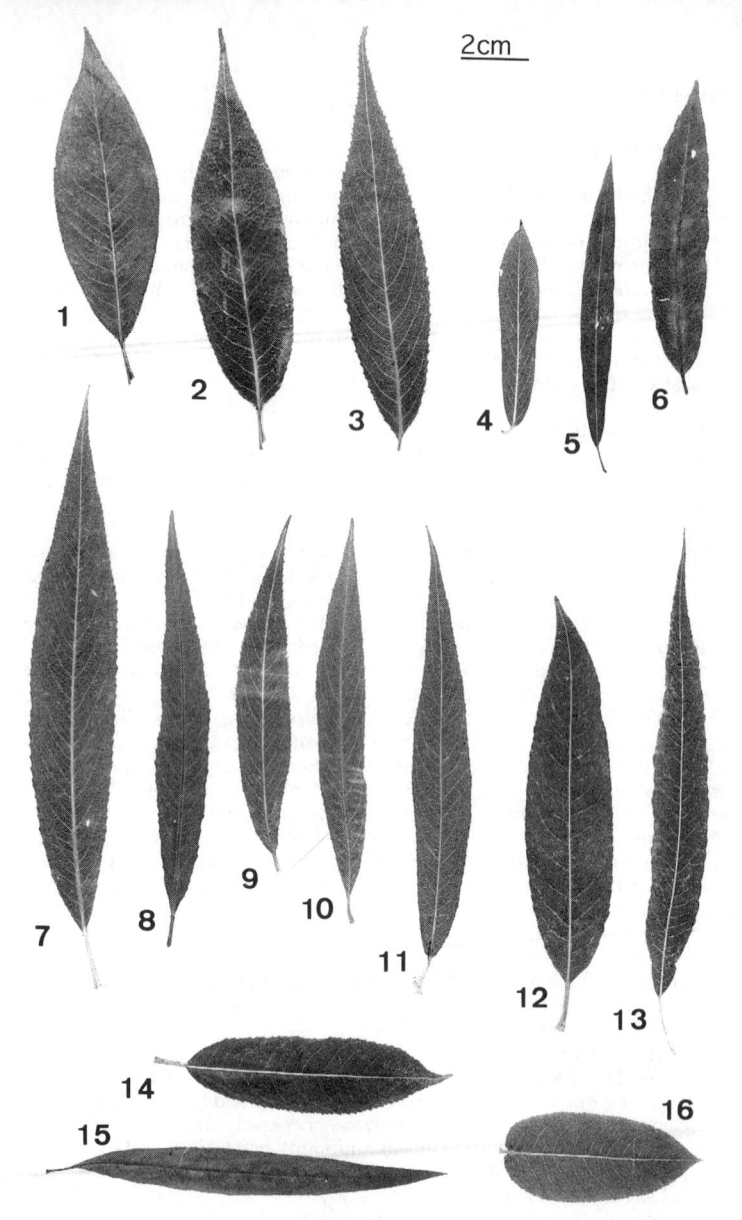

FIG 154 - Leaves of *Salix*. 1, *S. pentandra*. 2, *S. x meyeriana*. 3, *S. x ehrhartiana*. 4, *S. purpurea*. 5, *S. x rubra*. 6, *S. x forbyana*. 7, *S. fragilis*. 8, *S. x rubens*. 9, *S. x pendulina*. 10, *S. alba*. 11, *S. x sepulcralis*. 12, *S. daphnoides*. 13, *S. acutifolia*. 14, *S. triandra*. 15, *S. x mollissima*. 16, *S. eriocephala*.

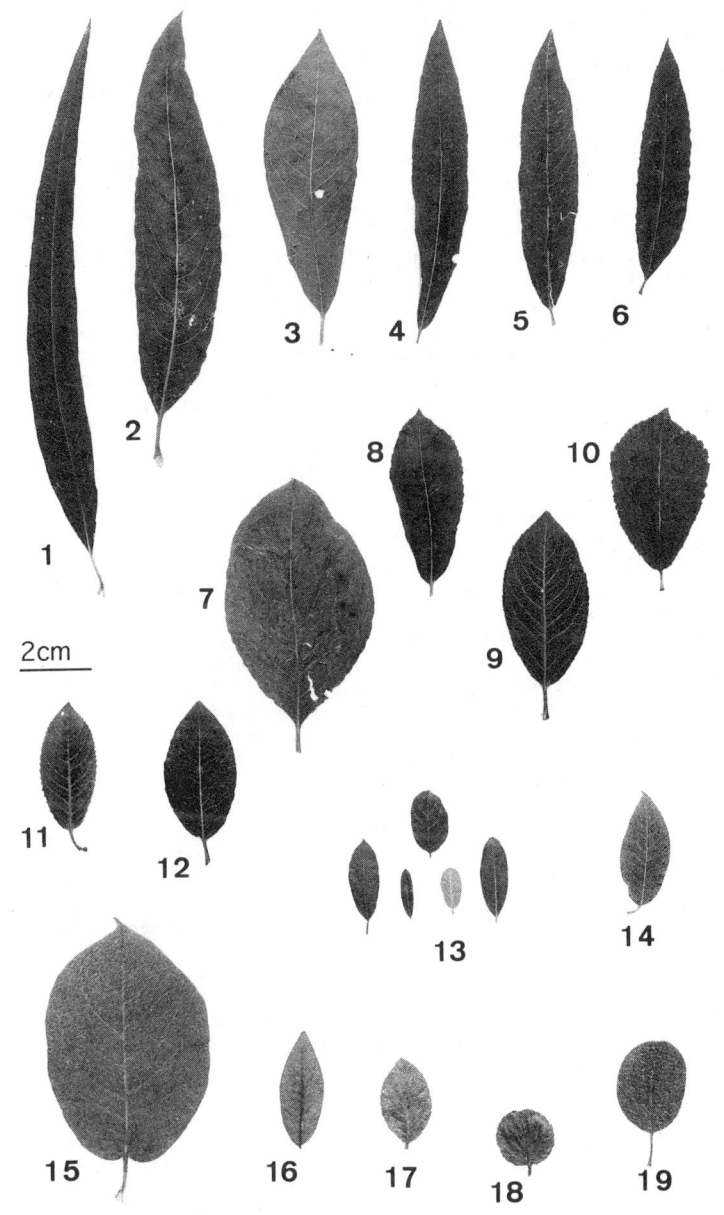

FIG 155 - Leaves of *Salix*. 1, *S. viminalis*. 2, *S. x sericans*. 3, *S. x calodendron*. 4, *S. x stipularis*. 5, *S. x smithiana*. 6, *S. x fruticosa*. 7, *S. caprea*. 8, *S. cinerea*. 9, *S. x laurina*. 10, *S. aurita*. 11, *S. myrsinifolia*. 12, *S. phylicifolia*. 13, *S. repens*. 14, *S. lapponum*. 15, *S. lanata*. 16, *S. arbuscula*. 17, *S. myrsinites*. 18, *S. herbacea*. 19, *S. reticulata*.

Key A - Shrubs <80(100)cm high
1 Stems usually <10cm; leaves not or scarcely longer than wide, rounded to emarginate at apex; catkins appearing when leaves ± mature, all or most from terminal buds 2
1 Stems usually >10cm; leaves longer than wide, usually obtuse to acute at apex; catkins appearing before leaves mature, mostly from lateral buds 3
  2 Leaves entire, with distinctly impressed veins on upperside; petioles mostly >10mm. Dwarf shrub <20cm. Native; wet rock-ledges and mountain-slopes; very local in C & N mainland Sc
Net-leaved Willow - **S. reticulata** L. 155
  2 Leaves crenate-serrate; veins not impressed on leaf upperside; petioles mostly <4mm. Dwarf shrub <10cm. Native; rock-ledges and rocky mountain-tops; locally common in Sc, N En and parts of Wa and Ir *Dwarf Willow* - **S. herbacea** L. 155
3 Leaves densely pubescent to tomentose at least until maturity at least on lowerside 4
3 Leaves glabrous to sparsely pubescent at maturity on both sides 7
  4 Stipules mostly large and ± persistent at maturity 5
  4 Stipules small and mostly soon dropping off 6
5 Leaves ± entire; catkins densely silky with yellow hairs. Low shrub to 1m. Native; damp mountain rock-ledges; local, highlands of C Sc
Woolly Willow - **S. lanata** L. 155
5 Leaves crenate-serrate; catkins with greyish to whitish hairs. Shrub to 2(3)m. Native; heathland, scrub and rocky hills on acid soils; throughout BI *Eared Willow* - **S. aurita** L. 155
  6 Leaves silky-pubescent on lowerside. Procumbent to erect shrub to 1.5(2)m, usually <1m. Native; acid heaths and moors, fens, dunes and dune-slacks; frequent *Creeping Willow* - **S. repens** L. 155
  6 Leaves tomentose on lowerside. Native; rocky mountain slopes and cliffs; N & C mainland Sc, rare in Westmorland and S Sc
Downy Willow - **S. lapponum** L. 155
7 Leaves cordate at base. Rhizomatous shrub to 2m. Intrd-natd; bogs; rare in En and Wa *Heart-leaved Willow* - **S. eriocephala** Michx. 154
7 Leaves cuneate at base 8
  8 Stipules mostly large and persistent to maturity 9
  8 Stipules mostly small and soon dropping off 10
9 Leaves shining green on both sides, not blackening when dried; ovary usually pubescent. Spreading shrub <50cm. Native; rocky ledges and slopes on mountains; local in N & C Sc, rare in S Sc
Whortle-leaved Willow - **S. myrsinites** L. 155
9 Leaves green on upperside, ± glaucous on lowerside, blackening when dried; ovary mostly ± glabrous. Native; by upland ponds and streams and in damp rocky places; frequent in N En and Sc, intrd in C & S Br *Dark-leaved Willow* - **S. myrsinifolia** Salisb. 155
  10 Leaves mostly <2.5 x 1.5cm. Shrub to 80cm. Native; damp rocky slopes and mountain ledges; locally abundant in C Sc
Mountain Willow - **S. arbuscula** L. 155

2. SALIX

   10 Leaves mostly >2.5 x 1.5cm. Native; by upland ponds and streams and in damp rocky places; frequent in N En and Sc, Co Leitrim, Co Sligo   ***Tea-leaved Willow*** - **S. phylicifolia** L.   **155**

Key B - Shrubs 1-1.5(2)m high
1   Extension growth malformed, with contorted, flattened and laterally fused (fasciated) branches. Shrub or small tree. Intrd-surv; where dumped or planted in wet ground; several sites NW En to C Sc
     Sachalin Willow - **S. udensis** Trautv. & C.A. Mey.   **339**
1   Extension growth not malformed and fasciated   **2**
    2 Leaves glabrous to sparsely pubescent at maturity on both sides   **3**
    2 Leaves densely pubescent to tomentose at least until maturity at least on lowerside   **5**
3   Leaves cordate at base   (see Key A, couplet 7) **S. eriocephala**
3   Leaves cuneate at base   **4**
    4 Stipules mostly large and persistent at maturity; leaves blackening when dried; ovary mostly ± glabrous, with pedicel c.1mm; twigs usually pubescent   (see Key A, couplet 9) **S. myrsinifolia**
    4 Stipules mostly small and soon dropping off; leaves not blackening when dried; ovary mostly pubescent, with 0 to very short pedicel; twigs usually ± glabrous
        (see Key A, couplet 10) **S. phylicifolia**
5   Leaves mostly >4x as long as wide, tomentose on lowerside; female only. Erect shrub to 5m. Native; often with parents but sometimes not; scattered over Br and Ir, especially in N & W (*S. viminalis* x *S. aurita*)
     Shrubby Osier - **S. x fruticosa** Döll   **155**
5   Leaves <4x as long as wide, or if more then appressed-pubescent on lowerside   **6**
    6 Stipules mostly large and ± persistent to maturity
        (see Key A, couplet 5) **S. aurita**
    6 Stipules small and mostly soon dropping off   **7**
7   Leaves silky-pubescent on lowerside; widespread
        (see Key A, couplet 6) **S. repens**
7   Leaves tomentose on lowerside; mountains of N Br only
        (see Key A, couplet 6) **S. lapponum**

Key C - Trees or shrubs >2m high; leaves closely and finely serrate; bracts yellowish (except dark-tipped in *S. daphnoides* and *S. acutifolia*)
1   Twigs dark purple, with conspicuous whitish bloom easily rubbed off; bracts dark-tipped   **2**
1   Twigs variously coloured, rarely dark purple, without bloom; bracts yellowish   **3**
    2 Leaves >5x as long as wide; ultimate branchlets ± pendent. Tall shrub or slender tree to 10m. Intrd-surv; widely planted (male only), rarely found in wild places; Br
     Siberian Violet-willow - **S. acutifolia** Willd.   **154**
    2 Leaves usually <5x as long as wide; branchlets not pendent. Tall shrub or slender tree to 10(18)m. Intrd-surv; widely planted (both

sexes), rarely found in wild places; Br
European Violet-willow - **S. daphnoides** Vill. 154

3 Branchlets pendent (weeping willows) 4
3 Branchlets erect or spreading 5
    4 Leaves pubescent to silky-pubescent ± until maturity; ovary little longer than subtending bract. Tree to 22m with distinctive weeping habit. Intrd-surv; much planted and often persisting in wild places; lowland BI (*S. alba* x *S. babylonica*)
Weeping Willow - **S. x sepulcralis** Simonk. 154
    4 Leaves glabrous to sparsely pubescent even when very young; ovary much longer than subtending bract. Tree to 15m with distinctive weeping habit. Intrd-surv; planted and found in wild places, less common than *S. x sepulcralis* (*S. fragilis* x *S. babylonica*)
Weeping Crack-willow - **S. x pendulina** Wender. 154
5 Leaves elliptic to ovate, 2-4x as long as wide; stamens ≥4. Tree to 10(18)m. Native; wet ground and by ponds and streams; N & C Br and Ir, planted further S   *Bay Willow* - **S. pentandra** L. 154
5 Leaves narrowly elliptic to lanceolate, >(3)4x as long as wide; stamens ≤4 6
    6 Stipules mostly large and ± persistent to maturity; bark smooth, flaking off in large patches or peels 7
    6 Stipules mostly 0 or small and soon dropping off; bark furrowed, not flaking off 8
7 Young leaves and twigs glabrous; twigs with strong ridges or angles. Small tree or shrub to 10m. Native; damp places; frequent in C & S En, much less so elsewhere   *Almond Willow* - **S. triandra** L. 154
7 Young leaves and twigs pubescent; twigs terete or obscurely ridged or angled. Shrub to 5m. Possibly native; damp places; scattered over Br and Ir, frequent in C & S En (*S. triandra* x *S. viminalis*)
Sharp-stipuled Willow - **S. x mollissima** Elwert 154
    8 Mature leaves silky-pubescent, or ± glabrous and dull on upperside 9
    8 Mature leaves glabrous and rather glossy on upperside 10
9 Leaves subglabrous on both sides at maturity. Tree to 30m. Native; frequent over most of lowland BI (*S. alba* x *S. fragilis*)
Hybrid Crack-willow - **S. x rubens** Schrank 154
9 Leaves ± silky-pubescent at least on lowerside at maturity. Possibly native; marshes, wet hollows and by streams and ponds; common over most of lowland BI   *White Willow* - **S. alba** L. 154
    10 Twigs and leaves glabrous from the first 11
    10 Twigs and leaves sparsely pubescent or leaves pubescent at margin at first 12
11 Leaves mostly with 3-6 serrations/cm; twigs very fragile at branches; buds shiny; male flowers with 2-3 stamens. Tree to 25m. Probably intrd-natd; common in damp places over most of lowland BI
Crack-willow - **S. fragilis** L. 154
11 Leaves mostly with 6-10 serrations/cm; twigs scarcely fragile at branches; buds scarcely shiny; male flowers mostly with 3-4 stamens.

2. SALIX

Tree to 15m. Native; damp areas and by rivers; frequent in N En, N Wa and N Ir (*S. pentandra* x *S. fragilis*)
   Shiny-leaved Willow - S. x meyeriana Willd.   154
12 Twigs yellowish-orange to reddish
   (see Key C, couplet 9) **S. x rubens**
12 Twigs brownish   13
13 Leaves mostly with 6-10 serrations/cm; male flowers mostly with 3-4 stamens; male only. Tree to 15(25)m. Probably intrd-surv; planted and often ± wild; scattered in En (*S. pentandra* x *S. alba*)
   Ehrhart's Willow - S. x ehrhartiana Sm.   154
13 Leaves mostly with 3-6 serrations/cm; male flowers with 2-3 stamens   (see Key C, couplet 11) **S. fragilis**

Key D - Trees or shrubs >2m high; leaves entire, obscurely crenate-serrate, or coarsely and irregularly serrate, <3x as long as wide; bracts dark or dark-tipped
1 Catkins developing with leaves, usually with small leaf-like bracts on peduncle   2
1 Catkins developing before leaves on bare twigs, with reduced or 0 bracts on peduncle   3
   2 Stipules mostly large and persistent at maturity; leaves blackening when dried; ovary mostly ± glabrous, with pedicel c.1mm; twigs usually pubescent   (see Key A, couplet 9) **S. myrsinifolia**
   2 Stipules mostly small and soon dropping off; leaves not blackening when dried; ovary mostly pubescent, with 0 or very short pedicel; twigs usually ± glabrous
      (see Key A, couplet 10) **S. phylicifolia**
3 Wood of twigs without ridges under bark   4
3 Wood of twigs with longitudinal ridges under bark   6
   4 Leaves glabrous, >2x as long as wide. Shrub 1.5-5m. Native; damp places; scattered throughout BI   **Purple Willow - S. purpurea** L.   154
   4 Leaves pubescent on lowerside, ≤2x as long as wide. Shrub or small tree to 10(19)m. Native (*Goat Willow - S. caprea* L.)   5   155
5 Leaves 5-12 x 2.5-8cm, densely pubescent on lowerside, irregularly undulate-serrate. Damp and rough ground, hedges and open woodland; locally common to abundant   **S. caprea ssp. caprea**
5 Leaves 3-7 x 1.5-4.5cm, densely appressed-pubescent on lowerside, entire to obscurely serrate. Damp ground on mountains; Sc
   S. caprea ssp. sphacelata (Sm.) Macreight
   6 Leaves very sparsely pubescent on lowerside (mostly on veins) at maturity; stipules small, soon falling off. Erect shrub to 6m. Native; frequent with the parents in N BI, scattered elsewhere (*S. cinerea* x *S. phylicifolia*)   **Laurel-leaved Willow - S. x laurina** Sm.   155
   6 Leaves distinctly pubescent on lowerside at maturity; stipules large, persistent to maturity   7
7 Leaves mostly >8cm; catkins mostly >3cm; female only. Erect shrub or small tree to 12m. Native; in damp places; frequent (*S. viminalis* x *S. caprea* x *S. cinerea*)   **Holme Willow - S. x calodendron** Wimm.   155

7 Leaves mostly <8cm; catkins mostly <3cm 8
   8 Leaves distinctly rugose, undulate at margin, without rust-coloured hairs on lowerside (see Key A, couplet 5) **S. aurita**
   8 Leaves scarcely rugose or undulate, usually with some rust-coloured hairs on lowerside. Shrub or small tree to 6(15)m. Native (*Grey Willow* - **S. cinerea** L.) 9
9 Twigs usually ± persistently pubescent; leaves mostly obovate or oblong, dull and ± pubescent on upperside, densely grey-pubescent on lowerside. Marshes and fens; predominant in E Anglia and Lincs, very scattered elsewhere **S. cinerea ssp. cinerea**
9 Twigs usually becoming glabrous; leaves mostly narrowly obovate or oblong, slightly glossy and usually nearly glabrous on upperside, grey-pubescent but with some stiff rust-coloured hairs on lowerside. Wet places, woods and marginal habitats; very common
                      S. cinerea ssp. oleifolia Macreight

Key E - Trees or shrubs >2m high; leaves entire, obscurely crenate-serrate, or coarsely and irregularly serrate, >3x as long as wide; bracts dark or dark-tipped (except yellowish in *S. x mollissima* and *S. elaeagnos*)
1 Extension growth malformed, with contorted, flattened and laterally fused (fasciated) branches (see Key B, couplet 1) **S. udensis**
1 Extension growth not malformed and fasciated 2
   2 Leaves glabrous to sparsely pubescent on lowerside at maturity 3
   2 Mature leaves pubescent to tomentose on lowerside 6
3 Leaves glabrous even when young, commonly opposite or subopposite; male flowers apparently with 1 stamen
                      (see Key D, couplet 4) **S. purpurea**
3 Leaves pubescent when young, alternate; male flowers with 2 free or only partially fused stamens 4
   4 Leaves glabrous at maturity. Erect shrub to 5m. Probably intrd-natd; wet places; scattered in C & S Br and N Ir (*S. purpurea* x *S. viminalis* x *S. cinerea*) *Fine Osier* - **S. x forbyana** Sm.
   4 Leaves sparsely pubescent on lowerside at maturity 5
5 Catkins appearing with leaves; bracts yellowish; male flowers with 2-3 free stamens (see Key C, couplet 7) **S. x mollissima**
5 Catkins appearing before leaves; bracts dark-tipped; male flowers with 2 free or partly united stamens. Shrub or small tree to 7m. Native; wet places; most of Br and Ir (*S. purpurea* x *S. viminalis*)
                  Green-leaved Willow - **S. x rubra** Huds.
   6 Leaves linear or nearly so, mostly >(6)10x as long as wide 7
   6 Leaves lanceolate to narrowly elliptic, mostly <6x as long as wide 8
7 Leaves densely silky-pubescent (hairs appressed) on lowerside; bracts reddish-brown, darker at apex; filaments free; ovary densely pubescent. Erect shrub or small tree to 6(-10)m. Native; mostly damp places; common in lowland BI *Osier* - **S. viminalis** L.
7 Leaves tomentose (hairs matted) on lowerside; bracts yellowish, not darker at apex; filaments fused at base, often ± to 1/2 way; ovary glabrous. Erect shrub to 6m. Intrd-surv; garden plant persisting

2. SALIX

where neglected or thrown out; scattered in En
Olive Willow - **S. elaeagnos** Scop. **339**
- 8 Wood of twigs without ridges under bark. Shrub or small tree to 9m. Native; usually with parents; common (*S. viminalis* x *S. caprea*)
Broad-leaved Osier - **S. x sericans** A. Kern. **155**
- 8 Wood of twigs with longitudinal ridges under bark   9
- 9 Leaves pubescent on lowerside at maturity, but not densely so or soft to touch. Erect shrub or small tree to 9m. Native; usually with parents; common (*S. viminalis* x *S. cinerea*)
Silky-leaved Osier - **S. x smithiana** Willd. **155**
- 9 Leaves densely pubescent, velvety or silky to touch   10
  - 10 Most leaves <4x as long as wide, not undulate at margin
  (see Key D, couplet 7) **S. x calodendron**
  - 10 Most leaves >4x as long as wide, strongly to slightly (or not) undulate at margin   11
- 11 Leaves usually slightly undulate at margin, velvety-tomentose on lowerside. Erect shrub or small tree to 10m. Native; damp places, often planted, scattered in N Br and N Ir (*S. viminalis* x *S. caprea* x *S. aurita*)   *Eared Osier* - **S. x stipularis** Sm. **155**
- 11 Leaves usually strongly undulate at margin, pubescent with appressed or crisped hairs on lowerside
(see Key B, couplet 5) **S. x fruticosa**

Hybrids are discussed under the genus description.

64. BRASSICACEAE - *Cabbage family*

Herbaceous annuals to perennials, rarely dwarf shrubs; easily recognized by the distinctive flowers with 4 sepals and 4 petals in decussate pairs, the (4-)6 stamens, and the characteristic 2-valved fruits.

It is difficult to construct keys without both ripe fruits and a knowledge of petal colour. Caution is needed when assessing flower colour of dried specimens, as sometimes yellow can fade to white and white discolour to yellow; the keys allow for this in most but perhaps not all cases. Hair-type is also very important, and a strong lens is often needed; both leaves and stems should be examined, as hair-type can differ on these 2 organs (branched hairs are commoner on leaves). Plants that are sterile for some reason (hybridity, lack of suitable pollen, genetic mutation) are often impossible to key out, or will key out wrongly due to mis-shapen fruits or fewer than normal seeds; these possibilities must always be borne in mind. In this work 16 groups of distinctive genera are first defined; some of these rarely produce ripe fruits and would not be identifiable with the main keys which, however, do cover all the genera. Fruit lengths include beak, style and stigma unless otherwise stated.

Distinctive genera

Stems woody for most of length
16. AUBRIETA, 18. ALYSSUM, 31. IBERIS

Fresh plants smelling of garlic when crushed **3. ALLIARIA, 30. THLASPI**

Flowers distinctly zygomorphic, 2 petals on 1 side much shorter than 2 on other **29. TEESDALIA, 31. IBERIS**

Erect plants with white scaly rhizome, pinnate basal leaves, pink flowers, and purple bulbils in leaf-axils **14. CARDAMINE**

Robust plants with strong tap-root, large *Rumex*-like leaves sometimes partly pinnately divided, and small white flowers in large panicles **13. ARMORACIA**

Small aquatic, often submerged, plants with leaves all subulate and in basal rosette; inflorescence with few small flowers or short swollen fruits **34. SUBULARIA**

Low straggly annuals with pinnate leaves, small inflorescences borne opposite the leaves, and bilobed 2-seeded fruits **33. CORONOPUS**

Erect plants with small yellow flowers in panicles, and pendent, winged, indehiscent, 1-seeded fruits **5. ISATIS**

Leaves succulent
23. COCHLEARIA, 37. BRASSICA, 43. CAKILE, 45. CRAMBE

Stem-leaves 0 **14. CARDAMINE, 21. DRABA, 22. EROPHILA, 29. TEESDALIA, 34. SUBULARIA, 41. COINCYA**

Petals 0 **14. CARDAMINE, 26. CAPSELLA, 32. LEPIDIUM, 33. CORONOPUS, 34. SUBULARIA, 35. CONRINGIA**

Petals bifid >1/3 way to base **19. BERTEROA, 22. EROPHILA**

Stamens 2-4 **4. ARABIDOPSIS, 14. CARDAMINE, 21. DRABA, 32. LEPIDIUM, 33. CORONOPUS**

Fruits mostly 1-seeded **5. ISATIS, 6. BUNIAS, 25. NESLIA, 43. CAKILE, 44. RAPISTRUM, 45. CRAMBE**

Fruits with stalk-like base >1mm above sepal-scars
1. SISYMBRIUM, 17. LUNARIA, 36. DIPLOTAXIS, 37. BRASSICA

Fruits with seed-containing beak beyond dehiscent lower segment
37. BRASSICA, 38. SINAPIS, 41. COINCYA, 42. HIRSCHFELDIA

General key

1 Fruit splitting longitudinally by 2 valves to release 1-many seeds on each side 2
1 Fruit indehiscent, or splitting transversely, or splitting longitudinally but not releasing seeds separately *Key A*
   2 Fruit >3x as long as wide 3
   2 Fruit ≤3x as long as wide 4
3 Petals pale yellow to golden *Key B*
3 Petals white, pink, purple, mauve or reddish, or rarely 0 *Key C*
   4 Fruit distinctly compressed, the septum at right angles to plane of compression *Key D*
   4 Fruit distinctly compressed, the septum parallel with the plane of compression, or fruit scarcely compressed *Key E*

64. BRASSICACEAE

Key A - Fruit indehiscent, or splitting transversely, or splitting longitudinally but not releasing seeds separately

1	Fruit pendent, flattened, winged, 1-seeded	**5. ISATIS**
1	Fruit not pendent, not winged	2
	2 Petals yellow	3
	2 Petals white to pink or purple	6
3	Fruit >12mm; fresh root smelling of radish	**46. RAPHANUS**
3	Fruit <12mm; fresh root not smelling of radish	4
	4 Fruit composed of 2 segments separated by transverse constriction	**44. RAPISTRUM**
	4 Fruit composed of 1 segment or with slight longitudinal constriction	5
5	Fruit irregularly warty, 5-8mm; at least some leaves pinnately lobed	**6. BUNIAS**
5	Fruit minutely tuberculate, <5mm; leaves entire to dentate	**25. NESLIA**
	6 Fruit clearly divided longitudinally into 2 1-seeded halves	7
	6 Fruit clearly divided transversely or not obviously divided	8
7	Leaves simple; inflorescence a terminal panicle	**32. LEPIDIUM**
7	Leaves pinnate; inflorescences mostly simple leaf-opposed racemes	**33. CORONOPUS**
	8 Plant with stiff hairs below; leaves not succulent; fresh root smelling of radish	**46. RAPHANUS**
	8 Plant glabrous; leaves succulent; fresh root not smelling of radish	9
9	Perennial with *Brassica*-like leaves; lower fruit-segment sterile, stalk-like; filaments of inner stamens toothed	**45. CRAMBE**
9	Annual with narrow, pinnately-lobed leaves; lower fruit-segment usually1-seeded; filaments of inner stamens not toothed	**43. CAKILE**

Key B - Fruits >3x as long as wide, splitting longitudinally from base into 2 valves; flowers pale yellow to golden

1	Stem-leaves distinctly clasping stem at their base	2
1	Stem-leaves not clasping stem, petiolate or narrowed to base	7
	2 Hairs present at least on lowest leaves, branched	**15. ARABIS**
	2 Hairs 0 or simple on all parts	3
3	Fruit with distinct stalk 0.5-1mm between valves and sepal-scars; at least lower flowers with lobed bracts	**40. ERUCASTRUM**
3	Fruit with base of valves immediately above sepal-scars; flowers with entire or toothed bracts or without bracts	4
	4 Fruit with broad-based beak ≥4mm	**37. BRASSICA**
	4 Fruit with narrowly cylindrical beak ≤4mm	5
5	Leaves all simple and entire	**35. CONRINGIA**
5	At least lower leaves pinnate or pinnately lobed	6
	6 Valves of fruit with prominent midrib; seeds in 1 row under each valve	**11. BARBAREA**
	6 Valves of fruit with midrib not or scarcely discernible; seeds ± in 2 rows under each valve	**12. RORIPPA**
7	Leaves linear, all in tight basal rosette	**21. DRABA**
7	Leaves rarely all linear, not in tight basal rosette	8

8	Fruit terminated by distinct beak >3mm; beak either wide at base and tapered distally or wide for most of length, sometimes with seeds	9
8	Fruit terminated by sessile stigmas or by short narrow cylindrical style ≤4mm	13
9	Seeds in 2 rows under each valve; petals pale yellow with conspicuous violet veins	**39. ERUCA**
9	Seeds in 1 row under each valve; petals without conspicuous violet veins	10
10	Fruit <2cm, closely appressed to stem	**42. HIRSCHFELDIA**
10	Fruit >2cm, not closely appressed to stem	11
11	Each valve of fruit with 1 prominent vein; beak long-conical	**37. BRASSICA**
11	Each valve of fruit with 3(-7) prominent parallel veins; beak long-conical or flat and sword-like	12
12	Sepals erect; beak of fruit flat and sword-like	**41. COINCYA**
12	Sepals patent (to erecto-patent); beak of fruit flat and sword-like or long-conical	**38. SINAPIS**
13	Hairs present at least below, at least some stellate or 2-3-armed (arms often closely appressed to plant)	14
13	Hairs 0, or present and all simple	16
14	Leaves 2-3x finely divided almost to midrib; petals c. as long as sepals	**2. DESCURAINEA**
14	Leaves entire to simply toothed or lobed; petals c.2x as long as sepals	15
15	Compact basal leaf-rosette present; basal leaves pinnately lobed	**15. ARABIS**
15	Basal leaf-rosette 0 or very ill-defined; lower leaves entire to toothed	**7. ERYSIMUM**
16	Valves of fruit with midrib not or scarcely discernible	**12. RORIPPA**
16	Valves of fruit with conspicuous midrib	17
17	Seeds in 2 rows under each valve in well developed fruits	**36. DIPLOTAXIS**
17	Seeds in 1 row under each valve	18
18	Each valve of fruit with 3 prominent parallel veins	**1. SISYMBRIUM**
18	Each valve of fruit with 1 prominent vein	19
19	Seeds ± globose; flowers all without bracts; valves of fruit rounded	**37. BRASSICA**
19	Seeds distinctly longer than wide; lower flowers usually with small bracts; valves of fruit keeled at midrib	**40. ERUCASTRUM**

Key C - Fruits >3x as long as wide, splitting longitudinally from base into 2 valves; flowers white, mauve, pink, purple or red

1	Fruit with flat, sword-like beak at apex; petals whitish with violet veins	**39. ERUCA**
1	Fruit with beak 0 or cylindrical; petals rarely whitish with violet veins	2
2	Basal and lower stem-leaves pinnate or ternate	3
2	Basal and lower stem-leaves entire to lobed, but lobes not	

64. BRASSICACEAE

	reaching midrib	4
3	Valves of fruit convex (fruit little wider than thick), opening by separating laterally from base	**12. RORIPPA**
3	Valves of fruit ± flat (fruit much wider than thick), opening by suddenly spiralling upwards from base	**14. CARDAMINE**
4	Petals bifid c.1/2 way to base; small annual with leaves confined to basal rosette (or ± so)	**2. EROPHILA**
4	Petals entire or notched to <1/4 way to base; stems usually bearing leaves	5
5	Plant glabrous or with simple hairs only	6
5	Plant pubescent; all or some hairs branched or stellate, sometimes sparse	11
6	Stem-leaves strongly clasping stem at base	**35. CONRINGIA**
6	Stem-leaves not or scarcely clasping stem	7
7	Basal leaves cordate at base; plant smelling of garlic when crushed	**3. ALLIARIA**
7	Basal leaves cuneate at base; plant not smelling of garlic	8
8	Petals >15mm; stigma deeply 2-lobed	**8. HESPERIS**
8	Petals <15mm; stigma capitate or shortly 2-lobed	9
9	Perennial of mountains of N & W; valves of fruit ± flat	**15. ARABIS**
9	Annual lowland weed; valves of fruit convex	10
10	Seeds in 1 row under each valve; petals shorter than stamens, white only after drying	**1. SISYMBRIUM**
10	Seeds in 2 rows under each valve in well developed fruits; petals longer than stamens, always white	**36. DIPLOTAXIS**
11	Stigma deeply 2-lobed, the lobes visible at apex of fruit (though often closed up together)	12
11	Stigma capitate or slightly notched	15
12	Seeds narrowly to broadly winged	13
12	Seeds not winged	14
13	Hairs branching from basal stalk in tree-like form; seeds broadly winged all round	**10. MATTHIOLA**
13	Hairs with 2 arms, both appressed to plant body, without common stalk; seeds narrowly winged, often not all round	**7. ERYSIMUM**
14	Spreading annual; most hairs with stalk 0 and 2-4 arms appressed close to plant surface	**9. MALCOLMIA**
14	Erect biennial or perennial; most hairs simple or with stalk and 2 arms, patent	**8. HESPERIS**
15	Fruit <8x as long as wide (excl. style), all or most >2mm wide	16
15	Fruit >8x as long as wide (excl. style), all or most <2mm wide	17
16	Flowers white; fruit flat, usually twisted; stems erect	**21. DRABA**
16	Flowers rarely white; fruit scarcely compressed, not twisted; stems procumbent to decumbent	**16. AUBRIETA**
17	Basal leaves pinnately lobed	**15. ARABIS**
17	Basal leaves entire to dentate	18
18	Annual; fruit circular to square in section	**4. ARABIDOPSIS**
18	Biennial to perennial; fruit strongly flattened	**15. ARABIS**

Key D - Fruits <3x as long as wide, obviously compressed, with septum at right angles to plane of compression, splitting longitudinally from base into 2 valves to release 1 or more seeds on each side

1	Corolla distinctly zygomorphic, with 2 petals on 1 side much shorter than 2 on other	2
1	Corolla actinomorphic, with 4 equal petals	3
	2 Seeds 2 under each valve; style ± 0; shorter petals c. as long as sepals	**29. TEESDALIA**
	2 Seed 1 under each valve; style distinct; shorter petals ≥2x as long as sepals	**31. IBERIS**
3	Fruit distinctly winged at edges (i.e. valve midribs)	4
3	Fruit sometimes keeled but not winged at edges (i.e. valve midribs)	6
	4 Basal leaves strongly cordate at base	**30. THLASPI**
	4 Basal leaves 0 or cuneate to rounded at base	5
5	Seeds 3-8 under each valve	**30. THLASPI**
5	Seeds 1(-2) under each valve	**32. LEPIDIUM**
	6 Fruit obtriangular	**26. CAPSELLA**
	6 Fruit orbicular to elliptic, oblong or obovate	7
7	Flowers appearing solitary, on long pedicels arising from axils of basal leaves	**28. JONOPSIDIUM**
7	Flowers in racemes, all or mostly without bracts	8
	8 Fruits distinctly keeled at edges (i.e. valve midribs)	**32. LEPIDIUM**
	8 Fruits obtuse to rounded at edges (i.e. valve midribs)	9
9	Basal leaves deeply pinnately lobed; flowers <2mm across	**27. HORNUNGIA**
9	Basal leaves entire to palmately angled or shallowly lobed; flowers >2mm across	**23. COCHLEARIA**

Key E - Fruits <3x as long as wide, either not compressed or obviously compressed and with septum parallel with plane of compression, splitting longitudinally to release 1 or more seeds on each side

1	Leaves all confined to basal rosette	2
1	At least some leaves borne on stem	5
	2 Petals yellow	**21. DRABA**
	2 Petals white	3
3	Aquatic, often submerged; leaves subulate	**34. SUBULARIA**
3	Not aquatic; leaves flat	4
	4 Annual; petals bifid c.1/2 way to base; fruit glabrous	**22. EROPHILA**
	4 Perennial; petals notched <1/4 way to base; fruit pubescent	**21. DRABA**
5	Petals yellow	6
5	Petals white to pink or purple	10
	6 Plant glabrous or with simple hairs	7
	6 Plant pubescent; all or many hairs branched	8
7	Fruit not or scarcely keeled at junction of valves	**12. RORIPPA**
7	Fruit strongly keeled (± winged) at junction of valves	**24. CAMELINA**
	8 Petals bifid nearly 1/2 way to base	**19. BERTEROA**
	8 Petals entire or notched <1/4 way to base	9

64. BRASSICACEAE

9	Fruit not or scarcely flattened; seeds >2 under each valve	**24. CAMELINA**
9	Fruit strongly flattened; seeds 2 under each valve	**18. ALYSSUM**
	10 Robust perennial with strong tap-root and large dock-like basal leaves (sometimes pinnately lobed)	**13. ARMORACIA**
	10 Plant without dock-like basal leaves	11
11	Fruits >2 x 1cm, very flat	**17. LUNARIA**
11	Fruits <2cm, <1cm wide, flat or not	12
	12 Plant glabrous or with simple hairs	13
	12 Plant with at least some branched or stellate hairs	14
13	Basal leaves abruptly contracted into long petiole	**23. COCHLEARIA**
13	Basal leaves ± sessile, gradually narrowed to base	**24. CAMELINA**
	14 Fruit scarcely compressed	15
	14 Fruit strongly compressed	16
15	Flowers rarely white; procumbent to decumbent much-branched perennial	**16. AUBRIETA**
15	Flowers white; erect little-branched annual	**24. CAMELINA**
	16 Petals bifid nearly 1/2 way to base	**19. BERTEROA**
	16 Petals entire to notched <1/4 way to base	17
17	Fruit c. as long as wide, notched at apex	**18. ALYSSUM**
17	Fruit longer than wide, rounded to obtuse at apex	18
	18 Seeds >2 under each valve; basal leaf-rosette present at flowering	**21. DRABA**
	18 Seeds 1 under each valve; basal leaf-rosette 0 at flowering	**20. LOBULARIA**

TRIBE 1 - SISYMBRIEAE (genera 1-6). Hairs branched or unbranched, sometimes glandular; petals yellow or white; fruit usually beakless, >3x as long as wide and with small stigma, sometimes 1-2-seeded and indehiscent.

1. SISYMBRIUM L. - *Rockets*

Annuals or perennials; basal leaves simple, entire to very deeply lobed; hairs unbranched; petals yellow, sometimes very pale; fruit beakless, >3x as long as wide, with convex valves; seeds in 1 row under each valve.

1 Leaves all entire to dentate   2
1 Lowest leaves lobed >1/4 way to midrib   4
   2 Annual; petals ≤3.5mm, not longer than sepals or stamens. Stems to 60cm. Intrd-casual; a fairly regular wool-alien in En
                                ***French Rocket* - S. erysimoides** Desf.
   2 Perennial; petals >5mm, longer than sepals and stamens   3
3 Leaves pubescent on lowerside, narrowly acute to acuminate at apex. Stems to 1.2m. Intrd-natd; walls, rough and waste ground; very few places in En   ***Perennial Rocket* - S. strictissimum** L.
3 Leaves ± glabrous on lowerside, subacute to apiculate at apex. Stems to 75cm. Intrd-natd; waste ground; widely scattered in En
                                Russian Mustard - S. volgense E. Fourn.
   4 Fruits ≤2cm, strongly appressed to stem. Stems to 1m. Native;

waste places, rough and cultivated ground, hedges and roadsides; very common   *Hedge Mustard* - **S. officinale** (L.) Scop.

4 At least most fruits >2cm, patent to ± erect but not appressed to stem   5
5 Upper stem-leaves pinnate with linear divisions. Stems to 1m. Intrd-natd; waste places; scattered over much of BI

Tall Rocket - **S. altissimum** L.

5 Upper stem-leaves variously divided but usually not to midrib and never with linear segments   6
   6 Pedicels c.1mm wide, >2/3 as wide as ripe fruit   7
   6 Pedicels c.0.3-0.6mm wide, <2/3 as wide as ripe fruit   8
7 Petals ≤3.5mm, not longer than sepals or stamens; fruit 2-5cm
(see couplet 2) **S. erysimoides**
7 Petals >6mm, much longer than sepals and stamens; fruit (2.5)5-12cm. Stems to 1m. Intrd-natd; waste places; scattered over much of BI

Eastern Rocket - **S. orientale** L.

   8 Lower part of stem with ± dense patent to reflexed stiff hairs >1mm. Stems to 1(1.5)m. Intrd-natd; waste places and tips; fairly frequent casual in S Br, natd in London area

False London-rocket - **S. loeselii** L.

   8 Lower part of stem glabrous to sparsely pubescent, or ± densely pubescent with hairs <0.5mm   9
9 Petals pale yellow, <5mm; anthers <1mm. Stems to 60cm. Intrd-natd; walls, roadsides and waste places; scattered in Br and Ir

London-rocket - **S. irio** L.

9 Petals bright yellow, mostly >5mm; anthers >1mm   10
   10 Rhizomatous perennial; most fruits >3cm (see couplet 3) **S. volgense**
   10 Annual; most fruits <3cm   (see couplet 8) **S. loeselii**

2. DESCURAINIA Webb & Berthel. - *Flixweed*

Annuals or biennials; basal leaves 2-3-pinnate or nearly so; hairs both unbranched and branched, ± patent; petals pale yellow; fruit beakless, >3x as long as wide, with convex valves; seeds in 1 row under each valve.

1 Stems to 1m. Possibly native; roadsides, rough and waste ground; throughout much of Br, mostly E, very local in Ir

Flixweed - **D. sophia** (L.) Prantl

3. ALLIARIA Fabr. - *Garlic Mustard*

Biennials; basal leaves simple, toothed; hairs unbranched; petals white; fruit beakless, >3x as long as wide, with angled valves; seeds in 1 row under each valve.

1 Stems to 1.2m. Native; rough ground, hedgerows and shady places; throughout most of BI

Garlic Mustard - **A. petiolata** (M. Bieb.) Cavara & Grande

4. ARABIDOPSIS (DC.) Heynh. - *Thale Cress*
Annuals; basal leaves simple; hairs both unbranched and branched; petals white; fruit beakless, >3x as long as wide, with convex to angled valves; seeds in 1 row under each valve.

1 Stems to 30(50)cm. Native; cultivated ground and bare places on banks, walls, rocks and waysides; throughout BI
Thale Cress - **A. thaliana** (L.) Heynh.

5. ISATIS L. - *Woad*
Biennials to perennials; basal leaves simple; hairs 0 or unbranched; petals yellow; fruit pendent, indehiscent, 1-seeded, winged; style 0.

1 Stems to 1.5m. Intrd-natd; on cliffs in E Gloucs and Surrey, infrequent casual elsewhere
Woad - **I. tinctoria** L.

6. BUNIAS L. - *Warty-cabbages*
Biennials to perennials; basal leaves simple; hairs mixed unbranched, branched and warty-glandular; petals yellow; fruit irregularly warty-ovoid, indehiscent, 1-2-seeded, with short style.

1 Stems to 1.2m. Intrd-natd; waste and rough grassy places; scattered over most of Br and CI
Warty-cabbage - **B. orientalis** L.

TRIBE 2 - HESPERIDEAE (genera 7-10). Hairs branched (sometimes closely appressed to plant), sometimes unbranched, sometimes glandular; petals usually white to pink or purple, sometimes yellow to red; fruit beakless, >3x as long as wide, with usually deeply (sometimes shallowly) 2-lobed stigma.

7. ERYSIMUM L. - *Wallflowers*
Annuals to perennials; leaves simple; hairs 2-3-branched, stalkless, the arms tightly appressed to plant; petals yellow to red, brown or purple; fruit flattened to 4-angled, with strong midribs; stigma ± capitate to 2-lobed; seeds in 1(-2) rows under each valve, not winged or with narrow wing.

1 Petals <1cm; fruits 1-3cm, erecto-patent. Stems to 60(100)cm. Intrd-natd; cultivated and waste ground; scattered over much of BI
Treacle-mustard - **E. cheiranthoides** L.
1 Petals >1cm; fruits 2.5-9cm, ± erect 2
  2 Petals orange; fruits square in section; stigma with 2 rounded closely appressed lobes. Stems to 50cm. Intrd-casual; fairly frequent on tips etc.; S En (*E. decumbens* (Willd.) Dennst. x *E. perofskianum* Fisch. & C.A. Mey.)
Siberian Wallflower - **E. x marshallii**
  2 Petals various colours; fruits distinctly flattened; stigma with 2 divergent lobes. Intrd-natd; walls, banks and other dry places; scattered throughout BI
Wallflower - **E. cheiri** (L.) Crantz

8. HESPERIS L. - *Dame's-violet*
Biennial to perennial; hairs mixed simple and branched, or 0; petals white to pink or purple; fruit ± cylindrical, slightly constricted, with ± strong midribs; stigma 2-lobed, seeds in 1 row under each valve, not winged.

1 Stems to 1.5m. Intrd-natd; garden escape of waste places, banks, grassland, hedges and verges; frequent *Dame's-violet* - **H. matronalis** L.

9. MALCOLMIA W.T. Aiton - *Virginia Stock*
Annuals; basal leaves simple; hairs branched, with 2-4 appressed arms and stalk 0; petals pink to purple; fruit constricted, with ± rounded valves with 3 veins; stigma 2-lobed; seeds in 1 row under each valve, not winged.

1 Stems to 50cm. Intrd-natd; common escape of tips and waste places; S & C Br and CI   *Virginia Stock* - **M. maritima** (L.) W.T. Aiton

10. MATTHIOLA W.T. Aiton - *Stocks*
Annuals to perennials; basal leaves simple; hairs branched, stalked; petals white to pink or purple; fruit not or sometimes ± constricted, with 1-veined flattened to ± rounded valves; stigma strongly 2-lobed, each lobe with dorsal horn-like or hump-like process; seeds in 1 row under each valve, broadly winged.

1 Fruit ± cylindrical, ± constricted between seeds, with apical horn-like processes (2)5-7(10)mm. Stems to 50cm. Intrd-casual; frequent on tips and waste places; En   *Night-scented Stock* - **M. longipetala** (Vent.) DC.
1 Fruit ± compressed, not constricted between seeds; with apical horn-like processes <3mm   2
    2 Leaves all entire; fruits without glands. Stems to 80cm. Possibly native on sea-cliffs in S En; garden escape, often natd on walls and banks etc. in C & S BI   *Hoary Stock* - **M. incana** (L.) W.T. Aiton
    2 Lower leaves sinuate to lobed; fruits with large yellow to black sessile glands. Stems to 1m. Native; sand-dunes and sea-cliffs; S Ir, S Wa, SW En and CI   *Sea Stock* - **M. sinuata** (L.) W.T. Aiton

TRIBE 3 - ARABIDEAE (genera 11-16). Hairs branched or unbranched, sometimes stellate, not glandular; petals yellow or white; fruit beakless, with small stigma, >3x or sometimes <3x as long as wide.

11. BARBAREA W.T. Aiton - *Winter-cresses*
Biennials or perennials; basal leaves pinnate; hairs 0 or few, unbranched; petals yellow; fruit >3x as long as wide, with angled valves; seeds in 1 row under each valve.

1 Uppermost stem-leaves simple, toothed or lobed to <1/2 way to midrib (ignore basal lobes); seeds 1-1.8mm   2
1 Uppermost stem-leaves ± pinnate to pinnately lobed to >1/2 way to midrib; seeds 1.6-2.4mm   3

2 Fruit with style (1.7)2-3.5(4)mm; flowers buds glabrous. Stems to 1m. Native; hedges, streamsides, roadsides and waste places; throughout most of BI   ***Winter-cress* - B. vulgaris** W.T. Aiton

2 Fruit with style 0.5-1.8(2.3)mm; flower buds with hairs at apex of sepals. Stems to 1m. Probably intrd-natd; similar places to *B. vulgaris*; very scattered in C & S Br

***Small-flowered Winter-cress* - B. stricta** Andrz.

3 At least some fruits >4cm; fresh petals >5.6mm. Stems to 1m. Intrd-natd; in waste, cultivated and open ground and by roads; scattered in most of BI   ***American Winter-cress* - B. verna** (Mill.) Asch.

3 Fruits <4cm; fresh petals ≤5.6(6.3)mm. Stems to 60cm. Intrd-natd; in waste, open and cultivated ground and by streams; scattered in most of BI   ***Medium-flowered Winter-cress* - B. intermedia** Boreau

12. RORIPPA Scop. - *Water-cresses*

Annuals to perennials; basal leaves simple and unlobed to pinnate; hairs 0 or unbranched; petals white or yellow; fruit >3x or <3x as long as wide, with convex valves and indistinct midribs; seeds in (1 or) 2 rows under each valve.

1 Petals white   2
1 Petals yellow   4

2 Seeds per fruit <5; pollen <40% with full contents. Stems to 1m. Native; in similar places to parents; scattered over most of BI (*R. nasturtium-aquaticum* x *R. microphylla*)

***Hybrid Water-cress* - R. x sterilis** Airy Shaw

2 Seeds per fruit >10; pollen >80% with full contents   3

3 Seeds in 2 rows under each valve, with c.6-12 cells showing across broadest width. Stems to 1m. Native; in and by streams, ditches and marshes; common   ***Water-cress* - R. nasturtium-aquaticum** (L.) Hayek

3 Seeds mostly in 1 row under each valve, with c.12-20 cells showing across broadest width. Stems to 1m. Native; similar distribution to *R. nasturtium-aquaticum* but less common in S and more common in N

***Narrow-fruited Water-cress* - R. microphylla** (Boenn.) Á. & D. Löve

4 Petals c. as long as sepals; stems not rooting   5
4 Petals ≥1.5x as long as sepals; stems often rooting when contacting ground or water   6

5 Fruit 2-3x as long as pedicels; sepals <1.6mm. Stems to 15(30)cm. Native; open pondsides and other damp places; very scattered in W Ir and W Br, mostly near sea

***Northern Yellow-cress* - R. islandica** (Gunnerus) Borbás

5 Fruit 0.8-2x as long as pedicels; sepals ≥1.6mm. Stems to 60cm. Native; open damp and waste ground; frequent to common

***Marsh Yellow-cress* - R. palustris** (L.) Besser

6 Stem-leaves with auricles at base, ± clasping stem   7
6 Stem-leaves with no or very small auricles, not clasping stem   9

7 Leaves toothed but not lobed; fruit (excl. style) <1.5x as long as wide. Stems to 1m. Intrd-natd; waste ground, and by roads and rivers;

scattered in En and S Wa

Austrian Yellow-cress - **R. austriaca** (Crantz) Besser

7 Leaves strongly lobed; fruit (excl. style) >1.5x as long as wide 8

   8 Petals mostly <3.5mm when fresh; stem-leaves c.4x as long as wide. Native; by R. Thames and Avon in S En (*R. palustris* x *R. amphibia*) *Thames Yellow-cress* - **R. x erythrocaulis** Borbás

   8 Petals mostly >3.5mm when fresh; stem-leaves 2-3x as long as wide. Native; damp waste ground; very scattered in Br (*R. sylvestris* x *R. austriaca*)
Walthamstow Yellow-cress - **R. x armoracioides** (Tausch) Fuss

9 Upper stem-leaves toothed but not lobed; fruit (excl. style) 2-2.5(3)x as long as wide. Stems to 1.2m. Native; in and by rivers, ponds and ditches; frequent in En, local in Wa and Ir, rare in S Sc
Great Yellow-cress - **R. amphibia** (L.) Besser

9 Upper stem-leaves lobed; fruit (excl. style) (2)2.5-7.5x as long as wide 10

   10 Stem-leaves with terminal segment <1/4 of total length; fruit 7-23 x 0.9-1.8mm, plus style ≤1.2mm. Stems to 60cm. Native; in damp places and disturbed ground; frequent
Creeping Yellow-cress - **R. sylvestris** (L.) Besser

   10 Stem-leaves with terminal segment >1/4 of total length; fruit 3-10 x 1.2-2.5mm, plus style usually ≥1.2mm 11

11 Fruiting pedicels mostly reflexed. Native; damp and waste ground and by water; scattered in En, Wa and Ir. The commonest hybrid yellow-cress, often fertile (*R. sylvestris* x *R. amphibia*)
Hybrid Yellow-cress - **R. x anceps** (Wahlenb.) Rchb.

11 Fruiting pedicels patent to erecto-patent
(see couplet 8) **R. x armoracioides**

Hybrids are not common but are effectively spread vegetatively and may occur in absence of both parents. Several occur in addition to those treated here. *R. sylvestris*, *R. amphibia* and *R. austriaca* are highly self-sterile and therefore often do not set seed; hence sterility is not diagnostic for hybridity in the yellow-flowered taxa.

13. ARMORACIA P. Gaertn., B. Mey. & Scherb. - *Horse-radish*

Perennials with deep strong roots; basal leaves simple; hairs ± 0; petals white; fruit not or very rarely ripening, the best developed being ± terete, <3x as long as wide, with few seeds in 2 rows under each valve.

1 Forming extensive patches; stems to 1.5m. Intrd-natd; grassy places and waste ground; frequent in En, Wa and CI, scattered in Ir and Sc
Horse-radish - **A. rusticana** P. Gaertn., B. Mey. & Scherb.

14. CARDAMINE L. - *Bitter-cresses*

Annuals to perennials; basal leaves pinnate or ternate; hairs 0 or unbranched; petals white to pink or purple, sometimes 0; fruit compressed, >3x as long as wide, with valves that spring open suddenly to release seeds; seeds in 1 row under each valve.

14. CARDAMINE

1 Upper stem-leaves with purple bulbils in axils. Stems to 75cm. Native; deciduous woodland; very local in C & SE En, rarely natd elsewhere   *Coralroot* - **C. bulbifera** (L.) Crantz
1 Stem-leaves without axillary bulbils   2
    2 All leaves with (1-)3 leaflets. Stems to 30cm. Intrd-natd; in shady places; scattered over En   *Trefoil Cress* - **C. trifolia** L.
    2 At least some lower leaves with ≥5 leaflets   3
3 Petals ≤5.5 x 2.8mm, white, sometimes 0   4
3 Petals ≥5.5 x 2.8mm, white to pink or purple   7
    4 Stem-leaves with small basal auricles clasping stem. Stems to 80cm. Native; damp woods and river-banks; locally frequent in Br N to S Sc, Westmeath   *Narrow-leaved Bitter-cress* - **C. impatiens** L.
    4 Stem-leaves without auricles   5
5 Most flowers with 4(5) stamens. Stems to 30cm. Native; cultivated and open ground, rocks and walls; common
                                                               Hairy Bitter-cress - **C. hirsuta** L.
5 Most flowers with 6 stamens   6
    6 Stems with 0-3 leaves, glabrous to sparsely pubescent; leaflets 3-7; petals 2-2.8mm wide. Stems to 10cm, often with 0 leaves and only 1 flower. Intrd-natd; paths, rockeries and pavement cracks; scattered in C & N Br and NE Ir
                                  New Zealand Bitter-cress - **C. corymbosa** Hook. f.
    6 Stems with (3)4-7(10) leaves, conspicuously pubescent; leaflets 5-17; petals 1-2(2.2)mm wide. Stems to 50cm. Native; marshes, streamsides and sometimes cultivated ground; common
                                               Wavy Bitter-cress - **C. flexuosa** With.
7 Plant with whitish succulent rhizome with succulent scale-leaves. Stems to 60cm. Intrd-natd; persistent relic in woodland; scattered in En   *Pinnate Coralroot* - **C. heptaphylla** (Vill.) O.E. Schulz
7 Rhizome 0 or present, not whitish or with succulent scale-leaves   8
    8 Anthers blackish-violet; petals usually white; stigma tapered from style. Stems to 60cm. Native; streamsides, marshes and flushes; local in Br and N Ir   *Large Bitter-cress* - **C. amara** L.
    8 Anthers yellow; petals usually pale to deep pink; stigma minutely capitate   9
9 All leaves large, with apical leaflet much wider than lateral ones. Stems to 70cm. Intrd-natd; in damp often shady places; very scattered in Br, mostly in N & W   *Greater Cuckooflower* - **C. raphanifolia** Pourr.
9 Upper leaves with much narrower leaflets than lower leaves, the apical leaflet not or scarcely wider than lateral ones. Stems to 60cm. Very variable; often sterile. Native; wet grassy places; common
                                                                        Cuckooflower - **C. pratensis** L.

C. flexuosa hybridises with *C. pratensis* and *C. hirsuta*, but rarely.

15. ARABIS L. - *Rock-cresses*
Annuals to perennials; basal leaves simple, deeply lobed to ± entire; hairs usually branched and unbranched, rarely 0 or all simple; petals white, pale

yellow or pinkish-purple; fruit compressed or 4-angled, >3x as long as wide; seeds in 1 or 2 rows under each valve.

1 Stem-leaves strongly clasping stem at base; auricles distinctly longer than stem-width 2
1 Stem-leaves not or scarcely clasping stem at base; auricles 0 or much shorter than stem-width 6
    2 Non-flowering shoots 0 or forming ± sessile rosettes in compact clump 3
    2 Non-flowering shoots elongated, some becoming stolons and mat-forming 5
3 Fruits >7cm, patent and curved downwards when ripe. Stems to 70cm. Intrd-natd; on old walls; Cambridge   *Tower Cress* - **A. turrita** L.
3 Fruits ≤7cm, erect 4
    4 Petals pale yellow; seeds in 2 rows under each valve. Stems to 1m. Native; dry grassy, rocky and waste places; very local in En, casual in Sc   *Tower Mustard* - **A. glabra** (L.) Bernh.
    4 Petals white; seeds in 1 row under each valve. Stems to 60cm. Native; limestone rocks and bare places in grassland, walls; locally common   *Hairy Rock-cress* - **A. hirsuta** (L.) Scop.
5 Lowest leaves with petiole c. as long as blade; stem-leaves with pointed to narrowly rounded auricles longer than wide. Stems to 40cm. Intrd-natd; on walls and limestone rocks; frequent in En and Wa, rare in Sc   *Garden Arabis* - **A. caucasica** Schltdl.
5 Lowest leaves with petiole much shorter than blade; stem-leaves with broadly rounded auricles c. as long as wide. Stems to 40cm. Native; rock-ledges at 820-850m; Skye (N Ebudes)
    Alpine Rock-cress - **A. alpina** L.
    6 Ripe fruits patent to erecto-patent, arising at ≤45°; basal leaves mostly deeply sinuate, with long petioles 7
    6 Ripe fruits erect or nearly so, arising at much <45°; basal leaves mostly shallowly sinuate with short petioles 8
7 Plant with branching rhizome; mountains of N & W BI. Stems to 25cm. Native; mountain rock-ledges, slopes and crevices; NW Wa, N & C Sc, very local in Ir   *Northern Rock-cress* - **A. petraea** (L.) Lam.
7 Plant with only root underground; alien in C & S En. Stems to 40cm. Intrd-casual; waste and other open ground; C & S En
    Sand Rock-cress - **A. arenosa** (L.) Scop.
    8 Basal leaves entire to slightly toothed; petals white; pedicels mostly <8mm   (see couplet 4) **A. hirsuta**
    8 Basal leaves sinuate-lobed; petals white, pink or yellow, if pure white then most pedicels >8mm 9
9 Petals pale yellow; pedicels mostly <7mm. Stems to 25cm. Native; limestone rock-crevices and rubble; near Bristol (N Somerset and W Gloucs)   *Bristol Rock-cress* - **A. scabra** All.
9 Petals white to pink; pedicels mostly >7mm. Stems to 30cm. Intrd-natd; on walls and banks; S & C Sc, N Somerset
    Rosy Cress - **A. collina** Ten.

16. AUBRIETA Adans. - *Aubretia*
Perennials; leaves simple, with few deep teeth; hairs unbranched and stellate; petals mauve to purple, fruit scarcely compressed, c.2-5x as long as wide; seeds in 2 rows under each valve.

1 Mat-forming, with numerous sterile shoots; flowering stems to 15(30)cm. Intrd-natd; on walls and rocky banks; scattered in Br and CI, mainly S **Aubretia - A. deltoidea** (L.) DC.

TRIBE 4 - ALYSSEAE (genera 17-22). Hairs unbranched or branched, often stellate, not glandular; petals yellow, white or pinkish-purple; fruit usually <3x, sometimes slightly >3x, as long as wide, with the septum parallel to plane of compression.

17. LUNARIA L. - *Honesty*
Biennials; basal leaves simple, long-stalked; hairs unbranched; petals pinkish-purple, sometimes white; fruit distinctive, large, flat, c.2x as long as wide; seeds in 2 rows under each valve.

1 Fruits 2.5-7.5cm. Stems to 1m. Intrd-natd; tips, roadsides, waste ground etc, occasionally persistent; scattered in Br and CI
Honesty - L. annua L.

18. ALYSSUM L. - *Alisons*
Annuals to perennials; hairs branched and unbranched, often ± stellate; leaves simple, ± entire; petals yellow; fruit <3x as long as wide, not or slightly inflated; seeds 2 under each valve.

1 Annual to biennial; basal leaves withered by flowering time; petals pale yellow, fading to white when dried; fruit pubescent, slightly inflated. Stems to 30cm. Intrd-natd; grassy and arable fields; ± confined to Suffolk **Small Alison - A. alyssoides** (L.) L.
1 Perennial; basal leaves still green at flowering time; petals usually bright yellow; fruit glabrous, ± flat. Stems to 45cm. Intrd-natd; escaping on to walls and dry banks; scattered in Br
Golden Alison - A. saxatile L.

19. BERTEROA DC. - *Hoary Alison*
Annuals to perennials; hairs stellate; leaves simple, ± entire; petals white, often discolouring yellow when dried, bifid nearly 1/2 way to base; fruit <3x as long as wide, slightly inflated; seeds 2-6 under each valve.

1 Plant grey-pubescent; stems to 60cm. Intrd-natd; rough grassy waste places, waysides; scattered in C & S Br, mainly casual
Hoary Alison - B. incana (L.) DC.

20. LOBULARIA Desv. - *Sweet Alison*
Annuals to perennials; hairs 2-armed with 0 stalk; leaves simple, entire;

petals white, sometimes purplish; fruit <3x as long as wide, slightly inflated; seeds 1 under each valve.

1 Plant grey-pubescent; stems to 30cm, much branched. Intrd-natd; on walls and other dry places, well natd on coastal sands in S Br and CI; scattered over most of BI   *Sweet Alison* - **L. maritima** (L.) Desv.

21. DRABA L. - *Whitlowgrasses*

Annuals to perennials; hairs unbranched or branched, often stellate; leaves simple; petals white or yellow; fruit <3x or sometimes slightly >3x as long as wide, ± not inflated; seeds in 2 rows under each valve.

1 Petals yellow; style ≥1mm in fruit; hairs all unbranched. Stems to 10(15)cm. Probably native; limestone rocks and walls near sea; Gower peninsula (Glam)   *Yellow Whitlowgrass* - **D. aizoides** L.
1 Petals white; style <1mm in fruit; at least some hairs branched and stellate   2
   2 Annual; stem-leaves <2x as long as wide, cordate and clasping stem at base. Stems to 40cm. Native; soil-pockets on limestone rocks and cliffs, also natd on walls and in gardens; scattered over Br and Ir, perhaps native only in SW & N En
Wall Whitlowgrass - **D. muralis** L.
   2 Perennial; at least some stem-leaves >2x as long as wide or stem-leaves 0, tapered to rounded at base and ± not clasping stem   3
3 Stems with 0-2(3) leaves; fruit 3-8mm, not twisted. Stems to 6cm. Native; bare places near mountain-tops, usually calcareous; very local in C & N Sc   *Rock Whitlowgrass* - **D. norvegica** Gunnerus
3 Stems normally with >3 leaves; fruit 5.5-12mm, usually twisted. Stems to 40cm. Native; rock-ledges and soil-pockets mainly on limestone, and sand-dunes in N & W Sc; very local in N & NW BI   *Hoary Whitlowgrass* - **D. incana** L.

22. EROPHILA DC. - *Whitlowgrasses*

Early flowering ephemerals; hairs unbranched or branched, often stellate; leaves confined to basal rosette, simple, entire or toothed; petals white, bifid ±1/2 way to base; fruit <3x or sometimes slightly >3x as long as wide, inflated or not; seeds in 2 rows under each valve; flowering in very early spring.

A much misunderstood genus. Many traditionally used characters, such as fruit size and shape and ovule number, do not correlate well with cytological characters or breeding behaviour and separate only pure-breeding lines.

1 Leaves and lower parts of stems densely grey-pubescent; petioles 1/5-1/2 as long as leaf-blades; seeds 0.3-0.5mm; petals bifid ≤1/2 way to base. Stems to 9cm. Native; all sorts of open, dry ground, especially on calcareous soils, but rarely or not on dunes; scattered in BI except N & W Sc   *Hairy Whitlowgrass* - **E. majuscula** Jord.

22. EROPHILA

1 Leaves and lower parts of stems subglabrous to moderately pubescent, green; petioles ≥1/2 as long as leaf-blades; seeds 0.5-0.8mm 2
- 2 Petioles 0.5-1x as long as leaf-blades; petals bifid 1/2-3/4 way to base; usually pubescent. Stems to 10(25)cm. Native; open, dry ground, especially calcareous, rocks, walls, open grassland and dunes; locally common ***Common Whitlowgrass* - E. verna** (L.) DC.
- 2 Petioles 1.5-2.5x as long as leaf-blades; petals bifid ≤1/2 way to base; usually subglabrous. Stems to 9cm. Native; habitat and distribution as in *E. verna*, but less common
***Glabrous Whitlowgrass* - E. glabrescens** Jord.

TRIBE 5 - LEPIDEAE (genera 23-34). Hairs 0, unbranched or (less often) branched; petals white, sometimes pink, yellow or 0; fruit ≤3x as long as wide, with septum at right angles to or sometimes parallel with plane of compression, sometimes indehiscent.

23. COCHLEARIA L. - *Scurvygrasses*

Annuals to perennials (mostly biennials); basal leaves simple, long-stalked; hairs 0 or unbranched; petals white or mauve; fruit ≤3x as long as wide, inflated or ± compressed and then with septum at right angles to plane of compression; seeds in 2 rows under each valve.

1 Pedicels >3x as long as ripe fruits; stems to 1(1.5)m. Intrd-natd; cultivated and waste ground; Nottingham (Notts)
***Tall Scurvygrass* - C. megalosperma** (Maire) Vogt
1 Pedicels <3x as long as ripe fruits; stems <50cm 2
- 2 Basal leaves cuneate at base; fruit compressed, with septum >3x as long as wide and at right angles to plane of compression. Stems to 40cm. Native; muddy shores and estuaries; coasts of most of BI but local in Ir ***English Scurvygrass* - C. anglica** L.
- 2 Basal leaves cordate to very broadly cuneate at base; fruit scarcely compressed, the septum <2(3)x as long as wide 3
3 All or all except extreme uppermost stem-leaves petiolate; flowers ≤5(6)mm across; fruits with ≤12(16) seeds. Stems to 25cm. Native; sandy and pebbly shores, banks and walls near sea, and by railways and salt-treated roads inland; coasts of most of BI and widespread inland ***Danish Scurvygrass* - C. danica** L.
3 Upper stem-leaves sessile, often clasping stem; flowers mostly >5mm across; fruits with ≤8 seeds 4
- 4 Perennial with ± woody base; at least some fruits acute at both ends and widest at or just below middle, with veins not or scarcely forming reticulation when dry. Stems to 20cm. Native; on micaceous schists above 800m; rare in C Sc
***Mountain Scurvygrass* - C. micacea** E.S. Marshall
- 4 Biennial to perennial, rarely woody at base; fruits rarely acute at both ends, if so widest above middle, with veins forming distinct reticulation when dry 5
5 Upper stem-leaves slightly or not clasping stem at base; flowers

5-8mm across. Stems to 30cm. Native; barish damp mostly base-rich soils, rocks and spoil-heaps in inland upland areas
(*Pyrenean Scurvygrass* - **C. pyrenaica** DC.)   6

5 Upper stem-leaves distinctly clasping stem at base; flowers (5)8-15mm across; maritime habitats and inland by some roads. Stems to 30(50)cm. Native (*Common Scurvygrass* - **C. officinalis** L.)   7

   6 Leaves not succulent; fruits cuneate at base. N En (Derbys to Cumberland), N Ebudes   **C. pyrenaica** ssp. **pyrenaica**

   6 Leaves usually succulent; fruits rounded at base. Often on less basic substrata; N Somerset, N Wa, W Ir, mountains in Sc (*C. alpina* (Bab.) H.C. Watson) **C. pyrenaica** ssp. **alpina** (Bab.) Dalby

7 Stems to 30(50)cm; basal leaves cordate at base, 1.5-3cm wide; petals mostly 4-8mm, usually white; fruits mostly 3-7mm. Salt-marshes, cliffs and other habitats by or near sea and by salt-treated roads inland; round the coasts of BI and by some roads in En and Wa
                                            C. officinalis ssp. **officinalis**

7 Stems to 10cm; basal leaves rounded to shallowly cordate at base, 0.6-1.6cm wide; petals mostly 2-4mm, often lilac; fruits mostly 2.5-3.5mm. Rocky and sandy coasts of Sc, N & W Ir, Man, SW En and perhaps Wa (*C. scotica* Druce)
                               C. officinalis ssp. **scotica** (Druce) P.S. Wyse Jacks.

C. officinalis hybridizes with *C. danica* and *C. anglica* more or less wherever they meet; both hybrids are intermediate and fertile.

24. **CAMELINA** Crantz - *Gold-of-pleasures*

Annuals or sometimes biennials; leaves simple, lanceolate to narrowly elliptic, entire or nearly so; hairs branched and unbranched; petals yellow; fruit <3x as long as wide, ± inflated, ± smooth, with septum parallel with plane of compression, with keeled to ± winged margin and long persistent style, on patent to erecto-patent pedicels; seeds in 2 rows under each valve.

1 Plant subglabrous or sometimes pubescent; petals 3.4-6mm; fruit 5.5-10.3 x (3.5)4-8mm (excl. style); style <1/3x as long as rest of fruit; seeds 1.2-2(2.5)mm. Stems to 70(100)cm. Intrd-casual; mostly on tips; scattered over Br and Ir   *Gold-of-pleasure* - **C. sativa** (L.) Crantz

1 Plant pubescent; petals 2.2-4.6mm; fruit 4-7(7.5) x (2)2.5-4.5(4.8)mm (excl. style); style >1/3x as long as rest of fruit; seeds 0.9-1.4mm. Stems to 70(100)cm. Intrd-casual; similar places to *C. sativa* but much less common; sporadic in Br   *Lesser Gold-of-pleasure* - **C. microcarpa** DC.

25. **NESLIA** Desv. - *Ball Mustard*

Annuals; leaves simple, narrowly ovate to elliptic, entire or nearly so, clasping stem at base; hairs ± stellate; petals yellow; fruit <3x as long as wide, indehiscent, reticulately ridged, unkeeled, with long persistent style, with septum parallel with plane of compression; seed usually 1.

1 Fruit 1.5-3.5mm. Stems to 80cm. Intrd-casual; tips and waste places;

25. NESLIA

sporadic in Br and Ir   *Ball Mustard* - **N. paniculata** (L.) Desv.

26. CAPSELLA Medik. - *Shepherd's-purses*
Annuals to biennials; basal leaves simple, entire to deeply pinnately lobed; stem-leaves clasping stem at base; hairs unbranched; petals white or sometimes red-tinged or 0; fruit <3x as long as wide, ± obtriangular, compressed, with septum at right angles to plane of compression; seeds in 2 rows under each valve.

1 Petals white, 1.5-3.5mm, sometimes 0; fruit 4-10 x 3.4-7.5mm, ± obtriangular, with straight to slightly convex sides. Native; cultivated and other open ground; common
   Shepherd's-purse - **C. bursa-pastoris** (L.) Medik.
1 Petals 1.5-2mm (scarcely longer than sepals), usually (like the sepals) red-tinged; fruit 5-7 x 4-6mm, with concave sides forming gradually tapered base. Intrd-casual; cultivated and waste ground; sporadic in Br and CI, mainly S & C En   *Pink Shepherd's-purse* - **C. rubella** Reut.

C. bursa-pastoris x *C. rubella* = **C. x gracilis** Gren. has been found sporadically in En with the parents

27. HORNUNGIA Rchb. - *Hutchinsia*
Annuals; leaves deeply pinnately lobed or ± pinnate; hairs stellate; petals white; fruit <3x as long as wide, ± compressed, with septum at right angles to plane of compression; seeds (1-)2 under each valve.

1 Early-flowering ephemeral; stems to 10(15)cm. Native; bare places becoming desiccated in summer, on calcareous rocks and dunes; very local in N & SW En, Wa and CI   *Hutchinsia* - **H. petraea** (L.) Rchb.

28. JONOPSIDIUM Rchb. - *Violet Cress*
Annuals; leaves simple, entire to 3-lobed; hairs 0; petals pink to purple, sometimes white; fruit <3x as long as wide, slightly compressed, with septum at right angles to plane of compression; seeds 2-5 under each valve.

1 Stems ± absent to short and congested; leaves appearing as if in a rosette, on long thin petioles. Intrd-casual; on tips, rough ground and roadsides; S En and Man   *Violet Cress* - **J. acaule** (Desf.) Rchb.

29. TEESDALIA W.T. Aiton - *Shepherd's Cress*
Annuals; leaves deeply pinnately lobed or ± pinnate; hairs unbranched or 0; petals white, the 2 abaxial c.2x as long as the 2 adaxial ones; fruit <3x as long as wide, compressed, with septum at right angles to plane of compression, ± keeled to very narrowly winged round edges; seeds 2 under each valve.

1 Most leaves in basal rosette; stems to 25cm. Native; open sand, gravel

or shingle; scattered through Br and CI, NE Ir
Shepherd's Cress - **T. nudicaulis** (L.) W.T. Aiton

30. THLASPI L. - *Penny-cresses*
Annuals to perennials, sometimes with rhizome; leaves simple, entire to dentate; hairs unbranched or 0; petals white; fruit <3x as long as wide, compressed, with septum at right angles to plane of compression, with narrow to broad wing round edges; seeds (1)3-8 under each valve.

1 Plant with strong rhizome; stem-leaves all petiolate. Stems to 40cm. Intrd-natd; woodland; Herts, N Somerset and Salop
                        Caucasian Penny-cress - **T. macrophyllum** Hoffm.
1 Plant without rhizome; stem-leaves sessile, clasping stem   2
  2 Fruit ≥9mm, with wing >1mm wide at midpoint. Stems to 60cm. Possibly native; weed of waste and arable land; scattered over most of BI   *Field Penny-cress* - **T. arvense** L.
  2 Fruit ≤10mm, with wing <1mm wide at midpoint   3
3 Biennial to perennial, usually with non-flowering leaf-rosettes; style equalling or exceeding apical notch of fruit. Stems to 40cm. Native; bare or sparsely grassed stony places mainly on limestone contaminated with lead or zinc; extremely local in Br from N Somerset to C Sc   *Alpine Penny-cress* - **T. caerulescens** J. & C. Presl
3 Annual; style c.1/2 as long as apical notch of fruit   4
  4 Plant glabrous, not smelling of garlic; stem-leaves with rounded auricles. Stems to 25cm. Native; bare limestone stony ground; E Gloucs, N Wilts and Oxon, rarely intrd elsewhere in En
                        Perfoliate Penny-cress - **T. perfoliatum** L.
  4 Plant smelling of garlic when crushed; stem sparsely pubescent at base; stem-leaves with acute auricles. Stems to 60cm. Intrd-natd; weed of arable fields and borders; E Kent and S Essex
                        Garlic Penny-cress - **T. alliaceum** L.

31. IBERIS L. - *Candytufts*
Annuals to perennials; leaves simple, entire to deeply lobed; hairs unbranched or 0; inflorescence a corymb at flowering, often elongate in fruit; petals white to purple or mauve, the 2 abaxial much longer than the 2 adaxial; fruit <3x as long as wide, compressed, with septum at right angles to plane of compression, with fairly narrow to broad wing round edges; seeds 1 under each valve.

1 Sprawling perennial; stems often >1m, woody at base; leaves evergreen. Intrd-natd; persistent relic or throwout, occasionally self-sown; mostly S Br, rare in N   *Perennial Candytuft* - **I. sempervirens** L.
1 Annual with herbaceous stem   2
  2 Inflorescence elongating in fruit; fruits mostly or all 3-6mm. Stems to 35cm. Native; bare places in grassland and arable fields on dry calcareous soils; CS En N to Cambs   *Wild Candytuft* - **I. amara** L.
  2 Inflorescence remaining corymbose in fruit; fruits mostly or all

31. IBERIS

7-10mm. Stems to 70cm. Intrd-casual; on tips and in waste places; common in Br and CI   *Garden Candytuft* - **I. umbellata** L.

32. LEPIDIUM L. - *Pepperworts*

Annuals to perennials; leaves simple to 2-3-pinnate; hairs unbranched or 0; inflorescence a raceme or panicle; petals usually white, sometimes reddish, yellowish or 0; fruit <3x as long as wide, compressed, with septum at right angles to plane of compression, sometimes ± indehiscent, strongly keeled to winged round edges; seeds 1(2) under each valve but sometimes not developing.

1 Fruit not or scarcely dehiscent; inflorescence a ± corymbose panicle. Stems to 60cm. Intrd-natd; waste ground, by roads, railways, paths and arable land, sandy ground near sea (*Hoary Cress* - **L. draba** L.)   2
1 Fruit readily dehicent; inflorescence a raceme or a racemose to pyramidal panicle   3
    2 Leaves greyish-green; fruit at least as wide as long, truncate to cordate at base, usually reticulately ridged. Throughout much of BI   **L. draba ssp. draba**
    2 Leaves brighter green; fruit usually longer than wide, rounded to broadly cuneate at base, usually smooth. Rare casual, natd in S En   **L. draba ssp. chalepense** (L.) Thell.
3 Lower and upper stem-leaves strikingly different, the lower finely 2-3-pinnate, the upper ovate, entire, ± encircling stem; petals yellow. Stems to 45cm. Intrd-casual; on tips and by docks; very scattered in Br   *Perfoliate Pepperwort* - **L. perfoliatum** L.
3 Lower and upper stem-leaves often different but with ± gradual transition; petals white, reddish or 0   4
    4 Fruit ≥(4)4.5mm, usually at least as long as pedicel; basal 1/2 of style usually fused with wings of fruit   5
    4 Fruit ≤4.2mm, usually shorter than pedicel; style free from wings of fruit or wings 0   7
5 Lower and middle stem-leaves very deeply pinnately lobed to 2-pinnate, not clasping stem. Stems to 50(100)cm. Intrd-casual; on tips and waysides; frequent in much of BI   *Garden Cress* - **L. sativum** L.
5 Lower and middle stem-leaves simple, toothed to very shallowly lobed, clasping stem at base   6
    6 Fruit covered with scale-like vesicles; style not exceeding apical notch of fruit. Stems to 60cm. Native; open grassland, banks, walls, waysides and arable fields; scattered in much of Br, mostly in S, intrd in Ir   *Field Pepperwort* - **L. campestre** (L.) W.T. Aiton
    6 Fruit with few or 0 vesicles; style exceeding apical notch of fruit. Stems to 50cm. Native; similar places to *L. campestre*; locally common in much of BI, mostly in W   *Smith's Pepperwort* - **L. heterophyllum** Benth.
7 Perennial; fruit unwinged, rounded to subacute at apex, with style projecting beyond   8
7 Normally annual or biennial; fruit usually winged at least apically,

notched at apex, with style not or scarcely projecting beyond notch 9
  8 Upper stem-leaves linear; fruit 2-4.2mm, in lax branched racemes.
     Stems to 60cm. Intrd-casual; mostly near docks; very sporadic
     in S Br   *Tall Pepperwort* - **L. graminifolium** L.
  8 Upper stem-leaves elliptic to narrowly so; fruit 1.5-2.7mm, in
     congested panicles. Stems to 1.5m. Native; damp barish ground
     near sea; on or near coasts of S Br, Guernsey, intrd on waste
     land elsewhere   *Dittander* - **L. latifolium** L.
9 Middle and upper stem-leaves deeply pinnately lobed to 2-pinnate.
   Stems to 60cm. Intrd-casual; on tips and in fields; very scattered in
   Br, mostly S En   *Argentine Pepperwort* - **L. bonariense** L.
9 Middle and upper stem-leaves entire to dentate   10
   10 Fruit 2.5-4mm wide; apical notch >2mm deep, c.1/10 of fruit
      length. Stems to 50cm. Intrd-casual; on tips and waste ground;
      sporadic in Br and Ir   *Least Pepperwort* - **L. virginicum** L.
   10 Fruit 1.4-2.3mm wide; apical notch <2mm deep, <1/10 of fruit
      length   11
11 Upper stem-leaves entire, rounded to subacute at apex; basal leaves
   (gone before fruiting) pinnate to 2-pinnate. Stems to 45cm. Probably
   intrd-natd; waste places, waysides and tips, natd in open ground
   especially near sea; locally common in E & SE En, scattered
   elsewhere   *Narrow-leaved Pepperwort* - **L. ruderale** L.
11 Upper stem-leaves entire to dentate, acuminate to subacute at apex;
   basal leaves (gone before fruiting) dentate to deeply pinnately lobed.
   Stems to 45cm. Intrd-casual; on tips and in fields; very scattered in Br
                                  African Pepperwort - **L. hyssopifolium** Desv.

33. CORONOPUS Zinn - *Swine-cresses*

Annuals or biennials with inflorescences mostly opposite leaves; leaves all deeply pinnately lobed; hairs unbranched or 0; petals white or 0; fruit <3x as long as wide, only slightly compressed, with septum at right angles to plane of compression, indehiscent or breaking into 2 halves; seed 1 under each valve.

1 Not scented; petals 1-2mm, longer than sepals; fertile stamens 6;
   pedicels shorter than fruit; fruit 2-3.5mm, with style protruding,
   indehiscent. Stems to 30cm. Probably native; waste ground, paths
   and round trodden gateways; throughout much of BI
                                      Swine-cress - **C. squamatus** (Forssk.) Asch.
1 Strong smelling when crushed; petals 0 or c.0.5mm, shorter than
   sepals; fertile stamens 2(-4); pedicels longer than fruit; fruit 1.2-
   1.7mm, with very short included style, breaking into 2 halves. Stems
   to 40cm. Intrd-natd; cultivated and waste ground; frequent in S BI,
   scattered N to C Sc   *Lesser Swine-cress* - **C. didymus** (L.) Sm.

34. SUBULARIA L. - *Awlwort*

Aquatic annuals or biennials; leaves confined to basal rosette, subulate, entire; hairs 0; petals white, rarely 0; fruit <3x as long as wide, ± inflated, with septum parallel with plane of compression; seeds 2-7 in 2 rows under each valve.

1 Leaves ± erect, to 4(7)cm; flowering stems to 8(12)cm. Native; in stony or gravelly base-poor lakes, usually totally submerged; very local in Wa, NW En, N & W Ir, Sc *Awlwort* - **S. aquatica** L.

TRIBE 6 - BRASSICEAE (genera 35-46). Hairs 0 or unbranched; petals yellow, white or pink to purple or mauve; fruit usually >3x as long as wide and many-seeded, beaked or unbeaked, usually dehiscing longitudinally to release seeds, sometimes dehiscing transversely but not releasing seeds separately, sometimes indehiscent, sometimes only 1-4-seeded and transversely dehiscent.

35. CONRINGIA Fabr. - *Hare's-ear Mustard*
Annuals; leaves simple, entire, basal ones petiolate, upper ones sessile, with large auricles clasping stem; petals greenish- to yellowish-white, rarely 0; fruit >3x as long as wide, unbeaked, longitudinally dehiscent; seeds in 1 row under each valve.

1 Glabrous, glaucous; stems to 60cm. Intrd-casual; in arable land and waste places, often near sea; scattered in Br and CI
Hare's-ear Mustard - **C. orientalis** (L.) Dumort.

36. DIPLOTAXIS DC. - *Wall-rockets*
Annuals to perennials; leaves deeply pinnately lobed, sometimes not so in *D. erucoides*, strongly smelling when crushed; petals yellow, rarely white; fruit >3x as long as wide, unbeaked, longitudinally dehiscent; seeds in 2 rows under each valve in well-developed fruits; valves with 1 strong vein.

1 Petals white; fruit with beak 2-4(6)mm. Stems to 50cm. Intrd-casual; waste ground and by paths; occasional in En
White Wall-rocket - **D. erucoides** (L.) DC.
1 Petals yellow; fruit with beak 1-3(3.5)mm 2
   2 Fruit with distinct stalk (0.5-6.5mm) between sepal-scars and base of valves; petals 8-15mm. Stems to 80cm. Possibly native; dry waste places, bare ground, banks and walls; scattered through Br and CI N to C Sc, locally common in S
Perennial Wall-rocket - **D. tenuifolia** (L.) DC.
   2 Fruit with base of valves immediately above sepal-scars; petals 4-8(8.5)mm. Stems to 60cm. Intrd-natd; dry waste places, rocks, walls and arable land; similar distribution to *D. tenuifolia*, also scattered in Ir *Annual Wall-rocket* - **D. muralis** (L.) DC.

37. BRASSICA L. - *Cabbages*
Annuals to perennials; leaves crenate to deeply pinnately lobed; petals yellow; fruit >3x as long as wide, beaked or unbeaked, longitudinally dehiscent; seeds in 1 row under each valve; valves with 1 strong vein.

1 Stem-leaves distinctly clasping stem at base 2
1 Stem-leaves not clasping stem, petiolate or narrowed to base 7
  2 Sepals erect in flower; flowering part of inflorescence elongated, the buds greatly overtopping open flowers; plant glabrous. Stems to 2m. Possibly native on sea-cliffs scattered round Br, mostly in S; common casual throughout BI *Cabbage* - **B. oleracea** L.
  2 Sepals erecto-patent to patent in flower; flowering part of inflorescence scarcely elongated, the buds slightly overtopping or overtopped by uppermost open flowers; lowest leaves usually with some hairs 3
3 Buds slightly overtopping open flowers, forming convex 'dome'; petals mostly >11mm, bright pale yellow. Stems to 1.5m. Intrd-natd; frequent relic of cultivation and from seed importation; throughout BI (**B. napus** L.) 4
3 Buds overtopped by open flowers, forming concave 'bowl'; petals mostly <12mm, bright deep yellow. Stems to 1.5m (**B. rapa** L.) 5
  4 Root swollen into a yellow-fleshed tuber
                          Swede - **B. napus** ssp. **rapifera** Metzg.
  4 Root slender   *Oil-seed Rape* - **B. napus** ssp. **oleifera** (DC.) Metzg.
5 Root swollen into a white-fleshed tuber. Intrd-casual; frequent relic of cultivation; throughout BI   *Turnip* - **B. rapa** ssp. **rapa**
5 Root slender 6
  6 Seeds <1.6mm, grey to blackish. Possibly native; occasional by streams and rivers, often with *B. nigra*; S & C Br
                Wild Turnip - **B. rapa** ssp. **campestris** (L.) A.R. Clapham
  6 Seeds mostly >1.6mm, red-brown. Intrd-casual; a birdseed or oil-processing alien on tips etc.; scattered in Br
                Turnip-rape - **B. rapa** ssp. **oleifera** (DC.) Metzg.
7 Fruit terminated by distinct ± conical beak ≥(4)5mm and sometimes with 1(-3) seeds 8
7 Fruit terminated by slender beak ≤4mm not or scarcely wider at base and seedless 10
  8 Lowest leaves with >3 pairs of lateral lobes; fruit with beak ≥10mm, some with 1(-3) seeds. Stems to 50cm. Intrd-casual; a rather frequent wool-alien; sporadic in Br
                Pale Cabbage - **B. tournefortii** Gouan
  8 Lowest leaves with ≤3 pairs of lateral lobes; fruit with seedless beak ≤10mm 9
9 Lowest leaves with 1-3 pairs of lateral lobes; fruit with beak (4)5-9(12)mm; sepals 4.5-7mm. Stems to 1m. Intrd-casual; a frequent birdseed-alien, sometimes from other sources; scattered in BI
                Chinese Mustard - **B. juncea** (L.) Czern.
9 Lowest leaves with 0-1 pairs of lateral lobes; fruit with beak 2.5-6(7)mm; sepals 7-10mm. Stems to 1m. Intrd-casual; occasional birdseed-alien; sporadic in En   *Ethiopian Rape* - **B. carinata** A. Braun
  10 Fruit closely appressed to stem; pedicels 3-8mm. Stems to 2m. Probably native; sea-cliffs, river banks, rough ground and waste

37. BRASSICA

places; frequent in Br and CI N to S Sc, very scattered in Ir
Black Mustard - **B. nigra** (L.) W.D.J. Koch
10 Fruit erecto-patent; pedicels (6)8-18mm 11
11 Fruit (1)1.5-2.5mm wide, with distinct stalk (1)1.5-5mm between sepal-scars and base of valves, with style ≤2(3)mm. Stems to 1m. Intrd-casual; occasional in waste places; sporadic in Br
Long-stalked Rape - **B. elongata** Ehrh.
11 Fruit 3-9mm wide, with base of valves within 1mm of sepal-scars, with style ≥2.5mm (see couplet 9) **B. carinata**

B. napus x *B. rapa* = **B. x harmsiana** O.E. Schulz occurs rarely in crops of *B. napus*.

38. SINAPIS L. - *Mustards*

Annuals; leaves crenate to deeply pinnately lobed; sepals patent; petals yellow; fruit >3x as long as wide, longitudinally dehiscent, with a distinct beak usually >1/3 as long as valves; seeds in 1 row under each valve; valves with 3(-7) strong veins.

1 Leaves lobed or not, if so the terminal lobe much the largest; petals 7.5-17mm; fruit 2.2-5.7cm, with 4-24 seeds, with 0-1-seeded conical beak 7-16mm and 1/3-3/4 as long as valves. Stems to 1(1.5)m. Probably native; arable and waste land, tips and roadsides; throughout BI *Charlock* - **S. arvensis** L.
1 Leaves deeply pinnately lobed petals 7.5-14mm; fruit 2-4.2cm, with 2-8 seeds, with 0-1-seeded strongly flattened beak 10-24(30)mm and 1-1.5x as long as valves. Stems to 70(100)cm
(*White Mustard* - **S. alba** L.) 2
   2 Leaves deeply pinnately lobed, the terminal lobe much the largest; fruit usually hispid. Intrd-natd; in arable or waste land and on waysides and tips, especially on calcareous soils; scattered over most of BI but absent from much of N **S. alba ssp. alba**
   2 Leaves 2-pinnately lobed, the terminal lobe little larger than largest laterals; fruit glabrous to slightly pubescent. Intrd-casual; on waste land; sporadic in Br **S. alba ssp. dissecta** (Lag.) Bonnier

39. ERUCA Mill. - *Garden Rocket*

Annuals; leaves deeply pinnately lobed; sepals erect; petals white to pale yellow with conspicuous violet veins; fruit >3x as long as wide, longitudinally dehiscent, with a distinct beak usually >1/3 as long as valves; seeds in 2 rows under each valve; valves with 1 strong vein.

1 Stems to 1m; fruit 1.2-3.5(4)cm, with seedless strongly flattened beak 4-11mm. Intrd-casual; on waste land; scattered over Br and CI
Garden Rocket - **E. vesicaria** (L.) Cav.

40. ERUCASTRUM C. Presl - *Hairy Rocket*

Annuals to perennials; leaves deeply pinnately lobed; sepals erect to ±

patent; petals yellow; fruit >3x as long as wide, beakless or shortly beaked, longitudinally dehiscent, slightly constricted between seeds; seeds in 1 row under each valve; valve with 1 strong vein.

1 Stems to 60cm; fruit (1.6)2-4.5cm, with seedless beak (1.5)2-4mm. Intrd-casual; arable and waste land, rarely persisting; scattered in BI, mainly S **Hairy Rocket - E. gallicum** (Willd.) O.E. Schulz

41. COINCYA Rouy - *Cabbages*
Annuals to perennials; leaves pinnately lobed to pinnate; sepals erect; petals yellow; fruit >3x as long as wide, longitudinally dehiscent, with distinct beak usually 1/5-1/3 as long as valves and with (0)1-4(5) seeds; seeds in 1 row under each valve; valves with 3 strong veins.

1 Biennials to perennials, often woody near base, pubescent over all or most of stem incl. inflorescence; basal leaves with 2-5(6) pairs of lateral lobes, the terminal lobe much larger than laterals; fruit 2-8cm incl. beak 7-16mm. Stems to 1m. Native; cliffs and slopes; Lundy Island (N Devon) **Lundy Cabbage - C. wrightii** (O.E. Schulz) Stace
1 Annuals to perennials, subglabrous to pubescent below; basal leaves with 3-9 pairs of lateral lobes, the terminal lobe not much larger than laterals; fruit (2.5)3.5-8(8.5)cm incl. beak 5-24(34)mm (**C. monensis** (L.) Greuter & Burdet) 2
  2 Stems procumbent to ascending, to 60cm; glabrous to sparsely hispid below; seeds 1.3-2mm. Native; sandy ground near sea; Man and W Br from S Lancs to Clyde Is, rare casual elsewhere
                            Isle of Man Cabbage **- C. monensis ssp. monensis**
  2 Stems usually erect, to 1m, hispid to rather sparsely so below; seeds 0.8-1.6mm. Intrd-natd; sandy ground, waste places and roadsides; casual in SW Br, natd in Mons and Jersey (*C. monensis* ssp. *recurvata* (All.) Leadlay) **Wallflower Cabbage - C. monensis ssp. cheiranthos** (Vill.) Aedo, Leadlay & Muñoz Garm.

42. HIRSCHFELDIA Moench - *Hoary Mustard*
Annuals to short-lived perennials; lower leaves pinnate to deeply pinnately lobed; sepals ± erect; petals yellow; fruit >3x as long as wide, longitudinally dehiscent, with distinct beak usually c.1/2 as long as valves with (0)1(-2) seeds; seeds in 1 row under each valve; valves with 1-3 ± strong veins.

1 Stems erect, to 1.3m; fruit 6-17mm, appressed to stem, with beak 3-6.5mm, swollen round seeds and abruptly narrowed distally. Intrd-natd; waste places and waysides; BI, especially S
                               Hoary Mustard - H. incana (L.) Lagr.-Foss.

43. CAKILE Mill. - *Sea Rocket*
Glabrous annuals; leaves entire to pinnately lobed; sepals erect; petals mauve to pink or white; fruit breaking transversely into 2 portions, the proximal (0-)1-seeded, the distal 1(-2)-seeded, longer and wider, both keeled

43. CAKILE

laterally and with prominent veins and margin.

1 Stems to 50cm; leaves ± glaucous, ± succulent; fruit 12-25mm; proximal segment 4-9mm; distal segment 8-20mm. Native; near sea drift-line on sand and sometimes shingle; around coasts of BI
Sea Rocket - **C. maritima** Scop.

44. RAPISTRUM Crantz - *Cabbages*
Annuals to perennials; leaves dentate to deeply pinnately lobed; sepals erecto-patent; petals yellow; fruit breaking transversely into 2 ± equal-lengthed portions, the proximal 0-1(3)-seeded, the distal 1-seeded, narrowed at apex into persistent style and variously ribbed or wrinkled.

1 Fruit 3-12mm; distal segment abruptly narrowed into (0.8)1-3.5(5)mm style, ribbed and rugose; proximal segment usually much narrower than distal, mostly 0(1)-seeded. Stems to 1m. Intrd-natd; waste and arable land, tips, waysides and open grassland; S & C BI
Bastard Cabbage - **R. rugosum** (L.) J.P. Bergeret
1 Fruit 5-10mm; distal segment gradually narrowed into 0.5-1.2mm style, longitudinally ribbed; proximal segment usually similar in size to distal but less or not ribbed, mostly 1-seeded. Intrd-natd; similar places to *R. rugosum* but much rarer; scattered in S & C Br
Steppe Cabbage - **R. perenne** (L.) All.

45. CRAMBE L. - *Sea-kale*
Large perennials with thick long roots; leaves large and irregularly lobed or toothed; sepals ± patent; petals white; fruit breaking transversely into 2 portions, the proximal small, sterile and stalk-like, the distal 1-seeded.

1 Glabrous, glaucous, densely branched cabbage-like plant to 75cm; basal leaves undulate at margins; stem-leaves many, similar; distal portion of fruit 6-14 x 6-11mm. Native; on sand, rocks and cliffs but mostly shingle, by sea; coasts of BI N to C Sc *Sea-kale* - **C. maritima** L.
1 Sparsely pubescent erect perennial to 2m; basal leaves plane at margins; stem-leaves very few, much smaller; distal portion of fruit 4-6 x 2.6-6.8mm. Intrd-natd; garden throwout with persistent roots, rarely self-sown; scattered in En *Greater Sea-kale* - **C. cordifolia** Steven

46. RAPHANUS L. - *Radishes*
Annuals to perennials with distinctive radish-like smell when crushed; leaves shallowly pinnately lobed to pinnate; sepals erect; petals white, mauve or yellow, often with darker veins; main part (upper segment) of fruit indehiscent or transversely dehiscent into 1-10 1-seeded portions (mericarps), with short, inconspicuous, 0-1-seeded lower segment and with long persistent narrow beak.

1 Root usually swollen, often reddish; fruit not or scarcely constricted between seeds, indehiscent. Stems to 1m. Intrd-casual; fields,

gardens and tips; sporadic throughout BI *Garden Radish* - **R. sativus** L.
1 Root slender; fruit strongly constricted between seeds, at least partly transversely dehiscent (**R. raphanistrum** L.) 2
  2 Fruit with cylindrical or oblong mericarps usually longer than wide, with beak (2.5)3-6x as long as most apical mericarp. Stems to 75cm. Probably intrd-natd; cultivated and rough ground, waste places and tips; frequent throughout BI
Wild Radish - **R. raphanistrum ssp. raphanistrum**
  2 Fruit with ± globose mericarps c. as long as wide, with beak 1-3(4)x as long as most apical mericarp 3
3 Leaves with crowded lateral lobes; petals (14)15-22(25)mm when fresh. Stems to 80cm. Native; sea-shores and maritime cliffs and waste places; coasts of BI, absent from most of E Br
Sea Radish - **R. raphanistrum ssp. maritimus** (Sm.) Thell.
3 Leaves with ± distant lateral lobes; petal 8-15mm when fresh. Stems to 80cm. Intrd-casual; in waste places; sporadic in En and Wa
Mediterranean Radish - **R. raphanistrum ssp. landra** (DC.) Bonnier & Layens

65. RESEDACEAE - *Mignonette family*

Herbaceous annuals to perennials; easily recognized by the zygomorphic flowers with open-topped ovary and 4-6 white or yellowish deeply lobed petals.

1. RESEDA L. - *Mignonettes*

1 At least upper and middle leaves deeply pinnately lobed 2
1 All leaves entire to minutely toothed, or a few with 1-2 lateral lobes 3
  2 Carpels 3; petals yellowish; filaments falling after flowering; seeds smooth. Stems decumbent to erect, to 75cm. Native; disturbed, waste and arable land, especially on calcareous soils; throughout much of BI *Wild Mignonette* - **R. lutea** L.
  2 Carpels 4; petals white; filaments persistent until fruit ripe; seeds tuberculate. Stems erect to ascending, to 75cm. Intrd-natd; waste ground, often near sea, occasionally persisting; S Br and CI
White Mignonette - **R. alba** L.
3 Sepals and petals 4; fruits crowded, stiffly erecto-patent, <7mm, on pedicels <4mm; seeds smooth. Stems erect, to 1.5m. Native; open grassland, disturbed, waste and arable land mostly on base-rich soils; throughout most of BI *Weld* - **R. luteola** L.
3 Sepals and petals 6; fruits well spaced, pendent, ≥7mm, on pedicels >5mm; seeds rugose 4
  4 Capsules 7-11mm; sepals ≤5mm at fruiting. Stems procumbent to ascending, to 30cm. Intrd-casual; tips and waste places; occasional in S Br *Garden Mignonette* - **R. odorata** L.
  4 Most mature capsules 11-15mm; many sepals >5mm at fruiting.

1. RESEDA

Stems procumbent to ascending, to 30cm. Intrd-natd; waste ground, cornfields and field margins; few places in S En
Corn Mignonette - **R. phyteuma** L.

66. EMPETRACEAE - *Crowberry family*

Dwarf, *Erica*-like, evergreen shrubs; distinctive in the narrow leaves with inrolled margins, very inconspicuous, 3-merous flowers and black fruits.

1. EMPETRUM L. - *Crowberry*

1 Leaves strongly revolute obscuring abaxial surface, 3-7 x 1-2mm; fruit 4-8mm across. Native; peaty and rocky moors, bogs and mountain-tops (*Crowberry* - **E. nigrum** L.) 2
    2 Stems to 1.2m, procumbent, slender, rooting along length; leaves ± parallel-sided, mostly 3-5x as long as wide; flowers dioecious, rarely bisexual. Frequent in suitable places to c.800m in Br and Ir NW of line from Devon to NE Yorks   **E. nigrum ssp. nigrum**
    2 Stems to 50cm, less procumbent, stiff, not rooting; leaves with curved sides, mostly 2-4x as long as wide; flowers bisexual (remains of stamens usually visible at base of some fruits). Usually at >650m in dry rocky places; Caerns, Lake District, highlands of Sc   **E. nigrum ssp. hermaphroditum** (Hagerup) Böcher

67. ERICACEAE - *Heather family*

Deciduous or evergreen trees or dwarf shrubs; very variable in superficial flower characters, but usually recognizable by the woody often evergreen habit, usually fused petals, stamens borne on receptacle (not on corolla), and anthers usually opening by pores. All our spp., except *Arbutus unedo*, *Erica terminalis* and *E. x darleyensis*, are calcifuges.

1 Ovary inferior; fruit a berry with persistent calyx-lobes at apex
    13. VACCINIUM
1 Ovary superior; fruit various, if succulent then calyx deciduous or persistent at base of fruit 2
    2 Petals free, white; leaves tomentose with rust-coloured hairs on lowerside   **1. LEDUM**
    2 Petals fused at least at base; leaves tomentose or not, if so then hairs not rust-coloured 3
3 Most leaves opposite or whorled; anthers with 0 or basal appendages 4
3 Leaves alternate or spiral; anthers with 0 or terminal appendages 7
    4 Sepals and petals 4; stamens 8; corolla persistent around ripe fruit 5
    4 Sepals and petals 5; stamens 5 or 10; corolla falling before fruiting 6
5 Leaves opposite; corolla shorter than calyx, divided >1/2 way to base
    11. CALLUNA

5 Leaves mostly in whorls of 3-4(5); corolla longer than calyx, normally divided <1/2 way to base **12. ERICA**
  6 Leaves <15mm, with strongly revolute margins; stamens 5; corolla divided c.1/2 way to base, without pouches **3. LOISELEURIA**
  6 Leaves >15mm, with flat margins; stamens 10; corolla divided much <1/2 way to base, with 10 small pouches on inside near base **4. KALMIA**
7 Corolla widened distally, >15mm, slightly zygomorphic **2. RHODODENDRON**
7 Corolla narrowed distally, <15mm, actinomorphic 8
  8 Fruit succulent or surrounded by a succulent calyx 9
  8 Fruit a dry capsule with a dry calyx 12
9 Leaves spine-tipped **8. GAULTHERIA**
9 Leaves without spines 10
  10 Calyx becoming succulent and surrounding fruit when ripe **8. GAULTHERIA**
  10 Calyx remaining dry and small at base of succulent fruit 11
11 Erect tree or shrub flowering in late autumn; fruit orange to red, very warty **9. ARBUTUS**
11 Procumbent shrub flowering in summer; fruit red or black, smooth, ±glossy **10. ARCTOSTAPHYLOS**
  12 Petals and sepals 4; leaves white-tomentose on lowerside **6. DABOECIA**
  12 Petals and sepals 5; leaves glabrous but often white on lowerside 13
13 Calyx and pedicels with glandular hairs; corolla purple; anthers without appendages **5. PHYLLODOCE**
13 Calyx and pedicels glabrous; corolla pink; anthers with horn-like appendages at apex **7. ANDROMEDA**

1. LEDUM L. - *Labrador-tea*

Leaves alternate, evergreen; flowers in dense terminal racemes, petals 5, free, white; stamens (5)6-8(10); anthers without appendages; ovary superior; fruit a capsule.

1 Well-branched ± upright shrub to 1.2m; leaves 1.5-5cm, with strongly revolute margins. Intrd-natd; well natd in bogs and other wet peaty ground; scattered in Br N to C Sc
Labrador-tea - **L. palustre** L. ssp. **groenlandicum** (Oeder) Hultén

2. RHODODENDRON L. - *Rhododendrons*

Leaves alternate, deciduous or evergreen; flowers in dense terminal racemes; petals 5, fused to form bell-shaped lobed corolla; stamens 5 or 10; anthers without appendages; ovary superior; fruit a capsule.

1 Leaves entire, evergreen, glabrous; flowers mauvish-purple, c.4-6cm across; stamens 10. Intrd-natd; extensively natd by seeding and suckering on sandy and peaty soils and on rocks both in woods and in open; throughout BI *Rhododendron* - **R. ponticum** L.

2. RHODODENDRON

1 Leaves shallowly serrate, deciduous, slightly pubescent; flowers yellow, c.5cm across; stamens 5. Intrd-natd; natd in woods by suckering and seeding; scattered in Br *Yellow Azalea* - **R. luteum** Sweet

3. LOISELEURIA Desv. - *Trailing Azalea*

Leaves opposite, evergreen; flowers 1-several in terminal apparent umbels; petals 5, pink, fused to form bell-shaped lobed corolla; stamens 5; anthers without appendages; ovary superior; fruit a capsule.

1 Densely branched domed or trailing shrub to 25cm high; leaves <1cm; flowers 3-6mm aross. Native; rocky and peaty moors and mountains above 400m; locally frequent in C & N Sc
Trailing Azalea - **L. procumbens** (L.) Desv.

4. KALMIA L. - *Sheep-laurels*

Leaves alternate, whorled or opposite, evergreen; flowers in small corymbose racemes; petals 5, pink, fused to form saucer-shaped slightly lobed corolla with 10 small pouches on inside near base; stamens 10, enclosed in corolla-pouches before dehiscence; anthers without appendages; ovary superior; fruit a capsule.

The frequently used character 'inflorescence terminal or lateral' is misleading; in all cases the inflorescence arises from leaf-axils near the apex of the previous year's growth, which might or might not become overtopped by a current season's shoot. The former is normal in only *K. angustifolia*, but even there is not constant.

1 Leaves alternate or irregularly arranged, finely acute; flowers 20-25mm across. Erect shrub (or tree) to 3(12)m. Intrd-natd; in a few wet acid places; SE En, Man *Mountain-laurel* - **K. latifolia** L.
1 Leaves opposite or in whorls of 3, obtuse to subacute; flowers 6-16mm across 2
  2 Leaves sessile or with petioles <4mm; pedicels glabrous; flowers 10-16mm across; inflorescence apparently terminal. Straggling to erect shrub to 70cm. Intrd-natd; in wet peaty bogs and moors; scattered in SE & N En and C Sc *Bog-laurel* - **K. polifolia** Wangenh.
  2 Leaves with petioles c. 4-8mm; pedicels very shortly pubescent; flowers 6-12mm across; inflorescence usually becoming overtopped by current season's shoot and appearing lateral. Erect to ascending shrub to 1m. Intrd-natd; in similar places to *K. polifolia*; scattered in En *Sheep-laurel* - **K. angustifolia** L.

5. PHYLLODOCE Salisb. - *Blue Heath*

Leaves alternate, evergreen; flowers few on long pedicels in subterminal clusters; petals 5, mauvish-purple, fused to form tubular corolla narrowed and shortly lobed distally; stamens 10; anthers without appendages; ovary superior; fruit a capsule.

1 Domed shrub to 20cm high; leaves <15mm. Native; rocky moorland

at 680-840m; very local in Westerness and M Perth

Blue Heath - **P. caerulea** (L.) Bab.

6. DABOECIA D. Don - *St Dabeoc's Heath*

Leaves alternate, evergreen; flowers in lax terminal racemes; petals 4, pinkish-purple, fused to form tubular corolla narrowed and shortly lobed distally; stamens 8; anthers without appendages; ovary superior; fruit a capsule.

1 Straggly or loosely domed shrub to 50(70)cm; leaves <15mm.
  Native; peaty and rocky moorland; locally common in Connemara

St Dabeoc's Heath - **D. cantabrica** (Huds.) K. Koch

7. ANDROMEDA L. - *Bog-rosemary*

Leaves alternate, evergreen; flowers in small terminal umbel-like clusters; petals 5, pale pink, fused to form tubular corolla narrowed and shortly lobed distally; stamens 10; anthers each with 2 long terminal appendages; ovary superior; fruit a capsule.

1 Straggly glabrous shrub; stems to 35cm; leaves 1-4cm. Native; wet
  peaty places; locally common, C En and S Wa to C Sc, Ir

Bog-rosemary - **A. polifolia** L.

8. GAULTHERIA L. - *Shallons*

Leaves alternate, evergreen; flowers solitary and axillary or in terminal and subterminal racemes; petals 5, white to pink, fused to form tubular corolla narrowed and shortly lobed distally; stamens 10; anthers with 4 short terminal appendages; ovary superior; fruit a capsule surrounded by succulent, berry-like swollen calyx, or a succulent berry with calyx remaining small and dry at its base.

1 Leaves ≤2cm, spine-tipped; fruit a succulent berry with calyx
  remaining small and dry at its base; functionally dioecious. Erect or
  spreading suckering shrub to 1.5m. Intrd-natd; in open woodland
  and shrubberies on sandy soil; scattered in Br and Ir

Prickly Heath - **G. mucronata** (L.f.) Hook. & Arn.
1 Leaves ≥2cm, not spine-tipped; fruit a capsule surrounded by
  succulent berry-like swollen calyx; plants bisexual   2
  2 Leaves cuneate at base; flowers solitary in leaf-axils; fruit bright
    red. Dwarf ground-covering shrub to 15cm. Intrd-natd; woodland;
    few places in Sc and S En   *Checkerberry* - **G. procumbens** L.
  2 Leaves rounded to cordate at base; flowers in terminal and
    subterminal racemes; fruit purplish-black. Thicket-forming shrub
    to 1.5m. Intrd-natd; woodland and shrubberies especially on
    sand and peat; scattered through BI   *Shallon* - **G. shallon** Pursh

G. shallon x *G. mucronata* = **G. x wisleyensis** D.J. Middleton has been found in S Hants.

9. ARBUTUS L. - *Strawberry-tree*
Leaves alternate, evergreen; flowers in terminal panicles; petals 5, white or pink-tinged, fused to form tubular corolla narrowed and shortly lobed distally; stamens 10; anthers with 2 long terminal appendages; ovary superior; fruit a warty, globose berry.

1 Shrub or tree to 5(11)m; leaves 4-11cm. Native; rocky ground in scrub and young woodland; S & N Kerry, W Cork and Sligo; rarely natd (bird-sown) on mostly chalk or limestone slopes in En and Wa *Strawberry-tree* - **A. unedo** L.

10. ARCTOSTAPHYLOS Adans. - *Bearberries*
Leaves alternate, deciduous or evergreen; flowers 1-few in terminal clusters; petals 5, white, pink- or green-tinged, fused to form tubular corolla narrowed and shortly lobed distally; stamens 10; anthers with 2 terminal appendages; ovary superior; fruit a smooth, globose, berry-like drupe.

1 Leaves 1-3cm, entire, evergreen; stamens with reflexed appendages c. equalling anthers; fruit c.8-10mm across, bright red. Stems procumbent, to 1.5m. Native; peaty and rocky moorland in lowlands and mountains; locally common in Sc, N En and N & W Ir
Bearberry - **A. uva-ursi** (L.) Spreng.
1 Leaves 1-2.5cm, serrate, dying in autumn but persistent until next spring; stamens with erect appendages much shorter than anthers; fruit c.6-10mm across, black. Stems procumbent, to 60cm. Native; mountain moorland; local in N & NW Sc
Arctic Bearberry - **A. alpinus** (L.) Spreng.

11. CALLUNA Salisb. - *Heather*
Leaves opposite, evergreen, sessile; flowers in long usually terminal racemes or panicles; petals 4, pink (or white), fused for basal 1/4 or less; stamens 8; anthers with 2 basal appendages; ovary superior; fruit a capsule dehiscing along line of fusion of carpels.

1 Decumbent to erect shrub to 60(150)cm; leaves 2-3.5mm, sessile. Native; heaths, moors, rocky places, bogs and open woodland, mainly on sandy or peaty soil; common *Heather* - **C. vulgaris** (L.) Hull

12. ERICA L. - *Heaths*
Leaves in whorls of 3-4(5), evergreen, shortly petiolate; flowers in various terminal and/or axillary clusters; petals 4, pink, reddish-purple or white, fused for basal 1/2 or more; stamens 8, anthers with 0 or 2 basal appendages; ovary superior; fruit a capsule dehiscing between lines of fusion of carpels.

1 Anthers at least partly exserted from corolla 2
1 Anthers included in corolla 4
   2 Pedicels longer than calyx; corolla-lobes divergent distally; summer-flowering. Straggly shrub to 80cm. Native; dry heaths;

W Cornwall, sometimes natd elsewhere, well natd or perhaps native in Fermanagh **Cornish Heath - E. vagans** L.
2 Pedicels shorter than calyx; corolla-lobes ± parallel distally; winter- to spring-flowering 3
3 Stems erect, to 1.2(-2)m, with well-developed main stems; young twigs with flanges of tissue running <1/2 way from leaf-bases to next lower node; flowering Mar-Jun. Native; in usually well-drained parts of bogs; W Galway and W Mayo **Irish Heath - E. erigena** R. Ross
3 Stems to 60cm, without well-developed main stems; young twigs with flanges of tissue running >1/2 way from leaf-bases to next lower node; flowering Nov-Jun. Intrd-natd; spreading vegetatively where planted; Surrey and W Cornwall (*E. erigena* x *E. carnea* L.)
Darley Dale Heath - E. x darleyensis Bean
4 Revolute leaf-margins meeting closely under leaf, entirely obscuring lowerside; flowers usually in panicles 5
4 Revolute leaf-margins not meeting or meeting only distally under leaf, revealing at least proximal part of lowerside; flowers in racemes or apparent umbels 7
5 Some bracteoles borne near apex of pedicel, overlapping calyx; straggly shrub to 60cm, summer-flowering. Native; usually dry heaths and moors; throughout BI **Bell Heather - E. cinerea** L.
5 Bracteoles borne only on proximal part of pedicel, not overlapping calyx; shrub usually >60cm, spring-flowering 6
6 Corolla 4-5mm; stigma red; all hairs smooth (microscope). Erect shrub to 2(-3)m. Intrd-natd; garden escape on heaths and railway-banks; E & W Cornwall and Dorset
Portuguese Heath - E. lusitanica Rudolphi
6 Corolla 2.5-4mm; stigma white; some hairs on young twigs with rough surface (microscope). Erect shrub to 2(-3)m. Intrd-natd; hedgerows and open woodland; Surrey, Scillies and Man, less common than *E. lusitanica* **Tree Heath - E. arborea** L.
7 Flowers in terminal ± elongated racemes; anthers with 0 basal appendages. Straggly shrub to 60cm. Native; heaths, often damp; very locally frequent in SC & SW En and W Galway
Dorset Heath - E. ciliaris L.
7 Flowers in terminal umbel-like clusters; anthers with basal appendages 8
8 Lowerside of leaves green; sepals glabrous or with only short hairs; anthers with triangular appendages. Bushy or erect shrub to 1m. Intrd-natd; on sand-dunes; Magilligan (Co Londonderry)
Corsican Heath - E. terminalis Salisb.
8 Lowerside of leaves whitish; sepals with long hairs; anthers with linear appendages 9
9 Sepals and uppersides of leaves usually glabrous except for long hairs; most of leaf lowerside exposed; ovary and fruit glabrous. Straggly to compact shrub to 60cm. Native; peaty bogs; very local in W & NW Ir **Mackay's Heath - E. mackaiana** Bab.
9 Sepals and uppersides of leaves usually with dense short hairs as

well as long hairs; most of leaf lowerside obscured; ovary and fruit pubescent. Straggly shrub to 70cm. Native; bogs and usually wet heaths and moors; throughout BI   *Cross-leaved Heath* - **E. tetralix** L.

E. tetralix forms sterile hybrids with *E. ciliaris*, *E. mackaiana* and *E. vagans* in mixed populations.

13. VACCINIUM L. - *Bilberries*
Leaves alternate, deciduous or evergreen; flowers solitary or clustered, terminal or axillary; petals 4-5, fused at base or for most part to form variously shaped corolla; stamens 8 or 10; anthers with or without terminal appendages; ovary inferior; fruit a berry.

1 Corolla divided >3/4 way to base; pedicels erect, filiform; leafy stems procumbent for most part 2
1 Corolla divided <2/3 way to base; pedicels not erect or filiform; leafy stems erect to decumbent 4
   2 Leaves narrowly oblong, at least some >1cm; bracteoles above middle of pedicel, mostly >1mm wide; leafy shoot continuing growth beyond flower cluster in same year. Procumbent shrub with stems to 1m. Intrd-natd; in peaty places; scattered in Br from S En to W Sc   *American Cranberry* - **V. macrocarpon** Aiton
   2 Leaves ovate-elliptic or narrowly so, rarely >1cm; bracteoles at or below middle of pedicel, <0.5mm wide; flowers in terminal groups of 1-c.5 3
3 Pedicels minutely pubescent. Procumbent shrub with stems to 30(80)cm. Native; bogs and very wet heaths; locally frequent in Br and Ir, but absent from most of S En, S Ir and N Sc
Cranberry - **V. oxycoccos** L.
3 Pedicels glabrous or almost so. Procumbent shrub with stems to 30cm. Native; bogs; C & N mainland Sc, S Northumb and Cheviot
Small Cranberry - **V. microcarpum** (Rupr.) Schmalh.
   4 Some leaves >3cm (to 8cm). Erect shrub to 1(-2.5)m. Intrd-natd; on heathland from bird-sown seeds; S Hants and Dorset
Blueberry - **V. corymbosum** L.
   4 Leaves <3cm 5
5 Leaves deciduous, serrate to serrulate; stems acutely angled. Erect to ascending shrub to 50(100)cm. Native; heaths, moors and woods; common throughout most of BI but absent from CI and C & E En
Bilberry - **V. myrtillus** L.
5 Leaves entire to obscurely crenulate; stems terete 6
   6 Leaves evergreen; flowers in terminal racemes; corolla widest at mouth, divided c.1/2 way to base; fruit red. Erect to decumbent shrub to 30cm. Native; moors and open peaty woods; locally abundant in Br from S Wa and C En northwards, scattered in Ir
Cowberry - **V. vitis-idaea** L.
   6 Leaves deciduous; flowers in axillary clusters of 1-4; corolla narrowed at mouth, divided ≤1/4 way to base; fruit bluish-black.

Erect to ascending shrub to 50(80)cm. Native; moors; locally common in C & N Sc, very local in S Sc and N En, natd in S Somerset   **Bog Bilberry - V. uliginosum** L.

V. vitis-idaea x *V. myrtillus* = ***V. x intermedium*** Ruthe occurs very locally with the parents in Staffs, Derbys and Yorks; it is slightly fertile..

68. PYROLACEAE - *Wintergreen family*

Herbaceous rhizomatous perennials; plants with ± white petals and with flowers similar in structure to those of Ericaceae.

1 Flowers solitary **3. MONESES**
1 Flowers in terminal raceme 2
    2 Flowers all turned to 1 side of axis; anther pores borne on main body of anther; petioles <2cm **2. ORTHILIA**
    2 Flowers facing all directions; anther pores borne at end of short tubular anther outgrowths; longest petioles >2cm **1. PYROLA**

1. PYROLA L. - *Wintergreens*
Flowers in terminal raceme, facing all directions; anthers with pores borne on very short tubes; pollen-grains released in tetrads. Leaf characters do not reliably separate any of the taxa in this genus.

1 Style ± straight; flowers ± globose 2
1 Style strongly curved; flowers saucer- to cup-shaped. Flowering stem erect, to 20(30)cm (***Round-leaved Wintergreen* - P. rotundifolia** L.) 3
    2 Style 1-2mm, included in flower, not widened below stigma. Flowering stem erect, to 20(30)cm. Native; on leaf-mould in woods in S, on damp rock-ledges and peaty moors in N, rarely on sand-dunes; scattered over Br and Ir   *Common Wintergreen* **- P. minor** L.
    2 Style 4-6mm, just exserted from flower, widened immediately below stigma-lobes. Flowering stem erect, to 20(30)cm. Native; humus-rich moors and woods; frequent in C & N Sc, very local in N Ir, S Sc and N En, Worcs
            Intermediate Wintergreen **- P. media** Sw.
3 Scale-leaves on stems above true leaves 1-2; pedicels 4-8mm; sepals triangular-lanceolate, acute; anthers 2.2-2.8mm; style 6-10mm. Native; damp rock-ledges, woods, bogs and fens; very local in En, Sc and C Ir   **P. rotundifolia ssp. rotundifolia**
3 Scale-leaves 2-5; pedicels 2-5mm; sepals oblong-lanceolate, obtuse; anthers 1.9-2.4mm; style 4-6mm. Native; damp hollows in sand-dunes; W coast of Br N to Cumberland, Co Wexford, uncertain in E Br   **P. rotundifolia ssp. maritima** (Kenyon) E.F. Warb.

2. ORTHILIA Raf. - *Serrated Wintergreen*
Stem-leaves alternate; flowers in terminal raceme, all turned to 1 side;

anthers with pores borne on main body of anther; pollen-grains released singly.

1 Flowering stem erect, often curved at top, to 10(20)cm. Native; woods and damp rock-ledges; local from N En to N Sc, very local in Wa and N & C Ir *Serrated Wintergreen* - **O. secunda** (L.) House

3. MONESES Gray - *One-flowered Wintergreen*
Stem-leaves opposite or in whorls of 3; flowers single, terminal, pendent or turned to 1 side; anthers with pores borne on short tubes; pollen-grains released in tetrads.

1 Flowering stem erect, to 10(15)cm. Native; on leaf-litter in pinewoods; very local in NE Sc *One-flowered Wintergreen* - **M. uniflora** (L.) A. Gray

69. MONOTROPACEAE - *Bird's-nest family*

Saprophytic, ± chlorophyll-less, brownish-yellow herbaceous perennials with scale-like leaves; flowers in terminal raceme, with 4-5 free petals and sepals and 8 or 10 stamens opening by longitudinal slits.

1. MONOTROPA L. - *Yellow Bird's-nest*

1 Stems to 30cm, pendent at apex in flower, erect in fruit. Native; on leaf-litter in woods (especially *Pinus* and *Fagus*) and on sand-dunes (*Yellow Bird's-nest* - **M. hypopitys** L.) 2
  2 Flowers ≤11; petals 9-13mm; stamens, carpels and inside of petals pubescent; style equalling or longer than ovary. Very scattered in En **M. hypopitys ssp. hypopitys**
  2 Flowers ≤8; petals 8-10mm; ovary glabrous; stamens, style and inside of petals glabrous or pubescent; style equalling or shorter than ovary. Local in Br and Ir
  M. hypopitys ssp.hypophegea (Wallr.) Holmboe

70. DIAPENSIACEAE - *Diapensia family*

Cushion-like, evergreen dwarf shrub; easily recognisable by its habit, 5 white petals fused to c.1/2 way, 5 stamens and 3-locular ovary.

1. DIAPENSIA L. - *Diapensia*

1 Plant dome-shaped, to 6cm high; leaves 5-10mm. Native; exposed mountain at c.760m; on hill NW of Fort William (Westerness)
Diapensia - **D. lapponica** L.

71. PRIMULACEAE - *Primrose family*

Herbaceous annuals or perennials; very variable in superficial flower characters, but usually recognizable by the herbaceous habit, fused petals, and 1-celled ovary with free-central placentation.

1	Leaves pinnate; plant a submerged aquatic	**2. HOTTONIA**
1	Leaves simple; plant not submerged	2
	2 All leaves basal	3
	2 Some or all leaves on stems	4
3	Corolla-lobes patent to erecto-patent; underground corm 0	
		1. PRIMULA
3	Corolla-lobes strongly reflexed; underground corm present	
		3. CYCLAMEN
	4 Corolla yellow	**4. LYSIMACHIA**
	4 Corolla white to red, blue or purple, or 0	5
5	All or most leaves in single apparent whorl at top of stem; corolla-lobes mostly 6-7	**5. TRIENTALIS**
5	Leaves opposite or alternate along stem; corolla-lobes 5	6
	6 Ovary 1/2-inferior; corolla white, longer than calyx	**8. SAMOLUS**
	6 Ovary superior; corolla usually coloured or 0 or shorter than calyx	7
7	Corolla present; calyx-lobes ± free; capsule dehiscing transversely	
		6. ANAGALLIS
7	Corolla 0; calyx-lobes fused for >1/4; capsule dehiscing longitudinally	
		7. GLAUX

1. PRIMULA L. - *Primroses*

Perennials; leaves all basal, simple; calyx-tube longer than lobes; corolla-lobes 5, patent to erecto-patent; corolla-tube c. as long as lobes or longer; capsule dehiscing by 5 teeth or valves. Most spp. are usually heterostylous, some plants having stigmas higher than anthers (pin-eyed) and others vice versa (thrum-eyed).

1	Pale mealy coating present on various parts of leaves, scape, pedicels and/or flowers	2
1	Mealy coating 0	6
	2 Corolla ≥(12)15mm across, yellow	3
	2 Corolla ≤15mm across, lilac to purple, rarely white	5
3	Leaves with narrow whitish border with short-stalked glands, often mealy or minutely pubescent on upperside. Scape to 15cm. Intrd-natd; on rock-ledge in Caenlochan Glen (Angus), also in MW Yorks	
	Auricula - **P. auricula** L.	
3	Leaves green to margin, glabrous, not mealy	4
	4 Leaves cuneate at base. Scape to 60cm. Intrd-natd; mountain rocks; Caerns *Sikkim Cowslip* - **P. sikkimensis** Hook. f.	
	4 Leaves truncate to cordate at base. Scape to 60cm. Intrd-natd; by ponds and streams and in marshes; scattered in N En and Sc *Tibetan Cowslip* - **P. florindae** Kingdon-Ward	

5 Flowers heterostylous; corolla-lobes usually lilac, with gaps between
at least near base. Scape to 15cm. Native; damp grassy, stony or
peaty ground on limestone; locally frequent in N En
Bird's-eye Primrose - **P. farinosa** L.
5 Flowers homostylous (anthers and stigma ± at same level); corolla-
lobes usually purple, overlapping or contiguous. Scape to 10cm.
Native; damp grassy places near sea on cliffs, dunes and pastures; W
Sutherland, Caithness and Orkney *Scottish Primrose* - **P. scotica** Hook.
   6 Flowers borne in 2 or more whorled tiers up scape; corolla
purplish-red or white. Intrd-natd; shady moist places; scattered
in En, W Sc and W Ir *Japanese Cowslip* - **P. japonica** A. Gray
   6 Flowers borne singly from base of plant or in a single umbel on
scape; corolla usually yellow, rarely white or purplish-red   7
7 Pedicels with long shaggy hairs; ripe capsules lying near or on
ground, with sticky seeds; flowers usually borne singly from base
of plant on pedicels to 12cm. Native; woods, hedgebanks and moist
grassland, often on heavy soils; throughout BI, often common
Primrose - **P. vulgaris** Huds.
7 Pedicels with short fine hairs; ripe capsules held ± erect, with dry
seeds; flowers borne in umbel on scape   8
   8 Corolla usually <15mm across, with folds in throat; calyx
uniformly pale green, with acute or obtuse and apiculate teeth;
capsule enclosed in calyx. Scape to 30cm. Native; grassy places
usually on light base-rich soils; most of BI   *Cowslip* - **P. veris** L.
   8 Corolla usually >15mm across, without folds in throat; calyx pale
green with dark green midribs, with ± acuminate teeth; capsule
c. as long as or longer than calyx. Scape to 30cm. Native; very
locally abundant in woods on clay in E Anglia, 2 small areas
in Bucks   *Oxlip* - **P. elatior** (L.) Hill

P. vulgaris x *P. elatior* = **P. x digenea** A. Kern., *P. vulgaris* x *P. veris* = **P. x polyantha** Mill., *P. vulgaris* x *P. elatior* x *P. veris* = **P. x murbeckii** Lindq. and *P. elatior* x *P. veris* = **P. x media** Peterm. all occur where the parents meet, the first 2 commonly. The first is highly and the others slightly fertile.

2. HOTTONIA L. - *Water-violet*

Perennials with submerged vegetative parts and emergent inflorescences; leaves ± whorled, pinnate with linear lobes; calyx divided nearly to base; corolla lobes 5, lilac with yellow throat, patent, longer than tube; capsule dehiscing by 5 valves.

1 Stems to 1m or more, floating under water; flowers in tiered whorls
on erect peduncle. Native; shallow ponds and ditches; scattered in
En and Wa, very locally natd in Ir   *Water-violet* - **H. palustris** L.

3. CYCLAMEN L. - *Sowbreads*

Perennials with large underground corm; leaves all basal, simple, with long petiole; flowers borne singly on long pedicels from corm; calyx-tube shorter than lobes; corolla-lobes 5, purplish- to pale pink (or white), strongly reflexed, longer than tube; capsule dehiscing by 5 valves which become reflexed and on pedicel which becomes tightly spiralled at maturity.

1 Flowers appearing in late summer and autumn, usually before the leaves; corolla with conspicuous flaps or bulges at point of reflexion of lobes from tube; corm rooting mainly from upperside. Intrd-natd; woods and hedgerows; very scattered in CI, Man and Br N to S Sc
Sowbread - **C. hederifolium** Aiton
1 Flowers appearing in spring with the leaves; corolla without lobes or bulges at point of reflexion; corm rooting only from lowerside 2
 2 Corolla-lobes 7-15mm, with prominent violet blotch at base; leaves usually only undulate or toothed. Intrd-natd; roadside verges and grassy places; scattered in S En
Eastern Sowbread - **C. coum** Mill.
 2 Corolla-lobes 15-30mm, without violet blotch at base; leaves usually angled or shallowly lobed. Intrd-natd; roadside verges, woods and grassy places; scattered in SW & SE En
Spring Sowbread - **C. repandum** Sibth. & Sm.

4. LYSIMACHIA L. - *Loosestrifes*

Perennials; leaves opposite or whorled, simple; calyx divided nearly to base; corolla-lobes 5-7, patent to ± erect, longer than tube, yellow; capsule opening by 5 valves.

1 Stems procumbent or decumbent; flowers borne singly in axils of normal leaves 2
1 Stems erect; at least some main leaf-axils with >1 flower, or flowers borne in terminal or axillary racemes 3
 2 Leaves obtuse to rounded at apex, dotted with usually black glands; calyx-lobes ovate. Stems to 60cm, procumbent. Native; damp places, often in shade; throughout most of BI but often intrd *Creeping-Jenny* - **L. nummularia** L.
 2 Leaves acute to subacute at apex, not glandular; calyx-lobes subulate to linear. Stems to 40cm, procumbent to decumbent. Native; woods and copses; throughout most of BI
Yellow Pimpernel - **L. nemorum** L.
3 Flowers all borne in axils of much reduced bracts in terminal raceme; bulbils up to 2cm x 2mm usually produced in leaf-axils late in season. Stems to 80cm, erect. Intrd-natd; damp places on shore of Lake Windermere (Westmorland) *Lake Loosestrife* - **L. terrestris** (L.) Britton
3 At least lower flower clusters or racemes borne in axils of ± normal leaves; bulbils 0 4
 4 Flowers <10mm across, borne in dense axillary racemes; petals linear. Stems to 70cm, erect. Native; wet places in marshes and by ditches and canals; scattered in N En and C & S Sc, E Donegal, rarely natd in C & S En *Tufted Loosestrife* - **L. thyrsiflora** L.
 4 Flowers >15mm across, borne in few-flowered axillary clusters or

         in terminal panicles 5
5 Pedicels >2cm; calyx and leaves glabrous. Stems to 1.2m, erect.
   Intrd-natd; rough ground and damp or shady places; scattered in
   Br mainly N   *Fringed Loosestrife* - **L. ciliata** L.
5 Pedicels ≤2cm; calyx glandular-pubescent; leaves pubescent 6
   6 Corolla-lobes glandular-pubescent at margins; calyx-teeth
      uniformly green. Stems to 1.2m, erect. Intrd-natd; rough ground
      and damp places; common   *Dotted Loosestrife* - **L. punctata** L.
   6 Corolla-lobes glabrous at margins; calyx-teeth with conspicuous
      orange margin. Stems to 1.5m, erect. Native; ditches, marshes and
      by lakes and rivers; scattered through most of BI except N Sc
                                  Yellow Loosestrife - **L. vulgaris** L.

5. TRIENTALIS L. - *Chickweed-wintergreens*
Glabrous perennials; leaves mostly in an apparent whorl at apex of stem, simple; calyx divided almost to base; corolla-lobes (5)6-7(9), erecto-patent, longer than tube, white; capsule opening by 5 valves.

1 Stems to 20(25)cm, simple, erect. Native; on humus in open pine-woods and heather-moors; N Br, E Suffolk
                         Chickweed-wintergreen - **T. europaea** L.

6. ANAGALLIS L. - *Pimpernels*
Glabrous annuals or perennials; leaves opposite or alternate, simple; calyx divided ± to base; corolla-lobes 5, erect to patent, scarcely longer to much longer than tube; capsule opening by transverse line of dehiscence.

1 Upper leaves alternate; corolla divided <3/4 way to base, <2mm, white to pink, much shorter than calyx-lobes. Stems erect to decumbent, to 5(8)cm. Native; bare damp sandy ground on heaths and in woodland rides; scattered over most of BI
                      Chaffweed - **A. minima** (L.) E.H.L. Krause
1 Leaves all opposite; corolla divided almost to base, the lobes >3mm, c. as long as or longer than calyx-lobes 2
   2 Stems procumbent, rooting at nodes, to 20cm; corolla-lobes >2x as long as calyx-lobes, pale pink; leaves suborbicular. Native; bogs and damp peaty ground; scattered over BI, common in parts of W, absent from much of E   *Bog Pimpernel* - **A. tenella** (L.) L.
   2 Stems decumbent to ascending, not rooting at nodes; corolla-lobes <2x as long as calyx-lobes; leaves ovate. Stems to 40cm. Native; arable and waste land and open ground (**A. arvensis** L.) 3
3 Corolla-lobes usually red, sometimes variously pink or white or blue, entire to crenulate, with numerous minute hairs with 3 cells (incl. basal one), the most distal globose and glandular. Most of BI, common in S, rare in N Sc   *Scarlet Pimpernel* - **A. arvensis ssp. arvensis**
3 Corolla-lobes blue, crenulate to denticulate, with sparse minute hairs with 4 cells (incl. basal one), the most distal ellipsoid and glandular.

Much rarer than ssp. *arvensis*; usually in arable land; mostly C & S En
(*A. arvensis* ssp. *caerulea* Hartm., invalid name)
Blue Pimpernel - A. arvensis ssp. **foemina** (Mill.) Schinz & Thell.

7. GLAUX L. - *Sea-milkwort*
Glabrous slightly succulent rhizomatous perennials; leaves opposite, simple; calyx divided c.1/2 way to base into 5 lobes, white to pink; corolla 0; capsule opening by 5 valves.

1   Stems procumbent to suberect, to 30cm. Native; saline sandy, muddy, rocky or grassy places; round coasts of BI and in a few inland salt-marshes ***Sea-milkwort*** **- G. maritima** L.

8. SAMOLUS L. - *Brookweed*
Glabrous perennials; leaves in basal rosette and alternate up stem, simple; calyx-lobes ± free but fused to ovary; corolla-lobes 5, erect, longer than tube, white; capsule opening by 5 teeth.

1   Stems to 45cm, erect. Native; wet places, especially by streams and flushes near the sea; coasts of BI, rare inland except in E En
Brookweed **- S. valerandi** L.

72. PITTOSPORACEAE - *Pittosporum family*

Shrubs; easily recognized by the evergreen entire leaves, purple petals and 1-celled, 2-4-carpellary ovary.

1. PITTOSPORUM Gaertn. - *Pittosporums*

1   Leaf upperside dark green, lowerside white-tomentose, with revolute margin; flowers male and female mixed in terminal clusters; capsule with 3(-4) valves. Dense shrub or tree to 5(8)m; resembles *Olearia traversii* when sterile but leaves are alternate and 1st-year twigs ± terete. Intrd-natd; planted as screen or windbreak by sea; W Cornwall and Jersey ***Karo*** **- P. crassifolium** A. Cunn.
1   Leaves mid-green on both surfaces, ± glabrous when mature, with undulate margin; flowers bisexual, solitary, axillary; capsule with 2 valves. Shrub or tree to 5(10)m; resembles *Olearia paniculata* when sterile but latter has thicker leaves with white lowerside. Intrd-natd; planted for ornament or screen by sea; W Cornwall
Kohuhu **- P. tenuifolium** Gaertn.

73. HYDRANGEACEAE - *Mock-orange family*

Deciduous shrubs; the only opposite-leaved shrubs with free petals, 9-numerous stamens and 1/2-inferior or inferior ovary.

73. HYDRANGEACEAE

1 Flowers in corymbs, at least the outer ones with much-enlarged sepals and sterile **3. HYDRANGEA**
1 Flowers in racemose or paniculate cymes, all similar and fertile 2
  2 Petals and sepals 4; styles 4, united >1/2 way; stamens ≥20
      1. PHILADELPHUS
  2 Petals and sepals 5; styles usually 3, ± free; stamens 10 **2. DEUTZIA**

1. PHILADELPHUS L. - *Mock-oranges*

Flowers fragrant, in raceme-like cymes terminal on lateral shoots, all similar and fertile; sepals 4; petals 4 (or *flore pleno*), white or creamy-white; stamens c.20-30; ovary 4-celled with 4 styles united ≥1/2 way to apex; fruit a 4-valved capsule.

1 Leaves glabrous to sparsely pubescent on lowerside; calyx glabrous; flowers with ± patent petals. Shrub to 3m. Intrd-natd; relic in hedges and copses but rarely (ever?) self-sown; scattered in En, Man
    Mock-orange - **P. coronarius** L.
1 Leaves pubescent on lowerside; calyx pubescent; flowers cup-shaped. Shrub to 5m. Intrd-surv; relic in hedges etc., now commoner than *P. coronarius*; scattered in Br *Hairy Mock-orange* - **P. x virginalis** Rehder

2. DEUTZIA Thunb. - *Deutzia*

Flowers not fragrant, in raceme-like cymes terminal on lateral shoots, all similar and fertile; sepals 5; petals 5 (or *flore pleno*), white or tinged with pink; stamens 10; ovary 3-celled with 3 ± free styles; fruit a 3-valved capsule.

1 Shrub to 3m. Intrd-natd; persistent as relics and sometimes regenerating; scattered in Br N to C Sc *Deutzia* - **D. scabra** Thunb.

3. HYDRANGEA L. - *Hydrangeas*

Flowers in terminal corymbs, at least the outer ones with much-enlarged sepals and sterile; sepals 4-5; petals 4-5, usually varying shades of red to blue or white; stamens 9-20; ovary 2-4-celled with 2-4 short but free styles; fruit a capsule with 2-4 teeth.

1 Soft-wooded shrub to 1m. Intrd-surv; persistent where thrown out or neglected; scattered in S & W Br, CI
    Hydrangea - **H. macrophylla** (Thunb.) Ser.

74. GROSSULARIACEAE - *Gooseberry family*

Shrubs; distinguishable by shrubby habit, 5 sepals, petals and stamens arising from hypanthium, and inferior 1-celled ovary with 2 parietal placentas.

74. GROSSULARIACEAE

1 Leaves evergreen, not lobed; fruit a capsule; petals longer than sepals
1. ESCALLONIA
1 Leaves deciduous, palmately lobed; fruit a berry; petals shorter than sepals **2. RIBES**

1. ESCALLONIA L.f. - *Escallonia*
Leaves evergreen, simple, serrate; flowers in terminal racemes or panicles; petals much longer than calyx, red, with distinct claw; fruit a capsule.

1 Shrub to 3m (rarely more); young growth glandular-sticky. Intrd-natd; planted for hedging and ornament near sea; persistent relic in SW En, Wa, Man, W Ir and CI, also Dunbarton, rarely self-sown
Escallonia - **E. macrantha** Hook. & Arn.

2. RIBES L. - *Gooseberries*
Leaves deciduous, palmately lobed, variously toothed; flowers solitary or in racemes, on short lateral branches; petals shorter than sepals, not forming a tube; fruit a berry.

1 Spines present on branches; flowers solitary or in short racemes of 2(-3). Spiny shrub to 1(-1.5)m. Probably native; hedges, scrub and open woods; most of BI, often only escape *Gooseberry* - **R. uva-crispa** L.
1 Spines 0; flowers in racemes of >4   2
  2 Flowers bright pink to red, bright yellow, or (rarely) white; hypanthium tubular, longer than wide   3
  2 Flowers green to yellowish-green, sometimes tinged purplish; hypanthium disk- to cup-shaped, wider than long   4
3 Flowers bright pink to red, rarely white; leaves pubescent, scented when crushed. Shrub to 2.5m. Intrd-natd; relic, sometimes self-sown, in hedges and scrub; scattered throughout BI
Flowering Currant - **R. sanguineum** Pursh
3 Flowers bright yellow; leaves glabrous, not scented. Shrub to 2.5m. Intrd-natd; relic or self-sown in hedgerows, roadsides and scrub; scattered in Br and Man   *Buffalo Currant* - **R. odoratum** H.L. Wendl.
  4 Leaves with sessile orange glands on lowerside, scented when crushed; fruit black. Shrub to 2m. Probably intrd-natd; woods, hedges and shady streamsides; throughout most of BI
Black Currant - **R. nigrum** L.
  4 Leaves with mostly stalked reddish glands, not scented; fruit red or rarely whitish   5
5 Dioecious; bracts >4mm. Shrub to 2m, sometimes pendent on rock-faces. Native; limestone woods, often on rocks or cliffs, also an escape in other shady places; native for certain only N Wa and N En, widespread as escape   *Mountain Currant* - **R. alpinum** L.
5 Flowers bisexual; bracts <2mm   6
  6 Hypanthium cup-shaped; anther-lobes contiguous. Shrub to 2m. Native; woods on limestone; very local in N En and Sc, rarely natd further S   *Downy Currant* - **R. spicatum** E. Robson

6 Hypanthium saucer-shaped; anther-lobes distinctly separated
by connective. Shrub to 2m. Probably intrd-natd; woods, hedges
and scrub; throughout most of BI *Red Currant* - **R. rubrum** L.

75. CRASSULACEAE - *Stonecrop family*

Annual to perennial herbs or rarely woody; easily recognized by the free (or ± free) carpels as many as sepals and petals, stamens as many or 2x as many as petals, and usually succulent leaves.

1 Petals fused to form tube for >1/2 their length; basal leaves peltate
  2. UMBILICUS
1 Petals free or fused only at base; leaves not peltate 2
  2 Stamens as many as petals; leaves opposite **1. CRASSULA**
  2 Stamens 2x as many as petals; leaves usually alternate or spiral 3
3 Flowers with 4-5 petals and sepals **5. SEDUM**
3 Flowers with 6 or more petals and sepals 4
  4 Leaves about as thick as wide **5. SEDUM**
  4 Leaves distinctly wider than thick, distinctly flat on upperside 5
5 Petals yellow **4. AEONIUM**
5 Petals dull pink to purplish **3. SEMPERVIVUM**

1. CRASSULA L. - *Pigmyweeds*

Aquatic or terrestrial annuals to perennials, glabrous or nearly so; leaves opposite, often fused in pairs at base, succulent or ± so, entire; flowers <5mm, 3-5-merous; petals free or ± so, white to pink; stamens as many as petals.

1 Flowers sessile or ± so (pedicels <1mm) 2
1 Flowers on ≥2mm pedicels 3
  2 Leaves 1-2mm; petals mostly 3, shorter than sepals. Procumbent
    to ascending, to 5cm. Native; open sandy or gravelly ground;
    S En, E Anglia, Notts, CI *Mossy Stonecrop* - **C. tillaea** Lest.-Garl.
  2 Leaves 3-5mm; petals 4, longer than sepals. Stems procumbent
    to decumbent, to 5cm. Probably native; muddy pool-margin,
    Westerness *Pigmyweed* - **C. aquatica** (L.) Schönland
3 Terrestrial; petals shorter than to c. as long as sepals; stems to 12cm,
  decumbent to ascending. Intrd-natd; weed in damp sandy bulbfields
  and tracksides; Scillies, occasional wool-alien in S En
  Scilly Pigmyweed - **C. decumbens** Thunb.
3 Aquatic or on mud; petals longer than sepals; stems to 30cm, trailing
  in water or ascending from it or decumbent in mud. Intrd-natd; in
  ponds; well natd in many places in S En and CI, scattered N to C Sc,
  Co Down *New Zealand Pigmyweed* - **C. helmsii** (Kirk) Cockayne

2. UMBILICUS DC. - *Navelwort*

Glabrous perennials; leaves alternate on stem and in basal rosette, succulent, the lower and basal ones peltate, crenate; flowers >5mm, 5-merous; petals fused >1/2 way from base; stamens 2x as many as petals, fused to corolla-tube.

1 Stem usually erect, to 30(50)cm. Native; rocks, walls and stony hedgebanks; frequent in Ir, CI and W Br N to C Sc, rare or absent in E & C Br *Navelwort* - **U. rupestris** (Salisb.) Dandy

3. SEMPERVIVUM L. - *House-leeks*

Glandular-pubescent perennials; leaves narrow, in dense basal rosette and alternate up stem, succulent, entire; flowering stems erect, arising from centre of mature rosette which then dies, with cymes of flowers at apex; flowers >5mm, 8-18(mostly 13)-merous; petals ± free, dull pink to purplish, narrow; stamens 2x as many as petals.

1 Rosettes mostly >3cm across; stems to 50cm; leaves 2-4cm, glabrous; flowers 15-30mm across. Intrd-natd; grown on wall-tops and roofs, rarely sand-dunes, very persistent but rarely well natd; scattered over Br *House-leek* - **S. tectorum** L.
1 Rosettes mostly <2cm across; stems to 12cm; leaves 0.7-1.2cm, pubescent and with apical tuft of long web-like hairs matted over rosette; flowers 12-20mm across. Intrd-surv; on barn roof and walls; 1 site in W Norfolk *Cobweb House-leek* - **S. arachnoideum** L.

4. AEONIUM Webb & Berthel. - *Aeonium*

Almost glabrous perennials, sometimes woody below; leaves in large dense rosette and alternate up stem, succulent, ± entire; flowers >5mm, 8-11-merous; petals ± free, yellow; stamens 2x as many as petals.

1 Leaf-rosettes near ground, saucer-shaped, to 50cm across; flowering stems to 80(120)cm; flowers 1-2cm across. Intrd-surv; on walls; persistent in Scillies *Aeonium* - **A. cuneatum** Webb & Berthel.

5. SEDUM L. - *Stonecrops*

Annuals or (usually) perennials; leaves alternate, sometimes crowded and ± in a rosette, succulent, entire or toothed; flowers >3mm, 4-9-merous; petals free, various in colour; stamens 2x (or c.2x) as many as petals.

1 Leaves ± flat and distinctly dorsiventral 2
1 Leaves ± terete, or rounded on lowerside and flattened on upperside 12
   2 Rhizome thick, succulent, scaly; flowers dioecious, usually 4-merous. Glabrous perennial; stems erect, to 35cm; petals greenish-yellow. Native; mountain rocks and sea cliffs; Wa, Ir, Sc, N En, rare escape elsewhere *Roseroot* - **S. rosea** (L.) Scop.
   2 Rhizome 0, or not scaly and non-succulent; flowers bisexual, 5(-6)-merous 3
3 Petals yellow 4
3 Petals pink to purplish-red, rarely white 7

    4   Leaves conspicuously white-bloomed, forming flat dense rosette at stem apex. Glabrous procumbent perennial with rooting stems to 15cm. Intrd-natd; garden throwout or relic; rare and scattered in En and Man   *Colorado Stonecrop* - **S. spathulifolium** Hook.
    4   Leaves not white-bloomed, not forming flat dense rosette at stem apex   5
5   Stems thin, annual. Glabrous perennial with ascending non-rooting stems to 30cm. Intrd-natd; garden relic or throwout; scattered in En
                    Kamchatka Stonecrop - **S. kamtschaticum** Mast.
5   Stems stout, woody at least at base   6
    6   Leaves mostly >4.5cm, 3-4x as long as wide; petals c.4x as long as wide. Bushy, glabrous, evergreen shrub to 75cm. Intrd-natd; on cliffs and banks; CI and S Devon
                    Greater Mexican-stonecrop - **S. praealtum** A. DC.
    6   Leaves mostly <4.5cm, c.2x as long as wide; petals c.3x as long as wide. Bushy, glabrous, evergreen shrub to 40cm. Intrd-natd; on banks; Guernsey and W Cornwall
                    Lesser Mexican-stonecrop - **S. confusum** Hemsl.
7   Stamens distinctly longer than petals and sepals. Glabrous perennial; stems erect, to 50cm. Intrd-natd; persistent relic or throwout; S & C En, natd in woodland in N Wilts *Butterfly Stonecrop* - **S. spectabile** Boreau
7   Stamens shorter than to c. as long as petals   8
    8   Petals 3-5mm; stems rooting only near base   9
    8   Petals 5-12mm; stems rooting along length   11
9   Leaves entire; non-flowering shoots procumbent; roots not tuberous. Glabrous perennial; stems ascending, to 25cm. Intrd-natd; garden relic or throwout; sporadic in S & C En
                    Love-restoring Stonecrop - **S. anacampseros** L.
9   Leaves toothed, non-flowering shoots ± erect; roots tuberous. Glabrous perennial; stems erect, to 60cm. Native; woods, hedgebanks and rocky places; local throughout most of Br, escape in Ir and parts of Br (*Orpine* - **S. telephium** L.)   10
    10 Follicles with groove on back; leaves usually sessile, tapering to ± truncate base   **S. telephium ssp. telephium**
    10 Follicles not grooved; leaves tapering to cuneate base, the lower often petiolate   **S. telephium ssp. fabaria** Syme
11 Petals mostly >8mm; leaves narrowed to base but scarcely petiolate; flowers mostly pedicellate. Glabrous perennial; stems decumbent or procumbent, rooting along length, to 20cm. Intrd-natd; very persistent as escape, relic or throwout; very scattered throughout BI   *Caucasian-stonecrop* - **S. spurium** M. Bieb.
11 Petals mostly <8mm; leaves with distinct petiole; flowers ± sessile. Like *S. spurium* but smaller. Intrd-natd; similar places to *S. spurium* but much rarer; S En and CI
                    Lesser Caucasian-stonecrop - **S. stoloniferum** S.G. Gmel.
    12 Petals yellow   13
    12 Petals white to pink or red   17
13 Most leaves >7mm, acute or apiculate; ripe follicles erect; flowers

	5-9-merous	14
13	Leaves <7mm, obtuse; ripe follicles ± patent; flowers 5-merous	16

    14 Sterile shoots with terminal tassel-like cluster of living leaves and persistent dead ones below; leaves flattened, abruptly apiculate; filaments and follicles smooth; sepals subacute to obtuse. Like *S. rupestre* but less robust. Native; rocks and screes, either dry in open or wet in woods; local in Wa and SW En, also natd as for *S. rupestre*   ***Rock Stonecrop* - S. forsterianum** Sm.

    14 Sterile shoots with long terminal region of living leaves; dead leaves not persistent; leaves subterete, acute to acuminate; base of filaments and inner side of follicles minutely papillose; sepals acute   15

15 Inflorescence erect in bud; leaves mostly >3mm wide. Like *S. rupestre* but stems to 50cm. Intrd-natd; dry sunny banks; by road in W Kent   ***Pale Stonecrop* - S. nicaeense** All.

15 Inflorescence pendent in bud; leaves mostly <3mm wide. Glabrous perennial; stems erect to ascending, to 35cm. Intrd-natd; on walls, rocks and stony banks; locally common over BI

    ***Reflexed Stonecrop* - S. rupestre** L.

    16 Leaves ovoid, broadest near base, with acrid taste when fresh. Glabrous perennial; stems procumbent, rooting, with ascending to erect flowering stems to 10cm. Native; walls, rocks, open grassland and maritime sand and shingle; throughout most of BI

    ***Biting Stonecrop* - S. acre** L.

    16 Leaves ± cylindrical, ± parallel-sided, not acrid. Like *S. acre* but flowering stems to 25cm. Intrd-natd; on walls and rocks; scattered in En and Wa   ***Tasteless Stonecrop* - S. sexangulare** L.

17	Leaves, pedicels and sepals with small glandular hairs	18
17	Plant glabrous	20

    18 Leaves mostly opposite. Glandular-pubescent perennial; stems ascending, rooting at base, to 10cm. Intrd-natd; on walls and rocks; very scattered in En, Wa and Ir

    ***Thick-leaved Stonecrop* - S. dasyphyllum** L.

    18 Leaves mostly alternate   19

19 Petals 5, pink; ripe follicles erect. Glandular-pubescent (rarely glabrous) biennial to perennial; stems erect to ascending, rooting near base, to 10(15)cm. Native; streamsides and stony flushes in hilly areas; N En, C & S Sc   ***Hairy Stonecrop* - S. villosum** L.

19 Petals mostly 6-7, white, with pink midrib; ripe follicles patent. Glandular-pubescent, usually perennial; stems decumbent to ± erect, to 10cm. Intrd-natd; on walls and stony ground; rare in S & C En

    ***Spanish Stonecrop* - S. hispanicum** L.

20	Sepals free to base	21
20	Sepals fused into short calyx-tube at base	22

21 Leaves ± ovoid; petals narrowly acute to acuminate. Glabrous perennial; stems procumbent, rooting, with ascending to erect flowering stems to 10cm. Native; rocks, sand and shingle; common in much of CI, Ir and W Br, very local and mainly coastal in C &

E Br   *English Stonecrop* - **S. anglicum** Huds.
21 Leaves semicylindric; petals obtuse to subacute (var. *glabratum*
    Rostrup)   (see couplet 19) **S. villosum**
   22 Petals c.2x as long as sepals, narrowly acute; inflorescence usually
      <20-flowered. Glabrous perennial; stems procumbent, rooting,
      with ascending to erect flowering stems to 25cm. Intrd-natd;
      garden outcast or escape; few places in En, Sc and Ir
                                   Least Stonecrop - **S. lydium** Boiss.
   22 Petals ≥3x as long as sepals, obtuse to subacute; inflorescence
      usually >20-flowered. Glabrous perennial; stems procumbent,
      rooting, with ascending to erect flowering stems to 20cm.
      Probably intrd-natd; walls, rocks and stony ground; scattered
      through most of BI   *White Stonecrop* - **S. album** L.

76. SAXIFRAGACEAE - *Saxifrage family*

Annual to perennial herbs, rarely woody at base; very variable in vegetative and floral characters, but distinguishable by the 2 carpels fused only at base or for varying distances to apex; *Parnassia* is distinct in its staminodes.

1 Leaves compound, or simple and divided >1/2 way to base   2
1 Leaves simple, divided <1/2 way to base   4
   2 Leaves (incl. petioles) <5cm; inflorescence few-flowered, <5cm
                                                           5. SAXIFRAGA
   2 Leaves >10cm; inflorescence many-flowered, >5cm   3
3 Leaves palmate, with 5-9 leaflets   **2. RODGERSIA**
3 Leaves ternate to pinnate, the main divisions ternate to pinnate
                                                                 1. ASTILBE
   4 Leaves peltate   **4. DARMERA**
   4 Leaves not peltate (petiole joining leaf-blade at edge)   5
5 Petals 0; sepals 4   **9. CHRYSOSPLENIUM**
5 Petals present; sepals 5   6
   6 Five large divided staminodes present, alternating with 5
       stamens; flowering stems with 1 flower and 1 leaf **10. PARNASSIA**
   6 Staminodes 0; flowering stems with >1 flower and/or >1 leaf   7
7 Stamens as many as sepals or fewer   8
7 Stamens 2x as many as sepals   9
   8 Petals 4, brown; stamens 3   **7. TOLMIEA**
   8 Petals 5, pink to red; stamens 5   **6. HEUCHERA**
9 Petals fringed with long narrow lobes   **8. TELLIMA**
9 Petals entire to minutely toothed   10
   10 Thick surface rhizome present; at least some leaves >10cm;
       petals pink to red   **3. BERGENIA**
   10 Rhizome 0 or thin; all leaves <10cm (if petals pink or red then
       leaves <1cm)   **5. SAXIFRAGA**

1. ASTILBE D. Don - *False-buck's-beards*

Perennials; leaves ternate to pinnate, the primary divisions ternate to pinnate; inflorescence a many-flowered terminal panicle; flowers bisexual or unisexual (dioecious to variously arranged), ± hypogynous; sepals 5; petals 0 or 5; stamens (5-)10; carpels 2(-3), ± free to united at base to form 2-celled ovary with axile placentation.

Often confused with *Spiraea* (always shrubs) or *Aruncus* (carpels 3, stamens >10) (both Rosaceae), but differs in floral details. Female plants have short sterile stamens.

1 Petals 0; plant usually >1m. Stems to 1.6m. Intrd-natd; by streams and tracks in forest plantations; Kintyre and Argyll
Tall False-buck's-beard - **A. rivularis** D. Don
1 Petals 5, longer than sepals; plant <1m 2
   2 Stems with short whitish to brown hairs; petals white to pale pink. Stems to 80cm. Intrd-natd; usually in damp places; very scattered in Br, mostly in N En and Sc
False-buck's-beard - **A. japonica** (C. Morren & Decne.) A. Gray
   2 Stems with dense long shaggy brown hairs; petals pink to red. Stems to 80cm. Intrd-natd; similar situations to *A. japonica*; very scattered in En and Sc (?*A. chinensis* (Maxim.) Franch. & Savat. x *A. japonica*) *Red False-buck's-beard* - **A. x arendsii** Arends

2. RODGERSIA A. Gray - *Rodgersia*

Perennials with short stout rhizome; leaves palmate with 5-9 simple leaflets; inflorescence a many-flowered terminal panicle; flowers bisexual, shallowly perigynous; sepals 5; petals 5, yellowish-white; stamens 10; carpels 2, united for most part to form 2-celled ovary with axile placentation.

1 Stems to 1.3m, erect, pubescent; leaves with long petiole and leaflets each up to 30cm; inflorescence up to 25cm. Intrd-natd; damp places by ponds and rivers, sometimes persistent and spreading vegetatively (rarely sets seed); very scattered in Br
Rodgersia - **R. podophylla** A. Gray

3. BERGENIA Moench - *Elephant-ears*

Glabrous perennials with stout scaly rhizome usually on soil surface; leaves large, simple, thick, serrate-crenate; inflorescence a many-flowered panicle on erect leafless stem; flowers perigynous with cup-shaped hypanthium; sepals 5; petals 5, pink; stamens 10; carpels 2, united only at base to form 2-celled ovary with axile placentation.

1 Leaves ovate to obovate, 6-20cm with petiole c.1/2 as long, cuneate to subcordate at base; flowering stem 10-40cm; flowers 15-25mm across. Intrd-natd; very persistent garden relic or throwout; scattered in Br and CI *Elephant-ears* - **B. crassifolia** (L.) Fritsch

4. DARMERA Post & Kuntze - *Indian-rhubarb*

Pubescent perennials with stout rhizome; leaves simple, peltate, palmately lobed and sharply serrate; inflorescence a ± corymbose panicle on erect leafless stem; flowers almost hypogynous; sepals 5; petals 5, pink to whitish; stamens 10; carpels 2, ± free.

1 Leaves 5-40cm across, on petioles up to 1m; flowering stem to 1(1.5)m; flowers 10-15mm across. Intrd-natd; damp places, natd where planted or outcast; scattered over Br and Ir
Indian-rhubarb - **D. peltata** (Benth.) Post & Kuntze

5. SAXIFRAGA L. - *Saxifrages*

Pubescent annuals or perennials; leaves simple to almost compound; flowers in simple or compound cymes, sometimes solitary; ovary superior to ± completely inferior; hypanthium ± 0; sepals 5; petals 5; stamens 10; carpels 2, fused at least at base to form 2-celled ovary with axile placentation.

1 Leaves opposite; petals purple. Procumbent mat-forming perennial; stems to 25cm. Native; damp mountain rocks and scree; local in Wa, N & W Ir, NW En and Sc *Purple Saxifrage* - **S. oppositifolia** L.
1 Leaves alternate (spiral); petals yellow or white 2
   2 Some or all flowers replaced by reddish bulbils. Perennial with basal leaf-rosette bearing axillary bulbils; stems to 15cm. Native; basic mountain rocks above 900m; very rare in C Sc
Drooping Saxifrage - **S. cernua** L.
   2 Bulbils in inflorescence 0 3
3 Petals bright yellow 4
3 Petals white to cream 6
   4 Annuals; leaves ± orbicular, abruptly contracted to petiole. Stems decumbent to suberect, to 20cm. Intrd-natd; weed of shady places; scattered in Br, Co Antrim
Celandine Saxifrage - **S. cymbalaria** L.
   4 Perennials; leaves linear to oblanceolate, gradually contracted to petiole or ± sessile 5
5 Ovary superior; flowers 1(-3) per stem. Stoloniferous perennial; stems leafy, to 20cm. Native; wet places on moors; very local in N En, N Ir and S & C Sc *Marsh Saxifrage* - **S. hirculus** L.
5 Ovary semi-inferior; flowers usually >3 per stem. Perennial, with sterile and fertile decumbent to ascending stems to 25cm. Native; wet rocks and streamsides in mountains, down to sea-level on dunes in N Sc; locally common in N En and N & C Sc, rare in N Ir
Yellow Saxifrage - **S. aizoides** L.
   6 Flowers strongly zygomorphic, the lower petals >2x as long as 3 upper ones. Perennial with long thin stolons producing new plants at apex; panicles on leafless stems to 60cm. Intrd-natd; on shady walls; Cornwall *Strawberry Saxifrage* - **S. stolonifera** Curtis
   6 Flowers actinomorphic or ± so 7

7 Ovary superior, the sepals arising from underneath it   8
7 Ovary partly inferior, the sepals arising from its side or top   16
    8 Leaves ± sessile or with petiole <1/2 as long as blade. Stoloniferous perennial with basal leaf-rosette and ± leafless erect stem to 20cm. Native; wet rocks and stony places, in flushes and by streams in mountains; frequent in N Wa, N En and Sc, local in Ir   *Starry Saxifrage* - **S. stellaris** L.
    8 Basal leaves with petiole almost as long as to longer than blade   9
9 Stems leafy; sepals erecto-patent. Perennial with basal leaf-rosette and erect leafy stem to 40cm. Intrd-natd; by shady streams; N En and Sc   *Round-leaved Saxifrage* - **S. rotundifolia** L.
9 Stems leafless; sepals reflexed   10
    10 Petals without red spots; leaf-blades entire for ≥ basal 1/3 of margin. Habit of *S. hirsuta* but stem to 30cm. Intrd-natd; on rocks and old walls; very scattered in N & W Br
       Lesser Londonpride - **S. cuneifolia** L.
    10 Petals usually with small red spots; leaf-blades toothed for ≥ apical 3/4 of margin   11
11 Petioles subterete; leaf-blades sparsely pubescent over both surfaces, cordate at base. Stoloniferous perennial with basal leaf-rosette and leafless erect stem to 40cm. Native; damp rocks in mountains; locally common in N & S Kerry and W Cork, natd in S & C Sc, N En and Man   *Kidney Saxifrage* - **S. hirsuta** L.
11 Petioles distinctly flattened; leaf-blades glabrous or nearly so at least on lowerside   12
    12 Petioles densely pubescent on lateral margins; glabrous on upperside. Habit of *S. hirsuta*. Intrd-natd; on shady limestone rocks; rare from N En to C Sc   *Pyrenean Saxifrage* - **S. umbrosa** L.
    12 Petioles ± glabrous to rather sparsely pubescent   13
13 Leaf-blades with acute teeth; petioles subglabrous. Habit of *S. hirsuta*. Native; damp rocks in mountains; locally common in Ir, mostly W & SW   *St Patrick's-cabbage* - **S. spathularis** Brot.
13 Leaf-blades with subacute to rounded teeth, and/or petioles distinctly pubescent on margins   14
    14 Petioles usually c. as long as to slightly longer than leaf-blades; leaf-blades glabrous. Habit of *S. hirsuta*. Intrd-natd; natd in waste places, in woods, by streams and on walls and rocks; throughout BI (*S. umbrosa* x *S. spathularis*) **Londonpride** - **S. x urbium** D.A. Webb
    14 Petioles usually much longer than leaf-blade; leaf-blades sparsely pubescent to glabrous   15
15 Leaves mostly c. as long as wide, with scarcely visible translucent border (<0.2mm) and rather sharp teeth. Habit of *S. hirsuta*. Native; with or without parents SW & W Ir, also natd garden escape in S Wa, N En and C Sc (*S. spathularis* x *S. hirsuta*)
       False Londonpride - **S. x polita** (Haw.) Link
15 Many leaves distinctly longer than wide, with conspicuous translucent border (≥0.2mm) and low, blunt teeth. Habit of *S. hirsuta*.

Intrd-natd; in shady and damp, often rocky places; W & N Br
(*S. umbrosa* x *S. hirsuta*) **Scarce Londonpride - S. x geum** L.

16 Leaves all in basal rosette or just reduced ones on stem 17
16 Normal leaves present on stem 18
17 Leaves sessile, finely serrate, encrusted with lime on margins. Shortly stoloniferous perennial with ± hemispherical basal leaf-rosettes and erect stems to 40cm. Intrd-surv; in crack in limestone scar in MW Yorks **Livelong Saxifrage - S. paniculata** Mill.
17 Leaves petiolate, crenate-serrate, not lime-encrusted. Perennial with basal leaf-rosette and leafless erect stem to 15cm. Native; mountain rocks and cliffs; very local in N Wa, NW En, NW Ir and Sc
Alpine Saxifrage - S. nivalis L.
   18 Annual, without perennating organs. Stems erect, to 10(16)cm. Native; bare dry ground on walls, rocks and sand, mostly calcareous; locally common **Rue-leaved Saxifrage - S. tridactylites** L.
   18 Perennial, with sterile rosettes, stolons, rhizomes or basal bulbils 19
19 Basal bulbils usually present; stolons terminating in leafy rosettes 0 20
19 Bulbils 0; procumbent stolons terminating in dense leafy rosettes present 21
   20 Flowers mostly >3 per stem; petals >(6)9mm; basal leaves mostly ≥7-lobed. Perennial with basal leaf-rosette bearing axillary bulbils; stems erect, leafy, to 50cm. Native; moist base-rich grassland; locally common in most of Br, very rare in E Ir, natd garden escape elsewhere **Meadow Saxifrage - S. granulata** L.
   20 Flowers 1-3 per stem; petals <6mm; basal leaves 3-7-lobed. Stoloniferous perennial usually with basal bulbils; stems ascending, to 12cm, with few leaves. Native; wet mountain rocks above 900m; rare in C & N Sc **Highland Saxifrage - S. rivularis** L.
21 Leaf-lobes acuminate to narrowly acute, apiculate to aristate; flower-buds ± pendent. Stoloniferous laxly mat-forming perennial with leaf-rosettes and erect nearly leafless stems to 20cm. Native; damp rock-ledges, boulders and dunes and by mountain streams and flushes; locally common in N & W Br, very local in Ir, rare garden escape elsewhere **Mossy Saxifrage - S. hypnoides** L.
21 Leaf-lobes rounded, obtuse or acute, shortly mucronate or not; flower-buds erect 22
   22 Leaf-lobes rounded, obtuse or subacute; petals dull creamy- or greenish-white. Stoloniferous cushion-forming perennial with basal leaf-rosettes and erect nearly leafless stems to 10cm. Native; mountain rocks above 600m; rare in C Sc and N Wa
Tufted Saxifrage - S. cespitosa L.
   22 Leaf-lobes subacute to acute; petals pure white. Stoloniferous cushion- or mat-forming perennial with leaf-rosettes and erect nearly leafless stems to 20cm. Native; damp cliffs, rocks and streamsides on mountains (***Irish Saxifrage* - S. rosacea** Moench) 23
23 Leaves with glandular and non-glandular hairs, many of which exceed 0.5mm. Locally common in S & W Ir, 1 place in Caerns; natd in N Somerset **S. rosacea ssp. rosacea**

23 Leaves with hairs all glandular and <0.5mm; more robust. Arranmore
Island (W Donegal) **S. rosacea ssp. hartii** (D.A. Webb) D.A. Webb

S. hypnoides x *S. rosacea* is intermediate in habit and leaf characters and occurs with the parents in S Tipperary and Co Clare; some persistent garden escapes in Br might be of the same parentage. Certain examples of *S. hypnoides* in N Wa also suggest past hybridization with *S. rosacea*, and its hybrid with *S. tridactylites* was once found in MW Yorks.

6. HEUCHERA L. - *Coralbells*
Pubescent perennials; leaves all basal, palmately lobed, serrate; inflorescence a panicle on erect leafless stem; flowers bright pinkish-red, 1/2- or more-inferior with bell-shaped hypanthium above; sepals 5; petals 5; stamens 5; carpels 2, fused to form 1-celled ovary with parietal placentation.

1 Leaves 2-6cm, broadly ovate to orbicular, cordate, on petioles 7-15cm; flowering stem to 50cm. Intrd-surv; on waste ground and tips; scattered in SE En *Coralbells* - **H. sanguinea** Engelm.

7. TOLMIEA Torr. & A. Gray - *Pick-a-back-plant*
Pubescent perennials; leaves mostly basal, palmately lobed, serrate, producing plantlets at junction with petiole in moist conditions; inflorescence a simple raceme on ± leafy stem; flowers brown, zygomorphic, perigynous, with tubular hypanthium; sepals 5; petals 4(-5), filiform, the lowest usually missing; stamens 3, opposite the 3 upper sepals; carpels 2, fused for most part to form 1-celled ovary with parietal placentation.

1 Leaves 4-10cm, broadly ovate to ± orbicular, cordate, on petioles 10-30cm; flowering stems to 70cm. Intrd-natd; in damp shady places and on tips and waste ground; scattered through most of Br
Pick-a-back-plant - **T. menziesii** (Pursh) Torr. & A. Gray

8. TELLIMA R. Br. - *Fringecups*
Pubescent perennials; leaves mostly basal, palmately lobed, serrate; inflorescence a simple raceme on ± leafy stem; flowers green, usually pink- or red-tinged, 1/4-1/2-inferior, with bell-shaped hypanthium above; sepals 5; petals 5, broad and fringed with long narrow lobes; stamens 10; carpels 2, fused for most part to form 1-celled ovary with parietal placentation.

1 Leaves 4-10cm, broadly ovate to orbicular, cordate, on petioles 5-20cm; flowering stems to 70cm. Intrd-natd; in woods and damp hedgerows; scattered through most of Br, Co Longford
Fringecups - **T. grandiflora** (Pursh) Lindl.

9. CHRYSOSPLENIUM L. - *Golden-saxifrages*
Sparsely pubescent perennials with procumbent sterile and erect leafy flowering stems; leaves orbicular, crenate, petiolate; inflorescence of dichotomous subcorymbose cymes; flowers golden-yellow, epigynous;

hypanthium 0; sepals 4; petals 0; stamens 8; carpels 2, fused to form 1-celled ovary with parietal placentation.

1 Sterile shoots leafy; flowering stems to 15cm; leaves opposite, cuneate to rounded at base, with petiole up to as long as blade. Native; wet places by streams, in flushes and boggy woods, on mountain ledges; throughout BI *Opposite-leaved Golden-saxifrage* - **C. oppositifolium** L.
1 Sterile shoots with only scale-leaves; flowering stems to 20cm; leaves alternate, cordate at base, the lowest with petiole much longer than blade. Native; similar places to *C. oppositifolium*, often with it; local over most of Br *Alternate-leaved Golden-saxifrage* - **C. alternifolium** L.

10. PARNASSIA L. - *Grass-of-Parnassus*
Glabrous perennials; leaves mostly basal, simple, entire, petiolate; flower solitary, terminal, hypogynous; sepals 5; petals 5, white; stamens 5, alternating with 5 large much divided staminodes; carpels 4, fused to form 1-celled ovary with parietal placentation at least below.

1 Flowering stems to 30cm, often much less, with 1 sessile leaf; flowers 15-30mm across. Native; marshes, damp grassland and dune-slacks; local in much of Br and Ir, rare in S *Grass-of-Parnassus* - **P. palustris** L.

77. ROSACEAE - *Rose family*

Trees, shrubs or annual to perennial herbs; extremely variable in most characters, but usually recognizable by the alternate stipulate leaves (stipules often present only in young state or on leader shoots in woody spp.); the perigynous or epigynous flowers with usually 5 free petals, numerous stamens, and 1-many free or fused 2-ovuled carpels each with a separate style; and the 1-2-seeded fruits often aggregated into false-fruits. Exceptions to virtually all the above occur.

General key
1 Herbs; stems annual, sometimes woody at base *Key C*
1 Trees or shrubs; stems woody, biennial to perennial 2
   2 Flowers hypogynous to perigynous; hypanthium variously developed and often enclosing carpels, but not fused with them; fruit dry (but surrounded by succulent hypanthium in *Rosa*) or 1-several drupes *Key A*
   2 Flowers epigynous; hypanthium completely enclosing carpels, becoming succulent or ± so and fused with them at fruiting *Key B*

Key A - Woody plants with hypogynous to perigynous flowers
1 Fruits a cluster of small drupes on strongly convex receptacle with ± flat hypanthium around **8. RUBUS**
1 Fruits dry (but surrounded by succulent hypanthium in *Rosa*), or if drupes then hypanthium cup-shaped (and drupe usually 1) 2

	2 Leaves simple	3
	2 Leaves pinnate	9
3	Petals yellow	**7. KERRIA**
3	Petals white to bright pink	4
	4 Petals >6; fruit terminated by long feathery appendage (style)	**14. DRYAS**
	4 Petals normally 5; fruit without feathery appendage	5
5	Fruit 1-5 succulent drupes	6
5	Fruit dry achenes, legumes or follicles	7
	6 Carpel, style and drupe 1; leaves serrate to crenate	**22. PRUNUS**
	6 Carpels and styles 5; drupes 1-5; leaves entire	**23. OEMLERIA**
7	Carpels with 1-2 ovules; fruits indehiscent, 1-seeded (achene)	**5. HOLODISCUS**
7	Carpels with >2 ovules; fruits dehiscent, >1-seeded (follicle or legume)	8
	8 Stipules 0; carpels free; fruitss not inflated, dehiscing along 1 side (follicle)	**3. SPIRAEA**
	8 Stipules present; carpels fused at base; fruits inflated, dehiscing along 2 sides (legume)	**2. PHYSOCARPUS**
9	Stems spiny; fruits enclosed in succulent hypanthium	**21. ROSA**
9	Stems not spiny; fruits dry, not enclosed in hypanthium	10
	10 Flowers in dense globose heads; petals 0; hypanthium usually with spines at maturity	**18. ACAENA**
	10 Flowers not in dense globose heads; petals present; hypanthium never spiny	11
11	Flowers <15mm across, many in each group; petals white; leaflets >9; fruit follicles	**1. SORBARIA**
11	Flowers >15mm across, 1-few in each group; petals yellow; leaflets <9; fruit achenes	**9. POTENTILLA**

Key B - Woody plants with epigynous flowers

1	Walls of carpels (within hypanthium) becoming stony at fruiting	2
1	Walls of carpels (within hypanthium) becoming cartilaginous at fruiting	5
	2 Flowers ≥2cm across; sepals ≥1cm; fruit ≥2cm, brown	**34. MESPILUS**
	2 Flowers <2cm across; sepals <1cm; fruit <2cm, orange to red or purple to black	3
3	Stems not spiny; leaves entire	**32. COTONEASTER**
3	Stems usually spiny; leaves toothed or lobed	4
	4 Leaves evergreen; stipules minute, falling early	**33. PYRACANTHA**
	4 Leaves deciduous; stipules persistent at least on leading shoots	**35. CRATAEGUS**
5	Carpels with ≥4 ovules, later usually with >2 seeds	6
5	Carpels with 1-2 ovules and seeds	7
	6 Leaves entire, dull green on upperside, pubescent on lowerside; styles free	**24. CYDONIA**
	6 Leaves serrate, shiny green on upperside, glabrous; styles fused below	**25. CHAENOMELES**

7	Flowers in racemes	**30. AMELANCHIER**
7	Flowers in umbels or corymbs	8
	8 Inflorescence a simple umbel or corymb, rarely producing >3 fruits	9
	8 Inflorescence a branched corymb, usually producing >3 fruits	10
9	Flesh (hypanthium) of fruit with gritty groups of stone-cells; styles free; fruit often pear-shaped	**26. PYRUS**
9	Flesh of fruit without stone-cells; styles fused at base; fruit usually apple-shaped	**27. MALUS**
	10 Leaves entire, evergreen; hypanthium not enclosing apex of carpels, easily separated from carpels at maturity	**31. PHOTINIA**
	10 Leaves serrate to pinnate, deciduous; hypanthium wholly enclosing and strongly adherent to carpels at maturity	11
11	Styles 2-4(5); leaves simple or pinnate, the main veins running straight to leaf margin	**30. SORBUS**
11	Styles 5; leaves simple, the main veins curved near apex and running parallel to (not reaching) leaf margin	**29. ARONIA**

Key C - Herbaceous plants

1	Petals 0	2
1	Petals present	6
	2 Leaves simple or palmate	3
	2 Leaves pinnate	5
3	Annuals; flowers in leaf-opposed clusters; stamens 1(-2)	**20. APHANES**
3	Perennials; flowers in terminal clusters; stamens 4-5(10)	4
	4 Carpel 1; leaves palmate or palmately lobed	**19. ALCHEMILLA**
	4 Carpels 5-12; leaves ternate	**10. SIBBALDIA**
5	Hypanthium usually with 4 spines at apex; stamens 2; plant with at least some procumbent stems	**18. ACAENA**
5	Hypanthium not spiny; stamens 4-many; stems erect	**17. SANGUISORBA**
	6 Petals >6	**14. DRYAS**
	6 Petals 4-6	7
7	Epicalyx 0	8
7	Calyx-like epicalyx present behind true sepals	11
	8 Petals yellow; fruit with hooked bristles	**15. AGRIMONIA**
	8 Petals white to red or purple; fruit without bristles	9
9	Leaves 2-3-pinnate; stipules 0 ; carpels 3, several-seeded (or plant male)	**4. ARUNCUS**
9	Leaves simple, 1-pinnate or 1-ternate; stipules present; carpels >5, 1-seeded (or plant male)	10
	10 Flowers numerous in inflorescence; fruit a head of achenes	**6. FILIPENDULA**
	10 Flowers 1-6(10) in inflorescence; fruit a head of drupes	**8. RUBUS**
11	Leaves pinnate	12
11	Leaves palmate or ternate	14
	12 Carpels and achenes enclosed in hypanthium; stamens 5-10	**16. AREMONIA**

12 Hypanthium not covering carpels and achenes; stamens
10-numerous 13
13 Styles strongly hooked, persistent in fruit **13. GEUM**
13 Styles not hooked, deciduous before fruiting **9. POTENTILLA**
  14 Receptacle becoming red and succulent in fruit (or not forming
  fruits) 15
  14 Receptacle remaining dry in fruit 16
15 Petals yellow; epicalyx segments serrate at apex **12. DUCHESNEA**
15 Petals white to pinkish; epicalyx segments entire **12. FRAGARIA**
  16 Petals ≤2mm; leaflets with (usually) 3 teeth at extreme apex
  10. SIBBALDIA
  16 Petals ≥(1.5)2mm, if <2.5mm then leaflets serrated round apical
  ≥1/2 of margin **9. POTENTILLA**

SUBFAMILY 1 - SPIRAEOIDEAE (genera 1-5). Stipules present or 0; shrubs, sometimes herbs; hypanthium cup- or saucer-shaped, not enclosing carpels and fused to them only at base; epicalyx 0; stamens >10; flowers 5-merous; carpels 3-5, often >2-seeded; fruit dry, a group of follicles or sometimes of achenes or legumes or a capsule.

1. SORBARIA (DC.) A. Braun - *Sorbarias*

Deciduous shrubs; leaves pinnate with ≥11 leaflets, stipulate; flowers crowded in large terminal panicles; petals white; carpels 5; fruit several-seeded follicles (dehiscing along 1 margin). Spp. often mis-identified.

1 Panicle rather dense; follicles pubescent, on erect stalks. Stems to
  2.5m, ± erect. Intrd-natd; extensively suckering, on walls and waste
  ground; SE En *Sorbaria* - **S. sorbifolia** (L.) A. Braun
1 Panicle rather diffuse; follicles glabrous, on recurved stalks 2
  2 Styles arising from apex of carpel; longest stamens c. as long as
  petals. Stems to 6m, spreading. Intrd-natd; on walls and in scrub;
  C & S Sc, C & S En
  Himalayan Sorbaria - **S. tomentosa** (Lindl.) Rehder
  2 Style arising from well below apex of carpel; longest stamens
  nearly 2x as long as petals. Stems to 6m, spreading. Intrd-natd;
  on walls and in scrub; SE En, Jersey
  Chinese Sorbaria - **S. kirilowii** (Regel) Maxim.

2. PHYSOCARPUS (Cambess.) Raf. - *Ninebark*

Deciduous shrubs; leaves simple, stipulate when young; flowers crowded in ± corymbose racemes; petals white to pinkish; carpels (3)4-5; fruit 2- to several-seeded inflated legumes (dehiscing along both margins).

1 Stems to 2(3)m, erect to arching. Intrd-natd; relic of cultivation in
  shrubberies, rough ground and by streams; C & N En and Sc,
  Tyrone *Ninebark* - **P. opulifolius** (L.) Maxim.

3. SPIRAEA L. - *Brideworts*

Deciduous shrubs; leaves simple, without stipules; flowers in dense or fairly dense panicles or corymbs; petals white to pinkish-purple; carpels 5; fruit several-seeded follicles (dehiscing along 1 margin).

1 Inflorescence terminating each leafy shoot corymbose (hemispherical to flat-topped, at least as wide as long) 2
1 Inflorescence terminating each leafy shoot cylindrical to conical, longer than wide, sometimes comprised of a number of lateral corymbs 5
  2 Inflorescences simple corymbs; petals white 3
  2 At least the larger inflorescences compound corymbs 4
3 Leaves cuneate at base, obovate, at least some 3(-5)-lobed distally; petals longer than stamens. Stems arching, to 2m. Intrd-natd; in hedges, etc.; scattered in En and Man (*S. cantoniensis* Lour. x *S. trilobata* L.)   *Van Houtte's Spiraea* - **S. x vanhouttei** (Briot) Carrière
3 Leaves rounded at base, ovate, not distinctly lobed; petals shorter than stamens. Stems arching, to 2m. Intrd-natd; in river gorge; Angus   *Elm-leaved Spiraea* - **S. chamaedryfolia** L.
  4 Petals white; branchlets strongly angled; leaves mostly <2cm; stamens <1.2x as long as petals; follicles pubescent. Stems arching, to 2m. Intrd-natd; in scrub; occasional in S Br
        Himalayan Spiraea - **S. canescens** D. Don
  4 Petals pink; branchlets ± terete; leaves mostly >2cm; stamens >1.2x as long as petals; follicles glabrous. Stems ± erect, to 1.5m. Intrd-natd; in hedges, etc.; occasional in Br, Man and CI
        Japanese Spiraea - **S. japonica** L. f.
5 Petals much longer than stamens, white. Stems arching, to 2m. Intrd-natd; in shrubberies, rough ground and by paths and roads; scattered in En and Sc, Man (*S. thunbergii* Blume x *S. x multiflora* Zabel)   *Bridal-spray* - **S. x arguta** Zabel
5 Petals shorter than stamens or ± as long, pink or white 6
  6 Leaves glabrous or nearly so when mature 7
  6 Leaves tomentose to slightly pubescent on lowerside even when mature 10
7 Inflorescence axis glabrous; leaves serrate only in apical 1/2; sepals reflexed in fruit. Stems arching, to 2m; petals white. Intrd-natd; by road and in open woodland; Westmorland
    Russian Spiraea - **S. media** Schmidt
7 Inflorescence axis pubescent; leaves serrate for most of length; sepals erect in fruit 8
  8 Panicles broadly conical, with long branches near base; leaves mostly widest above mid-way; petals usually white; stamens c. as long as petals. Habit as for *S. salicifolia*. Intrd-natd; in hedges and rough ground; frequent in Sc, Man, N Ir and probably N En   *Pale Bridewort* - **S. alba** Du Roi
  8 Panicles ± cylindrical, with short branches; leaves mostly widest below mid-way; petals usually pink; stamens usually distinctly

| | longer than petals | 9 |

9 Panicle-branches usually pubescent; leaves usually lanceolate; petals usually bright pink; pollen >90 per cent fertile. Strongly suckering; stems ± erect, to 2m. Intrd-natd; in hedges and rough ground; throughout BI but over-recorded **Bridewort - S. salicifolia** L.

9 Panicle-branches usually sparsely pubescent; leaves usually narrowly ovate; petals usually pale pink; pollen <20% fertile. Habit as for *S. salicifolia*. Intrd-natd; in hedges and rough ground; throughout Br (*S. salicifolia* x *S. alba*) **Intermediate Bridewort - S. x rosalba** Dippel

  10 Panicles usually both terminal and lateral, the terminal ones broadly conical, c. as long as wide; leaves of flowering shoots mostly <4cm. Stems ± erect, to 2m. Intrd-natd; in hedges and rough ground; Sc (*S. douglasii* x *S. canescens*)
Lange's Spiraea - S. x brachybotrys Lange

  10 Panicles terminal only, much longer than wide; leaves of flowering shoots mostly >4cm 11

11 Leaves tomentose on lowerside; petals pink 12
11 Leaves pubescent on lowerside, the leaf surface showing through 14

  12 Leaves ovate, grey- to buff-tomentose on lowerside, serrate for distal ≥2/3 of margin; follicles pubescent. Habit as for *S. salicifolia*. Intrd-natd; in hedges and rough ground; rare in En
Hardhack - S. tomentosa L.

  12 Leaves oblong, white- to pale grey-tomentose on lowerside, ± entire in proximal c.1/2 of margin; follicles glabrous. Habit as for *S. salicifolia*. Intrd-natd; in hedges and rough ground; commoner than *S. salicifolia* (**Steeple-bush - S. douglasii** Hook.) 13

13 Leaves whitish- or pale greyish-tomentose on lowerside. Common in Br, Man **S. douglasii** ssp. **douglasii**
13 Leaves subglabrous to pubescent on veins on lowerside. Rare
S. douglasii ssp. **menziesii** (Hook.) Calder & Roy L. Taylor

  14 Leaves oblong to narrowly so, ± entire in proximal c.1/2 of margin; pollen >90 per cent fertile (see couplet 12) **S. douglasii**

  14 Leaves ovate to elliptic to narrowly so, serrate for distal ≥2/3 of margin; pollen <20 per cent fertile 15

15 Panicles narrowly conical; petals often very pale pink. Habit as for *S. salicifolia*. Intrd-natd; in hedges and rough ground; rare in Br and Man (*S. alba* x *S. douglasii*) **Billard's Bridewort - S. x billardii** Hérincq

15 Panicles subcylindrical; petals pink. Habit as for *S. salicifolia*. Intrd-natd; in hedges and rough ground; throughout Br (*S. salicifolia* x *S. douglasii*) **Confused Bridewort - S. x pseudosalicifolia** Silverside

4. ARUNCUS L. - *Buck's-beard*

Herbaceous perennials; leaves 2-3-pinnate, without stipules; flowers ± dioecious, in very dense terminal panicles; petals white; carpels 3; fruit 2- to several-seeded follicles (dehiscing along 1 margin).

See *Astilbe* (Saxifragaceae) for differences.

1 Stems to 2m, erect; basal leaves to 1m; inflorescence up to 50cm.

Intrd-natd; very persistent in woodland and by water, but ± never seeding; scattered throughout En and Sc, especially N

Buck's-beard - **A. dioicus** (Walter) Fernald

5. HOLODISCUS (K. Koch) Maxim. - *Oceanspray*

Deciduous shrubs; leaves simple, without stipules; flowers in dense terminal panicles; petals creamy-white; carpels 5; fruit 1-seeded indehiscent achenes.

1 Stems to 4m, erect to arching; panicles arching to pendent, up to 30cm. Intrd-natd; in hedges and scrub and on walls; scattered in En and Sc *Oceanspray* - **H. discolor** (Pursh) Maxim.

SUBFAMILY 2 - ROSOIDEAE (genera 6-21). Stipulate herbs, sometimes shrubs; hypanthium variably concave, sometimes flat, not or sometimes enclosing carpels but fused to them only at base; epicalyx present or 0; stamens 1-numerous; flowers mostly 4-6-merous; carpels 1-numerous, 1-2-seeded; fruit a head of achenes or drupes, sometimes borne on succulent receptacle or surrounded by succulent or dry hypanthium.

6. FILIPENDULA Mill. - *Meadowsweets*

Herbaceous perennials; leaves pinnate or reduced to terminal lobe only; flowers in terminal ± flat-topped panicles, 5-6-merous; epicalyx 0; hypanthium ± flat to saucer-shaped; stamens numerous; carpels 4-12; fruit a head of achenes each with 1-2 seeds.

1 Basal leaves with terminal and 8-30 pairs of main leaflets all 0.5-2cm; petals usually 6. Stems erect, to 50(100)cm; petals creamy-white. Native; calcareous grassland; locally frequent in Br N to C Sc, sometimes natd elsewhere *Dropwort* - **F. vulgaris** Moench
1 Basal leaves with terminal and 0-5 pairs of main leaflets all >2cm; petals usually 5 2
  2 Carpels 6-10, spirally twisted, glabrous; leaves with 2-5 pairs of large lateral leaflets. Stems erect, to 1.2m; petals white. Native; all sorts of wet and damp places; common

Meadowsweet - **F. ulmaria** (L.) Maxim.
  2 Carpels 4-6, straight, pubescent on edges; leaves without large lateral leaflets. Stems erect, to 3m; petals white, pink or red. Intrd-natd; in damp places; very scattered in Br, especially Sc and N En *Giant Meadowsweet* - **F. kamtschatica** (Pall.) Maxim.

7. KERRIA DC. - *Kerria*

Deciduous shrubs; leaves simple; flowers solitary, terminal on lateral branches, 5-merous, yellow, often *flore pleno*; epicalyx 0; hypanthium ± flat; stamens numerous; carpels 5-8; fruit a head of achenes.

1 Stems to 2.5m, erect, bright green above. Intrd-natd; persistent in neglected shrubberies and old garden sites, self-sown in Middlesex;

very scattered in Br, Man   *Kerria* - **K. japonica** (L.) DC.

8. RUBUS L. - *Brambles*

Deciduous or semi-evergreen shrubs, often spiny, or herbaceous perennials; leaves simple, ternate, pinnate or palmate; flowers solitary or in racemes or panicles, usually 5-merous; epicalyx 0; hypanthium flat, with receptacle usually extended upwards from centre; stamens numerous; carpels usually numerous; fruit a head of (1)2-many 1-seeded drupes.

Most of the taxa of subg. *Rubus* form an extremely complex, largely apomictic group (sect. *Glandulosus*), often known collectively as **R. fruticosus** L. agg. Two other sections (*Rubus* and *Corylifolii*) are often included within this aggregate, but they are probably derived from ancient and some recent hybrids between it and *R. idaeus* and *R. caesius* respectively, and are here treated separately. *R. caesius* forms a 4th section. 330 microspp. are currently recognized in BI in these 4 sections together. In this work the microspp. of sects. *Rubus*, *Corylifolii* and *Glandulosus* are not treated in full but 11 rather ill-defined series, representing the main nodes in the spectrum of variation, are recognized in sect. *Glandulosus*. These are keyed out in couplets 15-24, but most of the characters used are relative and a high level of success will be achieved only after much experience.

1	Leaves simple	2
1	Leaves pinnate, palmate or ternate	5
	2 Stipules arising direct from stem; receptacle strongly convex	3
	2 Stipules fused to petiole proximally; receptacle flat	4
3	Stems herbaceous, erect, to 20cm; flowers solitary, dioecious. Fruit orange. Native; peaty moors and bogs on mountains; Br N from N Wa and Derbys, Tyrone *Cloudberry* - **R. chamaemorus** L.	
3	Stems arching or procumbent, trailing to several m, rooting at tips; flowers in clusters, bisexual. Intrd-natd; spreading in hedges and shrubberies; very scattered in En *Chinese Bramble* - **R. tricolor** Focke	
	4 Petals white. Stems to 2m, erect; fruit red. Intrd-natd; in rough ground; very scattered in Br, N Kerry *Thimbleberry* - **R. parviflorus** Nutt.	
	4 Petals reddish-purple. Stems to 3m, erect; fruit red. Intrd-natd; in rough ground; very scattered in S En *Purple-flowered Raspberry* - **R. odoratus** L.	
5	Stipules arising direct from stem; stems all annual, producing flowers in 1st year; receptacle flat	6
5	Stipules fused to petiole proximally; stems at least biennial, producing flowers in 2nd year; receptacle conical	7
	6 Stoloniferous, without rhizomes; petals <6mm, white. Flowering stems to 40cm, erect; fruit red. Native; woods, screes and mountain slopes on basic soils; rather scattered in Ir, Sc, Wa and C, N & W En *Stone Bramble* - **R. saxatilis** L.	
	6 Rhizomatous, without stolons; petals >6mm, pink. Stems to 30cm, erect; fruit dark red. Probably native; several records from highlands of Sc, the last in 1841 *Arctic Bramble* - **R. arcticus** L.	

7 Stems densely covered with white bloom, to 5m, erect and arching. Flowers purplish; fruit black. Intrd-natd; in hedges and rough ground; very scattered in En and Sc
White-stemmed Bramble - **R. cockburnianus** Hemsl.
7 Stems green to red, sometimes glaucous but not with dense white bloom 8
   8 Stems usually not rooting at tips; leaves ternate or pinnate; fruit usually red to orange, usually separating from receptacle at maturity 9
   8 Stems usually rooting at tips; leaves ternate or palmate; fruit usually black, sometimes red, coming away with extension of receptacle at maturity. *R. ulmifolius* is a sexual diploid with 2n=14, but all the other 329 microspp. are polyploid facultative apomicts. Abundant and characteristic in much of C & S BI, but much less common in Sc and not native in Orkney or Shetland (**R. subg. Rubus**, with 4 sections) 12
9 At least lower leaves pinnate with 5(-7) leaflets; petals white 10
9 Leaves normally ternate; petals pink to purple 11
   10 Stems ± erect, to 1.5(2.5)m; fruit separating from receptacle at maturity, red, rarely yellow or white; stem with weak prickles. Native; woods, heaths and marginal ground; frequent
Raspberry - **R. idaeus** L.
   10 Stems arching and eventually rooting at tips, to several m; fruit coming away with extension of receptacle at maturity, dark purplish-red; stem with moderate prickles. Intrd-natd; bird-sown or escaped in hedges and waste places; scattered throughout Br and CI *Loganberry* - **R. loganobaccus** L.H. Bailey
11 Leaves ± glabrous; flowers >2cm across, 1-few per group, appearing in spring; glandular hairs 0. Stems to 2m, ± erect; fruit orange. Intrd-natd; in woods and hedgerows; scattered throughout Br, Man and Ir
Salmonberry - **R. spectabilis** Pursh
11 Leaves white-tomentose on lowerside; flowers <1.5cm across, many per group, appearing in summer; glandular bristles dense on stems and flower-stalks. Stems to 2(3)m, erect and spreading; fruit red. Intrd-natd; in rough places and scrub; scattered in S Br, Lanarks
Japanese Wineberry - **R. phoenicolasius** Maxim.
   12 Fruit with glaucous bloom, composed of rather few, large, only loosely coherent drupes; leaflets 3, the lateral ones ± sessile. Flowers few, in corymbs, white, large (2-3cm across). Native; disturbed ground, grassland, scrub and sand-dunes, often on clayey or basic soils; common throughout C & S BI but local in W and in Sc (**R. subg. Rubus sect. Caesii** Lej. & Courtois, with 1 sp.) *Dewberry* - **R. caesius** L.
   12 Fruit without glaucous bloom, usually composed of many tightly coherent drupes; leaflets often >3 13
13 Leaflets ± overlapping, the basal pair ± sessile; stipules lanceolate; inflorescence usually a ± simple corymb. Variously intermediate between *R. caesius* and *R. fruticosus* agg. Native; open places; 24

microspp. almost throughout BI **R. subg. Rubus sect. Corylifolii** Lindl.
13 Leaflets mostly ± not overlapping, the basal pair usually stalked; stipules linear; inflorescence compound or ± racemose 14
  14 Stems suberect, usually not rooting at tips; suckers often produced from roots; fruits often red to purple, sometimes black. Variously intermediate between *R. idaeus* and *R. fruticosus* agg. Native; sunny and partly shaded places; 20 microspp. almost throughout BI, mostly En and Wa **R. subg. Rubus sect. Rubus**
  14 Stems procumbent to arching, usually rooting at tips; suckers not produced; fruits black. Native; all sorts of habitats both natural and man-made, but much less common on calcareous soils; 277 microspp. almost throughout BI
(**R. subg. Rubus sect. Glandulosus** Wimm. & Grab., with 11 series) (***Bramble*** - **R. fruticosus** L., agg.) 15
15 Stalked glands 0 or rare and inconspicuous on 1st-year stems 16
15 Stalked glands on 1st-year stems obvious 18
  16 Leaflets variously pubescent on lowerside but not, or only the upper ones, tomentose. 60 microspp. almost throughout BI. The alien **R. laciniatus** Willd., with distinctive dissected leaves, is grown for its fruits and is frequently natd over most of BI
R. sect. Glandulosus series Sylvatici (P.J. Müll.) Focke
  16 Leaflets all tomentose on lowerside 17
17 Leaflets greyish-white-tomentose on lowerside; few stalked glands sometimes present in inflorescence. 42 microspp. almost throughout BI **R. sect. Glandulosus series Rhamnifolii** (Bab.) Focke
17 Leaflets chalky-white-tomentose on lowerside; stalked glands 0. 11 microspp. almost throughout BI, incl. the alien **R. armeniacus** Focke, much grown for its fruit as **'Himalayan Giant'**. R. ulmifolius Schott is the only diploid and only wholly sexual sp. of subg. *Rubus* in BI. It is widespread, especially on chalk and clay where few other spp. occur, and hence is the commonest sp. in several areas incl. CI. It is easily recognized by its stems with a whitish bloom and very short pubescence but otherwise only large subequal prickles, rather small elliptic to obovate leaflets white-tomentose on lowerside, flowers produced relatively late (from Aug) in long narrow inflorescences and with rounded pink petals, and small fruit
R. sect. Glandulosus series Discolores (P.J. Müll.) Focke
  18 Stamens shorter than styles; stalked glands usually few. 4 microspp. in Br N to SW Sc, S Ir, CI
R. sect. Glandulosus series Sprengeliani Focke
  18 Stamens as long as or longer than styles; stalked glands usually many 19
19 Main prickles ± confined to angles of stem, distinct from smaller pricklets and acicles 20
19 Main prickles occurring all round stem, grading into pricklets and acicles 23
  20 Stems conspicuously pubescent, the stalked glands less

conspicuous. 22 microspp. throughout BI except most of Sc
R. sect. Glandulosus R. series Vestiti (Focke) Focke
- 20 Stems glabrous to pubescent, the stalked glands more conspicuous than hairs 21
- 21 Terminal leaflet obovate, often broadly so, with short cuspidate apex and serrulate margin. 11 microspp. throughout most of Br and Ir **R. sect. Glandulosus series Mucronati** (Focke) H.E. Weber
- 21 Terminal leaflet usually ovate to elliptic, usually less abruptly narrowed at apex, usually more serrate on margin 22
    - 22 Pricklets, acicles and stalked glands on stems of similar lengths. 43 microspp. almost throughout BI
    R. sect. Glandulosus R. series Radulae (Focke) Focke
    - 22 Pricklets, acicles and stalked glands on stems of different lengths. 30 microspp. local in En, Wa and CI, very local in Ir and Sc
    R. sect. Glandulosus R. series Micantes Bouvet
- 23 Prickles and stalked glands in very variable quantities on same plant. 20 microspp. almost throughout BI
R. sect. Glandulosus series Anisacanthi H.E. Weber
- 23 Prickles and stalked glands not markedly variable in quantity on same plant 24
    - 24 Prickles strong; pricklets often more numerous than stalked glands. 35 microspp. almost throughout Br, Ir and Man
    R. sect. Glandulosus series Hystrices Focke
    - 24 Prickles weak; stalked glands and acicles often very numerous. 7 microspp. scattered in C & S Br, 1 N En, 1 in Ir
    R. sect. Glandulosus series Glandulosi (Wimm. & Grab.) Focke

Hybrids within sect. *Rubus* are common, especially involving the wholly sexual *R. ulmifolius*; some of these are fertile but others sterile. Sections *Rubus* and *Corylifolii* probably arose from hybrids between *R. fruticosus* agg. and *R. idaeus* or *R. caesius* respectively, and more recent hybrids with these two parentages are not rare, again either fertile or sterile. In addition, *R. idaeus* x *R. caesius* = **R. x pseudoidaeus** (Weihe) Lej. is very sparsely scattered in En and in N Tipperary; it resembles *R. caesius* in habit and stem characters and *R. idaeus* in leaf characters. It is largely sterile, with undeveloped fruits or partially developed reddish-black ones.

9. POTENTILLA L. - *Cinquefoils*
Herbaceous (annuals to) perennials or rarely deciduous shrubs; leaves pinnate, ternate or palmate, or upper ones simple; flowers solitary or few in cymes, (4-)5-merous; epicalyx present; hypanthium flat to saucer-shaped, with receptacle slightly to strongly convex; stamens ≥(5)10; carpels 4-numerous; fruit a head of achenes.

- 1 Lower leaves pinnate 2
- 1 Lower leaves ternate or palmate 5
    - 2 Shrub; leaflets entire; achenes pubescent. Deciduous erect or spreading shrub to 1m. Native; rock-ledges, river- and lake-

	margins in full sun; extremely local in 2 areas of N En and in W Ir, also garden relic ***Shrubby Cinquefoil* - P. fruticosa** L.
2	Herb; leaflets toothed; achenes glabrous 3
3	Petals purple; plant with long woody rhizome. Herbaceous perennial with ascending stems to 50cm. Native; fens, marshes and bogs; common over most of Bl but very local in S & C Br and CI ***Marsh Cinquefoil* - P. palustris** (L.) Scop.
3	Petals yellow or white; plant without long rhizome 4
	4 Petals white; flowers in terminal cyme. Perennial with erect flowering stems to 60m arising from leaf-rosette. Native; on basic rocks; Monts, Rads and E Sutherland, also occasional escape ***Rock Cinquefoil* - P. rupestris** L.
	4 Petals yellow; flowers solitary, axillary. Perennial with long procumbent stolons and terminal leaf-rosettes. Native; waste places, waysides, pastures and sand-dunes; common ***Silverweed* - P. anserina** L.
5	Petals white; achenes pubescent on 1 side. Perennial with procumbent stolons and terminal leaf-rosettes. Native; wood-margins and clearings, scrub and hedgebanks; common ***Barren Strawberry* - P. sterilis** (L.) Garcke
5	Petals yellow; achenes glabrous (but receptacle often pubescent) 6
	6 Leaves grey- to white-tomentose on lowerside 7
	6 Leaf-surface visible through pubescence on lowerside 8
7	Leaves grey-tomentose on lowerside, with flat margin; upper part of stem with mixed straight and woolly hairs; petals 5-7mm. Perennial with erect to ascending stems to 50cm arising from leaf-rosette. Intrd-casual; waste places; occasional in Br, mainly SE En ***Grey Cinquefoil* - P. inclinata** Vill.
7	Leaves white-tomentose on lowerside, with revolute margin visible from lowerside as narrow green edge; upper part of stem with dense woolly hairs; petals 4-5mm. Perennial with decumbent to ascending stems to 30cm arising from leaf-rosette. Native; sandy grassland and waste ground; local in CI and Br N to C Sc, mainly E En ***Hoary Cinquefoil* - P. argentea** L.
	8 At least some flowers with 4 petals and sepals 9
	8 All flowers with 5 petals and sepals 12
9	Plant highly sterile (with 0-few achenes per flower); petioles of stem-leaves >1cm, all ± same length. Native; frequent, and commoner than *P. anglica* (*P. erecta* x *P. reptans* and *P. anglica* x *P. reptans*) ***Hybrid Cinquefoil* - P. x mixta** Rchb.
9	Plant fertile (many achenes per flower); stem-leaves ± sessile to petiolate, if latter petioles decreasing markedly in size towards stem apex 10
	10 Carpels >20, lower stem-leaves with stalks >10mm; some leaves with 4-5 leaflets; stems rooting at nodes late in season; some flowers 5-merous. Perennial, with persistent basal leaf-rosette and decumbent to procumbent stems to 80cm. Native; wood-

borders, heaths and dry banks; scattered over most of BI
Trailing Tormentil - **P. anglica** Laichard.
- 10 Carpels <20; stem-leaves sessile or with stalks <5mm; all or nearly all leaves ternate (ignore stipules); stems not rooting at nodes; flowers ± all 4-merous. Perennial with basal leaf-rosette (often withered by flowering) and erect to procumbent flowering stems to 45cm. Native; grassland and dwarf-scrub on heaths, moors, bogs, mountains, roadsides and pastures, mostly on acid soils but sometimes on limestone (*Tormentil* - **P. erecta** (L.) Raeusch.) 11
- 11 Stems to 25cm; stem-leaves serrate in distal 1/2 only, with teeth <1.5mm, the uppermost leaf c.6-16mm; petals 2.5-4.5mm; fruiting pedicels 6-30mm. Common, mostly in lowlands **P. erecta ssp. erecta**
- 11 Stems 15-45cm; stem-leaves serrate for most of length, with teeth >1.5mm, the uppermost leaf 12-30mm; petals 4-6mm; fruiting pedicels (12)20-50mm. Upland areas of BI N from S Devon and S Ir **P. erecta ssp. strictissima** (Zimmeter) A.J. Richards
  - 12 Flowers solitary in leaf-axils; main stems procumbent and rooting at nodes. Perennial with persistent basal leaf-rosette and procumbent flowering stems to 1m. Native; rough ground, hedgebanks, sand-dunes and open grassland; common in BI N to S Sc, very local further N *Creeping Cinquefoil* - **P. reptans** L.
  - 12 Flowers (often few) in terminal cymes; main stems not rooting at nodes 13
- 13 Flowering stem arising laterally from side of terminal leaf-rosette, usually <2mm wide, with only reduced (usually simple) leaves 14
- 13 Flowering stem arising from centre of leaf-rosette (the latter often withered by flowering-time), usually >2mm wide, with several well- developed leaves 15
  - 14 Vegetative stems long, procumbent, often rooting, mat-forming; free part of stipules of basal leaves linear-triangular; flowers mostly <l5mm across. Perennial with terminal leaf-rosettes and ascending flowering stems to 10cm. Native; dry basic grassland and rocky slopes; very local in Br

  Spring Cinquefoil - **P. neumanniana** Rchb.
  - 14 Vegetative stems short, not rooting or mat-forming; free part of stipules of basal leaves narrowly ovate; flowers mostly >15mm across. Perennial with terminal leaf-rosettes and ascending flowering stems to 20cm. Native; sparse basic grassland, rocky places and crevices on mountains; very local in Sc, N Wa and N En S to Derbys *Alpine Cinquefoil* - **P. crantzii** (Crantz) Fritsch
- 15 Petals >6mm, longer than sepals. Perennial with erect stems to 10cm arising from leaf-rosette. Intrd-natd; waste ground, banks and grassy places; scattered over C & S En and CI *Sulphur Cinquefoil* - **P. recta** L.
- 15 Petals <5mm, shorter than to as long as sepals 16
  - 16 Petals c.1/2 as long as sepals; stamens 5-10; achenes <0.8mm, smooth. Annual to biennial with erect to ascending stems to

50cm. Intrd-natd; by pool; Salop
Brook Cinquefoil - **P. rivalis** Torr. & A. Gray
16 Petals c.3/4-1x as long as sepals; stamens c.20; achenes >0.8mm, minutely rugose 17
17 Leaves ± all ternate; epicalyx-segments longer than sepals in fruit. Annual to short-lived perennial with erect to ascending stems to 50cm. Intrd-natd; casual or sometimes natd in waste places; very scattered in En and Wa, rare casual in Sc and N Ir
Ternate-leaved Cinquefoil - **P. norvegica** L.
17 Most lower leaves with 5 leaflets; epicalyx-segments shorter than sepals in fruit. Biennial or perennial with erect to ascending stems to 50cm arising from leaf-rosette. Intrd-natd; waste and grassy places, rarely natd; scattered in En *Russian Cinquefoil* - **P. intermedia** L.

P. erecta x *P. anglica* = **P. x suberecta** Zimmeter occurs frequently with the parents in BI. It resembles *P. erecta* in habit but the stems may rarely root at nodes late in the season and it is intermediate in leaflet- and petal-number and petiole-length and is partially fertile. *P. crantzii* x *P. neumanniana* = **P. x beckii** Murr might be the identity of intermediate plants in N En and Sc.

10. SIBBALDIA L. - *Sibbaldia*
Herbaceous perennials; differ from *Potentilla* in leaves ternate; flowers 5-merous, in compact heads; petals ≤2mm or 0; stamens (4)5(-10); carpels 5-12.

1 Leaves in basal rosette with obovate-obtriangular leaflets mostly with 3 apical teeth; flowering stems 1-5cm, procumbent to ascending. Native; grassy and rocky slopes and rock-crevices above 470m; frequent in C & N Sc *Sibbaldia* - **S. procumbens** L.

11. FRAGARIA L. - *Strawberries*
Herbaceous perennials, usually stoloniferous; leaves ternate; flowers in cymes on ± leafless stems arising from axils of leaf-rosette, 5-merous; epicalyx present, with entire segments; hypanthium ± flat, with strongly convex receptacle; petals white or flushed pink; stamens and carpels numerous (sometimes flowers ± dioecious); fruit a head of achenes borne on outside of enlarged, red, succulent receptacle.

1 Leaflets glabrous or nearly so on upperside; most fruiting heads >15mm wide, usually with sepals appressed; achenes sunk into surface of ripe receptacle. Flowering stems about as long as rosette-leaves, to 30cm. Intrd-natd; frequent in waste places and field-borders; scattered throughout BI
Garden Strawberry - **F. ananassa** (Duchesne) Duchesne
1 Leaflets pubescent to sparsely so on upperside; fruiting heads <15mm wide, with sepals not appressed; achenes prominent from surface of ripe receptacle 2
   2 Flowers bisexual; uppermost pedicel in each cyme with apically

directed hairs at fruiting; leaves rather glossy on upperside when fresh. Flowering stems about as long as rosette-leaves, to 30cm. Native; woods, scrub and hedgerows; common
Wild Strawberry - **F. vesca** L.

2 Flowers functionally dioecious; uppermost pedicel in each cyme with many patent hairs at fruiting; leaves dull on upperside when fresh. Flowering stems longer than leaves, to 40cm. Intrd-natd; scrub and hedgerows; rare throughout Br (*F. muricata* auct.)
Hautbois Strawberry - **F. moschata** (Duchesne) Weston

12. DUCHESNEA Sm. - *Yellow-flowered Strawberry*

Stoloniferous perennials; differ from *Fragaria* in solitary flowers; epicalyx-segments 3-toothed at apex, much larger than sepals; petals yellow.

1 Stolons to 50cm, slender, bearing solitary axillary flowers on erect pedicels 3-10cm; fruit ± globose, 8-16mm across, not juicy, tasteless. Intrd-natd; shady places; very scattered in Br
Yellow-flowered Strawberry - **D. indica** (Jacks.) Focke

13. GEUM L. - *Avens*

Herbaceous perennials; leaves pinnate; flowers in terminal cymes on stems arising from leaf-rosette, rarely solitary, 5(rarely more)-merous; epicalyx present; hypanthium ± flat to saucer-shaped, with strongly convex receptacle; petals yellow to purple; stamens and carpels numerous; fruit a head of achenes with long styles terminating in hook (after apical segment has fallen off).

1 Petals creamy-pink to purplish, with long claw at base; ripe achenes carried up from flower centre on stalk >5mm. Stems to 50cm. Native; marshes, streamsides, mountain rock-ledges and open woodland; throughout Br and Ir but very local in S, garden escape in non-native areas *Water Avens* - **G. rivale** L.
1 Petals yellow, scarcely clawed at base; ripe achenes ± sessile in flower centre 2
   2 Achenes <150, in globose head, the receptacle with dense hairs >1mm. Stems to 70cm. Native; woods and hedgerows; common almost throughout BI *Wood Avens* - **G. urbanum** L.
   2 Achenes >150, in ovoid head, the receptacle with sparse hairs <1mm. Stems to 1m. Intrd-natd; in woods and by paths; Cards and scattered in Sc *Large-leaved Avens* - **G. macrophyllum** Willd.

G. rivale x *G. urbanum* = *G. x intermedium* Ehrh. is common wherever the parents meet. It is intermediate in all respects and highly fertile, forming a complete spectrum between the parents.

14. DRYAS L. - *Mountain Avens*

Perennials, woody at base, herbaceous above; leaves ± evergreen, simple, bluntly serrate; flowers solitary, axillary, 7-10-merous; epicalyx 0;

hypanthium saucer-shaped, with convex receptacle; petals white, mostly 8; stamens and carpels numerous; fruit a head of achenes with long feathery styles.

1 Stems to 50cm, procumbent, rooting; leaves dark glossy green on upperside, white-tomentose on lowerside. Native; base-rich rock-crevices and -ledges on mountains; very local in NW Wa, N Ir and N En, local in MW Ir and N & W Sc   *Mountain Avens* - **D. octopetala** L.

15. AGRIMONIA L. - *Agrimonies*

Herbaceous perennials; leaves pinnate; flowers in long terminal racemes, 5-merous; epicalyx 0; hypanthium deeply concave, narrow at mouth, surrounding carpels; petals yellow; stamens 5-20; carpels 2; fruit of 1-2 achenes enclosed in woody hypanthium which has ring of hooked bristles distally.

Vegetative differences between the 2 spp. are often exaggerated; *A. eupatoria* is also fragrant when crushed and is often glandular on leaf lowersides, though glands may be concealed by denser pubescence.

1 Fruiting hypanthium obconical, deeply grooved ± to apex, with outermost bristles patent to erecto-patent. Stems to 1m. Native; grassy places in fields and hedgerows; throughout most of BI
Agrimony - **A. eupatoria** L.
1 Fruiting hypanthium bell-shaped, with grooves extending <3/4 way to apex and outermost bristles reflexed. Stems to 1m. Native; same habitats and distribution as *A. eupatoria* but much more scattered
Fragrant Agrimony - **A. procera** Wallr.

A. eupatoria x *A. procera* = **A. x wirtgenii** Asch. & Graebn. occurs rarely in En.

16. AREMONIA Nestl. - *Bastard Agrimony*

Herbaceous perennials; leaves pinnate; flowers in small terminal cymes, 5-merous; epicalyx of 5 lobes, surrounded by 8-12 fused bracts; hypanthium deeply concave, surrounding carpels; petals yellow; stamens 5-10; carpels 2; fruit of 1-2 achenes enclosed in woody pubescent hypanthium without hooked bristles but concealed by bracts.

1 Stems to 30cm, decumbent, scarcely or not longer than basal leaves; fruiting hypanthium 5-6mm, ± globose. Intrd-natd; woods and shady roadsides   *Bastard Agrimony* - **A. agrimonioides** (L.) DC.

17. SANGUISORBA L. - *Burnets*

Herbaceous perennials; leaves pinnate; flowers in very dense terminal spikes, all bisexual or bisexual and unisexual mixed; sepals 4; epicalyx 0; hypanthium deeply concave, surrounding carpels; petals 0; stamens 4 or numerous; carpels 1-2; stigmas papillate or tasselled; fruit 1-2 achenes enclosed in hard hypanthium.

1 Upper flowers female, others bisexual or male; stamens numerous;
    stigmas 2, tasselled. Stems to 50(80)cm (**S. minor** Scop.) 2
1 All flowers bisexual; stamens 4; stigma 1, papillate 3
    2 Fruiting hypanthium 3-4.5mm, with thickened but scarcely
      winged ridges on angles, the faces distinctly reticulate with finer
      ridges. Native; calcareous or sometimes neutral grassland and
      rocky places; locally common throughout BI N to C Sc
                              Salad Burnet - **S. minor ssp. minor**
    2 Usually more robust and leafy with more deeply and sharply
      toothed leaflets; Fruiting hypanthium 3-5mm, with often undulate
      wings on angles, the faces smooth to irregularly rugose. Intrd-
      natd; grassy places; very scattered in BI, frequent only in C &
      S En   *Fodder Burnet* - **S. minor ssp. muricata** (Gremli) Briq.
3 Flower-heads 1-3cm; sepals and stamens dull purplish; stamens
    c.4mm. Stems to 1.2(1.7)m. Native; damp unimproved grassland;
    locally frequent in Br N to C Sc, N Ir   *Great Burnet* - **S. officinalis** L.
3 Flower-heads 3-16cm; sepals green and white; stamens white,
    c.10mm. Stems to 2m. Intrd-natd; rough and marginal ground; very
    scattered in Sc and En, frequent in C Sc   *White Burnet* - **S. canadensis** L.

18. ACAENA L. - *Pirri-pirri-burs*

Perennials, with stems woody at base but herbaceous distally; leaves
pinnate; flowers in globose heads, bisexual; sepals 4; epicalyx 0;
hypanthium deeply concave, surrounding carpels; petals 0; stamens 2;
carpels 1-2; stigmas long-papillate; fruit 1-2 achenes enclosed in dry
hypanthium which usually develops barbed spines at apex.

1 Apical pair of leaflets c. as long as wide; spines on hypanthium 0, or
    imperfect, or not barbed. Habit as *A. novae-zelandiae* but herbaceous
    stems to 6cm. Intrd-natd; barish ground; very scattered in Sc, Carms
                              Spineless Acaena - **A. inermis** Hook. f.
1 Apical pair of leaflets 1.2-2.5x as long as wide; spines on hypanthium
    2-4, barbed at apex 2
    2 Apical pair of leaflets each with (11)17-23 teeth; hypanthium with
      2 spines. Habit as in *A. novae-zelandiae*. Intrd-natd; barish ground;
      very few places from S En to N Sc, SE Ir
                              Two-spined Acaena - **A. ovalifolia** Ruiz & Pav.
    2 Apical pair of leaflets each with 5-12(15) teeth; hypanthium with
      up to 4 spines 3
3 Leaflets glossy green on upperside; apical pair of leaflets 1.8-2.5x as
    long as wide. Woody stems procumbent, mat-forming; herbaceous
    stems ascending, to 15cm. Intrd-natd; barish ground; S & E Br N to
    SE Sc, very scattered in Ir   *Pirri-pirri-bur* - **A. novae-zelandiae** Kirk
3 Leaflets matt green on upperside, often edged and veined with
    brown; apical pair of leaflets 1.2-2x as long as wide. Habit as in *A.
    novae-zelandiae*. Intrd-natd; barish ground; very few places from
    S En to C Sc, N Ir
                Bronze Pirri-pirri-bur - **A. anserinifolia** (J.R. & G. Forst.) Druce

19. ALCHEMILLA L. - *Lady's-mantles*

Herbaceous perennials; leaves palmate or simple and palmately lobed; flowers in terminal compound cymes, bisexual; sepals 4; epicalyx present; hypanthium deeply concave, surrounding carpel; petals 0; stamens 4; carpel 1; fruit an achene enclosed in dry hypanthium.

All our spp. are obligate apomicts and many differ only by the small characters often diagnosing agamospecies. The most important characters are degree and distribution of pubescence; shape of leaves, especially number, shape and toothing of leaf-lobes; and shape of sinuses between leaf-lobes and at base of leaf-blade. The terms small, medium and large are relative and often help to distinguish taxa in 1 locality; in general 'small' means stems usually <20cm, leaves usually <3cm; 'medium' means stems usually <50cm, leaves usually <5cm; 'large' means stems up to 60(80)cm, leaves up to 7(10)cm. The terms leaves and petioles refer to those of the basal rosette, excluding the first-formed ones. Numbers of teeth refer to those of the mid-lobe of each leaf.

1 Leaves palmately divided >1/2 way to base, densely silver-silky-pubescent on lowerside 2
1 Leaves palmately divided <1/2 way to base, variously pubescent but never silver-silky-pubescent. Stems to 60(80)cm. Damp rich grassland, woodland margins and rides, rock-ledges; throughout most of Br and Ir but rare or absent in most of SE En (*Lady's-mantle* - **A. vulgaris** L. agg.) 3
   2 Leaves divided ± to base, the leaflets mostly <6mm wide. Stems to 20cm. Native; grassland, scree and rock-crevices on mountains; Lake District, C & N Sc, Kerry and Co Wicklow, intrd in Derbys *Alpine Lady's-mantle* - **A. alpina** L.
   2 Leaves divided 3/5-4/5 way to base, the lobes mostly >6mm wide. Stems to 30cm. Intrd-natd; walls and rocky places; scattered in N & C Br *Silver Lady's-mantle* - **A. conjuncta** Bab.
3 Petioles and stem glabrous, subglabrous or with appressed or subappressed hairs 4
3 Petioles and lower part of stem with erecto-patent, patent or reflexed hairs 7
   4 Stems up to and including 1st inflorescence-branches pubescent; leaves pubescent on upperside 5
   4 Stems pubescent only on lowest 2 or 3 internodes; leaves glabrous on upperside 6
5 Hypanthium tapered to cuneate base. Plant medium. Native; very local in S Northumb (*A. gracilis* auct.) **A. micans** Buser
5 Hypanthium rounded at base. Plant medium. Native; local in N En and Sc **A. glomerulans** Buser
   6 Sinuses between leaf-lobes toothed to base; teeth in middle of each side of leaf-lobes larger than those above or below. Plant large. Native; almost throughout range of agg., but almost absent from En S of Peak District **A. glabra** Neygenf.
   6 Sinuses between leaf-lobes with toothless region at base c.2x as

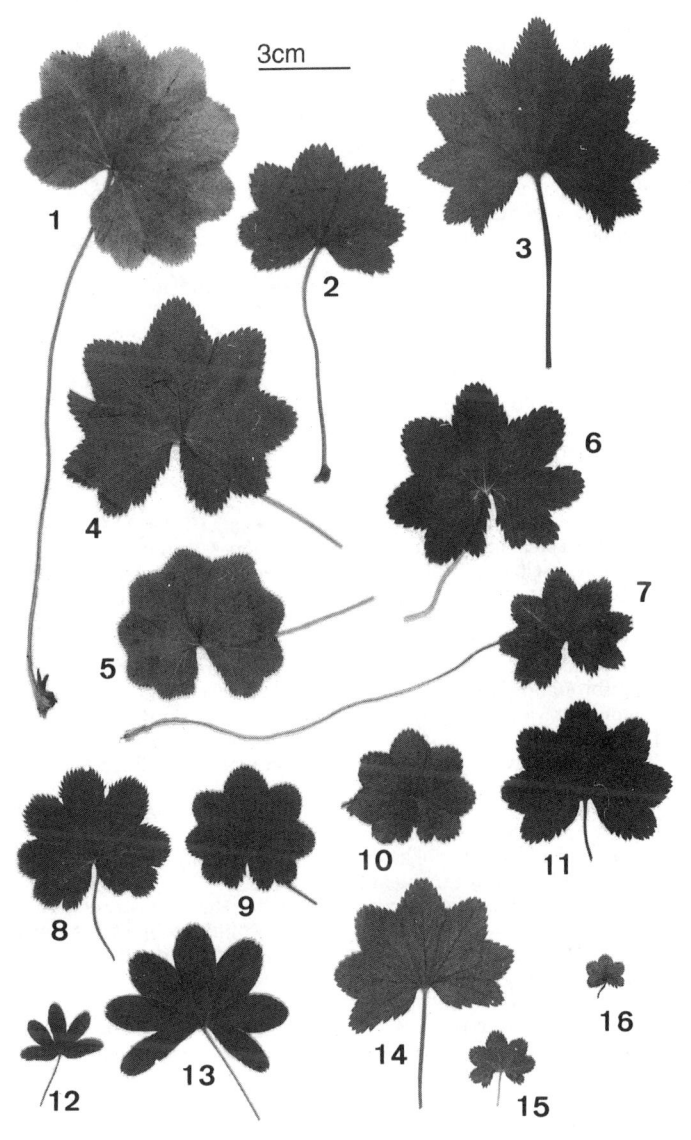

FIG 233 - Leaves of *Alchemilla*. 1, *A. mollis*. 2, *A. xanthochlora*. 3, *A. acutiloba*. 4, *A. glomerulans*. 5, *A. tytthantha*. 6, *A. subcrenata*. 7, *A. micans*. 8, *A. wichurae*. 9, *A. monticola*. 10, *A. glaucescens*. 11, *A. glabra*. 12, *A. alpina*. 13, *A. conjuncta*. 14-15, *A. filicaulis*. 16, *A. minima*.

　　　　long as a tooth; teeth on leaf-lobes subequal. Plant rather small.
　　　　Native; local in N En and Sc, intrd in NW Wa
　　　　　　　　　　　　　　　　　　　　　　A. wichurae (Buser) Stefánsson
7　Epicalyx-segments c. as long as sepals; hypanthium much shorter
　　than mature achene. Plant large to very large. Intrd-natd; very
　　scattered nearly throughout Br　　　　　　　**A. mollis** (Buser) Rothm.
7　Epicalyx-segments distinctly shorter than sepals; hypanthium as
　　long as mature achene　　　　　　　　　　　　　　　　　　　　8
　　　8　Pedicels and hypanthia both pubescent, sometimes sparsely so　9
　　　8　Pedicels glabrous; hypanthia usually glabrous　　　　　　　11
9　Leaf-lobes usually 5, with sinuses between them with toothless
　　region at base ≥2x as long as a tooth; stems rarely >5cm. Plant small.
　　Native; very local on 2 hills in MW Yorks　　　**A. minima** Walters
9　Leaf-lobes (5)7-9, with sinuses between them toothed to base; stems
　　usually >5cm　　　　　　　　　　　　　　　　　　　　　　10
　　　10　Base of stems and petioles tinged wine-red; leaf basal sinus open
　　　　 (≥45°). Plant small to medium. Native; distribution of agg., the
　　　　 commonest taxon in genus
　　　　　　　　　　　　A. filicaulis Buser **ssp. vestita** (Buser) M.E. Bradshaw
　　　10　Base of stems and petioles brownish; leaf basal sinus very narrow
　　　　 (<30°) to closed. Plant small. Native; very local in N En, NW Sc,
　　　　 NW Ir, rare escape elsewhere in En　　　**A. glaucescens** Wallr.
11　Leaves glabrous on upperside (or with very sparse hairs in folds)　12
11　Leaves pubescent on upperside, sometimes only in folds　　　　13
　　　12　Leaf-lobes rounded, with subequal teeth. Plant medium. Native;
　　　　 ± throughout range of agg., common in N, very local in SE En
　　　　 and S Ir　　　　　　　　　　　　　　**A. xanthochlora** Rothm.
　　　12　Leaf-lobes ± straight-sided then rounded to subtruncate at apex,
　　　　 with teeth in middle of each side larger than those above or
　　　　 below. Plant large. Native; quite widespread in Co Durham,
　　　　 natd in Lanarks　　　　　　　　　　　　　**A. acutiloba** Opiz
13　Leaves ± densely pubescent on both surfaces　　　　　　　　　14
13　Leaves rather sparsely pubescent on upperside, with hairs usually
　　± confined to folds (or frequent only there)　　　　　　　　　16
　　　14　Hypanthium tapered to cuneate base; stem-hairs mostly erecto-
　　　　 patent　　　　　　　　　　　　　(see couplet 5) **A. micans**
　　　14　Hypanthium rounded at base; stem-hairs patent to reflexed　15
15　Flowers mostly >2.5mm across; petioles and stems with patent hairs.
　　Plant medium. Native; very local in NW Yorks and Co Durham,
　　rare escape elsewhere in En　　　　　　　　**A. monticola** Opiz
15　Flowers mostly <2.5mm across; petioles and stems with many
　　reflexed hairs. Plant medium. Intrd-natd; C & S Sc　**A. tytthantha** Juz.
　　　16　Leaf-lobes ± straight-sided then rounded to subtruncate at apex,
　　　　 with teeth in middle of each side larger than those above or
　　　　 below　　　　　　　　　　　　(see couplet 12) **A. acutiloba**
　　　16　Leaf-lobes ± rounded, with teeth ± equal to unequal but not with
　　　　 large ones in middle of each side　　　　　　　　　　　17
17　Leaf-lobes usually 5, with sinuses between them with toothless

region at base ≥2x as long as a tooth; stems rarely >5cm
(see couplet 9) **A. minima**

17 Leaf-lobes (5)7-9, with sinuses between them toothed to base; stems usually >5cm 18

   18 Hypanthium often with some hairs; stems and petioles with patent hairs, tinged wine-red at base. Plant small to medium. Native; rather scattered from Wa and C En to N Sc, NW Ir
A. filicaulis Buser ssp. **filicaulis** 233

   18 Hypanthium glabrous; stems and petioles usually with some reflexed hairs, brownish at base. Plant medium. Native; very local in NW Yorks and Co Durham **A. subcrenata** Buser 233

20. APHANES L. - *Parsley-pierts*

Annuals; leaves deeply palmately lobed; flowers in small dense, leaf-opposed clusters, bisexual; sepals 4; epicalyx present; hypanthium deeply concave, surrounding carpel; petals 0; stamen 1(-2); carpel 1; fruit an achene enclosed in dry hypanthium.

1 Stipules at fruiting nodes fused into a leaf-like cup with ovate-triangular teeth at apex c.1/2 as long as entire portion; fruiting hypanthium 2-2.6mm incl. erect sepals c.0.6-0.8mm, with a slight constriction where hypanthium and sepals meet, reaching ± to apex of stipules at maturity. Native; cultivated and other bare ground on well-drained soils; frequent *Parsley-piert* - **A. arvensis** L.

1 Stipule-teeth at fruiting nodes ovate-oblong, c. as long as entire portion; fruiting hypanthium 1.4-1.9mm incl. convergent sepals c.0.3-0.5mm, the sepals continuing curved outline of hypanthium, falling well short of apex of stipules at maturity. Native; similar places to *A. arvensis* but rarely on base-rich soils and less often in arable ground; frequent (*A. inexspectata* W. Lippert)
Slender Parsley-piert - **A. australis** Rydb.

21. ROSA L. - *Roses*

Deciduous or rarely evergreen shrubs with spiny stems and petioles; leaves pinnate; flowers solitary or in few-(many-)flowered corymbs, usually 5-merous; epicalyx 0; hypanthium deeply concave, surrounding carpels, narrowed at apex; stamens and carpels numerous; fruit (actually a false fruit) a head of achenes enclosed by succulent hypanthium.

An extremely complex genus, much hybridized and selected in cultivation. Half of our spp. (sect. *Caninae* DC.: couplets 14-24, plus *R. ferruginea* and *R. stylosa*) contribute unbalanced gametes, the male ones having 7 chromosomes and the female ones 21, 28, 35 or 42 (from plants with 2n=28, 35, 42, 56 respectively). The ripe fruits and leaves provide the most important characters, but a collection from the same plant at flowering is often also desirable. *Disc* refers to the thickened rim at top of the hypanthium at fruiting, in centre of which is the *orifice* through which the styles (which are sometimes united into a *column*) project. Only the spp. are keyed here, but even for these access to accurately named material is often a

77. ROSACEAE

prerequisite for successful determination; in some areas hybrids are commoner than either parent, but can often be determined if the parents present in the area are known.

1 Styles exserted and fused into a column, sometimes becoming free at fruiting 2
1 Styles exserted or not, free (may appear fused in dried material) 6
    2 Leaflets 3(-5). Scrambler, to 5m. Intrd-natd; in scrub; Jersey and Guernsey **Prairie Rose - R. setigera** Michx.
    2 Leaflets 5-9 3
3 Styles pubescent; semi-evergreen; stems ± procumbent. Semi-evergreen, ± procumbent, to 4m. Intrd-natd; in open places, low scrub and beaches mostly near sea; very scattered in Br and CI
                                                Memorial Rose - R. luciae Crép.
3 Styles glabrous; deciduous; stems trailing to strongly arching 4
    4 Flowers 2-3cm across, in groups of >(6)10; stipules lobed >1/2 way to petiole. Scrambler, to 5m. Intrd-natd; in hedges and copses; scattered in CI and Br
                                         Many-flowered Rose - R. multiflora Thunb.
    4 Flowers mostly 3-5cm across, in groups of 1-6(10); stipules not lobed or lobed <1/2 way to petiole 5
5 Styles as long as stamens; top of hypanthium flat; inner sepals entire, outer with very few lobes. Stems weakly trailing, to 1(2)m, with rather few and slender curved prickles. Native; low scrub, hedgerows, woods and open places; frequent in En, Wa and much of Ir, very rare in Man and Sc **Field-rose - R. arvensis** Huds.
5 Styles shorter than stamens; top of hypanthium conical; sepals pinnately lobed. Stems arching, to 3(4)m, with hooked prickles. Native; hedges, scrub, wood-borders; scarce in Br and Ir S of a line from Co Dublin to E Suffolk **Short-styled Field-rose - R. stylosa** Desv.
    6 Sepals entire, ± tapering to apex, erect or suberect and persistent until after fruit ripe 7
    6 At least some sepals lobed or with strongly expanded tips, if ± entire then patent to reflexed and/or falling before fruit ripe 10
7 Fruit blackish when ripe; flowers all solitary, without bract. Strongly suckering; stems erect, to 50(100)cm, with numerous slender prickles and acicles. Native; dry sandy places near sea, on inland heaths and limestone; round most coasts of BI, very local inland
                                               Burnet Rose - R. pimpinellifolia L.
7 Fruit red when ripe; flowers 1-several, with 1 or more much reduced leaves (bracts) at base of pedicels 8
    8 Stems and prickles ± glabrous; flowers 3-5cm across; pedicels glandular-pubescent (see couplet 24) **R. mollis**
    8 Stems and prickles tomentose; flowers 6-9cm across; pedicels tomentose 9
9 Fruits 1.5-2.5cm, usually wider than long; leaflets bullate, rather shiny. Strongly suckering; stems erect, to 1.5(2)m, tomentose, with numerous tomentose prickles and acicles. Intrd-natd; on dunes,

rough ground and banks, often mass-planted; scattered through much of BI *Japanese Rose* - **R. rugosa** Thunb.

9 Fruits 0.8-1.5cm, usually longer than wide; leaflets scarcely bullate, matt. Differs from *R. rugosa* in stems less stout and less tomentose. Intrd-natd; hedges, roadsides and waste ground; scattered in most of Br, mainly Sc *Dutch Rose* - **R. 'Hollandica'**

  10 Stems with many mixed prickles, pricklets and acicles; flowers often solitary. Strongly suckering; stems erect, to 1.5m, with slender prickles, acicles and glandular hairs. Intrd-natd; scrub and hedges; very scattered in En, Man and Guernsey
Red Rose (of Lancaster) - **R. gallica** L.

  10 Stems without acicles; flowers usually >1 per branch   **11**

11 Flowers 6-8cm across, *flore pleno* to some degree, usually white. Stems erect, to 2m, with slightly curved prickles. Intrd-natd; in hedges and other marginal sites; SE Yorks, Kirkcudbrights and Man *White Rose (of York)* - **R. x alba** L.

11 Flowers (2)3-5(6)cm across, very rarely *flore pleno*   **12**

  12 Sepals entire or some with few very narrow lateral lobes, some or all falling before fruit ripe   **13**

  12 Outer sepals on ± all flowers with lateral lobes; sepals falling early or persistent   **14**

13 Leaves strongly red-tinged; petals usually shorter than sepals; pedicels, fruits and sepals glabrous to very sparsely glandular-pubescent. Stems erect, to 3m, glabrous, with few prickles, with numerous acicles on suckers. Intrd-natd; bird-sown in rough and marginal ground; very scattered in Sc, Man and En
(*R. glauca* Pourr.) *Red-leaved Rose* - **R. ferruginea** Vill.

13 Leaves green; petals longer than sepals; pedicels, fruits and sepals densely glandular-pubescent. Variably suckering; stems ± erect, to 2m, with few curved prickles and 0 or few hairs and stalked glands. Intrd-natd; scrub and hedgerows; scattered in En and Wa
Virginian Rose - **R. virginiana** Herrm.

  14 Leaflets glabrous, sometimes with few stalked glands on midrib but without eglandular hairs   **15**

  14 Leaflets tomentose to pubescent with eglandular hairs on lowerside, at least on midrib   **16**

15 Orifice of disc c.1/5 its total width; styles glabrous, pubescent or woolly, forming ± loose group; sepals mostly patent to reflexed after flowering, falling before fruit ripe. Stems arching, to 3(4)m, with usually strongly curved to hooked prickles. Native; hedges, scrub, wood-borders; common throughout most of BI except parts of N. Extremely variable *Dog-rose* - **R. canina** L.

15 Orifice of disc c.1/3 its total width; styles woolly, forming dense mass ± obscuring disc; sepals mostly erect to erecto-patent after flowering, usually persistent until fruit ripe. Stems arching, to 2(3)m, with strongly curved to hooked prickles, often strongly red-coloured and glaucous. Native; hedges, scrub and wood-borders; throughout

most of N 3/4 of BI, rare and very scattered in S (*R. caesia* ssp. *glauca* (Nyman) G.G. Graham & Primavesi, invalid name)*Glaucous Dog-rose* - **R. caesia** Sm. **ssp. vosagiaca** (N.H.F. Desp.) D.H. Kent

16 Leaflets with prominent, ± sticky, sessile and short-stalked glands on lowerside, giving fresh leaf fruity smell when rubbed, pubescent to sparsely so with eglandular hairs on lowerside   17

16 Leaflets with 0 or ± inconspicuous glands on lowerside, the glands ± confined to veins or if over whole surface then with no or with resinous smell and leaves usually ± tomentose on lowerside   19

17 Pedicels glabrous; leaflets cuneate at base. Differs from *R. rubiginosa* in stems erect but somewhat flexuous, to 1.5(2)m; prickles ± equal. Native; scrub, mostly on calcareous soils; very scattered and mostly rare in Br and Ir N to S Sc   *Small-leaved Sweet-briar* - **R. agrestis** Savi

17 Pedicels glandular-pubescent; leaflets rounded at base   18

   18 Stems erect; prickles unequal; styles pubescent; sepals mostly erect to patent, persistent until fruit reddens. Stems erect, ± straight, to 2m, with hooked prickles. Native; mostly in scrub on calcareous soils; scattered throughout BI, ± common on chalk in SE En   *Sweet-briar* - **R. rubiginosa** L.

   18 Stems arching; prickles ± equal; styles glabrous or nearly so; sepals mostly reflexed, falling before fruit reddens. Differs from *R. rubiginosa* in stems arching, often scrambling, to 3m. Native; similar places to *R. rubiginosa* but often not on calcareous soils; scattered throughout BI, mostly En, Wa and S Ir

Small-flowered Sweet-briar - **R. micrantha** Sm.

19 Leaflets without glands or with few on midrib on lowerside, 1-2-serrate with teeth not or variably gland-tipped   20

19 Leaflets glandular on lowerside, at least on midrib and lateral veins, 2-serrate with gland-tipped teeth   22

   20 Orifice of disc c.1/3 its total width; styles woolly, forming dense mass ± obscuring disc; sepals mostly erect to erecto-patent after flowering, usually persistent until fruit ripe. Stems arching, to 2(3)m, with strongly curved to hooked prickles, green or somewhat red. Native; hedges, scrub and wood-borders; throughout most of N 3/4 of BI, rare and very scattered in S

Hairy Dog-rose - **R. caesia** Sm. **ssp. caesia**

   20 Orifice of disc c.1/5 its total width; styles glabrous, pubescent or woolly, forming ± loose group; sepals mostly patent to reflexed after flowering, falling before fruit ripe   21

21 Lobes on outer sepals narrow, usually entire; prickles moderately hooked, longer than width of base; leaves usually eglandular on lowerside   (see couplet 15) **R. canina**

21 Lobes on outer sepals broad, usually lobed or toothed; prickles strongly and abruptly hooked, c. as long as width of base; leaves usually glandular on lowerside of midrib. Stems arching, to 2(3)m. Native; hedges and scrub; En, Wa and Ir, frequent only in C En

Round-leaved Dog-rose - **R. obtusifolia** Desv.

22 Orifice of disc c.1/5 its total width; styles glabrous to pubescent	23
22 Orifice of disc c.1/3-1/2 its total width; styles woolly	24

23 Prickles strongly and abruptly hooked, c. as long as width of base; pedicels 5-15mm; pedicels, fruits and sepals glabrous or sparsely glandular-pubescent (see couplet 21) **R. obtusifolia**

23 Prickles ± straight to arched, longer than width of base; pedicels (10)15-25mm; pedicels, fruits and sepals glandular-pubescent to densely so. Stems arching, to 3m. Native; hedges, scrub and open woods; frequent in BI except very rare in C & N Sc
Harsh Downy-rose - **R. tomentosa** Sm.

24 Sepals erect or suberect, entire or with few lateral lobes, persistent until fruit decays; orifice of disc c.2/5-1/2 its total width; prickles ± all straight. Differs from *R. tomentosa* and *R. sherardii* in more compact habit with erect stems to 1.5(2)m. Native; similar habitats and range to *R. sherardii*, but S to only Glam and Derbys *Soft Downy-rose* - **R. mollis** Sm.

24 Sepals erect to erecto-patent, with lateral lobes, falling from ripe fruit; orifice of disc c.1/3 its total width; at least some prickles curved. Differs from *R. tomentosa* in more compact habit, to 1.5(2)m; prickles more slender. Native; scrub, hedges and open woods, throughout most of Br and Ir, common in Sc, rare in SE En *Sherard's Downy-rose* - **R. sherardii** Davies

Hybrids are very common (in many areas commoner than their parents) and, because of the unequal donation of genetic material by the 2 sexes, reciprocal crosses often produce offspring of very different appearance, the hybrid more closely resembling its female parent. Hybrids potentially arise between any 2 of the 12 native species; so far 50 of the 66 possible combinations have been recorded, and probably some involving 3 or even more spp occur. In addition *R. rugosa* occasionally forms spontaneous hybrids with native spp., but the latter sp. is often very difficult to determine. Hybrids are mostly fertile to some, often a high, degree, but are sometimes ± sterile (especially in hybrids of *R. arvensis*).

SUBFAMILY 3 - AMYGDALOIDEAE (genera 22-23). Stipulate trees or shrubs; hypanthium concave, not enclosing carpels and fixed to them only at base; epicalyx 0; stamens >10; flowers 5-merous; carpels 1(or 5), 1-2-seeded; fruit 1(-5) drupes.

22. PRUNUS L. - *Cherries*

Trees or shrubs; leaves simple, serrate to crenate; flowers solitary or in racemes, corymbs or umbels; carpel 1; fruit a drupe.

1 Flowers usually >10 in elongated racemes	2
1 Flowers solitary or 2-c.10 in umbels or corymbs	5
2 Leaves coriaceous, evergreen; racemes without leaves at base; fruit conical-acute at apex	3
2 Leaves herbaceous, deciduous; racemes usually with 1-few leaves	

3   (often reduced) near base; fruit rounded to apiculate at apex   4
3 Petioles and 1st-year stems deep red; leaves serrate; racemes mostly longer than leaves. Evergreen shrub or tree to 12m. Intrd-natd; woods, shrubberies and waste land; scattered in Br N to C Sc
Portugal Laurel - **P. lusitanica** L.
3 Petioles and 1st-year stems green; leaves crenate to obscurely serrate; racemes mostly shorter than leaves. Evergreen shrub or tree to 10m. Intrd-natd; woods and shrubberies; throughout most of BI
Cherry Laurel - **P. laurocerasus** L.
   4 Petals >5mm; sepals falling before fruit ripe. Deciduous shrub or tree to 19m. Native; woods and scrub; Br from C En and S Wa to N Sc, very scattered in Ir, natd in S & C En *Bird Cherry* - **P. padus** L.
   4 Petals <5mm; sepals persistent until fruit ripe. Deciduous shrub or tree to 20m. Intrd-natd; woods and commons; S En and Wa
Rum Cherry - **P. serotina** Ehrh.
5 Ovary and fruit pubescent; leaves mostly lanceolate to oblanceolate   6
5 Ovary and fruit glabrous; leaves mostly ovate to obovate   7
   6 Drupe becoming dry and splitting at maturity, much wider than thick, with pitted stone. Deciduous tree to 8m. Intrd-surv; street- and park-tree frequent on tips and waste places; S Br
Almond - **P. dulcis** (Mill.) D.A. Webb
   6 Drupe becoming very succulent and not splitting at maturity, subglobose, with deeply grooved stone. Deciduous tree to 6m. Intrd-surv; frequent on tips and waste ground in towns, from discarded stones, but mostly not reaching maturity; S & C En
Peach - **P. persica** (L.) Batsch
7 Flowers in short subcorymbose racemes with green bracts on proximal part of axis. Deciduous shrub or small tree to 6(10)m. Intrd-natd; by railways, in grassland and in woods; several places in S En   *St Lucie Cherry* - **P. mahaleb** L.
7 Flowers solitary or in umbels, or rarely very short corymbose racemes without green bracts on axis   8
   8 Flowers (1)2-6(10) together, with group of large (>5mm) often green or reddish bud-scales at base of cluster (not in *P. pensylvanica*); ripe fruit not pruinose, with longer pedicel (*cherries*)   9
   8 Flowers 1-3 together, with 0 or small (<3mm) brown bud-scales at base of cluster; ripe fruit pruinose, with usually shorter pedicel (*plums*)   13
9 Leaves with acuminate to aristate teeth   10
9 Leaves with acute to obtuse teeth   11
   10 Leaves <8cm, with acuminate teeth; flowers not pendent. Deciduous shrub or small tree to 6(10)m. Intrd-natd; in oakwoods; Chinnor Hill (Oxon)   *Fuji Cherry* - **P. incisa** Thunb.
   10 Some leaves >8cm, with aristate teeth; flowers usually pendent. Deciduous tree to 12m. Intrd-surv; much planted by roads and in parks, and often found as relic in wild places; much of BI
Japanese Cherry - **P. serrulata** Lindl.
11 Bud-scales at base of inflorescence falling before flowers fully open;

22. PRUNUS

  flowers 1-2cm across. Deciduous tree to 12m. Intrd-natd; self-sown trees ≤9m in and by woodland; Surrey *Pin Cherry* - **P. pensylvanica** L.f.
11 Bud-scales at base of inflorescence persistent; flowers 2-3.5cm across 12
 12 Hypanthium cup-shaped, not constricted at opening; some bud-scales at base of flowers usually green and leaf-like; never a large tree. Deciduous shrub or small tree to 8m. Intrd-natd; hedges and copses; through most of Bl N to C Sc *Dwarf Cherry* - **P. cerasus** L.
 12 Hypanthium bowl-shaped, constricted at opening; bud-scales not green and leaf-like; often a large tree. Deciduous tree to 31m. Native; hedgerows, wood-borders and copses; throughout Bl
              Wild Cherry - **P. avium** (L.) L.
13 1st-year twigs green, shiny, glabrous. Deciduous, sometimes spiny shrub or tree to 8(12)m. Intrd-natd; common in hedges and spreading by suckers, often not fruiting; Cl and Br N to C Sc
            Cherry Plum - **P. cerasifera** Ehrh.
13 1st-year twigs brown to grey, dull, often pubescent 14
 14 Fruit <2cm, blue-black; flowers appearing before leaves; petals 5-8mm; twigs very spiny. Deciduous, dense, spiny shrub to 4m. Native: hedges, scrub and woods; common almost throughout Bl
             Blackthorn - **P. spinosa** L.
 14 Fruit usually >2cm, blue-black, red or yellow-green; flowers appearing with leaves; petals 7-12mm; twigs not or sparsely spiny. Deciduous large shrub or tree to 8(12)m. Intrd-natd; hedges, copses, scrub and waste ground; throughout most of Bl. Very variable   *Wild Plum* - **P. domestica** L.

P. spinosa x *P. domestica* = ***P. x fruticans*** Weihe occurs in hedges sporadically throughout En; it is intermediate, fertile and variable. *P. spinosa* var. *macrocarpa* Wallr. probably belongs here.

23. OEMLERIA Rchb. - *Osoberry*
Deciduous shrubs; leaves simple, entire; flowers dioecious or partially so, in racemes; carpels 5; fruit a cluster of 1-5 drupes.

1 Suckering shrub with erect stems to 2(3)m; flowers greenish-white. Intrd-natd; on rough ground; scattered in S Br
       Osoberry - **O. cerasiformis** (Hook. & Arn.) J.W. Landon

SUBFAMILY 4 - MALOIDEAE (genera 24-35). Trees or shrubs with stipules; hypanthium concave, enclosing carpels and fused to them all round; epicalyx 0; stamens >10; flowers 5-merous; carpels 1-5, 1-2(several)-seeded; fruit consisting of fused carpels surrounded by usually succulent hypanthium.

24. CYDONIA Mill. - *Quince*
Leaves simple, entire, deciduous; flowers solitary; stamens 15-25; carpels 5, with free styles and numerous ovules, the walls cartilaginous in fruit.

1 Spineless shrub or tree to 3(6)m; flowers 4-5cm across, white or pink; fruit up to 12cm, yellow. Intrd-surv; very persistent in hedges and woods; scattered in S Br, Man   **Quince - C. oblonga** Mill.

25. CHAENOMELES Lindl. - *Japanese Quinces*
Leaves simple, serrate, deciduous, flowers in clusters of 1-4; stamens 40-60; carpels 5, with styles fused at base and numerous ovules, the walls cartilaginous in fruit.

1 Leaves 4-10cm, finely toothed; flowers 3.5-4.5cm across, usually red; fruit up to 7.5cm. Open ± spiny shrub to 3m. Intrd-surv; very persistent in hedges and woods; scattered in CI and S En
   Chinese Quince - **C. speciosa** (Sweet) Nakai
1 Leaves 2.5-6cm, coarsely toothed; flowers 3.5-4cm across, usually pink; fruit up to 5cm. Open ± spiny shrub to 1m. Intrd-surv; less persistent than *C. speciosa*; rare in S En
   Japanese Quince - **C. japonica** (Thunb.) Spach

26. PYRUS L. - *Pears*
Leaves simple, serrate to crenate, deciduous; flowers in simple corymbs; petals white; stamens 20-30; carpels 2-5, with free styles and 2 ovules, the walls cartilaginous in fruit; fruit with groups of gritty stone-cells in hypanthium.

1 Fruit >(5)6cm, soft and sweet when mature, usually pear-shaped. Usually non-spiny tree to 20m. Intrd-natd; found in the wild in hedges and waste ground from old trees or discarded seeds; frequent   ***Pear*** - **P. communis** L.
1 Fruit <5cm, hard and sour even when mature, not pear-shaped   2
   2 Fruit with calyx falling early, <1.5(2)cm; petals 6-10mm; inflorescence rhachis >1cm. Spiny shrub to 8m. Probably native; hedges and wood-margins; 2 sites in S Devon, 3 sites in W Cornwall   ***Plymouth Pear*** - **P. cordata** Desv.
   2 Fruit with persistent calyx, 1.5-4cm; petals 10-17mm; inflorescence rhachis <1cm. Usually ± spiny shrub or tree to 15m. Intrd-natd; hedges and wood-margins; scattered through S & C Br, CI and Ir   ***Wild Pear*** - **P. pyraster** (L.) Burgsd.

27. MALUS Mill. - *Apples*
Leaves simple, serrate, deciduous; flowers in simple corymbs; stamens 15-50; carpels 3-5, with styles fused below and with 2 ovules, the walls cartilaginous in fruit; fruit without groups of gritty stone-cells.

1 Leaves purplish-green to purple; fruit dark reddish-purple, c. as long as pedicel. Tree to 11m, not spiny. Intrd-natd; grown in parks and by roads, rarely self-sown; W Kent, perhaps elsewhere (*M. niedzwetzkvana* Dieck x *M. atrosanguinea* (Späth) C.K. Schneid.)
   Purple Crab - **M. x purpurea** (E. Barbier) Rehder

27. MALUS

1 Leaves green; fruits green to yellow or red, longer than pedicel　2
　2　Leaves glabrous when mature; pedicels and outside of calyx glabrous. Tree to 10m, often spiny. Native; woods, hedges and scrub; throughout BI　***Crab Apple* - M. sylvestris** (L.) Mill.
　2　Leaves pubescent on lowerside; pedicels and outside of calyx pubescent. Tree to 10(20)m, not spiny. Intrd-natd; hedges, scrub and waste ground; throughout BI　***Apple* - M. domestica** Borkh.

28. SORBUS L. - *Whitebeams*

Leaves pinnate, or simple and serrate to pinnately lobed, deciduous; flowers in compound corymbs; petals white; stamens 15-25; carpels 2-4(5), with styles free or fused below and 2 ovules, the walls cartilaginous in fruit.

A difficult genus, consisting of several well-defined but variable sexual species and a number of apomictic ones, many of which are of hybrid origin. The apomicts mostly fall into 3 groups: those similar to *S. aria* (*S. aria* agg.); those intermediate between either *S. torminalis* or *S. aria* agg. and *S. aucuparia* (*S. intermedia* agg.); and those intermediate between *S. aria* agg. and *S. torminalis* (*S. latifolia* agg.). Most diagnostic characters concern leaves and fruits, both of which are required for identification by beginners. 'Leaves' refers to the broader leaves of the short-shoots (not those on leading shoots). It is important that plants are surveyed as fully as possible and that the means of each character-state are used for identification.

1　Leaves pinnate, at least proximally, with ≥1 pair of completely free leaflets　2
1　Leaves not to deeply lobed, but without free leaflets　5
　2　Leaves completely pinnate, with ≥4 pairs of leaflets and a single terminal leaflet　3
　2　Leaves pinnate only proximally, with 1-2(3) pairs of leaflets and a distal lobed part of leaf (if ≤5 pairs of leaflets, probably hybrids of *S. aucuparia* with *S. intermedia* and *S. aria*; these have (1)2-3(5) pairs of free well-separated leaflets and a total of 10-12 pairs of lateral veins. The former hybrid has scarlet fruits with no seed and sterile pollen; the latter has brownish-red fruits with some good seed and some fertile pollen)　4
3　Buds glabrous, sticky, greenish to pale brown; stipules 6-14mm, forked at or below 1/2 way, soon falling; fruits mostly >2cm across, greenish-brown sometimes red-tinged. Tree to 23m; trunk deeply and closely fissured. Probably native; on limestone sea-cliffs in Glam (stunted, to 3(5)m, rarely fruiting); planted and sometimes self-sown in a few places in S En, odd specimens possibly native in W Gloucs　***Service-tree* - S. domestica** L.
3　Buds tomentose, not sticky, brown; stipules 4-8mm, simple or fan-shaped, persistent on long-shoots; fruits <15mm across, scarlet (to yellow). Tree to 18m; trunk smooth or shallowly and sparsely fissured. Native; woods, moors, rocky places except on heavy soils; throughout Br and Ir, intrd in CI　***Rowan* - S. aucuparia** L.
　4　Leaves 5.5-8.5cm, rather thinly grey-tomentose on lowerside, with

mostly 1 pair of free leaflets; fruit longer than wide; petals c.4mm. Tree to 7m; leaves differing from those of 2 hybrids in couplet 2 in having total of 7-9(10) pairs of lateral veins and 1(-2) pairs of broader ± overlapping free leaflets. Native; steep granite streambank; Glen Catacol in Arran (Clyde Is)
Arran Service-tree - **S. pseudofennica** E.F. Warb.

4 Leaves 7.5-10.5cm, densely whitish-grey-tomentose on lowerside, with mostly 2 pairs of free leaflets; fruit globose; petals c.6mm. Tree to 10m; leaves differing from those of 2 hybrids in couplet 2 in having total of 8-10 pairs of lateral veins and (1-)2(-3) pairs of free well-separated leaflets. Intrd-natd; frequently grown in gardens and parks and sometimes self-sown; W Kent and N Aberdeen *Swedish Service-tree* - **S. hybrida** L.

5 Mature leaves sparsely pubescent on lowerside, at least some lobed ≥1/3 way to midrib proximally; fruit brown. Tree to 27m. Native; woods, scrub and hedgerows mostly on clay or limestone; local in most of En and Wa *Wild Service-tree* - **S. torminalis** (L.) Crantz

5 Mature leaves tomentose on lowerside (wearing off with age), if lobed ≥1/3 way to midrib then fruit red 6
  6 Leaves white-tomentose on lowerside; fruit red
(***Common Whitebeam*** - ***S. aria*** agg.) 7
  6 Leaves grey- to yellowish-tomentose on lowerside; fruit red, orange or brown 15

7 Fruit mostly 15-22mm, longer than wide 8
7 Fruit mostly 8-15mm, often globose or wider than long 9
  8 Leaves (1.3)1.5-1.7(1.9)x as long as wide; fruit scarlet. Shrub to 3m. Native; carboniferous limestone crags; 2 areas in Brecs
S. leptophylla E.F. Warb.
  8 Leaves 1.1-1.4(1.7)x as long as wide; fruit crimson. Shrub or small tree to 6m. Native; rocky carboniferous limestone woodland; Avon Gorge and Wye Valley (N Somerset, W Gloucs, Mons and Herefs) **S. eminens** E.F. Warb.

9 Leaves with (9)10-14(15) pairs of lateral veins; fruit scarlet, 8-15mm. Tree to 15(23)m. Native; woods, scrub, rocky places, mostly on calcareous soils; probably native only in En N to Derbys, and in Galway, but commonly planted and ± natd throughout most of Br and Ir. Sexual and very variable, incl. several cultivars, overlapping to some extent with most other spp. in the agg. **S. aria** (L.) Crantz
9 Leaves with (6)7-10(11) pairs of lateral veins, or (Ir only) if with >10 pairs then fruit crimson or pinkish-scarlet and c.15mm 10
  10 Fruit longer than wide 11
  10 Fruit ± globose or wider than long 12
11 Fruit 10-13mm, crimson; leaves with sides curved ± throughout. Shrub or small tree to 6m. Native; rocky carboniferous limestone woodland and scrub; Avon Gorge (N Somerset and W Gloucs)
S. wilmottiana E.F. Warb.
11 Fruit 12-15mm, scarlet; leaves with sides ± straight in lower 1/3-1/2. Small tree to 6m. Native; rocky woods near coast of Bristol Channel

28. SORBUS

(N Devon and S Somerset), not on limestone **S. vexans** E.F. Warb. 246
- 12 Fruit 8-12mm. Shrub or small tree to 5m. Native; rocky limestone woods; SW Br from S Devon to Brecs **S. porrigentiformis** E.F. Warb. 246
- 12 Fruit 12-15mm 13
- 13 Leaves mostly 1.2-1.5x as long as wide, curved-sided throughout, with margins entire for basal ≤1/5. Tree to 6m. Native; rocky carboniferous limestone scrub; Ir, mostly C **S. hibernica** E.F. Warb. 246
- 13 Leaves mostly 1.5-2.1x as long as wide, straight-sided for basal 1/3-1/2, with margins entire for basal c.1/4-1/2 14
  - 14 Leaf-margin straight-sided and entire for basal c.1/4-1/3; most leaves ≤1.8x as long as wide. Shrub or small tree to 5m. Native; rocky scrub and woodland, usually on carboniferous limestone; W Lancs and Westmorland **S. lancastriensis** E.F. Warb. 246
  - 14 Leaf-margin straight-sided and entire for basal c.1/3-1/2; most leaves ≥1.8x as long as wide. Shrub or small tree to 6(10)m. Native, rocky woodland, scrub and cliffs, usually on limestone; scattered over Br and Ir but not in E, SE or SC En **S. rupicola** (Syme) Hedl. 246
- 15 Fruit red (*Swedish Whitebeam - S. intermedia* agg.) 16
- 15 Fruit orange to brown (*Broad-leaved Whitebeam - S. latifolia* agg.) 20
  - 16 Leaves mostly with 7-8 pairs of lateral veins. Tree to 7.5m. Native; steep granite stream-banks; Arran (Clyde Is) **S. arranensis** Hedl. 247
  - 16 Leaves mostly with 8-10 pairs of lateral veins 17
- 17 Fruit 12-15mm, longer than wide. Tree to 10(18)m. Intrd-natd; much planted and frequently self-sown in copses and on rough ground; scattered over most of Br and Ir **S. intermedia** (Ehrh.) Pers. 247
- 17 Fruit ± globose or wider than long 18
  - 18 Leaves mostly lobed 1/3-1/2 way to midrib. Shrub to 3m, rarely tree to 12m. Native; carboniferous limestone crags; near Merthyr Tydfil (Brecs) **S. leyana** Wilmott 247
  - 18 Leaves mostly lobed 1/6-1/3 way to midrib 19
- 19 Petals mostly c.4mm; fruit 6-8mm, scarlet. Shrub to 3m. Native; carboniferous limestone crags; near Crickhowell (Brecs) **S. minima** (Ley) Hedl. 247
- 19 Petals mostly c.6mm; fruit 7-12mm, crimson or crimson-scarlet. Shrub to 3m. Native; woods and rocky places mostly on carboniferous limestone; very local in Wa, SW En and N Kerry. A variant from near Llangollen (Denbs) with more narrowly cuneate leaf-bases might be a separate sp. **S. anglica** Hedl. 247
  - 20 Fruit orange-brown, sometimes turning brown 21
  - 20 Fruit orange 23
- 21 Fruit mostly longer than wide, with few, small lenticels. Occasional trees, both fertile and sterile, near the parents in N Somerset, W Gloucs, Mons and Herefs (*S. aria* x *S. torminalis*) **S. x vagensis** Wilmott 246
- 21 Fruit mostly subglobose, with many large lenticels 22
  - 22 Leaves divided 1/8(-1/6) way to midrib, 1.3-1.6(1.8)x as long as wide. Tree to 15m. Native; woods and hedges on well-drained soils; widespread in Devon, very local in E Cornwall, S Somerset, Man and SE & NE Ir **S. devoniensis** E.F. Warb. 247

FIG 246 - Leaves of *Sorbus*. 1, *S. aria*. 2, *S. leptophylla*. 3, *S. eminens*. 4, *S. hibernica*. 5, *S. wilmottiana*. 6, *S. lancastriensis*. 7, *S. rupicola*. 8, *S. vexans*. 9, *S. porrigentiformis*. 10, *S. pseudofennica*. 11, *S. hybrida*. 12, *S. x thuringiaca*. 13, *S. x vagensis*.

FIG 247 - Leaves of *Sorbus*. 1, *S. latifolia*. 2, *S. decipiens*. 3, *S. bristoliensis*.
4, *S. devoniensis*. 5, *S. croceocarpa*. 6, *S. leyana*. 7, *S. intermedia*.
8, *S. minima*. 9, *S. subcuneata*.
10, *S. anglica* (Llangollen). 11, *S. arranensis*. 12, *S. anglica* (typical).

77. ROSACEAE

22 Leaves divided 1/6-1/4(1/3) way to midrib, (1.5)1.6-1.9(2.5)x as
long as wide. Tree to 10m. Native; open rocky *Quercus* woods
near coast; N Devon and S Somerset **S. subcuneata** Wilmott
23 Leaves divided ≤1/6 way to midrib at largest lobes 24
23 Leaves divided ≥1/6 way to midrib at largest lobes 25
   24 Anthers pink; fruit 9-11mm, mostly longer than wide. Tree to
     10m. Native; rocky woods and scrub on carboniferous limestone;
     Avon Gorge (N Somerset and W Gloucs) **S. bristoliensis** Wilmott
   24 Anthers cream; fruit 11-22mm, mostly subglobose. Tree to 21m.
     Intrd-natd; frequently planted, natd in Br N to C Sc. Leaves
     least deeply lobed in the agg. **S. croceocarpa** P.D. Sell
25 Leaves 1.3-1.8(2.5)x as long as wide, with 10-13 pairs of veins. Tree
to 10m. Intrd-natd; woods and scrub; Avon Gorge (W Gloucs) and
Achnashellach (W Ross), planted elsewhere
                                        S. decipiens (Bechst.) Irmisch
25 Leaves 1.1-1.3x as long as wide, with 7-9 pairs of veins. Tree to 20m.
Intrd-natd; frequently planted, natd in Br N to C Sc
                                                      S. latifolia (Lam.) Pers.

The 3 widespread sexual natives, *S. aria*, *S. aucuparia* and *S. torminalis*, occasionally hybridise with each other and with some of the apomictic spp. 6 combinations are known; the 3 most frequent are in the above key.

29. ARONIA Medik. - *Chokeberries*

Leaves simple, serrate, deciduous; flowers in compound corymbs; petals white; stamens c.20; carpels 5, with 5 styles fused near base, the walls cartilaginous at fruiting. Very few records, but likely to increase in future.

1 Leaves narrowly obovate to elliptic, densely pubescent on lowerside;
fruit red. Suckering shrub to 3m. Intrd-natd; in woodland on sandy
soil; Surrey   **Red Chokeberry** - **A. arbutifolia** (L.) Pers.
1 Leaves obovate, subglabrous; fruit black. Suckering shrub to 1.5m.
Intrd-natd; in boggy areas; Dorset, Caerns and S Lancs
                         Black Chokeberry - **A. melanocarpa** (Michx.) Elliott

A. arbutifolia x *A. melanocarpa* = *A. x prunifolia* (Marshall) Rehder probably also occurs but is overlooked.

30. AMELANCHIER Medik. - *Juneberry*

Leaves simple, serrate, deciduous; flowers in racemes; petals white; stamens 10-20; carpels 5, with 5 styles fused near base, the walls cartilaginous at fruiting.

1 Tree to 10m; fruit red then purplish-black. Intrd-natd; on mainly
sandy soils in woodland and scrub; frequent in SC & SE En, very
sparse elsewhere in En, Wa and Jersey
                          Juneberry - **A. lamarckii** F. G. Schroed.

31. PHOTINIA Lindl. - *Stranvaesia*
Leaves simple, entire, evergreen; flowers in compound corymbs; petals white; stamens c.20; carpels 5, with 5 styles fused to ≥1/3 way, the walls cartilaginous at fruiting; hypanthium not quite reaching apex of carpels at fruiting.

1 Shrub or tree to 3(8)m; fruit scarlet. Intrd-natd; bird-sown in rough ground; extremely scattered in En, Man and Sc

Stranvaesia - **P. davidiana** (Decne.) Cardot

32. COTONEASTER Medik. - *Cotoneasters*
Leaves simple, entire; flowers in compound corymbs, small clusters, or solitary; stamens 10-20; carpels 1-5, with 1-5 ± free styles, the walls stony at fruiting.

The genus parallels *Sorbus* in that it contains both very variable sexual spp. and much less variable (in this case ± invariable) apomictic spp.; the precise extent of apomixis is unknown but perhaps c.95% of spp. exhibit it. Much misidentification has occurred of both garden and wild plants.

A large genus becoming increasingly natd via bird-sown seed from garden or roadside ornamentals. Until familiar with the genus, flowers, ripe fruit and summer leaves (after flowering but before fruit ripe) are necessary for determination, and a knowledge of the degree of leaf retention in winter is also desirable. Fruit colours are often diagnostic, but are difficult to describe, so only 6 colours are defined here: dark purple to black; yellow (to yellowish-orange); orange-red; bright red; crimson (a deep red mildly tinted with blue); and maroon (often becoming brownish). The number of carpels ('stones') per fruit should be counted in at least 5 fruits; closely adherent stones are counted as separate. Leaf-sizes and pubescence refer to those of fully grown summer leaves; at flowering they may be much smaller and more densely pubescent in deciduous spp. All the species occur in rough and waste ground, in developing scrub (especially on chalk), and on roadsides, banks and walls.

The following keys first use 8 characters to separate 12 groups in a multi-access key. 10 of these 12 groups are then dealt with by dichotomous keys; the other 2 groups each contain only *C. cooperi*.

Multi-access general key
Petals patent, usually white	A
Petals erect to erecto-patent, usually pink	B
Leaves deciduous or mostly so, most dropped by Jan	C
Leaves evergreen	D
Fruits dark purple to black	E
Fruits yellow to bright red, crimson or maroon	F
Veins deeply impressed on leaf upperside	G
Veins not or slightly impressed on leaf upperside	H
Summer leaves densely pubescent to tomentose on lowerside, largely or wholly obscuring surface	I
Summer leaves glabrous to pubescent on lowerside, leaving most of	

surface exposed   J
Hypanthium and calyx glabrous to sparsely pubescent   K
Hypanthium and calyx pubescent   L
Hypanthium and calyx densely pubescent to tomentose   M
Flowers mostly 1-2(3) together   N
Flowers mostly 3-10(12) together   O
Flowers mostly (10)12-many together   P
   Anthers white (to pale yellow)   Q
   Anthers pigmented (pink, mauve, violet, purple or blackish)   R

ACEH(IJ)(KLM)(OP)(QR)   ***Key A***
ACFH(IJ)KPR   ***Key B***
ACF(GH)(IJ)(LM)PR   ***Key C***
ADEH(IJ)KPR   (see Key A, couplet 7) **C. cooperi**
ADF(GH)(IJ)M(OP)R   ***Key D***
ADFH(IJ)(KL)(NO)R   ***Key E***
ADFH(IJ)KPR   (see Key A, couplet 7) **C. cooperi**
BCE(GH)(IJ)(KLM)(NOP)Q   ***Key F***
BCF(GH)I(KLM)(OP)Q   ***Key G***
BCFGJ(KL)PQ   ***Key H***
BCFHJ(KL)(NO)Q   ***Key I***
BDF(GH)I(KLM)O(QR)   ***Key J***

Key A - Petals patent; leaves deciduous; fruits dark purple to black
1   Leaves <5(6)cm; flowers mostly <20 per inflorescence; anthers white
    or purplish-black   2
1   Some leaves >6cm; flowers mostly >(15)20 per inflorescence; anthers
    pink or red to violet or mauve   4
    2   Anthers purplish-black; fruits 8-11mm, oblong-ellipsoid, maroon;
        leaves c.1.25-1.5x as long as wide. Deciduous shrub to 3m with
        arching branches. Intrd-natd; on chalk in Beds and W Kent
        (*C. multiflorus* auct.)   ***One-stoned Cotoneaster -***
              C. monopyrenus (W.W. Sm.) Flinck & B. Hylmö
    2   Anthers white; fruits 6-9mm, subglobose, bluish-black; leaves
        c.1-1.25x as long as wide   3
3   Shrub to 2m; leaves 1.5-4cm; flowers in loose inflorescence;
    hypanthium and calyx pubescent. Deciduous shrub with rigid
    close branches. Intrd-natd; SE En
            Circular-leaved Cotoneaster - **C. hissaricus** Pojark.
3   Shrub or tree to 6m; leaves 2.5-5(6)cm; flowers in compact
    inflorescence; hypanthium and calyx tomentose. Deciduous shrub or
    tree with erect to arching branches. Intrd-natd; SE En (*C. ellipticus*
    Loudon, invalid name)   ***Lindley's Cotoneaster -*** **C. insignis** Pojark.
    4   Hypanthium and calyx densely pubescent to tomentose; leaves
        sparsely to densely pubescent on lowerside in summer; midrib
        lowersides, peduncles and pedicels pubescent to densely so at
        fruiting   5
    4   Hypanthium and calyx sparsely pubescent to pubescent

32. COTONEASTER

(sometimes hypanthium densely so); leaves subglabrous on lowerside in summer; midrib lowersides, peduncles and pedicels very sparsely pubescent at fruiting   **6**

5   Flowers mostly <15 per inflorescence; fruiting inflorescences mostly <3 x 3cm. Deciduous shrub or tree to 8m with erect to arching branches. Intrd-natd; Br N to Cheshire
Purpleberry Cotoneaster - **C. affinis** Lindl.   252

5   Flowers mostly >15 per inflorescence; fruiting inflorescences mostly >3 x 3cm. Deciduous shrub or tree to 6m with erect to arching branches. Intrd-natd; SE En
Black-grape Cotoneaster - **C. ignotus** G. Klotz   252

   6   Larger leaves mostly <2x as long as wide, very obtuse to rounded and apiculate at apex. Deciduous shrub or tree to 5m with erect to arching branches. Intrd-natd; SE En
Dartford Cotoneaster - **C. obtusus** Lindl.   252

   6   Larger leaves mostly ≥2x as long as wide, acuminate to obtuse and apiculate at apex   **7**

7   Fruits red, turning to dark purple when fully ripe, not exposing stones at apex; inflorescences 5-20-flowered. Deciduous to evergreen shrub to 8m. Intrd-natd; roadside in S Devon
Cooper's Cotoneaster - **C. cooperi** C. Marquand   252

7   Fruits soon becoming purplish- to brownish-black when ripe, usually exposing stones at apex; inflorescences 15-30-flowered   **8**

   8   Fruits with dense whitish bloom, with very open apex exposing stones. Deciduous shrub or tree to 5m, with widely arching branches. Intrd-natd; SE En
Open-fruited Cotoneaster - **C. bacillaris** Lindl.   252

   8   Fruits without or with sparse whitish bloom, with scarcely open apex ± not exposing stones. Deciduous shrub or tree to 5m, with widely arching branches. Intrd-natd; scattered in S En
Godalming Cotoneaster - **C. transens** G. Klotz   252

Key B - Petals patent; leaves deciduous to ± so; fruits yellow to bright red; hypanthium and calyx glabrous to sparsely pubescent

1   Leaves pubescent to densely so on lowerside, <2x as long as wide; anthers purplish-black   (see Key A, couplet 2) **C. monopyrenus**

1   Leaves sparsely pubescent to subglabrous on lowerside, ≥2x as long as wide; anthers mauve   **2**

   2   Fruit eventually brownish-black, with stones slightly showing at apex; inflorescences 5-20-flowered
(see Key A, couplet 8) **C. transens**

   2   Fruit eventually purplish-black, with stones not showing at apex; inflorescences 15-30-flowered   (see Key A, couplet 7) **C. cooperi**

Key C - Petals patent; leaves deciduous to ± so; fruits yellow to bright red or crimson; hypanthium and calyx pubescent to tomentose

1   Leaves flat on upperside, very dull, the veins not impressed   **2**
1   Leaves with slightly to strongly impressed veins on upperside, usually somewhat shiny   **3**

FIG 252 - Leaves of *Cotoneaster*. 1, *C. monopyrenus*. 2, *C. tomentellus*. 3, *C. insignis*. 4, *C. hissaricus*. 5, *C. ignotus*. 6, *C. affinis*. 7, *C. obtusus*. 8, *C. bacillaris*. 9, *C. transens*. 10, *C. cooperi*. 11, *C. frigidus*. 12, *C. x watereri*. 13, *C. salicifolius*. 14, *C.* 'Hybridus Pendulus'. 15, *C. henryanus*. 16, *C. hylmoei*. 17, *C. dammeri*. 18, *C. x suecicus*. 19, *C. pannosus*.

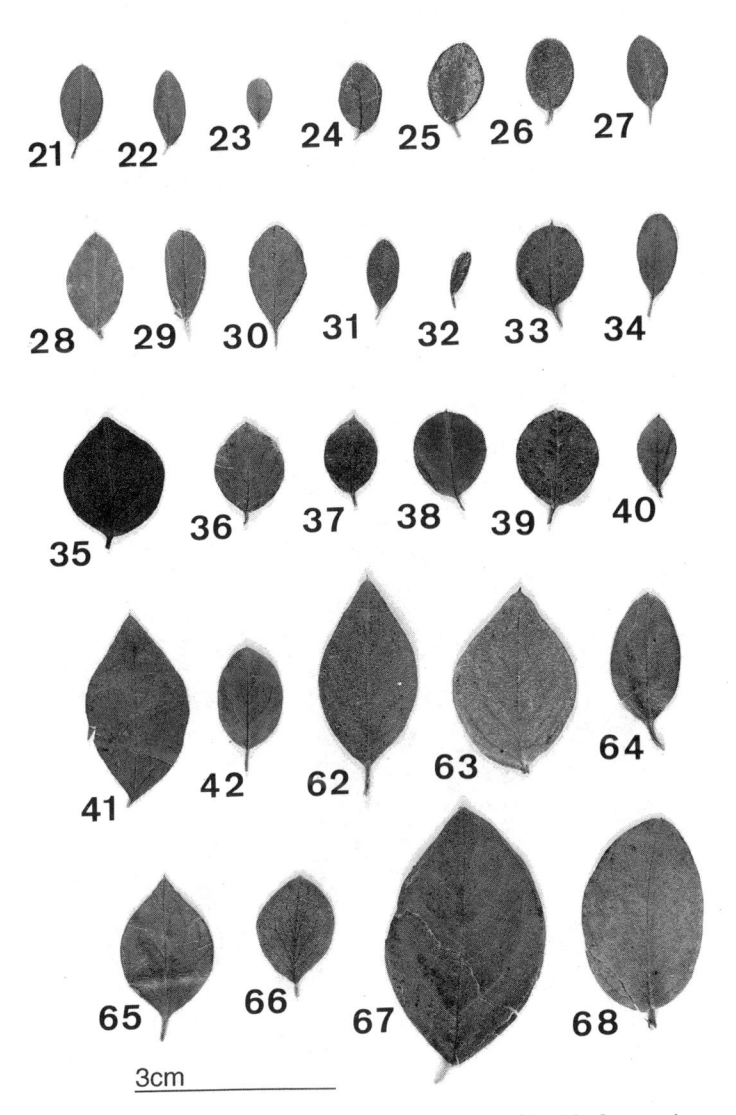

FIG 253 - Leaves of *Cotoneaster*. 21, *C. microphyllus*. 22, *C. conspicuus*. 23, *C. astrophoros*. 24, *C. sherriffii*. 25, *C. rotundifolius*. 26, *C. cochleatus*. 27, *C. cashmiriensis*. 28, *C. prostratus*. 29, *C. marginatus*. 30, *C. congestus*. 31, *C. integrifolius*. 32, *C. linearifolius*. 33, *C. nitidus*. 34, *C. adpressus*. 35, *C. apiculatus*. 36, *C. nanshan*. 37, *C. horizontalis*. 38, *C. atropurpureus*. 39, *C. hjelmqvistii*. 40, *C. ascendens*. 41, *C. divaricatus*. 42, *C. nitens*. 62, *C. vilmorinianus*. 63, *C. insculptus*. 64, *C. amoenus*. 65, *C. dielsianus*. 66, *C. splendens*. 67, *C. zabelii*. 68, *C. fangianus*.

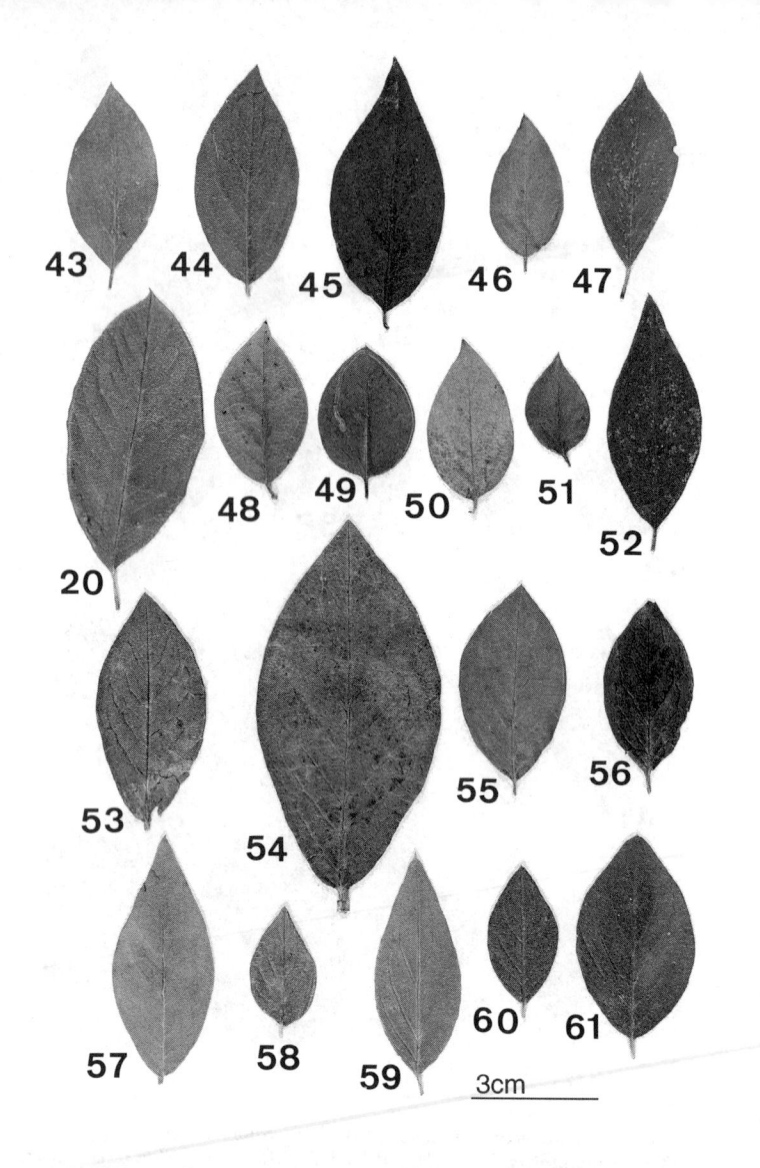

FIG 254 - Leaves of *Cotoneaster*. 20, *C. lacteus*. 43, *C. lucidus*.
44, *C. villosulus*. 45, *C. laetevirens*. 46, *C. pseudoambiguus*. 47, *C. hummelii*.
48, *C. hsingshangensis*. 49, *C. cambricus*. 50, *C. mucronatus*. 51, *C. simonsii*.
52, *C. tengyuehensis*. 53, *C. bullatus*. 54, *C. rehderi*. 55, *C. boisianus*.
56, *C. obscurus*. 57, *C. moupinensis*. 58, *C. franchetii*.
59, *C. wardii*. 60, *C. mairei*. 61, *C. sternianus*.

32. COTONEASTER

2 Leaves 2-5cm, obovate to suborbicular; fruits 8-13mm, mostly with 1 stone; inflorescences c.6-20-flowered. Deciduous shrub to 5m with spreading branches. Intrd-natd; pathside; S Lancs
Short-felted Cotoneaster - **C. tomentellus** Pojark. 252

2 Leaves 6-15cm, elliptic-oblong; fruits 4-6mm, with 2 stones; inflorescences mostly >20-flowered. Deciduous to semi-evergreen, erect shrub or strong tree to 8(18)m. Intrd-natd; frequent in BI *Tree Cotoneaster* - **C. frigidus** Lindl. 252

3 Leaves mostly 2.5-3x as long as wide; flowers ≥20 per inflorescence; fruits with 2(-5) stones, 5-8mm. Usually semi-evergreen, erect shrub to 8m. Intrd-natd; many parts of BI (*C. frigidus* x *C. salicifolius*). [*Henry's Cotoneaster* - **C. henryanus** (C.K. Schneid.) Rehder & E.H. Wilson would also key out here; it is a semi-evergreen shrub to 5m which comes very close to some variants of *C. x watereri* that are closer to *C. salicifolius* than to *C. frigidus*, and might be involved in its parentage. Considerable experience is necessary for certain determination. Intrd-natd; in a few places in S En, C Sc and N Ir]
Waterer's Cotoneaster - **C. x watereri** Exell 252

3 Leaves mostly 2-2.5x as long as wide; flowers <15(20) per inflorescence; fruits with 1-2 stones, 6-10mm
(see Key A, couplet 5) **C. affinis**

Key D - Petals patent; leaves evergreen or ± so; fruits yellow to bright red or crimson; hypanthium and calyx densely pubescent to tomentose

1 Leaves most or all <3cm; flowers <15 per inflorescence. Stems erect, long, slender, slightly arching, to 3(5)m; fruits often with conspicuous erect sepals as in *C. amoenus*. Intrd-natd; W Kent
Silverleaf Cotoneaster - **C. pannosus** Franch. 252

1 Leaves most or all >3cm; flowers >15 per inflorescence 2
  2 Leaves flat on upperside, the veins not impressed, usually semi-deciduous (see Key C, couplet 2) **C. frigidus**
  2 Leaves with somewhat to strongly impressed veins on upperside, usually ± evergreen 3

3 Leaves obtuse to rounded at apex; fruits usually with 2 stones. Evergreen, spreading shrub to 5m. Intrd-natd; Br N to C Sc, planted as field-hedges in E Anglia *Late Cotoneaster* - **C. lacteus** W.W. Sm. 254

3 Leaves acute to subacute at apex; fruits with 2-5 stones 4
  4 Leaves sparsely pubescent on lowerside. Procumbent (in the wild), evergreen shrub with branches to 1m. Intrd-natd; Glam, probably overlooked(*C. salicifolius* x *C. dammeri*)
Weeping Cotoneaster - **C. 'Hybridus Pendulus'** 252
  4 Leaves densely pubescent to tomentose on lowerside 5

5 Leaves oblanceolate to narrowly elliptic; fruits mostly 4-5mm, often slightly wider than long, with 3-5 stones. Erect, arching or ± procumbent, evergreen shrub to 5m. Intrd-natd; frequent in BI
Willow-leaved Cotoneaster - **C. salicifolius** Franch 252

5 Leaves narrowly obovate to elliptic; fruits mostly 5-8mm, often slightly longer than wide, with 2-3(5) stones 6

6 Petals pink; leaves with 5-7 pairs of lateral veins. Erect, arching, evergreen shrub to 3m. Intrd-natd; on rough ground; Lanarks and Offaly   *Hylmö's Cotoneaster* - **C. hylmoei** Flinck & J. Fryer
6 Petals white; larger leaves with >7 pairs of lateral veins
(see Key C, couplet 3) **C. x watereri** & **C. henryanus**

Key E - Petals patent; leaves evergreen; fruits bright red to crimson or orange; hypanthium and calyx glabrous to sparsely pubescent or pubescent

1 Inflorescences with >10 flowers   2
1 Inflorescences with <10 flowers   3
  2 Erect shrub to 8m; fruits with 2 stones
(see Key A, couplet 7) **C. cooperi**
  2 Procumbent shrub; fruits with 3-5 stones
(see key D, couplet 4) **C. 'Hybridus Pendulus'**
3 Some leaves >1.5cm; fruits with 2-5 stones   4
3 Leaves all <1(1.5)cm, or if >1.5mm then fruits with 2(-3) stones   6
  4 Leaves all <2.5cm; at least some stems erect or arching; fruits with 2-4 stones. Stems arching to 60cm high, trailing to 2m. Intrd-natd; Br N to C Sc (?*C. dammeri* x *C. conspicuus*)
Swedish Cotoneaster - **C. x suecicus** G. Klotz
  4 Some leaves >2.5cm; stems all procumbent; fruits often with 5 stones   5
5 Leaf apex acute to acuminate
(see Key D, couplet 4) **C. 'Hybridus Pendulus'**
5 Leaf apex obtuse to rounded. Procumbent, evergreen shrub with branches to 3m. Intrd-natd; Br N to C Sc
Bearberry Cotoneaster - **C. dammeri** C.K. Schneid.
  6 Fruits orange, with 1(-2) stones. Evergreen, stiffly erect to spreading shrub to 2m. Intrd-natd; scattered sites in Br
Sherriff's Cotoneaster - **C. sherriffii** G. Klotz
  6 Fruits bright red to crimson, with 2(-3) stones   7
7 Flowers mostly 2-5(7) per cyme   8
7 Flowers mostly 1(-2) per cyme except at ends of branches   9
  8 Leaves 10-25mm, ± matt on upperside, densely pubescent on lowerside. Evergreen, erect to procumbent shrub to 3m. Intrd-natd; numerous sites in Br N to Westmorland
Fringed Cotoneaster - **C. marginatus** (Loud.) Schltdl.
  8 Leaves 5-12mm, shiny on upperside, rather sparsely pubescent on lowerside. Evergreen, procumbent to ascending shrub to 1m. Intrd-natd; several places in C Ir, not confirmed in Br
Small-leaved Cotoneaster - **C. microphyllus** Lindl.
9 Leaves mostly ≥2x as long as wide   10
9 Leaves <2x as long as wide   12
  10 Fruits bright red, shiny; leaves mid-green and slightly shiny on upperside. Evergreen, stiffly erect to spreading shrub to 1.5m. Intrd-natd; Br N to C Sc
Tibetan Cotoneaster - **C. conspicuus** C. Marquand

32. COTONEASTER

10 Fruits crimson, dull; leaves dark green and very shiny on upperside 11
11 Leaves 7-15 x 3-6mm with petioles 2-4mm; fruits 8-10mm; flowers c.11mm across. Procumbent to arching, evergreen shrub to 1m. Intrd-natd; frequent over much of BI
Entire-leaved Cotoneaster - **C. integrifolius** (Roxb.) G. Klotz ... 253
11 Leaves 4-7 x 1-3mm with petioles 0.5-1.5mm; fruits 4-5mm; flowers c.5mm across. Evergreen, procumbent or arching shrub to 0.6m. Intrd-natd; S En, W Ir and M Ebudes
Thyme-leaved Cotoneaster - **C. linearifolius** (G. Klotz) G. Klotz ... 253
12 Leaves mid-green and matt on upperside. Tightly branched, evergreen shrub to 0.7m. Intrd-natd; W Kent and Man
Congested Cotoneaster - **C. congestus** Baker ... 253
12 Leaves medium- to dark-green on upperside 13
13 Larger leaves >15mm; shrub to 2m, often ± erect. Evergreen, stiffly erect to spreading shrub. Intrd-natd; scattered sites in Br N to Westmorland *Round-leaved Cotoneaster* - **C. rotundifolius** Lindl. ... 253
13 All leaves <15mm; procumbent to arching shrub rarely >0.5m 14
14 Leaves densely pubescent to tomentose on lowerside. Evergreen, arching and spreading shrub to 0.3m. Intrd-natd; by railway; Lanarks *Starry Cotoneaster* - **C. astrophoros** J. Fryer & E.C. Nelson ... 253
14 Leaves sparsely to very sparsely pubescent on lowerside 15
15 Fruits 4-6mm. Evergreen, procumbent shrub to 0.3m. Intrd-natd; E Kent *Kashmir Cotoneaster* - **C. cashmiriensis** G. Klotz ... 253
15 Fruits 6-10mm 16
16 Some branches arching above ground to 50cm high; fruits bright red. Evergreen, procumbent shrub to 0.5m. Intrd-natd; W Kent and N Somerset (*C. buxifolius* auct.)
Procumbent Cotoneaster - **C. prostratus** Baker ... 253
16 Branches all flat to ground, rarely >20cm high; fruits crimson-red. Evergreen, procumbent shrub to 0.2m. Intrd-natd; Midlothian
Yunnan Cotoneaster - **C. cochleatus** (Franch.) G. Klotz ... 253

Key F - Petals erect to erecto-patent; leaves deciduous; fruits dark purple to black
1 Leaves <2cm, obtuse at apex; flowers 1-3(4) together; stamens 10. Erect, densely branched, deciduous shrub to 3.5m. Intrd-natd; S & SE En *Few-flowered Cotoneaster* - **C. nitens** Rehder & E.H. Wilson ... 253
1 Leaves all or nearly all >2cm, acute to acuminate at apex; flowers 3-30 together; stamens 15-20 2
2 Leaves strongly bullate with deeply impressed veins on upperside; inflorescences mostly 12-30-flowered; fruits with 3-5 stones 3
2 Leaves not bullate but veins slightly to strongly impressed on upperside; inflorescences mostly 3-15-flowered; fruits with 2-3 stones 4
3 Leaves 3-5cm, acute to shortly acuminate at apex; fruits maroon, with 3-5 stones. Erect, deciduous shrub to 2m. Intrd-natd; in hedges;

S Lancs and S Hants

Obscure Cotoneaster - **C. obscurus** Rehder & E.H. Wilson

3 Leaves 4-12cm, long-acuminate at apex; fruits purplish-black, with 5 stones. Erect, deciduous shrub to 3m. Intrd-natd; scattered in En & Wa

Moupin Cotoneaster - **C. moupinensis** Franch.
- 4 Leaves 2-5.5(7)cm, most or all <5cm ... 5
- 4 Leaves 3-11cm, most or many >5cm ... 6

5 Calyx and hypanthium glabrous or nearly so on outside at flowering, glabrous at fruiting; leaves shiny on upperside. Spreading to erect, deciduous shrub to 3m. Intrd-natd; scattered places in En

Shiny Cotoneaster - **C. lucidus** Schltdl.

5 Calyx and hypanthium pubescent at flowering, sparsely pubescent at fruiting; leaves matt on upperside. Erect, deciduous shrub to 5m. Intrd-natd; Westmorland

Kangting Cotoneaster - **C. pseudoambiguus** J. Fryer & B. Hylmö
- 6 Leaves tapering-acuminate ... 7
- 6 Leaves acute to shortly and ± abruptly acuminate ... 8

7 Leaves with not or scarcely impressed lateral veins on upperside, pubescent on lowerside; inflorescences with 3-9(15) flowers; fruits mostly with 2 stones. Erect, deciduous shrub to 5m. Intrd-natd; S Hants

Ampfield Cotoneaster - **C. laetevirens** (Rehder & E.H. Wilson) G. Klotz

7 Leaves with distinctly impressed lateral veins on upperside, very sparsely pubescent on lowerside; inflorescences with 7-15 flowers; fruits mostly with 3 stones. Erect, deciduous shrub to 5m. Intrd-natd; S Hants

Hummel's Cotoneaster - **C. hummelii** J. Fryer & B. Hylmö

8 Leaves with apex abruptly acuminate to fine point; calyx-lobes pubescent only at base; fruits 10-11mm, globose, with 2-3 stones. Erect, deciduous shrub to 5m. Intrd-natd; S Hants and Westmorland

Hsing-Shan Cotoneaster - **C. hsingshangensis** J. Fryer & B. Hylmö

8 Leaves acute to shortly and gradually acuminate apex; calyx-lobes tomentose; fruits 8-10mm, broadly obovoid, with 2(-3) stones. Erect, deciduous shrub to 5m. Intrd-natd; scattered places in Br

Lleyn Cotoneaster - **C. villosulus** (Rehder & E.H. Wilson) Flinck & B. Hylmö

Key G -Petals erect to erecto-patent; leaves deciduous; fruits bright red to orange-red; leaves densely pubescent to tomentose on lowerside

1 Hypanthium and calyx glabrous or nearly so. Irregularly branched, spreading, deciduous shrub to 1.5m. Native; very few plants on limestone of Great Orme's Head (Caerns) (*C. integerrimus* auct.)

Wild Cotoneaster - **C. cambricus** J. Fryer & B. Hylmö

1 Hypanthium and calyx pubescent to tomentose, the pubescence persisting on calyx until fruit ripe ... 2
- 2 Fruits pendent, with 2 stones; some leaves usually >2.5cm ... 3
- 2 Fruits held stiffly, with 3-4 stones; usually ± all leaves <2.5cm ... 4

3 Fruits strongly obovoid to almost pear-shaped; often some leaves

32. COTONEASTER

>3cm. Erect, deciduous shrub to 3m. Intrd-natd; W Kent, Surrey and Man *Cherryred Cotoneaster* - **C. zabelii** C.K. Schneid. 253

3 Fruits globose to ellipsoid or slightly obovoid; leaves ± all <3cm. Erect, deciduous shrub to 3m. Intrd-natd; S Lancs
Fang's Cotoneaster - **C. fangianus** T.T. Yu 253

   4 Fruits bright red, 6-8mm, ± globose; branches arching-erect, long and whip-like; plant usually >1.5m. Deciduous shrub to 2(3)m. Intrd-natd; rather frequent in Br and Ir
Diels' Cotoneaster - **C. dielsianus** Diels 253

   4 Fruits orange-red, 8-10mm, usually slightly longer than wide; branches stiffly spreading; plant usually <1.5m. Widely spreading, deciduous shrub to 1(2)m. Intrd-natd; SW & SE En
Showy Cotoneaster - **C. splendens** Flinck & B. Hylmö 253

Key H - Petals erect; leaves deciduous; fruit bright red to orange-red or maroon; leaves pubescent to sparsely so on lowerside, bullate or at least with veins deeply impressed on upperside

1 Leaves 1.5-3cm; inflorescences with 3-7(15) flowers. Erect, semi-evergreen shrub to 3m. Intrd-natd; S Devon, Caerns and S Lancs
Maire's Cotoneaster - **C. mairei** H. Lév. 254

1 Most or all leaves >3cm; inflorescences with 9-30 flowers   2
   2 Fruits maroon; leaves 3-5cm   (see Key F, couplet 3) **C. obscurus**
   2 Fruits bright red to orange-red; leaves 3-15cm   3

3 Fruits orange-red, with 3-4(5) stones; leaves 3-6cm. Erect, deciduous shrub to 3m. Intrd-natd; N Hants and Lanarks
Bois's Cotoneaster - **C. boisianus** G. Klotz 254

3 Fruits bright red, with (4-)5 stones; leaves 3.5-15cm   4
   4 Leaves 3.5-7cm, ± bullate, pubescent on lowerside at flowering; fruits mostly <8mm. Arching, deciduous shrub to 4m. Intrd-natd; frequent in Br and Ir, Man
Hollyberry Cotoneaster - **C. bullatus** Bois 254

   4 Leaves 5-12cm, strongly bullate, sparsely pubescent on lowerside at flowering; fruits mostly >8mm. Arching, deciduous shrub to 5m. Intrd-natd; frequent in Br and Ir
Bullate Cotoneaster - **C. rehderi** Pojark. 254

Key I - Petals erect to erecto-patent; leaves deciduous; fruit bright red to orange-red; leaves pubescent to sparsely so on lowerside, not bullate and with veins not or scarcely impressed on upperside

1 Leaves most or all >1.3x as long as wide, acute or acuminate to obtuse at apex   2

1 Leaves most or all ≤1.3x as long as wide, mostly rounded to broadly obtuse (sometimes subacute or apiculate) at apex   5

   2 Fruits parallel-sided, oblong in side view (sausage-shaped); stamens10(-15). Deciduous shrub to 2m with wide spreading or arching branches. Intrd-natd; scattered in En, C Sc
Spreading Cotoneaster - **C. divaricatus** Rehder & E.H. Wilson 253

   2 Fruits subgloboid to broadly ellipsoid or obovoid, with curved

77. ROSACEAE

	sides; stamens (15-)20	3
3	Fruits orange-red; leaves 1.5-3cm. Erect, deciduous shrub to 3(4)m. Intrd-natd; rather common throughout most of BI	
	Himalayan Cotoneaster - **C. simonsii** Baker	2
3	Fruits bright red; leaves 2-5.5cm, most >3cm	4
4	Leaves ± tomentose on lowerside; flowers in groups of 1-4; fruits with 2-3 stones. Erect, deciduous shrub to 4m. Intrd-natd; S Br	
	Mucronate Cotoneaster - **C. mucronatus** Franch.	2
4	Leaves pubescent on lowerside; flowers in groups of (3)5-7(9); fruits with 3-4(5) stones. Erect, evergreen to semi-evergreen shrub to 2.5m. Intrd-natd; W Kent	
	Tengyueh Cotoneaster - **C. tengyuehensis** J. Fryer & B. Hylmö	2
5	Shrub to 3m; flowers pendent, with pedicels c.5mm; stamens (15-)20. Erect, evergreen or semi-evergreen shrub to 3m, with wide-spreading branches. Intrd-natd; S Hants	
	Distichous Cotoneaster - **C. nitidus** Jacques	2
5	Shrub to 1(-2)m; flowers usually erect, with pedicels to c.2mm; stamens10-13	6
6	Leaves matt or ± so on upperside, ± undulate; fruits bright red	7
6	Leaves shiny on upperside, flat or less often undulate; fruits orange-red to bright red	8
7	Leaves mostly 1-2.5cm; flowers 2-4 together; fruits mostly 10-12mm. Spreading, deciduous shrub to 0.5(1)m. Intrd-natd; W Kent	
	Dwarf Cotoneaster - **C. nanshan** Mottet	2
7	Leaves mostly 0.5-1.5cm; flowers 1(-2) together; fruits mostly 6-7mm. Usually procumbent, deciduous shrub to 0.3m, with irregular branching. Intrd-natd; W Kent and Lanarks	
	Creeping Cotoneaster - **C. adpressus** Bois	2
8	Petals red with purplish-black base and narrow white fringe. Arching to horizontal, deciduous shrub to 1(3)m, with less regular branching than in *C. horizontalis*. Intrd-natd; S Br and C Ir	
	Purple-flowered Cotoneaster - **C. atropurpureus** Flinck & B. Hylmö	2
8	Petals pink and red and/or white	9
9	Sepals glabrous on outer surface; leaves 1-2cm	10
9	Sepals pubescent on outer surface; leaves 0.6-1.5cm	11
10	Leaves apiculate to very shortly acuminate at apex; fruits 10-12mm, bright red. Procumbent to ascending, deciduous shrub to 1m. Intrd-natd; Renfrews and Lanarks	
	Apiculate Cotoneaster - **C. apiculatus** Rehder & E.H. Wilson	2
10	Leaves obtuse to rounded or mucronate at apex; fruits 6-8mm, orange-red. Arching to horizontal, deciduous shrub to 1.5(4.5)m, with stronger growth and less regular branching than in *C. horizontalis*. Intrd-natd; Br N to C Sc	
	Hjelmqvist's Cotoneaster - **C. hjelmqvistii** Flinck & B. Hylmö	2
11	Branches forming very regular herring-bone pattern; fruits orange-red. Arching to horizontal, deciduous shrub to 1(3)m, often vertical on walls. Intrd-natd; common in much of BI	
	Wall Cotoneaster - **C. horizontalis** Decne.	2

11 Branches irregular, not forming herring-bone pattern; fruits bright red. Deciduous shrub to 1(2)m with ascending branches. Intrd-natd; open woodland; W Lancs
Ascending Cotoneaster - **C. ascendens** Flinck & B. Hylmö   253

Key J - Petals erect to erecto-patent; leaves evergreen; fruits orange-red to bright red
1 Most leaves >2cm   2
1 Most leaves ≤2cm   7
   2 Leaves pubescent on lowerside; anthers white
       (see Key I, couplet 4) **C. tengyuehensis**
   2 Leaves tomentose on lowerside; anthers white or pigmented   3
3 Most leaves >3cm   4
3 Most leaves <3cm   5
   4 Anthers pale mauve; fruits with 2 stones. Erect, evergreen shrub to 2.5m. Intrd-natd; Offaly   *Ward's Cotoneaster* - **C. wardii** W.W. Sm.   254
   4 Anthers white; fruits with (2-)3 stones. Erect, evergreen shrub to 3m. Intrd-natd; scattered in Br and Ir
       Stern's Cotoneaster - **C. sternianus** (Turrill) Boom   254
5 Anthers white; inflorescences mostly with 3-7 flowers; leaves semi-evergreen   (see Key H, couplet 1) **C. mairei**
5 Anthers pink to mauve or pale purple; inflorescences with 5-15 flowers; leaves fully evergreen   6
   6 Fruits orange-red, with (2-)3 stones; leaves with veins deeply impressed on upperside, silvery- to yellowish-tomentose on lowerside. Erect or arching, evergreen shrub to 3m. Intrd-natd; scattered in Br and Ir   *Franchet's Cotoneaster* - **C. franchetii** Bois   254
   6 Fruits bright red, with 2-3 stones; leaves with veins slightly impressed on upperside, greyish-white-tomentose on lowerside. Evergreen shrub to 2m. Intrd-natd; scattered in En N to Westmorland   *Vilmorin's Cotoneaster* - **C. vilmorinianus** G. Klotz   253
7 Leaves 0.5-1.3cm, shiny on upperside, subglabrous on lowerside.
       (see Key I, couplet 5) **C. nitidus**
7 Leaves 1-2(2.5)cm, scarcely shiny on upperside, tomentose on lowerside   8
   8 Veins deeply impressed on leaf upperside; flowers 1-4 together; fruit with 3-4(5)stones; anthers white. Erect, evergreen shrub to 3m with spreading branches. Intrd-natd; Caerns
       Engraved Cotoneaster - **C. insculptus** Diels   253
   8 Veins not or scarcely impressed on leaf upperside; flowers 6-10 together; fruit with 2-3 stones; anthers pinkish-purple. Densely branched, evergreen shrub to 1.5m. Intrd-natd; S & SE En, Offaly. See note under *C. pannosus* (Key D, couplet 1)
       Beautiful Cotoneaster - **C. amoenus** E.H. Wilson   253

33. PYRACANTHA M. Roem. - *Firethorns*
Leaves simple, serrate, evergreen; flowers in compound corymbs; petals white; stamens 20; carpels 5, with 5 free styles, the walls stony at fruiting.

1   Petioles and inflorescence-stalks pubescent; leaves 2-7cm. Spiny shrub
    to 2(6)m. Intrd-natd; bird-sown plants on banks and walls and in
    rough ground; frequent in S & C Br   **Firethorn - P. coccinea** M. Roem.
1   Petioles and inflorescence-stalks ± glabrous; leaves up to 3(5)cm.
    Spiny shrub to 2(6)m. Intrd-natd; rare in SE En but probably
    overlooked   *Asian Firethorn* - **P. rogersiana** (A.B. Jacks.) Coltm.-Rog.

34. MESPILUS L. - *Medlar*

Leaves simple, ± entire, deciduous; flowers solitary; petals white; stamens 30-40; carpels 5, with 5 free styles, the walls stony at fruiting.

1   Shrub or tree to 9m, sometimes spiny; flowers 3-5cm across excl.
    projecting sepals; fruit subglobose, solitary, 2-3cm (to 6cm in
    cultivars). Intrd-natd; in hedges; local in CI and S Br, sporadic
    in C & N En   ***Medlar* - M. germanica** L.

34 x 35. MESPILUS X CRATAEGUS = X CRATAEMESPILUS E.G. Camus - *Haw-medlar*

1   Shrub or tree; leaves serrate, usually slightly lobed; fruits brown,
    obovoid, <2cm, with 2-3 styles and carpels, some usually 2-3 together.
    Probably intrd-natd; isolated trees of uncertain origin very scattered
    through Br (*M. germanica* x *C. laevigata*)
    [The graft-hybrid **+ Crataegomespilus dardarii** Bellair, which has
    very variable leaves on one plant and usually branches of the pure
    spp. here and there, and clustered fruits with sepals like those of
    Mespilus, may also occur planted in the wild]
                                Haw-medlar - **X C. grandiflora** (Sm.) E.G. Camus

35. CRATAEGUS L. - *Hawthorns*

Leaves simple, serrate, lobed or not, deciduous; flowers in corymbs; petals white to pink or red; stamens (5)10-20; carpels 1-5, with as many free styles, the walls stony at fruiting.

1   At least some leaves usually lobed >1/3 way to midrib; apices both of
    lobes and of sinuses between lobes on at least some leaves reached by
    major vein from midrib   2
1   Leaves not lobed or lobed usually <1/3 way to midrib; only apices of
    lobes or main teeth (not sinuses) reached by major lateral vein   5
    2   Leaves densely pubescent; styles and nutlets (3)4-5. Tree to 6m;
        spines very sparse. Intrd-natd; in hedges and on banks; En N to
        Derbys (*C. laciniata* auct.)
                                    ***Oriental Hawthorn* - C. orientalis** M. Bieb.
    2   Leaves glabrous to sparsely pubescent; styles and nutlets 1-3   3
3   Leaves varying from unlobed to deeply lobed on 1 tree, some
    narrowly oblong to oblanceolate and entire in basal 1/2. Tree to 6m;
    spines very sparse. Intrd-natd; urban woodland; Middlesex, Surrey
    and Midlothian   ***Various-leaved Hawthorn* - C. heterophylla** Flüggé

3 Leaves not strongly heterophyllous, none narrowly oblong, unlobed and entire in basal 1/2 4
    4 Styles and nutlets 1(-2); deepest sinus between leaf-lobes reaching >2/3 way to midrib; leaf-lobes (3)5-7, the lowest pair acute. Shrub or tree to 10(15)m; spines 1-2.5cm, strong to medium. Native; wood-borders, scrub and hedges; abundant
Hawthorn - **C. monogyna** Jacq.
    4 Styles and nutlets (1)2-3; deepest sinus between leaf-lobes reaching <2/3 way to midrib; leaf-lobes usually 3, the lateral pair obtuse. Tree to 10m, less spiny than *C. monogyna*. Native; woods, often well shaded, and hedges; common in C & SE En, scattered W to Wa and N to N En, natd in SW & NE Ir
Midland Hawthorn - **C. laevigata** (Poir.) DC.
5 Styles and nutlets mostly 2-3 6
5 Styles and nutlets mostly 4-5 8
    6 Leaves and inflorescences glabrous; nutlets without hollows on inner surfaces. Tree to 6m; spines 3-8cm, medium. Intrd-natd; hedges; S & C Br, over-recorded for *C. persimilis*
Cockspurthorn - **C. crus-galli** L.
    6 Leaves pubescent on lowerside veins; inflorescence-stalks pubescent; nutlets with hollows on inner surfaces 7
7 Stamens 10-15; leaves cuneate (<90°) at base. Tree to 6m; spines 3-8cm, medium. Intrd-natd; sometimes self-sown; S En
Broad-leaved Cockspurthorn - **C. persimilis** Sarg.
7 Stamens 15-20; leaves broadly cuneate (>90°) at base. Tree to 6m; spines 3-5cm, strong. Intrd-natd; woods and hedges; S En
Round-fruited Cockspurthorn - **C. succulenta** Schrad.
    8 Leaves pubescent on lowerside at first, only on veins later; inflorescence-stalks tomentose; stamens 10. Tree to 8m; spines 5-7cm, thin. Intrd-natd; self-sown in hedges and rough ground; W Kent and S Devon *Hairy Cockspurthorn* - **C. submollis** Sarg.
    8 Leaves glabrous to pubescent only on veins at first, glabrous later; inflorescence-stalks glabrous or with sparse shaggy hairs 9
9 Stamens 20; fruits subglobose; flowers >2cm across. Tree to 7m; spines 3-5cm, strong. Intrd-natd; natd as for *C. submollis*; Surrey and Yorks *Large-flowered Cockspurthorn* - **C. coccinioides** Ashe
9 Stamens 10; fruits ellipsoid to pear-shaped; flowers ≤2cm across. Tree to 7m; spines 3-5cm, medium. Intrd-natd; natd as for *C. submollis*; SE En and CI *Pear-fruited Cockspurthorn* - **C. pedicellata** Sarg.

C. monogyna x *C. laevigata* = ***C. x media*** Bechst. (*C. x macrocarpa* auct.) is common throughout the range of *C. laevigata* and even beyond. It is fertile and covers the whole spectrum of intermediacy. *C. monogyna* x *C. heterophylla* has arisen naturally in Surrey with both parents.

78. MIMOSACEAE - *Australian Blackwood family*

Suckering trees; at once recognizable by the racemes of small spherical pom-poms of flowers, of which the yellow stamens are the most conspicuous part, and the lanceolate to oblanceolate leaves with 3-5 longitudinal veins.

1. ACACIA Mill. - *Australian Blackwood*

1 Tree to 15m; leaves 6-13(20)cm; flower-heads c.10mm across. Intrd-natd; seaside scrub; S Devon and Scillies
Australian Blackwood - **A. melanoxylon** R. Br.

79. FABACEAE - *Pea family*

Annual to perennial herbs, shrubs or trees, sometimes spiny; leaves simple to palmate or pinnate, often with tendrils, usually stipulate. Flowers always like that of the pea in organization, with 5 petals - the upper (*standard*), 2 free laterals (*wings*) and 2 fused lower (*keel*); stamens 10, usually all fused into tube or the uppermost free and the 9 lower fused, rarely all 10 free; fruit basically a legume, but very variably modified, usually dehiscent along 2 sides but often a schizocarp (breaking transversely into 1-seeded units).

General key
1	Leaves simple, sometimes reduced to a tendril, spine or scale, sometimes 0	**Key A**
1	At least some leaves with at least 2 leaflets	2
2	Leaves with 1-many pairs of leaflets, with or without an odd terminal leaflet, if with then pairs of leaflets >1, if without then often with tendrils	**Key C**
2	Leaves ternate or palmate, without tendrils	**Key B**

Key A - Leaves simple, sometimes reduced to tendril, spine or scale, or 0
1	Herbaceous annuals or perennials	2
1	Woody shrubs	5
2	Fruit opening along 2 sides like a pea-pod	**21. LATHYRUS**
2	Fruit indehiscent, or breaking transversely between seeds	3
3	Fruit 1(-2)-seeded, enclosed in calyx	**11. ANTHYLLIS**
3	Fruit >2-seeded, exserted from calyx	4
4	Plant glabrous; fruit curved, smooth	**15. CORONILLA**
4	Plant slightly pubescent; fruit spiralled, with tubercles or weak spines	**18. SCORPIURUS**
5	Plant spiny, at least some spines branched	6
5	Spines 0 or simple	7
6	Upper calyx-lip with 2 short teeth; small bracteole present on either side of flower	**35. ULEX**
6	Upper calyx-lip divided >1/3 way to base; bracteoles 0	**34. GENISTA**
7	Flowers white	**32. CYTISUS**

79. FABACEAE

7	Flowers yellow to reddish	8
	8 Twigs strongly angled or grooved; upper calyx-lip with 2 short teeth (divided <1/5 way to base)	**32. CYTISUS**
	8 Twigs finely grooved; upper calyx-lip divided ≥1/4 way to base	9
9	Spines 0; upper calyx-lip divided nearly to base, the 2 halves inclined downwards near lower lip	**33. SPARTIUM**
9	Spines 0 or present; calyx with distinct upper and lower lips, the upper divided 1/4-3/4 way to base	**34. GENISTA**

Key B - Leaves ternate or palmate, without tendrils

1	Leaves palmate, with >4 leaflets	**30. LUPINUS**
1	Leaves ternate	2
	2 Woody trees or shrubs	3
	2 Herbaceous annuals or perennials	8
3	Stems spiny; corolla yellow	**35. ULEX**
3	Stems not spiny, or if spiny corolla not yellow	4
	4 Fruit spiral	**27. MEDICAGO**
	4 Fruit ± straight	5
5	Leaflets toothed; flowers usually pinkish-purple, rarely white	**24. ONONIS**
5	Leaflets entire; flowers usually yellow, rarely white	6
	6 Leaflets mostly >3cm; racemes usually pendent	**31. LABURNUM**
	6 Leaflets mostly <3cm; flowers 1-few or in ± erect racemes	7
7	Upper lip of calyx deeply bifid	**34. GENISTA**
7	Upper lip of calyx with 2 short teeth	**32. CYTISUS**
	8 Main lateral veins of leaflets running whole way to margin; leaflets often toothed	9
	8 Main lateral veins of leaflets not reaching margin; leaflets rarely toothed	17
9	Calyx with glandular (and often non-glandular) hairs; all 10 stamens fused into tube	**24. ONONIS**
9	Calyx without glandular hairs; 9 stamens fused into tube, the 10th free	10
	10 Flowers in elongated racemes; fruits ≤7mm, all with 1-2 seeds	11
	10 Flowers few, or in short dense heads, or if in elongated racemes then fruits >7mm and at least some with >2 seeds	12
11	Flowers yellow or white; fruits exserted from calyx-tube	**25. MELILOTUS**
11	Flowers cream or pink to purple; fruits included in calyx-tube	**28. TRIFOLIUM**
	12 Fruits spiralled into >1/2 complete coil, often spiny	**27. MEDICAGO**
	12 Fruits straight to curved (≤1/2 complete coil), never spiny	13
13	Fruits >3cm, plus beak >1cm	**26. TRIGONELLA**
13	Fruits <3cm, incl. beak <1cm	14
	14 Flowers yellow; fruits ± curved, >7mm, at least some >2-seeded	15
	14 Flowers not yellow, or if yellow then fruits straight, <7mm, 1-2-seeded	16
15	Ripe fruits pendent; plant annual	**26. TRIGONELLA**

15	Ripe fruits ± erect; plant perennial	**27. MEDICAGO**
	16 Fruits inflated, exserted from calyx and forming a compact naked head	**26. TRIGONELLA**
	16 Fruits not or scarcely inflated, usually at least partly covered by calyx or persistent corolla, not forming a compact naked head	**28. TRIFOLIUM**
17	Leaflets with small stipule-like outgrowths at base	18
17	Leaflets without stipule-like outgrowths at base	20
	18 Fruit erect to patent; stipules broadly ovate	**3. VIGNA**
	18 Fruit pendent; stipules triangular to narrowly so	19
19	Corolla c. as long as calyx; common peduncle 0 to very short; plant with brown patent hairs	**4. GLYCINE**
19	Corolla much longer than calyx; common peduncle long; plant glabrous to rather sparsely pubescent with whitish hairs	**2. PHASEOLUS**
	20 Petals blue to white; leaflets dentate	**5. PSORALEA**
	20 Petals yellow; leaflets entire	21
21	All 10 stamens free	**29. THERMOPSIS**
21	9(-10) stamens fused into a tube	22
	22 Fruit 1(-2)-seeded, enclosed in calyx	**11. ANTHYLLIS**
	22 Fruit >2-seeded, exserted	23
23	Fruit curved, breaking transversely between seeds at maturity	**15. CORONILLA**
23	Fruit ± straight, dehiscing longitudinally along 2 sides, sometimes tardily	24
	24 Fruit with 4 longitudinal wings	**13. TETRAGONOLOBUS**
	24 Fruit not winged	**12. LOTUS**

Key C - Leaves with 1-many pairs of leaflets, if with 1 then without odd terminal leaflet, often with tendrils

1	Leaves with even no. of leaflets, terminated by point or tendril	2
1	Leaves with odd no. of leaflets, terminated by single leaflet	6
	2 Stem winged, and/or leaflets parallel-veined	**21. LATHYRUS**
	2 Stem not or scarcely winged; leaflets pinnately veined	3
3	At least some stipules >2cm, larger than leaflets	**22. PISUM**
3	Stipules <2cm, smaller than leaflets	4
	4 Calyx-teeth equal, >2x as long as tube	**20. LENS**
	4 Calyx-teeth usually unequal, 2-5 of them <2x as long as tube	5
5	Style glabrous, or pubescent all round, or pubescent only on lowerside	**19. VICIA**
5	Style pubescent only on upperside	**21. LATHYRUS**
	6 Woody shrubs or trees	7
	6 Herbaceous, sometimes ± woody at base	10
7	Tree; corolla white	**1. ROBINIA**
7	Shrub; corolla pale yellow to orange	8
	8 Fruit strongly inflated, indehiscent or dehiscing longitudinally; flowers in racemes	**7. COLUTEA**
	8 Fruit ± not inflated, breaking transversely between seeds;	

79. FABACEAE

	flowers in umbels	9
9	Claws of wings and standard 2-3x as long as calyx; fruits 5-11cm; stems ridged or furrowed	**16. HIPPOCREPIS**
9	Claws of wings and standard 1-1.3x as long as calyx; fruits 1-5cm; stems terete	**15. CORONILLA**
	10 Flowers in racemes	11
	10 Flowers solitary or in umbels	14
11	Corolla pink to purple; fruits with 1 seed	**10. ONOBRYCHIS**
11	Corolla white, yellow or blue; fruits with ≥2 seeds	12
	12 Keel beaked at apex	**9. OXYTROPIS**
	12 Keel subacute to rounded at apex	13
13	Fruit not partitioned internally, terete, glabrous; uppermost stamen fused to stamen-tube for part of its length	**6. GALEGA**
13	Fruit longitudinally partitioned by internal membrane variously shaped but not terete and glabrous; uppermost stamen free	**8. ASTRAGALUS**
	14 Fruit with 2 longitudinal sutures, usually dehiscing along them, without transverse sutures	15
	14 Fruit breaking along transverse sutures between seeds	17
15	Fruits >3x as long as wide, with >3 seeds	**12. LOTUS**
15	Fruits <3x as long as wide, with 1-2 seeds	16
	16 Fruits longer than calyx; leaflets toothed	**23. CICER**
	16 Fruits enclosed within calyx; leaflets entire	**11. ANTHYLLIS**
17	Fruit-segments and seeds horseshoe-shaped	**16. HIPPOCREPIS**
17	Fruit-segments and seeds oblong-ellipsoid	18
	18 Strong perennials; flowers usually ≥10 per inflorescence, 8-15mm	**17. SECURIGERA**
	18 Annuals; flowers ≤8(12) per inflorescence, 3-9mm	**14. ORNITHOPUS**

TRIBE 1 - ROBINEAE (genus 1). Trees; leaves imparipinnate, with entire leaflets; flowers in pendent racemes; 9 stamens forming tube, 10th free; fruit longitudinally dehiscent, ≥3-seeded.

1. ROBINIA L. - *False-acacia*

1 Deciduous tree to 29m; twigs with stipular spines; flowers white; fruit 5-10cm. Intrd-natd; banks, scrub and woodland; scattered over most of BI but natd ± only in S *False-acacia* - **R. pseudoacacia** L.

TRIBE 2 - PHASEOLEAE (genera 2-4). Herbs; leaves ternate, with entire leaflets, with stipule-like outgrowths below each leaflet; flowers in axillary racemes or clusters; 9 stamens forming tube, 10th free or also part of tube; fruit longitudinally dehiscent, often tardily so, ≥2-seeded.

2. PHASEOLUS L. - *Beans*
Herbs, often climbing, subglabrous to shortly or appressed-pubescent; flowers in racemes with long common peduncle; corolla much longer than

calyx; fruit pendent, many-seeded.

1 Annual, with spirally climbing stem to 3m or (usually) not climbing; flowers ≤6, white, sometimes purplish; cotyledons borne above ground. Intrd-casual; tips and waste places; scattered throughout lowland Br **French Bean - P. vulgaris** L.
1 Perennial with tuberous roots but rarely surviving winter, with climbing stem to 5m; flowers ≤20, bright red, rarely white; cotyledons subterranean. Intrd-casual; casual as for *P. vulgaris*
Runner Bean - P. coccineus L.

3. VIGNA Savi - *Mung-bean*
Herbs, not climbing, with patent hairs; flowers with long common peduncle; corolla much longer than calyx; fruit erect to patent, (3)8-14-seeded.

1 Erect annual to 60(100)cm; flowers white to purple. Intrd-casual; waste land and near docks and factories; very scattered in S Br
Mung-bean - V. radiata (L.) Wilczek

4. GLYCINE Willd. - *Soyabean*
Herbs, not climbing, with long patent brown hairs; flowers with 0 or very short common peduncle; corolla c. as long as calyx; fruit pendent, 2-4-seeded.

1 Erect annual to 60(100)cm; flowers white to purple. Intrd-casual; waste land and near docks and factories; very scattered in S Br
Soyabean - G. max (L.) Merr.

TRIBE 3 - PSORALEEAE (genus 5). Herbs; leaves ternate, with dentate leaflets; flowers in erect axillary racemes; 10 stamens forming tube; fruit 1-seeded, indehiscent.

5. PSORALEA L. - *Scurfy Pea*

1 Erect perennial (but not surviving winter) to 50cm; flowers c.8mm, white tinged violet. Intrd-casual; birdseed-alien on tips; sporadic in S Br **Scurfy Pea - P. americana** L.

TRIBE 4 - GALEGEAE (genera 6-9). Herbs or shrubs; leaves imparipinnate, with entire leaflets; flowers in axillary racemes; 9 stamens forming tube, 10th free or fused to others for c.1/2 way; fruit longitudinally dehiscent or ± indehiscent, few- to many-seeded.

6. GALEGA L. - *Goat's-rue*
Flowers numerous in erect racemes; keel not beaked; 10th stamen partly fused to other 9; fruit ± terete, erect, not inflated, dehiscent.

1 Erect, glabrous or sparsely pubescent perennial to 1.5m; leaves with 9-17 leaflets; flowers white to bluish-mauve. Intrd-natd; on tips and in waste and grassy places; frequent in S & C Br
Goat's-rue - **G. officinalis** L.

7. COLUTEA L. - *Bladder-sennas*
Flowers c.2-8 in ± erect racemes; keel beaked or not; 10th stamen free; fruit greatly inflated, ± pendent.

1 Flowers pale to deep yellow, with beakless keel; fruit indehiscent. Deciduous shrub to 4m. Intrd-natd; much grown in gardens and natd in waste and grassy places, on roadsides and railway banks; frequent in S Br, especially SE En, Man *Bladder-senna* - **C. arborescens** L.
1 Flowers orange-bronze, with beaked keel; fruits splitting open at apex. Deciduous shrub to 4m. Intrd-natd; habitats as for *C. arborescens*; very scattered in S En (*C. arborescens* x *C. orientalis* Mill.)
Orange Bladder-senna - **C. x media** Willd.

8. ASTRAGALUS L. - *Milk-vetches*
Herbaceous perennials (rarely annuals); flowers in ± erect racemes; keel not beaked; 10th stamen free; fruit inflated to not so, variable.

1 Corolla blue to purple (rarely white) 2
1 Corolla whitish-cream to yellow 3
   2 Stipules free from each other; flowers patent to reflexed; fruit with dark, appressed hairs. Ascending perennial to 30cm. Native; grassy rocky places on mountains at 700-800m; very rare in C Sc *Alpine Milk-vetch* - **A. alpinus** L.
   2 Stipules fused below; flowers erect to erecto-patent; fruit with whitish, spreading hairs. Ascending perennial to 30cm. Native; short grass on calcareous well-drained soils; local in E Br from Beds to E Sutherland, extremely scattered elsewhere in En and Sc, and Aran Isles (Co Clare) *Purple Milk-vetch* - **A. danicus** Retz.
3 Leaflets mostly <17; fruits >2cm. Sprawling perennial to 1(1.5)m. Native; grassy places and scrub, mostly on calcareous soils; very scattered in Br N to Moray *Wild Liquorice* - **A. glycyphyllos** L.
3 Leaflets mostly >17; fruits <2cm 4
   4 Fruits ovoid-globose, 10-15mm; calyx 7-10mm; standard 14-16mm. Spreading to suberect perennial to 60(100)cm. Intrd-natd; hedgebank; Midlothian *Chick-pea Milk-vetch* - **A. cicer** L.
   4 Fruits oblong, compressed, 8-10mm; calyx 4-5mm; standard 9-12mm. Erect to ascending perennial to 30cm. Intrd-natd; in grassy places; very scattered in C & S En
Lesser Milk-vetch - **A. odoratus** Lam.

9. OXYTROPIS DC. - *Oxytropises*
Herbaceous perennials; flowers in erect racemes; keel beaked at apex; 10th stamen free; fruit elongated, grooved abaxially, slightly inflated, erect.

1 Flowers usually pale purple, rarely white; fruit 15-20(25)mm, pubescent, divided internally by septa from both adaxial and abaxial sutures and ± bilocular. Leaves and leafless peduncle to 30cm, arising from dense tuft. Native; grassy rocky places; very local in SW, C & N mainland Sc *Purple Oxytropis* - **O. halleri** W.D.J. Koch
1 Flowers pale yellow, often strongly tinged with purple; fruit 14-18mm, divided internally by septum from abaxial suture only and semi-bilocular. Habit as *O. halleri*. Native; cliffs and rock-ledges; rare in C and SW Sc *Yellow Oxytropis* - **O. campestris** (L.) DC.

TRIBE 5 - HEDYSAREAE (genus 10). Herbaceous perennials; leaves imparipinnate, with entire leaflets; flowers in axillary racemes; 9 stamens forming tube, 10th free; fruit indehiscent, 1-seeded.

10. ONOBRYCHIS Mill. - *Sainfoin*

1 Stems suberect, to 60(80)cm; flowers pinkish-red; fruit 5-8mm, reticulately ridged, toothed. Possibly native; grassland and bare places mostly on chalk and limestone; locally frequent in Br N to Yorks, scattered casual or natd elsewhere *Sainfoin* - **O. viciifolia** Scop.

TRIBE 6 - LOTEAE (genera 11-13). Annuals or herbaceous perennials; leaves ternate to imparipinnate, with entire leaflets; flowers solitary or in umbels; 9 stamens forming tube, 10th free or variably and loosely fused with others; fruit dehiscent or not, 1-many-seeded.

11. ANTHYLLIS L. - *Kidney Vetch*

Lower leaves with large terminal and 0-3 pairs of small lateral leaflets, grading variably to upper leaves with ≤15 ± equal leaflets; stipules small, falling early; 10th stamen variably and loosely fused with other 9; calyx inflated, enclosing indehiscent 1(-2)-seeded fruit.

1 Erect to procumbent perennial to 60cm; calyx densely white-pubescent; petals yellow to red (*Kidney Vetch* - **A. vulneraria** L.) 2
   2 Calyx (4.5)5-7mm wide, the lateral teeth not appressed to upper ones; upper leaves with large terminal and (0)1-4 pairs of smaller lateral leaflets 3
   2 Calyx 2-4(5)mm wide, the lateral teeth appressed to upper ones and ± obscured; upper leaves with 4-7 pairs of lateral leaflets scarcely smaller than terminal one 4
3 Calyx-hairs appressed, sparse. Corolla pale yellow. Intrd-natd; marginal and disturbed ground; scattered over Br, Ir and Man
A. vulneraria ssp. **carpatica** (Pant.) Nyman
3 Calyx-hairs ± patent. Corolla yellow. Native; banks, cliffs and rock-ledges; mostly mountainous areas of N En, W & NW Ir and Sc
A. vulneraria ssp. **lapponica** (Hyl.) Jalas
   4 Stem-hairs all patent. Corolla yellow. Native; sea-cliffs; Anglesey,

W Cornwall and CI (Sark and Guernsey)
A. vulneraria ssp. corbierei (C.E. Salmon & Travis) Cullen
4 At least upper part of stem with appressed hairs 5
5 Calyx usually with red tip; stems usually with appressed hairs throughout or semi-appressed ones at base. Corolla yellow to orange, rarely pink or red. Native; grassland, dunes, cliff-tops, waste ground, usually calcareous; locally common **A. vulneraria ssp. vulneraria**
5 Calyx usually without red tip; lower part of stems with patent hairs. Corolla pale yellow. Intrd-natd; grassy places; E Sc and S En
A. vulneraria ssp. polyphylla (DC.) Nyman

12. LOTUS L. - *Bird's-foot-trefoils*
Leaves with 5 leaflets, the lowest 2 at base of rhachis and resembling stipules; stipules minute, falling or withering early; 10th stamen free; calyx not enclosing fruit; fruit several-seeded, longitudinally dehiscent, not ridged or angled.

1 Annuals, with dense patent hairs; fruit mostly <2mm wide 2
1 Perennials, glabrous to variously pubescent; fruit mostly >2mm wide 3
   2 Fruit 12-30mm, ≥3x as long as calyx, with >12 seeds; keel with ± right-angled bend c.1/2 way along lower edge of limb. Stems procumbent to decumbent, to 30(80)cm. Native; dry grassy places near sea; very local in S En and CI
Slender Bird's-foot-trefoil - **L. angustissimus** L.
   2 Fruit 6-15mm, ≤3x as long as calyx, with ≤12 seeds; keel with obtuse-angled bend near base of lower edge of limb. Stems procumbent to decumbent, to 30(80)cm. Native; dry grassy places near sea; very local in SW En E to S Hants, extreme SW Wa, SW & SE Ir, CI *Hairy Bird's-foot-trefoil* - **L. subbiflorus** Lag.
3 Stem hollow; calyx-teeth recurved in bud, the upper 2 with acute sinus between them. Stems erect to ascending, to 1m. Native; damp grassy places, marshes and pondsides; frequent throughout most of BI *Greater Bird's-foot-trefoil* - **L. pedunculatus** Cav.
3 Stem solid or rarely hollow; calyx-teeth erect in bud, the upper 2 with obtuse sinus between them 4
   4 Leaflets of upper leaves mostly >4x as long as wide, acute to acuminate. Stems sprawling to suberect, to 90cm. Native; dry grassy places; scattered in Br N to C Sc, N & E Ir, CI
Narrow-leaved Bird's-foot-trefoil - **L. glaber** Mill.
   4 Leaflets of upper leaves mostly <3(4)x as long as wide, subacute to obtuse. Stems procumbent to ascending, to 50cm. Native; grassy and barish places, mostly on well-drained soils; common to abundant *Common Bird's-foot-trefoil* - **L. corniculatus** L.

13. TETRAGONOLOBUS Scop. - *Dragon's-teeth*
Leaves with 3 leaflets; stipules herbaceous, persistent; 10th stamen free; calyx not enclosing fruit; fruits several-seeded, tardily longitudinally dehiscent, square in section with wing c.1mm wide on each angle.

1   Sparsely pubescent perennial with decumbent stems to 30cm; flowers
    solitary, yellow with brownish streaks. Intrd-natd; rough calcareous
    grassland; S En, rare casual elsewhere in S & C Br
                               Dragon's-teeth - **T. maritimus** (L.) Roth

TRIBE 7 - CORONILLEAE (genera 14-18). Herbaceous annuals or
perennials or shrubs; leaves simple, ternate or imparipinnate, with entire
leaflets; flowers solitary or in umbels; 9 stamens forming tube, 10th free;
fruit dehiscing transversely between seeds or ± indehiscent, few- to many-
seeded.

14. ORNITHOPUS L. - *Bird's-foots*
Annuals; leaves imparipinnate with ≥3 pairs of lateral leaflets; fruits curved
to ± straight, beaked, not or slightly constricted between the 3-12 cylindrical
to oblong-ellipsoid segments.

1   Flower-heads without bract at base, or with minute scarious bracts.
    Stems to 50cm, decumbent to ascending. Native; short turf or open
    ground on sandy soil; CI (all islands) and Scillies
                         Orange Bird's-foot - **O. pinnatus** (Mill.) Druce
1   Flower-heads with leaf-like bract at base   2
    2 Corolla yellow; fruits not or scarcely constricted between
       segments. Stems to 50cm, procumbent to decumbent. Intrd-natd;
       barish sandy banks; W Kent, rare casual elsewhere in S Br and
       CI   *Yellow Serradella* - **O. compressus** L.
    2 Corolla white to pink; fruits distinctly constricted between
       segments   3
3   Corolla >5.5mm; bract c.1/2 as long as flowers. Stems to 70cm,
    procumbent to ascending. Intrd-natd; on china-clay waste;
    E Cornwall, rare casual elsewhere in S Br   *Serradella* - **O. sativus** Brot.
3   Corolla <5.5mm; bract at least as long as flowers. Stems to 30cm,
    procumbent to decumbent. Native; dry barish sandy and gravelly
    ground; locally common in BI, especially S & E, but absent from
    much of Ir and Sc   *Bird's-foot* - **O. perpusillus** L.

15. CORONILLA L. - *Scorpion-vetches*
Annuals or shrubs, with terete branches; leaves simple, ternate or
imparipinnate; fruits curved to ± straight, beaked, not or scarcely
constricted between the (1)2-11 ± cylindrical segments.

1   Shrub to 1m; leaves with 5-13 leaflets; flowers 4-8(12), 7-12(14)mm,
    yellow. Intrd-natd; on cliffs and banks by sea; S Devon and E Sussex
                          Shrubby Scorpion-vetch - **C. valentina** L.
1   Annual to 40cm; leaves with 1-3 leaflets, rarely all with 1; flowers
    2-5, 3-8mm, yellow. Intrd-casual; fairly frequent, mostly from
    birdseed; S Br   *Annual Scorpion-vetch* - **C. scorpioides** (L.) W.D.J. Koch

16. HIPPOCREPIS L. - *Horseshoe Vetches*
Herbaceous perennials or shrubs, with furrowed or ridged stems; leaves imparipinnate; fruits ± straight to curved, beaked, with horseshoe-shaped segments, strongly compressed.

1 Shrub to 1.5(2)m; leaves with 5-9 leaflets; flowers (12)14-20mm, yellow; fruits 5-11cm. Intrd-natd; roadsides and banks; En N to S Lincs *Scorpion Senna* - **H. emerus** (L.) Lassen
1 Procumbent to suberect, herbaceous perennial to 30(50)cm; leaves with 7-25 leaflets; flowers 5-10(14)mm, yellow; fruits 1-3cm. Native; dry calcareous grassland and cliff-tops; local in Br N to Westmorland, NW Jersey *Horseshoe Vetch* - **H. comosa** L.

17. SECURIGERA DC. - *Crown Vetch*
Herbaceous perennials with furrowed or ridged stems; leaves imparipinnate; fruits ± straight to slightly curved, beaked, not or scarcely constricted between the 3-8(12) cylindrical segments.

1 Stems sprawling, to 1.2m; leaves with 11-25 leaflets; flowers 8-15mm, white to pink or purple; fruits 2-6(8)cm. Intrd-natd; in grassy places and rough ground; scattered through Br N to C Sc, E Ir, Guernsey
Crown Vetch **S. varia** (L.) Lassen

18. SCORPIURUS L. - *Caterpillar-plant*
Annuals with furrowed or ridged stems; leaves simple; fruits much curved, spiralled or variously contorted, longitudinally ridged, variously ornamented with spines and tubercles, beaked, indehiscent or tardily dehiscent.

1 Stems procumbent to suberect, to 80cm; flowers 5-12mm, yellow, often red-tinged. Intrd-casual; birdseed- or wool-alien on tips, rough ground and in gardens and parks; C & S Br, mainly S En
Caterpillar-plant - **S. muricatus** L.

TRIBE 8 - FABEAE (genera 19-22). Herbaceous annuals or perennials; leaves usually paripinnate, often with tendril(s) at apex, rarely simple or reduced to a tendril, with usually entire (rarely dentate) leaflets; flowers solitary or in axillary racemes; 9 stamens forming tube, 10th free; fruit dehiscing longitudinally, (1)2-many seeded.

19. VICIA L. - *Vetches*
Stem ± not winged (often ridged); leaves paripinnate with (1)2-many pairs of pinnately-veined leaflets, usually with terminal tendril(s); stipules smaller than leaflets; at least 2 calyx-teeth <2x as long as tube; style glabrous, or pubescent all round, or pubescent only on lowerside.

1 All leaves without tendrils, terminated by small point 2
1 At least upper leaves terminated by tendril(s) 4

2	Perennial; flowers ≥6; peduncles >3cm; leaflets >5 pairs. Stems erect, to 60cm. Native; grassy and rocky places and scrub; scattered through W En, Sc, Wa and C & N Ir ***Wood Bitter-vetch* - V. orobus** DC.	
2	Annual; flowers ≤6; peduncles 0 or <2cm; leaflets <5 pairs	3
3	Flowers <1cm; leaflets <2cm; fruits <3cm. Procumbent to weakly climbing annual to 20cm. Native; maritime sand and inland sandy heaths; scattered over most of BI ***Spring Vetch* - V. lathyroides** L.	
3	Flowers >1cm; leaflets >2cm; fruit >5cm. Erect annual to 1m. Intrd-casual; tips and waste ground; throughout BI, mainly S Br ***Broad Bean* - V. faba** L.	
	4 Peduncle 0, or shorter than each flower	5
	4 Peduncle longer than each flower	13
5	Standard pubescent on back. Climbing annual to 60cm. Intrd-natd; waste places and roadside banks; frequent casual in En and Wa, natd in W Kent ***Hungarian Vetch* - V. pannonica** Crantz	
5	Standard glabrous on back	6
	6 Flowers 6-9mm, solitary; seeds tuberculate (see couplet 3) **V. lathyroides**	
	6 Flowers 9-25(30)mm, 1-several; flowers rarely cleistogamous and <9mm but then seeds smooth	7
7	Leaflets 1-3 pairs; stipules c.1cm	8
7	At least some leaves with >3 pairs leaflets; stipules <8mm	9
	8 Fruit pubescent all over, ≤1cm wide; leaflets >2x as long as wide. Climbing or scrambling annual to 60cm. Probably native; scrub, rough grassland and hedges; rare near coast in CI and Br N to Wigtowns, mainly S En ***Bithynian Vetch* - V. bithynica** (L.) L.	
	8 Fruit pubescent only along 2 sutures, ≥1cm wide; leaflets <2x as long as wide. Erect climbing annual to 60cm. Intrd-casual; on waste land from grain; scattered in S Br ***Narbonne Vetch* - V. narbonensis** L.	
9	Perennial; seeds with hilum >1/2 total circumference; lower calyx-teeth longer than upper but shorter than tube. Stems climbing or sprawling, to 60(100)cm. Native; grassy places, hedges, scrub and wood-borders, rarely sand-dunes; common ***Bush Vetch* - V. sepium** L.	
9	Annual; seeds with hilum <1/2 total circumference; all calyx-teeth equal, or unequal and lower longer than tube	10
	10 Calyx-teeth unequal (the lower longer); seeds with hilum 1/3-1/2 total circumference; corolla usually yellow. Procumbent to sprawling annual to 60cm. Native; maritime shingle and cliffs; very scattered round coasts of CI and Br N to S Sc, casual inland ***Yellow-vetch* - V. lutea** L.	
	10 Calyx-teeth ± equal; seeds with hilum 1/4-1/3 total circumference; corolla usually pink to purple (***Common Vetch* - V. sativa** L.)	11
11	Plant heterophyllous, the leaflets of upper leaves much (and abruptly) narrower than those of lower leaves; flowers ± concolorous, usually bright pinkish-purple. Climbing to sprawling annual to 75cm. Native; sandy banks, heathland, maritime sand	

and shingle; throughout most of BI **V. sativa ssp. nigra** (L.) Ehrh.
11 Plant ± isophyllous, the leaflets of upper leaves little (and gradually) narrower than those of lower leaves; flowers usually bicolorous, the standard much paler than wings 12
  12 Fruits smooth, usually glabrous, brown to black. Climbing annual to 1m. Probably intrd-natd; grassy and rough places and field-borders; common **V. sativa ssp. segetalis** (Thuill.) Gaudin
  12 Fruits slightly constricted between seeds, often pubescent, yellowish to brown. Robust climbing annual to 1.5m. Intrd-casual; waste places and field-borders; very scattered in Br
  V. sativa ssp. sativa
13 Flowers 2-8(9)mm, white to purple (not blue), 1-8 per raceme 14
13 Flowers 8-20mm, if <10mm then blue and >8 per raceme 16
  14 Calyx-teeth equal, all at least as long as tube; fruits usually 2-seeded. Scrambling annual to 80cm. Native; rough ground and grassy places; throughout lowland BI
  Hairy Tare - **V. hirsuta** (L.) Gray
  14 Calyx-teeth unequal, at least the upper shorter than tube; fruits 3-6-seeded 15
15 Flowers 1-2; seeds (3)4(-5), with hilum >2x as long as wide and c.1/5 seed circumference. Scrambling annual to 60cm. Native; grassy places; common in CI, En and Wa, very scattered elsewhere
  Smooth Tare - **V. tetrasperma** (L.) Schreb.
15 Flowers 1-4(5); seeds 4-6(8), with hilum little longer than wide and <1/3 seed circumference. Scrambling annual to 60cm. Native; grassy places; local in S En N to Hunts *Slender Tare* - **V. parviflora** Cav.
  16 Leaves with 2-3 pairs of leaflets; flowers 1-3
  (see couplet 8) **V. bithynica**
  16 Leaves with ≥4 pairs of leaflets; flowers usually >4 17
17 Standard with limb c.1/2 as long as claw; calyx very asymmetrical at base, with large bulge on upperside 18
17 Standard with limb c. as long as or longer than claw; calyx only slightly asymmetrical at base 19
  18 Corolla reddish-purple with blackish tip. Scrambling or climbing annual to 80cm. Intrd-casual; on tips and rough and waste ground; sporadic in C & S Br *Purple Vetch* - **V. benghalensis** L.
  18 Corolla blue to violet or purple, sometimes with white or yellow wings. Scrambling or climbing annual to 2m. Intrd-natd; grassy places, tips, rough and waste ground, mostly casual; throughout much of Br *Fodder Vetch* - **V. villosa** Roth
19 Lower lobe of stipules strongly toothed; corolla white with blue or purple veins; seeds with hilum >1/2 total circumference. Scrambling or climbing perennial to 2m. Native; open woods and wood-borders, scree, scrub, maritime cliffs and shingle; local in Br and Ir
  Wood Vetch - **V. sylvatica** L.
19 Lower lobe of stipules entire; corolla blue to purple or violet; seeds with hilum <1/2 total circumference 20
  20 Corolla 8-12(13)mm; limb of standard c. as long as claw; seeds

with hilum 1/4-1/3 total circumference. Scrambling or climbing perennial to 2m. Native; grassy and bushy places and hedgerows; common *Tufted Vetch* - **V. cracca** L.

20 Corolla (10)12-18mm; limb of standard longer than claw; seeds with hilum 1/5-1/4 total circumference. Scrambling or climbing perennial to 2m. Intrd-natd; grassy places and rough ground; scattered in En and Sc *Fine-leaved Vetch* - **V. tenuifolia** Roth

20. LENS Mill. - *Lentil*

Stems scarcely winged, markedly ridged; leaves paripinnate, with 3-8 pairs of pinnately-veined leaflets; simple tendril present on upper leaves; stipules smaller than leaflets; calyx-teeth all >2x as long as tube; style pubescent on upperside.

1 Weakly climbing annual to 40cm flowers 1-3, white to pale mauve; fruits 1-2(3)-seeded. Intrd-casual; fairly frequent grain-alien on tips and rough ground; scattered in S Br *Lentil* - **L. culinaris** Medik.

21. LATHYRUS L. - *Peas*

Stems angled or winged; leaves usually paripinnate with 1-many pairs of pinnately- or parallel-veined leaflets, sometimes reduced to a simple blade or a simple tendril; terminal tendril present or 0; stipules variable; calyx-teeth variable; style pubescent only on upperside.

1	Leaves reduced to single blade or tendril	2
1	Leaves with 1-many pairs of leaflets; terminal tendril present or 0	3
	2 Leaf reduced to single blade; stipules <3mm; flowers reddish. Erect or ascending annual to 90cm. Native; grassy places; local in En and S Wa N to N Lincs, rare casual elsewhere *Grass Vetchling* - **L. nissolia** L.	
	2 Leaf reduced to simple tendril; stipules >3mm, leaf-like; flowers yellow. Scrambling annual to 40(100)cm. Probably native; dry banks, grassy places and rough ground; local in S & SE En and CI, casual elsewhere *Yellow Vetchling* - **L. aphaca** L.	
3	Leaves (±) all with >1 pair of leaflets	4
3	Leaves (±) all with 1 pair of leaflets	7
	4 Stem angled, not winged	5
	4 Stem, at least above, with wings ≥1/2 as wide as stem	6
5	Stipules ovate-triangular; leaves usually with tendrils. Procumbent perennial to 90cm. Native; maritime shingle or rarely sand, very local on coasts of BI *Sea Pea* - **L. japonicus** Willd.	
5	Stipules lanceolate; leaves without tendrils. Erect perennial to 80cm. Intrd-natd; garden escape in grassy, rocky and scrubby places; rare in Sc and En *Black Pea* - **L. niger** (L.) Bernh.	
	6 Tendrils 0; fruit scarcely compressed. Erect perennial to 40cm. Native; wood-borders, hedgerows, scrub; local throughout Br and Ir *Bitter-vetch* - **L. linifolius** (Reichard) Bässler	
	6 Tendrils well developed; fruit strongly compressed. Climbing	

perennial to 1.2m. Native; fens and tall damp grassland; very
locally scattered in En, Wa, SW Sc and Ir *Marsh Pea* - **L. palustris** L.
7 Stem angled, not winged 8
7 Stem, at least above, with wings ≥1/2 as wide as stem 10
   8 Flowers yellow; leaflets acute; fruits strongly compressed.
      Climbing perennial to 1.2m. Native; grassy places and rough
      ground; common   *Meadow Vetchling* - **L. pratensis** L.
   8 Flowers pink to purple; leaflets obtuse to rounded or retuse at
      apex; fruits scarcely compressed 9
9 Plant minutely pubescent; flowers >25mm. Climbing perennial to
2m. Intrd-natd; garden escape in hedges, waste ground near old
gardens; scattered in Br N to C Sc, Alderney
         Two-flowered Everlasting-pea - **L. grandiflorus** Sm.
9 Plant glabrous to very sparsely pubescent; flowers <25mm.
Climbing perennial to 1.2m. Intrd-natd; cornfields, hedgerows and
roadsides; N Essex since 1859, scattered through Br N to C Sc but
mostly casual or shortly persisting only   *Tuberous Pea* - **L. tuberosus** L.
10 Perennials; flowers 3-many 11
10 Easily uprooted annuals; flowers 1-3(4) 13
11 Stipules <1/2 as wide as stem; all calyx-teeth shorter than tube.
Climbing or scrambling perennial to 2m. Native; scrub, wood-
borders, hedgerows, rough ground; scattered through Br N to C Sc,
but mostly intrd   *Narrow-leaved Everlasting-pea* - **L. sylvestris** L.
11 Stipules >1/2 as wide as stem (often as wide); lowest calyx-tooth as
long as or longer than tube 12
   12 Flowers 12-22mm; leaflets >4x as long as wide; ovules ≤15 per
      ovary. Climbing or scrambling perennial to 2m. Intrd-natd;
      grassy and scrubby places; W Norfolk and N Hants
         Norfolk Everlasting-pea - **L. heterophyllus** L.
   12 Flowers 15-30mm; leaflets <4x as long as wide; ovules ≥16 per
      ovary. Climbing or scrambling perennial to 2m. Intrd-natd;
      hedges, roadsides, railway banks and rough ground; scattered
      through Br   *Broad-leaved Everlasting-pea* - **L. latifolius** L.
13 Plant glabrous (or with glands on young fruits) 14
13 Plant pubescent at least on pedicels, calyx and fruits 15
   14 Corolla white, pink or blue; fruit with 2 narrow wings along
      dorsal suture. Scrambling or climbing annual to 1m. Intrd-casual;
      tips and waste places; sporadic in S En   *Indian Pea* - **L. sativus** L.
   14 Corolla yellow; fruit not winged. Scrambling or climbing annual
      to 1m. Intrd-casual; tips and waste places; sporadic in S En
         Fodder Pea - **L. annuus** L.
15 Flowers >20mm; leaflets ovate-oblong. Climbing annual to 2.5m.
Intrd-casual; tips and waste places; En and Wa, mostly S
         Sweet Pea - **L. odoratus** L.
15 Flowers <20mm; leaflets linear-oblong. Scrambling or climbing
annual to 1m. Intrd-natd; grassy and rough ground and waste
places, very scattered in En and S Sc, natd in Essex
         Hairy Vetchling - **L. hirsutus** L.

22. PISUM L. - *Garden Pea*
Stems not winged, ± terete; leaves paripinnate with 1-3 pairs of pinnately-veined leaflets, with terminal branched tendril; stipules larger than leaflets; calyx-teeth broad, ± leafy; style pubescent only on upperside, proximal part with reflexed margins.

1 Climbing or sprawling annual to 2m (usually much less); flowers 1-3, 15-35mm, white to purple. Intrd-casual; grown on field-scale and a common casual by roads and fields, in waste places and on tips; scattered throughout BI *Garden Pea* - **P. sativum** L.

TRIBE 9 - CICEREAE (genus 23). Annuals; leaves imparipinnate, with sharply serrate leaflets; tendrils 0; flowers solitary, axillary; 9 stamens forming tube, 10th free; fruit dehiscing longitudinally, 1-2-seeded.

23. CICER L. - *Chick Pea*

1 Erect, glandular-pubescent annual to 35(60)cm; corolla 10-12mm, white to pale purple. Intrd-casual; tips and waste places; scattered in Br, mainly En *Chick Pea* - **C. arietinum** L.

TRIBE 10 - TRIFOLIEAE (genera 24-28). Annual or perennial herbs or rarely shrubs; leaves ternate, main lateral veins of leaflets running whole way to often toothed margin; flowers solitary or in racemes (often greatly condensed); 9 stamens forming tube, 10th free, or all 10 forming tube; fruits indehiscent to longitudinally dehiscent, with 1-many seeds.

24. ONONIS L. - *Restharrows*
Flowers solitary or in terminal racemes; calyx with glandular hairs; all 10 stamens forming tube; fruits straight, 1-many seeded, dehiscing longitudinally.

1 Perennials with stems woody at least below, sometimes spiny 2
1 Herbaceous annuals, not spiny 4
   2 Corolla yellow, often streaked red; fruits 12-25mm, with 4-10 seeds. Dwarf shrub to 60cm, without spines. Intrd-natd; rough ground; Berks *Yellow Restharrow* - **O. natrix** L.
   2 Corolla pink, rarely white; fruits 5-10mm, with 1-2(4) seeds 3
3 Stems equally hairy all round, procumbent to ascending; leaflets <3x as long as wide, obtuse to emarginate. Rhizomatous perennial; stems woody at base, to 60cm, with or without spines. Native; rough grassy places on well-drained soils, especially coastal; locally common *Common Restharrow* - **O. repens** L.
3 Stems mainly hairy along 1 side or 2 opposite sides, ascending to erect; leaflets >3x as long as wide, acute to subacute. Usually spiny shrub to 70cm. Native; grassy places and rough ground on mostly well-drained soils; locally frequent in Br N to S Sc
 Spiny Restharrow - **O. spinosa** L.

 4 Fruits >8-seeded; flowers 5-10mm; pedicels >5mm. Erect to
procumbent annual to 15cm. Native; barish sand or limestone;
rare and very local in SW Br, Wigtowns and Alderney
Small Restharrow - **O. reclinata** L.
 4 Fruits 2-3(6)-seeded; flowers 9-17mm; pedicels <2mm 5
5 Racemes not leafy, their lower bracts with 0-1 leaflet, borne on bare
peduncles. Erect to ascending annual to 50cm. Intrd-casual;
occasional birdseed-alien on tips; sporadic in S En
Andalucian Restharrow - **O. baetica** Clemente
5 Racemes leafy, their lower bracts with 3 leaflets, borne on leafy
peduncles 6
 6 Calyx 6.5-9mm, with ovate-acuminate teeth; seeds 1.5-2mm,
tuberculate. Erect to procumbent annual to 60cm. Intrd-casual;
occasional birdseed-alien on tips; sporadic in S En
Mediterranean Restharrow - **O. mitissima** L.
 6 Calyx 10-12mm, with linear-lanceolate teeth; seeds 2-2.3mm,
smooth. Erect to procumbent annual to 75cm. Intrd-casual;
frequent birdseed-alien on tips and in waste places; sporadic in
S En (*O. baetica* auct.) *Salzmann's Restharrow* - **O. alopecuroides** L.

O. spinosa x *O. repens* = ***O. x pseudohircina*** Schur occurs with both parents
in a few places from Cambs to Durham; it is intermediate in leaflet-shape,
stem-pubescence and fruit-size, and fertile.

25. MELILOTUS Mill. - *Melilots*
Flowers in elongated racemes, calyx without glandular hairs; 9 stamens
forming tube, 10th free; fruits straight, 1-2-seeded, indehiscent or very
tardily dehiscent longitudinally.

1 Corolla white. Erect annual or biennial to 1.5m. Intrd-natd; open
grassland and rough ground; frequent in S & C En, scattered W to Ir
and N to C Sc, CI *White Melilot* - **M. albus** Medik.
1 Corolla yellow 2
 2 Fruits with strong concentric ridges; wings shorter than keel.
Usually erect annual to 40cm. Intrd-casual; tips and waste land;
sporadic in S En *Furrowed Melilot* - **M. sulcatus** Desf.
 2 Fruits with weak to strong transverse or reticulate ridges; wings
as long as or longer than keel 3
3 Flowers 2-3.5mm; fruits <3mm. Erect to ascending annual to 40cm.
Intrd-natd; rough ground and waste places, usually casual; frequent
in CI, scattered in BI N to C Sc *Small Melilot* - **M. indicus** (L.) All.
3 Flowers >4mm; fruits >3mm 4
 4 Fruits >5mm, mostly 2-seeded, black when ripe, pubescent; keel
± equalling wings. Erect biennial or perennial to 1.5m. Intrd-natd;
open grassland and rough ground; frequent in S & C En,
scattered W to Ir and N to C Sc *Tall Melilot* - **M. altissimus** Thuill.
 4 Fruits <5mm, mostly 1-seeded, brown when ripe, glabrous; keel
shorter than wings. Erect to decumbent biennial to 1.5m. Intrd-

natd; open grassland and rough ground; frequent in S & C En, scattered W to Ir and N to C Sc, CI

Ribbed Melilot - **M. officinalis** (L.) Pall.

26. TRIGONELLA L. - *Fenugreeks*

Annuals; flowers solitary in leaf-axils or in axillary racemes; calyx without glandular hairs; 9 stamens forming tube, 10th free; fruits straight to curved, (1)2-many-seeded, dehiscent longitudinally, often tardily.

1 Flowers >10mm, 1(-2) in leaf-axils; fruits >6cm, >9-seeded. Stems erect to spreading, to 50cm. Intrd-casual; tips and waste land; sporadic in C & S En *Fenugreek* - **T. foenum-graecum** L.
1 Flowers <8mm, in axillary racemes; fruits <2cm, <9-seeded 2
   2 Racemes elongated; corolla yellow; fruits >8mm, 4-8-seeded. Stems erect to procumbent, to 50cm. Intrd-casual; tips and waste ground; sporadic in S Br

Sickle-fruited Fenugreek - **T. corniculata** (L.) L.
   2 Racemes subcapitate; corolla white to blue; fruits <8mm, 1-3-seeded. Stems erect to decumbent, to 50cm. Intrd-casual; tips and waste land; very scattered in Br *Blue Fenugreek* - **T. caerulea** (L.) Ser.

27. MEDICAGO L. - *Medicks*

Flowers 1-many in axillary racemes; calyx without glandular hairs; 9 stamens forming tube, 10th free; fruits slightly curved to spiral with several complete turns, 1-many-seeded, indehiscent, often spiny.

Spp. in couplets 5-9 have quite distinct fruits but the differences are difficult to describe. Spine length is of virtually no value; in most spp. they can be very short to longer than the coil diameter, and rare spineless or tuberculate variants exist. The margin (outer edge) of each coil is occupied by a variously thickened vein or 'border'. Just inside this, on each face of the coil, is a variously thickened 'submarginal vein', and between these 2 thickened veins, in each face, is the (usually channelled) 'submarginal border'. The base of each spine straddles this border and originates from both the thickened veins. The 'face' of each coil between the coil centre and the submarginal border is often characteristically veined.

1 Fruits curved to spiralled, without spines or tubercles; if flowers yellow then fruits curved to spiralled in ≤1.5 turns 2
1 Fruits spiralled in >(1.5)2.5 turns, usually spiny, rarely tuberculate or smooth; flowers yellow 5
   2 Fruits <3mm in longest plane; flowers <4mm, yellow; seed 1. Procumbent to scrambling annual or short-lived perennial to 80cm. Native; grassy places and rough ground; common. See *Trifolium dubium* for differences *Black Medick* - **M. lupulina** L.
   2 Fruits >4mm in longest plane; flowers >6mm, yellow, white, green or purplish; seeds >1. Erect to decumbent perennial to 90cm (**M. sativa** L.) 3
3 Fruits spiralled in 2-3(4) complete turns; coils ± closed in centre;

flowers mauve to violet. Intrd-natd; relic of cultivation on field-
margins, roadsides, rough grassland and waste places; common
in CI and S & C En, scattered elsewhere   *Lucerne* - **M. sativa ssp. sativa**
3 Fruits curved or spiralled in ≤1.5 complete turns; coils ± open in
centre; flowers yellow or white to purple or green or blackish   4
   4 Fruits nearly straight to curved in arc ≤1/2 circle; flowers yellow.
Native; grassy places and rough or waste ground; local in E
Anglia, casual or sometimes natd elsewhere in Br N to C Sc
Sickle Medick - **M. sativa ssp. falcata** (L.) Arcang.
   4 Fruits curved or spiralled in 0.5-1.5 complete turns; flowers
yellow or other colours. Native; on sandy or rough ground,
arising *in situ* or intrd as hybrid seed; scattered in Br N to C Sc,
especially E Anglia, Co Dublin (ssp. *sativa* x ssp. *falcata*)
Sand Lucerne - **M. sativa ssp. varia** (Martyn) Arcang.
5 Leaflets usually each with dark blotch; fruit border grooved along
centre, hence (with the 2 submarginal borders) forming 3 grooves at
edge of each coil. Procumbent to scrambling annual to 60cm. Native;
grassy and barish places, especially near sea; CI, S & C Br, also casual
throughout BI in waste places   *Spotted Medick* - **M. arabica** (L.) Huds.
5 Leaflets not dark-blotched; fruit border not grooved, hence edge of
each coil with 0 or 2 grooves   6
   6 Fruit border thinner than submarginal veins, scarcely contributing
to origin of spines; spines ± not grooved at base; fruits with sparse
long hairs. Procumbent to scrambling annual to 50cm. Intrd-
casual; waste places; scattered and sporadic in Br
Strong-spined Medick - **M. truncatula** Gaertn.
   6 Fruit border as thick or thicker than submarginal veins,
contributing strongly to origin of spines; spines hence deeply
grooved at least at base; fruits usually ± glabrous   7
7 Fruit border 0.5-0.8mm thick in lateral view, at least as thick as rest
of each coil and ± obscuring it in lateral view. Procumbent annual
to 20cm. Intrd-casual; on tips and rough ground; sporadic
throughout most of Br   *Early Medick* - **M. praecox** DC.
7 Fruit border <0.4mm thick in lateral view, thinner than rest of each
coil and not completely obscuring it in lateral view   8
   8 Stipules entire to denticulate, with teeth much shorter than
entire part; leaves and stems ± densely pubescent. Procumbent
annual to 20(40)cm. Native; sandy heaths and dunes and shingle
by sea; very local in E En from Kent to Norfolk, Jersey; also
casual throughout most of Br   *Bur Medick* - **M. minima** (L.) Bartal.
   8 Stipules deeply incised, with teeth much longer than entire part;
leaves and stems glabrous to sparsely pubescent   9
9 Submarginal border broadly grooved, wider than border; fruit face
with curved veins anastomosing to form reticulum adjacent to
submarginal vein; wings longer than keel. Procumbent to
scrambling annual to 60cm. Native; local in sandy ground near
sea; CI, S En; also common casual   *Toothed Medick* - **M. polymorpha** L.
9 Submarginal border narrowly grooved, narrower than border; fruit

face with sigmoid veins not or scarcely anastomosing before joining submarginal vein; keel longer than wings. Procumbent annual to 40cm. Intrd-casual; tips and rough ground; sporadic throughout most of Br **Tattered Medick - M. laciniata** (L.) Mill.

28. TRIFOLIUM L. - *Clovers*

Flowers 1-many in axillary congested racemes; calyx without glandular hairs; 9 stamens forming tube, 10th free; fruits indehiscent, longitudinally dehiscent, or dehiscent by apex falling off, ± straight, 1-9-seeded, often partly or wholly enclosed in calyx or persistent corolla.

The keys require flowering and fruiting racemes, usually both present for a long period in the season. Calyx-tube vein-number is important; in spp. with densely pubescent calyx-tubes it is best observed by splitting the tube up one side and observing the veins from the inside.

General key
1 Racemes with sterile corolla-less flowers mixed with normal ones, becoming turned down and thrust into ground as fruit ripens. Decumbent to procumbent pubescent annual to 20(80)cm, often <10cm. Native; short turf and barish places on sandy soils, especially by sea; scattered in C & S Br, Man, CI, Co Wicklow; also frequent wool-alien in Br **Subterranean Clover - T. subterraneum** L.
1 Racemes wholly of sexual flowers with corollas, not becoming subterranean 2
   2 Calyx becoming greatly inflated in fruit, the inflation confined to upper lip with 2 teeth *Key A*
   2 Calyx not becoming inflated in fruit, or only moderately and symmetrically so 3
3 Calyx-tube with 5 veins; corolla yellow, ≤8mm; fruit stalked, 1-seeded *Key B*
3 Calyx-tube with (5)10-20 veins; corolla mostly white or pink to purple, if yellow then >8mm; fruit sessile, 1-9-seeded 4
   4 Throat of calyx (remove corolla) at least partly closed by a thickening or a ring of hairs; flowers without bracts, ± sessile *Key C*
   4 Throat of calyx glabrous and open; each flower with small bract at base, or the bracts variably fused; flowers sessile or distinctly pedicellate *Key D*

Key A - Calyx asymmetrically strongly inflated in fruit
1 Perennial; stems rooting at nodes; flowers with standard uppermost. Subglabrous perennial with procumbent stems to 30cm. Native; grassy places, often on heavy or brackish soils; locally common in BI N to S Sc. Vegetatively resembles *T. repens* but lateral veins of leaflets are thickened and recurved distally almost as in *T. scabrum* (not so in *T. repens*) **Strawberry Clover - T. fragiferum** L.
1 Annual; stems not rooting at nodes; corolla upside-down so that standard is lowermost 2
   2 Fruiting calyx densely woolly, with calyx-teeth ± completely

obscured; leaflets 4-12mm. Glabrous, procumbent annual to 15cm. Intrd-casual; waste places; sporadic mainly in S Br
Woolly Clover - **T. tomentosum** L.

2 Fruiting calyx sparsely to densely pubescent, with 2 prominent divergent calyx-teeth at apex; leaflets 10-20(25)mm. Glabrous, procumbent to suberect annual to 30cm. Intrd-natd; rather frequent casual in waste places, sometimes persisting; mainly S Br *Reversed Clover* - **T. resupinatum** L.

Key B - Calyx not inflated or moderately and symmetrically so, with 5 veins; corolla yellow, <1cm

1 Apical leaflet with stalk >0.5mm, distinctly longer than that of lateral leaflets 2
1 All leaflets with stalks <0.5mm 3
   2 Corolla 3-4mm, with standard folded longitudinally over fruit; racemes mostly <25-flowered. Procumbent to suberect very sparsely pubescent annual to 25cm. Native; grassy and open ground; very common. Differs from *Medicago lupulina* in glabrous (not pubescent) calyx, and subglabrous (not pubescent) leaflets without (not with) apical apiculus *Lesser Trefoil* - **T. dubium** Sibth.
   2 Corolla 4-7mm, with standard ± flat over fruit; racemes mostly >25-flowered. Procumbent to suberect sparsely pubescent annual to 30cm. Native; grassy and barish places; frequent
Hop Trefoil - **T. campestre** Schreb.
3 Corolla 5-8mm; pedicels c.1mm; racemes mostly >20-flowered. Erect to ascending, sparsely pubescent annual to 30cm. Intrd-natd; in grassy and rough ground; very local in Br, well natd ± only in C & S Sc *Large Trefoil* - **T. aureum** Pollich
3 Corolla 1.5-3mm; pedicels c.1.5mm; racemes <10-flowered. Procumbent to ascending subglabrous annual to 15cm. Native; short turf, especially close-cut lawns; common in CI and SE En, scattered and often intrd elsewhere *Slender Trefoil* - **T. micranthum** Viv.

Key C - Calyx not inflated or moderately and symmetrically so, with 10-20 veins, its throat ± closed by thickening or hairs; bracts 0

1 Firmly rooted perennials with habit of *T. pratense* or rhizomatous; flowers ≥12mm 2
1 Easily uprooted annuals; flowers often <12mm 5
   2 Free part of stipules of stem-leaves triangular-ovate, abruptly narrowed to brown bristle-like point. Decumbent to erect, pubescent, tufted perennial to 60cm. Native; grassy places, waste and rough ground; abundant *Red Clover* - **T. pratense** L.
   2 Free part of stipules of stem-leaves linear to lanceolate, green ± to apex 3
3 Corolla reddish-purple (rare albinos); lowest calyx-lobe c.1.5x as long as upper lobes and as calyx-tube. Ascending, sparsely pubescent, rhizomatous perennial to 50cm. Native; grassy places, hedgerows and wood-borders; frequent *Zigzag Clover* - **T. medium** L.

3 Corolla whitish-yellow, (?sometimes pale pink); lowest calyx-lobe
  ≥2x as long as upper lobes and as calyx-tube 4
   4 Flowers 15-18mm; leaflets elliptic to oblong or obovate; racemes
    ± subsessile. Ascending to erect, pubescent, tufted but shortly
    rhizomatous perennial to 50cm. Native; grassy places on heavy
    soils; very local in E Anglia W to Northants, casual elsewhere
    in En **Sulphur Clover - T. ochroleucon** Huds.
   4 Flowers 20-25mm; leaflets elliptic to linear-oblong; some racemes
    pedunculate. As *T. ochroleucon* but stems ± erect, to 60(80)cm.
    Intrd-natd; mainly as grass-seed contaminant, rarely ± natd;
    sporadic in S En **Hungarian Clover - T. pannonicum** Jacq.
5 Calyx-tube with c.20 veins 6
5 Calyx-tube with 10 veins 7
   6 Calyx-tube with many long hairs; corolla >10mm. Ascending
    pubescent annual to 35cm. Intrd-casual; tips and waste ground;
    sporadic in En **Rose Clover - T. hirtum** All.
   6 Calyx-tube without long hairs, usually glabrous; corolla <10mm.
    Erect or ascending sparsely pubescent annual to 35cm. Intrd-
    casual; tips and waste places; sporadic in S Br
    Bur Clover - T. lappaceum L.
7 Free part of stipules of stem-leaves ovate-triangular, abruptly
  narrowed to obtuse, acute or shortly apiculate apex 8
7 Free part of stipules of stem-leaves gradually tapered to long narrow
  apex 10
   8 Racemes ± globose; stipules sharply serrate. Erect pubescent
    annual to 20(30)cm. Intrd-natd; on shingle in W Sussex,
    infrequent casual elsewhere in S Br **Starry Clover - T. stellatum** L.
   8 Racemes cylindrical; stipules obscurely denticulate. Pubescent
    annual to (30)50cm (**T. incarnatum** L.) 9
9 Usually erect, with ± patent hairs; flowers crimson. Intrd-natd; rough
  and waste ground, grassy places; sporadic over CI, Man and Br
  Crimson Clover - T. incarnatum ssp. incarnatum
9 Usually decumbent to ascending to 20cm, with ± appressed hairs;
  flowers yellowish-white to pale pink. Native; short grassland near
  sea; Jersey, Lizard Peninsula (W Cornwall)
  Long-headed Clover - T. incarnatum ssp. molinerii (Hornem.) Ces.
  10 Racemes ± cylindrical in fruit, >2x as long as wide 11
  10 Racemes globose to ovoid in fruit, <1.5(2)x as long as wide 12
11 Flowers 10-13mm; calyx-lobes unequal, the lowest much the longest.
  Erect, appressed-pubescent annual to 50cm. Intrd-casual; tips and
  waste ground; sporadic in Br **Narrow Clover - T. angustifolium** L.
11 Flowers 3-6mm; calyx-lobes subequal. Erect or ascending pubescent
  annual to 20(40)cm. Native; barish ground on sandy soils; locally
  frequent in BI N to C Sc **Hare's foot Clover - T. arvense** L.
  12 At least the 2 uppermost leaves on each stem-branch exactly
    opposite; flowers (5)7-12mm 13
  12 All leaves alternate (sometimes 2 uppermost close together, but
    not exactly opposite); flowers 4-7mm 15

13 Calyx with ring of hairs in throat; fruit exserted from calyx-tube. Erect to ascending pubescent (to sparsely so) annual to 60cm. Intrd-casual; in newly sown grass by roads and in parks; Guernsey, rare in S En *Egyptian Clover* - **T. alexandrinum** L.

13 Calyx with bilateral swellings in throat, completely closed; fruit enclosed in calyx-tube 14

    14 Calyx-lobes with 1 vein, or with 3 veins just at base. Procumbent to erect, sparsely pubescent annual to 50cm. Intrd-casual; tips and rough ground; sporadic in En
Hedgehog Clover - **T. echinatum** M. Bieb.

    14 Calyx-lobes with 3 veins for >1/2 length. Erect to ascending rather sparsely pubescent annual to 40cm. Native; short, often brackish, turf by sea; very local in S Br, Guernsey
Sea Clover - **T. squamosum** L.

15 Leaflets with lateral veins thickened and arched-recurved at leaf-margin. Procumbent to erect pubescent annual to 20cm. Native; similar habitats and distribution to *T. striatum* (couplet 16), and often with it *Rough Clover* - **T. scabrum** L.

15 Leaflets with lateral veins thin and straight or slightly forward-curved at leaf-margin 16

    16 Leaflets ± glabrous on upperside; fruiting calyx-tube not inflated. Erect to ascending pubescent annual to 20(30)cm. Native; short turf near sea; Jersey, Lizard Peninsula (W Cornwall)
Twin-headed Clover - **T. bocconei** Savi

    16 Leaflets pubescent on upperside; fruiting calyx-tube inflated. Procumbent to erect pubescent annual to 30cm. Native; short grassland and open places on sandy ground, especially near sea; locally frequent through BI N to CE Sc
Knotted Clover - **T. striatum** L.

Key D - Calyx not inflated or slightly and symmetrically so, with (5)10(-12) veins, its throat open; bracts present

1 Flowers 1-4(5), with pedicels ≥1mm; seeds 5-9. Procumbent ± glabrous annual to 20cm. Native; sandy semi-open ground, mainly near sea; scattered round coasts of C & S BI, also Midlothian
Bird's-foot Clover - **T. ornithopodioides** L.

1 Flowers (3)10-many, if <10 then ± sessile; seeds 1-4 2

    2 Flowers ± sessile, not reflexed after pollination; fruit shorter than calyx (incl. lobes) 3

    2 Flowers with pedicels ≥1mm, strongly reflexed after pollination; fruit longer than calyx (incl. lobes) 5

3 Stipules serrate; teeth of leaflets and stipules gland-tipped. Erect to ascending glabrous annual to 15(25)cm. Native; rocky grassy places; Jersey, W Cornwall, Rads *Upright Clover* - **T. strictum** L.

3 Stipules usually entire, sometimes serrate; teeth of stipules and leaflets not gland-tipped 4

    4 Corolla 3-4mm, shorter than calyx (incl. lobes), whitish; many (often ± all) racemes congested at base of plant. Procumbent

subglabrous annual to 3(8)cm. Native; grassy places on sandy soil mostly near sea; CI, S & E coast of En N to SE Yorks, Man
Suffocated Clover - **T. suffocatum** L.

4   Corolla 4-7mm, longer than calyx (incl. lobes), purplish; racemes dispersed along stems. Procumbent to ascending subglabrous annual to 25cm. Native; grassy places on sandy soil mostly near sea; CI, S & E coast of En N to Norfolk
Clustered Clover - **T. glomeratum** L.

5   Flowers 4-5mm; upper racemes with peduncles <1cm, often subsessile. Procumbent to ascending subglabrous annual to 25cm. Intrd-casual; waste ground, sometimes persisting for few years; sporadic in C & S En   *Nodding Clover* - **T. cernuum** Brot.

5   Flowers (5)7-13mm; all racemes with peduncles >1cm; perennial   6

    6   Stems not rooting at nodes, usually erect to ascending. Erect to decumbent subglabrous perennial to 40(70)cm. Intrd-natd; grassy and rough ground   *Alsike Clover* - **T. hybridum** L.

    6   Stems procumbent, rooting at nodes   7

7   Leaflets usually >10mm, obovate, often with light or dark markings, with veins translucent when fresh; petioles glabrous; calyx-lobes triangular-lanceolate; standard rounded at apex. Subglabrous perennial with procumbent stems to 50cm rooting at nodes. Native; grassy and rough ground; abundant   *White Clover* - **T. repens** L.

7   Leaflets usually <10mm, suborbicular, without light or dark markings, with veins not translucent when fresh; petioles sparsely pubescent; calyx-lobes triangular-ovate; standard emarginate at apex. Subglabrous perennial with procumbent stems to 20cm rooting at nodes. Native; short turf by sea; CI, SW En, S Wa, Anglesey, E Ir   *Western Clover* - **T. occidentale** Coombe

TRIBE 11 - THERMOPSIDEAE (genus 29). Rhizomatous perennial herbs; leaves ternate, with entire leaflets; flowers in erect racemes with small ovate bracts; all 10 stamens free; fruits longitudinally dehiscent, with 2-7 seeds.

29. THERMOPSIS R. Br. - *False Lupin*

1   Stems erect, to 70(100)cm; leaflets up to 10cm; flowers yellow, 20-25mm. Intrd-natd; on old garden sites or rough grassy places; Fetlar (Shetland)   *False Lupin* - **T. montana** Torr. & A. Gray

TRIBE 12 - GENISTEAE (genera 30-35). Perennial or annual herbs or woody shrubs or trees; leaves simple, entire, ternate or palmate; flowers in axillary or terminal racemes, sometimes reduced to 1 or 2; all 10 stamens forming tube; fruits longitudinally dehiscent, with 2-many seeds.

30. LUPINUS L. - *Lupins*

Herbaceous annuals or perennials, sometimes woody at base; leaves palmate, with long petiole; flowers in terminal erect racemes, variously coloured; fruits several-seeded, erecto-patent.

1 Stems woody towards base, not dying down to ground in winter, erect, much branched, to 2m; corolla usually yellow, sometimes blue. Intrd-natd; in waste places and on maritime shingle or sand; very scattered mostly on coasts of BI N to C Sc *Tree Lupin* - **L. arboreus** Sims
1 Stems herbaceous, dying down to ground in winter 2
  2 Annuals, easily uprooted 3
  2 Tuft-forming perennials 4
3 Leaflets oblanceolate; upper lip of calyx very shallowly 2-lobed or ± entire; seeds mostly >7mm. Stems erect to ascending, sparingly branched, to 60(100)cm; corolla white, usually variably tinged bluish-violet. Intrd-casual; on tips, at docks and in waste places; sporadic in S En *White Lupin* - **L. albus** L.
3 Leaflets linear; upper lip of calyx deeply bifid; seeds <7mm. Stems erect, well branched, to 60(100)cm; corolla blue. Intrd-casual; docks and waste places; sporadic in Br, mainly S
  Narrow-leaved Lupin - **L. angustifolius** L.
  4 Basal leaves absent at flowering time; upper part of stem and petioles usually shaggy-hairy; lower lip of calyx 7-13mm. Stems erect, to 1m, usually with long shaggy hairs; corolla bluish-purple, sometimes whitish tinged purplish. Intrd-natd; on riverside shingle and moorland; C & N Sc from M Perth to Orkney, NW Ir *Nootka Lupin* - **L. nootkatensis** Sims
  4 Basal leaves present at flowering time; stem and petioles with rather sparse short hairs; lower lip of calyx 3-8mm 5
5 Stems unbranched, with 1 inflorescence; flowers blue; leaflets obtuse to acute; lower lip of calyx 3-6mm. Stems erect, to 1.2m. Intrd-natd; by rivers and railways, sometimes in waste places; scattered in Sc, much over-recorded *Garden Lupin* - **L. polyphyllus** Lindl.
5 Stems mostly branched, with >1 inflorescence; flowers various shades of blue, pink, purple, yellow or white; leaflets acute to acuminate; lower lip of calyx 5-8mm. Stems erect, to 1.5m. Native; on rough ground, banks of roads and railways; throughout lowland Br; garden origin, but spontaneous hybrids and perhaps backcrosses also occur in Moray, Wigtowns and Midlothian (*L. arboreus* x *L. polyphyllus*) *Russell Lupin* - **L. x regalis** Bergmans

L. nootkatensis hybridizes with *L. polyphyllus* (= **L. x pseudopolyphyllus** C.P. Sm.) and probably with *L. x regalis* in Moray and M Perth

31. LABURNUM Fabr. - *Laburnums*
Deciduous non-spiny trees; leaves ternate, with long petiole; flowers yellow, in pendent racemes on short-shoots; fruits several-seeded, pendent.

1 Young twigs, petioles, leaf lowerside, peduncle and young fruits appressed-pubescent, densely silvery so when very young; racemes mostly 15-30cm; flowers 17-23mm; fruits with narrow dorsal ridge thick and ± truncate in section. Tree to 8m. Intrd-natd; self-sown in rough ground and on banks by roads and railways;

scattered through BI *Laburnum* - **L. anagyroides** Medik.
1 Young twigs, petioles, leaf lowerside, peduncle and young fruits glabrous to sparsely pubescent; racemes mostly 25-35cm with more but less densely crowded flowers 13-18mm; fruits with distinctly winged dorsal ridge narrow and acute in section. Tree to 13m. Intrd-natd; similar occurrence to *L. anagyroides* but much confused, less common in S, perhaps commoner in N Br and Ir
Scottish Laburnum - **L. alpinum** (Mill.) J. Presl

L. anagyroides x *L. alpinum* = ***L. x watereri*** (Wettst.) Dippel is now more commonly planted than either sp. but much less or never self-sown. It has the longer racemes with more flowers of *L. alpinum*, the larger (15-21mm) more densely arranged flowers of *L. anagyroides*, and usually forms few good seeds; it has apparently been used for hedging in the W.

32. CYTISUS Desf. - *Brooms*
Non-spiny shrubs; leaves simple or ternate, sessile or petiolate; flowers yellow, in terminal or axillary racemes or 1-few in leaf-axils; calyx with upper lip 2-toothed or deeply bifid; fruits several- to many-seeded, erect to patent.

1 Flowers in terminal leafless racemes; all leaves ternate, most with petioles >1cm. Stems erect, to 1.5m. Intrd-natd; on waste ground by railway and in gravel-pits; Middlesex, E Kent and W Ross
Black Broom - **C. nigricans** L.
1 Flowers 1-few in lateral groups with (often reduced) leaves; upper leaves usually simple, lower with petioles <1cm 2
   2 Flowers white, 7-12mm; calyx ≤3mm. Stems erect or arching, to 2m. Intrd-natd; on banks of roads and railways; very scattered in Br *White Broom* - **C. multiflorus** (Aiton) Sweet
   2 Flowers yellow to red, 10-25mm; calyx >3mm 3
3 Fruits covered with dense white shaggy hairs; twigs c.10-angled, fragile. Stems erect or arching, to 3m. Intrd-natd; on roadside banks; very scattered in Br *Hairy-fruited Broom* - **C. striatus** (Hill) Rothm.
3 Fruits with long hairs on sutures, ± glabrous on faces; twigs c.5-angled, pliable (*Broom* - **C. scoparius** (L.) Link) 4
   4 Plant erect or arching, to 2.5m high. Native; calcifuge of heathland, sandy banks, open woodland, rough ground; throughout most of BI **C. scoparius ssp. scoparius**
   4 Plant ± procumbent, to 50cm high; young branches more densely silky-pubescent than in ssp. *scoparius*. Native; maritime cliffs and perhaps shingle; SW Ir, S & SW Br, CI
C. scoparius ssp. maritimus (Rouy) Heywood

33. SPARTIUM L. - *Spanish Broom*
Non-spiny shrubs; leaves simple, shortly petiolate to sessile; flowers yellow, in leafless terminal racemes; calyx with upper lip divided nearly to base; fruits many-seeded, erecto-patent.

1 Stems erect, to 3m, terete with wide soft pith, glabrous; flowers 20-28mm, yellow. Intrd-natd; on sandy roadside banks and rough ground; scattered in C & S Br **Spanish Broom - S. junceum** L.

34. GENISTA L. - *Greenweeds*
Spiny or non-spiny shrubs; leaves simple or ternate, sessile or shortly petiolate; flowers yellow, in terminal racemes or lateral clusters; calyx with upper lip divided 1/4-3/4 way to base; fruits 1-several seeded, erect to patent.

1 Leaves all ternate. Stems not spiny, erect, to 2(3)m. Intrd-natd; on banks, roadsides and rough ground; scattered in CI and S En
Montpellier Broom - **G. monspessulana** (L.) L.A.S. Johnson
1 Leaves all simple, or 0 due to early fall 2
  2 Plant with branched spines, erect, to 70cm. Intrd-natd; on sandy and rocky hills and roadsides; Cards, Salop, S En and Man
Spanish Gorse - **G. hispanica** L.
  2 Plant with 0 or simple spines 3
3 Seeds 1-2; leaves most or all fallen by flowering time. Stems not spiny, erect, to 5m. Intrd-natd; on waste ground; W Kent and Surrey *Mount Etna Broom* - **G. aetnensis** (Biv.) DC.
3 Seeds 3-12; leaves present at flowering time 4
  4 Calyx, corolla and leaves pubescent. Stems not spiny, procumbent or scrambling, to 50cm. Native; cliff-tops and heathland; extremely local in S Br *Hairy Greenweed* - **G. pilosa** L.
  4 Calyx, corolla and leaves ± glabrous 5
5 Flowers 7-10mm; fruits 12-20mm; plant usually (not always) spiny. Stems erect to spreading, to 1m. Native; sandy and peaty heaths and moors; throughout most of Br *Petty Whin* - **G. anglica** L.
5 Flowers 10-15mm; fruits (15)20-30mm; plant never spiny (*Dyer's Greenweed* - **G. tinctoria** L.) 6
  6 Stems erect to ascending, to 60(100)cm; leaves narrowly elliptic, mostly >4x as long as wide; fruits glabrous. Native; grassy places, banks and rough ground; locally common in Br N to S Sc, Jersey (not Cornwall) **G. tinctoria ssp. tinctoria**
  6 Stems procumbent, to 60cm; leaves elliptic-oblong, mostly <4x as long as wide; fruits pubescent or glabrous. Native; grassy cliff-tops; N Devon, Cornwall, SW Pembs
G. tinctoria ssp. littoralis (Corb.) Rothm.

35. ULEX L. - *Gorses*
Shrubs with branched spines; leaves ternate on young plants, simple and reduced to scales or weak spines on mature plants; flowers yellow, 1-few in lateral clusters, with small bracteole on each side between calyx-lips; calyx with upper lip shortly 2-toothed; fruits (1)2-6(8)-seeded, >1/2 enclosed in calyx.

1 Bracteoles 1.8-4.5 x 1.5-4mm, ≥2x as wide as pedicels. Densely spiny spreading shrub to 2(2.5)m; flowering mainly winter-spring (autumn

flowers accompanied by many buds) but often overlapping with
other 2 spp. in Sep-Oct; calyx 10-16(20)mm, with convergent teeth;
standard 12-18mm. Native; grassy places, heathland, open woods,
mostly on sandy or peaty soil; throughout BI *Gorse* - **U. europaeus** L.
1 Bracteoles <1.5 x 1mm, ≤2x as wide as pedicels 2
   2 Teeth of lower calyx-lip parallel to convergent; calyx
9-13(15)mm; standard (12)13-18(22)mm. Densely spiny spreading
shrub to 1.5(2)m; flowering mainly summer (autumn flowers
accompanied by many withered ones). Native; habitat as for
U. europaeus; mainly in CI, W 1/2 of En, Wa, Man, SW Sc, C & S
Ir, but scattered in E En and E Sc *Western Gorse* - **U. gallii** Planch.
   2 Teeth of lower calyx-lip divergent; calyx 5-9.5(10)mm; standard
7-12(13)mm. Densely spiny spreading shrub to 1(1.5)m; flowering
as in *U. gallii*. Native; heaths; mainly in S En from E Kent to
Dorset and Wilts, very scattered elsewhere and sometimes
natd *Dwarf Gorse* - **U. minor** Roth

U. europaeus x *U. gallii* occurs rarely in W Br and Ir where the 2 spp. overlap;
it is highly fertile and variously intermediate, and flowers in autumn and
winter.

80. ELAEAGNACEAE - *Sea-buckthorn family*

Deciduous or evergreen, usually spiny shrubs; easily recognized by the
entire, silver-scaly leaves, petal-less, 2- or 4-sepalled flowers, and succulent
1-seeded fruits.

1 Leaves linear-lanceolate; petiole <2mm; flowers dioecious,
appearing before leaves **1. HIPPOPHAE**
1 Leaves elliptic to broadly ovate; petiole >5mm; flowers bisexual,
appearing with leaves **2. ELAEAGNUS**

1. HIPPOPHAE L. - *Sea-buckthorn*
Flowers dioecious, wind-pollinated; sepals and stamens 2.

1 Spiny, spreading and suckering shrub to 3(9)m; leaves deciduous,
linear-lanceolate; flowers 3-4mm, early spring; fruits translucent-
orange, 6-10mm. Native; dunes and other sandy places by sea;
round coasts of Br and Ir, perhaps native only in E from E Sussex
to C Sc; sometimes natd inland *Sea-buckthorn* - **H. rhamnoides** L.

2. ELAEAGNUS L. - *Oleasters*
Flowers bisexual, insect-pollinated; sepals and stamens 4.

1 Leaves deciduous or ± so, elliptic to ovate or narrowly so; flowers
May-Jun; calyx-lobes <1/2 as long as tube; fruits scaly-red, 6-15mm.
Spiny shrub to 3(6)m. Intrd-natd; rough ground and by roads;

scattered in S Br and CI   *Spreading Oleaster* - **E. umbellata** Thunb.
1   Leaves evergreen, elliptic to ovate or broadly so; flowers Oct-Nov; calyx-lobes >1/2 as long as tube; fruits white-scaly, 15-20mm. Spiny shrub to 3m. Intrd-surv; grown as hedging (often on *E. umbellata* as stock) in Guernsey and Sark and long persistent, rarely self-sowing in S En   *Broad-leaved Oleaster* - **E. macrophylla** Thunb.

81. HALORAGACEAE - *Water-milfoil family*

Perennial, mainly subaquatic herbs with weak trailing stems; leaves simple and opposite, or in whorls of 3-6 and finely pinnate. The inconspicuous flowers with 4 sepals, 4 petals and 8 stamens, and fruits of 1 or 4 nutlets are diagnostic.

1   Leaves opposite, simple, not or slightly toothed   **1. HALORAGIS**
1   Leaves in whorls of 3-6, finely pinnate   **2. MYRIOPHYLLUM**

1. HALORAGIS J.R. & G. Forst. - *Creeping Raspwort*

Plant growing on peat surface; leaves opposite, simple, entire or weakly serrate; ovary 1-celled; fruit a single nutlet.

1   Stems decumbent, to 20cm; leaves broadly ovate to suborbicular, 3-10mm. Intrd-natd; on bare peat in bog; W Galway
            Creeping Raspwort - **H. micrantha** (Thunb.) Siebold & Zucc.

2. MYRIOPHYLLUM L. - *Water-milfoils*

Plants normally subaquatic, sometimes on mud; leaves in whorls of 3-6, finely pinnate; ovary 4-celled; fruit a group of ≤4 nutlets.

Vegetative characters, especially numbers of leaf-segments, are not sufficient for certain identification; in *M. verticillatum*, for example, there are usually 24-35 segments, but in some mud-growing plants there can be only 4.

1   Leaves many or mostly 5 in a whorl; uppermost bracts deeply serrate to pinnately dissected   2
1   Leaves (3)4(-5) in a whorl; uppermost bracts simple, entire or minutely serrate   3
    2   Emergent leaves with dense sessile glands; only female flowers present, whitish. Stems to 2m; leaves 4-6 in a whorl, usually with 8-30 segments. Intrd-natd; ponds and canals; scattered in Br N to S Sc, Jersey   *Parrot's-feather* - **M. aquaticum** (Vell.) Verdc.
    2   Emergent leaves with sparse sessile glands; each plant with upper flowers male, lower female, and usually some bisexual between; flowers reddish. Stems to 3m; leaves (4)5(-6) in a whorl, usually with 24-35 segments. Native; mostly base-rich ponds, lakes and slow rivers in lowlands; very scattered in CI, En, Wa and Ir   *Whorled Water-milfoil* - **M. verticillatum** L.

3   All flowers in whorls, reddish; leaves usually with 13-38 segments. Stems to 2.5m; leaves (3)4(-5) in a whorl, usually with 13-38 segments. Native; mostly base-rich ponds, lakes, slow rivers and ditches, mostly lowland; locally common
Spiked Water-milfoil - **M. spicatum** L.
3   Upper flowers opposite or alternate, all yellowish; leaves usually with 6-18 segments. Stems to 1.2m; leaves (3-)4 in a whorl, usually with 6-18 segments. Native; mostly base-poor lakes, ponds, slow streams and ditches, often upland; locally frequent, mostly in N & W   *Alternate Water-milfoil* - **M. alterniflorum** DC.

82. GUNNERACEAE - *Giant-rhubarb family*

Huge herbaceous perennials; stems wholly rhizomatous; unique in the huge rhubarb-like leaves and erect, compact, many-flowered inflorescences.

1. GUNNERA L. - *Giant-rhubarbs*

1   Leaves ≤2m across, cordate at base; petioles ≤1.5m, with pale bristles and weak spines; inflorescences ≤1m. Intrd-natd; self-sown by lakes; scattered in lowland BI   *Giant-rhubarb* - **G. tinctoria** (Molina) Mirb.
1   Leaves often >2m across, peltate; petioles ≤2m, with reddish bristles and spines; inflorescences ≤1.2m. Intrd-surv; persistent by lakes; scattered in lowland BI   *Brazilian Giant-rhubarb* - **G. manicata** André

83. LYTHRACEAE - *Purple-loosestrife family*

Annuals or herbaceous perennials; leaves opposite or in whorls of 3, or upper ones alternate, simple; distinguished by the perigynous flowers bearing 6 petals, sepals and epicalyx-segments near the hypanthium mouth.

1. LYTHRUM L. - *Purple-loosestrifes*

1   Petals >4mm; stigma or some stamens exceeding sepals; flowers trimorphic on separate plants   2
1   Petals <4mm; stigma and stamens not reaching apex of sepals; flowers monomorphic   3
  2   Petals >7mm; flowers clustered in whorls. Erect perennial to 1.5m; petals 8-10mm, purple. Native; by water and in marshes and fens; common   *Purple-loosestrife* - **L. salicaria** L.
  2   Petals <7mm; flowers 1-2 in each axil. Erect to decumbent annual (in Br) to 70cm; petals 5-6mm, purple. Intrd casual; birdseed-alien in parks and waste places; frequent in S & C Br
False Grass-poly - **L. junceum** Banks & Sol.
3   Leaves obovate-spathulate; hypanthium funnel- to cup-shaped; capsule subglobose. Procumbent annual to 25cm, rooting at nodes;

petals 0-6, c.1mm, purplish. Native; open or bare ground by or in
water, or in damp trackways; scattered throughout most of BI
Water-purslane - **L. portula** (L.) D.A. Webb

3 Leaves linear-oblong; hypanthium tubular; capsule cylindrical.
Erect to decumbent usually annual to 25cm; petals 2-3mm, pink.
Native; seasonally wet bare ground; extremely local in S En and
Jersey, rare casual elsewhere *Grass-poly* - **L. hyssopifolium** L.

84. THYMELAEACEAE - *Mezereon family*

Early-flowering, glabrous, poisonous shrubs; distinctive in the perigynous
flowers with 4 sepals, 0 petals and ovary with 1 ovule.

1. DAPHNE L. - *Mezereons*

1 Erect, deciduous shrub to 2m; leaves light green; flowers bright
pink, appearing before leaves, very fragrant; fruit bright red. Native;
calcareous woods; very local in Br N to Yorks and Lancs but often
only natd *Mezereon* - **D. mezereum** L.

1 Erect to decumbent evergreen shrub to 1.5m; leaves dark green;
flowers yellowish-green, not or slightly scented; fruit black. Native;
woods mostly on calcareous or clayey soils; locally frequent in En,
Wa and CI *Spurge-laurel* - **D. laureola** L.

D. mezereum x *D. laureola* = ***D. x houtteana*** Lindl. & Paxton has been found
rarely in En, but not recently.

85. MYRTACEAE - *Myrtle family*

Evergreen trees or shrubs; recognized by the combination of evergreen,
aromatic, entire leaves, inferior ovary, numerous stamens and 4-5 petals.

1 All leaves opposite; ovary (2-)3-celled; fruit a berry **3. LUMA**
1 Adult leaves alternate; ovary 4-5-celled; fruit a capsule 2
  2 Leaves <2cm; petals free; shrubs **1. LEPTOSPERMUM**
  2 Leaves >3cm; petals united into a hood covering flower in bud;
trees **2. EUCALYPTUS**

1. LEPTOSPERMUM J.R. & G. Forst. - *Tea-trees*
All leaves alternate, <2cm; flowers solitary, 0.8-1.8cm across; petals 5, white
or sometimes pink, free; ovary 5-celled; fruit a woody capsule, 6-10mm
wide.

1 Leaves 4-20 x 2-6mm, tapered to sharp point; sepals falling as soon
as flowers fade. ± glabrous shrub to 5m. Intrd-natd; self-sowing in
woods; Tresco (Scillies) *Broom Tea-tree* - **L. scoparium** J.R. & G. Forst.

1 Leaves 4-15 x 2-4mm, abruptly pointed and scarcely sharply so;
sepals, hypanthium and lowerside of leaf white-pubescent; sepals
persistent on fruit. Shrub to 5m. Intrd-natd; with *L. scoparium* on
Tresco but less well natd   **Woolly Tea-tree - L. lanigerum** (Aiton) Sm.

2. EUCALYPTUS L'Hér. - *Gums*

Leaves on adult shoots alternate, those on juvenile shoots opposite and of
very different shape, >3cm; flowers solitary or in small umbel-like clusters;
petals united into a hood covering flower in bud and breaking off
transversely; ovary mostly 4-celled; fruit a woody capsule.

1 Flowers and fruits solitary. Tree to 45m. Intrd-natd; planted for
  ornament and rarely forestry, sometimes self-sown; W Ir, Scillies
  and Man   **Southern Blue-gum - E. globulus** Labill.
1 Flowers and fruits in groups of 3 or more   2
  2 Flowers in groups of 5-12. Tree to 15m. Intrd-natd; planted for
    ornament in extreme SW En, self-sown in Scillies
                              White Peppermint-gum - E. pulchella Desf.
  2 Flowers in groups of 3   3
3 Juvenile leaves lanceolate to narrowly elliptic; fruit 5-6mm. Tree to
  50m. Intrd-natd; planted for small-scale forestry in Ir, sometimes
  self-sown   **Ribbon Gum - E. viminalis** Labill.
3 Juvenile leaves orbicular to elliptic or ovate; fruit ≥7mm   4
  4 Flower buds and fruit with 2-4 longitudinal ribs, sessile. Tree to
    50m. Intrd-natd; planted for small-scale forestry in Ir, sometimes
    self-sown   **Johnston's Gum - E. johnstonii** Maiden
  4 Flower buds and fruit ± terete, usually stalked   5
5 Mature leaves 4-7cm; flower buds 6-8mm; fruit 7-10mm. Tree to
  30m. Intrd-natd; planted widely in Ir and S & W Br, sometimes
  self-sown   **Cider Gum - E. gunnii** Hook. f.
5 Mature leaves 8-18cm; flower buds c.12mm; fruit c.17mm. Tree to
  30m. Intrd-natd; planted for small-scale forestry in Ir, sometimes
  self-sown   **Urn-fruited Gum - E. urnigera** Hook. f.

3. LUMA A. Gray (*Amomyrtus* auct.) - *Myrtles*

Leaves all opposite; flowers 1(-3) in leaf-axils; petals 4(-5), white, free; ovary
with (2-)3 cells; fruit a berry.

1 Shrub or tree rarely to 18m; leaves 1.5-3cm; fruit dark purple, globose,
  6-10mm. Intrd-natd; self-sown in semi-natural woodland; SW En,
  Guernsey, SW Ir, Man (*Amomyrtus luma* auct.)
                              Chilean Myrtle - L. apiculata (DC.) Burret

86. ONAGRACEAE - *Willowherb family*

Herbaceous annuals, biennials or perennials or rarely shrubs; distinguished
by the combination of epigynous flowers with 2 or 4 sepals, 2, 4 or 8

stamens, and (1)2-4-celled ovary.

1 Sepals 2; petals 2; stamens 2; ovary 1-2-celled; fruit with hooked
  bristles **7. CIRCAEA**
1 Sepals 4; petals 4 (rarely 0); stamens 4 or 8; ovary 4-celled; fruit
  without hooked bristles 2
    2 Shrub; fruit a berry **6. FUCHSIA**
    2 Herb; fruit a capsule 3
3 Petals 0 or caducous and <1mm; stamens 4 **3. LUDWIGIA**
3 Petals 4, >1mm, not caducous; stamens 8 4
    4 Petals yellow (sometimes streaked or tinged reddish);
      hypanthium ≥20mm **4. OENOTHERA**
    4 Petals pink, red or purple, sometimes white, never yellow;
      hypanthium ≤11mm 5
5 Seeds without hairy plume; hypanthium 2-11mm; all leaves
  alternate **5. CLARKIA**
5 Seeds with hairy plume; hypanthium 0-c.3mm; lower leaves often
  opposite 6
    6 At least lowest leaves opposite; flowers ± erect when open,
      actinomorphic **1. EPILOBIUM**
    6 All leaves alternate; flowers held horizontally, slightly
      zygomorphic **2. CHAMERION**

1. EPILOBIUM L. - *Willowherbs*

Perennial herbs; at least lower leaves opposite or rarely whorled; flowers in loose terminal racemes, actinomorphic; hypanthium very short; sepals 4; petals 4, pink to purple, rarely white; stamens 8; ovary 4-celled; fruit a linear capsule; seeds with a hairy plume.

Plants vary greatly in stature, leaf-size, and degree of branching and of pubescence, but the *type* of hairs and certain aspects of leaf-shape are relatively constant. Seed-coat ornamentation is highly diagnostic, as is the presence of a terminal appendage, but a high magnification (x≥20) is required.

1 Stigma 4-lobed 2
1 Stigma clavate 5
    2 Stems with many patent non-glandular (and some gland-tipped)
      hairs 3
    2 Stems with 0 or few patent hairs and those all gland-tipped 4
3 Petals 10-16mm, purplish-pink; leaves slightly clasping stem at
  base; white subterranean rhizomes produced. Stems erect, to 1.8m.
  Native; in all sorts of wet or damp places; common throughout
  lowland BI except N & C Sc *Great Willowherb* - **E. hirsutum** L.
3 Petals 5-9mm, paler; leaves not clasping stem at base; short, green
  surface stolons produced. Stems erect, to 75cm. Native; in all sorts
  of wet or damp places; frequent throughout lowland BI
  Hoary Willowherb - **E. parviflorum** Schreb.
    4 Lower leaves ovate, rounded at base, abruptly delimited from

petiole 2-6mm. Stems erect, to 75cm; perennating by subsessile leafy buds. Native; shady places, walls, rocks and cultivated ground; common **Broad-leaved Willowherb - E. montanum** L.

4 Lower leaves narrowly elliptic, cuneate at base, gradually narrowed to petiole 3-10mm. Stems erect, to 60cm; perennating as *E. montanum*. Native; waysides, walls and waste places; locally frequent in S & C Br and CI

Spear-leaved Willowherb - E. lanceolatum Sebast. & Mauri

5 Stems wholly procumbent, rooting at nodes; leaves suborbicular to broadly ovate-elliptic; flowers solitary in leaf axils 6
5 Stems not procumbent, or procumbent and rooting only near base; leaves lanceolate to ovate; flowers forming terminal raceme (rarely solitary and terminal) 8

6 Leaves with ± prominent veins and bronze-coloured on upperside; seeds obscurely reticulate, not papillose. Stems procumbent, to 20cm. Intrd-surv; barely natd garden weed; few places in En, Sc and N Ir

Bronzy Willowherb - E. komarovianum H. Lév.

6 Leaves with obscure veins and without bronze colour (sometimes reddish) on upperside; seeds minutely papillose 7

7 Leaves entire or nearly so, 3-7(10)mm; petals 2.5-4mm. Stems procumbent, to 20cm. Intrd-natd; all sorts of damp barish ground, especially gravelly hillsides, railway sidings, waste-tips; most of BI, especially N & W ***New Zealand***
***Willowherb* - E. brunnescens** (Cockayne) P.H. Raven & Engelhorn

7 Leaves acutely dentate, 3-10(14)mm; petals 3-5mm. Stems procumbent, to 20cm. Intrd-natd; weed of barish damp ground, by roads in C Sc and W Ir, garden weed in En and Sc

Rockery Willowherb - E. pedunculare A. Cunn.

8 At least stem apices, sepals, hypanthium, ovaries and young capsules with numerous patent gland-tipped and appressed glandless hairs; plant perennating by sessile dense leafy basal rosettes 9
8 Patent gland-tipped hairs 0 or few, if present then plant perennating by elongated leafy or leafless stolons 10

9 Petioles ≤4mm; leaves rounded to subcordate at base; seeds with papillae on well separated ridges, with rounded appendage c.0.05mm at hairy end. Stems erect, to 75(100)cm. Intrd-natd; waste and cultivated ground, by roads, rivers and railways, on walls; most of BI and the commonest sp. in S & C Br

American Willowherb - E. ciliatum Raf.

9 Petioles 4-15mm; leaves cuneate at base; seeds uniformly papillose, without appendage. Stems erect, to 75cm. Native; shady places, damp ground, cultivated and waste land; scattered throughout most of Br and Ir **Pale Willowherb - E. roseum** Schreb.

10 Seeds without appendage; stem apices, sepals, hypanthium, ovaries and young capsules densely appressed-pubescent 11
10 Seeds with appendage c.0.2mm at hairy end; stem apices, sepals,

1. EPILOBIUM

  hypanthium, ovaries and young capsules rather sparsely
  pubescent   12
11 Patent glandular hairs 0; plant perennating by ± sessile lax leaf-
rosettes; capsules (5.5)6.5-8(10)cm. Stems erect, to 75cm. Native;
hedgerows, open woods, by water, cultivated and waste ground;
locally common in CI, S & C Br, very scattered elsewhere
Square-stalked Willowherb - **E. tetragonum** L.
11 Patent glandular hairs present on hypanthium and sometimes
capsule; plant perennating by elongated leafy stolons; capsules
(3)4-6(6.5)cm. Stems erect, to 75cm. Native; same habitats as *E.
tetragonum*; frequent *Short-fruited Willowherb* - **E. obscurum** Schreb.
 12 Seeds uniformly papillose; plant erect, perennating by long
 slender stolons ending in tight bud. Stems erect, to 60cm.
 Native; marshes, fens and ditches, often with *E. parviflorum*;
 frequent   *Marsh Willowherb* - **E. palustre** L.
 12 Seeds obscurely reticulate, not papillose; plant decumbent to
 ascending, perennating by stolons ending in loose leafy bud 13
13 Leaves narrowly elliptic-oblong, entire to denticulate; stolons on
soil surface, with green leaves. Stems ascending to decumbent, to
20cm. Native; mountain flushes and streamsides; locally frequent
in N En and Sc   *Alpine Willowherb* - **E. anagallidifolium** Lam.
13 Leaves ovate to narrowly ovate, distinctly dentate; stolons below
soil surface, with yellowish scale-leaves. Stems (erect or) ascending
to decumbent, to 25(30)cm. Native; similar habitats and distribution
to *E. anagallidifolium*, but also Caerns and Co Leitrim
    Chickweed Willowherb - **E. alsinifolium** Vill.

Hybrids occur commonly where 2 or more spp. occur together, especially in
quantity for several years in disturbed ground. Hybrids are often
recognizable by their larger and more-branched stature, longer flowering
season, unusually large or small flowers markedly more darkly coloured at
petal-tips, and partially or entirely abortive fruits. Most seeds are abortive
but some are fertile; backcrossing and even triple-hybrids rarely occur. Most
hybrids are variously intermediate in diagnostic characters, notably stigma-
form and type of pubescence. Probably all possible combinations can occur;
45 each involving 2 parents have so far been reliably recorded

2. CHAMERION (Raf.) Raf. - *Rosebay Willowherb*

Perennial rhizomatous herbs; all leaves alternate; flowers in dense terminal
racemes, slightly zygomorphic; hypanthium ± 0; sepals 4; petals 4, pinkish-
purple, rarely white; stamens 8; ovary 4-celled; fruit a linear capsule; seeds
with a hairy plume.

1 Stems erect, to 1.5m; 2 upper petals wider than 2 lower ones; style
deflexed; stigma 4-lobed. Native; waste ground, woodland-
clearings, embankments, rocky places and screes on mountains;
common   *Rosebay Willowherb* - **C. angustifolium** (L.) Holub

3. LUDWIGIA L. - *Hampshire-purslane*

Annual to perennial ± aquatic herbs; all leaves opposite; flowers solitary in leaf-axils, actinomorphic; hypanthium 0; sepals 4; petals 0(-4) stamens 4; ovary 4-celled; fruit a short, cylindrical, scarcely dehiscent capsule retaining sepals; seeds not plumed.

1 Stems procumbent to decumbent or ascending, sometimes floating at ends, to 30(60)cm; flowers 2-5mm, inconspicuous; stigma capitate. Native; acid pools; extremely local in S Hants

Hampshire-purslane - **L. palustris** (L.) Elliott

Natd plants in E Sussex, Surrey, Dorset and S Essex, appear to be an aquarists' hybrid with no valid name (*L. x muellertii* hort.), possibly *L. palustris* x *L. natans* Elliott; the flowers have 1-4 cream caducous petals <1mm and 2 bracteoles 1-1.5mm.

4. OENOTHERA L. - *Evening-primroses*

Annual to biennial (rarely perennial) herbs; all leaves alternate; flowers in terminal racemes, ± actinomorphic; hypanthium 20-45mm, a narrow tube; sepals 4; petals 4, yellow, sometimes streaked or tinged red; stamens 8; ovary 4-celled; fruit a long ± cylindrical capsule; seeds not plumed.

A critical genus where species limits are a matter of opinion. Flower measurements refer to the lower (first opened) ones of the inflorescence (later ones may be much smaller); fruit pubescence should also be noted on the lower part of the inflorescence.

1 Capsule widest near apex, c.2-4mm wide near base; petals tinged reddish when withering; seeds not angled. Stems erect to ascending, to 1m; green parts of stems and fruits without red bulbous-based hairs; rhachis usually reddened towards tip. Intrd-natd; in sandy places mostly on coasts; locally frequent in CI and S Br

Fragrant Evening-primrose - **O. stricta** Link

1 Capsule widest near base, there c.6-8mm wide; petals always yellow; seeds sharply angled 2
    2 Petals 3-5cm; style longer than stamens, the stigmas held above anthers. Stems erect, to 1.8m; green parts of stems and fruits with many red bulbous-based hairs; rhachis red at tip. Intrd-natd; sand-dunes, waste ground, waysides; common in Bl except N & C Sc   *Large-flowered Evening-primrose* - **O. glazioviana** P. Micheli
    2 Petals 1.5-3cm; style shorter than stamens, the stigmas held ± at same level as anthers 3
3 Green parts of stems and fruits without red bulbous-based hairs 4
3 Green parts of stems and fruits with red bulbous-based hairs 5
    4 Petals 1.5-3cm, wider than long; capsules 2-3(3.5)cm, all with glandular hairs. Stems erect, to 1.5m; green parts of stems and fruits without red bulbous-based hairs but these present on red blotches on stems; rhachis green at tip. Intrd-natd; same habitats

as *O. glazioviana*; frequent in Br N to C Sc, ?CI
Common Evening-primrose - **O. biennis** L.
4 Petals 2-3cm, c. as wide as long; capsules 3-4(5)cm, only upper with glandular hairs. Stems erect, to 1.2m; green parts of stems and fruits usually with, sometimes without (but probably because of hybridization with *O. biennis*), red bulbous-based hairs; rhachis green at tip. Intrd-natd; in maritime sandy places, inland where intrd from coast; frequent by coast in En and Wa, CI
Small-flowered Evening-primrose - **O. cambrica** Rostanski
5 Sepals red-striped; rhachis reddish towards apex; capsules 2-3cm, all with glandular hairs. Stems erect, to 1.5m. Native; same habitats as *O. glazioviana*; scattered in Br N to C Sc, CI (*O. glazioviana* x *O. biennis*) *Intermediate Evening-primrose* - **O. x fallax** Renner
5 Sepals and rhachis green; capsules 3-4(5)cm, only upper with glandular hairs (see couplet 4) **O. cambrica**

Hybrids usually occur wherever *O. biennis*, *O. cambrica* and/or *O. glazioviana* occur together; reciprocal crosses often have different characters. The hybrids are fertile and may backcross or form the triple-hybrid.

5. CLARKIA Pursh - *Clarkias*

Annuals; all leaves alternate; flowers in loose terminal racemes, actinomorphic; hypanthium 2-11mm, narrowly tubular; sepals 4; petals 4, pink to purple, rarely white; stamens 8; ovary 4-celled; fruit a linear capsule; seeds not plumed.

1 Leaves glabrous; flower-buds pendent; hypanthium 2-5mm; sepals 10-16mm; petals 1-2cm, with claw c. as long as limb. Stems erect to ascending, to 50(80)cm. Intrd-casual; tips, parks and waste places; scattered in Br *Clarkia* - **C. unguiculata** Lindl.
1 Leaves minutely pubescent; flower-buds erect; hypanthium 5-30mm; sepals 12-30mm; petals 1-4(6)cm, with claw much shorter than limb. Stems erect to ascending, to 50(80)cm. Intrd-casual; occurs as for *C. unguiculata* *Godetia* - **C. amoena** (Lehm.) A. Nelson & J.F. Macbr.

6. FUCHSIA L. - *Fuchsias*

Deciduous shrubs; all leaves opposite or sometimes in whorls of 3-4; flowers solitary in leaf-axils, actinomorphic, pendent on long pedicels; hypanthium 5-16mm, broadly tubular; sepals 4; petals 4, pink to purple or violet (or white); stamens 8; ovary 4-celled; fruit a ± cylindrical black berry.

1 Leaves 2.5-5.5cm; sepals 12-24mm; petals 6-12mm. Spreading shrub to 1.5(3)m. Intrd-natd; planted as hedging in Ir, Man and W Br; natd in Ir, Man and W Br N to Orkney *Fuchsia* - **F. magellanica** Lam.
1 Leaves 3-8cm; sepals 25-40mm; petals 15-25mm. Spreading shrub to 1.5(3)m. Intrd-surv; relic of cultivation; Lleyn (Caerns) and Lundy Island (N Devon) *Large-flowered Fuchsia* - **F. 'Corallina'**

7. CIRCAEA L. - *Enchanter's-nightshades*
Perennial rhizomatous and/or stoloniferous herbs; all leaves opposite; flowers in loose terminal racemes, ± zygomorphic, ± horizontally held; hypanthium very short; sepals 2; petals 2, deeply 2-lobed, white or pinkish; stamens 2; ovary 1-2-celled, each cell with 1 seed; fruit a 1-2-seeded achene.

1 Open flowers crowded at inflorescence apex; pedicels, hypanthia and sepals glabrous; ovary 1-celled. Stems erect, to 30cm. Native; same habitats as *C.* x *intermedia*; very scattered in Wa, Lake District and Sc *Alpine Enchanter's-nightshade* - **C. alpina** L.
1 Open flowers on elongated raceme; pedicels, hypanthia and sepals with glandular hairs; ovary 2-celled 2
   2 Stolons produced from lower leaf-axils; petioles pubescent on upperside, subglabrous on lowerside; ovary with 1 large and 1 small cell; fruit not ripening. Stems erect, to 45cm. Native; woods and shady rocky places, often on mountains; locally frequent in N & W Br, very scattered in N & C Ir (*C. lutetiana* x *C. alpina*)
Upland Enchanter's-nightshade - **C. x intermedia** Ehrh.
   2 Stolons 0; petioles pubescent all round; ovary with 2 equal cells; fruit ripening. Stems erect, to 60cm. Native; woods, hedgerows and shady places; common *Enchanter's-nightshade* - **C. lutetiana** L.

87. CORNACEAE - *Dogwood family*

Deciduous or evergreen shrubs or perennial herbs; distinguished by the usually opposite leaves, inferior ovary with 1 ovule per cell, 4 or 5 free petals, and drupe.

1 Leaves deciduous, not thick and glossy; flowers bisexual **1. CORNUS**
1 Leaves evergreen, thick and glossy; dioecious 2
   2 Leaves opposite; petals dark purple, 4 **2. AUCUBA**
   2 Leaves alternate; petals yellowish-green, 5 **3. GRISELINIA**

1. CORNUS L. - *Dogwoods*
Perennial herbs or deciduous shrubs; leaves opposite, ± entire; flowers in corymbs or umbels, bisexual; sepals and petals 4; style 1; ovary 2-celled; fruit a drupe with 1 2-celled stone.

1 Rhizomatous herb; petals (not the white petal-like bracts) purple. Stems erect, to 20cm. Native; upland moors among low shrubs; extremely local in N En, locally frequent in W & C mainland Sc
Dwarf Cornel - **C. suecica** L.
1 Shrub; petals white to yellow 2
   2 Inflorescences appearing before leaves, with 4 yellow petal-like bracts at base. Shrub or small tree to 4(8)m. Intrd-natd; hedges and roadsides, often long persistent, sometimes bird-sown; scattered in S Br, rarely N to C Sc, Man *Cornelian-cherry* - **C. mas** L.

2 Inflorescences appearing after leaves, without petal-like bracts
at base; petals whitish 3
3 Fruit purplish-black; leaves rarely with >5 pairs of lateral veins;
petals 4-7mm. Shrub to 4m. Native; woods and scrub on limestone
or base-rich clays; common in most of S & C lowland Br, very local
in S Ir, escape elsewhere **Dogwood - C. sanguinea** L.
3 Fruit white to cream or pale blue; many larger leaves with 6(-7)
pairs of lateral veins; petals 2-4mm 4
4 Stone flattened-ellipsoid, longer than wide, tapered at both
ends; leaves shortly and abruptly acuminate to acute; fruit
whitish to pale blue. Shrub to 3m. Intrd-natd; occurs as for
C. sericea, but less well natd; very scattered in lowland Br
White Dogwood - C. alba L.
4 Stone flattened-subglobose, c. as long as wide, rounded at base;
leaves tapering-acuminate; fruit white to cream. Shrub to 3m.
Intrd-natd; roadsides and scrub, natd by suckers; scattered in
most of lowland BI **Red-osier Dogwood - C. sericea** L.

2. AUCUBA Thunb. - *Spotted-laurel*

Evergreen shrubs; leaves opposite, entire to remotely serrate; flowers dioecious, the male in erect terminal panicles, the female in small terminal clusters; sepals and petals 4; ovary 1-celled; style 1; fruit a 1-seeded drupe.

1 Shrub to 5m; leaves 8-20cm, dark green but often with yellow
blotches; drupes bright scarlet. Intrd-natd; shrubberies, rarely
self-sown; very scattered in Br, mainly W, N to MW Sc, Man
Spotted-laurel - A. japonica Thunb.

3. GRISELINIA G. Forst. - *New Zealand Broadleaf*

Evergreen shrubs; leaves alternate, entire; flowers dioecious, in axillary racemes or panicles; sepals and petals 5; ovary 1-2-celled; styles 3; fruit a 1-seeded drupe.

1 Shrub to 3m or rarely tree to 15m; leaves 3-10cm, rather yellowish-green; drupes dark purple. Intrd-natd; planted in S & W, especially
near sea, sometimes self-sown; very scattered in S & W Br N to C Sc,
Man, Co Down **New Zealand Broadleaf - G. littoralis** (Raoul) Raoul

88. SANTALACEAE - *Bastard-toadflax family*

Herbaceous perennials semi-parasitic on roots of various herbs; easily recognized by the narrow, 1-veined, yellowish-green leaves and the epigynous flowers with 5 tepals which are retained on the nut.

1. THESIUM L. - *Bastard-toadflax*

1 Stems procumbent to weakly ascending, to 20cm; flowers yellowish-

green, 2-3mm, to c.4mm in fruit. Native; chalk and limestone grassland; very local in En N to S Lincs and E Gloucs, Jersey and Alderney   *Bastard-toadflax* - **T. humifusum** DC.

89. VISCACEAE - *Mistletoe family*

Semi-parasitic yellowish-green evergreen dioecious shrubs growing on tree-branches; flowers inconspicuous, Feb-Apr; fruit a white 1-seeded berry.

1. VISCUM L. - *Mistletoe*

1   Stems green, divergently branching to form ± spherical loose mass to 2m across. Native; on many spp. of tree, especially *Malus* and *Populus*; En and Wa N to Yorks, mostly local but common in S part of En/Wa borders   *Mistletoe* - **V. album** L.

90. CELASTRACEAE - *Spindle family*

Evergreen or deciduous shrubs or woody climbers; the distinctive brightly coloured seeds are diagnostic; otherwise distinguished by the woody habit, ± hypogynous flowers with 4-5 sepals, petals and stamens, with 3-5 fused carpels, and with a large nectar-secreting disc at base.

1   Erect shrub with opposite leaves; capsule 4-5-celled   **1. EUONYMUS**
1   Woody climber with alternate leaves; capsule 3-celled   **2. CELASTRUS**

1. EUONYMUS L. - *Spindles*
Erect non-spiny shrubs with opposite leaves; ovary 4-5-celled; stigma capitate; capsule 4-5-celled, with 4-5 distinct rounded to winged lobes, pinkish-red or creamish-yellow on outside; aril orange.

1   Leaves leathery, evergreen; fruit with rounded lobes. Bushy shrub or small tree to 5(8)m; fruits pink or creamish-yellow. Intrd-natd; self-sown in extreme S and Man; frequent relic by sea in S & W Br N to S Wa, CI   *Evergreen Spindle* - **E. japonicus** L. f.
1   Leaves thin, deciduous; fruit with obtuse to winged lobes   2
    2   Terminal buds (Jul-Mar) <5mm; flowers mostly with 4 sepals and stamens; fruits with mostly 4 obtuse angles. Much-branched shrub or small tree to 5(8)m; fruits pinkish-red. Native; hedges, scrub and open woods on calcareous or base-rich soils; frequent in Br and Ir N to C Sc   *Spindle* - **E. europaeus** L.
    2   Terminal buds (Jul-Mar) >5mm; flowers mostly with 5 sepals and stamens; fruits with mostly 5 winged angles. Much-branched shrub or small tree to 5(8)m; fruits pinkish-red. Intrd-natd; hedges and scrub from bird-sown seeds; scattered in C & S En   *Large-leaved Spindle* - **E. latifolius** (L.) Mill.

2. CELASTRUS L. - *Staff-vine*

Woody climbers with alternate leaves each with 2 spinose stipules in young state; ovary 3-celled; stigma 3-lobed; capsule 3-celled, subglobose, brownish- to greenish-yellow on outside, golden-yellow inside; seeds with red aril.

1 Stems twining, climbing potentially to 12m. Intrd-natd; garden plant spreading into woodland; Surrey   *Staff-vine* - **C. orbiculatus** Thunb.

91. AQUIFOLIACEAE - *Holly family*

Evergreen trees or shrubs; usually at least some leaves with very spiny margins; flowers in small axillary cymes, usually dioecious; sepals, petals and stamens 4; fruit a red to yellow (2-)4-seeded drupe.

1. ILEX L. - *Hollies*

1 Leaves mostly ≥2x as long as wide, at least the lower undulate and strongly spinose at margins. Shrub or tree to 23m. Native; woods, hedges and scrub; common   *Holly* - **I. aquifolium** L.
1 Leaves mostly <2x as long as wide, ± flat, without lateral spines or with few ± forwardly-pointed ones. Intrd-natd; in hedges and woodland as relics or bird-sown plants; scattered in Br and Man (*I. aquifolium* x *I. perado* Aiton)
Highclere Holly - **I. x altaclerensis** (Loudon) Dallim.

92. BUXACEAE - *Box family*

Evergreen shrubs or small trees; recognized by the evergreen leaves, distinctively arranged unisexual flowers with 4 stamens and 2-3-celled ovary, and 2- or 3-horned fruits.

1 Erect shrub or tree with opposite, entire leaves; fruit 3-horned
1. BUXUS
1 Stoloniferous dwarf shrub with alternate, dentate leaves; fruit 2-horned   **2. PACHYSANDRA**

1. BUXUS L. - *Box*

Erect shrub or tree with opposite, entire, glabrous to sparsely pubescent leaves; inflorescences axillary, usually with 1 apical female and several lower male flowers; ovary 3-celled, with 3 styles; stigma bilobed; fruit a dry capsule with 3 horns.

1 Shrub or small tree to 5(11)m; leaves 1-2.5cm; flowers pale yellow. Native; woods and scrub on chalk and limestone; extremely local in S En, sometimes natd   *Box* - **B. sempervirens** L.

2. PACHYSANDRA Michx. - *Carpet Box*

Dwarf stoloniferous shrub with alternate glabrous leaves dentate distally; inflorescence terminal on previous year's growth, with terminal male and lower female flowers; ovary 2-celled, with 2 styles; stigma linear; fruit a ± succulent drupe with 2 horns.

1 Stems ascending, to 25cm; leaves 5-10cm; flowers white. Intrd-surv; much grown in public places as ground-cover, sometimes 'running wild'; W Kent *Carpet Box* - **P. terminalis** Siebold & Zucc.

93. EUPHORBIACEAE - *Spurge family*

Annual to perennial herbs or rarely woody annuals, often with white latex; 3 extremely distinct genera with superficially little in common, but all with monoecious or dioecious flowers with 2-3-celled ovary with 1 ovule per cell and 2-3 strongly papillose or branched stigmas.

1 Leaves palmately lobed **2. RICINUS**
1 Leaves not lobed, entire to serrate 2
  2 Plant with copious white latex, monoecious; ovary and fruit 3-celled; stamen 1 **3. EUPHORBIA**
  2 Plant with watery sap, usually dioecious; ovary and fruit 2-celled; stamens numerous **1. MERCURIALIS**

1. MERCURIALIS L. - *Mercuries*

Herbs with watery sap; leaves opposite, unlobed, serrate; flowers usually dioecious, the male in catkin-like ± erect axillary spikes, the female in smaller axillary clusters; tepals 3, green; stamens numerous, with free, simple filaments; ovaries 2-celled.

1 Rhizomatous perennial; stems and leaves pubescent; female flower-clusters on stalks usually >1cm. Stems to 40cm. Native; woods, hedgerows and shady places among rocks; common over much of Br, rare in CI and Ir *Dog's Mercury* - **M. perennis** L.
1 Annual with fibrous root system; stems and leaves glabrous or nearly so; female flowers subsessile. Stems to 40cm. Possibly native; cultivated ground and waste places; frequent in S En and CI, scattered and only intrd elsewhere *Annual Mercury* - **M. annua** L.

2. RICINUS L. - *Castor-oil-plant*

Annual herb or shrub with watery sap; leaves alternate, lobed, serrate; flowers monoecious, in branched axillary groups, the male below the female; tepals 3-5, membranous; stamens numerous, on branched filaments; ovary 3-celled.

1 Stems simple or branched, to 2(4)m; leaves long-petiolate, peltate, palmately lobed, up to 60cm. Intrd-casual; on tips and in waste

places; scattered in S Br and CI   *Castor-oil-plant* - **R. communis** L.

3. EUPHORBIA L. - *Spurges*
Annual, biennial or perennial herbs with white latex; leaves opposite, alternate or whorled, unlobed, entire or serrate; flowers monoecious, in distinctive small units composed of 1 female and few male flowers together in a cup-shaped *cyathium*, which has 4-5 conspicuous glands at top; cyathia solitary in leaf-axils, or (usually) in terminal compound cymes with paired or whorled branches each subtended by a bract which is leafy but often different in shape from the leaves; perianth 0; stamen 1, with jointed filament; ovary 3-celled.

1 Plants usually procumbent; stipules present; bracts and leaves similar, markedly unequal at base   2
1 Plants usually erect; stipules 0; bracts and leaves often different, ± equal at base   3
   2 Stems and capsules glabrous. Glabrous, often purplish, annual with procumbent stems to 10cm. Native; on sandy or shingly beaches; extinct since 1976   *Purple Spurge* - **E. peplis** L.
   2 Stems and capsules pubescent. Pubescent annual with procumbent stems to 50cm. Intrd-surv; scarcely natd weed of nurseries and in a quarry; very scattered in En, S Wa, Jersey   *Spotted Spurge* - **E. maculata** L.
3 Leaves on main stems opposite. Glabrous biennial; stems to 1m in 1st year, producing inflorescence from top and to 2m in 2nd year. Possibly native in shady places in S En, frequent casual or natd over much of Br and CI   *Caper Spurge* - **E. lathyris** L.
3 Leaves on main stems alternate   4
   4 Glands on cyathia rounded on outer edge   5
   4 Glands on cyathia concave on outer edge, prolonged into 2 points 11
5 Ovary and capsule smooth to granulose, but sometimes pubescent   6
5 Ovary and capsule conspicuously warty or papillose   7
   6 Ovary and capsule pubescent; leaves pubescent at least on lowerside, oblong to oblong-lanceolate. Pubescent perennial with 0 or weak rhizomes; stems erect, to 60cm. Intrd-natd; woods and hedgerows; W Sussex and N Somerset   *Coral Spurge* - **E. corallioides** L.
   6 Ovary and capsule glabrous; leaves glabrous, obovate. Glabrous (or ± so) erect annual to 50cm. Native; cultivated ground and waste places; common   *Sun Spurge* - **E. helioscopia** L.
7 Annuals with simple root system   8
7 Rhizomatous perennials   9
   8 Capsules with hemispherical papillae; umbel with 5 main branches, the bracts at that node similar to leaves below but markedly different from bracts at next higher node. Glabrous or pubescent erect annual to 80cm. Native; cultivated and rough ground; very local in S En   *Broad-leaved Spurge* - **E. platyphyllos** L.
   8 Capsules with cylindrical papillae; umbel with 2-5 main branches,

the bracts at that node intermediate between leaves below and bracts at next higher node. Glabrous erect annual to 80cm. Native; limestone woods; W Gloucs and Mons, natd rarely in S En *Upright Spurge* - **E. serrulata** Thuill.

9 Capsules with most papillae c. as long as wide (± hemispherical). Densely pubescent perennial with strong rhizomes; stems erect, to 90cm. Intrd-natd; grassy banks; S Hants, Wight and N Essex
Balkan Spurge - **E. oblongata** Griseb.

9 Capsules with many papillae c.2x as long as wide (± cylindrical)   10

  10 Stems without scales near base; capsule 5-6mm; bracts yellowish. Sparsely pubescent or rarely glabrous perennial with strong rhizomes; stems erect, to 60cm. Native; woods, hedgerows, grassy places and stream-banks; extremely local in SW En, frequent in SW Ir, natd in Cards   *Irish Spurge* - **E. hyberna** L.

  10 Stems with scales near base; capsule (2)3-4mm; bracts green. Sparsely pubescent to subglabrous perennial with strong rhizomes; stems erect, to 50cm. Intrd-natd; shady places; very scattered in Br   *Sweet Spurge* - **E. dulcis** L.

11 Opposite pairs of bracts fused at base; stems pubescent   12
11 Opposite pairs of bracts not fused at base; stems glabrous   15

  12 Capsules glabrous; primary branches of topmost whorl of inflorescence 4-12. Pubescent perennial; stems biennial, with inflorescences arising from tops in 2nd year, to 90cm (*Wood Spurge* - **E. amygdaloides** L.)   13

  12 Capsules densely pubescent; primary branches of topmost whorl of inflorescence 10-20. Densely pubescent, tufted perennial; stems biennial, with inflorescences arising from tops in 2nd year, to 1.5m. Intrd-natd; old garden sites and waste ground (*Mediterranean Spurge* - **E. characias** L.)   14

13 Rhizomes short or 0; leaves of 1st-year stems herbaceous, dull, pale- to mid-green, pubescent on lowerside and margins. Native; woods and shady hedgerows; common in CI and much of S Br, rare alien further N and in Ir   **E. amygdaloides ssp. amygdaloides**

13 Rhizomes long; leaves of lst-year stems ± coriaceous, ± shiny, dark green, ± glabrous. Intrd-natd; woods and other shady places; scattered in SE & C En   **E. amygdaloides ssp. robbiae** (Turrill) Stace

  14 Glands on cyathia dark reddish-brown, with short points. Surrey   **E. characias ssp. characias**

  14 Glands on cyathia yellowish, with long points. N Somerset (ssp. *wulfenii* (W.D.J. Koch) Radcl.-Sm.)
E. characias ssp. veneta (Willd.) Litard.

15 Annuals to perennials; rhizomes 0 (stems sometimes buried in sand)   16
15 Rhizomatous perennials   19

  16 Annuals with thin leaves, rarely on maritime sands; bracts and leaves similar   17

  16 Biennials to perennials with ± succulent leaves, on maritime sands; bracts and leaves markedly different   18

17 Leaves linear to narrowly oblong, sessile. Glabrous erect annual to

20(30)cm. Probably native; arable land, rarely elsewhere; common in S & E En, scattered elsewhere in most of BI   *Dwarf Spurge* - **E. exigua** L.

17 Leaves ovate to obovate, petiolate. Glabrous erect annual to 30(40)cm. Native; cultivated and waste ground; common
Petty Spurge - **E. peplus** L.

   18 Midrib prominent on leaf lowerside; seeds pitted. Glabrous, erect biennial to perennial to 40(50)cm. Native; maritime sand-dunes; rather local on coasts of Ir, CI and S & W Br N to Kintyre *Portland Spurge* - **E. portlandica** L.

   18 Midrib obscure on leaf lowerside; seeds smooth. Glabrous erect annual to 60cm. Native; maritime sand-dunes, often with *E. portlandica*; rather local on coasts of Ir, CI and Br N to W Norfolk and Wigtowns   *Sea Spurge* - **E. paralias** L.

19 Leaves ≤2(3)mm wide, often linear, those of lateral shoots crowded and conifer-like. Glabrous, rhizomatous perennial; stems erect, to 50cm. Possibly native in chalk grassland in E Kent and perhaps elsewhere in SE En, natd in rough grassland and waste places scattered throughout BI   *Cypress Spurge* - **E. cyparissias** L.

19 Some or all leaves ≥(2)3mm wide, not linear, those on lateral shoots not conifer-like   20

   20 Leaves oblanceolate to oblong-oblanceolate, widest above middle, attenuate at base   21

   20 Leaves linear- to oblong-lanceolate, widest at or below middle, not attenuate at base   22

21 Leaves ≤4mm wide. Glabrous, rhizomatous perennial; stems erect, to 60cm. Intrd-natd; hedgerows and grassy and waste places; scattered in S Br (*E. esula* x *E. cyparissias*)
Figert's Spurge - **E. x pseudoesula** Schur

21 At least some leaves >4mm wide. Glabrous, rhizomatous perennial; stems erect, to 60cm. Intrd-natd; hedgerows and grassy and waste places; scattered in Br, most frequent in C Sc   *Leafy Spurge* - **E. esula** L.

   22 Leaves rounded to broadly cuneate-rounded at base. Glabrous, rhizomatous perennial; stems erect, to 1m. Intrd-natd; hedgerows and grassy and waste places; in scattered places in En and N to Lanarks   *Waldstein's Spurge* - **E. waldsteinii** (Soják) Czerep.

   22 Leaves abruptly narrowed to cuneate base   23

23 Leaves mostly c.2-3mm wide. Glabrous, rhizomatous perennial; stems erect, to 60cm. Intrd-natd; hedgerows and grassy and waste places; Cumberland and Brecs (*E. waldsteinii* x *E. cyparissias*)
Gáyer's Spurge - **E. x gayeri** Boros & Soó

23 Leaves mostly c.4-5mm wide. Glabrous, rhizomatous perennial; stems erect, to 1m. Intrd-natd; hedgerows and grassy and waste places; frequently natd in Br N to C Sc, Jersey. Easily the commonest of the *E. esula* agg. (couplets 19-23) in Br (*E. waldsteinii* x *E. esula*)
Twiggy Spurge - **E. x pseudovirgata** (Schur) Soó

E. portlandica x *E. paralias* has been found rarely with the parents in Wa and Ir.

94. RHAMNACEAE - *Buckthorn family*

Evergreen or deciduous shrubs or small trees; recognizable by the shrubby habit, simple stipulate leaves, small 4-5-merous perigynous flowers, 2-4-celled ovary with 1 basal ovule per cell, and black berry.

1 Leaves serrate; winter buds with scales **1. RHAMNUS**
1 Leaves entire; winter buds without scales **2. FRANGULA**

1. RHAMNUS L. - *Buckthorns*
Evergreen or deciduous shrubs with alternate or subopposite serrate leaves; flowers 4-5-merous; style divided into 3 or 4 distally.

1 Deciduous, usually spiny shrub to 8m; leaves mostly with 2-4(5) pairs of major lateral veins; sepals and petals mostly 4. Native; hedgerows, scrub and open woods on peat and base-rich soils; frequent in En, scattered in Wa and Ir *Buckthorn* - **R. cathartica** L.
1 Evergreen, non-spiny shrub to 5m; leaves mostly with 3-6 pairs of major lateral veins; sepals 5; petals 0. Intrd-natd; in scrub near sea; Caerns and Denbs *Mediterranean Buckthorn* - **R. alaternus** L.

2. FRANGULA Mill. - *Alder Buckthorn*
Deciduous shrubs with alternate entire leaves; flowers 5-merous; style not divided, with 2-3-lobed stigma.

1 Non-spiny shrub to 5m; leaves mostly with 6-10 pairs of major lateral veins. Native; scrub, bogs and open woods usually on damp peaty soils; locally common in En and Wa, very scattered in Ir and Sc *Alder Buckthorn* - **F. alnus** Mill.

95. VITACEAE - *Grape-vine family*

Deciduous woody climbers with leaf-opposed tendrils; leaves simple and palmately lobed or palmate; differs from Rhamnaceae also in hypogynous flowers with fused sepals in leaf-opposed cymes.

1 Leaves simple; tendrils not ending in discs; petals fused distally, falling as flowers open **1. VITIS**
1 Leaves palmate or simple, if simple then tendrils ending in discs; petals free **2. PARTHENOCISSUS**

1. VITIS L. - *Grape-vine*
Leaves simple, palmately lobed; petals fused distally, forming cap in bud which drops as flowers open.

1 Woody vine potentially >10m; tendrils branched, lacking discs; leaves orbicular, cordate, with 5-7 palmate lobes. Intrd-natd;

hedges and scrub and by tips; scattered in CI, S En and S & WC Wa
Grape-vine - **V. vinifera** L.

2. PARTHENOCISSUS Planch. - *Virginia-creepers*
Leaves simple and palmately lobed or palmate; petals free, remaining for while after flowers open.

1   At least some leaves simple, 3-lobed, often some ternate. Woody vine potentially >20m; tendril-branches ending in discs. Intrd-natd; similar places to *P. quinquefolia* but rarer; scattered in S En, Jersey
Boston-ivy - **P. tricuspidata** (Siebold & Zucc.) Planch.
1   All leaves palmate, most or all with 5 leaflets   2
    2   Tendrils with 5-8(12) branches each ending in adhesive disc. Woody vine potentially >20m. Intrd-natd; on old walls and tips and in hedges and scrub; scattered in Br N to SW Sc, CI
Virginia-creeper - **P. quinquefolia** (L.) Planch.
    2   Tendrils with 3-5 branches not ending in adhesive disc. Woody vine potentially >20m. Intrd-natd; similar places to *P. quinquefolia* but rarer; scattered in S Br
False Virginia-creeper - **P. inserta** (A. Kern.) Fritsch

96. LINACEAE - *Flax family*

Herbaceous annuals or perennials; distinguished by the 4-5 free sepals, petals and stamens, entire leaves without stipules, and 8- or 10-valved capsule with 8 or 10 seeds.

1   Sepals, petals and stamens 5; sepals entire to minutely serrate at apex; capsule with 10 valves   **1. LINUM**
1   Sepals, petals and stamens 4; sepals deeply 2-4-toothed at apex; capsule with 8 valves   **2. RADIOLA**

1. LINUM L. - *Flaxes*
Glabrous annuals to perennials; leaves opposite or alternate; sepals, petals and stamens 5; sepals entire to minutely serrate at apex; petals white or blue, much longer than sepals; capsule with 10 valves.

1   Leaves opposite; petals white, <7mm. Annual; stems erect, to 25cm. Native; dry calcareous or sandy soils, also moorland and mountains; frequent   *Fairy Flax* - **L. catharticum** L.
1   Leaves alternate; petals usually blue, >7mm   2
    2   Sepals c.1/2 as long as ripe capsule, at least the 2 inner rounded and apiculate at apex; stigmas capitate, either higher or lower than anthers. Perennial; stems >1, to 60cm; petals 13-20mm. Native; calcareous grassland; very local from N Essex to Durham and Kirkcudbrights, mainly E En   *Perennial Flax* - **L. perenne** L.
    2   Sepals c. as long as ripe capsule, all abruptly acuminate at apex;

	stigmas elongate-clavate, c. as high as anthers	3
3	Stems usually >1; sepals and capsules 4-6mm. (Annual,) biennial or perennial; stems to 60cm; petals 8-12mm. Native; dry grassy places; local in S BI N to Man *Pale Flax* **- L. bienne** Mill.	
3	Stem usually 1; sepals and capsules 6-9mm. Annual; stem to 85cm; petals 12-20mm. Intrd-natd; from birdseed on tips and as relic in fields; throughout most of BI *Flax* **- L. usitatissimum** L.	

2. RADIOLA Hill - *Allseed*

Annuals; leaves opposite; sepals, petals and stamens 8; sepals deeply 2-4-toothed at apex; petals white, c. as long as sepals; capsule with 8 valves.

1 Stems much branched, extremely slender, ± erect, to 6(10)cm; sepals and petals c.1mm; capsule 0.7-1mm. Native; seasonally damp, bare, peaty or sandy, acid ground; scattered over most BI but mostly near coast *Allseed* **- R. linoides** Roth

97. POLYGALACEAE - *Milkwort family*

Small herbaceous perennials often woody at base; the strange zygomorphic flowers, with 3 petals (the lower one dissected distally) and 8 stamens fused together, are unique; flowers various shades of blue, pink or white, often mixed in 1 population.

1. POLYGALA L. - *Milkworts*

1 Leaves near base of stems smaller than those above, ± acute, not congested into a rosette; inner sepals with veins anastomosing around edges 2
1 Leaves near base of stems larger than those above, ± obtuse, congested into a rosette; inner sepals with veins not anastomosing or sparingly so and not around edges 4
   2 Lower stem-leaves (sometimes lost by fruiting - see scars left) opposite. Stems not or scarcely woody at base, procumbent to scrambling, to 25cm. Native; acid grassland and heathland; frequent *Heath Milkwort* **- P. serpyllifolia** Hosé
   2 All leaves alternate. Stems woody at base, procumbent or scrambling to erect, to 30cm. Native; calcareous or acid grassland, heathland and dunes (*Common Milkwort* **- P. vulgaris** L.) 3
3 Inner sepals 6-8.5 x 3.5-5mm, c. as wide as capsule, with 6-20 inter-veinlet areolae; style c. as long as fruit apical notch. Frequent throughout BI **P. vulgaris ssp. vulgaris**
3 Inner sepals 4-6 x 2-3.5mm, distinctly narrower than capsule, with 8-16(22) inter-veinlet areolae; style longer than fruit apical notch. Scattered throughout much of Br but distribution very uncertain
  P. vulgaris ssp. collina (Rchb.) Borbás
   4 Flowers 6-7mm; stems with ± leafless portion below leaf-rosette.

Stems woody at base, procumbent below leaf-rosette then erect
to ascending, to 20cm. Native; chalk and limestone grassland;
local in S En N to S Lincs *Chalk Milkwort* - **P. calcarea** F.W. Schultz

4 Flowers 2-5mm; stems with leaf-rosette at or very near base.
Plants bitter-tasting; stems ± woody at base, erect to ascending,
to 10(16)cm. Native; chalk and limestone grassland; very local
in E & W Kent and NE En *Dwarf Milkwort* - **P. amarella** Crantz

P. vulgaris has been found to hybridize with *P. calcarea* and *P. amarella* in a few places in S En.

98. STAPHYLEACEAE - *Bladdernut family*

Deciduous shrubs; vegetatively very like *Sambucus nigra*, but without the characteristic smell to the crushed leaves and with flowers in pendent panicles and with diagnostic bladder-like 2-3-lobed capsules.

1. STAPHYLEA L. - *Bladdernut*

1 Shrub to 5m; leaflets (3)5(-7), glabrous; fruit 2.5-4cm, subglobose.
Intrd-natd; hedges and banks; rare in S En *Bladdernut* - **S. pinnata** L.

99. SAPINDACEAE - *Pride-of-India family*

Deciduous trees or shrubs; resembles Staphyleaceae in its pinnate leaves and inflated capsule, but leaves are alternate and flowers are yellow and have 4 petals and 8 stamens (see also *Colutea*, Fabaceae).

1. KOELREUTERIA Laxm. - *Pride-of-India*

1 Tree to 16m; leaves with 9-15 ovate, serrate leaflets; flowers
numerous, 10-15mm across; fruits 3-5cm, ovoid-conical. Intrd-natd;
waste land; few sites in SE En *Pride-of-India* - **K. paniculata** Laxm.

100. HIPPOCASTANACEAE - *Horse-chestnut family*

Deciduous trees; the only trees with opposite, palmate leaves; the fruits, inflorescences and flowers are also unique. Resemblance of the fruits to those of *Castanea* (Fagaceae) is purely superficial; in the latter the prickly husk is a cupule containing fruits (nuts).

1. AESCULUS L. - *Horse-chestnuts*

1 Petals 4, white, to varying degrees tinged and marked with red, pink
and yellow; stamens exceeding petals by c.2cm; leaflets with stalks

c.1cm; fruits obovoid to pear-shaped. Tree to 19m. Intrd-natd; rarely self-sown in parks and by roads; SE En
Indian Horse-chestnut - **A. indica** (Cambess.) Hook.
1 Petals 4-5, always some flowers with 5; stamens exceeding petals by ≤1cm; leaflets sessile or with stalks <1cm; fruits globose   2
  2 Petals pink to red; fruits with 0 or few blunt protuberances. Tree to 28m. Intrd-natd; rarely self-sown in grassy places, copses and rough ground; SE En   *Red Horse-chestnut* - **A. carnea** J. Zeyh.
  2 Petals predominantly white; fruits with many conical pointed protuberances. Wide-spreading tree to 39m. Intrd-natd; self-sown in grassy places, copses and rough ground; throughout lowland BI   *Horse-chestnut* - **A. hippocastanum** L.

101. ACERACEAE - *Maple family*

Deciduous trees; the 2-seeded, 2-winged fruit is diagnostic; the opposite, palmately lobed leaves of all spp. except *A. negundo* occur elsewhere in only *Viburnum opulus* (Caprifoliaceae).

1. ACER L. - *Maples*

1 Leaves ternate or pinnate; trees dioecious. Tree to 17m. Intrd-natd; in parks and by roads and railways; sometimes self-sown where both sexes occur in SE En   *Ashleaf Maple* - **A. negundo** L.
1 Leaves simple, palmately lobed; trees usually bisexual   2
  2 Leaves white on lowerside; flowers in small stiff compact clusters. Tree to 31m. Intrd-natd; in parks and by roads; rarely seeding, self-sown in London area   *Silver Maple* - **A. saccharinum** L.
  2 Leaves green on lowerside; flowers in erect or pendent panicles   3
3 Leaf-lobes serrate; panicles pendent. Tree to 35m. Intrd-natd; fully natd and 1 of the most abundant trees in wide range of habitats throughout BI   *Sycamore* - **A. pseudoplatanus** L.
3 Leaf-lobes entire to irregularly dentate; flowers in stiff ± corymbose panicles   4
  4 Leaf-lobes obtuse; body of fruit convex. Tree to 25m. Native; woods, scrub and hedgerows on calcareous or clay soils; common in En and Wa N to Co Durham, natd elsewhere
Field Maple - **A. campestre** L.
  4 Leaf-lobes acuminate; body of fruit flat   5
5 Leaf-lobes entire. Tree to 26m. Intrd-natd; in parks and by roads; self-sown and extensively suckering in SE En, rarely N to MW Yorks   *Cappadocian Maple* - **A. cappadocicum** Gled.
5 Leaf-lobes with few acuminate teeth or sub-lobes. Tree to 30m. Intrd-natd; self-sown in rough grassland, scrub, hedges and woodland; throughout lowland BI   *Norway Maple* - **A. platanoides** L.

102. ANACARDIACEAE - *Sumach family*

Deciduous shrubs; distinct in the thick, pithy, pubescent, little-branched stems with large pinnate leaves and large reddish inflorescences.

1. RHUS L. - *Stag's-horn Sumach*

1 Shrub to 5(10)m; twigs thick, densely pubescent; leaflets (7)11-15(21), 5-12cm; inflorescence stiffly erect, 10-20cm. Intrd-natd; on verges and banks by roads and railways, extensively suckering but very rarely or never self-sown; S Br scattered N to Cumberland
(*R. hirta* (L.) Sudw.) *Stag's-horn Sumach* - **R. typhina** L.

103. SIMAROUBACEAE - *Tree-of-heaven family*

Deciduous trees; leaves and fruits resemble those of *Fraxinus* (Oleaceae), but leaves are alternate and fruits are usually >1 per flower and with seeds in middle (not at base) of wing.

1. AILANTHUS Desf. - *Tree-of-heaven*

1 Tree to 26m; leaves up to 90cm with up to 41 leaflets; panicles 10-20cm, greenish-white; achenes pendent, 3-4cm. Intrd-natd; parks, roadsides and by rivers; extensively suckering and self-sown; SE En, especially London area *Tree-of-heaven* - **A. altissima** (Mill.) Swingle

104. RUTACEAE - *Rue family*

Deciduous or evergreen shrubs; the 4-5 sepals, free attractive petals and carpels, the lobed disc, the 8 or 10 stamens, and the 4-5-lobed capsule are diagnostic.

1 Leaves ternate, glabrous; flowers erect; petals white   **1. CHOISYA**
1 Leaves simple, tomentose on lowerside; flowers pendent; petals greenish-yellow   **2. CORREA**

1. CHOISYA Kunth - *Mexican Orange*

Evergreen shrub; leaves ternate; flowers several to many in erect corymbose cymes; petals 5, free, white; stamens 10; carpels 5, each 2-lobed and with 2 ovules; stigma 5-lobed.

1 Shrub to 2(3)m; leaflets 3-7.5cm, glabrous, entire; flowers sweetly scented, 2-3cm across. Intrd-natd; in shrubberies and on estates; SE En and Man; Mexico.   *Mexican Orange* - **C. ternata** Kunth

2. CORREA J. Kenn. - *Tasmanian-fuchsia*

Evergreen shrub; leaves simple; flowers 1-3 in pendent groups; petals 4, free at base and apex (4-lobed corolla), but fused into cylindrical tube in middle part, greenish-yellow; stamens 8; carpels 4, each 1-lobed and with 2 ovules; stigma 4-lobed.

1 Shrub to 2(3)m; leaves 1.5-3cm, ± glabrous on upperside, densely tomentose on lowerside, entire; flowers c.2.5cm long. Intrd-natd; in woods; Tresco (Scillies) *Tasmanian-fuchsia* - **C. backhouseana** Hook.

105. OXALIDACEAE - *Wood-sorrel family*

Perennial, rarely annual, often slightly succulent herbs, often with bulbs and/or rhizomes; the ternate or less often palmate leaves and conspicuous actinomorphic flowers are diagnostic.

1. OXALIS L. - *Wood-sorrels*

Most bulbous spp. are trimorphic and self-incompatible, and reproduce mainly or wholly vegetatively. The different clones may show morphological differences, but those present in BI represent only a small part of their whole range and a relatively broad view of sp. limits is taken here. Many spp. are only marginally natd, occurring ± wholly in cultivated ground, but can be very persistent. Corolla colours given are those in the fresh state; after drying the red/pink colours often fade or become more bluish.

1 Petals yellow 2
1 Petals red, pink, mauve or white 8
  2 3 sepals cordate; leaves succulent. Stems erect to ascending, subterranean and aerial, to 20cm, thick and succulent, ± unbranched. Intrd-natd; on walls and banks; Scillies, 1 place in W Cornwall *Fleshy Yellow-sorrel* - **O. megalorrhiza** Jacq.
  2 No sepals cordate; leaves thin, not succulent 3
3 Aerial stem 0; bulbils present at or below soil level. Leaves and flowers arising at soil level among group of bulblets at top of short underground stem arising from deep-seated main bulb. Intrd-natd; weed of arable land; Scillies, rare in S Devon and CI
Bermuda-buttercup - **O. pes-caprae** L.
3 Aerial stems present; bulbils 0 4
  4 Stems procumbent, rooting freely at nodes 5
  4 Stems decumbent to erect, not or very sparsely rooting 6
5 Inflorescences always 1-flowered; capsules 3-4.5mm, with 3-4 seeds per cell; usually 5 stamens with and 5 without anthers. Bulbs 0; stems mostly procumbent, to 20cm, rooting at nodes, much-branched. Intrd-natd; similar places and distribution to *O. corniculata* *Least Yellow-sorrel* - **O. exilis** A. Cunn.
5 At least most inflorescences 2-8(12)-flowered; capsules (4)8-20mm,

1. OXALIS

with >4 seeds per cell; usually all 10 stamens with anthers. Bulbs 0; stems mostly procumbent, to 50cm, rooting at nodes, often subterranean, much-branched. Intrd-natd; pernicious weed of gardens, paths, walls and waste ground; common in most of C & S BI, very scattered in Sc and Ir *Procumbent Yellow-sorrel* - **O. corniculata** L.

6 Capsules <2x as long as wide; petals 10-15mm, with purple veins. Bulbs 0; stems erect to decumbent, to 30cm, not or little branched, rather succulent. Intrd-natd; garden weed; very scattered in C & S En
Chilean Yellow-sorrel - **O. valdiviensis** Barnéoud

6 Capsules >3x as long as wide; petals 5-11mm, not purple-veined 7

7 Pedicels (but not capsules) patent or reflexed in fruit; inflorescence an umbel; vegetative parts with only white simple hairs. Stems erect to decumbent, to 20cm, little-branched. Intrd-natd; weed in sandy arable fields; Herm (CI) *Sussex Yellow-sorrel* - **O. dillenii** Jacq.

7 Pedicels erect in fruit; inflorescence cymose; vegetative parts with translucent septate hairs as well as white simple ones. As *O. dillenii* but stems to 40cm. Intrd-natd; weed of gardens and arable fields; scattered over most of BI *Upright Yellow-sorrel* - **O. stricta** L.

8 Leaves with 5-10 leaflets. Leaves and peduncles arising from scaly bulb which produces many bulblets at the ends of thin rhizomes to 15cm. Intrd-surv; weed in public flower-beds; Man
Ten-leaved Pink-sorrel - **O. decaphylla** Kunth

8 Leaves with 3-4 leaflets 9

9 Stem aerial, ± erect 10

9 Stem 0 or a rhizome at or below soil level 11

10 Bulbs 0; inflorescences >1-flowered. Bulbs 0; stems erect to ascending, to 25cm, branched or not. Intrd-natd; garden weed; Jersey and W Cornwall *Annual Pink-sorrel* - **O. rosea** Jacq.

10 Stem arising from bulb and producing axillary aerial bulbs; inflorescences 1-flowered. Bulb producing annual, erect, branched stem to 20cm with axillary sessile bulblets. Intrd-natd; weed of cultivated ground, walls and banks; frequent in SW En and CI, scattered elsewhere *Pale Pink-sorrel* - **O. incarnata** L.

11 Bulbs 0; stem a rhizome 12

11 Leaves arising from bulb at or below soil level; bulb often producing thin rhizomes 13

12 Flowers solitary; rhizome slender, with distant succulent scales. Stems horizontal thin rhizomes with distant succulent scale leaves. Native; woods, hedgebanks, shady rocks, often on humus; common *Wood-sorrel* - **O. acetosella** L.

12 Flowers in umbels; rhizome thick, with dense papery scales. Stem a thick brown-scaly horizontal to oblique rhizome. Intrd-natd; garden escape in waste and stony and sandy ground on roadsides, banks and seashores; frequent in SW En and CI, scattered in C & S Br and Ir *Pink-sorrel* - **O. articulata** Savigny

13 Leaves with 4 leaflets. As *O. debilis* but bulblets often formed at end of rhizomes to 6cm. Intrd-natd; weed of gardens and arable land;

Jersey and Scillies   *Four-leaved Pink-sorrel* - **O. tetraphylla** Cav.
13 Leaves with 3 leaflets   14
 - 14 Leaflets widest about the middle, with submarginal orange or dark dots on lowerside. Leaves and peduncles arising from scaly bulb which produces many sessile bulblets and 0-2 swollen succulent roots. Intrd-natd; weed of gardens and other open ground; frequent in S En and CI, scattered elsewhere in C & S Br and E Ir   *Large-flowered Pink-sorrel* - **O. debilis** Kunth
 - 14 Leaflets widest at or near apex, without submarginal dots. Differs from *O. debilis* in bulblets often formed at end of rhizomes to 3cm. Intrd-natd; habitat and distribution of *O. debilis*   *Garden Pink-sorrel* - **O. latifolia** Kunth

106. GERANIACEAE - *Crane's-bill family*

Herbaceous annuals to perennials, sometimes woody below; the distinctive 5-seeded fruits elongated distally into a sterile *column* and the usually conspicuous actinomorphic (or nearly so) flowers are diagnostic.

1 Corolla zygomorphic, the 2 upper petals much wider than others; uppermost sepal with spur tightly fused to pedicel (but spur inconspicuous)   **4. PELARGONIUM**
1 Corolla actinomorphic to weakly zygomorphic, the petals scarcely different in width; sepals not spurred   2
 - 2 Stamens 15, all with anthers, fused to c.1/2 way into 5 groups of 3 each   **2. MONSONIA**
 - 2 Stamens 10, or 5 alternating with anther-less staminodes, free   3
3 Beaks of mericarps curved or loosely spirally twisted for 1-2 turns at maturity; leaves palmate or palmately lobed, or rarely ternate or ternately lobed   **1. GERANIUM**
3 Beaks of mericarps becoming tightly spiralled along their own axis at maturity; leaves pinnate or pinnately lobed or rarely ternate or ternately lobed   **3. ERODIUM**

1. GERANIUM L. - *Crane's-bills*

Annuals to perennials; leaves simple and palmately or rarely ternately lobed, or palmate or rarely ternate; flowers actinomorphic or ± so; stamens 10, free, sometimes the outer 5 anther-less; fruits dispersing variously, but only in *G. phaeum* with the mericarp separating whole from the column and its beak twisting spirally, and then the spirals only 1-2 and in large loops.

'Leaves' refers to lower leaves with long petioles. Most taxa occur as rare white-petalled mutants; this aspect of variation is not mentioned for each sp.

1 Petals narrowed at base to distinct claw 1/2 as long as limb to longer than limb   2
1 Petals without claw or with claw <1/2 as long as limb   8

1. GERANIUM

- 2 Most petals >14mm — **3**
- 2 Most petals <14mm — **4**
- 3 Perennial with thick rhizome. Shortly rhizomatous perennial with ± erect stems to 50cm. Intrd-natd; on walls and banks and in grassy places; very scattered in Br   *Rock Crane's-bill* - **G. macrorrhizum** L.
- 3 Biennial with thin fibrous roots. Erect, reddish, faintly scented biennial to 60cm due to long petioles and large leaves, but with stem usually much shorter. Intrd-natd; on rough ground; Guernsey and Man   *Greater Herb-Robert* - **G. rubescens** Yeo
  - 4 Leaves divided ≤3/4 to base — **5**
  - 4 Leaves divided ± to base (ternate to palmate) — **7**
- 5 Leaves glossy green to purple, very sparsely pubescent; sepals strongly keeled on back. Shining, erect to ascending annual to 40cm. Native; bare ground, rocks, walls and stony banks, mostly on calcareous ground; locally common in most of BI
Shining Crane's-bill - **G. lucidum** L.
- 5 Leaves grey-green, pubescent; sepals rounded on back — **6**
  - 6 Mericarps appressed-pubescent, smooth; outer 5 stamens lacking anthers. Decumbent to ascending annual to 40cm. Native; cultivated and waste land and barish places among grass; scattered throughout most of BI
Small-flowered Crane's-bill - **G. pusillum** L.
  - 6 Mericarps (excluding beaks) glabrous, usually ridged; all stamens with anthers. Decumbent to ascending annual to 40cm. Native; similar places to *G. pusillum*; common
Dove's-foot Crane's-bill - **G. molle** L.
- 7 Anthers orange or purple (pale yellow in albinos); petals 8-14mm; mericarps with sparse fine ridges and 0-1(2) deep collar-like ridges at apex. Procumbent to erect, strong-smelling, annual to biennial to 50cm. Native; woods, hedgerows, banks, scree and maritime shingle; common   *Herb-Robert* - **G. robertianum** L.
- 7 Anthers yellow; petals 5-9mm; mericarps with dense wrinkle-like ridges and 2-3(4) deep collar-like ridges at apex. Procumbent to erect, strong-smelling, annual to biennial to 50cm. Native; rocky and stony places, hedgebanks and on shingle and cliffs, usually near sea; very local in SW Br E to W Sussex, SW Ir, CI
Little-Robin - **G. purpureum** Vill.
  - 8 Annuals to biennials, or sometimes perennials ± without rhizome; petals mostly ≤10mm, rarely to 22mm — **9**
  - 8 Perennials with distinct thick and/or elongated rhizome; petals >10mm, rarely less — **17**
- 9 Petals ≥15mm — **10**
- 9 Petals ≤10mm — **11**
  - 10 Plant usually <60cm; flowers ≤3cm across; petal-claw 6-7mm (see couplet 3) **G. rubescens**
  - 10 Plant usually >60cm; flowers >3cm across; petal-claw 2-2.5mm. Erect perennial to 1(2)m, like a giant *G. rubescens* with stem up to

5cm wide. Intrd-natd; on cliffs in dense low vegetation; Scillies

Giant Herb-Robert - **G. maderense** Yeo
- 11 Seeds smooth　12
- 11 Seeds pitted or reticulately ridged　14
  - 12 Mericarps (excl. beaks) glabrous, usually ridged.

   (see couplet 6) **G. molle**
  - 12 Mericarps pubescent, smooth　13
- 13 Petals <5mm; sepals <3mm; outer 5 stamens lacking anthers

   (see couplet 6) **G. pusillum**
- 13 Petals >6mm; sepals >3mm; all 10 stamens with anthers. Perennial without rhizome, erect to ascending, to 60cm. Possibly native; hedgerows, grassy places and rough ground; locally frequent in BI but absent from much of N & W

   Hedgerow Crane's-bill - **G. pyrenaicum** Burm. f.
  - 14 Petals ± rounded at apex; leaves divided <3/4 way to base, sepals with apiculus <0.5mm. Erect to decumbent annual to 40cm. Native; banks, walls and stony ground; local in C & S BI

     Round-leaved Crane's-bill - **G. rotundifolium** L.
  - 14 Petals distinctly notched at apex; leaves divided ≥3/4 way to base; sepals with apiculus >0.5mm　15
- 15 Mericarps glabrous to sparsely pubescent; most pedicels >2.5cm. Erect to ascending or scrambling annual to 60cm. Native; grassy places, banks and scrub, mostly on calcareous soils; locally frequent in much of BI, rare in N　*Long-stalked Crane's-bill* - **G. columbinum** L.
- 15 Mericarps pubescent; pedicels ≤2.5cm　16
  - 16 Leaves divided almost to base; stalked glands frequent on upper parts of plant. Erect to procumbent annual to 60cm. Native; grassy and stony ground, waste places and cultivated ground; common　*Cut-leaved Crane's-bill* - **G. dissectum** L.
  - 16 Leaves divided c.3/4-7/8 way to base; stalked glands 0. Perennial without rhizome, decumbent to ascending, to 60cm. Intrd-natd; hedgerows and grassy and waste places; Guernsey and Alderney, rare casual in Jersey

     Alderney Crane's-bill - **G. submolle** Steud.
- 17 Flowers all solitary on pedicel + peduncle. Shortly rhizomatous, erect to procumbent perennial to 40cm. Native; grassland, rocky places, sand-dunes, open woods on calcareous soils; local in N & W Br and C Ir, natd elsewhere　*Bloody Crane's-bill* - **G. sanguineum** L.
- 17 At least most flowers in pairs, with 2 pedicels on a common peduncle 18
  - 18 Mericarps pointed at base, with 2-4 collar-like ridges at apex; petals not notched, usually apiculate or with small triangular point at apex, sometimes ruffled or subentire　19
  - 18 Mericarps rounded at base, with 0-1 collar-like ridge at apex; petals rounded to notched at apex, sometimes with small point in notch　20
- 19 Petals c. as long as wide, patent to slightly reflexed, usually purplish-black. Shortly rhizomatous to compact, ± erect perennial to 80cm. Intrd-natd; shady places in hedges and wood-borders;

1. GERANIUM

scattered throughout most of Br, very local in Ir

Dusky Crane's-bill - **G. phaeum** L.

19 Petals c.1.5x as long as wide, strongly reflexed, never purplish-black. Habit of *G. phaeum*. Intrd-natd; roadside verges; E Sussex and Midlothian (*G. phaeum* x *G. reflexum* L.)

Munich Crane's-bill - **G. x monacense** Harz

- 20 Stalked glands 0 or <0.3mm long   21
- 20 Sepals and pedicels, and usually peduncles and upperparts of stems, with stalked glands >0.3mm long   26

21 Base of mericarps without tuft of bristles on inside; petals violet-blue. Shortly rhizomatous to compact ± erect perennial to 50cm. Intrd-natd; grassy places; Cards and E Lothian

Caucasian Crane's-bill - **G. ibericum** Cav.

21 Base of mericarps with tuft of apically pointed bristles on inside directed on to seed or into cavity; petals whitish to bright- or purplish-pink   22

- 22 Hairs on pedicels, peduncles and upper parts of stems <0.2mm, appressed; main leaf-lobes toothed, with teeth up to c.5mm long. Shortly rhizomatous ± erect perennial to 50cm. Intrd-natd; hedgerows and woodland; very scattered in C & S Br

Knotted Crane's-bill - **G. nodosum** L.

- 22 Pedicels, peduncles and upper parts of stem with patent hairs >0.5mm; main leaf-lobes with sub-lobes or deep teeth >(5)10mm long   23

23 Petals with veins darker than ground-colour; beaks of mericarps with hairs c.0.1mm   24

23 Petals with veins paler than or same colour as ground-colour; beaks of mericarps with some hairs >(0.2)0.5mm   25

- 24 Petals curved outwards at apex (flowers trumpet-shaped), with white to very pale pink ground-colour. Shortly rhizomatous ± erect perennial to 60cm. Intrd-natd; grassy places and waste ground; scattered in C & S Br, Ir and CI

Pencilled Crane's-bill - **G. versicolor** L.

- 24 Petals not curved outwards at apex (flowers funnel-shaped), ground-colour pink. Extensively rhizomatous ± erect perennial to 70cm. Intrd-natd; grassy places and waste ground; very scattered in BI (*G. endressii* x *G. versicolor*)

Druce's Crane's-bill - **G. x oxonianum** Yeo

25 Fruit with style (between tip of column and base of stigmas) 2.5-3(4)mm; petals deep bright pink. Extensively rhizomatous ± erect perennial to 70cm. Intrd-natd; grassy places and waste ground; scattered over most of Br, rare in Ir

French Crane's-bill - **G. endressii** J. Gay

25 Fruit with style (3)4-6mm; petals usually mid-pink, often variable on 1 plant   (see couplet 24) **G. x oxonianum**

- 26 Stalked glands sparse, confined to floral parts and pedicels   Return to 25
- 26 Stalked glands abundant on pedicels, peduncles and upper parts

            of stems   27
27 Petals conspicuously notched; base of mericarps (if ripening)
    without tuft of bristles on inside   28
27 Petals rounded at apex; base of mericarps with tuft of apically
    pointed bristles on inside directed on to seed or into cavity   29
    28 Basal leaves divided c.1/2 way to base, their main lobes broadest
        near apex; flowers pointing horizontally; seeds ripening. Shortly
        rhizomatous to compact ± erect perennial to 40cm. Intrd-natd;
        grassy places; Dunbarton
                    Glandular Crane's-bill - **G. platypetalum** Fisch. & C.A. Mey.
    28 Basal leaves divided c.2/3-7/8 way to base, their main lobes
        broadest well below apex; flowers pointing upwards; seeds
        not ripening. Shortly rhizomatous to compact ± erect perennial
        to 75cm. Intrd-natd; grassy places, by roads and on waste land;
        scattered over much of Br, Guernsey (*G. ibericum* x
        G. platypetalum)   *Purple Crane's-bill* - **G. x magnificum** Hyl.
29 Petals blue to violet-blue, with white base; flowers and immature
    fruits pointing sideways or ± downwards; fruits with styles >4mm   30
29 Petals pinkish-purple to magenta, with black or white base; flowers
    and immature fruits pointing obliquely or vertically upwards;
    fruits with styles <4mm   31
    30 Sepals with point >1/5 as long as main part; leaves divided
        >5/6 way to base. Compact ± erect perennial to 1m. Native;
        meadows, roadsides, open woodland, often in damp places;
        common in much of Br, absent from N Sc and parts of Wa and
        S En, very local in N Ir, natd elsewhere
                                Meadow Crane's-bill - **G. pratense** L.
    30 Sepals with point <1/5 as long as main part; leaves divided
        <5/6 way to base. Extensively rhizomatous ± erect perennial to
        60cm. Intrd-natd; grassy places; extremely scattered in En and Sc,
        perhaps overlooked for *G. pratense* or *G.* x *magnificum*
                            Himalayan Crane's-bill - **G. himalayense** Klotzsch
31 Petals magenta with black base, >16mm, with few hairs on either
    side at base. Compact ± erect perennial to 120cm. Intrd-natd; grassy
    places; extremely scattered in En and Sc
                                Armenian Crane's-bill - **G. psilostemon** Ledeb.
31 Petals pinkish-purple with white base, <16mm, with abundant
    hairs at base on upperside. Compact ± erect perennial to 70cm.
    Native; woods and hedges in lowlands, rock-ledges and meadows
    in uplands; locally common in N Br S to Yorks, very local in C En,
    Wa and N Ir, natd elsewhere   *Wood Crane's-bill* - **G. sylvaticum** L.

G. robertianum x *G. purpureum* has been found in E & M Cork.

2. MONSONIA L. - *Dysentery-herbs*

Annuals; leaves simple, pinnately toothed; flowers actinomorphic, stamens
15, all with anthers, with filaments fused for c.1/2 their length into 5 groups
of 3; seeds dispersed inside mericarps with beaks attached, the beaks tightly

2. MONSONIA

spiralled (twisted) on their own axes.

1 Sparsely pubescent, branched, ascending annual to 40cm; flowers 1-3 on common peduncle; petals 5-8mm, bluish; fruits 22-30mm. Intrd-casual; wool-alien in fields and waste places; scattered in En
Short-fruited Dysentery-herb - **M. brevirostrata** R. Knuth

3. ERODIUM L'Hér. - *Stork's-bills*

Annuals to perennials, leaves simple and pinnately to ternately lobed or pinnate to ternate; flowers actinomorphic or slightly zygomorphic; stamens 5, free, alternating with staminodes; fruits dispersing as in *Monsonia*, with 2 pits at apex of mericarp, 1 either side of base of beak.

1 Leaves pinnate   2
1 Leaves simple, shallowly to deeply lobed, or if ± compound then ternate   5
   2 Primary leaflets divided ≤3/4 way to midrib; apical pits of mericarp with sessile glands. Suberect to procumbent annual to 60cm. Native; rough ground and barish places in short grassland, mainly near sea; coasts of CI and S & W Br, probably intrd in Ir, elsewhere as casual   *Musk Stork's-bill* - **E. moschatum** (L.) L'Hér.
   2 Primary leaflets divided nearly to base; apical pits of mericarp glandless   3
3 Petals 15-20mm; bracts green, fused into cupule; fruits with column 4-7cm. Perennial with peduncle and leaves arising from base. Intrd-natd; hedgerows and open ground; W Kent
Garden Stork's-bill - **E. manescavii** Coss.
3 Petals 4-12mm; bracts brown, several, free; fruits with column 1-4cm   4
   4 Apical pits of mericarp separated from main part of mericarp by sharp ridge and groove, not overarched by hairs; flowers 3-7 per peduncle, mostly >10mm across. Suberect to procumbent annual to 60cm. Native; barish places in grassland, coastal dunes, waste and rough ground, on sandy or chalky soils, also common wool-alien; scattered over most of BI, especially coasts
Common Stork's-bill - **E. cicutarium** (L.) L'Hér.
   4 Apical pits of mericarp not delimited by sharp ridge and groove, overarched by hairs from main part of mericarp; flowers 2-4(5) per peduncle, mostly <10mm across. Suberect to procumbent annual to 15(25)cm. Native; barish places on fixed dunes; coasts of Br from E Kent to Wigtowns, Ir, CI
Sticky Stork's-bill - **E. lebelii** Jord.
5 Beak of fruit 0.8-1cm; petals not exceeding sepals or 0. Procumbent to decumbent annual to 10(20)cm. Native; fixed maritime dunes and barish places in short grassland, rarely inland; coasts of CI, E & S Ir, W Br N to Wigtowns   *Sea Stork's-bill* - **E. maritimum** (L.) L'Hér.
5 Beak of fruit >1.5cm; petals exceeding sepals   6
   6 Petals pinkish-purple; fruits with column ≤4cm; apical pits of mericarps delimited by very distinct groove and with

       conspicuous sessile glands, or with no groove and no glands   7
- 6 Petals blue to violet-purple; fruits with column >4cm; apical pits of mericarps delimited by distinct groove, without glands   8
- 7 Hairs on sepals and pedicels eglandular, appressed; apical pits of mericarp without glands, not delimited by groove. Suberect to ascending annual to 40cm. Intrd-casual; infrequent wool-alien; scattered in En and Wa   **Three-lobed Stork's-bill - E. chium** (L.) Willd.
- 7 Most hairs on sepals and pedicels glandular, patent; apical pits of mericarp with sessile glands, delimited by very distinct groove. Suberect to ascending annual to 40cm. Intrd-casual; infrequent wool-alien; scattered in En and Wa
       Soft Stork's-bill - E. malacoides (L.) L'Hér.
- 8 Lower leaves pinnately lobed, with ≥2 basal pairs of lobes not very different in size   9
- 8 Lower leaves ternately lobed, with 1 pair of basal lobes greatly exceeding all others   10
- 9 Apical pits of mericarp with sparse bristles, bounded below by 1 blunt-rimmed groove. Suberect to ascending annual to 50cm. Intrd-casual; frequent wool-alien; scattered in En
       Hairy-pitted Stork's-bill - E. brachycarpum (Godr.) Thell.
- 9 Apical pits of mericarps completely glabrous, bounded below by (1)2-3 sharp-rimmed grooves. Suberect to ascending annual to 50cm. Intrd-casual; common wool-alien; scattered in Br
       Mediterranean Stork's-bill - E. botrys (Cav.) Bertol.
- 10 Sepals and pedicels with many patent glandular hairs. Habit as *E. crinitum*. Intrd-casual; frequent wool-alien; very scattered in En
       Western Stork's-bill - E. cygnorum Nees **ssp. glandulosum** Carolin
- 10 Sepals and pedicels without glandular hairs   11
- 11 Pedicels glabrous or with hairs near apex only; sepals with only ± appressed hairs <0.3mm. Habit and distribution as for ssp. *glandulosum*   **Western Stork's-bill - E. cygnorum** Nees **ssp. cygnorum**
- 11 Pedicels sparsely pubescent along length; sepals with short appressed and some longer ± patent hairs >0.5mm. Decumbent to ascending annual with thick ± succulent root (perennial in native area) to 50cm. Intrd-casual; frequent wool-alien; scattered in En
       Eastern Stork's-bill - E. crinitum Carolin

E. cicutarium x *E. lebelii* = **E. x anaristatum** Andreas is known from coastal dunes in Wa and S Lancs; it is intermediate and sterile.

4. PELARGONIUM Aiton - *Geraniums*

Perennials, woody or ± so at base; leaves simple, tomentose, deeply palmately lobed; flowers zygomorphic, with the upper 2 petals wider than the others and the upper sepal with a backward-directed spur tightly fused to pedicel; stamens 10, free, but only 7 with anthers; fruits dispersing as in *Monsonia*.

- 1 Scrambling to 1m; leaves peppermint-scented; flowers up to 20 in

4. PELARGONIUM

umbel; petals whitish-pink, red at base. Intrd-natd; open woodland; Tresco (Scillies) *Peppermint-scented Geranium* - **P. tomentosum** Jacq.

107. LIMNANTHACEAE - *Meadow-foam family*

Annual, slightly succulent, ± glabrous herbs; the buttercup-like flowers, but with quite different stamens and carpels, are distinctive; differs from Geraniaceae in lacking stipules, ovary with 1 ovule per cell, and cells of fruit remaining closed.

1. LIMNANTHES R.Br. - *Meadow-foam*

1 Stems erect to ascending, to 35cm; flowers 16-25mm across, petals white with conspicuous yellow basal part; beak of fruit 5-9mm. Intrd-casual; on tips and by roads, seashores and lakes, rarely briefly persisting; scattered in Br *Meadow-foam* - **L. douglasii** R. Br.

108. TROPAEOLACEAE - *Nasturtium family*

Annual or perennial, sometimes climbing, somewhat succulent herbs with peltate leaves; the zygomorphic flowers with 5 free petals and sepals, 8 stamens and 3-celled ovary are unmistakable.

1. TROPAEOLUM L. - *Nasturtiums*

1 Rhizomatous perennial to several m; leaves deeply 5-6-lobed, almost palmate; flowers scarlet, c.18-25mm across. Intrd-natd; clambering through bushes and in hedges; very scattered in Br and Ir *Flame Nasturtium* - **T. speciosum** Poepp. & Endl.
1 Annual remaining dwarf or ± climbing to 2m; leaves slightly angled, otherwise entire; flowers various shades of yellow or red, usually orange, c.25-60mm across. Intrd-casual; on tips and waste ground; scattered in Br and CI *Nasturtium* - **T. majus** L.

109. BALSAMINACEAE - *Balsam family*

Glabrous, somewhat succulent, annual herbs; the zygomorphic flowers, with spurred lower sepal and 2 others, fused pairs of lateral petals, and distally fused stamens, are unique.

1. IMPATIENS L. - *Balsams*

1 Flowers deep pinkish-purple to white; leaves opposite or in whorls. Stems erect, to 2m. Intrd-natd; banks of rivers and canals, damp

places and waste ground; locally common throughout most of BI

Indian Balsam - **I. glandulifera** Royle

1 Flowers pale yellow to orange   2
- 2 Flowers (incl. spur) <1.5(2)cm; larger leaves with >20 teeth on each side. Stems erect, to 60(100)cm. Intrd-natd; damp shady places in woods and hedges and disturbed or cultivated ground; scattered through most of Br   *Small Balsam* - **I. parviflora** DC.
- 2 Flowers (incl. spur) all or many >2cm; larger leaves with <20 teeth on each side   3
3 Flowers yellow; sepal-spur held at c.90° to rest of sepal in fresh state. Stems erect, to 1m. Native; damp places in woods; very local in Br N to C Sc, frequent in Lake District

Touch-me-not Balsam - **I. noli-tangere** L.

3 Flowers orange; sepal-spur held ± parallel to rest of sepal in fresh state. Stems erect, to 1.5m. Intrd-natd; banks of rivers and canals; locally frequent in Br N to SW Yorks

Orange Balsam - **I. capensis** Meerb.

110. ARALIACEAE - *Ivy family*

Evergreen, woody climbers, herbaceous perennials or evergreen or deciduous shrubs; the umbellate inflorescence and 2-5-celled berries are diagnostic; the growth-forms of each of the genera are unmistakable.

1 Deciduous shrub or herbaceous; leaves 1-2-pinnate   **3. ARALIA**
1 Woody evergreen; leaves simple, usually palmately lobed   2
- 2 Upright shrub; bisexual flowers with 5 free styles; most or all leaves >25cm across   **2. FATSIA**
- 2 Climbing or scrambling; flowers with 1 style; leaves ≤25cm across   **1. HEDERA**

1. HEDERA L. - *Ivies*

Evergreen, non-spiny, woody climbers with numerous short roots borne along climbing stems; leaves ≤25cm, simple, usually palmately lobed; inflorescence a simple umbel or group of umbels; flowers autumn; fruits spring; style 1; fruits black unless otherwise stated.

In all spp. leaves on creeping or climbing stems differ from those on non-rooting, flowering stems; the former are more lobed, and are those referred to in key and descriptions as they and their hairs are the diagnostic ones.

1 Hairs on leaves and young stems semi-peltate, orange-brown, with rays mostly 8-25 and fused for 1/4-1/2 their length; leaves often >10cm wide, scarcely or not lobed or with 3 rather shallow lobes   2
1 Hairs on leaves and young stems stellate, white to yellowish-brown, with rays mostly 4-8(10) and fused only at extreme base; at least some leaves well lobed, rarely >10cm wide. Stems variously creeping and climbing, often to great height. Native; on trees,

1. HEDERA

banks, rocks and sprawling over the ground (**H. helix** L.)   3
2   Leaves with matt surface, with ± cordate base, usually not or scarcely lobed, mostly >15cm across; hairs with mostly 15-25 rays. Stems creeping, scrambling and weakly climbing to c.10m. Intrd-natd; often long-persistent and spreading in shrubberies and on walls; scattered over Br, Man
Persian Ivy - **H. colchica** (K. Koch) K. Koch
2   Leaves with glossy surface, with truncate to rounded base, usually some 3-lobed, mostly <15cm across; hairs with mostly 8-18 rays. Stems variously creeping and climbing to c.5m. Intrd-natd; sometimes persisting and spreading on patios and walls; scattered in S & W Br N to CW Sc, Man (*H. canariensis* auct.)
Algerian Ivy - **H. algeriensis** Hibberd
3   Hairs whitish, with rays lying parallel to leaf surface and also projecting away from it; leaves usually <8cm across, often lobed >1/2 way to base. Common over E, C & N Br, uncertain further W
Common Ivy - **H. helix** ssp. **helix**
3   Hairs often pale yellowish-brown, with rays ± all lying parallel to leaf surface; leaves often >8cm across, usually lobed <1/2 way to base. The commoner taxon in CI, Ir, and W & SW Br N to SW Sc, but with much overlap
Atlantic Ivy - **H. helix** ssp. **hibernica** (G. Kirchn.) D.C. McClint.

2. FATSIA Decne. & Planch. - *Fatsia*

Evergreen, non-spiny shrubs, without roots along stems; leaves simple, mostly >25cm, deeply palmately lobed; inflorescence a large (20-40cm) panicle of umbels, the lateral ones male; flowers autumn; fruits spring; styles 5; fruits black.

1   Shrub to 3(-5)m; leaves 15-40cm, palmately divided >1/2 way into 7-9 lobes. Intrd-natd; surviving as relics in SE En, relics and self-sown plants in Man   *Fatsia* - **F. japonica** (Thunb.) Decne. & Planch.

3. ARALIA L. - *Angelica-trees*

Herbaceous perennials or deciduous, spiny shrubs, without roots along stems; leaves 1-2-pinnate, mostly >50cm; inflorescence a large (20-45cm) panicle of umbels; flowers summer; fruits autumn; styles 5; fruits dark purplish to black.

1   Spineless herbaceous perennial to 2m. Intrd-natd; shrubberies and open woodland; Salop   *American-spikenard* - **A. racemosa** L.
1   Spiny deciduous shrub or small tree often >2m   2
2   Inflorescence a ± conical panicle; leaves glabrous on lowerside except on veins. Suckering shrub to 3(6)m, with thick spiny branches. Intrd-natd; planted on banks and in shrubberies and spreading by suckers; very scattered in Br
Chinese Angelica-tree - **A. chinensis** L.
2   Inflorescence a ± umbrella-shaped panicle; leaves pubescent on

lowerside. Shrub or tree to 5(10)m, with spiny branches. Intrd-natd; occurrence as for *A. chinensis*, also Jersey

Japanese Angelica-tree - **A. elata** (Miq.) Seem.

111. APIACEAE - *Carrot family*

Herbaceous annuals to perennials, rarely shrubs; most spp. are unmistakable 'umbellifers', with compound (less often simple) umbels and distinctive fruits; genera not so conforming (*Eryngium, Hydrocotyle*) are individually highly distinctive.

Flowers in terminal and lateral umbels, the umbels sometimes simple but usually compound, in *Hydrocotyle* often with whorls of flowers below main umbel, the 5 petals actinomorphic or sometimes zygomorphic (those away from centre of umbel larger); sepals 5, usually represented by small teeth near top of ovary, or 0; styles 2, often arising from swelling (*stylopodium*) on top of ovary; fruit a dry 2-celled schizocarp, the 2 mericarps usually separating from the central sterile *carpophore* (or carpophore lacking) but each remaining indehiscent at maturity.

Fully ripe fruits are vital for identification. Important points are the degree and pattern of longitudinal ridging and the shape and position of sub-surface oil-bodies on the mericarps. The face of the mericarps where they join is the *commissure*, and the outer face is dorsal. Fruits may be compressed to varying degrees: either dorsally (with commissure as wide as fruit) or laterally (with commissure through short axis). Bracteoles (when present) subtend the flowers at the base of the pedicels; bracts (when present) subtend the main branches (*rays*) of the compound umbel. Fruit length excludes the stylopodium; width is that in widest view.

In this work 11 distinctive groups are first defined, followed by the keys which cover all taxa except for male plants of *Trinia*. The list of distinctive taxa and then the keys should be tackled successively.

Distinctive genera
Fresh plants smelling of aniseed when crushed
8. MYRRHIS, 21. FOENICULUM, 22. ANETHUM
Plant with subterranean swollen tubers (not swollen tap-roots)
11. BUNIUM, 12. CONOPODIUM, 19. OENANTHE
Base of stem sheathed by mass of fibres (remains of old petioles)
3. ASTRANTIA, 13. PIMPINELLA, 18. SESELI, 23. SILAUM, 24. MEUM, 28. TRINIA, 38. CARUM, 44. PEUCEDANUM
Plant entirely male or female **28. TRINIA**
Stem procumbent; leaves ± orbicular, shallowly lobed, stipulate
1. HYDROCOTYLE
Plant with subaquatic leaves much more finely divided than aerial ones
15. SIUM, 19. OENANTHE, 30. APIUM
Leaves all simple, not divided or divided <1/2 way to base
1. HYDROCOTYLE, 4. ERYNGIUM, 10. SMYRNIUM, 27. BUPLEURUM
Leaves divided >1/2 way to base but at most 1-ternate to 1-palmate or

111. APIACEAE

ternately to palmately lobed, the leaflets/lobes toothed to shallowly lobed
 2. SANICULA, 3. ASTRANTIA, 4. ERYNGIUM, 14. AEGOPODIUM, 37. FALCARIA, 40. LIGUSTICUM, 44. PEUCEDANUM
Inflorescence with lobed or divided bracts
 2. SANICULA, 4. ERYNGIUM, 16. BERULA, 28. TRINIA, 29. CUMINUM, 31. TRACHYSPERMUM, 32. PETROSELINUM, 36. AMMI, 38. CARUM, 44. PEUCEDANUM, 49. DAUCUS
Fruits with hooked or barbed bristles **2. SANICULA, 4. ERYNGIUM, 6. ANTHRISCUS, 48. TORILIS, 49. DAUCUS**
Fruits tuberculate (not smooth, scaly, spiny or pubescent)
 27. BUPLEURUM, 31. TRACHYSPERMUM, 48. TORILIS

General key
1 Leaves all simple and entire   **27. BUPLEURUM**
1 Leaves simple to compound; if simple at least toothed   2
 2 Stem-leaves spiny   **4. ERYNGIUM**
 2 Leaves not spiny   3
3 Leaves stipulate, all simple, ± orbicular, shallowly lobed, with long petiole   **1. HYDROCOTYLE**
3 Leaves without stipules, if all simple then either lobed >1/2 way to base or sessile   4
 4 Leaves all simple, with small teeth only   **10. SMYRNIUM**
 4 Leaves simple to compound, if all simple then divided >1/2 way to base   5
5 Basal leaves all ternately or palmately lobed almost to base   6
5 Basal leaves ternate, palmate or pinnate, often compoundly so   7
 6 Flowers subsessile, with inconspicuous bracteoles; fruits with hooked bristles   **2. SANICULA**
 6 Flowers distinctly pedicellate, with bracteoles at least as long; fruits covered with bifid scales   **3. ASTRANTIA**
7 Fruit with apical beak >2x as long as seed-bearing part   **7. SCANDIX**
7 Fruit beakless or with beak shorter than seed-bearing part   8
 8 Stem and often basal leaves with white, flexuous subterranean part arising from brown tuber   9
 8 Stem and basal leaves (if present) arising from roots at ground level, or from rhizome, or plant aquatic   10
9 Stem hollow at fruiting; fruit with ± erect styles gradually narrowed from stylopodium   **12. CONOPODIUM**
9 Stem solid at fruiting; fruit with reflexed styles suddenly contracted from stylopodium   **11. BUNIUM**
 10 Petals yellow   *Key A*
 10 Petals white to pink or purplish, or greenish-white   11
11 Fruits with spines, bristles, hairs or conspicuous tubercles   *Key B*
11 Fruits glabrous, ± smooth   12
 12 Fruits strongly compressed dorsally, distinctly wider in dorsal view than in lateral view   *Key C*
 12 Fruits not compressed dorsally or scarcely so   13
13 Base of stem sheathed by mass of fibres (remains of old petioles)   14

13 Basal mass of petiole-fibres 0   17
    14 Basal leaves 1-pinnate, each lobe palmately divided ± to base into filiform segments appearing as if whorled   **38. CARUM**
    14 Basal leaves 1-4-pinnate, if 1-pinnate then lobes not divided ± to base into filiform segments   15
15 Fruit 4-10mm; basal leaves 3-4-pinnate with filiform ultimate lobes   **24. MEUM**
15 Fruit 2-4mm; basal leaves 1-3-pinnate with ovate to linear ultimate lobes   16
    16 Plant usually dioecious; all leaf-segments linear; umbels with <10 rays   **28. TRINIA**
    16 Plant bisexual; leaf-segments variable, some often ovate, very rarely all linear; umbels usually with ≥10 rays   **13. PIMPINELLA**
17 Fruits >(2)2.5x as long as widest width   *Key D*
17 Fruits ≤2x as long as widest width   18
    18 Sepals 0, or minute teeth, or vestigial rim, not or scarcely visible at top of fruit   *Key F*
    18 Sepals ≥0.2mm, distinctly visible at top of fruit   *Key E*

Key A - Petals yellow
1 Fruits strongly compressed dorsally, i.e. distinctly wider in dorsal view than in lateral view   2
1 Fruits not compressed dorsally, or scarcely so   5
    2 Bracts >3; bracteoles fused at base   **42. LEVISTICUM**
    2 Bracts 0-3; bracteoles 0 or free   3
3 Leaves 1-pinnate, with ovate lobes   **45. PASTINACA**
3 Leaves 2-several times pinnate or ternate, with linear lobes   4
    4 Stem surrounded by fibrous remains of petioles at base; stem-leaves with narrow petiole; fruits 5-8mm   **44. PEUCEDANUM**
    4 Stem not surrounded by fibrous remains of petioles; stem-leaves with wide sheathing petiole; fruits 10-16mm   **43. FERULA**
5 Ultimate leaf-lobes filiform, <0.5mm wide   6
5 Ultimate leaf-lobes flat, linear to ovate, >0.5mm wide   8
    6 Firmly rooted perennials   **21. FOENICULUM**
    6 Easily uprooted annuals   7
7 Fruits dorsally compressed, with lateral wings; fresh plant smelling of aniseed   **22. ANETHUM**
7 Fruits laterally compressed, not winged; fresh plant not smelling of aniseed   **33. RIDOLFIA**
    8 Fruit laterally compressed, c.2x as wide in lateral view as in dorsal view   9
    8 Fruit not or scarcely compressed, c. as wide in lateral view as in dorsal view   10
9 Bracts and bracteoles 0-2, very short; mericarps with 3 acute dorsal ridges   **19. SMYRNIUM**
9 Bracts 1-3, often lobed; bracteoles >3; mericarps with 3 rounded dorsal ridges   **32. PETROSELINUM**
    10 Leaves succulent; bracts >4; mericarps not winged   **17. CRITHMUM**

111. APIACEAE

    10 Leaves not succulent; bracts 0-3; mericarps with narrow lateral wings **23. SILAUM**

Key B - Petals not yellow; fruits with spines, bristles, hairs or conspicuous tubercles
1. Fruits strongly compressed dorsally, i.e. distinctly wider in dorsal view than in lateral view (incl. projections)   2
1. Fruits not compressed dorsally, or scarcely so   3
    2 Stems 1.5-5.5m; umbels with >30 rays; fruits >8mm **46. HERACLEUM**
    2 Stems <1.5m; umbels with <20 rays; fruits <7mm **47. TORDYLIUM**
3. Fruits with conspicuous tubercles **31. TRACHYSPERMUM**
3. Fruit with spines, bristles or hairs   4
    4 Fruits with hairs or weak or minute bristles   5
    4 Fruits with usually stout, terminally hooked or barbed, spines   7
5. Fruits >3x as long as wide; fresh plant smelling of aniseed **8. MYRRHIS**
5. Fruits <3x as long as wide; fresh plant not smelling of aniseed   6
    6 Slender annual without basal fibres; bracts and bracteoles each <5; fruits >3.5mm **29. CUMINUM**
    6 Perennial with base of stem sheathed by mass of fibres; bracts and bracteoles each >5; fruits <3.5mm **18. SESELI**
7. Bracts deeply pinnately or ternately divided **49. DAUCUS**
7. Bracts 0 or simple   8
    8 Fruits without beak, with spines up to base of stylopodium, with persistent sepals **48. TORILIS**
    8 Fruits with spine-less but ridged beak below stylopodium; sepals 0 **6. ANTHRISCUS**

Key C - Petals not yellow; fruits without spines, bristles, hairs or conspicuous warts, strongly dorsally compressed (see also *Levisticum* and *Pastinaca*; petals yellow, Key A)
1. Leaves and/or stems pubescent to hispid **46. HERACLEUM**
1. Leaves and main parts of stems glabrous or nearly so, sometimes coarsely papillose   2
    2 Easily uprooted annuals   3
    2 Firmly rooted perennials   4
3. Fruits ovoid, with strong ridges; outer petals <1.5mm; sepals 0 **20. AETHUSA**
3. Fruits ± globose, scarcely ridged; outer petals >2mm; sepals persistent **9. CORIANDRUM**
    4 Leaf-lobes <2 x 1cm   5
    4 Most leaf-lobes >2 x 1cm   6
5. Stems solid; bracts 0 or few and soon falling; bracteoles not or weakly reflexed **39. SELINUM**
5. Stems hollow; bracts >3, reflexed; bracteoles reflexed **43. PEUCEDANUM**
    6 Larger leaves 2-3-pinnate (smaller ones sometimes 2-3-ternate or pinnate-ternate) **41. ANGELICA**

111. APIACEAE

     6   All leaves 1-2-ternate   7
7   Umbels with <20 rays; bracts 1-5; fruits with 3 prominent acute dorsal ridges   **40. LIGUSTICUM**
7   Umbels with >20 rays; bracts 0(-2); fruits with low, obtuse dorsal ridges   **44. PEUCEDANUM**

Key D - Petals not yellow; fruits without spines, bristles, hairs or conspicuous warts, not or scarcely dorsally compressed, >(2)2.5× as long as wide

1   Sepals 0 or minute or a vestigial rim, not or scarcely visible at top of fruit   2
1   Sepals ≥0.2mm, distinctly visible at top of fruit   4
    2   Fruit not ridged in mid or basal regions; stems hollow   **6. ANTHRISCUS**
    2   Fruit ridged along length; stems solid or hollow   3
3   Stems solid; fruits with low, rounded ridges; fresh plant not aniseed-scented   **5. CHAEROPHYLLUM**
3   Stems hollow; fruits with sharp, prominent ridges; fresh plant aniseed-scented   **8. MYRRHIS**
    4   Umbels with 1-5 rays; at least some bracts >1/2 as long as rays   **29. CUMINUM**
    4   Umbels with ≥6 rays, if fewer then bracts 0 or much <1/2 as long as rays   5
5   Fruits laterally compressed; lobes of lower leaves linear-lanceolate, >5cm, with regularly and sharply serrate margins; not aquatic   **37. FALCARIA**
5   Fruits not compressed or slightly dorsally so; leaf-lobes various, but if linear or lanceolate and >5cm then entire to distantly and irregularly toothed; often aquatic   **19. OENANTHE**

Key E - Petals not yellow; fruits without spines, bristles, hairs or conspicuous warts, not or scarcely dorsally compressed, ≤2× as long as wide; sepals ≥0.2mm, distinctly visible at top of fruit

1   Fruits subglobose, in lateral view c. as wide as long   2
1   Fruits in lateral view distinctly longer than wide   4
    2   Easily uprooted annuals; mericarps remaining fused even at maturity   **9. CORIANDRUM**
    2   Firmly rooted perennials; mericarps splitting apart at maturity   3
3   Stems solid; fruits >2.5mm; petioles not widened at base, nor sheathing lateral stems   **25. PHYSOSPERMUM**
3   Stems hollow; fruits <2.5mm; petioles widened at base and sheathing lateral stems   **35. CICUTA**
    4   Not aquatic; lobes of lower leaves linear-lanceolate, >5cm, with regularly and sharply serrate margins   **37. FALCARIA**
    4   Often aquatic; leaf-lobes various, but if linear or lanceolate and >5cm then entire to distantly or irregularly toothed   5
5   Fruits laterally compressed; lobes of lower leaves mostly >4 × 2cm   **15. SIUM**

111. APIACEAE

5 Fruits not compressed or slightly dorsally so; lobes of lower leaves <4 x 2cm **19. OENANTHE**

Key F - Petals not yellow; fruits without spines, bristles, hairs or conspicuous warts, not or scarcely dorsally compressed, ≤2x as long as wide, with sepals 0 or scarcely visible

1 Lower leaves simply pinnate, the lobes not divided as far as midrib 2
1 Lower leaves 2-4-pinnate or 1-2-ternate 8
   2 Bracts 0(-2), if constantly present then stem mostly procumbent with ± only leaves and peduncles erect 3
   2 Bracts 2-c.8; at least apical part of stem erect to ascending 5
3 Plant often aquatic, at least lower part of stem procumbent and rooting **30. APIUM**
3 Plant ± never aquatic; stem erect, not rooting 4
   4 Fruit <2mm; plant glabrous **30. APIUM**
   4 Fruit ≥2mm, or if <2mm then at least rays hispid **13. PIMPINELLA**
5 Styles at fruiting at least as long as stylopodium 6
5 Styles at fruiting much shorter than stylopodium 7
   6 Lowest leaves with 2-5 pairs of leaflets each 3-6cm; all bracts <1/2 as long as all rays **34. SISON**
   6 Lowest leaves with 4-12 pairs of leaflets each 0.5-3.5cm; longest bracts >1/2 as long as shortest rays **32. PETROSELINUM**
7 Stems solid; bracts divided to base into linear to filiform lobes, the longest >1/2 as long as rays; upper leaves >1-pinnate **36. AMMI**
7 Stems hollow; bracts lobed but usually not to base, <1/2 as long as rays; all leaves 1-pinnate **16. BERULA**
   8 Bracts ≥4 9
   8 Bracts 0-2(3) 10
9 Bracts <1/4 as long as rays, undivided; stems hollow; fruits with prominent ± wavy-edged ridges **26. CONIUM**
9 Bracts >1/2 as long as rays, deeply divided; stems solid; fruits with low smooth ridges **36. AMMI**
   10 Plant in water or on mud; stems procumbent and rooting at least near base; styles at fruiting much shorter than stylopodium **30. APIUM**
   10 Plant not aquatic; stems erect but sometimes rhizomes produced; styles at fruiting at least as long as stylopodium 11
11 Plant rhizomatous; leaves 1-2-ternate **14. AEGOPODIUM**
11 Plant not rhizomatous; leaves 2-3-pinnate 12
   12 Easily uprooted annuals; bracteoles long, strongly reflexed **20. AETHUSA**
   12 Firmly rooted biennials or perennials; bracteoles 0 or scarcely reflexed 13
13 Usually some umbels with 1-2(3) bracts; all ultimate leaf-lobes linear to linear-lanceolate; styles ± appressed to stylopodium **38. CARUM**
13 Bracts 0; usually at least some leaf-lobes ovate, if all linear to linear-lanceolate then styles not appressed to stylopodium **13. PIMPINELLA**

111. APIACEAE

SUBFAMILY 1 - HYDROCOTYLOIDEAE (genus 1). Leaves simple, ± orbicular, not or shallowly lobed, stipulate; flowers in simple axillary umbels, often with whorls below; fruit with a woody inner wall, but no carpophore and no oil-bodies.

1. HYDROCOTYLE L. - *Pennyworts*
Perennials with thin procumbent rooting stems; leaves on thin, usually erect petioles as long as to much longer than axillary peduncles; flowers very small, dull; fruits strongly laterally compressed, ± orbicular in lateral view.

1 Plant in water or bogs or on mud, glabrous at least on leaf uppersides and on stems; fruits orbicular or slightly wider than long 2
1 Plant in turf or on banks, with pubescent leaves and stems; fruits longer than wide 3
  2 Leaves peltate, scarcely lobed. Stems to 30cm, but usually much less; petioles 1-25cm. Native; in bogs, fens and marshes and at sides of lakes; locally common *Marsh Pennywort* - **H. vulgaris** L.
  2 Leaves not peltate, with deep basal sinus, lobed ±1/2 way to base. Similar to *H. vulgaris* in size; stems often floating. Intrd-natd; in rivers, canals and ponds; several places in SE En
    Floating Pennywort - **H. ranunculoides** L. f.
3 Fruits ≥2 x 1.8mm; leaf-lobes crenate. Similar to small-sized *H. vulgaris*. Intrd-natd; on lawns and golf-courses; W Cornwall and Angus *New Zealand Pennywort* - **H. novae-zeelandiae** DC.
3 Fruits ≤1.6 x 1.5mm; leaf-lobes serrate. Similar to small-sized *H. vulgaris*. Intrd-natd; in lawns and on grassy banks; Valencia Island (S Kerry), E Cornwall, E Sussex and Ayrs
    Hairy Pennywort - **H. moschata** G. Forst.

SUBFAMILY 2 - SANICULOIDEAE (genera 2-4). Leaves mostly simple, palmately or ternately lobed, rarely pinnately lobed and then spiny, without stipules; flowers in simple umbels or capitula; fruit with soft inner wall and with oil-bodies, but without carpophore.

2. SANICULA L. - *Sanicle*
± glabrous perennials; most leaves in basal rosette; leaves palmately lobed almost to base; petioles long; umbels with pedicellate male and sessile bisexual flowers; bracteoles several, shorter than flowers; fruits laterally compressed, covered with long, hooked bristles; sepals persistent.

1 Stems erect, to 40(60)cm. Native; in deciduous woods on leaf-mould; locally common throughout Br and Ir *Sanicle* - **S. europaea** L.

3. ASTRANTIA L. - *Astrantia*
± glabrous perennials; most leaves in basal rosette; leaves ternately or palmately lobed almost to base; petioles long; umbels with pedicellate male and bisexual flowers; bracteoles numerous, lanceolate to oblanceolate, at least as long as flowers; fruits scarcely compressed, covered with swollen

scales; sepals persistent.

1 Stems erect, to 80(100)cm. Intrd-natd; grassy and shady places; scattered in Br, mostly in N & W   *Astrantia* - **A. major** L.   340

4. ERYNGIUM L. - *Sea-hollies*

± glabrous perennials; basal leaves petiolate, simple to ± pinnate; stem-leaves various, at least the upper spiny; flowers sessile, in ± globose to ovoid capitula with ± leaf-like spiny bracts at base; bracteoles 3-lobed or entire, with 1-3 spines, longer than flowers; fruits scarcely compressed, densely scaly or bristly; sepals persistent.

1 Basal leaves and lower stem-leaves pinnately or ternately divided or lobed almost to midrib   2
1 Basal leaves and lower stem-leaves not lobed or lobed ≤1/2 way to midrib   3
   2 Basal leaves with petiole unwinged, c. as long as its leaf; plant not blue. Stems erect, to 75cm. Probably native in Kent, doubtfully elsewhere; grassland or open places, especially calcareous, mostly near sea; very local in extreme SE, S & SW Br, rarely intrd elsewhere   *Field Eryngo* - **E. campestre** L.   340
   2 Basal leaves with petiole broadly winged, much shorter than its leaf; plant strongly blue-tinged. Stems erect, to 45cm. Intrd-natd; on dunes in Caerns, rare relic elsewhere
    Italian Eryngo - **E. amethystinum** L.   340
3 Upper stem-leaves and bracts palmate or palmately lobed almost to base. Stems erect, to 50(100)cm. Intrd-natd; waste places; scattered in S Br, Lanarks   *Blue Eryngo* - **E. planum** L.   340
3 Upper stem-leaves toothed, or lobed <1/2 way to base   4
   4 Basal leaves and at least lower stem-leaves shallowly toothed, not spiny. Stems erect, to 1.2m. Intrd-natd; on waste ground; near Otley (MW Yorks)   *Tall Eryngo* - **E. giganteum** M. Bieb.   340
   4 Basal leaves and stem-leaves with deep, strongly spiny teeth. Stems erect, to 60cm. Native; on maritime sand and shingle; coasts of BI, now rare in most of Sc and NE En
    Sea-holly - **E. maritimum** L.   340

SUBFAMILY 3 - APIOIDEAE. (genera 5-48). Leaves various, often much divided, not spiny, without stipules; flowers usually in compound umbels; fruit with soft inner wall, carpophore and usually oil-bodies.

5. CHAEROPHYLLUM L. - *Chervils*

Biennials or perennials; stems solid; leaves 2-3-pinnate; bracts 0 or present, entire; bracteoles present; sepals ± 0; petals white or pink, actinomorphic; fruits slightly compressed laterally, >3x as long as wide, glabrous, with low, wide, rounded ridges.

1 Petals minutely pubescent at margin; styles suberect, forming angle

111. APIACEAE

    <45°; petals usually pinkish. Erect, ± softly pubescent perennial to 1m. Intrd-natd; by roads and rivers; Westmorland and Lanarks
                                              Hairy Chervil - **C. hirsutum** L.

1   Petals glabrous; styles ± divergent, forming angle >45°; petals white   2
    2   Fruits 8-10mm; leaf-segments acute. Erect, softly to rather roughly pubescent perennial to 1.2m. Intrd-natd; in grassy places; very scattered in En (mainly N) and Sc, rare casual elsewhere in Br   *Golden Chervil* - **C. aureum** L.
    2   Fruits 4-6.5mm; leaf-segments obtuse or abruptly contracted to acute apex. Erect, roughly pubescent biennial to 1m. Native; grassy places, hedgerows and wood-borders; common in En and Wa, sparse in CI, Sc and Ir   *Rough Chervil* - **C. temulum** L.

6. ANTHRISCUS Pers. - *Chervils*

Annuals to perennials; stems hollow; leaves 2-3-pinnate; bracts 0(-1); bracteoles present; sepals ± 0; petals white, actinomorphic to zygomorphic; fruits slightly compressed laterally, >3x as long as wide, glabrous, ridged only near apex.

1   Fruits <5mm, with abundant hooked bristles. Erect to decumbent, sparsely pubescent annual to 70cm. Native; waste places, open ground and open hedgerows on sandy or shingly soils, especially near sea; common in parts of E En, scattered elsewhere, rare in N
                                              Bur Chervil - **A. caucalis** M. Bieb.
1   Fruits >5mm, without bristles   2
    2   Annual; rays pubescent; fruits with well differentiated beak 1-4mm. Erect to spreading, sparsely pubescent annual to 70cm. Intrd-natd; on rock-face in Herefs, decreasing casual in waste and marginal places in S & C Br and Guernsey
                                      Garden Chervil - **A. cerefolium** (L.) Hoffm.
    2   Perennial; rays ± glabrous; fruits with scarcely differentiated beak ≤1mm. Erect, pubescent perennial to 1.5m. Native; grassy places, hedgerows and wood-margins; abundant
                                              Cow Parsley - **A. sylvestris** (L.) Hoffm.

7. SCANDIX L. - *Shepherd's-needle*

Annuals; stems hollow at fruiting; leaves 2-4-pinnate; bracts 0 or umbels simple; bracteoles present; sepals very small (<0.5mm) but persistent; petals white, zygomorphic; fruit ± not compressed, many times longer than wide, with beak 3-5x as long as seed-bearing part, glabrous, with wide low rounded ridges.

1   Stems usually erect, to 50cm; leaf-segments linear; fruits 3-7cm. Possibly native; weed of arable land and waste places; rare and ± only in C & S En   *Shepherd's-needle* - **S. pecten-veneris** L.

8. MYRRHIS Mill. - *Sweet Cicely*

Perennials smelling of aniseed when crushed; stems hollow; leaves 2-4-

8. MYRRHIS

pinnate; bracts 0; bracteoles present; sepals ± 0; petals white, slightly zygomorphic; fruit slightly laterally compressed, >3x as long as wide, glabrous or with minute bristles, with acute well developed ridges.

1 Stems erect, to 1.8m; fruits 15-25mm. Intrd-natd; banks, pathsides, waste and grassy places; common in Br N from Derbys and Monts, rare in S Wa, S En and N & C Ir   *Sweet Cicely* - **M. odorata** (L.) Scop.   340

9. CORIANDRUM L. - *Coriander*
Annuals; stems solid; leaves simple to 3-pinnate; bracts 0-1(2), entire; bracteoles present; sepals conspicuous, persistent; petals white to purplish, zygomorphic; fruits ± globose, glabrous, the mericarps not separating at maturity, with low ± rounded ridges.

1 Stems erect, to 50cm, ± unridged. Intrd-casual; tips and waste places; scattered in Br   *Coriander* - **C. sativum** L.   340

10. SMYRNIUM L. - *Alexanders*
Glabrous biennials (to perennials?); stems solid; leaves simple to 3-pinnate or -ternate; bracts and bracteoles 0-few, entire; sepals minute; petals yellow, actinomorphic; fruits laterally compressed, not or scarcely longer than wide, glabrous, with prominent sharp ridges.

1 Stem-leaves (1)2-3-pinnate or -ternate, petiolate; fruit 6.5-8mm, c. as long as wide or slightly longer, blackish when ripe. Stems erect, to 1.5m. Intrd-natd; on cliffs and banks, by roads and ditches and in waste places, mostly near sea; common on coasts of BI N to C Sc, very scattered inland   *Alexanders* - **S. olusatrum** L.   340
1 Stem-leaves simple, deeply cordate, sessile; fruits 2-3.5mm, wider than long, dark brown when ripe. Stems erect, to 60(100)cm. Intrd-natd; in grassy places and flower-borders; few places in SE En, Co Durham   *Perfoliate Alexanders* - **S. perfoliatum** L.   340

11. BUNIUM L. - *Great Pignut*
Glabrous perennials; stems solid, arising from subterranean ± globose tuber; leaves 2-3-pinnate, with linear lobes; bracts and bracteoles several, entire; sepals ± 0; petals white, actinomorphic; fruits slightly laterally compressed, <2x as long as wide, glabrous, with low, rounded ridges.

1 Stems erect, to 50(80)cm; leaves mostly withered by flowering time. Native; chalk grassland and banks; very local in Herts, Bucks, Beds and Cambs   *Great Pignut* - **B. bulbocastanum** L.   340

12. CONOPODIUM W.D.J. Koch - *Pignut*
Glabrous perennials; stems hollow after flowering, arising from subterranean ± globose tuber; leaves 2-3-pinnate, with linear lobes; bracts 0(-2); bracteoles several; sepals 0; petals white, actinomorphic; fruits slightly laterally compressed, <2x as long as wide, glabrous, with low, rounded

ridges.

1 Stems erect, to 40(75)cm; basal leaves mostly withered by fruiting
 time. Native; grassland, hedgerows, woods; common
 Pignut - **C. majus** (Gouan) Loret 3

13. PIMPINELLA L. - *Burnet-saxifrages*
Perennials; stems hollow or solid; leaves 1-2(3)-pinnate; bracts and bracteoles 0; sepals 0; petals white or pinkish-white, actinomorphic; fruits slightly laterally compressed, slightly longer than wide, glabrous, with narrow low ridges.

1 Stems erect, to 1(2)m, hollow, ± glabrous; basal leaves simply
 pinnate; fruits 3-4mm. Native; grassland, hedgerows, wood-
 borders; locally common in C and S En and S & WC Ir, Guernsey,
 rare casual in Wa and Sc *Greater Burnet-saxifrage* - **P. major** (L.) Huds. 3
1 Stems erect, to 70(100)cm, solid, glabrous to densely minutely
 pubescent; basal leaves 1-2-pinnate; fruits 2-3mm. Native; grassland
 and open rocky places; common *Burnet-saxifrage* - **P. saxifraga** L. 3

14. AEGOPODIUM L. - *Ground-elder*
Rhizomatous glabrous perennials; stems hollow; leaves 1-2-ternate with large, ovate, serrate lobes; bracts and bracteoles 0; sepals 0; petals white, ± actinomorphic; fruits laterally compressed, <2x as long as wide, glabrous, with narrow ridges.

1 Stems erect, to 1m; rhizomes slender, far-creeping; fruits 3-4mm.
 Probably intrd-natd; waste places and cultivated and other open
 ground; common *Ground-elder* - **A. podagraria** L. 3

15. SIUM L. - *Greater Water-parsnip*
Glabrous perennials; stems hollow; aerial leaves pinnate with ovate, serrate leaflets; bracts and bracteoles several, entire; sepals conspicuous, persistent; petals white, actinomorphic; fruits laterally compressed, distinctly longer than wide, glabrous, with thick, prominent ridges.

1 Stems erect, to 2m; submerged leaves 2-3-pinnate in spring; fruits
 2.5-4mm. Native; in ditches and fens; very local in En and C Ir,
 Jersey *Greater Water-parsnip* - **S. latifolium** L. 3

16. BERULA W.D.J. Koch - *Lesser Water-parsnip*
Glabrous stoloniferous perennials; stems hollow; lower leaves pinnate with ovate, serrate leaflets, if submerged then scarcely different; bracts and bracteoles several, often lobed; sepals distinct but not persistent; petals white, actinomorphic; fruits slightly laterally compressed, scarcely longer than wide, glabrous, with low, slender ridges.

1 Stems decumbent to erect, to 1m; fruits 1.3-2mm. Native; in and by

16. BERULA

water in ditches, marshes, lakes and rivers; frequent over much of
Ir and Br N to C Sc   *Lesser Water-parsnip* - **B. erecta** (Huds.) Coville   341

17. CRITHMUM L. - *Rock Samphire*
Glabrous perennials; stems solid, ± woody near base; leaves 2-3-pinnate, with linear, succulent lobes; bracts and bracteoles several, entire; sepals minute; petals yellowish-green, actinomorphic; fruits not compressed, each mericarp ± triangular in section, slightly longer than wide, spongy when fresh, glabrous, with thick, prominent ridges.

1   Stems erect to decumbent, to 45cm. Native; cliffs, rocks and less
    often sand and shingle by sea; coasts of BI but N only to E Suffolk
    on E coast   *Rock Samphire* - **C. maritimum** L.   341

18. SESELI L. - *Moon Carrot*
Minutely pubescent biennial to monocarpic perennial; stems solid; leaves 2-3-pinnate, with rather narrow lobes; bracts and bracteoles numerous, entire; sepals small, sometimes ± persistent; petals white, actinomorphic; fruits not compressed, slightly longer than wide, minutely pubescent, with thick, prominent ridges.

1   Stems erect, to 60cm, with dense sheath of fibres at base. Native;
    grassland or rough ground on chalk; E Sussex, Herts, Cambs and
    Beds   *Moon Carrot* - **S. libanotis** (L.) W.D.J. Koch   341

19. OENANTHE L. - *Water-dropworts*
Glabrous annuals to perennials, often with tuberous roots; stems hollow or solid; leaves 1-4-pinnate; bracts 0-several, entire; bracteoles usually numerous; sepals conspicuous, persistent; petals white, ± actinomorphic to slightly zygomorphic; fruits not or very slightly dorsally or laterally compressed, up to c.2.5x as long as wide, glabrous, with obscure to rather prominent ridges.

1   Some umbels leaf-opposed, with peduncles shorter than rays; styles
    ≤1/4 as long as mature fruit   2
1   All umbels terminal, with peduncles longer than rays; styles >1/4 as
    long as mature fruit   3
    2   Fruit ≤4.5mm; stems ascending to erect, often terrestrial. Annual
        to biennial to 1.5m; root-tubers 0 at maturity; stems hollow.
        Native; ditches and ponds, often drying up in summer; scattered
        in Br and Ir N to SE Sc
                        Fine-leaved Water-dropwort - **O. aquatica** (L.) Poir.   341
    2   Fruit ≥5mm; stems usually floating at least at base. Erect to
        ascending or floating perennial to 1m; root-tubers 0 at maturity;
        stems hollow. Native; in slow rivers; scattered in S & E En N
        to MW Yorks, C & E Ir, Denbs
                        River Water-dropwort - **O. fluviatilis** (Bab.) Coleman   341
3   Ultimate clusters of ripe fruits globose; all fruits sessile; all leaves

usually with petioles longer than divided part. Erect, stoloniferous perennial to 80cm; root-tubers spindle-shaped; stems hollow. Native; marshes, ditches and other wet places; locally frequent in BI N to C Sc **Tubular Water-dropwort - O. fistulosa** L.

3 Ultimate clusters of ripe fruits not globose; some fruits stalked; all leaves usually with petioles shorter than divided part 4

  4 Fruits ≥4mm; segments of mid stem-leaves ovate to ± orbicular, <2x as long as wide. Erect perennial to 1.5m; root-tubers spindle-shaped; stems hollow. Native; ditches, pondsides and other wet places; locally common in BI
Hemlock Water-dropwort - O. crocata L.

  4 Fruits ≤3.5mm; segments of mid stem-leaves ± linear, >3x as long as wide 5

5 Bracts 0 at least on most umbels; rays ≥1mm thick at fruiting; stems at maturity hollow, straw-like, with walls c.0.5mm thick. Erect perennial to 1m; root-tubers spindle-shaped. Native; marshes, dykes and ditches; scattered in C & S En N to SE Yorks
Narrow-leaved Water-dropwort - **O. silaifolia** M. Bieb.

5 Bracts (0)1-c.5; rays <1mm thick at fruiting; stems at maturity solid to hollow with walls >0.5mm thick 6

  6 Rays and pedicels thickened at fruiting, the pedicels >0.5mm thick; root-tubers ellipsoid, the proximal part of the root not thickened. Erect perennial to 1m. Native; in dry or damp grassland, often by ditches or ponds; very local in C & S Br, Co Clare *Corky-fruited Water-dropwort* - **O. pimpinelloides** L.

  6 Rays and pedicels scarcely thickened at fruiting, the pedicels <0.5mm thick; root-tubers cylindrical to spindle-shaped, gradually widening ± from base of root. Erect perennial to 1m. Native; ditches, marshes and dykes, mostly near sea and often brackish; coasts of BI except N & E Sc, scattered inland in En
Parsley Water-dropwort - **O. lachenalii** C.C. Gmel.

20. AETHUSA L. - *Fool's Parsley*

Glabrous annuals; stems hollow; leaves 2-3-pinnate; bracts usually 0; bracteoles usually 3-4; sepals 0; petals white, ± actinomorphic; fruits slightly dorsally compressed, slightly longer than wide, glabrous, with prominent, wide, keeled ridges.

1 Stems erect, to 1(1.5)m; bracteoles strongly reflexed. Cultivated, arable and waste ground; throughout BI except much of C & N Sc, intrd in Ir (*Fool's Parsley* - **A. cynapium** L.) 2

  2 Stems to 1m; longest pedicels mostly <1/2 as long as bracteoles, mostly c.2x as long as fruits. Native; common
A. cynapium ssp. cynapium

  2 Stems to 20cm; longest pedicels c. as long as bracteoles, mostly shorter than fruits. Intrd; arable land; probably only in S Br and CI **A. cynapium ssp. agrestis** (Wallr.) Dostál

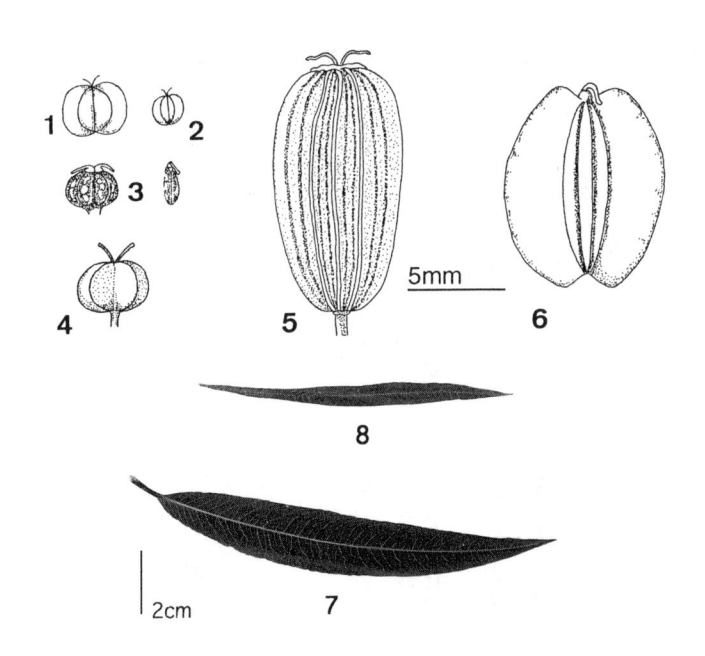

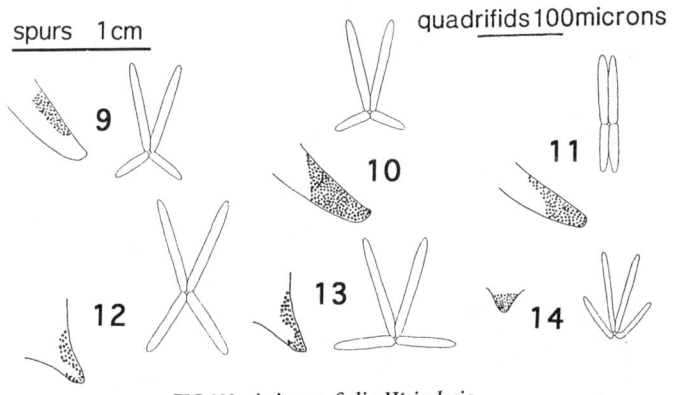

FIG 339 - **Apiaceae**, *Salix*, *Utricularia*.
1-6. Fruits of **Apiaceae**. 1, *Hydrocotyle vulgaris*.
2, *H. novae-zeelandiae*. 3, *H. moschata*. 4, *H. ranunculoides*.
5, *Ferula communis*. 6, *Angelica pachycarpa*.
1, lateral & dorsal views. 2-4, lateral view. 5-6, dorsal view.
7-8. Leaves of *Salix*. 7, *Salix udensis*. 8, *S. elaeagnos*.
9-14. Flower spur showing gland distribution and bladder quadrifids
of *Utricularia*. 9, *U. vulgaris*. 10, *U. australis*. 11, *U. intermedia*.
12, *U. stygia*. 13, *U. ochroleuca*. 14, *U. minor*.
1-6, drawings by Hilli Thompson. 9-14, courtesy of G. Thor.

FIG 540 - Fruits (mostly with lateral and dorsal views) of **Apiaceae**.
1, *Eryngium amethystinum*. 2, *E. giganteum*. 4, *Sanicula europaea*.
5, *Coriandrum sativum*. 6, *Scandix pecten-veneris*. 7, *Myrrhis odorata*.
8, *Conopodium majus*. 9, *Bunium bulbocastanum*. 10, *Astrantia major*.
11, *Chaerophyllum hirsutum*. 12, *C. aureum*. 13, *C. temulum*.
14, *Anthriscus cerefolium*. 15, *A. sylvestris*. 16, *A. caucalis*.
17, *Eryngium planum*. 18, *E. campestre*. 19, *E. maritimum*.
20, *Smyrnium perfoliatum*. 21, *S. olusatrum*. Drawings by Hilli Thompson.

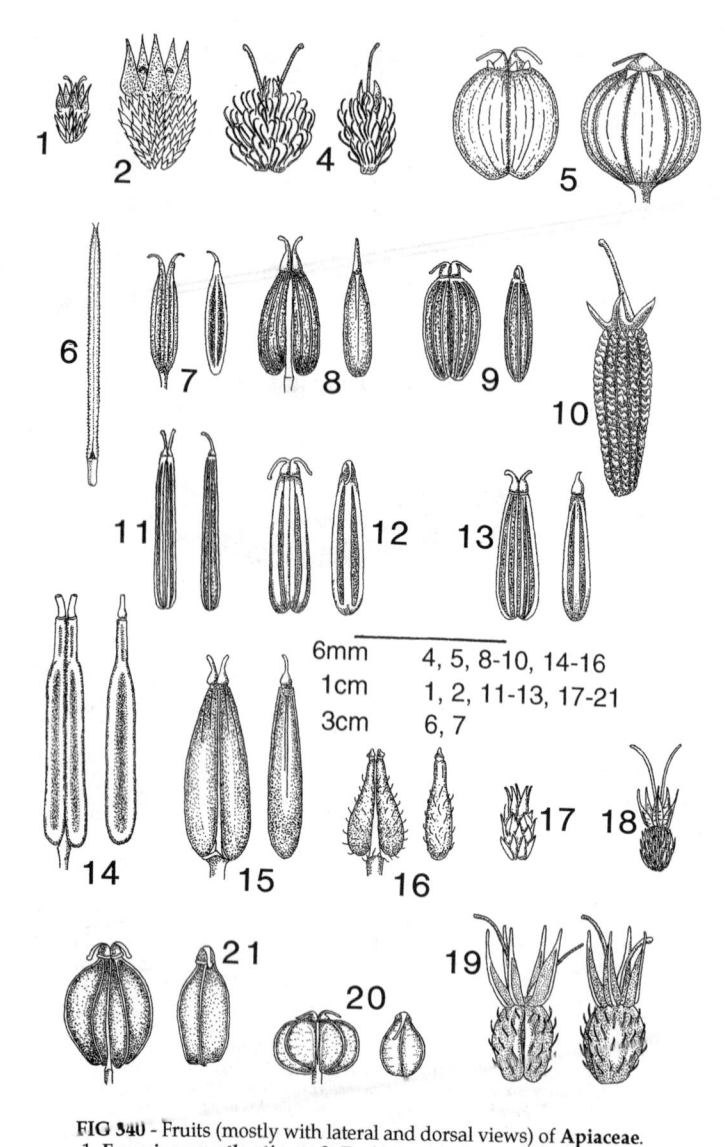

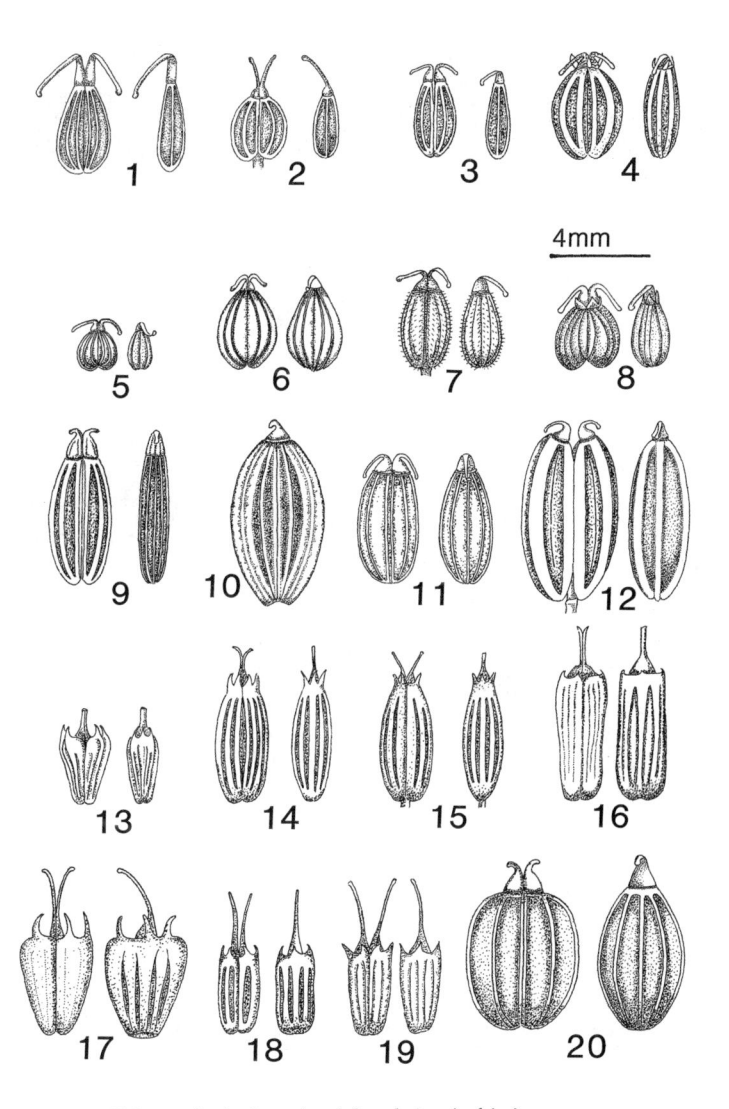

FIG 341 - Fruits (lateral and dorsal views) of **Apiaceae**.
1, *Aegopodium podagraria*. 2, *Pimpinella saxifraga*. 3, *P. major*.
4, *Sium latifolium*. 5, *Berula erecta*. 6, *Aethusa cynapium*.
7, *Seseli libanotis*. 8, *Physospermum cornubiense*. 9, *Foeniculum vulgare*.
10, *Anethum graveolens*. 11, *Silaum silaus*. 12, *Meum athamanticum*.
13, *Oenanthe lachenalii*. 14, *O. fluviatilis*. 15, *O. aquatica*. 16, *O. crocata*.
17, *O. fistulosa*. 18, *O. pimpinelloides*. 19, *O. silaifolia*.
20, *Crithmum maritimum*. Drawings by Hilli Thompson.

FIG 342 - Fruits (mostly with lateral and dorsal views) of **Apiaceae**.
1, *Bupleurum rotundifolium*. 2, *B. subovatum*. 3, *B. falcatum*.
4, *B. fruticosum*. 5, *B. baldense*. 6, *B. tenuissimum*. 7, *Ammi visnaga*.
8, *A. majus*. 9, *Trinia glauca*. 10, *Cuminum cyminum*.
11, *Conium maculatum*. 12, *Trachyspermum ammi*. 13, *Ridolfia segetum*.
14, *Apium nodiflorum*. 15, *A. inundatum*. 16, *A. repens*. 17, *A. graveolens*.
18, *Sison amomum*. 19, *Petroselinum crispum*. 20, *P. segetum*.
21, *Cicuta virosa*. Drawings by Hilli Thompson.

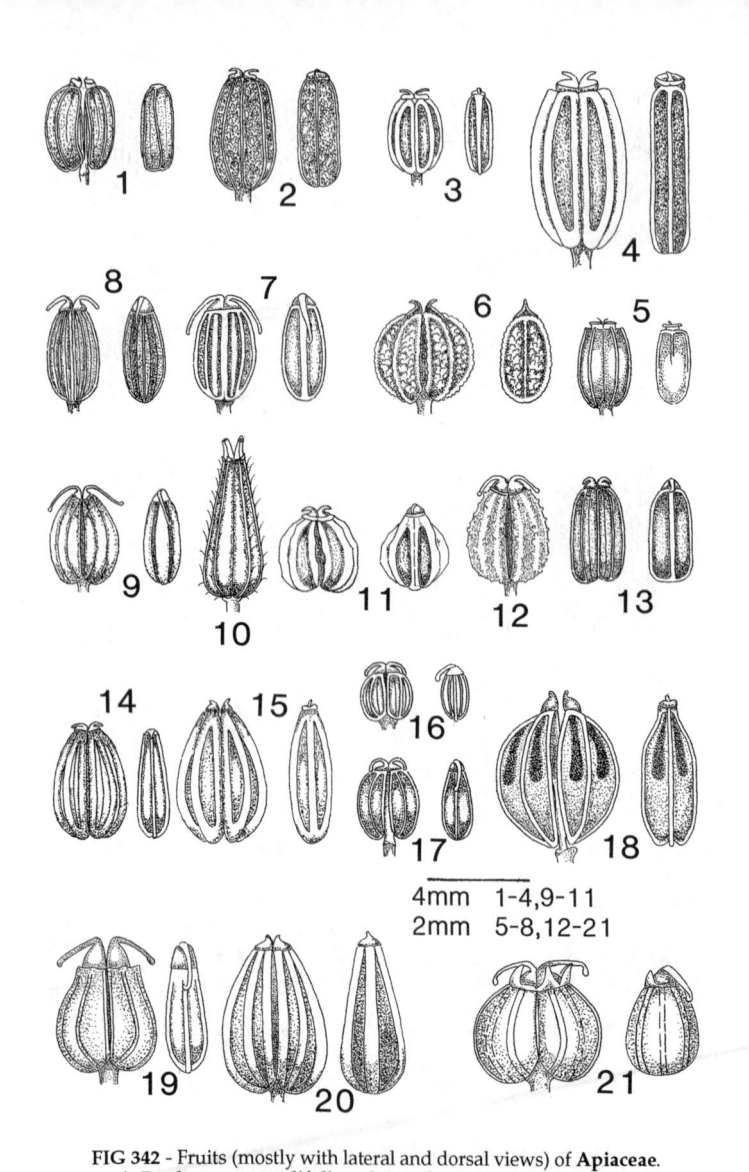

FIG 343 - Fruits (some with lateral and dorsal views) of **Apiaceae**.
1, *Falcaria vulgaris*. 2, *Selinum carvifolia*. 3, *Carum verticillatum*.
4, *C. carvi*. 5, *Ligusticum scoticum*. 6, *Levisticum officinale*.
7, *Pastinaca sativa*. 8, *Peucedanum officinale*.
9, *P. palustre*. 10, *P. ostruthium*. 11, *Angelica sylvestris*. 12, *A. archangelica*.
13, *Tordylium maximum*. 14, *Torilis nodosa*. 15, *T. arvensis*. 16, *T. japonica*.
17, *Heracleum sphondylium*. 18, *H. mantegazzianum*.
19-20, *Daucus*, lateral view shows 1 mericarp only. 19, *D. carota*.
20, *D. glochidiatus*. Drawings by Hilli Thompson.

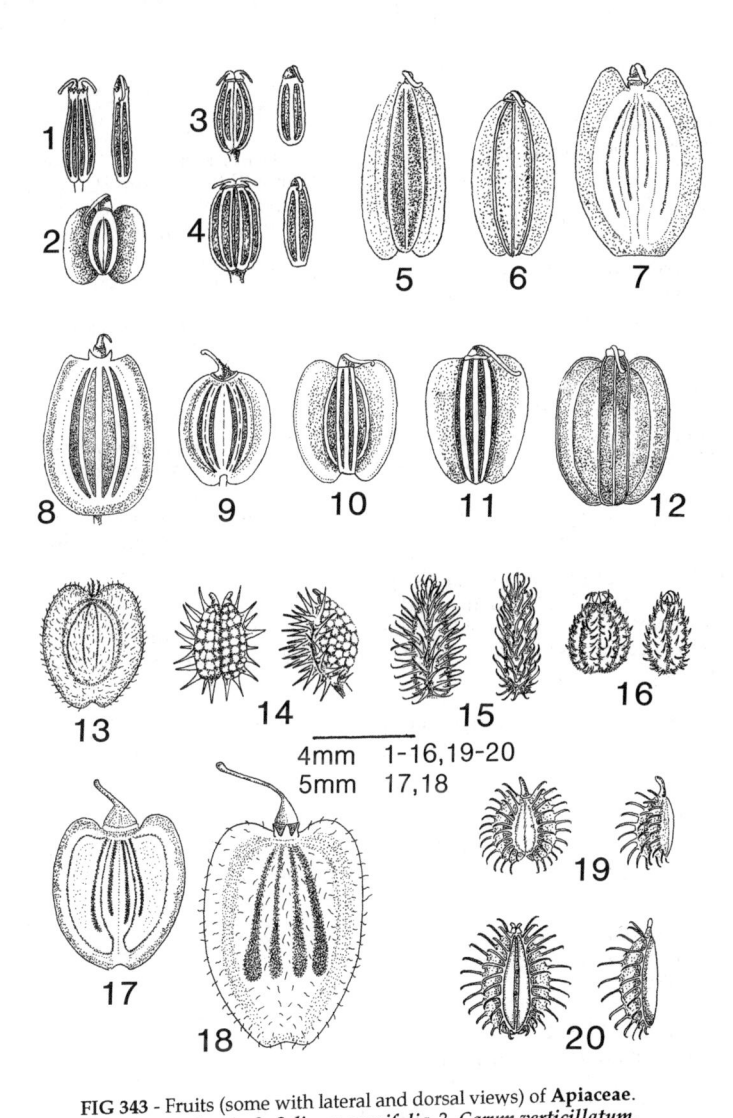

111. APIACEAE

21. FOENICULUM Mill. - *Fennel*
Glabrous perennials smelling strongly of aniseed; stems solid at first, becoming ± hollow; leaves 3-4-pinnate, with long filiform segments; bracts 0; bracteoles 0; sepals 0; petals yellow, actinomorphic; fruits scarcely compressed, c.2-4x as long as wide, glabrous, with prominent thick ribs.

1 Stems erect, ± glaucous, to 2.5m. Probably intrd-natd; open ground and waste places, especially near the coast; Bl N to N Ir, Man and Yorks, casual further N *Fennel* - **F. vulgare** Mill.

22. ANETHUM L. - *Dill*
Glabrous annuals smelling strongly of aniseed; stems hollow; leaves as in *Foeniculum*; bracts 0; bracteoles 0; sepals 0; petals yellow, actinomorphic; fruits strongly dorsally compressed, c.2x as long as wide, glabrous, with prominent slender dorsal ridges and conspicuously winged lateral ones.

1 Stems erect, to 60cm. Intrd-casual; waste and cultivated ground; scattered in Br *Dill* - **A. graveolens** L.

23. SILAUM Mill. - *Pepper-saxifrage*
Glabrous perennials; stems solid; leaves 1-4-pinnate; bracts 0-3, entire; bracteoles numerous; sepals 0; petals yellowish, actinomorphic; fruits scarcely compressed, c.2x as long as wide, glabrous, with prominent slender ridges.

1 Stems erect, to 1m. Native; grassy places; locally frequent in Br N to C Sc, mainly in E *Pepper-saxifrage* - **S. silaus** (L.) Schinz & Thell.

24. MEUM Mill. - *Spignel*
Glabrous perennials; stems hollow; leaves 3-4-pinnate; bracts 0-few, entire; bracteoles several; sepals 0; petals white to pinkish, actinomorphic; fruits scarcely compressed, <2x as long as wide, glabrous, with prominent, rather narrow ridges.

1 Plant sweetly aromatic; stems erect, to 60cm, with dense sheath of fibres at base. Native; mountain grassland; local in N En, N Wa, C & S Sc *Spignel* - **M. athamanticum** Jacq.

25. PHYSOSPERMUM Juss. - *Bladderseed*
Almost glabrous perennials; stems solid; basal leaves 2-ternate; stem-leaves simple to 1-ternate; bracts and bracteoles several, entire; sepals conspicuous, persistent; petals white, actinomorphic; fruits ± inflated, wider than long, glabrous, with narrow low ridges.

1 Stems erect, to 1.2m. Native; arable fields, hedgebanks, scrub and woods; very local in E Cornwall and S Devon, natd in Bucks
Bladderseed - **P. cornubiense** (L.) DC.

26. CONIUM L. - *Hemlock*
Glabrous biennials; stems hollow; leaves 2-4-pinnate; bracts and bracteoles several, entire; sepals 0; petals white, actinomorphic; fruits scarcely compressed, c. as wide as long, glabrous, with very prominent, narrow, ± undulate ridges.

1 Stems erect, to 2.5m, usually purple-spotted. Native; damp ground, roadsides, ditches, waste ground; common over most of BI except W & C Sc *Hemlock* - **C. maculatum** L. 342

27. BUPLEURUM L. - *Hare's-ears*
Annual to perennial herbs, rarely shrubs; stems hollow or solid; leaves simple, entire; bracts 0 to several, entire; bracteoles several; sepals 0; petals yellow, actinomorphic; fruits slightly laterally compressed, 1-1.5x as long as wide, glabrous, sometimes papillose, strongly to scarcely ridged.

1 Upper leaves fused right around stem; bracts 0 2
1 Leaves not fused around stem; bracts present 3
  2 Umbels with 4-8 rays; fruit smooth between ridges; leaves mostly <2x as long as wide. Erect annual to 30cm. Intrd-casual; formerly common in cornfields in most of En, especially C & S, extinct since 1960s except as rare casual *Thorow-wax* - **B. rotundifolium** L. 342
  2 Umbels with 2-3 rays; fruit strongly papillose between ridges; leaves mostly >2x as long as wide. Erect annual to 30cm. Intrd-casual; common birdseed-alien over much of BI *False Thorow-wax* - **B. subovatum** Spreng. 342
3 Firmly rooted perennial; at least some leaves often >1cm wide 4
3 Easily uprooted annual; leaves all <1cm wide 5
  4 Evergreen shrub; leaves with strong midrib and many lateral veins. Evergreen shrub to 2.5m, but often much less due to annual frost-damage. Intrd-natd; on roadside and railway banks; scattered in S En *Shrubby Hare's-ear* - **B. fruticosum** L. 342
  4 Herb, often woody at extreme base; leaves with ≥3 equally strong ± parallel main veins. Herbaceous ± erect perennial to 1m. Possibly native; damp roadsides, ditches, hedgebanks and field-borders; 1 locality in S Essex, rarely natd in C & N En *Sickle-leaved Hare's-ear* - **B. falcatum** L. 342
5 Bracteoles linear, not concealing flowers or fruits; fruit tuberculate. Erect to procumbent slender annual to 50cm. Native; grassy or barish brackish ground; sparse on coasts of S & C Br *Slender Hare's-ear* - **B. tenuissimum** L. 342
5 Bracteoles lanceolate, concealing flowers and fruits; fruits smooth. Erect annual to 25cm, usually much <10cm. Native; barish ground on fixed dunes and cliffs by sea; very local in CI, S Devon and E Sussex *Small Hare's-ear* - **B. baldense** Turra 342

28. TRINIA Hoffm. - *Honewort*
Glabrous, dioecious biennials to monocarpic perennials; stems solid; leaves

1-3-pinnate; bracts 0-1, 3-lobed; bracteoles 0-several, entire to 3-lobed; sepals 0; petals white, actinomorphic; fruits scarcely compressed, <1.5x as long as wide, glabrous, with prominent, wide ridges.

1 Stems erect, to 20cm, with dense sheath of fibres at base; female plants with longer, more unequal rays and fewer, longer-pedicelled flowers than male plants. Native; dry limestone turf; very local in S Devon, N Somerset and W Gloucs **Honewort - T. glauca** (L.) Dumort.

29. CUMINUM L. - *Cumin*

Glabrous annuals; stems solid; leaves 2-ternate; bracts 2-4, entire to 3-lobed with long filiform lobes; bracteoles usually 3; sepals conspicuous, persistent; petals white to pinkish, actinomorphic; fruits slightly dorsally compressed, c.2x as long as wide, glabrous or more usually bristly-pubescent, with prominent, narrow ridges.

1 Stems erect, to 50cm. Intrd-casual; birdseed- and spice-alien on tips and in waste places; scattered in S En **Cumin - C. cyminum** L.

30. APIUM L. - *Marshworts*

Glabrous biennials to perennials; stems hollow or solid; leaves pinnate or lower ones 2-3-pinnate; bracts and bracteoles 0-several, entire; sepals 0; petals white, actinomorphic; fruits laterally compressed, wider than long to longer than wide, glabrous, with prominent, slender to thick ridges.

1 Bracts and bracteoles 0; fresh plant smelling of celery. Usually erect biennial to 1m; stems solid. Native; damp barish usually brackish places usually near sea; coasts of BI N to S Sc, very scattered inland
Wild Celery - A. graveolens L.
1 Bracts 0-7; bracteoles 3-7; fresh plant not smelling of celery 2
  2 Lower leaves 2-3-pinnate, with ± filiform segments if submerged; styles shorter than stylopodium in fruit. Stems decumbent to procumbent, to 50cm, hollow, often largely submerged, rooting at lower nodes. Native; in still, usually shallow water and on bare mud nearby; scattered over most of BI
Lesser Marshwort - A. inundatum (L.) Rchb.f.
  2 All leaves 1-pinnate, even if submerged; styles longer than stylopodium in fruit 3
3 Bracts 0(-2); peduncles shorter than rays and adjacent petioles; leaflets distinctly longer than wide; fruits longer than wide. Stems to 1m, suberect to procumbent, rooting at lower nodes, hollow. Native; ditches, marshes and by lakes and rivers; common in BI except very local in C & N Sc **Fool's-water-cress - A. nodiflorum** (L.) Lag.
3 Bracts (1)3-7; peduncles longer than rays and adjacent petioles; leaflets c. as long as wide; fruits slightly wider than long. Differs from dwarfest plants of *A. nodiflorum* in stems procumbent, rooting at ± all nodes. Native; open wet places; perhaps now only in Oxon **Creeping Marshwort - A. repens** (Jacq.) Lag.

30. APIUM

A. nodiflorum x *A. repens* has been found with *A. nodiflorum* in some of the present and past sites of *A. repens* in En and Sc. *A. nodiflorum* x *A. inundatum* = *A. x moorei* (Syme) Druce occurs with the parents scattered over most of Ir and locally in CE En.

31. TRACHYSPERMUM Link - *Ajowan*
Almost glabrous annuals; stems hollow; leaves mostly 2-pinnate; bracts several, entire or lobed; bracteoles several; sepals small, not persistent; petals white, actinomorphic; fruits laterally compressed, slightly longer than wide, densely tuberculate, with rather obscure, wide ridges.

1 Stems erect, to 30cm; leaves with filiform segments to 2cm; fruits 1.5-2mm, very characteristically tuberculate. Intrd-casual; on tips and waste ground; scattered in En   *Ajowan* - **T. ammi** (L.) Sprague   342

32. PETROSELINUM Hill - *Parsleys*
Glabrous annuals or biennials; stems solid; leaves 1-3-pinnate, with wide lobes; bracts 1-several, entire or lobed; bracteoles several; sepals 0 or very small; petals white or yellow, actinomorphic; fruits laterally compressed, somewhat longer than wide, glabrous, with prominent thick to narrow ridges.

1 Erect biennial to 75cm, with characteristic parsley smell; lower leaves 3-pinnate; petals yellow; fruits with styles ± as long as stylopodium. Intrd-natd; escape on tips and in waste places; scattered in BI
 Garden Parsley - **P. crispum** (Mill.) A.W. Hill   342
1 Erect annual or biennial to 1m; lower leaves 1-pinnate; petals white; fruits with styles much shorter than stylopodium. Native; barish or grassy places in arable fields, pastures and hedgerows and on banks; local in S & C Br, Jersey   *Corn Parsley* - **P. segetum** (L.) W.D.J. Koch   342

33. RIDOLFIA Moris - *False Fennel*
Glabrous annuals; stems solid; leaves 3-4-pinnate, with filiform segments; bracts 0; bracteoles 0; sepals 0; petals yellow, actinomorphic; fruits laterally compressed, longer than wide, glabrous, with slender, rather low ridges.

1 Stems erect, to 1m; leaves with filiform segments, resembling those of *Anethum* but lacking aniseed smell. Intrd-casual; tips and waste ground; very scattered in S En   *False Fennel* - **R. segetum** (Guss.) Moris   342

34. SISON L. - *Stone Parsley*
Glabrous biennials, stems solid; leaves 1-2-pinnate, the lower with wide lobes; bracts 2-4 and bracteoles 2-4, entire; sepals 0; petals white, actinomorphic; fruits laterally compressed, somewhat longer than wide, glabrous, with narrow prominent ridges.

1 Fresh plant smelling rather like petrol when crushed; stems erect, to 1m. Native; hedgebanks, grassland, roadsides; locally frequent

in Br N to Cheshire and SE Yorks, rare casual further N
Stone Parsley - **S. amomum** L.

35. CICUTA L. - *Cowbane*
Glabrous perennials; stems hollow; leaves 2-3-pinnate; bracts 0; bracteoles numerous; sepals conspicuous, persistent; petals white, actinomorphic; fruits laterally compressed but each mericarp ± globose, wider than long, glabrous, with wide inconspicuous ridges.

1 Stems erect, to 1.5m; leaves with narrowly elliptic to linear-lanceolate segments, up to 5(9)cm. Native; ditches, marshy fields, pondsides; very local in N & C Ir and scattered areas of Br *Cowbane* - **C. virosa** L.

36. AMMI L. - *Bullworts*
Glabrous annuals to biennials; stems solid; leaves 1-3(4)-pinnate; bracts several, mostly pinnately divided with linear to filiform lobes; bracteoles numerous; sepals 0; petals white, very slightly zygomorphic; fruits slightly laterally compressed; somewhat longer than wide, glabrous, with slender prominent ridges.

1 Lower leaves with elliptic to narrowly elliptic, serrate lobes; rays remaining slender and bracts not becoming strongly reflexed in fruit; fruits 1.5-2mm. Stems erect, to 1m. Intrd-casual; on tips and waste ground and in fields; very scattered in CI and C & S Br
Bullwort - **A. majus** L.
1 Lower leaves with linear to filiform segments; rays becoming rigid, thick and erect, and bracts becoming strongly reflexed, in fruit; fruits 2-2.8mm. Stems erect, to 1m. Intrd-casual; on tips and waste ground; S En, N to SW Yorks *Toothpick-plant* - **A. visnaga** (L.) Lam.

37. FALCARIA Fabr. - *Longleaf*
Glabrous, glaucous, perennial; stems solid; leaves 1-2-ternate; bracts and bracteoles numerous, entire; sepals conspicuous, persistent; petals white, actinomorphic; fruits laterally compressed, 2-4x as long as wide, glabrous, with wide, low ridges.

1 Stems erect, to 60cm; leaf-segments c.10-30cm, linear-lanceolate, serrate. Intrd-natd; in grassy and waste places and scrub; very local in CI and S & C Br *Longleaf* - **F. vulgaris** Bernh.

38. CARUM L. - *Caraways*
Glabrous biennials or perennials; stems hollow; leaves 1-3-pinnate; bracts and bracteoles 0-numerous; sepals 0 or minute, not persistent; petals white, actinomorphic or nearly so; fruits laterally compressed, 1.2-1.7x as long as wide, glabrous, with narrow, ± prominent ridges.

1 Erect biennial to 60cm, without fibres at base; leaves 2-3-pinnate, the lower with linear to lanceolate segments. Intrd-natd; roadsides, fields

38. CARUM

and waste places; sparsely scattered in BI *Caraway* - **C. carvi** L. 343
1 Erect perennial to 60cm, with dense sheath of fibres at base; leaves 1-pinnate, each leaflet deeply palmately divided to base into filiform segments. Native; marshes, damp meadows and streamsides; local in Ir and W Br *Whorled Caraway* - **C. verticillatum** (L.) W.D.J. Koch 343

39. SELINUM L. - *Cambridge Milk-parsley*

Glabrous perennials; stems solid; leaves 2-3-pinnate; bracts 0-few, entire, soon falling; bracteoles numerous; sepals ± 0; petals white, actinomorphic; fruits dorsally compressed, c. as long as wide, glabrous, with conspicuous winged ridges.

1 Stems erect, to 1m; lower leaves with lanceolate to ovate, deeply lobed segments <10mm. Native; fens and damp meadows; very local in Cambs *Cambridge Milk-parsley* - **S. carvifolia** (L.) L. 343

40. LIGUSTICUM L. - *Scots Lovage*

Glabrous perennials; stems hollow; leaves 1-2-ternate, with stalked, wide leaflets; bracts 1-several, entire; bracteoles several; sepals conspicuous, persistent; petals greenish-white, actinomorphic; fruits strongly dorsally compressed, c.2x as long as wide, glabrous, with prominent, ± winged ridges.

1 Stems erect, to 60(90)cm; leaves with ovate-trullate, serrate segments >2 x 1.5cm. Native; cliffs and rocky places near sea; coasts of Sc, local in N & W Ir *Scots Lovage* - **L. scoticum** L. 343

41. ANGELICA L. - *Angelicas*

Minutely pubescent to ± glabrous biennials to often monocarpic perennials; stems hollow; leaves 2-3-pinnate, the upper with very strongly inflated petioles; bracts 0-few, entire, soon falling; bracteoles numerous; sepals very small; petals white, greenish-white or pinkish-white, actinomorphic; fruits strongly dorsally compressed, slightly longer than wide, glabrous, with low dorsal and conspicuously winged lateral ridges.

1 Fruits with thin membranous wings, 4-5mm. Stems erect, to 2.5m, often much less, usually somewhat purplish. Native; damp grassy places, fens, marshes, by streams, ditches and ponds, in damp open woods; common *Wild Angelica* - **A. sylvestris** L. 343
1 Fruits with thick corky wings, 5-10mm 2
   2 Fruits 5-6mm; leaflets acute to acuminate. Stems erect, to 2m, usually green. Intrd-natd; on river-banks and waste places; scattered in Br *Garden Angelica* - **A. archangelica** L. 343
   2 Fruits 8-10mm; leaflets acute to subacute. Stems erect, to 1m. Intrd-natd; in hedgebank; Guernsey
 Portuguese Angelica - **A. pachycarpa** Lange 339

111. APIACEAE

42. LEVISTICUM Hill - *Lovage*
Almost glabrous perennials smelling of celery when crushed; stems hollow; leaves 2-3-pinnate; bracts and bracteoles numerous, entire, the latter fused at base; sepals ± 0; petals yellow, actinomorphic; fruits strongly dorsally compressed, <2x as long as wide, glabrous; with very prominent dorsal and winged lateral ridges.

1 Stems erect, to 2.5m; leaves with trullate to rhombic, sparsely but deeply serrate leaflets. Intrd-natd; in rough ground, by walls and paths; very scattered in Br, mostly N **Lovage - L. officinale** W.D.J. Koch   34

43. FERULA L. - *Giant Fennel*
Erect, glabrous perennials; stems solid; leaves 4-6-pinnate; bracts 0; bracteoles few, soon falling; sepals minute; petals yellow, actinomorphic; fruits strongly dorsally compressed, longer than wide, glabrous, with low dorsal and winged lateral ridges.

1 Stems erect, to 3m; leaves with ultimate segments 10-50mm, linear, with very conspicuous, wide, sheathing petioles. Intrd-natd; on roadside verges; W Suffolk and Northants, occasionally elsewhere in En   **Giant Fennel - F. communis** L.   33

44. PEUCEDANUM L. - *Hog's Fennels*
Erect, glabrous to nearly glabrous biennials to perennials; stems hollow or solid; leaves 1-6-ternate or 2-4-pinnate; bracts 0-several, entire; bracteoles few to several; sepals ± 0 to conspicuous, persistent or not; petals white or yellow, actinomorphic; fruits strongly dorsally compressed, somewhat longer than wide, glabrous, with low dorsal and winged lateral ridges.

1 Petals yellow; stems solid, with dense sheath of fibres at base; ultimate leaf-segments >3cm, linear. Stems to 2m; leaves 3-6-ternate. Native; rough brackish grassland, banks of creeks and pathsides near sea; extremely local in E Kent and N Essex
    Hog's Fennel - P. officinale L.   34
1 Petals white; stems hollow, without basal fibres; ultimate leaf-segments not linear, or if so <2cm   2
    2 Leaves 2-4-pinnate, with linear to narrowly oblong-lanceolate ultimate segments <1.5 x 0.5cm; bracts ≥3, reflexed. Stems to 1.5m. Native; fens and marshes; local in En N to SE Yorks
        Milk-parsley - P. palustre (L.) Moench   34
    2 Leaves 1-2-ternate, with ovate ultimate segments >2 x 1cm; bracts 0(-2). Stems to 1m; leaves 1-2-ternate. Intrd-natd; in grassy places, marshy fields and riversides; scattered in N Ir and Br N from S Lincs and Staffs   **Masterwort - P. ostruthium** (L.) W.D.J. Koch   34

45. PASTINACA L. - *Parsnips*
Somewhat pubescent biennials with strong characteristic smell; stems hollow or solid; leaves 1-pinnate, with large ovate leaflets; bracts and

45. PASTINACA

bracteoles 0-2, entire, soon falling; sepals 0; petals yellow, actinomorphic; fruits strongly dorsally compressed, somewhat longer than wide, glabrous, with low dorsal and winged lateral ridges.

1 Stems erect, to 1.8m. Native; grassland, roadsides, rough ground, especially on chalk and limestone; common in SE 1/2 of En, rare elsewhere in En, Wa and Ir. Var. *hortensis* Gaudin (*Parsnip*), more glabrous and with a swollen root, is sometimes found as a relic of cultivation **Wild Parsnip - P. sativa** L. 343

46. HERACLEUM L. - *Hogweeds*
Erect, pubescent biennials to often monocarpic perennials; stems hollow; leaves simple and pinnately or ternately divided, or 1(-2)-pinnate, or ternate; bracts 0-several, entire; bracteoles several; sepals minute or conspicuous and persistent; petals white to purplish or greenish-white, zygomorphic to scarcely so; fruits strongly dorsally compressed, slightly to somewhat longer than wide, glabrous or pubescent, with very low dorsal and winged lateral ridges.

1 Stems to 5.5m, rather softly pubescent; leaves pinnate to ternate, or simple and ternately to pinnately lobed; bracts several; fruits 9-14mm, glabrous or pubescent, with persistent sepals, with conspicuous oil-bodies swollen to 0.6-1mm wide at proximal end. Intrd-natd; waste ground, roadside and riverside banks, rough grassland; scattered throughout BI
Giant Hogweed - H. mantegazzianum Sommier & Levier 343
1 Stems to 2(3)m, hispid; leaves 1(-2)-pinnate; bracts 0-few; fruits 4-10mm, glabrous, without persistent sepals, with linear oil-bodies scarcely widened (<0.4mm wide) at proximal end. Native; grassy places, rough ground and roadsides (*Hogweed* - **H. sphondylium** L.) 2 343
  2 Petals white or pinkish-white to purplish, the outer ones on outermost flowers of umbel bilobed and c.2x inner unlobed ones. Common **H. sphondylium ssp. sphondylium**
  2 Petals greenish-white, scarcely zygomorphic. NE parts of E Norfolk (?intrd) (ssp. *sibiricum* Simonk.)
H. sphondylium ssp. flavescens (Willd.) Soó

H. sphondylium x *H. mantegazzianum* is scattered in En, Ir and Sc, especially SE Sc and the London area; it is intermediate in size, pubescence, leaf-shape and fruit characters, and has very low fertility

47. TORDYLIUM L. - *Hartwort*
Hispid annuals or biennials; stems hollow or ± solid; leaves 1-pinnate (basal ± simple but gone by flowering); bracts and bracteoles several, entire; sepals conspicuous, persistent; petals white, zygomorphic; fruits strongly dorsally compressed, scarcely longer than wide, hispid, with very low dorsal and broadly whitish-winged lateral ridges.

111. APIACEAE

1 Stems erect, to 1m; lower leaves with 2-5 pairs of lanceolate, coarsely serrate leaflets. Possibly native; rough scrubby grassland; near R. Thames in S Essex *Hartwort* - **T. maximum** L.

48. TORILIS Adans. - *Hedge-parsleys*

Hispid annuals (rarely biennials); stems solid; leaves 1-3-pinnate; bracts 0-numerous, entire; bracteoles several; sepals persistent but inconspicuous at fruiting due to spines; petals white to purplish-white, not or slightly zygomorphic; fruits ± not compressed, somewhat longer than wide, variously furnished with curved or hooked spines.

1 Fruits with dimorphic mericarps, 1 with spines, 1 tuberculate; peduncles ≤1cm; rays <5mm, ± hidden by flowers or fruits. Stems procumbent to ascending, to 50cm. Native; arable and barish ground, especially near sea; rather scattered in BI N to SE Sc, mostly in E En *Knotted Hedge-parsley* - **T. nodosa** (L.) Gaertn.
1 Both mericarps with spines; peduncles >1cm; rays >5mm, conspicuous 2
  2 Bracts 0-1; fruits 3-4mm (excl. spines), with ± straight spines minutely hooked at end. Stems erect, with wide-spreading branches, to 50cm. Probably intrd-natd; weed of arable land; now rare and mainly casual in S En
Spreading Hedge-parsley - **T. arvensis** (Huds.) Link
  2 Bracts >2; fruits 2-2.5mm (excl. spines), with curved spines not hooked at end. Stems erect, to 1.2m. Native; grassy places, hedgerows, wood-borders and -clearings; frequent except in N & NW Sc *Upright Hedge-parsley* - **T. japonica** (Houtt.) DC.

49. DAUCUS L. - *Carrots*

Glabrous annuals to biennials with strong characteristic smell (especially in crushed root); stems solid; leaves (1)2-3-pinnate; bracts numerous, usually longer than rays, pinnately divided into filiform lobes; bracteoles numerous; sepals small, scarcely visible in fruit; umbel often with 1 dark purple central flower; other petals white, slightly zygomorphic; fruits strongly dorsally compressed, somewhat longer than wide; mericarps each with 2 lateral and 2 secondary dorsal ridges each with row of terminally barbed spines, the 3 primary dorsal ridges alternating with the secondaries and bearing only short weak bristles.

1 Slender annual; rays <8, very uneven lengthed, slender, each with ≤6 flowers, or sometimes umbels simple; fruits 3-5mm, with dense, rather slender spines. Stems erect, ± glabrous, to 40cm. Intrd-casual; rather frequent wool-alien; very scattered in En *Australian Carrot* - **D. glochidiatus** (Labill.) Fisch., C.A. Mey. & Avé-Lall.
1 Usually biennial; rays >10, stout, each with >6 flowers; fruits 2-3mm, with stout, ± straight spines. Stems erect to procumbent, scarcely branched to strongly branched with widely spreading branches, often hispid, to 1m (**D. carota** L.) 2

49. DAUCUS

2 Umbels convex to slightly concave in fruit. Native; cliffs, dunes and rocky places by sea; coasts of CI, S & SE Ir, S & W Br
Sea Carrot - D. carota ssp. gummifer (Syme) Hook. f.
2 Umbels very contracted in fruit, very concave 3
3 Root swollen in 1st year, usually orange; leaves usually bright green. Intrd-natd; casual in waste places and tips and a relic where planted; scattered over BI **Carrot - D. carota ssp. sativus** (Hoffm.) Arcang.
3 Root not swollen, whitish; leaves usually grey-green. Native; grassy and rough ground, mostly on chalky soils and near sea (there often very stunted); throughout BI but mainly coastal in N & W Br
Wild Carrot - D. carota ssp. carota

112. GENTIANACEAE - *Gentian family*

Glabrous annuals to herbaceous perennials; easily recognized by the opposite, entire, glabrous leaves, 4-5(8) fused petals with as many stamens borne on corolla-tube, and 1-celled ovary with many ovules forming 2-celled capsule.

1 Petals purple to blue, rarely white, very rarely pinkish; style and stigmas persistent in fruit 2
1 Petals pink, yellow or white; stigmas and sometimes style falling before fruiting 3
   2 Corolla with small inner lobes alternating with main ones; distal part of calyx-tube membranous between the calyx-lobe origins **6. GENTIANA**
   2 Corolla without small inner lobes but with long fringes; calyx-tube without membranous part **5. GENTIANELLA**
3 Pairs of stem-leaves fused at base; flowers 6-8-merous **4. BLACKSTONIA**
3 Stem-leaves not fused in pairs; flowers 4-5-merous 4
   4 Calyx-lobes shorter than calyx-tube; corolla yellow **1. CICENDIA**
   4 Calyx-lobes longer than calyx-tube; corolla pink or white 5
5 Calyx-lobes (4-)5, keeled; corolla-lobes (4-)5, ≥2mm; anthers twisting after flowering **3. CENTAURIUM**
5 Calyx-lobes 4, not keeled; corolla-lobes 4, <2mm; anthers not becoming twisted **2. EXACULUM**

1. CICENDIA Adans. - *Yellow Centaury*
Annuals; flowers 4-merous; calyx-lobes triangular, shorter than calyx-tube; corolla yellow; anthers not becoming twisted; style simple; stigma 1, peltate.

1 Stems erect, very slender, to 10(18)cm; leaves few, linear, ≤6mm; corolla 3-7mm. Native; damp sandy and peaty barish ground mostly near coast; very local in W Ir, S & SW Br N to Caerns and E to E Sussex **Yellow Centaury - C. filiformis** (L.) Delarbre

2. EXACULUM Caruel - *Guernsey Centaury*
Annuals; flowers 4-merous; calyx-lobes linear, flat, longer than calyx-tube; corolla pale pink; anthers not becoming twisted; style 1, divided near apex; stigmas 2.

1 Stems procumbent to ascending, very slender, to 4(10)cm; leaves linear, ≤7mm; corolla 3-6mm. Native; with *Cicendia* in short ± open turf in dune-slacks; few places in Guernsey
Guernsey Centaury - **E. pusillum** (Lam.) Caruel

3. CENTAURIUM Hill - *Centauries*
Annuals, biennials or perennials; flowers (4-)5-merous; calyx-lobes linear, keeled, longer than calyx-tube; corolla pink or white; anthers becoming twisted at fruiting; style 1, divided near apex; stigmas 2.

1 Perennials with procumbent to decumbent non-flowering stems; corolla-lobes >7mm. Flowering stems to 30cm, ascending. Native; grassy cliff-tops and dunes by sea; Pembs and W Cornwall, natd as lawn-weed in SE En *Perennial Centaury* - **C. scilloides** (L.f.) Samp.
1 Annuals to biennials without procumbent to decumbent non-flowering stems; corolla-lobes ≤7mm   2
    2 Usually biennials, normally with basal leaf-rosette at flowering; flowers with 1-2 bracts at base of calyx, the stalk between 0-1mm   3
    2 Usually annuals, normally without basal leaf-rosette at flowering; flowers with stalks 1-4mm between base of calyx and bracts   4
3 Stem-leaves narrowly oblong-elliptic, almost parallel-sided, obtuse to rounded at apex; calyx usually >3/4 as long as corolla-tube; stigmas broadly rounded to nearly flat at apex. Erect biennial to 26cm. Native; coastal dunes and sandy turf; local in Br N from S Wa and NE Yorks, Co Londonderry
Seaside Centaury - **C. littorale** (Sm.) Gilmour
3 Stem-leaves ovate to elliptic, acute to subacute at apex; calyx usually <3/4 as long as corolla-tube; stigmas narrowly rounded to nearly conical at apex. Erect biennial (rarely annual) to 50cm. Native; grassy and rather open ground on well-drained soils; frequent throughout BI, more local in Sc *Common Centaury* - **C. erythraea** Rafn
    4 Main stem with 2-4 internodes; all branches arising at c.30-45° and forming rather open inflorescence; corolla usually pink. Erect annual to 20cm but often <6cm. Native; woodland rides, damp grassy or ± open ground, especially near sea; local in En & Wa, CI, S & E Ir *Lesser Centaury* - **C. pulchellum** (Sw.) Druce
    4 Main stem with 5-9 internodes; upper branches arising at c.20-30° and forming rather dense inflorescence; corolla usually white. Erect annual to 35cm. Native; damp grassy places near sea; very rare in Dorset
Slender Centaury - **C. tenuiflorum** (Hoffmanns. & Link) Fritsch

C. erythraea x *C. littorale* = ***C. x intermedium*** (Wheldon) Druce occurs with

3. CENTAURIUM

the parents on coastal dunes in S & W Lancs, Anglesey and Merioneth; it has low fertility. *C. erythraea* x *C. pulchellum* occurs with the parents on the coast in N Somerset, S Essex and W Lancs; it is highly fertile

4. BLACKSTONIA Huds. - *Yellow-wort*
Annuals; flowers 6-8-merous; calyx divided almost to base into linear, flat lobes; corolla yellow; anthers not becoming twisted; style 1, divided near apex; stigmas 2.

1 Stems erect, to 50cm; stem-leaves triangular-ovate, pairs fused round stem at base, glaucous. Native; calcareous grassland, bare chalk and dunes; locally frequent in C & S BI
Yellow-wort - **B. perfoliata** (L.) Huds.

5. GENTIANELLA Moench - *Gentians*
Annuals to biennials; flowers 4-5-merous; calyx-lobes shorter or longer than tube; corolla blue or dark- to whitish-purple, rarely pink, the lobes with long fringes at margins or at base on inner face; anthers not becoming twisted; style scarcely distinct; stigmas 2.
 Spp. in couplets 3-7 are variable in size and habit. The existence of hybrids and of diminutive annuals (with smaller than normal floral parts) in normally biennial taxa can make determination difficult.

1 Corolla-lobe with long narrow fringes along sides, not at base on inner side. Erect biennial to 30cm; corolla 25-40mm, blue. Possibly native; chalk grassland; 1 place in Bucks
Fringed Gentian - **G. ciliata** (L.) Borkh.
1 Corolla-lobes with long narrow fringes at base on inner side, not along sides 2
  2 Flowers 4-merous; calyx with 2 lobes several times wider than other 2 lobes. Erect annual or biennial to 30cm; corolla 15-25(30)mm, bluish-purple to ± white. Native; grassland and dunes; scattered over Br and Ir but rare or absent in S
Field Gentian - **G. campestris** (L.) Börner
  2 Flowers 4-5-merous (often both on same plant); calyx with 4-5 lobes, the widest ≤2x as wide as others 3
3 Corolla (15)25-35mm, ≥2x as long as calyx; plant with 9-15 internodes. Erect (annual or) biennial to 40cm; corolla bright bluish-purple. Native; chalk grassland, mostly sheltered or among scrub; very local in SC En *Chiltern Gentian* - **G. germanica** (Willd.) Börner
3 Corolla 9-22mm, ≤2x as long as calyx; plant with 0-9(11) internodes 4
  4 Internodes (2)4-9(11); apical pedicel <1/4 total plant height to pedicel apex. Erect (annual or) biennial to 30cm. Native; basic pastures and dunes (*Autumn Gentian* - **G. amarella** (L.) Börner) 5
  4 Internodes 0-3; apical pedicel usually ≥1/2 total plant height to pedicel apex 7
5 Corolla creamy-white, suffused purplish-red on outside, (12)14-17mm. Stems with 2-7 internodes. Locally frequent in Sc N

from N Aberdeen and M Perth, Cheviot, MW Yorks
G. amarella ssp. septentrionalis (Druce) N.M. Pritch.
5 Corolla usually dull purple, rarely pale blue, pink or whitish, (14)16-22mm
   6 Corolla (14)16-18(20)mm. Stems with 4-9(11) internodes. Locally frequent in Br N to Angus and S Ebudes **G. amarella ssp. amarella**
   6 Corolla (17)19-22mm. Stems with 7-11 internodes. Frequent over most of Ir **G. amarella ssp. hibernica** N.M. Pritch.
7 Stem-leaves lanceolate to oblong- or linear-lanceolate; calyx-lobes ± equal, appressed to corolla. Erect (annual or) biennial to 20cm, usually much less, with (0)2-3(4) internodes. Native; chalk and limestone grassland and fixed dunes; local in C & S En from S Devon to S Lincs and E Kent, Pembs
Early Gentian - **G. anglica** (Pugsley) E.F. Warb.
7 Stem-leaves ovate to narrowly ovate; some calyx-lobes usually distinctly longer and wider than others, somewhat divergent from corolla. Erect annual or biennial to 15cm, with 0-2(3) internodes. Native; coastal dunes and dune-slacks; Glam, Carms, Pembs and S Ebudes
Dune Gentian - **G. uliginosa** (Willd.) Börner

G. germanica x *G. amarella* = ***G. x pamplinii*** (Druce) E.F. Warb. is ≥50% fertile; it occurs near most populations of *G. germanica*. *G. amarella* x *G. anglica* = ***G. x davidiana*** T.C.G. Rich (*G. anglica* ssp. *cornubiensis* N.M. Pritch.) occurs with the parents in S En and is fertile. *G. amarella* x *G. uliginosa* occurs with most colonies of *G. uliginosa* in S Wa and is fertile.

6. GENTIANA L. - *Gentians*

Annuals to perennials; flowers 5-merous; calyx-lobes shorter than to c. as long as tube, with small membranous connexion at base of sinuses; corolla blue, rarely white or pink, the lobes not fringed, with 5 small lobes alternating with the 5 large ones; anthers, styles and stigmas as in *Gentianella*.

1 Corolla-tube >1cm wide, widening distally; leaves all or most >15(20)mm 2
1 Corolla-tube <8mm wide, ± cylindrical; leaves all <15mm 4
   2 Leaves crowded in basal rosette, few reduced ones up stem. Perennial; stems erect, to 6cm, each with 1 flower. Intrd-natd; in chalk grassland in 3 nearby places in Surrey
Trumpet Gentian - **G. clusii** E.P. Perrier & Songeon
   2 Leaves spread ± evenly up stem 3
3 Leaves linear, <1cm wide, with 1 vein. Perennial; stems erect, to 40cm, with 1-10(28) flowers. Native; wet heathland; very local in Br from Dorset and E Sussex to NE Yorks and Westmorland
Marsh Gentian - **G. pneumonanthe** L.
3 Leaves lanceolate (to ovate), >1cm wide, with 3-5 veins. Perennial; stems erect, to 60cm, with 1-30 flowers. Intrd-natd; by streams and

6. GENTIANA

in shady places; E & W Sussex, MW Yorks and C Sc
Willow Gentian - **G. asclepiadea** L.

4 Rhizomatous perennial with several rosettes of leaves; corolla lobes >8mm. Stems erect, 0.5-7cm, each with 1 flower. Native; grassland on limestone, calcareous glacial drift and fixed dunes; extremely local in N En and W Ir, rarely natd elsewhere
Spring Gentian - **G. verna** L.

4 Annual, with or without 1 basal leaf-rosette; corolla-lobes <6mm. Stems erect, to 15cm, with 1-10 flowers. Native; rock-ledges above 730m; very local and rare in M Perth and Angus
Alpine Gentian - **G. nivalis** L.

113. APOCYNACEAE - *Periwinkle family*

Slightly woody perennials; the opposite, evergreen, entire leaves and blue 5-lobed flowers of characteristic shape are diagnostic.

1. VINCA L. - *Periwinkles*

1 Margin of leaves and calyx-lobes minutely pubescent. Vegetative stems ascending-arching then often procumbent, to 2m; flowering stems as in *V. difformis*. Intrd-natd; in hedgebanks, shrubberies and rough ground; most of BI *Greater Periwinkle* - **V. major** L.   2
1 Margin of leaves and calyx-lobes glabrous
  2 Corolla-tube 9-11mm; corolla-limb 25-30mm across; calyx-lobes 3-4(5)mm. Vegetative stems procumbent to arching, rooting at tips, to 1m; flowering stems erect to ascending, to 20cm. Probably intrd-natd; woods, hedgebanks and other shady places; scattered over most of BI *Lesser Periwinkle* - **V. minor** L.
  2 Corolla-tube 12-18mm; corolla-limb 30-45mm across; calyx-lobes 5-14mm. Similar to *V. minor* in habit but more robust; vegetative stems to 2m; flowering stems to 30cm. Intrd-natd; on bank in W Kent and (white-flowered) W Cornwall
  Intermediate Periwinkle - **V. difformis** Pourr.

114. SOLANACEAE - *Nightshade family*

Annuals to herbaceous perennials or shrubs; distinguished from all but the very distinctive *Verbascum* (Scrophulariaceae) in the ± actinomorphic 5-merous flowers with fused calyx and corolla and usually 2-celled ovary with many ovules.

1 Open flowers with anthers touching laterally, forming cone-shaped group around style   2
1 Open flowers with anthers separated laterally, not forming cone-shaped group around style   4

2   Stamens opening by apical pores; fruit succulent; if corolla
    yellow then plant spiny **10. SOLANUM**
2   Stamens opening by longitudinal slits; if fruit very succulent
    then corolla yellow and plant not spiny 3
3 Leaves simple; corolla white to purple; fruit rather dry **8. CAPSICUM**
3 Leaves pinnate; corolla yellow; fruit succulent **9. LYCOPERSICON**
    4   Woody shrub; corolla blue to purple, rarely white; fruits red or
        yellowish-green 5
    4   Stems herbaceous, or if woody towards base then flowers and
        fruits both whitish 6
5 Stems usually spiny; corolla <2cm, divided c.1/2 way; stamens
    exserted **2. LYCIUM**
5 Stems not spiny; corolla >2cm, divided much <1/2 way; stamens
    not exserted **3. IOCHROMA**
    6   Calyx toothed, the teeth <1/4 total length at flowering 7
    6   Calyx lobed, ≥1 lobe ≥1/3 total length at flowering (often not so
        at fruiting) 8
7 Calyx in fruit funnel-shaped, persistent, c.2x as long as fruit; capsule
    opening by lid; corolla with network of dark veins **5. HYOSCYAMUS**
7 Calyx in fruit tubular to bell-shaped, withering, <2x as long as fruit;
    capsule opening by longitudinal valves; corolla without network
    of dark veins **11. DATURA**
    8   Fruit a capsule; corolla tubular to trumpet-shaped 9
    8   Fruit a berry; corolla cup-, bowl-, bell- or star-shaped 10
9 Calyx-teeth ≥3/4 total length of calyx; flowers solitary in leaf-axils
    13. PETUNIA
9 Most or all calyx teeth ≤2/3 total length of calyx; flowers opposed
    to or in axils of much reduced bracts **12. NICOTIANA**
    10 Fruit ± completely enclosed in enlarged calyx 11
    10 Fruit well exposed from calyx 12
11 Fruiting calyx with lobes much longer than tube; ovary and fruit
    3-5-celled **1. NICANDRA**
11 Fruiting calyx with tube much longer than lobes; ovary and fruit
    2-celled **7. PHYSALIS**
    12 Corolla brownish- to greenish-purple, >20mm; largest leaves
        >5cm **4. ATROPA**
    12 Corolla whitish, <12mm; largest leaves <5cm **6. SALPICHROA**

1. NICANDRA Adans. - *Apple-of-Peru*
Subglabrous annuals; leaves simple, toothed to ± lobed; flowers solitary, axillary; calyx deeply 5-lobed, later enlarging and enclosing fruit; corolla bell-shaped, shallowly lobed; ovary 3-5-celled; fruit a rather dry berry.

1   Stems erect, to 80cm; corolla 25-40mm long and across, blue to
    mauve. Intrd-natd; frequent casual in waste and cultivated ground
    and on tips, sometimes persistent; scattered in CI and Br, mainly S
    Apple-of-Peru - **N. physalodes** (L.) Gaertn.

2. LYCIUM L. - *Teaplants*
Almost glabrous, usually spiny, deciduous shrubs with arching branches; leaves simple, entire; flowers axillary, 1-few together; calyx irregularly 2-lipped, not enclosing fruit; corolla funnel-shaped, purplish, rather deeply lobed; ovary 2-celled; fruit a red berry.

1 Leaves narrowly elliptic, widest near middle; calyx c.4mm; corolla 10-15mm, divided to c.1/2 way. Intrd-natd; in rough ground, hedges and on walls; frequent in CI, En, Wa and Man, very scattered in Ir and Sc   *Duke of Argyll's Teaplant* - **L. barbarum** L.
1 Leaves lanceolate to narrowly ovate, widest below middle; calyx c.3mm; corolla 10-15mm, divided to >1/2 way. Intrd-natd; similar places to *L. barbarum* and much confused with it
                                       Chinese Teaplant - **L. chinense** Mill.

3. IOCHROMA Benth. - *Argentine-pear*
Pubescent, non-spiny, deciduous shrubs; leaves simple, entire; flowers axillary, (1)2-5 together; calyx bell-shaped, with 5 lobes, enlarging but not enclosing fruit; corolla narrowly bell-shaped, with 5 shallow lobes; ovary 2-celled; fruit a berry.

1 Shrub to 5m; leaves narrowly elliptic, 4-8cm; corolla 25-45cm, deep blue, sometimes white. Intrd-natd; waste places; Nottingham (Notts)   *Argentine-pear* - **I. australe** Griseb.

4. ATROPA L. - *Deadly Nightshade*
Nearly glabrous to glandular-pubescent perennials; leaves simple, entire; flowers solitary, axillary; calyx rather deeply 5-lobed, slightly enlarging later but not enclosing fruit; corolla bell-shaped, shallowly lobed; ovary 2-celled; fruit a berry.

1 Stems erect, to 2m; corolla 24-30mm, greenish- or brownish-purple; fruit globose to depressed-globose, 15-20mm across, shiny black. Native; woods, scrub, rough and cultivated ground especially on chalk and limestone; locally frequent in C & S En, scattered in Ir and elsewhere in Br   *Deadly Nightshade* - **A. belladonna** L.

5. HYOSCYAMUS L. - *Henbane*
Glandular-pubescent stinking annuals to biennials; leaves simple, toothed to ± lobed; flowers solitary, axillary, forming 2 rows on 1 side of stem; calyx funnel-shaped, enlarging later and becoming swollen at base to accommodate fruit, broadly 5-toothed; corolla broadly funnel-shaped, rather deeply lobed; ovary 2-celled; fruit a capsule dehiscing by lid.

1 Stems erect, to 80cm; corolla 2-3cm, yellowish with strong purple reticulate venation, 2-3cm across; capsule enclosed by calyx. Native; maritime sand and shingle, inland rough and waste ground; scattered in Br and Ir, mainly C & S   *Henbane* - **H. niger** L.

114. SOLANACEAE

6. SALPICHROA Miers - *Cock's-eggs*
Pubescent perennials, somewhat woody below; leaves simple, entire, with petiole c. as long; flowers solitary, axillary; calyx cup-shaped, divided nearly to base, not enlarging; corolla bowl-shaped, with rather short lobes; ovary 2-celled; fruit an ovoid berry.

1 Stems much-branched, sprawling, to 1.5m; corolla 6-10mm, whitish, with reflexed lobes; berry 10-15mm, whitish. Intrd-natd; rough ground and open places; few places on S & SE coasts of En, Guernsey *Cock's-eggs* - **S. origanifolia** (Lam.) Thell.

7. PHYSALIS L. - *Japanese-lanterns*
Subglabrous to pubescent annuals to perennials; leaves simple, entire to coarsely dentate; flowers solitary, axillary; calyx bell-shaped, 5-lobed, the calyx-tube later enlarging to enclose fruit; corolla broadly bell-shaped to funnel-shaped, shallowly to rather deeply lobed; ovary 2-celled; fruit a globose berry.

1 Fruiting calyx and berry red to orange; corolla whitish. Rather sparsely pubescent, erect, rhizomatous perennial to 60cm; fruit not filling calyx. Intrd-natd; waste land, roadsides and in shrubberies; scattered in En and Wa *Japanese-lantern* - **P. alkekengi** L.
1 Fruiting calyx green to yellowish-green; berry yellow, green or purple; corolla yellowish 2
  2 Sparsely pubescent; leaves cuneate to ± rounded at base. Erect annual to 60cm; fruit green to purple, completely filling calyx. Intrd-casual; mostly as wool-alien; scattered in En and Wa
Tomatillo - **P. ixocarpa** Hornem.
  2 Densely pubescent; leaves cordate at base. Erect, rhizomatous perennial to 1m; fruit yellow, not filling calyx. Intrd-natd; casual on tips, natd in Herts; occasional in S Br and E Ir
Cape-gooseberry - **P. peruviana** L.

8. CAPSICUM L. - *Sweet Pepper*
Glabrous annuals; leaves simple, entire; flowers solitary, axillary; calyx bell-shaped, shallowly toothed, slightly enlarging; corolla star-shaped, deeply lobed; ovary 2-3(5)-celled; fruit a rather dry ovoid berry, with large cavities when mature.

1 Stems erect, to 60cm; corolla white to purplish, 2-3cm across; fruit green, yellow or red, (1)3-15(25)cm. Intrd-casual; tips and sewerage works; scattered in S En *Sweet Pepper* - **C. annuum** L.

9. LYCOPERSICON Mill. - *Tomato*
Glandular-pubescent annuals; leaves pinnate with mixed large and small leaflets; flowers in leaf-opposed cymes; calyx star-shaped, lobed nearly to base, slightly enlarged and lobes reflexed at fruiting; corolla star-shaped, deeply lobed; ovary 2-3(5)-celled; fruit a succulent, depressed-globose to

9. LYCOPERSICON

globose berry.

1 Stems erect to decumbent or scrambling, to 2m; corolla yellow, 18-25mm across; fruit usually red, rarely yellow to orange, 2-10cm across. Intrd-natd; on tips and in sewerage works and waste places; frequent throughout BI **Tomato - L. esculentum** Mill.

10. SOLANUM L. - *Nightshades*
Annual to perennial herbs or shrubs; leaves simple and entire to pinnate; flowers in axillary or leaf-opposed cymes or solitary; calyx star- to cup-shaped, usually deeply lobed; corolla star-shaped, deeply to scarcely lobed; ovary 2(-4)-celled; fruit a succulent to dry berry.

1 Stems and leaves with strong spines 2
1 Spines 0 4
  2 Corolla yellow; one anther longer than 4 others. Similar in appearance to *S. sisymbriifolium* but stems to 60cm. Intrd-casual; in arable fields, tips and waste places; scattered in En, Wa and CI **Buffalo-bur - S. rostratum** Dunal
  2 Corolla whitish to bluish-purple; 5 anthers of equal length 3
3 Annual; berry red; leaves lobed mostly >1/2 way to midrib, with toothed or lobed lobes. Stems erect, to 1m, with branched and glandular simple hairs. Intrd-casual; waste places and cultivated ground; scattered in S Br **Red Buffalo-bur - S. sisymbriifolium** Lam.
3 Rhizomatous perennial; berry yellow; leaves lobed <1/2 way to midrib, with ± entire lobes. Stems erect, to 80(100)cm, with branched hairs. Intrd-casual; tips and waste places; S En
Horse-nettle - S. carolinense L.
  4 Perennials with stems ± woody below 5
  4 Annuals to perennials with entirely herbaceous stems 7
5 Stems scrambling to procumbent; many inflorescences >10-flowered; at least some leaves with 2 small leaflets at base. Stems to 3(7)m, glabrous to pubescent. Native; walls, hedges, woods, ditches, fens, pondsides, rough ground and shingle beaches; common
Bittersweet - S. dulcamara L.
5 Stems erect to spreading; 0 or few inflorescences >10-flowered; leaves simple, entire to laciniate 6
  6 Corolla purple; fruits yellow to orange; plant subglabrous. Shrub to 1(2)m. Intrd-natd; rough ground, tips and maritime sand, mostly casual; W Cornwall, S Somerset and CI
Kangaroo-apple - S. laciniatum Aiton
  6 Corolla white; fruits purplish-black; plant appressed-pubescent. Stems to 1.6m, woody at least below. Intrd-natd; on rough ground; Guernsey, Jersey, London area, very rare casual elsewhere
Tall Nightshade - S. chenopodioides Lam.
7 Perennial with subterranean stem-tubers; leaves pinnate 8
7 Annual; leaves entire to deeply pinnately lobed 9
  8 Leaves sparsely pubescent on lowerside; stems green to slightly

114. SOLANACEAE

      purple-tinged; corolla white to purple. Stems erect to decumbent, to 1m. Intrd-natd; casual and often persistent on tips, waste ground and in fields; scattered through BI **Potato - S. tuberosum** L.
8 Leaves grey-tomentose on lowerside; stems strongly purple-blotched; corolla purplish-violet. Differs from *S. tuberosum* in stems to 1(2)m, strongly purple-blotched. Intrd-surv; weed in flower-beds and by paths; Reading (Berks)
                                ***Purple Potato* - S. vernei** Bitter & Wittm.
9 Leaves pinnately lobed >3/4 way to base. Diffusely branched, sparsely pubescent annual to 60cm. Intrd-natd; cultivated and rough ground; scattered in En and Wa, rarely natd
                              ***Small Nightshade* - S. triflorum** Nutt.
9 Leaves entire to toothed <1/2 way to base   10
  10 Anthers brownish-yellow; stems scrambling, with weak spine-like outgrowths. Differs from *S. nigrum* in stems usually scrambling, sparsely pubescent, to 1m. Intrd-natd; on tips, sewerage farms and fields spread with sludge; S En, natd in Herts   **Garden Huckleberry - S. scabrum** Mill.
  10 Anthers bright yellow; stems erect to decumbent, without spine-like outgrowths (but sometimes with dentate angles)   11
11 Ripe berries yellow to red. Habit of *S. nigrum*. Intrd-natd; casual from wool, birdseed and oilseed, very rarely natd
(***Red Nightshade* - S. villosum** Mill.)   12
11 Ripe berries green, black or brownish-purple   13
  12 Many glandular, patent hairs present; stems with rounded, entire ridges. Infrequent casual in S Br, natd in Nottingham (Notts)   **S. villosum ssp. villosum**
  12 Hairs eglandular, mostly appressed, often sparse; stems with angled, slightly dentate ridges. Infrequent casual in S Br
                            S. villosum ssp. miniatum (Willd.) Edmonds
13 Plant without gland-tipped hairs. Sparsely to densely pubescent, erect to decumbent annual to 70cm. Native; waste and cultivated ground; common in CI, most of En and S Wa, sparse elsewhere
                        ***Black Nightshade* - S. nigrum** L. **ssp. nigrum**
13 Plant with many gland-tipped hairs   14
  14 Calyx not enlarging in fruit, with obtuse teeth; berries usually black, sometimes green, without groups of stone-cells in flesh. Intrd-natd; waste and cultivated ground, locally natd; sporadic in En, mainly SE   **S. nigrum ssp. schultesii** (Opiz) Wessely
  14 Calyx enlarging in fruit, with acute teeth; berries green to purplish-brown, with ≥2 groups of stone-cells in flesh   15
15 Calyx-lobes ≥3mm in flower, >5mm in fruit, usually at least as long as berry; petals 5-7mm wide; fruits with >50 seeds. Annual with erect to decumbent stems to 2m, to 60cm tall. Intrd-natd; tips and waste ground in S En, natd in S Essex
                      ***Leafy-fruited Nightshade* - S. sarachoides** Sendtn.
15 Calyx-lobes ≤2mm in flower, <4mm in fruit, usually shorter than berry; petals 2-4mm wide; fruits with <30 seeds. Annual with

10. SOLANUM

decumbent stems to 1(2)m, to 40cm tall. Intrd-natd; cultivated and waste ground; C & S En and CI, casual elsewhere
Green Nightshade - **S. physalifolium** Rusby

S. nigrum x *S. physalifolium* = *S. x procurrens* A.C. Leslie occurs with the parents in cultivated ground in S En; the black berries have 0-few seeds.

11. DATURA L. - *Thorn-apples*
Glabrous to sparsely pubescent annuals; leaves simple, coarsely toothed to lobed; flowers solitary, axillary; calyx tubular, with 5 teeth; corolla trumpet-shaped, very shallowly lobed; ovary 2-celled towards apex, 4-celled towards base; fruit a usually spiny capsule dehiscing by 4 valves.

1 Calyx 3-5cm, with teeth (3)5-10mm; corolla 5-10cm, white (rarely purple); capsule (2.5)3.5-7cm incl. spines, with slender spines 2-15mm (rarely spineless). Stems erect, to 1(1.5)m. Intrd-natd; on tips and waste and cultivated ground, mostly casual; sporadic ± throughout BI *Thorn-apple* - **D. stramonium** L.
1 Calyx 2.5-4cm, with teeth 3-5mm; corolla 4-6cm, white; capsule 5-8cm incl. spines, with stout spines 1-3cm, some broad-based and >2cm. Intrd-casual; on tips and in fields; occasional in En and Wa *Angels'-trumpets* - **D. ferox** L.

12. NICOTIANA L. - *Tobaccos*
Glandular pubescent, ± sticky annuals; leaves simple, ± entire; flowers in terminal panicle- or raceme-like cymes; calyx tubular, with 5 unequal lobes c.1/4-2/3 total calyx length; corolla tubular with expanded limb to trumpet-shaped; ovary 2-celled; fruit a capsule with 2 short valves, each 2-lobed at apex.

1 Petioles not winged. Erect annual to 1.5m; corolla-limb 9-16mm across, scarcely lobed, greenish-yellow. Intrd-casual; occasional on tips; scattered in S En *Wild Tobacco* - **N. rustica** L.
1 Petioles broadly winged, the wing clasping stem 2
   2 Inflorescence a cymose panicle; corolla-tube with distal wider part c.1/4 total length. Erect annual to 2(3)m; corolla-limb 20-30mm across, lobed c.1/2 way, whitish to dingy red. Intrd-casual; on tips and as relic, once much commoner; scattered in S En *Tobacco* - **N. tabacum** L.
   2 Inflorescence a simple raceme-like cyme; corolla-tube with distal wider part ≤1/8 total length 3
3 Corolla-limb red to purple on upperside; filaments inserted in basal 1/2 of corolla-tube. Erect annual to 1m; corolla-limb 25-40mm across. Intrd-casual; grown and found as for *N. alata*
Red Tobacco - **N. forgetiana** Hemsl.
3 Corolla-limb white on upperside; filaments inserted in apical 1/2 of corolla-tube. Erect annual to 1.5m; corolla-limb 35-60mm across. Intrd-casual; much grown for ornament and frequent on tips and

rough ground; scattered in S Br   *Sweet Tobacco* - **N. alata** Link & Otto

N. alata x *N. forgetiana* = **N. x sanderae** W. Watson is grown in gardens like its parents and similarly occurs as a casual; it is intermediate in all characters, with a range of flower colours.

13. PETUNIA Juss. - *Petunia*
Glandular-pubescent, ± sticky annuals; leaves simple, ± entire; flowers solitary, axillary; calyx divided ≥3/4 way to base into 5 narrow lobes; corolla trumpet-shaped; ovary 2-celled; fruit a capsule with 2 valves, each slightly notched.

1 Stems procumbent to erect, to 60cm; corolla 5-12cm, often equally wide distally, white to red, mauve or purple. Intrd-casual; on tips and rough ground, sometimes self-sown; S Br (*P. axillaris* (Lam.) Britton, Sterns & Poggenb. x *P. integrifolia* (Hook.) Schinz & Thell.)
Petunia - **P. x hybrida** (Hook.) Vilm.

115. CONVOLVULACEAE - *Bindweed family*

Annuals to herbaceous perennials, with twining or procumbent stems; distinguishable by the large funnel- to trumpet-shaped corolla, usually 2-carpellary ovary with usually 2 stigmas and 2 ovules per cell, and often twining stems.

1 Bracteoles ovate, often pouched, partly or wholly obscuring sepals **2. CALYSTEGIA**
1 Bracteoles linear, distant from sepals 2
  2 Stigmas linear; sepals obtuse to retuse, rarely acuminate
                                                    1. CONVOLVULUS
  2 Stigmas ± globose; sepals acute to acuminate   **3. IPOMOEA**

1. CONVOLVULUS L. - *Field Bindweed*
Rhizomatous perennials with trailing or climbing stems; leaves triangular or ovate-oblong to linear, hastate to sagittate at base; flowers 1-few, with 2 linear bracteoles some way below; corolla usually scarcely divided, much longer than calyx; style 1; stigmas 2, linear; ovary and fruit unlobed.

1 Stems to 1(2)m, often much less; corolla 10-25mm, white, pink or striped pink-and-white, rarely deeply 5-lobed. Native; waste and cultivated ground, waysides, banks and rough or short grassland; common   *Field Bindweed* - **C. arvensis** L.

2. CALYSTEGIA R. Br. - *Bindweeds*
Rhizomatous perennials with trailing or climbing stems; leaves triangular and sagittate, or reniform; flowers usually 1, with 2 ovate, often ± pouched bracteoles partly or wholly concealing calyx; corolla usually scarcely

2. CALYSTEGIA

divided, rarely deeply 5-lobed, much longer than calyx; style 1; stigmas 2, ellipsoid; ovary and fruit not lobed.

1 Leaves reniform; stems not or weakly climbing. Stems trailing or weakly climbing, to 1m; corolla pink with 5 white stripes, yellowish in centre. Native; on sand-dunes and sometimes shingle, by sea; coasts of BI N to C Sc **Sea Bindweed - C. soldanella** (L.) R. Br.
1 Leaves triangular and sagittate; stems usually strongly climbing 2
   2 Bracteoles 10-18mm wide when flattened, not or little overlapping at edges, not or little obscuring sepals in lateral view; ratio of midrib- to-midrib to edge-to-edge distances in natural condition 0.4-1.1. Stems strongly climbing, to 2(3)m. Native; hedges, ditches, fens, marshes, by water and on rough and waste ground (*Hedge Bindweed* - **C. sepium** (L.) R. Br.) 3
   2 Bracteoles 18-45mm wide when flattened, strongly overlapping at edges, completely or nearly obscuring sepals in lateral view; ratio of midrib-to-midrib to edge-to-edge distance in natural condition 1.1-2.2 4
3 Glabrous; corolla 3-5(5.5)cm, white, or pink with 5 white stripes; stamens 15-25mm. Throughout BI, local in N **C. sepium ssp. sepium**
3 Stems, petioles and pedicels sparsely short-pubescent; corolla 4-5.5cm, pink with 5 white stripes; stamens 17-25mm. Local near W coast of Br and Ir, CI, occasionally intrd elsewhere
                                              C. sepium ssp. roseata Brummitt
   4 Corolla pink or pink-and-white striped; pedicels shortly pubescent (often only sparsely or partly so), usually with narrow wavy-edged wing near apex. Stems strongly climbing to 3(5)m. Intrd-natd; in hedges and on rough and waste ground; scattered ± throughout BI *Hairy Bindweed* - **C. pulchra** Brummitt & Heywood
   4 Corolla white (sometimes narrowly pink-striped on outside only); pedicels glabrous; without wing. Stems strongly climbing, to 3(5)m. Intrd-natd; as for *C. pulchra*, but much commoner
                                     Large Bindweed - **C. silvatica** (Kit.) Griseb.

C. sepium x *C. silvatica* = **C. x lucana** (Ten.) G. Don is intermediate between the parents and highly fertile; scattered over C & S Br and CI, frequent in SE En, rare further N. *C. pulchra* also rarely hybridises with *C. sepium* and *C. silvatica*.

3. IPOMOEA L. - *Morning-glories*

Annuals with usually strongly climbing stems; leaves ovate, cordate, entire to deeply 3-lobed; flowers 1-few, with 2 linear bracteoles some way below; corolla scarcely to shallowly lobed, much longer than calyx, funnel- to trumpet-shaped; style 1; stigma 1, 2-3-lobed; ovary and fruit not lobed.

1 Flower-stalks shorter than petioles, with 0-few patent to forwardly directed hairs; corollas usually white (rarely purple). Stems to 2m; leaves entire or rather shallowly 3-lobed. Intrd-casual; a rather

constant casual from soyabean waste; very local and sporadic in
S En   *White Morning-glory* - **I. lacunosa** L.
1 Flower-stalks longer than petioles, with reflexed hairs; corollas
  usually blue, fading or drying to pinkish-purple (rarely white)   2
    2 Corolla <5cm; sepals abruptly long-acuminate; most leaves
      deeply 3-lobed. Stems to 2m. Intrd-casual; occurrence as for
      I. purpurea but less common
                                 Ivy-leaved Morning-glory - **I. hederacea** Jacq.
    2 Corolla ≥5cm; sepals acute; most leaves entire. Stems to 3m.
      Intrd-casual; tips and waste places; sporadic in S En
                                 Common Morning-glory - **I. purpurea** Roth

116. CUSCUTACEAE - *Dodder family*

Herbaceous, annual to perennial parasites; the very thin, chlorophyll-less, rootless stems with haustoria and small globose inflorescences are unique.

1. CUSCUTA L. - *Dodders*

1 Stigmas capitate (style ending in distinct knob); stems yellowish.
  Intrd-natd; on a range of cultivated plants, especially carrot;
  scattered in En and Wa   *Yellow Dodder* - **C. campestris** Yunck.
1 Stigmas linear (style scarcely thickened distally); stems reddish   2
    2 Styles + stigmas shorter than ovary; corolla-scales bifid,
      minute, or 0; sepals obtuse; stamens included. Native; on a range
      of hosts, mostly *Urtica dioica*, often near water; scattered and
      rare in En N to Northants   *Greater Dodder* - **C. europaea** L.
    2 Styles + stigmas longer than ovary; corolla-scales not bifid,
      reaching to base of filaments; sepals acute; stamens ± exserted.
      Native; on a wide range of hosts, most commonly *Ulex* spp. and
      Calluna on heathland; frequent in S Br and CI, scattered
      elsewhere in BI N to C Sc   *Dodder* - **C. epithymum** (L.) L.

117. MENYANTHACEAE - *Bogbean family*

Glabrous, stoloniferous, aquatic or semi-aquatic perennials with easily recognized showy white to pink or yellow flowers with 5 fringed petals.

1 Leaves ternate; corolla white to pink   **1. MENYANTHES**
1 Leaves simple; corolla yellow   **2. NYMPHOIDES**

1. MENYANTHES L. - *Bogbean*
Leaves ternate, all alternate, held above water level; flowers monomorphic, in erect racemes; corolla white to pink, with many long fringes on inner side of lobes; capsule dehiscing by 2 valves.

1 Stems procumbent or floating, to 1.5m; racemes ≤30cm; flowers
1.5-2cm across. Native; in shallow water, bogs and fens; local but
throughout Br and Ir *Bogbean* - **M. trifoliata** L.

2. NYMPHOIDES Ség. - *Fringed Water-lily*
Leaves simple, alternate on vegetative stems, opposite on flowering stems, cordate, floating on water; flowers dimorphic, in small axillary groups, on long pedicels; corolla yellow, with fringes on margins of lobes; capsule dehiscing irregularly. The flowers are heterostylous, as in *Primula*, with pin and thrum morphs.

1 Stems floating, to 1.5m; pedicels ≤8cm; flowers 3-4cm across.
Possibly native (presumably only where dimorphic); in ponds
and slow rivers; fens of E Anglia and Thames basin, natd
elsewhere *Fringed Water-lily* - **N. peltata** Kuntze

118. POLEMONIACEAE - *Jacob's-ladder family*

Herbaceous perennials; told from other families with 5 fused sepals and petals and actinomorphic, hypogynous flowers (except the distinctive Diapensiaceae) by the 3-celled ovary and 3 stigmas.

1 Leaves pinnate; corolla-tube much shorter than -lobes, with long-
exserted stamens **1. POLEMONIUM**
1 Leaves simple; corolla-tube longer than -lobes, with anthers at
apex of corolla-tube **2. PHLOX**

1. POLEMONIUM L. - *Jacob's-ladder*
Leaves pinnate, petiolate, with 6-15 pairs of entire leaflets; corolla-lobes much longer than -tube; stamens ± equal-lengthed, well exserted, with hairy base to filaments.

1 Stems erect, to 1m; flowers blue (or white), 2-3cm across. Native;
limestone grassland, scree, rock-ledges, wood-borders; locally
frequent in Peak District, Yorkshire Dales, 1 place in S Northumb,
garden escape elsewhere in Br *Jacob's-ladder* - **P. caeruleum** L.

2. PHLOX L. - *Phlox*
Leaves simple, ± sessile, entire; corolla-lobes shorter than tube, patent; stamens with anthers at different heights, the longest at mouth of corolla-tube, glabrous.

1 Stems erect, woody at base, to 1.5m; flowers white to pink, purple
or mauve, 2-3cm across. Intrd-natd; on rough and waste ground;
sporadic in En *Phlox* - **P. paniculata** L.

119. HYDROPHYLLACEAE - *Phacelia family*

Annuals; similar to Boraginaceae in its flowers in scorpioid cymes, but with 2-valved capsule and deeply divided style.

1. PHACELIA Juss. - *Phacelia*

1 Pubescent, erect to ascending annual to 70(100)cm; corolla blue or pale mauve, 6-10mm. Intrd-natd; casual on tips, waste ground and among crops and new grass, rarely persistent; very scattered in En and Wa **Phacelia - P. tanacetifolia** Benth.

120. BORAGINACEAE - *Borage family*

Annual to perennial herbs, often hispid or scabrid; leaves alternate, simple, entire or ± so; like Verbenaceae and Lamiaceae in its 4-celled ovary with 1 ovule per cell and a fruit of 4 nutlets, but differing from both in usually alternate leaves, and spiralled cymose inflorescence.

Much value is placed in many keys on the presence or absence of folds, scales or bands of hairs at the throat of the corolla-tube; these are often difficult to make out and very little use is made of them here. When they are well developed they may meet in the centre or around the style and the corolla-tube appears closed.

1 Style bifid at apex; flowers distinctly zygomorphic, with unequal stamens and corolla-lobes **2. ECHIUM**
1 Style simple; flowers actinomorphic to weakly zygomorphic, with equal stamens and ± equal corolla-lobes 2
  2 All anthers completely exserted 3
  2 All anthers completely included or only tips exserted 4
3 Annual; filaments glabrous; anthers longer than filaments; calyx divided nearly to base **9. BORAGO**
3 Rhizomatous perennial; filaments pubescent; anthers shorter than filaments; calyx divided c.1/2 way **10. TRACHYSTEMON**
  4 Calyx-lobes with some small teeth between 5 main ones, enlarging greatly in fruit and forming 2-lipped covering **14. ASPERUGO**
  4 Calyx-lobes 5, entire, not or slightly enlarging in fruit 5
5 Nutlets with hooked or barbed bristles 6
5 Nutlets smooth to warty, ridged or pubescent 7
  6 Flowers and fruits all or mostly without bract; nutlets >4.5mm **18. CYNOGLOSSUM**
  6 Flowers and fruits all or mostly with bract; nutlets <4.5mm **16. LAPPULA**
7 Plant glabrous, often very glaucous **11. MERTENSIA**
7 Plant bristly to (sometimes appressed-)pubescent, not or scarcely glaucous 8

120. BORAGINACEAE

8	At least lower leaves opposite	**13. PLAGIOBOTHRYS**
8	All leaves alternate (rarely uppermost pair opposite in *Myosotis*)	9
9	Open flowers pendent, with exserted stigma	**4. SYMPHYTUM**
9	Open flowers erect, with stigma included or at throat of corolla-tube	10
10	Ripe nutlets smooth (sometimes pubescent or with keel round edge)	11
10	Ripe nutlets tuberculate to strongly warty and/or with variously branched ridges	15
11	Basal and all or most stem-leaves petiolate	**17. OMPHALODES**
11	All or most stem-leaves sessile	12
12	Corolla-tube plus -lobes >10mm	13
12	Corolla-tube plus -lobes <10mm	14
13	Calyx-lobes divided nearly to base	**1. LITHOSPERMUM**
13	Calyx-lobes fused ≥1/2 way	**3. PULMONARIA**
14	Corolla-tube longer than -lobes; calyx-hairs straight; throat of corolla partially closed by hairy folds	**1. LITHOSPERMUM**
14	Corolla-tube usually shorter than -lobes, if longer then calyx-hairs hooked; throat of corolla closed by glabrous or papillate scales	**15. MYOSOTIS**
15	Basal leaves strongly cordate at base	**5. BRUNNERA**
15	Basal leaves gradually to abruptly cuneate at base	16
16	Leaves ovate to obovate, at least most basal ones >5cm wide	17
16	Leaves lanceolate to oblanceolate or linear-oblong, <5cm wide	18
17	Corolla-lobes rounded; corolla-scales closing throat of corolla-tube; nutlets stalked	**8. PENTAGLOTTIS**
17	Corolla-lobes acute; corolla-scales not closing throat of corolla-tube; nutlets sessile	**9. BORAGO**
18	Nutlets tuberculate to strongly warty, not ridged apart from marginal keel, without collar-like base	19
18	Nutlets tuberculate and with strong branching ridges, with distinct collar-like base at point of attachment	20
19	Corolla yellow to orange; nutlets coarsely warty	**12. AMSINCKIA**
19	Corolla white to bluish-purple; nutlets minutely tuberculate	**1. LITHOSPERMUM**
20	Corolla-tube longer than -limb	**6. ANCHUSA**
20	Corolla-tube shorter than -limb	**7. CYNOGLOTTIS**

1. LITHOSPERMUM L. - *Gromwells*

Pubescent to hispid annuals to perennials; leaves lanceolate to narrowly elliptic, narrowed to base; flowers solitary in leaf-axils, congested at flowering, distant at fruiting; calyx divided ± to base; corolla actinomorphic, with narrow tube at least as long as expanded limb, purplish-blue or white to yellowish; stamens equal, included; style simple, included; nutlets smooth to warty, without collar-like base.

1 Corolla purplish-blue, 11-16mm. Rhizomatous perennial with procumbent sterile and erect flowering stems to 60cm. Native; scrub and wood-margins on chalk and limestone; very local in

SW En, S & N Wa   *Purple Gromwell* - **L. purpureocaeruleum** L.
1   Corolla usually white to yellowish, <10mm   2
   2   Leaves with lateral veins apparent on lowerside; nutlets white, smooth. Shortly rhizomatous perennial with erect stems to (80)100cm. Native; grassy and bushy places, hedgerows and wood-borders mostly on basic soils; locally frequent in En, very local in Wa and Ir   *Common Gromwell* - **L. officinale** L.
   2   Leaves without lateral veins apparent; nutlets brown, tuberculate. Erect annual to 50(80)cm. Native; arable fields, rough ground and open grassy places; locally frequent in En, very scattered and often casual elsewhere   *Field Gromwell* - **L. arvense** L.

2. ECHIUM L. - *Viper's-buglosses*

Plants hispid, monocarpic (usually biennials); leaves lanceolate to oblanceolate or the lower ovate, tapered to base; cymes terminal and lateral, forming compound narrow panicle; calyx divided nearly to base; corolla zygomorphic, pink, purple or blue with tube shorter than limb, the latter with unequal lobes; stamens unequal, at least some exserted; apex of style bifid and exserted; nutlets without collar-like base, warty to ridged.

1   Monocarpic shrubs with unbranched woody stem to 75 x 3-5cm and with terminal panicle up to 3.5m; leaves up to 50cm, crowded below panicle. Stems erect, to 4m. Intrd-natd; self-sown garden escapes on rough ground in CI, Man, Caerns, W Cornwall and Scillies
                 Giant Viper's-bugloss - **E. pininana** Webb & Berthel.
1   Stems herbaceous or ± so, ≤1m (incl. panicle) x 1cm; leaves scattered up stem, the panicle not sharply delimited from rest of stem   2
   2   Corolla pubescent on veins and margins only; usually 2 stamens exserted. Stems erect to ascending, to 75cm. Native; disturbed or open grassy, sandy ground near sea; frequent in Jersey, very local in Scillies and W Cornwall, rare casual elsewhere in S En   *Purple Viper's-bugloss* - **E. plantagineum** L.
   2   Corolla ± uniformly pubescent on outside; usually 3-5 stamens exserted   3
3   Corolla blue when fully open; stems stiffly erect, with narrow rather dense inflorescence with bracts scarcely exceeding cymes. Stems usually erect, to 1m. Native; open grassy places, cliffs, dunes, shingle, rough ground, usually on light, often calcareous soils; locally frequent in BI, especially S & E En   *Viper's-bugloss* - **E. vulgare** L.
3   Corolla pinkish-violet when fully open; stems ascending, with rather loose inflorescence with conspicuous leafy bracts. Stems ascending, to 75cm. Intrd-natd; waste ground; Barry Docks (Glam)
                 Lax Viper's-bugloss - **E. rosulatum** Lange

3. PULMONARIA L. - *Lungworts*

Pubescent to slightly hispid tufted perennials; leaves lanceolate to ovate, abruptly to gradually contracted at base; flowers in rather dense terminal clusters of cymes; calyx divided <1/2 way to base; corolla actinomorphic,

3. PULMONARIA

blue, red or purple, with tube slightly shorter to slightly longer than limb; stamens equal, included; style simple, included; nutlets smooth, sparsely pubescent, with collar-like base.

Leaf-characters must be observed on basal leaves that develop during the flowering season and reach maturity during the summer. There are 5 conspicuous hair-tufts at the throat of the corolla; pubescence of inside of corolla-tube refers to region below these tufts. The flowers are heterostylous, as in *Primula*, with pin and thrum morphs.

1 Basal leaves developing at flowering cordate to broadly cuneate at base, abruptly contracted into petiole 2
1 Basal leaves developing at flowering gradually cuneate at base, tapered into petiole 4
  2 Corolla bright red when open; inside of corolla-tube pubescent below hair-tufts; basal leaves abruptly cuneate to rounded at base. Stems erect, to 50cm; basal leaves usually unspotted. Intrd-natd; in grassy places, hedges and scrub; scattered in N En, C & S Sc **Red Lungwort - P. rubra** Schott
  2 Corolla reddish- to bluish-violet when open; inside of corolla-tube glabrous below hair-tufts; basal leaves cordate at base 3
3 Basal leaves with large white spots. Stems erect, to 30cm. Intrd-natd; on banks and in scrub, woods and rough ground; scattered throughout Br **Lungwort - P. officinalis** L.
3 Basal leaves unspotted or with faint pale green spots. Stems erect, to 30cm. Native; woods and woodland rides and clearings; 3 sites in E Suffolk **Suffolk Lungwort - P. obscura** Dumort.
  4 Stalked glands 0 to very sparse in inflorescence; leaves hispid on upperside, the basal ones white-spotted; flowers pin or thrum. Stems erect, to 40cm. Native; woods and scrub; extremely local in Dorset, S Hants and Wight
      Narrow-leaved Lungwort - P. longifolia (Bastard) Boreau
  4 Stalked glands very frequent in inflorescence; leaves softly pubescent on upperside, unspotted; flowers always thrum. Stems ascending to erect, to 30cm. Intrd-natd; shady places; very scattered in S & C En, S & C Sc
      Mawson's Lungwort - P. 'Mawson's Blue'

4. SYMPHYTUM L. - *Comfreys*

Hispid perennials; leaves ovate-elliptic, subcordate to broadly cuneate at base, the basal long-petiolate; flowers in rather dense cymes forming terminal panicle; calyx lobed c.1/5-9/10 way to base; corolla actinomorphic, various colours, with limb little wider than tube and c. as long, the limb with short lobes; stamens equal, included, alternating with 5 long corolla-scales; style simple, exserted; nutlets smooth to granulate, sometimes also ridged, with collar-like base.

1 Plant with decumbent to procumbent leafy stolons 2
1 Plant without stolons 3

2   Corolla predominantly blue or pink when open; larger flowering
stems branched. Ascending to erect flowering stems to 50(100)cm
and procumbent to decumbent stolons arising from rhizomes.
Intrd-natd; in hedges and woodland; very scattered in Br,
Guernsey (*S. grandiflorum* x ?*S. x uplandicum*)
Hidcote Comfrey - **S. 'Hidcote Blue'**

2   Corolla pale yellow when open, often flushed reddish on outside;
flowering stems unbranched. Habit of *S.* 'Hidcote Blue' but
flowering stems to 40cm. Intrd-natd; in woods and hedges;
scattered in C & S Br, rarely N to C Sc
Creeping Comfrey - **S. grandiflorum** DC.

3   Nutlets ± smooth, shining; stem-leaves strongly decurrent, forming
wings on stem extending down for >1 internode. Stems erect, well-
branched, to 1.5m, from thick, vertical root; corolla purplish or
cream, rarely white. Native; by streams and rivers, in fens and
marshy places, also roadsides and rough ground; locally frequent
in BI
Common Comfrey - **S. officinale** L.

3   Nutlets minutely tuberculate, dull; stem-leaves not to moderately
decurrent, the wings rarely extending for >1 internode   4
    4   Corolla pink, purple or blue   5
    4   Corolla pale yellow to white   7

5   Calyx divided <1/2 way to base. Stems erect, to 60cm, from thick,
branched roots; corolla blue. Intrd-natd; in hedgerows and other
shady places; very scattered in SE En, Flints
Caucasian Comfrey - **S. caucasicum** M. Bieb.

5   Calyx divided ≥1/2 way to base   6
    6   Calyx-hairs almost all broad-based whitish bristles, with some
much finer and smaller hairs; upper stem-leaves shortly petiolate,
not decurrent or clasping stem. Habit as in *S. officinale*; corolla sky
blue when open. Intrd-natd; in rough and waste ground; very
scattered over Br, Co Sligo   *Rough Comfrey* - **S. asperum** Lepech.
    6   Calyx-hairs a mixture of broad-based bristles and finer and
smaller hairs and all intermediates; upper stem-leaves sessile,
shortly decurrent or clasping stem. Habit as in *S. officinale*;
corolla blue to violet or purplish when open. Intrd-natd;
roadsides, rough and damp ground, wood-borders; frequent
over most of BI (*S. officinale* x *S. asperum*)
Russian Comfrey - **S. x uplandicum** Nyman

7   Corolla-scales exserted for >1mm. Stems simple, erect, to 50cm, from
rhizome with subglobose tubers; corolla pale yellow. Intrd-natd; in
woods and by streams; very scattered in C & S Br
Bulbous Comfrey - **S. bulbosum** K.F. Schimper

7   Corolla-scales included   8
    8   Calyx divided <1/2 way to base; corolla pure white. Stems erect,
little-branched, to 70cm, from thick branched roots. Intrd-natd;
in hedgerows and other shady places; frequent in E & S En
and C Sc, scattered elsewhere   *White Comfrey* - **S. orientale** L.
    8   Calyx divided >1/2 way to base; corolla yellow to pale yellow   9

9 Stems not or little branched; middle and upper stem-leaves sessile, the upper shortly decurrent; rhizomes with swollen tubers present. Stems erect, little or not branched, to 60cm, from rhizomes with thick swollen regions. Native; damp woods, ditches and river banks; frequent in lowland Sc, scattered in En, Wa and Ir
Tuberous Comfrey - **S. tuberosum** L.
9 Stems well-branched; middle stem-leaves petiolate, uppermost ones sessile but none decurrent; rhizomes 0. Stems erect, well-branched, to 60cm, from thick, vertical root. Intrd-natd; on hedgebank; Cambs
Crimean Comfrey - **S. tauricum** Willd.

S. x uplandicum x *S. tuberosum* occurs near the parents in very scattered localities in En and Sc. *S. asperum* x *S. caucasicum* was recorded from Midlothian in 1994 in the absence of both parents.

5. BRUNNERA Steven - *Great Forget-me-not*
Appressed-pubescent densely tufted perennials; basal and lower stem-leaves ovate-cordate, petiolate; flowers in bractless, dense cymes in terminal subcorymbose panicles; calyx divided nearly to base; corolla actinomorphic, blue, with tube shorter than limb, the latter with patent lobes; stamens equal, included; style simple, included; nutlets ridged, with collar-like base.

1 Stems erect, to 50cm, the basal leaves plus petioles often not much shorter; corolla c.3-4mm across, resembling a *Myosotis*. Intrd-natd; in woods and on rough ground and tips; scattered in Br
Great Forget-me-not - **B. macrophylla** (Adams) I.M. Johnst.

6. ANCHUSA L. - *Alkanets*
Hispid annuals to perennials; leaves lanceolate to oblanceolate, sessile or narrowed to short petiole; flowers in terminal, branched, spiralled cymes; calyx divided c.1/3 way to nearly wholly to base; corolla actinomorphic to slightly zygomorphic, blue to purple, or yellow, with tube longer than limb; stamens equal, included; style simple, included; nutlets ridged and tuberculate, with collar-like base.

1 Corolla with curved tube and 5 slightly unequal lobes. Hispid, procumbent to erect annual to 50cm. Native; weed of arable and rough ground on light, acid or calcareous soils; locally common throughout lowland BI *Bugloss* - **A. arvensis** (L.) M. Bieb.
1 Corolla with straight tube and 5 equal lobes 2
  2 Corolla yellow; calyx-lobes obtuse to rounded. Rather softly pubescent, erect perennial to 50cm. Intrd-natd; on rough ground; Phillack Towans (W Cornwall), rare escape elsewhere
Yellow Alkanet - **A. ochroleuca** M. Bieb.
  2 Corolla blue to purple; calyx-lobes acute 3
3 Calyx divided c.1/2 way to nearly to base; nutlets <5mm in longest plane. Hispid, erect perennial to 1.5m. Intrd-natd; on rough and

waste ground and tips, rarely natd; scattered over lowland Br

Alkanet - **A. officinalis** L.

3 Calyx divided nearly to base; nutlets >5mm in longest plane. Habit as for *A. officinalis*. Intrd-casual; on waste and rough ground and tips; scattered over lowland Br   *Garden Anchusa* - **A. azurea** Mill.

A. ochroleuca x *A. officinalis* = **A. x baumgartenii** (Nyman) Gusul. occurs in W Cornwall with both parents; it has pale blue and/or greyish-yellow corollas and forms backcrosses.

7. CYNOGLOTTIS (Gusul.) Vural & Kit Tan - *False Alkanet*

Hispid perennials; differ from *Anchusa* in calyx divided >1/2 way to base; corolla actinomorphic, blue, with tube much shorter than limb.

1 Plant appressed-hispid, erect, to 70cm; corolla-tube 1-2mm; corolla-limb 7-10mm across. Intrd-natd; persistent escape in rough and waste ground; very scattered in S En

False Alkanet - **C. barrelieri** (All.) Vural & Kit Tan

8. PENTAGLOTTIS Tausch - *Green Alkanet*

Hispid perennials with deep, thick roots; leaves ovate, abruptly contracted at base; flowers in terminal and lateral dense cymes; calyx divided >3/4 way to base; corolla actinomorphic, blue, with tube shorter than limb; stamens equal, included; style simple, included; nutlets ridged, with knob-like, stalked base.

1 Stems erect, to 1m; corolla 8-10mm across. Intrd-natd; in hedges and wood-borders and on rough ground; frequent over much of BI   *Green Alkanet* - **P. sempervirens** (L.) L.H. Bailey

9. BORAGO L. - *Borages*

Hispid annuals or perennials; leaves lanceolate to ovate or obovate, the lower abruptly tapered to petiole; cymes terminal, rather lax; calyx divided nearly to base; corolla actinomorphic, blue (rarely white), with lobes longer than rest of limb plus tube; stamens equal, exserted or included; style simple, included in corolla or between anthers; nutlets with collar-like base, ridged.

1 Erect annual to 60cm; calyx 7-15mm at flowering, up to 20mm in fruit; corolla with patent to reflexed lobes 7-15mm. Intrd-natd; persistent on tips, rough ground and waysides; scattered over much of BI   *Borage* - **B. officinalis** L.
1 Decumbent to ascending perennial to 60cm; calyx 4-6mm at flowering, up to 8mm in fruit; corolla with ± erect lobes 5-8mm. Intrd-natd; on heathy ground in Jethou (CI), less permanent on rough ground and by paths in very scattered places in Wa and S En

Slender Borage - **B. pygmaea** (DC.) Chater & Greuter

10. TRACHYSTEMON D. Don - *Abraham-Isaac-Jacob*

Hispid, rhizomatous perennials; leaves ovate, the basal long-petiolate and cordate to rounded at base; cymes dense, several in terminal panicle; calyx divided c.1/2 way to base; corolla actinomorphic, blue, white near base, with lobes longer than rest of limb plus tube; stamens equal, completely exposed; style simple, included between closely appressed stamens; nutlets with collar-like base, ridged.

1 Stems erect, to 40cm; corolla with patent and revolute lobes 9-12mm. Intrd-natd; on shady banks and in dampish woods; scattered in Br N to C Sc   *Abraham-Isaac-Jacob* - **T. orientalis** (L.) G. Don

11. MERTENSIA Roth - *Oysterplants*

Glabrous, usually glaucous perennials; leaves papillose, obovate or elliptic to oblanceolate, the middle and lower ones with winged petioles, the upper sessile; flowers in terminal, rather dense cymes; calyx divided ± to base; corolla actinomorphic, blue or blue and pink, pink in bud, with tube shorter than limb, the latter bell-shaped and lobed c.1/2 way to base; stamens equal, included or slightly exserted; style simple, included; nutlets slightly flattened, smooth, succulent then papery on outside.

1 Stems usually decumbent, to 60cm; corolla 4-6mm, c.6mm across. Native; on bare shingle or shingly sand by sea; local on coasts of N Ir, Man and N Br S to Denbs   *Oysterplant* - **M. maritima** (L.) Gray

12. AMSINCKIA Lehm. - *Fiddlenecks*

Hispid annuals; leaves linear to oblong or lanceolate, sessile; flowers in terminal, spiralled, bracteate or bractless cymes; calyx divided nearly to base; corolla actinomorphic, yellow to orange, with tube longer than limb; stamens equal, included; style simple, included; nutlets keeled, warty, without collar-like base.

1 Fruiting calyx (6)8-11(15)mm; corolla yellow to orange, 5-8mm, with pubescent scales at apex of tube visible from above, with stamens inserted c.1/2 way up tube or just below. Intrd-natd; on rough ground, especially on sandy soils; very scattered in En, especially NE   *Scarce Fiddleneck* - **A. lycopsoides** Lehm.
1 Fruiting calyx 5-6(9)mm; corolla yellow, 3-5mm, without hairs or scales inside, with stamens borne on upper 1/2 of tube. Intrd-natd; arable land and sandy rough ground; frequent in E En and E Sc N to E Ross, rare in C & W Br   *Common Fiddleneck* - **A. micrantha** Suksd.

13. PLAGIOBOTHRYS Fisch. & C.A. Mey. - *White Forget-me-not*

Appressed-pubescent annuals; leaves linear-oblong, the lower ones opposite, sessile; flowers in terminal, semi-bractless, spiralled cymes; calyx divided >1/2 way to base; corolla actinomorphic, white, with tube shorter than limb, the latter divided c.1/2 way into 5 rounded lobes; stamens equal, included; style simple, included; nutlets keeled, ridged, minutely tuberculate, without collar-like base.

1 Stems erect to decumbent, to 20cm, often well-branched; flowers resembling a white *Myosotis*. Intrd-casual; contaminant with grass-seed; N Sc and S En

White Forget-me-not - P. scouleri (Hook. & Arn.) I.M. Johnst.

14. ASPERUGO L. - *Madwort*

Hispid annuals; leaves lanceolate to oblanceolate, tapered to base; flowers 1-2 in leaf-axils; calyx with small teeth between 5 main ones, at fruiting much enlarged and forming 2-lipped structure around nutlets; corolla actinomorphic, purplish-blue, with tube shorter than limb; stamens equal, included; style simple, included; nutlets strongly compressed, without collar-like base, with dense low tubercles.

1 Stems procumbent to scrambling, to 60cm; corolla 1.5-3mm, inconspicuous. Intrd-natd; arable fields, waste and rough ground; very scattered over Br, especially Sc **Madwort - A. procumbens** L.

15. MYOSOTIS L. - *Forget-me-nots*

Pubescent annuals to perennials; leaves mostly narrowly oblong to oblanceolate, the basal ones tapered to petiole-like base, the upper ones sessile; flowers in terminal spiralled cymes; calyx divided c.1/4 way to nearly to base; corolla actinomorphic, blue or sometimes white or yellow, usually pink in bud, with tube shorter to longer than limb, with limb divided ± to base; stamens equal, included; style simple, usually included; nutlets slightly compressed, with distinct keel, smooth.

Maximum corolla-size is of diagnostic value, but larger-flowered spp. often produce flowers with unusually small corollas.

1 Calyx with all hairs ± straight and closely appressed 2
1 Calyx with some hairs patent and distally hooked or at least strongly curved 7
  2 Nutlets ≤1mm, shining olive-brown; annual without sterile shoots; calyx divided <1/2 way to base at flowering. Erect to decumbent, appressed-pubescent annual to 20cm. Native; damp grassland and by pond; very local in Jersey

Jersey Forget-me-not - M. sicula Guss.
  2 Nutlets ≥1.2mm, mid-brown to black, shining or not, or if <1.2mm then plants with axillary stolons; calyx often divided ≥1/2 way to base at flowering 3
3 Style longer than calyx-tube and often exceeding calyx-lobes at flowering; calyx divided <1/2 way to base at flowering, with broad teeth forming equilateral triangle. Erect to ascending, appressed or sometimes partly patent-pubescent perennial to 70cm, with rhizomes and/or stolons. Native; by or in edges of ponds and rivers, in damp fields; common throughout most of BI

Water Forget-me-not - M. scorpioides L.

3 Style shorter than calyx-tube at flowering; calyx often divided ≥1/2
way to base at flowering, with narrow teeth forming isosceles
triangle with base shorter than sides 4
    4 Lower part of stem with ± patent hairs 5
    4 Stem with only appressed hairs 6
5 Pedicels eventually 2.5-5x as long as fruiting calyx; nutlets <2mm.
Erect to ascending annual to perennial to 50cm, with stolons, with
hairs appressed above but patent below. Native; wet, often acidic
places by streams and in pools and bogs; common in N & W BI, rare
or absent in C & E En *Creeping Forget-me-not* - **M. secunda** Al. Murray
5 Pedicels <2x as long as fruiting calyx; nutlets usually >2mm. Erect,
± rhizomatous, patent-pubescent perennial to 25cm. Native;
mountain slopes and ledges, 700-1200m; very local in N En
and C Sc   *Alpine Forget-me-not* - **M. alpestris** F.W. Schmidt
    6 Stolons produced from lower nodes; leaves rarely >3x as long
as wide. Erect, appressed-pubescent perennial to 20(30)cm.
Native; wet flushes and streamsides on hills; local in N En and
S Sc   *Pale Forget-me-not* - **M. stolonifera** (DC.) Leresche & Levier
    6 Stolons 0; larger leaves >(3)4x as long as wide. Erect to ascending,
appressed-pubescent annual to biennial. Native; same places as
M. scorpioides and often with it; fairly common
                                        Tufted Forget-me-not - **M. laxa** Lehm.
7 Perennial, with sterile basal shoots at fruiting 8
7 Annual, without sterile shoots at fruiting 10
    8 Fruiting calyx narrowed and acute to subacute at base; nutlets
obtuse to rounded at apex; only >700m (see couplet 5) **M. alpestris**
    8 Fruiting calyx rounded to broadly obtuse at base; nutlets acute
to subacute at apex; lowland or upland 9
9 Corolla ≤8mm across, with ± flat limb; calyx-teeth erecto-patent,
exposing ripe nutlets. Erect to ascending, tufted, patent-pubescent
perennial to 50cm. Native; woods, scree and rock-ledges; locally
common in C & N Br, very local in S Br, frequent garden escape
elsewhere   *Wood Forget-me-not* - **M. sylvatica** Hoffm.
9 Corolla ≤5mm across, with saucer-shaped limb; calyx-teeth erect,
± appressed and concealing ripe nutlets (often squashed open when
pressed). Erect to ascending, tufted, patent-pubescent annual to
perennial to 40cm. Native; open, well-drained ground in many
habitats, including gardens; common
                                    Field Forget-me-not - **M. arvensis** (L.) Hill
10 Pedicels 1.2-2x as long as fruiting calyx (see couplet 9) **M. arvensis**
10 Pedicels shorter than to c. as long as fruiting calyx 11
11 Corolla blue (rarely white) from start, with tube shorter than calyx.
Erect to decumbent, patent-pubescent annual to 25cm. Native; dry
open places on sandy or limestone soils; locally common over
most of lowland BI, especially En and CI
                                      Early Forget-me-not - **M. ramosissima** Rochel
11 Corolla cream to yellow at first (rarely white), with tube eventually
longer than calyx. Erect, patent-pubescent annual to 25cm. Native;

similar places to *M. ramosissima* but common also in Sc, Wa and Ir, and in N Br also in damp places, marshes and dune-slacks

Changing Forget-me-not - **M. discolor** Pers.

M. scorpioides x *M. laxa* = ***M. x suzae*** Domin has been found with the parents very scattered in En, Wa and Ir; it is partially fertile and intermediate.

16. LAPPULA Moench - *Bur Forget-me-not*

Pubescent annuals or biennials; leaves linear-lanceolate to oblong-lanceolate, the basal ones tapered to base, the upper ones sessile; flowers in terminal, spiralled cymes; calyx divided nearly to base; corolla actinomorphic, blue, with tube shorter than limb, the latter 5-lobed, nearly flat; stamens equal, included; style simple, included; nutlets covered with hook-tipped spines.

1 Stems erect, to 50cm, with dense appressed and patent white hairs; corolla 2-4mm across. Intrd-casual; in waste places, rough ground and tips; very scattered in C & S Br and E Ir

Bur Forget-me-not - **L. squarrosa** (Retz.) Dumort.

17. OMPHALODES Mill. - *Blue-eyed-Mary*

Rhizomatous and stoloniferous, rather sparsely pubescent perennials; leaves ovate, rounded to cordate at base, all or nearly all long-petiolate; flowers in few-flowered terminal cymes; calyx divided nearly to base; corolla actinomorphic, blue, with tube shorter than limb, the latter 5-lobed, nearly flat; stamens equal, included; style simple, included; nutlets smooth, pubescent.

1 Stems erect to ascending, to 25cm; corolla 10-15mm across. Intrd-natd; persistent relic or throwout mostly in woods; very scattered over Br, mainly in N and Wa, Man *Blue-eyed-Mary* - **O. verna** Moench

18. CYNOGLOSSUM L. - *Hound's-tongues*

Pubescent biennials; leaves ovate to linear-lanceolate, the lower petiolate, the upper sessile; flowers in spiralled terminal and lateral cymes; calyx divided nearly to base; corolla actinomorphic, reddish-purple, with tube about as long as funnel-shaped limb; stamens equal, included; style simple, included; nutlets large (5-9mm in longest plane), covered with hooked spines.

1 Leaves grey-green, densely pubescent; pedicels <5mm; corolla 6-10mm across; nutlets with distinct thickened rim, uniformly spiny. Native; rather open ground mostly on sand, shingle or limestone, and waste ground; locally frequent in S, C & E Br, CI and E Ir *Hound's-tongue* - **C. officinale** L.
1 Leaves green, sparsely pubescent; pedicels mostly >5mm; nutlets without thickened rim, but spines longer and denser at edges. Native; woods and hedgerows; rare in E Gloucs, Oxon, Bucks and

Surrey   *Green Hound's-tongue* - **C. germanicum** Jacq.

121. VERBENACEAE - *Vervain family*

Herbaceous perennials or rarely annuals; told from Boraginaceae by square-sectioned stems, opposite leaves and 4 stamens, and from Lamiaceae by scarcely lobed ovary, terminal style and capitate stigma.

1. VERBENA L. - *Vervains*

1 At least lower stem-leaves petiolate, deeply pinnately lobed; spikes long, slender. Stems erect, to 75cm. Native; barish ground and rough grassy places, on well-drained often calcareous soils; locally common in S Br, scattered N to N En, intrd in C Ir   *Vervain* - **V. officinalis** L.
1 Stem-leaves sessile, at most sharply serrate; spikes short, very dense, forming subcorymbose panicle   2
   2 Bracts shorter than to as long as calyx; corolla mostly <2x as long as calyx. Stems erect, to 1(1.5)m. Intrd-casual; on tips and waste land; very scattered in En   *Argentinian Vervain* - **V. bonariensis** L.
   2 Bracts longer than calyx; corolla mostly >2x as long as calyx. Stems erect, to 50cm. Intrd-casual; garden escape on tips; occasional in En and Sc   *Slender Vervain* - **V. rigida** Spreng.

122. LAMIACEAE - *Dead-nettle family*

Herbaceous annuals to perennials or dwarf shrubs, often aromatic; stems usually square in section; like Verbenaceae and Boraginaceae in its 4-celled ovary with 1 ovule per cell and fruit of 4 nutlets, but differing as under those 2 families. Some Scrophulariaceae resemble Lamiaceae vegetatively, but have totally different ovaries.

Several genera in both subfamilies may produce male-sterile flowers on same plant as or on different plants from functionally bisexual flowers; such male-sterile flowers usually have smaller corollas and much reduced, pollen-less stamens, often included in the corolla-tube. They are not covered in the following keys, but usually occur with bisexual flowers or plants. Corolla-lengths refer to length from base of calyx to tip of longest corolla-lobe on fresh flowers; considerable shortening occurs on drying.

8 groups of distinctive genera are first defined, but all are included in the main keys which follow.

Distinctive genera
Plants annual   1. STACHYS, 5. LAMIUM, 6. GALEOPSIS, 11.TEUCRIUM, 12. AJUGA, 18. CLINOPODIUM, 26.SALVIA
Shrubs >40cm high   7. PHLOMIS, 17. SATUREJA, 19. HYSSOPUS, 21. THYMUS, 24. LAVANDULA, 25. ROSMARINUS

Flowers with only 2 fertile stamens
    22. LYCOPUS, 25. ROSMARINUS, 26. SALVIA
Corolla yellow or yellowish
  1. STACHYS, 4. LAMIASTRUM, 6. GALEOPSIS, 7. PHLOMIS,
   11. TEUCRIUM, 12. AJUGA, 15. PRUNELLA, 16. MELISSA
Calyx-teeth spiny tipped   **1. STACHYS, 3. LEONURUS,**
   6. GALEOPSIS, 7. PHLOMIS, 22. LYCOPUS
Calyx-teeth 10, hooked at apex   **9. MARRUBIUM**
Leaves lobed >1/2 way to midrib   **3. LEONURUS,**
  11. TEUCRIUM, 12. AJUGA, 15. PRUNELLA, 22. LYCOPUS
Stamens longer than corolla (incl. lips)**19. HYSSOPUS, 20. ORIGANUM,**
  21. THYMUS, 22. LYCOPUS, 23. MENTHA, 25. ROSMARINUS

General key
1 Corolla with well-developed lower lip; upper lip 0 or represented by 1-2 short lobes   2
1 Corolla with upper and lower lips well developed, or ± actinomorphic (4-5-lobed)   3
  2 Corolla with ring of hairs inside tube; lower lip 3-lobed (central lobe often ± bifid); upper lip of 1-2 short lobes   **12. AJUGA**
  2 Corolla without ring of hairs inside tube; lower lip 5-lobed (central lobe sometimes slightly bifid); upper lip 0   **11. TEUCRIUM**
3 Stamens 2   4
3 Stamens 4 (often very reduced in female flowers)   6
  4 Shrub; leaves entire   **25. ROSMARINUS**
  4 Herbaceous annual or perennial; leaves crenate to pinnately lobed, rarely some entire   5
5 Calyx and corolla both distinctly 2-lipped   **26. SALVIA**
5 Calyx with 5 equal lobes; corolla with 4 subequal lobes (the uppermost slightly wider and emarginate)   **22. LYCOPUS**
  6 Calyx with 2 entire lips, the upper with a dorsal outgrowth   **10. SCUTELLARIA**
  6 Calyx with 5-10 lobes or teeth, often 3 forming upper and 2 lower lip, without dorsal outgrowth   7
7 Calyx-teeth 10, hooked at apex   **9. MARRUBIUM**
7 Calyx-teeth (4-)5, not hooked at apex   8
  8 Corolla ± actinomorphic, indistinctly 2-lipped, or distinctly 2-lipped with upper lip ± flat; stamens (except in female flowers) usually fully exposed from front view of flower, sometimes longer than corolla   *Key A*
  8 Corolla distinctly 2-lipped, with upper lip distinctly hooded and usually at least partially concealing stamens from front view   *Key B*

Key A - Calyx with 5 teeth or lobes; stamens 4, not included within corolla-tube (but often very reduced in female flowers); corolla ± actinomorphic to zygomorphic and 2-lipped with upper lip ± flat and not concealing stamens from front view.

1	Corolla ± actinomorphic, with 4 ± equal lobes or 1 lobe shortly bifid **23. MENTHA**
1	Corolla strongly 2-lipped, or weakly zygomorphic with upper lip of 2 and lower lip of 3 lobes 2
	2 Upper calyx-tooth with widened apical appendage; shrub with flowers crowded in long-stalked spikes **24. LAVANDULA**
	2 Upper calyx-tooth without appendage; if a shrub, flowers not crowded in long-stalked spikes 3
3	Calyx-teeth ± equal, not forming 2 lips 4
3	Calyx-teeth unequal, the upper 3 differing markedly in length and/or breadth from lower 2 9
	4 Evergreen shrubs up to c.60cm; leaves entire 5
	4 Herbaceous perennials; leaves often crenate to serrate 6
5	Stamens divergent, the longer 2 longer than corolla; corolla bluish-violet (very rarely white) **19. HYSSOPUS**
5	Stamens convergent, shorter than corolla; corolla pinkish-purple (very rarely white) **17. SATUREJA**
	6 Flowers few in axillary clusters; plant with long procumbent stolons **14. GLECHOMA**
	6 Flowers numerous in terminal inflorescences; plant without stolons 7
7	Leaves entire or nearly so **20. ORIGANUM**
7	Leaves conspicuously crenate to serrate 8
	8 Calyx 15-veined; stem-leaves many pairs **13. NEPETA**
	8 Calyx 5-10-veined; stem-leaves ≤4(5) pairs **1. STACHYS**
9	Corolla ≥25mm; calyx with upper lip 2-3-toothed, with lower lip with 2 much deeper but scarcely narrower lobes **8. MELITTIS**
9	Corolla ≤22mm; calyx with upper lip 3-lobed, with lower lip with 2 much narrower lobes 10
	10 Stems woody, either erect to ascending or filiform and procumbent to decumbent **21. THYMUS**
	10 Stems herbaceous, usually ± erect 11
11	Corolla-tube curved; stigmas ± equal; fresh plant lemon-scented **16. MELISSA**
11	Corolla-tube straight; stigmas distinctly unequal; fresh plant variously scented but not of lemon **18. CLINOPODIUM**

Key B - Calyx with 5 teeth or lobes; stamens 4, not included within corolla-tube (but often very reduced in female flowers); corolla distinctly 2-lipped, with upper lip distinctly hooded and often at least partially concealing stamens from front view.

1	Plant with stolons >10cm 2
1	Plant without stolons 5
	2 Corolla yellow **4. LAMIASTRUM**
	2 Corolla variously bluish, purplish or white 3
3	Flowers few in axillary clusters; bracts all leaf-like **14. GLECHOMA**
3	Flowers in dense whorls forming terminal inflorescence; at least upper bracts much reduced 4

4	Lateral lobes of lower lip of corolla obscure or pointed, if rounded much <1/2 as large as terminal lobe; terminal lobe bifid for ≥1/3 length; carpels and mericarps truncate at apex **5. LAMIUM**
4	Lateral lobes of lower lip of corolla conspicuous, rounded, usually c.1/2 as large as terminal lobe; terminal lobe not bifid or bifid for <1/3 length; carpels and mericarps rounded at apex **1. STACHYS**
5	Calyx-teeth unequal, the upper 3 differing markedly in length and/or breadth from lower 2 6
5	Calyx-teeth ± equal, not forming 2 lips 8
6	Corolla >2cm; calyx ≥12mm **8. MELITTIS**
6	Corolla <2cm; calyx <12mm 7
7	Flowers forming dense terminal head; upper lip of calyx nearly entire, with 3 small teeth; fresh plant not strongly scented **15. PRUNELLA**
7	Flowers in axillary whorls, only the uppermost of which merge; upper lip of calyx with 3 distinct lobes; fresh plant strongly lemon-scented **16. MELISSA**
8	Lower leaves lobed >1/2 way to midrib **3. LEONURUS**
8	Leaves entire to serrate 9
9	Calyx-teeth and associated bracteoles spine-tipped; lower lip of corolla with conical projection at base of each of 2 lateral lobes **6. GALEOPSIS**
9	Calyx-teeth and bracteoles not spine-tipped, or former sometimes so; lower lip of corolla without 2 conical projections 10
10	Stigmas distinctly unequal; leaves whitish-tomentose on lowerside **7. PHLOMIS**
10	Stigmas ± equal; leaves rarely white-tomentose on lowerside 11
11	Lateral lobes of lower lip of corolla obscure or pointed, if rounded much <1/2 as large as terminal lobe; terminal lobe bifid for ≥1/3 length; carpels and mericarps truncate at apex **5. LAMIUM**
11	Lateral lobes of lower lip of corolla conspicuous, rounded, usually c.1/2 as large as terminal lobe; terminal lobe not bifid or bifid for <1/3 length; carpels and mericarps rounded at apex 12
12	Upper lip of corolla scarcely hooded; calyx 15-veined **13. NEPETA**
12	Upper lip of corolla strongly hooded; calyx 5-10-veined 13
13	Calyx-teeth c. as long as wide; calyx-tube trumpet-shaped (cylindrical or ± so for most part but conspicuously expanded in distal 1/2) **2. BALLOTA**
13	Calyx-teeth usually distinctly longer than wide; calyx-tube cylindrical to obconical, not conspicuously expanded in distal 1/2 **1. STACHYS**

SUBFAMILY 1 - LAMIOIDEAE (genera 1-12). Plants mostly not pleasantly scented; male-sterile plants rather rare; upper lip of corolla usually conspicuously hooded (not in *Marrubium*) or 0; stamens 4, shorter than corolla; cotyledons usually at least as long as wide; seeds with endosperm.

1. STACHYS L. - *Woundworts*

Herbaceous annuals or perennials; leaves crenate to serrate; calyx with 5 ± equal acute (sometimes ± spine-tipped) lobes; corolla yellow, pink to purple, or white, with hooded upper lip, with 3-lobed lower lip; stamens 4, shorter than upper lip of corolla; whorls distant in leaf-axils, or congested in axils of reduced bracts.

1 Corolla cream to yellow 2
1 Corolla purplish, very rarely white 3
    2 Flowers ≤6 per node; corolla 10-16mm. Erect, shortly pubescent annual to 30cm, with or without glandular hairs. Intrd-casual; waste ground; rather rare in C & S Br
Annual Yellow-woundwort - **S. annua** (L.) L.
    2 Flowers usually ≥6 per node; corolla 15-20mm. Erect to ascending, shortly pubescent biennial to perennial to 70cm, without rhizomes, without glandular hairs. Intrd-natd; waste ground; Barry Docks (Glam) *Perennial Yellow-woundwort* - **S. recta** L.
3 Corolla <10mm. Erect to ascending, pubescent annual to 25cm, without glandular hairs. Native; arable ground on non-calcareous soils; scattered over much of BI *Field Woundwort* - **S. arvensis** (L.) L.
3 Perennial; corolla >10mm 4
    4 Bracteoles 0 or minute 5
    4 Bracteoles mixed in with flowers, at least as long as calyx 7
5 Middle and upper stem-leaves sessile. Erect, slightly-smelling, pubescent perennial to 1m, with stalked glands in upper parts, with strong ± surface rhizomes developing slight swellings at tips late in year. Native; damp places, by rivers and ponds, and on rough ground; common *Marsh Woundwort* - **S. palustris** L.
5 All leaves up to 1st inflorescence node petiolate 6
    6 Petioles of middle and upper stem-leaves 1/10-1/5 total leaf + petiole length; few or 0 fruits ripening. Intermediate between parents. Native; in habitats of either parent, often in absence of 1 or both; scattered over BI (*S. sylvatica* x *S. palustris*)
Hybrid Woundwort - **S. x ambigua** Sm.
    6 Petioles of middle and upper stem-leaves 1/4-2/5(1/2) total leaf + petiole length; all or most fruits ripening. Erect, strong-smelling, harshly pubescent perennial to 1m, with stalked glands in upper parts, with strong ± surface rhizomes. Native; woods, hedgerows, rough ground; common
Hedge Woundwort - **S. sylvatica** L.
7 Corolla-tube longer than calyx; calyx very sparsely pubescent to glabrous on outside; sterile leaf-rosettes present at flowering. Erect, sparsely pubescent perennial to 75cm. Native; hedgebanks, grassland, heaths, avoiding heavy soils; common in En and Wa, local in Jersey, extremely so in Sc and Ir *Betony* - **S. officinalis** (L.) Trevis.
7 Corolla-tube shorter than calyx; calyx densely pubescent on outside; sterile leaf-rosettes 0 at flowering 8
    8 Upper part of stem and calyx with stalked glands as well as eglandular hairs. Erect, softly pubescent perennial to 1m, with stalked glands in upper parts, without rhizomes. Native; open woods; Denbs and W Gloucs *Limestone Woundwort* - **S. alpina** L.

8   Stem and calyx with only eglandular hairs   9
9   Leaves cuneate at base; stem and both leaf-surfaces densely white-tomentose. Erect to ascending perennial to 75cm, with thick surface rhizomes. Intrd-natd; on tips and waste ground; scattered in Br N to C Sc   *Lamb's-ear* - **S. byzantina** K. Koch
9   At least lower leaves cordate at base; green upper leaf-surface showing through pubescence. Erect, densely white-pubescent to -subtomentose biennial to 80(100)cm, without rhizomes. Native; rough open grassland, hedgebanks and wood-borders on calcareous soils; Oxon   *Downy Woundwort* - **S. germanica** L.

2. BALLOTA L. - *Black Horehound*

Herbaceous perennials; leaves serrate; calyx with 5 ± equal acuminate lobes; corolla reddish-mauve, with hooded upper lip, with 3-lobed lower lip; stamens 4, shorter than upper lip of corolla; whorls distant, in leaf-axils.

1   Erect perennial to 1m, with unpleasant smell when bruised; corolla 9-15mm. Native; hedgerows, waysides, rough ground; common in C & S Br and CI, very local in Sc and Ir   *Black Horehound* - **B. nigra** L.

3. LEONURUS L. - *Motherwort*

Herbaceous perennials; lower and middle stem-leaves and basal leaves deeply palmately to ternately lobed; calyx with 5 ± equal spine-tipped patent lobes; corolla pinkish-purple, with hooded upper lip, with 3-lobed lower lip; stamens 4, shorter than upper lip of corolla; whorls distant in leaf-axils.

1   Rhizomatous; stems erect, to 1.2m; corolla 8-12mm. Intrd-natd; waste places and waysides; thinly scattered over much of BI

Motherwort - **L. cardiaca** L.

4. LAMIASTRUM Fabr. - *Yellow Archangel*

Stoloniferous herbaceous perennials; leaves serrate; calyx with 5 ± equal triangular-acuminate lobes; corolla yellow, with hooded upper lip, with 3-lobed lower lip; stamens 4, shorter than upper lip of corolla; whorls distant in leaf-axils. Differs from *Lamium* in glabrous (not pubescent) anthers.

1   Flowering stems erect, to 60cm; stolons leafy, rooting at nodes, often >1m (*Yellow Archangel* - **L. galeobdolon** (L.) Ehrend. & Polatschek)   2
    2   Most leaves with large conspicuous whitish blotches for whole year; fruiting calyx ≥(11.5)12mm; upper lip of corolla ≥(7.5)8mm wide. Intrd-natd; shrubberies and waysides; scattered over BI

L. galeobdolon ssp. argentatum (Smejkal) Stace
    2   Leaves without whitish blotches, or some with white marbling early in year; fruiting calyx ≤12(12.5)mm; upper lip of corolla ≤8(8.5)mm wide   3
3   Bracts 1-2(2.2)x as long as wide, obtusely serrate; flowers ≤8(9) per node, at ≤4(5) nodes; flowering stems with hairs ± confined to 4

angles. Native; woods, wood-borders and hedgerows; very local
in small area of N Lincs **L. galeobdolon ssp. galeobdolon**

3 Bracts (1.5)1.7-3.6x as long as wide, acutely serrate; flowers ≥(7)10 per node, at ≥(3)4 nodes; flowering stems with hairs on faces as well as angles. Native; woods, wood-borders and hedgerows; common over most of En and Wa, very local in SW Sc, E Ir and CI
L. galeobdolon ssp. montanum (Pers.) Ehrend. & Polatschek

5. LAMIUM L. - *Dead-nettles*

Annuals or herbaceous perennials; leaves serrate to deeply so, rarely ± entire, calyx with 5 ± equal narrowly triangular-acuminate lobes; corolla white, or pink to purple or mauve, with hooded upper lip, with ± 1-lobed lower lip of which lateral lobes are much reduced and pointed or rounded; stamens 4, shorter than upper lip of corolla; whorls distant in leaf-axils, or ± congested in axils of modified leaves.

1 Perennials with rhizomes and/or stolons; corolla-tube curved 2
1 Annuals; corolla-tube straight 3
   2 Corolla white; leaves never blotched whitish; lower lip of corolla with 2-3 teeth each side. Rhizomatous or sometimes stoloniferous perennial to 60cm. Native; hedgebanks, waysides, rough ground; common in most of lowland Br, rare in CI and Ir
White Dead-nettle - **L. album** L.
   2 Corolla usually pinkish-purple; leaves usually blotched whitish; lower lip of corolla with 1 tooth on each side. Rhizomatous and/or stoloniferous perennial to 60cm. Intrd-natd; rough ground and tips; scattered in Br and CI
Spotted Dead-nettle - **L. maculatum** (L.) L.
3 Plant with ± all leaves petiolate 4
3 Middle and upper leaves subtending whorls sessile 5
   4 Leaves subtending whorls serrate to crenate-serrate, with teeth <2mm long. Annual to 25cm. Native; cultivated and waste ground; common   *Red Dead-nettle* - **L. purpureum** L.
   4 Leaves subtending whorls deeply serrate, with many teeth >2mm long. Annual to 25cm. Native; cultivated and waste ground, scattered over lowland BI
Cut-leaved Dead-nettle - **L. hybridum** Vill.
5 Calyx 5-7mm at flowering, densely white- ± patent-pubescent, the teeth erect to convergent at fruiting; lower lip of corolla <3mm. Annual to 25cm. Native, open, cultivated and waste ground; throughout BI, commoner in E *Henbit Dead-nettle* - **L. amplexicaule** L.
5 Calyx 8-12mm at flowering, ± appressed-pubescent, the teeth divergent at fruiting; lower lip of corolla >3mm. Annual to 25cm. Native; cultivated and waste ground; local near coast in N, W & E Sc, Man, very scattered in Ir *Northern Dead-nettle* - **L. confertum** Fr.

6. GALEOPSIS L. - *Hemp-nettles*

Annuals; leaves usually serrate; calyx with 5 ± equal, weakly spine-tipped lobes; corolla variously white, pinkish-purple or yellow, with hooded upper lip, with 3-lobed lower lip; stamens 4, shorter than upper lip of corolla; whorls somewhat congested, in axils of reduced leaves.

1 Stems with soft hairs, not swollen at nodes   2
1 Stems with rigid bristly hairs, swollen at nodes   3
    2 Corolla predominantly pale yellow, (20)25-30mm; leaves and calyx densely silky- or velvety-pubescent. Stems erect, to 50cm. Native; casual or sporadic in arable and waste ground; scattered in En and Wa   *Downy Hemp-nettle* - **G. segetum** Neck.
    2 Corolla predominantly reddish-pink, rarely white, 14-25mm; leaves and calyx variously pubescent but not densely silky or velvety. Stems erect, to 50cm. Native; arable land, open ground mostly on calcareous soils or maritime sand or shingle; very scattered in SC En N to SE Yorks
          Red Hemp-nettle - **G. angustifolia** Hoffm.
3 Corolla (22)27-35mm, yellow with purple blotch on lower lip; corolla-tube c.2x as long as calyx (incl. teeth). Stems erect to ascending, to 1m. Native; arable land, often on peaty soil with root-crops, and waste places; locally common in C & N Br and N Ir, scattered elsewhere   *Large-flowered Hemp-nettle* - **G. speciosa** Mill.
3 Corolla 13-20(25)mm, variously coloured; corolla-tube rarely >1.5x as long as calyx (incl. teeth)   4
    4 Terminal lobe of lower lip of corolla entire to very slightly emarginate, ± flat. Stems erect, to 1m. Native; arable land, rough ground, woodland clearings, damp places; common over most of BI   *Common Hemp-nettle* - **G. tetrahit** L.
    4 Terminal lobe of lower lip of corolla clearly emarginate, convex (with revolute sides). Stems erect, to 1m. Native; in similar places to *G. tetrahit* and as common   *Bifid Hemp-nettle* - **G. bifida** Boenn.

G. tetrahit x *G. bifida* = **G. x ludwigii** Hausskn. is intermediate in corolla shape and has 20-70 per cent pollen fertility and low seed-set; it has been recorded from Wa and C En.

7. PHLOMIS L. - *Sages*

Perennial herbs or shrubs; leaves entire to crenate, white-tomentose; calyx with 5 ± equal, spine-tipped teeth; corolla yellow, with strongly hooded upper lip, with 3-lobed lower lip; stamens 4, shorter than upper lip of corolla; whorls somewhat congested, in axils of reduced leaves.

1 Herb to 1m; basal leaves cordate at base, 6-20cm, with longer petiole; calyx 20-25mm; corolla yellow. Intrd-natd; banks by roads and railways, rough ground; very scattered in En and Sc
          Turkish Sage - **P. russeliana** (Sims) Benth.
1 Evergreen shrub to 1.3m; basal leaves 0; lower leaves cuneate to truncate at base, 3-9cm, with shorter petiole; calyx 10-20mm; corolla yellow. Intrd-natd; on sea-cliffs, banks and rough ground; S En,

Man and Ir   *Jerusalem Sage* - **P. fruticosa** L.

8. MELITTIS L. - *Bastard Balm*
Perennial herbs; leaves serrate; calyx with 2 lips, the upper with 2-3 short lobes, the lower with 2 deeper lobes; corolla white, or pink to mauve or purple, with flat or slightly hooded upper lip, with 3-lobed lower lip; stamens 4, shorter than upper lip of corolla; whorls distant, in leaf-axils.

1   Stems erect, to 70cm; lower leaves ovate, cordate to rounded at base, petiolate; corolla 25-40mm. Native; woods and hedgerows; very local in SW Wa, SW & S En E to W Sussex, rare escape elsewhere
Bastard Balm - **M. melissophyllum** L.

9. MARRUBIUM L. - *White Horehound*
Perennial herbs; leaves serrate; calyx with 10 equal teeth, each hooked at end and patent at fruiting; corolla white, with erect, flat, bilobed upper lip, with 3-lobed lower lip; stamens 4, all included in corolla-tube; whorls distant, in leaf-axils.

1   Plant whitish-tomentose or densely pubescent; stems erect or ascending, to 60cm. Native; short grassland, open or rough ground and waste places; sparsely scattered in BI to Moray
White Horehound - **M. vulgare** L.

10. SCUTELLARIA L. - *Skullcaps*
Perennial herbs; leaves entire to serrate; calyx 2-lipped, both lips entire, the upper with a dorsal outgrowth; corolla pinkish to blue or purple, with hooded upper lip, with rather obscurely 3-4-lobed lower lip; stamens 4, shorter than to c. as long as upper lip of corolla; flowers 2 at each node, in leaf-axils.

1   Flowers in axils of bracts markedly smaller than foliage leaves; lower leaves 5-15cm, with petioles >1cm. Stems erect, to 80cm. Intrd-natd; in hedgerows and wood borders; N Somerset and Surrey
Somerset Skullcap - **S. altissima** L.
1   Flowers in axils of foliage leaves; leaves very rarely >5cm, with petioles <1cm   2
  2   Corolla 6-10mm, pale pinkish-purple, with nearly straight tube. Stems erect to decumbent, to 25cm. Native; wet heaths and open woodland on acid soils; locally frequent in S & W Br and S Ir, rare elsewhere in En and Jersey   *Lesser Skullcap* - **S. minor** Huds
  2   Corolla 10-20mm, blue, with strongly bent tube. Stems erect to decumbent, to 50cm. Native; fens, wet meadows, by ponds and rivers; locally common   *Skullcap* - **S. galericulata** L.

S. galericulata x *S. minor* = **S. x hybrida** Strail occurs locally near the parents in S En and S Ir; it is intermediate and highly sterile.

11. TEUCRIUM L. - *Germanders*

Annual to perennial herbs or very low shrubs; leaves serrate to deeply lobed; calyx rather unequally 5-lobed, not 2-lipped; corolla pinkish-purple or greenish-cream, with 5-lobed lower lip and 0 upper lip; stamens 4, shorter than lower lip; whorls distant in leaf-axils to congested in axils of reduced bracts.

1 Leaves divided much >1/2 way to midrib. Erect annual or biennial to 30cm. Native; bare chalk and chalky fallow fields on downs; N Hants, W Kent, Surrey and E Gloucs
Cut-leaved Germander - **T. botrys** L.
1 Leaves serrate much <1/2 way to midrib; perennial 2
 2 Corolla greenish-cream; upper calyx-tooth much wider than other 4 . Stems herbaceous, erect, to 50cm. Native; woods, hedgerows, hilly areas, fixed shingle and dunes, on acidic or alkaline usually well-drained soils; common in suitable places
Wood Sage - **T. scorodonia** L.
 2 Corolla pinkish-purple (rarely white); upper calyx-tooth scarcely different from other 4 3
3 Very dwarf evergreen shrub; whorls forming a terminal inflorescence. Stems suberect to decumbent, to 40cm. Native; chalk grassland at 1 site in E Sussex, also natd on old walls and dry banks; very scattered in BI   *Wall Germander* - **T. chamaedrys** L.
3 Entirely herbaceous; whorls spaced down stem. Stems herbaceous, ascending to decumbent, to 50cm. Native; fens, dune-slacks and riverbanks on calcareous soils; N Devon and Cambs, locally frequent in WC Ir   *Water Germander* - **T. scordium** L.

12. AJUGA L. - *Bugles*

Annual or perennial herbs; leaves subentire to deeply divided; calyx ± equally 5-lobed; corolla pink, blue, white or yellow, with 3-lobed lower lip (terminal lobe often bifid), with very short 1-2-lobed upper lip; stamens 4, shorter than lower lip; whorls distant to slightly congested, in leaf-axils.

1 Leaves divided much >1/2 way to midrib; annual; corolla yellow. Stems erect to decumbent, usually branched low down, to 20cm. Native; bare chalk and chalky arable fields on downs; local in S & SE En N to W Suffolk   *Ground-pine* - **A. chamaepitys** (L.) Schreb.
1 Leaves subentire or serrate much <1/2 way to midrib; perennial with stolons and/or rhizomes; corolla blue, pink or white 2
 2 Plant with stolons; upper bracts shorter than flowers; upper part of stem hairy only on 2 opposite sides. Flowering stems erect, to 30cm. Native; woods, shady places, damp grassland; common   *Bugle* - **A. reptans** L.
 2 Plant without stolons, with rhizomes; all bracts longer than flowers; stem hairy all round. Flowering stems erect, to 30cm. Native; rock crevices in hilly areas; very local in N, NW & S Sc, WC & NE Ir, Westmorland   *Pyramidal Bugle* - **A. pyramidalis** L.

A. reptans x *A. pyramidalis* = ***A. x pseudopyramidalis*** Schur has occurred near the parents in Co Clare, W Sutherland, E Ross and Orkney; it is sterile and intermediate, forming stolons late in the season.

SUBFAMILY 2 - NEPETOIDEAE (genera 13-26). Plants mostly pleasantly scented; male-sterile plants very common; upper lip of corolla usually ± flat (hooded in *Prunella*, *Rosmarinus* and *Salvia*) or corolla ± actinomorphic; stamens 2 or 4, sometimes exceeding corolla; cotyledons usually wider than long; seeds without endosperm.

13. NEPETA L. - *Cat-mints*
Perennial herbs; leaves serrate; calyx with 5 subequal teeth; corolla white to blue, with flat to slightly hooded upper lip, with 3-lobed lower lip; stamens 4, shorter than upper lip of corolla.

1 Corolla white with small purple spots, the tube shorter than calyx; leaves ovate, cordate at base. Stems erect, to 1m. Probably native; open grassland, waysides, rough ground on calcareous soils; rather scattered in En and Wa, rare in Ir   *Cat-mint* - **N. cataria** L.
1 Corolla blue, the tube longer than calyx; leaves narrowly ovate-oblong, mostly truncate at base. Stems ascending, to 1.2m. Intrd-natd; tips and rough ground; scattered in En, Wa and CI (*N. mussinii* Henckel x *N. nepetella* L.)   *Garden Cat-mint* - **N. x faassenii** Stearn

14. GLECHOMA L. - *Ground-ivy*
Perennial herbs with long trailing stolons; leaves crenate-serrate; calyx with 5 subequal teeth; corolla blue, rarely pink or white, as in *Nepeta*; stamens as in *Nepeta*; flowers only 2-4 per node, the whorls ± distant in leaf-axils.

1 Stolons often >1m; flowering stems suberect, to 30cm; leaves broadly ovate to orbicular, cordate at base, those on stolons with long petioles. Native; woods, hedgerows, rough ground, often on heavy soils; common   *Ground-ivy* - **G. hederacea** L.

15. PRUNELLA L. - *Selfheals*
Perennial herbs; leaves entire to divided ± to midrib; calyx 2-lobed, the upper lip ± truncate with 3 very short teeth, the lower lip with 2 long teeth; corolla yellow, blue, pink or white, with 3-lobed lower lip, with ± entire, strongly hooded upper lip; stamens 4, shorter than upper lip of corolla; whorls congested, in axils of strongly modified bracts.

1 Leaves entire or shallowly toothed; corolla 10-15mm, bluish-violet, rarely pink or white. Stems erect to decumbent, to 30cm. Native; grassland, lawns, wood-clearings, rough ground; common
    Selfheal - **P. vulgaris** L.
1 Leaves variably shaped, often some entire but at least the upper deeply divided ± to midrib; corolla usually 15-17mm, creamy-yellow or -white, rarely pale blue. Stems erect to decumbent, to 30cm.

Possibly native; calcareous grassland; scattered in S & C En N to
N Lincs   *Cut-leaved Selfheal* - **P. laciniata** (L.) L.

P. vulgaris x *P. laciniata* = **P. x intermedia** Link occurs scattered in S & C En wherever *P. laciniata* occurs or has occurred; it is variably intermediate in leaf-shape and corolla-colour.

16. MELISSA L. - *Balm*
Perennial herbs; leaves serrate; calyx 2-lipped, the upper lip ± truncate with 3 short teeth, the lower lip with 2 long teeth; corolla pale yellow, becoming whitish or pinkish, with 3-lobed lower lip, with 2-lobed flat or slightly hooded upper lip; stamens 4, shorter than upper lip of corolla; whorls distant, in leaf-axils.

1 Plant lemon-scented when fresh; stems erect, to 1m; leaves ovate, the lower with cordate to truncate base. Intrd-natd; rough ground and waste places; scattered in C & S BI   *Balm* - **M. officinalis** L.

17. SATUREJA L. - *Winter Savory*
Evergreen low shrubs; leaves entire; calyx with 5 subequal teeth; corolla pinkish-purple, rarely white, with 3-lobed lower lip, with shallowly 2-lobed ± flat upper lip; stamens 4, shorter than upper lip; stigmas ± equal; flowers in contracted cymes in axils of reduced leaves, the whorls slightly congested.

1 Stems erect to ascending, to 50cm; leaves linear-lanceolate, finely acute. Intrd-natd; on old walls; N Somerset and S Hants, less persistent elsewhere in S Br   *Winter Savory* - **S. montana** L.

18. CLINOPODIUM L. - *Calamints*
Herbaceous annuals or perennials; leaves subentire to serrate, calyx 2-lipped, the lower lip with 2 longer and narrower teeth than the 3 teeth of upper lip; corolla pinkish-purple or violet to pale lilac or almost white, with short 3-lobed lower lip, with shallowly 2-lobed flat upper lip; stamens 4, shorter than corolla; stigmas markedly unequal; whorls distant to congested, in axils of reduced leaves.

1 Axillary flower-clusters very dense, without common stalk; calyx-tube asymmetrically curved (upper side convex, lower side concave near apex); flowers pinkish-purple to violet (rarely white)   2
1 Axillary flower-clusters in contracted cymes with common stalk; calyx-tube straight or slightly curved symmetrically on upper and lower sides; corolla very pale lilac to mauvish-pink   3
   2 Most or all whorls with >8 flowers; calyx-tube not or scarcely swollen. Stems erect, to 75cm. Native; hedgerows, wood-borders and scrubby grassland on light soils; frequent in Br N to C Sc, intrd in Ir, Alderney   *Wild Basil* - **C. vulgare** L.
   2 Whorls with ≤6(8) flowers; calyx-tube strongly swollen near base

on lower side, especially in fruit. Usually annual; stems erect to decumbent, to 25cm. Native; bare or rocky ground, arable fields on dry, usually calcareous soils; rather local in Br N to C Sc, very local in C & SE Ir **Basil Thyme - C. acinos** (L.) Kuntze

3 Teeth of lower calyx-lobe 1-2mm, with hairs all <0.1mm or very few longer; hairs in throat of calyx protruding beyond tube. Stems erect, to 60cm. Native; dry banks and rough grassland, usually calcareous; local in CE and SE En, W to Pembs, CI
Lesser Calamint - **C. calamintha** (L.) Stace

3 Teeth of lower calyx-lobe 2-4mm, with many hairs ≥0.2mm; hairs in throat of calyx usually entirely included 4

    4 Teeth of lower calyx-lobe 3-4mm; corolla 15-22mm; leaves often >4cm, with 6-10 teeth on each side. Stems erect, to 60cm. Native; scrubby laneside bank on chalk; Wight
Wood Calamint - **C. menthifolium** (Host) Stace

    4 Teeth of lower calyx-lobe 2-3(3.5)mm; corolla 10-16mm; leaves rarely >4cm, with 3-8 teeth on each side. Stems erect, to 60cm. Native; dry banks and rough grassland, usually calcareous; local in C & S BI *Common Calamint* - **C. ascendens** (Jord.) Samp.

19. HYSSOPUS L. - *Hyssop*

Evergreen low shrubs; leaves entire; calyx with 5 equal teeth; corolla blue, rarely white, with 3-lobed lower lip and shallowly 2-lobed ± flat upper lip; stamens 4, longer than corolla; styles ± equal; whorls distant below, fairly congested above, in axils of reduced leaves.

1 Stems erect to ascending, to 60cm; leaves narrowly oblong-elliptic to ± linear; corolla 7-12mm. Intrd-natd; on old walls; very scattered in S En *Hyssop* - **H. officinalis** L.

20. ORIGANUM L. - *Wild Marjoram*

Herbaceous perennials; leaves entire or remotely crenate-denticulate; calyx with 5 subequal teeth; corolla reddish-purple, rarely white, with short 3-lobed lower lip, with shallowly 2-lobed flat upper lip; stamens 4, longer than corolla in bisexual flowers; inflorescence a mass of dense cymes forming corymbose panicle, with large purple bracteoles.

1 Stems erect, to 50(80)cm; leaves ovate, 1-4cm; corolla 4-7mm. Native; dry grassland, hedgebanks and scrub, usually on calcareous soils; locally common in BI N to C Sc *Wild Marjoram* - **O. vulgare** L.

21. THYMUS L. - *Thymes*

Dwarf evergreen shrubs; leaves entire; calyx 2-lipped, the upper lip with 3 short teeth, the lower lip with 2 long teeth; corolla pinkish-purple or mauve to white, with 2 ill-defined lips, the upper of 1 emarginate lobe, the lower of 3 lobes; stamens 4, longer than corolla in bisexual flowers; whorls crowded into dense terminal heads or the lower ones more distant, in axils of reduced leaves.

1 Leaf margin revolute so that leaves are linear to narrowly oblong-elliptic in outline; plant without procumbent stems rooting at nodes. Stems erect to decumbent, to40cm. Intrd-natd; on old walls and stony banks; very scattered in CI and S En
Garden Thyme - **T. vulgaris** L.
1 Leaf margins not or scarcely revolute; leaves elliptic or elliptic-oblong to narrowly so; plant usually with procumbent stems rooting at nodes 2
  2 Lower internodes of flowering stems with hairs all or nearly all on the 4 angles. Vegetative stems procumbent to ascending, rooting but not forming dense mats; flowering stems suberect to decumbent, to 25cm. Native; short turf or barish places in coarser turf on chalky or sandy soils; locally frequent in S & C En, scattered N to Sc, very rare in Ir *Large Thyme* - **T. pulegioides** L.
  2 Lower internodes of flowering stems with hairs mainly on 2 or 4 faces 3
3 Lower internodes of flowering stems with hairs on all faces ± evenly distributed. Closely resembles small *T. polytrichus*. Native; sandy heaths; W Suffolk and W Norfolk *Breckland Thyme* - **T. serpyllum** L.
3 Lower internodes of flowering stems with hairs on 2 opposite faces, the 2 other faces glabrous or nearly so. Vegetative stems procumbent, abundantly rooting, usually forming dense mats; flowering stems decumbent to ascending, to 10cm. Native; short fine turf or open sandy or rocky places; common *Wild Thyme* - **T. polytrichus** Borbás

22. LYCOPUS L. - *Gypsywort*

Herbaceous perennials; leaves sharply serrate to deeply and acutely lobed; calyx with 5 equal teeth; corolla white with small purple dots, nearly actinomorphic, with 4 subequal lobes, the uppermost usually wider and shallowly bifid; stamens 2, longer than corolla; whorls remote, in leaf-axils.

1 Stems erect, to 1m; lower leaves partly divided >1/2 way to midrib into very acute lobes; flowers densely clustered, 3-5mm. Native; fens, wet fields, by lakes and rivers; common in En and Wa, scattered elsewhere *Gypsywort* - **L. europaeus** L.

23. MENTHA L. - *Mints*

Herbaceous rhizomatous and/or stoloniferous perennials with characteristic scents when fresh; leaves entire to serrate; calyx with 5 equal to rather unequal teeth; corolla pinkish to bluish-mauve or white, nearly actinomorphic, with 4 subequal lobes, the uppermost usually wider and shallowly bifid; stamens 4, longer than corolla in bisexual flowers of most fertile taxa; whorls all distant in leaf-axils, or the upper congested in axils of reduced leaves.

Taxonomically difficult due to well marked plasticity, widespread hybridization, and the clonal propagation of mutants and nothomorphs by the strongly developed rhizomes. The stamens are typically exserted, but are included in *M. requienii*, some plants of *M. pulegium*, female flowers of all other spp., and most (but not all) hybrids. With practice the scent of fresh

23. MENTHA

plants is very helpful, but difficult to describe.

Taxa in couplets 10-13 (*M. spicata* group), involving *M. spicata*, *M. suaveolens* and their hybrids with each other and with the non-British *M. longifolia*, are particularly difficult, especially owing to the great variation of *M. spicata*, which is itself derived from *M. longifolia* x *M. suaveolens*. *M. x villosa* and *M. x villosonervata* are sterile, but *M x rotundifolia* and the species are fertile in bisexual plants or in female plants open to a pollen source. Pubescent plants that are not *M. suaveolens* are often impossible to name for certain. Characters of *M. suaveolens* often seen in its hybrids are the broad, obtuse, very rugose leaves with teeth partly folded under the margin and with patchy or clumped indumentum on lowerside; of *M. longifolia* are the lanceolate-oblong, acute, flat leaves with sharp patent teeth and with felted grey indumentum; and of *M. spicata* are the lanceolate to ovate, acute, not to slightly rugose leaves with usually forward-directed teeth and relatively coarse pubescence. *M. spicata* is the most variable; very broad-leaved plants, and strongly rugose-leaved plants occur, but this is the only sp. of the 3 that can be glabrous to sparsely pubescent and the only sp. that can smell of spearmint. Hybrids (especially *M. x villosa*) can be very variable, showing many combinations of characters not always connected by intermediates, but they do not exactly duplicate the combinations shown by any of the spp.

1 Glabrous to sparsely pubescent plant with pungent scent; stems to 12cm, filiform, procumbent, rooting, mat-forming; flowers <6 per node. Intrd-natd; on damp paths and rocky places; very scattered in S Br and Ir **Corsican Mint - M. requienii** Benth.
1 Stems sometimes procumbent but not rooting along length and mat-forming; flowers usually >6 per node 2
    2 Whorls usually all axillary, the axis terminated by leaves, or by a reduced whorl; bracts like the leaves but reduced 3
    2 Upper whorls contracted into terminal long or rounded head; upper bracts much reduced, unlike leaves 7
3 Calyx with hairs in throat; lower 2 calyx-teeth narrower and slightly longer than 3 upper. Plant (often sparsely) pubescent, with pungent scent, resembling a small *M. arvensis*; stems erect to procumbent, to 30cm. Native; damp grassy or heathy places and by ponds; very local in S En, S Wa, Man, Jersey and Ir, sometimes natd elsewhere
Pennyroyal - M. pulegium L.
3 Calyx without hairs in throat; calyx-teeth ± equal 4
    4 Calyx 1.5-2.5mm, incl. triangular teeth ≤0.5mm, pubescent all over; usually fertile. Plant pubescent, with sickly scent; stems erect to decumbent, to 60cm. Native; arable fields, damp places in wood- clearings, fields and pondsides; fairly common
Corn Mint - M. arvensis L.
    4 Calyx 2-4mm, incl. narrowly triangular to subulate teeth 0.5-1.5mm, glabrous or pubescent; usually sterile 5
5 Calyx 2-3.5mm, incl. teeth usually ≤1mm, the tube <2x (usually c.1.5x) as long as wide. Plant glabrous to pubescent, usually with spearmint scent; stems erect to scrambling, to 90cm. Possibly native;

damp places and waste ground, usually escape or throwout; scattered ± throughout BI (*M. arvensis* x *M. spicata*)
Bushy Mint - M. x gracilis Sole

5 Calyx 2.5-4mm, incl. teeth usually ≥1mm, the tube c.2x as long as wide 6

  6 Plant subglabrous; calyx mostly >3.5mm; stamens usually exserted. Plant subglabrous, usually red-tinged, with spearmint scent; stems erect to scrambling, to 1.5m. Probably intrd-natd; damp places and waste ground, usually escape or throwout; scattered ± throughout Br and Ir, mostly S (*M. arvensis* x *M. aquatica* x *M. spicata*)   **Tall Mint - M. x smithiana** R.A. Graham

  6 Plant pubescent; calyx mostly <3.5mm; stamens usually included. Similar in habit and scent to *M. arvensis* but more robust, to 90cm. Native; similar places to both parents; frequent (*M. arvensis* x *M. aquatica*)   **Whorled Mint - M. x verticillata** L.

7 Leaves distinctly petiolate; flower-head 12-25mm across 8
7 Leaves sessile or ± so; flower-head 5-15mm across 10

  8 Leaves and calyx-tube glabrous to very sparsely pubescent. Plant glabrous or pubescent, often red-tinged, variously scented, often of peppermint; stems erect, to 90cm. Native; damp ground and waste places, escape or throwout when glabrous, usually spontaneous when pubescent; scattered throughout BI (*M. aquatica* x *M. spicata*)   **Peppermint - M. x piperita** L.

  8 Leaves and calyx-tube pubescent 9

9 Leaves ovate to narrowly ovate; flowers in rounded heads; plant normally fertile. Plant subglabrous to pubescent, with strong pleasant (not spearmint-like) scent; stems erect, to 90cm. Native; marshes, ditches, wet fields and by ponds; common
Water Mint - M. aquatica L.

9 Leaves lanceolate to narrowly ovate; flowers usually in elongate ± pyramidal heads; plant normally sterile   (see couplet 8) **M. x piperita**

  10 Leaves suborbicular to ovate, strongly rugose, pubescent, obtuse to ± rounded at apex, with teeth bent under and hence appearing as crenations from above; corolla whitish; fresh plant with sickly scent. Plant pubescent; stems erect, to 1m. Native; ditches and other damp places, waysides; local in W & S Wa, SW En and CI, natd elsewhere   **Round-leaved Mint - M. suaveolens** Ehrh.

  10 Leaves lanceolate to ovate-oblong, rugose or not, glabrous to densely pubescent, acute to subobtuse at apex, with teeth not bent over and hence appearing acute from above; if leaves close to those of *M. suaveolens* then corolla pinkish and fresh plant with sweet scent 11

11 Plant subglabrous to densely pubescent; corolla white to pinkish; leaves with acute, usually forwardly-directed teeth unless leaves broadly ovate to suborbicular 12

11 Plant pubescent to densely so; corolla pinkish; leaves with often acuminate teeth that curve outwards and become patent, especially near leaf-base, never broadly ovate to suborbicular 13

  12 Plant normally fertile; leaves lanceolate to broadly ovate, rarely

23. MENTHA

both rugose and pubescent. Plant glabrous to tomentose, usually with characteristic spearmint scent; stems erect, to 90cm. Intrd-natd; rough and waste ground; scattered throughout most of BI
Spear Mint - **M. spicata** L.

12 Plant sterile; leaves usually ovate to suborbicular, often both rugose and pubescent. Very variable sterile plants, varying in form from 1 parent to other. Intrd-natd; rough and waste ground; scattered ± throughout BI (*M. spicata* x *M. suaveolens*)
Apple-mint - **M. x villosa** Huds.

13 Plant normally fertile; leaves oblong-lanceolate to -ovate, with nearly parallel sides and broad rounded base. Close to some variants of *M. x villosa* but always pubescent. Intrd-natd; damp places and rough ground; very scattered in Br, mainly Sc (*M. longifolia* x *M. suaveolens*) *False Apple-mint* - **M. x rotundifolia** (L.) Huds.

13 Plant sterile; leaves lanceolate-elliptic, broadest near middle, narrowed into rounded to subcuneate base. Sterile plants resembling pubescent, lanceolate-leaved *M. spicata*. Intrd-natd; rough and waste ground; sparsely scattered in Br (*M. spicata* x *M. longifolia*) *Sharp-toothed Mint* - **M. x villosonervata** Opiz

M. aquatica x *M. suaveolens* = **M. x suavis** Guss. occurs near the parents in W Cornwall, N Devon and Jersey.

24. LAVANDULA L. - *Lavenders*

Evergreen shrubs; leaves entire, lanceolate to oblanceolate or linear; calyx with 5 subequal teeth, the uppermost with obcordate appendage at apex; corolla purple, weakly zygomorphic, with short 3-lobed lower lip, with short shallowly 2-lobed flat upper lip; stamens 4, included within corolla-tube; flowers on long peduncles, congested in spikes with much shorter bracts.

1 Erect shrub to 1m, with characteristic scent; leaves 2-4cm, white-tomentose when young; peduncles up to 40cm, simple or with 3 branches. Intrd-natd; self-sown on walls and banks or persistent throwout; sporadic in S En *Garden Lavender* - **L. angustifolia** Mill.

25. ROSMARINUS L. - *Rosemary*

Evergreen shrubs; leaves linear, entire, with revolute margins; calyx 2-lipped, with large ± entire upper lip, with narrower 2-lobed lower lip; corolla pale to deep mauvish-blue, strongly zygomorphic, with 2-lobed upper lip, with 3-lobed lower lip; stamens 2, longer than corolla; whorls ± distant in leaf-axils.

1 Usually erect shrub to 2m, with characteristic scent; leaves 1.5-4cm, tomentose on (mostly hidden) lowerside. Intrd-natd; self-sown on walls and rough ground; sporadic in En, Wa and CI
Rosemary - **R. officinalis** L.

26. SALVIA L. - *Claries*

Herbaceous annuals to perennials or rarely small shrubs; leaves simple, serrate to crenate; calyx 2-lipped, the lower with 2 longer teeth, the upper entire or with 3 shorter teeth; corolla blue to purple or pink, or yellow, rarely red or white, strongly zygomorphic, with shortly bilobed strongly hooded upper lip and 3-lobed lower lip; stamens 2, shorter than upper lip of corolla; whorls of 1-many flowers in axils of modified bracts in terminal interrupted spikes.

1 Corolla ≥18mm 2
1 Corolla ≤18mm 4
  2 Corolla yellow with reddish markings, >30mm; leaves often hastate at base. Erect perennial to 1m, very glandular above. Intrd-natd; woods, hedges and on road- and river-banks; scattered in En and Sc *Sticky Clary* - **S. glutinosa** L.
  2 Corolla blue to violet, rarely pink or white, <30mm; leaves never hastate at base 3
3 Bracts green, often tinged with violet-blue, much shorter than flowers. Erect perennial to 80cm, glandular above. Native; calcareous grassland, scrub and wood-borders; very local in S En, Mons, natd elsewhere in C & S Br *Meadow Clary* - **S. pratensis** L.
3 Bracts pink or white, at least as long as flowers. Erect biennial or perennial to 1m, glandular above. Intrd-natd; relic or throwout on tips and rough ground, sometimes established on walls; scattered in S Br *Clary* - **S. sclarea** L.
  4 Flowers (8)15-30 in each whorl. Erect perennial to 80cm, with eglandular hairs and sessile glands. Intrd-natd; casual or sometimes natd in rough ground and by roads and railways; scattered in C & S Br, Lanarks *Whorled Clary* - **S. verticillata** L.
  4 Flowers 1-6(8) in each whorl 5
5 Inflorescence with conspicuous tuft of green or coloured flower-less bracts at apex. Erect glandular or eglandular annual to 50cm. Intrd-natd; casual or rarely natd on tips and waste ground; scattered in SE En *Annual Clary* - **S. viridis** L.
5 Inflorescence without terminal tuft of conspicuous flower-less bracts 6
  6 Eglandular; stem-leaves usually >3 pairs, linear to lanceolate-oblong, crenate; calyx with hairs <0.2mm. Erect annual to 60cm. Intrd-casual; waste and rough ground; scattered in En and S Sc *Mintweed* - **S. reflexa** Hornem.
  6 Glandular above; stem-leaves ≤3 pairs, ovate to ovate-oblong, the lower doubly serrate to pinnately lobed and serrate; calyx with some hairs >0.5mm 7
7 Longest hairs on calyx white, eglandular; corolla with 0 or few glandular hairs; lower leaves often distinctly lobed. Erect perennial to 80cm, glandular above. Native; dry grassy and barish rough ground, roadsides, dunes; rather frequent in S & E Bl N to C Sc, N Wa and E Ir *Wild Clary* - **S. verbenaca** L.
7 Longest hairs on calyx brownish, glandular; corolla with many

26. SALVIA

glandular hairs; leaves at most strongly doubly serrate
(see couplet 3) **S. pratensis**

123. HIPPURIDACEAE - *Mare's-tail family*

Rhizomatous, perennial, aquatic or mud-dwelling herbs, with conspicuous air-cavities in stems and very inconspicuous flowers; the only aquatic or mud-plant with >4 (6-12) linear simple entire leaves in a whorl.

1. HIPPURIS L. - *Mare's-tail*

1 Usually aquatic, then with stems to 1(2)m, the apical emergent part bearing flowers, and with leaves up to 8cm; sometimes on mud, then much smaller. Native; in ponds and slow-flowing rivers, especially base-rich; locally frequent throughout BI   *Mare's-tail* - **H. vulgaris** L.

124. CALLITRICHACEAE - *Water-starwort family*

Annual or perennial herbs, aquatic or on mud, with filiform stems; distinguished from other aquatics and mud-dwellers by the opposite, thin, narrow, often notched leaves, single stamen and distinctive fruits.

1. CALLITRICHE L. - *Water-starworts*

Vegetatively very variable and often shy-fruiting; difficult to identify certainly without a strong lens or microscope. The following 3 keys should be used only on material with mature fruits, preferably with some stamens available as well. Plants normally reaching water surface but recently flooded can be told by their terminal rosettes of leaves different in shape from those further back. Fruit-wings are whitish or translucent when viewed in transmitted light. Leaf-shape is notoriously variable and misleading.

General key
1 Fruiting plants terrestrial or on wet mud   *Key C*
1 Fruiting plants mainly or completely submerged   2
  2 Fruiting plants completely submerged   *Key A*
  2 Fruiting plants mainly submerged but with terminal leaf rosette at water surface   *Key B*

Key A - Fruiting plants completely submerged; terminal leaves not in distinct rosette; pollen-grains with ± colourless unsculptured wall (beware re-submerged plants with terminal leaf-rosette).
1 Leaves transparent, all with 1 vein, without stomata; minute sessile hairs present only in leaf-axils, composed of a ± irregular cell-mass   2
1 Leaves opaque, often some with >1 vein, often some with stomata; minute sessile hairs present on leaves, stems and in leaf-axils,

composed of cells radiating from central point 3
2 Fruits ± orbicular in side view, c.1.4-2.2(3.3)mm, with conspicuous wing 0.1-0.5mm wide; leaves mostly ≥1cm, conspicuously notched at apex, usually pale to mid green. Submerged annual to 50cm. Native; lakes and rivers; scattered in BI N from S Lincs, Worcs, Brecs and W Cork, S Wilts

Autumnal Water-starwort - **C. hermaphroditica** L.

2 Fruits wider than long, c.1-1.2 x 1.4-1.6mm, not winged nor sharply keeled; leaves usually <1cm, truncate to shallowly notched at apex; usually dark green. Submerged annual to 50cm. Native; ponds, rivers and canals; very local in En, mostly S & E of range of *C. hermaphroditica*, Co Wexford, Anglesey

Short-leaved Water-starwort - **C. truncata** Guss.

3 Leaves expanded suddenly just at apex with deep notch ± like a bicycle-spanner; fruits ± orbicular in side view, 1.2-1.5mm, with wing usually <0.1mm wide. Submerged, floating or terrestrial annual or perennial. Native; in usually acid ponds, rivers and ditches, less often drying out than for *C. brutia*; probably frequent throughout Bl

Intermediate Water-starwort - **C. hamulata** W.D.J. Koch

3 Leaves not expanded suddenly just at apex, with variable, often shallow or asymmetric notch; fruits often longer than wide, 1-1.4 x 1-1.2mm, with wing usually ≥0.1mm wide. Submerged, floating or terrestrial usually annual. Native; shallow water often drying up in summer; S & W Br N to Kintyre and E to E Kent, W & N Ir, ?CI

Pedunculate Water-starwort - **C. brutia** Petagna

Key B - Fruiting plants aquatic but with terminal rosette of leaves at water surface

1 Flowers submerged; pollen grains with ± colourless unsculptured wall; styles usually persistent, reflexed and appressed to sides of fruit 2
1 Flowers aerial; pollen-grains yellow, with sculptured wall; styles, if persistent, erect to patent 3
   2 Submerged leaves expanded at apex with deep notch ± like a bicycle-spanner; fruits ± orbicular in side view, 1.2-1.5mm, with wing usually <0.1mm wide (see Key A, couplet 3) **C. hamulata**
   2 Submerged leaves not expanded at apex, with variable, often shallow or asymmetric notch; fruits often longer than wide, 1-1.4 x 1-1.2mm, with wing usually ≥0.1mm wide

(see Key A, couplet 3) **C. brutia**

3 Fruits without wing, rounded to obtuse at edges, distinctly longer than wide. Aquatic or terrestrial perennial, but not fruiting if submerged. Native; in and by ponds and streams; rather scattered in BI N to S Sc and Co Antrim

Blunt-fruited Water-starwort - **C. obtusangula** Le Gall

3 Fruits with wing 0.07-0.25mm wide, often nearly orbicular in side view 4
   4 Fruits with wing 0.12-0.25mm wide; stamens c.2mm at anthesis, with anthers c.0.5mm wide; pollen grains ± globose. Aquatic or

1. CALLITRICHE

terrestrial annual or perennial, but not fruiting if submerged. Native; in ponds, rivers, ditches and muddy places; common
Common Water-starwort - C. stagnalis Scop.
4 Fruits with wing 0.07-0.1mm wide; stamens c.4mm at anthesis, with anthers c.1mm wide; many pollen grains mis-shapen or ellipsoid. Aquatic or terrestrial annual or perennial, but not fruiting if submerged. Native; more often in flowing water than *C. stagnalis*; probably widespread in lowlands of BI
Various-leaved Water-starwort - C. platycarpa Kütz.

Key C - Fruiting plants terrestrial or on wet mud
1 Fruits without wing, rounded to obtuse at edges
  (see Key B, couplet 3) **C. obtusangula**
1 Fruits with wing 0.07-0.25mm wide 2
  2 Fruits with a stalk 2-10mm (see Key A, couplet 3) **C. brutia**
  2 Fruits sessile or with stalk <2mm 3
3 Styles usually persistent, reflexed and appressed to sides of fruit; pollen grains with ± colourless unsculptured wall
  (see Key A, couplet 3) **C. hamulata**
3 Styles, if persistent, erect to patent; pollen grains yellow, with sculptured wall 4
  4 Fruits with wing 0.12-0.25mm wide; stamens c.2mm long, with anthers c.0.5mm wide; pollen grains ± globose
    (see Key B, couplet 4) **C. stagnalis**
  4 Fruits with wing 0.07-0.1mm wide; stamens c.4mm at anthesis, with anthers c.1mm wide; many pollen grains mis-shapen or ellipsoid (see Key B, couplet 4) **C. platycarpa**

125. PLANTAGINACEAE - *Plantain family*

Annual to perennial herbs, very rarely dwarf shrubs; easily distinguished by the brownish spike-like inflorescence (or long-stalked solitary male flowers in *Littorella*) and leaves usually all in a basal rosette.

1 Flowers bisexual, in compact spikes; fruit a 2- to many-seeded dehiscent capsule **1. PLANTAGO**
1 Flowers unisexual, the male solitary on long stalks; fruit a 1-seeded nut **2. LITTORELLA**

1. PLANTAGO L. - *Plantains*
Stolons 0; flowers bisexual, in compact spikes, 4-merous; stamens borne on corolla-tube; ovary 2- or 4-celled, each cell with 2-many axile (and basal) ovules; capsule transversely dehiscent.

1 Spikes borne in axils of opposite stem-leaves 2
1 Spikes borne on leafless scapes 3
  2 Inflorescence and peduncles with many glandular hairs; spikes

with all bracts similar. Erect to decumbent pubescent annual to 30(50)cm. Intrd-casual; tips and waste places; scattered throughout much of Br **Glandular Plantain - P. afra** L.

2 Inflorescence and peduncles with 0 glandular hairs; spikes with lower bracts strongly differing from upper bracts. Erect to decumbent pubescent annual to 30(50)cm. Intrd-natd; open and rough ground on sandy soil, casual or sometimes natd; very scattered in S Br **Branched Plantain - P. arenaria** Waldst. & Kit.

3 Scapes strongly furrowed. Glabrous to pubescent perennial with 1-several rosettes. Native; grassy places; abundant
Ribwort Plantain - P. lanceolata L.

3 Scapes not furrowed 4

    4 Corolla-tube glabrous on outside; leaves narrowly elliptic to broadly ovate, at most bluntly and distantly toothed 5

    4 Corolla-tube pubescent on outside; leaves linear to linear-elliptic, often sharply toothed or deeply and finely lobed 7

5 Capsule usually 4-seeded; seeds >2mm; stamens exserted >5mm; anthers >1.5mm. Pubescent perennial with 1-few rosettes. Native; neutral and basic grassland; locally common in En, very local in Ir (intrd), Wa, CI and S & E Sc **Hoary Plantain - P. media** L.

5 Capsule usually >4-seeded; seeds <2mm; stamens exserted <5mm; anthers <1.5mm. Native (**Greater Plantain - P. major** L.) 6

    6 Leaves mostly with 5-9 veins, usually obtuse at apex, subcordate to rounded at base and subentire; capsules mostly with 4-15 seeds; seeds (1)1.2-1.8(2.1)mm. Open and rough ground, either cultivated or grassy, and on lawns; abundant **P. major ssp. major**

    6 Leaves mostly with 3-5 veins, usually subacute at apex, broadly cuneate at base and ± undulate-toothed near base; capsules mostly with (9)14-25(36) seeds; seeds (0.6)0.8-1.2(1.5)mm. Damp, usually slightly saline places near sea and less often inland; probably scattered through much of BI
P. major ssp. intermedia (Gilib.) Lange

7 Capsule 2-celled, with 1-2 seeds per cell; seeds >1.5mm; corolla-lobes with conspicuous brown midrib; leaves usually entire, sometimes slightly toothed. Usually glabrous perennial with 1-many rosettes. Native; in salt-marshes, rock-crevices and short turf near sea, wet rocky places on mountains; common on coasts of BI, local on mountains, rare in inland salty places **Sea Plantain - P. maritima** L.

7 Capsule 3(-4)-celled, with 1-2 seeds per cell; seeds <1.5mm; corolla-lobes with 0 or inconspicuous midrib; leaves usually deeply and narrowly lobed, sometimes toothed, rarely all entire. Usually pubescent annual to perennial with 1-many rosettes. Native; barish places or in very short turf on sandy or gravelly soils, sometimes on rocks; common on coasts of BI, inland in scattered lowland places in C & S Br **Buck's-horn Plantain - P. coronopus** L.

2. LITTORELLA P.J. Bergius - *Shoreweed*

Stoloniferous; flowers unisexual, the male solitary on long scapes and 3-4-

2. LITTORELLA

merous, the female 1-few, sessile at base of male scape and 2-4-merous; ovary 1-celled, with 1(-2) ovules; fruit a 1-seeded nut.

1 Leaves in rosette, semi-cylindrical, usually subulate, 1.5-10(25)cm; male scapes up to as long as leaves. Native; in shallow (rarely down to 4m) water at lake edges, often on exposed shore (only then flowering), on sandy or gravelly acid soils; scattered through much of BI but very local in lowlands   *Shoreweed* - **L. uniflora** (L.) Asch.

126. BUDDLEJACEAE - *Butterfly-bush family*

Deciduous or semi-evergreen shrubs; recognized by the numerous small but brightly coloured flowers with 4 fused sepals and petals, 2-celled ovary and 4 stamens.

1. BUDDLEJA L. - *Butterfly-bushes*

1 Leaves alternate; flowers mauve, borne in small clusters along previous year's wood. Large shrub to 8m with long thin arching branches. Intrd-natd; persistent and sometimes self-sown in woodland, hedges or banks; very scattered in En
   Alternate-leaved Butterfly-bush - **B. alternifolia** Maxim.
1 Leaves opposite; flowers borne in panicles on current year's growth   2
   2 Flowers orange, in dense globose stalked clusters c.2cm across arranged in large open panicles. Shrub to 5m with stiff erect to spreading branches. Intrd-natd; surviving on roadsides and rough ground, sometimes self-sown; very scattered in C & S Br, Man
   Orange-ball-tree - **B. globosa** Hope
   2 Flowers in long dense narrowly pyramidal panicles, often interrupted mostly below and sometimes the segments ± globose   3
3 Flowers dull yellow to greyish- or purplish-yellow. Habit more of *B. davidii* but inflorescence more interrupted. Intrd-surv; status as for *B. globosa* but not self-sowing; scattered in S En (*B. davidii* x *B. globosa*)
   Weyer's Butterfly-bush - **B. x weyeriana** Weyer
3 Flowers lilac to purple or white. Shrub to 5m with long ± arching branches. Intrd-natd; waste ground, walls, banks and scrub; common in S BI, decreasing N to C Sc   *Butterfly-bush* - **B. davidii** Franch.

127. OLEACEAE - *Ash family*

Trees, shrubs or woody trailers; distinguished by the woody habit, opposite leaves, and flowers with 4 fused sepals, 4-6 fused petals (0 or free in *Fraxinus*), 2-celled ovary and 2 stamens.

1 Corolla and calyx 0 (rarely both present and then petals free); flowers

often unisexual; leaves usually pinnate; fruit a winged achene
3. FRAXINUS

1 Corolla showy; calyx present; petals fused; flowers bisexual; leaves simple or with 1 main leaflet plus 1 or 2 small basal ones; fruit a capsule or berry 2
   2 Flowers yellow, appearing before leaves 3
   2 Flowers white, or lilac to mauve or red, appearing after leaves 4
3 Corolla-lobes mostly 4; flowers spring; leaves simple or with 1 main leaflet plus 1 or 2 small basal ones; fruit a capsule **1. FORSYTHIA**
3 Corolla-lobes (5-)6; flowers autumn to spring; leaves ternate; fruit a usually 2-lobed berry **2. JASMINUM**
   4 Fruit a capsule; leaves truncate to cordate at base **4. SYRINGA**
   4 Fruit a berry; leaves cuneate to rounded at base 5
5 Corolla <1cm across, white; leaves simple **5. LIGUSTRUM**
5 Corolla >1cm across; leaves ternate to pinnate, or if simple then corolla red to pink **2. JASMINUM**

1. FORSYTHIA Vahl - *Forsythias*
Deciduous shrubs; leaves simple or with 2 small leaflets at base; petals 4(-6), united into 4(-6)-lobed tube, yellow, appearing before leaves; fruit a capsule.

1 Erect to arching shrub to 5m; flowers in small clusters on old wood, 2-3.5cm. Intrd-natd; relic or throwout in rough ground or on roadsides and walls; few places in S En, Mons, Man (*F. suspensa* (Thunb.) Vahl x *F. viridissima* Lindl.) **Forsythia - F. x intermedia** Zabel

2. JASMINUM L. - *Jasmines*
Deciduous scrambling shrubs, rooting along stems; leaves simple, ternate or pinnate; petals 4-6, united into 4-6-lobed tube; fruit a usually 2-lobed black berry.

1 Flowers yellow, produced well before leaves, not fragrant. Stems scrambling to 5m; leaves ternate. Intrd-natd; stem-rooting on walls and marginal ground; S En **Winter Jasmine - J. nudiflorum** Lindl.
1 Flowers white to red, produced after leaves, fragrant 2
   2 Leaves simple; corolla red to pink, mostly 6-lobed. Erect or scrambling shrub to 2m. Intrd-natd; on walls and in marginal ground, sometimes self-sown; En N to Cheshire
      Red Jasmine - J. beesianum Forrest & Diels
   2 Leaves pinnate; corolla white, often tinged pinkish-purple, mostly 4-5-lobed. Stems scrambling to 10(20)m. Intrd-natd; on walls and in marginal ground, sometimes self-sown; S En
      Summer Jasmine - J. officinale L.

3. FRAXINUS L. - *Ashes*
Deciduous trees; leaves normally pinnate (rare form with 1 leaflet exists); petals 0 (rarely present, free, white); fruit an achene with a single long wing.

3. FRAXINUS

1 Thick-twigged tree to 37m; flowers in dense axillary or terminal panicles appearing before leaves. Native; woods, scrub and hedgerows, especially on damp or base-rich soils; common
Ash - F. excelsior L.

4. SYRINGA L. - *Lilac*

Deciduous shrubs; leaves simple; petals usually 4, united into usually 4-lobed tube, white, or lilac to mauve or purple, appearing after leaves; fruit a capsule.

1 Erect shrub to 7m; leaves cordate to truncate at base, entire; flowers in large terminal panicles, sweetly scented. Intrd-natd; relic or throwout in hedges and road- and railway-banks, spreading by suckers; scattered in most of BI, rarely self-sown **Lilac - S. vulgaris** L.

5. LIGUSTRUM L. - *Privets*

Deciduous to evergreen shrubs; leaves simple; petals 4, united into 4-lobed tube, white, appearing after leaves; fruit a black berry.

1 One-year-old stems and panicle branches densely but minutely pubescent; corolla-tube c. as long as -limb. Native; hedgerows and scrub, especially on base-rich soils; throughout most of BI
Wild Privet - L. vulgare L.

1 One-year-old stems and panicle branches glabrous; corolla-tube distinctly longer than lobes. Erect, nearly evergreen shrub to 3(5)m. Intrd-natd; in hedges and rough ground, rarely self-sown; scattered in most of BI **Garden Privet - L. ovalifolium** Hassk.

128. SCROPHULARIACEAE - *Figwort family*

Herbaceous annuals to perennials or very rarely small shrubs or trees; distinguished from Lamiaceae, Verbenaceae and Boraginaceae by the totally different ovary and fruit, and from other families with zygomorphic flowers and 2-celled ovary by the herbaceous habit (woody in *Paulownia*, *Hebe* and *Phygelius*), 2 or 4 stamens (5 in *Verbascum*), and non-spiny bracts. See also Orobanchaceae.

1 Stamens 2   2
1 Stamens 4 or 5   4
    2 Corolla 2-lipped, the lower lip large and inflated, yellow to reddish-brown; stamens included **7. CALCEOLARIA**
    2 Corolla 4-lobed, without inflated portion, white to blue or pink; stamens exserted   3
3 Stems woody; leaves evergreen, entire **18. HEBE**
3 Herbaceous plants; stems not woody or woody only at base and then with serrate leaves **17. VERONICA**
    4 Deciduous tree; flowers purplish-blue to violet **4. PAULOWNIA**

4	Herbaceous, or if a woody shrub then flowers red	5
5	Flowers 2.5-4cm, red, pendent in terminal panicles; stems woody	**3. PHYGELIUS**
5	Flowers often <2cm, if >2cm and red then in racemes; stems herbaceous or ± so	6
6	Corolla with conspicuous basal spur or pouch on lowerside	7
6	Corolla not spurred or pouched at base	13
7	Leaves palmately veined and lobed	8
7	Leaves with single midrib, often also with lateral pinnate veins, entire to serrate	9
8	Plant glandular-pubescent; corolla ≥3cm, yellow with purple veins, pouched at base	**11. ASARINA**
8	Plant glabrous to minutely pubescent, not glandular; corolla ≤3cm, mauve to purple, often with yellow centre, spurred at base	**12. CYMBALARIA**
9	Corolla-tube with broad, rounded pouch (wider than long) at base	10
9	Corolla-tube with narrow, often pointed, spur (longer than wide)	11
10	Calyx-lobes ± equal, all shorter than corolla-tube; corolla >2.5cm	**8. ANTIRRHINUM**
10	Calyx-lobes distinctly unequal, all longer than corolla-tube; corolla <2cm	**10. MISOPATES**
11	Leaves ovate to obovate, rounded to cordate at base; capsule opening by detachment of 2 oblique lids leaving large pores	**13. KICKXIA**
11	Leaves linear to lanceolate or oblanceolate, narrowed to base, rarely ovate to obovate and rounded at base and then capsule without detachable lids	12
12	Mouth of corolla completely closed by boss-like swelling on lower lip	**14. LINARIA**
12	Mouth of corolla incompletely closed by small swelling	**9. CHAENORHINUM**
13	Fertile stamens 5, at least in most flowers	**1. VERBASCUM**
13	Fertile stamens 4, sometimes with a sterile staminode representing fifth	14
14	Leaves all in basal rosettes, linear to spathulate	**6. LIMOSELLA**
14	At least some leaves borne on stems	15
15	Stems procumbent, rooting at nodes; leaves reniform, long-stalked; flowers solitary in leaf-axils	**19. SIBTHORPIA**
15	Stems not procumbent and rooting at nodes; leaves not reniform; flowers in terminal inflorescences	16
16	Calyx 5-lobed or -toothed	17
16	Calyx 4-lobed or -toothed, rarely 2-5-lobed with toothed lobes and inflated tube	20
17	Leaves opposite	18
17	Leaves alternate	19
18	Calyx-tube shorter than -lobes; corolla purplish-brown to dull yellowish-green, with tube scarcely longer than wide	**2. SCROPHULARIA**
18	Calyx-tube longer than -teeth; corolla bright yellow, often with	

128. SCROPHULARIACEAE

　　　　red spots or blotches, with tube much longer than wide
　　　　　　　　　　　　　　　　　　　　　　　　　　5. MIMULUS
19 Corolla distinctly zygomorphic, with tube >2x as long as calyx, lobes
　　not strongly patent　　　　　　　　　　　　　　**15. DIGITALIS**
19 Corolla scarcely zygomorphic, with tube <2x as long as calyx, with
　　strongly patent lobes　　　　　　　　　　　　　　**16. ERINUS**
　　20 Calyx irregularly 2-5-lobed, with toothed lobes; leaves divided
　　　　almost to base, the lobes toothed　　　　**26. PEDICULARIS**
　　20 Calyx regularly 4-lobed, with entire lobes; leaves entire to
　　　　simply toothed up to c.1/2 way to base　　　　　　　　21
21 Calyx-tube inflated, especially at fruiting; seeds discoid, with
　　marginal wing　　　　　　　　　　　　　　　**25. RHINANTHUS**
21 Calyx-tube not inflated; seeds not discoid, without marginal wing　22
　　22 Lower lip of corolla with 3 distinctly emarginate lobes
　　　　　　　　　　　　　　　　　　　　　　　21. EUPHRASIA
　　22 Lower lip of corolla with 3 entire lobes, or sometimes middle
　　　　lobe slightly emarginate　　　　　　　　　　　　　　23
23 Mouth of corolla partially closed by boss-like swellings on lower
　　lip; capsules with 1-4 seeds　　　　　　　　**20. MELAMPYRUM**
23 Mouth of corolla open; lower lip without swellings; capsules with
　　>4 seeds　　　　　　　　　　　　　　　　　　　　　　24
　　24 Corolla yellow　　　　　　　　　　　　　　　　　　25
　　24 Corolla pink to dark purple, rarely white　　　　　　　26
25 Leaves serrate; seeds c.0.5mm, ± smooth　　　**24. PARENTUCELLIA**
25 Leaves entire or ± so; seeds >1mm, ridged and grooved
　　　　　　　　　　　　　　　　　　　　　　　22. ODONTITES
　　26 Perennial; corolla dark purple, >12mm　　　　**23. BARTSIA**
　　26 Annual; corolla pink to reddish-purple, rarely white, <12mm
　　　　　　　　　　　　　　　　　　　　　　　22. ODONTITES

TRIBE 1 - SCROPHULARIEAE (genera 1-4). Herbaceous annuals or perennials or rarely shrubs; leaves alternate or opposite; flowers nearly actinomorphic to zygomorphic; corolla not to obscurely 2-lipped, not spurred or pouched at base; stamens 4-5.

1. VERBASCUM L. - *Mulleins*
Herbaceous biennials, less often annuals or perennials, with tall, erect, terete to ridged stems; leaves alternate; corolla yellow, purplish or white, with 5 ± equal lobes and short tube; stamens 5, rarely 4 in some flowers, all fertile, at least the upper 3 with very hairy filaments.

1　Anthers all reniform, symmetrical, placed transversely on filaments　2
1　Anthers of 3 upper stamens as above, of 2 lower stamens
　　asymmetrical, placed obliquely or ± longitudinally on filaments
　　and often ± decurrent on them　　　　　　　　　　　　　8
　　2　All hairs on filaments yellow or white　　　　　　　　3
　　2　All or many hairs on filaments violet　　　　　　　　5
3　Stems and leaves uniformly and persistently densely pubescent.

Biennial to 2m. Intrd-natd; escape in waste and rough ground;
W Kent, W Norfolk and E Suffolk

Hungarian Mullein - **V. speciosum** Schrad.

3 Stems and leaves unevenly mealy- or powdery-pubescent at first,
becoming less pubescent to glabrous later   4

    4 Corolla usually white (less often yellow); all or most pedicels
>6mm; leaves sparsely pubescent to glabrous and green on
upperside. Biennial to 1.5m. Native; barish ground on chalky
soils; local in S En from Devon to E Kent, infrequent escape
elsewhere in Br   *White Mullein* - **V. lychnitis** L.

    4 Corolla ± always yellow; all or most pedicels <6mm; leaves
whitish-pubescent on upperside. Biennial to 1.5m. Native;
barish ground on chalky soils; E Anglia (mostly Norfolk), rare
escape elsewhere in Br   *Hoary Mullein* - **V. pulverulentum** Vill.

5 Flowers 1 per node in axil of bract; bracteoles 0   6
5 Flowers 2-several per node in axil of bract; each pedicel with 2 small
bracteoles   7

    6 Corolla violet to purple; hairs all simple, mostly glandular;
inflorescence usually simple. Perennial to 1m. Intrd-casual; on
tips and waste ground; very scattered in S & C Br

Purple Mullein - **V. phoeniceum** L.

    6 Corolla yellow; many hairs stellate; inflorescence much branched.
Perennial to 1.5m. Intrd-natd; waste and rough ground; scattered
in S En   *Caucasian Mullein* - **V. pyramidatum** M. Bieb.

7 Basal leaves truncate to rounded at base; pedicels all similar length,
c. as long as calyx. Biennial or perennial to 1m. Intrd-natd; escape
on waste or rough ground; very scattered in S Br

Nettle-leaved Mullein - **V. chaixii** Vill.

7 Basal leaves cordate at base; pedicels variable in length, many c. as
long as calyx, the longest ≥2x as long as calyx. Biennial or perennial
to 1.2m. Native; waste and rough ground, open places on banks and
in grassland, mostly on soft limestone; locally common in CI and
C & S Br, sparse casual elsewhere   *Dark Mullein* - **V. nigrum** L.

    8 Simple stalked glands present; branched hairs present or 0   9
    8 Simple stalked glands 0; branched hairs always present   10

9 Flowers 1 per node; pedicels mostly longer than calyx; plant usually
with stalked glands only in upper parts. Annual to biennial to 1m.
Intrd-natd; waste and rough ground; scattered in C & S Br

Moth Mullein - **V. blattaria** L.

9 Flowers usually >1 per node in lower parts of inflorescence; pedicels
mostly shorter than calyx; plant usually with stalked glands ±
throughout. Biennial to 1m. Native; fields, waste places and dry
banks; locally frequent in Devon and Cornwall, casual elsewhere
in Br   *Twiggy Mullein* - **V. virgatum** Stokes

    10 Upper and middle stem-leaves distinctly decurrent   11
    10 Stem-leaves not decurrent   12

11 Stigma capitate. Biennial to 2m. Native; waste and rough ground,
banks and grassy places, mostly on sandy or chalky soils; common

1. VERBASCUM

in C & S Br and CI, local elsewhere   *Great Mullein* - **V. thapsus** L.
11 Stigma elongated, spathulate. Biennial to 2m. Intrd-natd; waste and rough ground; scattered in C & S En
Dense-flowered Mullein - **V. densiflorum** Bertol.
    12 Two lower stamens with filaments pubescent in lower 1/2, with anthers ≤4mm. Biennial to 2m. Intrd-natd; waste and rough ground; scattered in S En and Guernsey
Broussa Mullein - **V. bombyciferum** Boiss.
    12 Two lower stamens with glabrous to subglabrous filaments, with anthers ≥4mm. Biennial to 2m. Intrd-natd; waste and rough ground; frequent in C & S Br and CI, rare in N Br
Orange Mullein - **V. phlomoides** L.

Hybrids arise frequently where 2 or more spp. occur together; they are usually highly but often not completely sterile. 12 combinations have been recorded in the wild. Characters may be as in 1 or other parent or intermediate. In hybrids between spp. with violet stamen-hairs and spp. with white stamen-hairs all the stamens may have all violet hairs, all have all pale violet hairs, all have mixed white and violet hairs, or the upper 3 have white and the lower 2 have violet hairs. Oblique-asymmetrical anthers of 1 parent may or may not appear in the hybrid. Hybrids between white-flowered *V. lychnitis* and yellow-flowered spp. are yellow-flowered.

2. SCROPHULARIA L. - *Figworts*
Herbaceous perennials, less often biennials, with erect, usually 4-angled stems usually square in section; leaves opposite; corolla dull purplish, brownish or greenish-yellow, with 5 nearly equal lobes and short tube, the lobes usually obscurely organised into 2-lobed upper and 3-lobed lower lips; fertile stamens 4, with the 5th (uppermost) usually represented by a sterile staminode.

1   Staminode 0; corolla-lobes equal, not organized into upper and lower lips. Biennial or perennial to 50(80)cm. Intrd-natd; waste and rough ground, hedges and woodland clearings; scattered throughout much of Br   *Yellow Figwort* - **S. vernalis** L.
1   Staminode present, larger than anthers; corolla-lobes unequal, distinctly forming 2 weak lips   2
    2   Leaves and stems pubescent. Perennial to 1m. Native; grassy field-borders and hedgerows; very local, mostly near coast; Devon and Cornwall, CI, natd in S Wa   *Balm-leaved Figwort* - **S. scorodonia** L.
    2   Leaves and stems ± glabrous   3
3   Stem-angles not or scarcely winged; sepals with narrow scarious border <0.3mm wide. Perennial to 1m. Native; damp open or shady places and in hedgerows; common   *Common Figwort* - **S. nodosa** L.
3   Stem-angles distinctly winged; sepals with conspicuous scarious border >0.3mm wide   4
    4   Staminode ± orbicular, entire; leaves broadly cuneate to subcordate at base. Perennial to 1.2m. Native; places similar to or

wetter than those of *S. nodosa*; common throughout En, Wa and CI, frequent in Ir, rare in Sc and Man **Water Figwort - S. auriculata** L.

4 Staminode bifid or with 2 divergent lobes at apex; leaves cuneate to rounded at base. Perennial to 1.2m. Native; damp shady places; very locally scattered in Br and Ir

Green Figwort - **S. umbrosa** Dumort.

3. PHYGELIUS Benth. - *Cape Figwort*

Evergreen shrubs with 4-angled stems; leaves opposite; corolla red, trumpet-shaped, with 5 nearly equal lobes not organized into 2 lips and with tube much longer than lobes; stamens 4, exserted.

1 Glabrous shrub to 1.5m; flowers pendent, in large terminal erect panicles, 2.5-4cm. Intrd-natd; by rivers and other wet places; Co Wicklow, less well natd in Dunbarton, E & W Kent and W Galway

Cape Figwort - **P. capensis** Benth.

4. PAULOWNIA Siebold & Zucc. - *Foxglove-tree*

Deciduous tree with spreading crown; leaves opposite; corolla purplish-blue to violet, trumpet-shaped, with 5 nearly equal lobes scarcely organized into 2 lips, with tube much longer than lobes; stamens 4, included.

1 Tree to 26m; leaves broadly ovate, cordate at base, often shallowly lobed; flowers erect, in large terminal erect panicles, 3.5-6cm. Intrd-natd; self-sown in rough ground; Middlesex

Foxglove-tree - **P. tomentosa** (Thunb.) Steud.

TRIBE 2 - GRATIOLEAE (genera 5-6). Herbaceous annuals to perennials; leaves opposite (at least below) or all in basal rosette; flowers nearly actinomorphic to strongly zygomorphic; corolla not to strongly 2-lipped, not spurred or pouched at base; stamens 4.

5. MIMULUS L. - *Monkeyflowers*

Perennials with leafy stolons; leaves opposite; flowers strongly zygomorphic, with well-defined upper and lower lips, showy, yellow to red.

M. guttatus and *M. luteus* agg. (incl. *M. variegatus* Lodd., invalid name and *M. cupreus*) form a difficult, interfertile group; hybrids are frequent and at least 3 occur in the absence of either parent. Hybrids between *M. guttatus* and *M. luteus* agg. are sterile (pollen <40% full; seeds 0 or few per capsule), but hybrids within *M. luteus* agg. are fertile (pollen >50% full; seeds many per capsule). In this account *M. cupreus* is considered distinct from *M. luteus*, but *M. variegatus* is included in the latter, as in most American Floras.

All taxa occur in wet or damp places by rivers, streams, ponds and ditches, etc.

1 Calyx-teeth ± equal; plant glandular-pubescent ± all over; corolla <2.5cm. Stems decumbent to ascending, to 40cm. Intrd-natd;

5. MIMULUS

scattered over most of Br and Ir   *Musk* - **M. moschatus** Lindl.

1 Upper calyx-tooth distinctly longer than lower 4; plant glabrous below, often glandular pubescent above; corolla >2.5cm   **2**

    2 Inflorescence with abundant stalked glands; small simple hairs present at least on keels of calyx, often also elsewhere in inflorescence; plant fertile or sterile   **3**

    2 Inflorescence glabrous or with sparse stalked glands; small simple hairs 0 except inside calyx; plant fertile   **5**

3 Corolla copper-coloured, often also spotted or blotched red or purplish; calyx sometimes petaloid. Stems erect to ascending, to 50cm. Intrd-natd; W & N Br N from S Somerset and Yorks (*M. guttatus* x *M. cupreus* Dombrain)
                       Coppery Monkeyflower - **M. x burnetii** S. Arn.(*)

3 Corolla yellow, often spotted or blotched with orange, red or purplish; calyx not petaloid   **4**

    4 Throat of corolla ± closed by 2 boss-like swellings on lower lip; corolla wholly yellow or with red spots in throat, but with unmarked lobes; plant usually fertile. Stems erect to ascending, to 75cm. Intrd-natd; scattered and locally common over most of BI, commonest lowland taxon   *Monkeyflower* - **M. guttatus** DC.(**)

    4 Throat of corolla ± open, the boss-like swellings low or inconspicuous; corolla lobes with orange, red or purplish spots or blotches; plant usually sterile. Stems erect to ascending, to 50cm. Intrd-natd; locally common over much of Br and Ir, mostly in N & W, commonest upland taxon (*M. guttatus* x *M. luteus*)
                     Hybrid Monkeyflower - **M. x robertsii** Silverside

5 Leaves with even, triangular, flat teeth; corolla yellow, usually with coppery-orange spots or blotches, or mainly coppery. Stems decumbent to ascending, to 50cm. Sc N from Peebless (*M. luteus* x *M. cupreus*)   *Scottish Monkeyflower* - **M. x maculosus** T. Moore

5 Leaves with irregular, oblong, often twisted teeth; corolla yellow, with 1 or more dark red to purplish-brown blotches. Stems decumbent to ascending, to 50cm. Intrd-natd; rather uncommon in N Br S to Durham   *Blood-drop-emlets* - **M. luteus** L.

(*) The rarer *M. guttatus* x *M. luteus* x *M. cupreus* would also key out here.
(**) *M. guttatus* is fertile; sterile or partially sterile plants keying out here are hybrids of *M. guttatus* otherwise ± indistinguishable from it.

6. **LIMOSELLA** L. - *Mudworts*

Glabrous annuals (usually) with thin leafless stolons; leaves ± erect in basal rosette; flowers ± actinomorphic, small and inconspicuous, solitary on pedicels from leaf-rosettes.

1 Leaves (incl. petioles) to 6(10)cm, mostly with narrowly elliptic blades up to 2cm; flowers 2.5-3mm; calyx longer than corolla-tube; corolla white to pale mauve. Native; wet sandy mud by ponds, often dried out in summer; very scattered in Br and Ir   *Mudwort* - **L. aquatica** L.

1 Leaf-blades and petioles not differentiated, subulate, to 4cm; flowers

3.5-4mm; calyx shorter than corolla-tube; corolla white. Native; similar places to *L. aquatica*; extremely local in Glam, Merioneth and Caerns **Welsh Mudwort - L. australis** R.Br.

L. aquatica x *L. australis* occurs with the parents in Glam; it is sterile, with intermediate leaves and calyx length but more vigorous than either parent.

TRIBE 3 - CALCEOLARIEAE (genus 7). Herbaceous annuals; leaves opposite; flowers strongly zygomorphic; corolla strongly 2-lipped, not spurred or pouched at base, with greatly incurved 'slipper-like' lower lip; stamens 2.

7. CALCEOLARIA L. - *Slipperwort*

1 Stems well-branched, to 40cm, glandular-pubescent; leaves pinnate, with serrate leaflets; corolla yellow, 10-14mm. Intrd-natd; tips, waste places, cultivated ground; scattered in Br N to S Sc, especially E Anglia **Slipperwort - C. chelidonioides** Kunth

TRIBE 4 - ANTIRRHINEAE (genera 8-14). Herbaceous annuals to perennials, rarely ± woody at base; leaves usually opposite below, often mostly alternate; flowers strongly zygomorphic; corolla strongly 2-lipped, with boss-like swelling on lower lip and conspicuous spur or pouch at base of tube; stamens 4.

8. ANTIRRHINUM L. - *Snapdragon*

Tufted, perennating by means of basal shoots; leaves entire, with single midrib; calyx-lobes ± equal, shorter than corolla-tube; corolla with broad, rounded pouch at base, with mouth closed by boss-like swelling on lower lip; capsule opening by 3 apical pores.

1 Stems erect to ascending, glabrous or glandular-pubescent above, to 1m; corolla 3-4.5cm, usually pink to purple, sometimes white, yellow, orange or combinations. Intrd-natd; rough ground, walls, rocks, buildings; throughout most of BI **Snapdragon - A. majus** L.

9. CHAENORHINUM (Duby) Rchb. - *Toadflaxes*

Annuals, or tufted perennials with basal new shoots; leaves entire, with single midrib; calyx with slightly unequal lobes >1/2 as long to c. as long as corolla-tube; corolla with narrow conical spur at base, with mouth not completely closed by low boss-like swelling on lower lip; capsule opening by irregular large apical pores or tears.

1 Lower leaves elliptic; corolla (incl. spur) 8-15(20)mm, bluish-mauve with pale yellow boss. Stems decumbent to erect, to 30cm. Intrd-natd; on old walls; West Malling (W Kent), rare and impermanent elsewhere in En **Malling Toadflax - C. origanifolium** (L.) Kostel.
1 Lower leaves oblong-oblanceolate; corolla (incl. spur) 6-9mm, pale

purple with pale yellow boss. Stems erect, to 25cm. Native; arable land, waste places, railway tracks, in open ground; frequent over most of Br and (intrd) Ir *Small Toadflax* - **C. minus** (L.) Lange

10. MISOPATES Raf. - *Weasel's-snouts*

Annuals; leaves entire, with single midrib; calyx-lobes distinctly unequal, all longer than corolla-tube; corolla and capsule as in *Antirrhinum*.

1 Stems usually glandular-pubescent above, to 50cm; corolla 10-17mm, usually bright pink, rarely white. Probably native; weed of cultivated ground; locally frequent in S Br and CI, very scattered elsewhere *Weasel's-snout* - **M. orontium** (L.) Raf.
1 Stems usually glabrous, to 50cm; corolla 18-22mm, pale pink to white. Intrd-casual; tips and waste ground; rather scarce in S En
Pale Weasel's-snout - **M. calycinum** Rothm.

11. ASARINA Mill. - *Trailing Snapdragon*

Stoloniferous perennials; leaves crenate to shallowly lobed, with palmate main veins; calyx with slightly unequal lobes much shorter than corolla-tube; corolla as in *Antirrhinum*; capsule opening by 2 apical pores.

1 Stems procumbent, glandular-pubescent, to 60cm; corolla 3-3.5cm, pale yellow with pinkish-purple veins, the boss deep yellow. Intrd-natd; dry banks, walls and cliffs; scattered in Br N to Dumfriess
Trailing Snapdragon - **A. procumbens** Mill.

12. CYMBALARIA Hill - *Toadflaxes*

Stoloniferous perennials; leaves subentire to shallowly and obtusely lobed, with palmate main veins; calyx with unequal lobes shorter than corolla-tube; corolla with narrow cylindrical to conical spur at base, with mouth completely closed by boss-like swelling on lower lip; capsule opening by irregular apical longitudinal slits.

1 Corolla 9-15mm, incl. spur 1.5-3mm; stems trailing, often >20cm. Intrd-natd
 (*Ivy-leaved Toadflax* - **C. muralis** P. Gaertn., B. Mey. & Scherb.) 2
1 Corolla 15-25mm, incl. spur 4-9mm; stems usually not trailing, <20cm 3
 2 Plant glabrous, or sparsely pubescent on calyx and young parts. Walls, pavements, rocky or stony banks; frequent over most of BI **C. muralis ssp. muralis**
 2 Plant pubescent on ± all parts. On waste ground; Surrey
C. muralis ssp. visianii (Ják.) D.A. Webb
3 Stems, leaves, petioles and calyx shortly and densely pubescent. Intrd-natd; on walls, shingle and stony places; scattered in Br, mainly N En and Sc *Italian Toadflax* - **C. pallida** (Ten.) Wettst.
3 Plant glabrous or nearly so. Intrd-natd; near gardens and nurseries; very scattered in En and Sc
Corsican Toadflax - **C. hepaticifolia** (Poir.) Wettst.

13. KICKXIA Dumort. - *Fluellens*

Annuals; leaves ovate to elliptic, ± entire to remotely dentate, with pinnate venation; calyx with equal lobes c. as long as corolla-tube; corolla with narrowly conical spur at base, with mouth completely closed by boss-like swelling on lower lip; capsule opening by 2 large oblique lids.

1 Pedicels pubescent only immediately below flower; leaves hastate; corolla 7-12mm, with violet upper lip. Stems procumbent to suberect, to 50cm. Probably native; arable fields on light, usually calcareous soils; locally common in Br N to N Wa and N Lincs, CI, intrd in S & W Ir *Sharp-leaved Fluellen* - **K. elatine** (L.) Dumort.
1 Pedicels pubescent; leaves rounded at base; corolla 8-15mm, with purple upper lip. Stems procumbent to suberect, to 50cm. Probably native; similar places to *K. elatine*, often with it; SE Br NW to N Lincs and S Wa *Round-leaved Fluellen* - **K. spuria** (L.) Dumort.

14. LINARIA Mill. - *Toadflaxes*

Annuals or perennials, sometimes rhizomatous; leaves entire, with single midrib; calyx with usually unequal lobes shorter than corolla-tube; corolla with narrow conical spur at base, with mouth closed (or ± so) by boss-like swelling on lower lip; capsule opening by irregular apical longitudinal slits.

1 Spur longer than rest of corolla. Erect annual to 50cm; corolla very varied in colour. Intrd-casual; tips and waste places; scattered in Br and CI, mainly S En *Annual Toadflax* - **L. maroccana** Hook. f.
1 Spur much shorter than to nearly as long as rest of corolla   2
  2 Whole plant glandular-pubescent; corolla (incl. spur) 4-7mm. Erect annual to 15cm; corolla yellowish with yellowish to violet spur. Intrd-natd; semi-fixed dunes; Braunton Burrows (N Devon) *Sand Toadflax* - **L. arenaria** DC.
  2 Plant glabrous below, glabrous to glandular-pubescent above; corolla (incl. spur) usually ≥8mm   3
3 Corolla predominantly yellow, sometimes very pale or with purplish tinge (if with violet veins probably *L. vulgaris* x *L. repens*)   4
3 Corolla predominantly mauve, violet, purple or pink, sometimes very pale but then with darker veins, sometimes with yellow to orange boss   6
  4 Annual; stems decumbent, with conspicuous region below inflorescence bare of leaves. Procumbent to decumbent annual to 20cm; corolla pale yellow with yellowish-orange boss. Possibly native; sandy and waste ground; Cornwall, Devon and Carms, rare casual elsewhere *Prostrate Toadflax* - **L. supina** (L.) Chaz.
  4 Perennial; stems normally erect, with leaves ± up to inflorescence; rhizomes often present   5
5 Seeds disc-like, with broad wing round circumference; plant often glandular-pubescent above; leaves linear to narrowly elliptic-oblanceolate, cuneate at base. Erect to ascending perennial to 80cm; corolla yellow with orange boss. Native; rough and waste ground,

14. LINARIA

stony places, banks, open grassland; common over most of BI
Common Toadflax - **L. vulgaris** Mill.
5 Seeds angular, scarcely winged; plant always glabrous; at least some leaves lanceolate to ovate and subcordate at base. Erect perennial to 80cm; corolla yellow. Intrd-natd; waste places, waysides and by railways; scattered in S En   *Balkan Toadflax* - **L. dalmatica** (L.) Mill.
  6 Annual; capsule shorter than calyx; seeds disc-like, with broad wing round circumference. Erect annual to 30cm; corolla purplish-violet with whitish boss. Native; rough ground, rocky places and hedgebanks; last seen in 1955 in Jersey, rare casual in C & S Br   *Jersey Toadflax* - **L. pelisseriana** (L.) Mill.
  6 Perennial; capsule longer than calyx; seeds angular, not winged   7
7 Spur <1/2 as long as rest of corolla, straight, subacute to rounded at tip; corolla with orange patch on boss. Decumbent to erect perennial to 80cm; corolla whitish to pale mauve with violet thin veins. Native; stony places, rough ground, banks and walls; scattered over much of BI, frequent only in S & W Br
Pale Toadflax - **L. repens** (L.) Mill.
7 Spur ≥1/2 as long as rest of corolla, usually curved, acute at tip; corolla without orange (sometimes with white) patch on boss. Erect perennial to 1m; corolla mauve with heavy purplish-violet veins or wholly purplish-violet, rarely pink. Intrd-natd; rough ground, walls, banks; frequent through much of BI
Purple Toadflax - **L. purpurea** (L.) Mill.

L. vulgaris x *L. repens* = ***L. x sepium*** G.J. Allman and *L. purpurea* x *L. repens* = ***L. x dominii*** Druce are frequent within the range of *L. repens*; both are fertile and can form widely varying hybrid swarms. *L. repens* x *L. supina* = ***L. x cornubiensis*** Druce has been found twice.

TRIBE 5 - DIGITALEAE (genera 15-19). Herbaceous annuals to perennials or sometimes shrubs; leaves opposite or alternate; flowers nearly actinomorphic to zygomorphic; corolla not to weakly 2-lipped, not spurred or pouched at base; stamens 2 or 4 (rarely 3 or 5).

15. DIGITALIS L. - *Foxgloves*
Herbaceous biennials to perennials; leaves alternate, with pinnate venation; flowers in terminal racemes; corolla showy, weakly 2-lipped, with 5 lobes shorter than tube; stamens 4.

1 Erect usually densely pubescent biennial to short-lived perennial to 2m; leaves ± rugose; corolla 40-55mm, pink to purple with dark spots inside, sometimes white. Native; open places, especially woodland clearings, heaths and mountainsides, also waste ground, on acid soils; common   *Foxglove* - **D. purpurea** L.
1 Erect glabrous to slightly pubescent usually perennial to 1m; leaves ± smooth; corolla 9-25mm, pale yellow. Intrd-natd; waste ground, roadsides and walls; scattered in S En   *Straw Foxglove* - **D. lutea** L.

16. ERINUS L. - *Fairy Foxglove*

Herbaceous perennials; leaves alternate, with pinnate venation; flowers in terminal raceme; corolla showy, not 2-lipped, with 5 ± equal, patent, emarginate lobes little shorter than tube; stamens 4.

1 Stems to 20cm, ascending to suberect, pubescent; corolla-tube 3-7mm, corolla-limb 6-9mm across, purple, sometimes white. Intrd-natd; on walls and in stony places; scattered in BI, mainly N En to C Sc

Fairy Foxglove - **E. alpinus** L.

17. VERONICA L. - *Speedwells*

Herbaceous annuals to perennials, sometimes woody at base; leaves opposite, at least below, with pinnate or sometimes ± palmate venation; flowers solitary in leaf-axils or in terminal or axillary racemes; corolla showy to inconspicuous, not 2-lipped, with 4 subequal lobes much longer than tube; stamens 2.

General key
1 Flowers (at least mostly) in axillary racemes (include plants with a single raceme in 1 of the most apical leaf-axils) **Key A**
1 Flowers in terminal racemes or solitary in leaf-axils 2
  2 Flowers in terminal racemes; bracts all or at least the upper very different from the foliage leaves, but the lower sometimes similar to them **Key B**
  2 Flowers solitary in axils of leaves closely resembling the foliage leaves, though the upper often smaller than them **Key C**

Key A - Flowers in axillary racemes
1 Leaves and stems glabrous, except stems sometimes glandular-pubescent in inflorescence 2
1 Leaves and stems pubescent 6
  2 Racemes 1 per node; capsule dehiscing into 2 valves. Stems decumbent to scrambling-erect, to 60cm. Native; bogs, marshes, wet meadows, by ponds and lakes; locally frequent throughout BI *Marsh Speedwell* - **V. scutellata** L.
  2 Racemes mostly 2 per node; capsule dehiscing into 4 valves 3
3 Leaves all shortly petiolate; flowering stems procumbent or decumbent (to ascending). Stems to 60cm. Native; streams, ditches, marshes, pondsides and river-banks; common
Brooklime - **V. beccabunga** L.
3 Upper leaves sessile; flowering stems usually erect (to ascending) 4
  4 Plants sterile or slightly fertile, with racemes increasing in length through season. Native; frequent with parents or sometimes without, especially in S & E En (*V. anagallis-aquatica* x *V. catenata*)
Hybrid Water-speedwell - **V. x lackschewitzii** J.B. Keller
  4 Plants normally fully fertile, with racemes ceasing growth as fruits ripen 5
5 Corolla usually pale blue; pedicels erecto-patent in fruit; capsule ±

17. VERONICA

orbicular, not or shallowly notched at apex. Stems erect, to 50cm. Native; by ponds and streams, in marshes and wet meadows; scattered and locally common
Blue Water-speedwell - **V. anagallis-aquatica** L.

5 Corolla usually pinkish; pedicels patent in fruit; capsule wider than long, deeply notched at apex. Stems erect, to 50cm. Native; mostly in open muddy places with little or no flowing water; locally frequent in En, Wa and Ir, rare elsewhere
Pink Water-speedwell - **V. catenata** Pennell

6 Stems pubescent along 2 opposite lines only; capsule shorter than calyx. Stems erect to ascending, to 50cm. Native; woods, hedgerows, grassland in damper areas; common
Germander Speedwell - **V. chamaedrys** L.

6 Stems pubescent all round; capsule longer than calyx   7

7 Petioles >6mm; capsules >6mm wide. Stems procumbent to ascending, to 40cm. Native; dampish woods; scattered and locally frequent through most of Br and Ir   *Wood Speedwell* - **V. montana** L.

7 Petioles <6mm; capsules <6mm wide   8

   8 Leaves linear-oblong to -lanceolate; pedicels >6mm in fruit, longer than bracts   (see couplet 2) **V. scutellata**

   8 Leaves lanceolate-oblong to ovate or elliptic; pedicels <6mm in fruit, up to as long as bracts   9

9 Calyx-lobes usually 5, 1 much shorter than other 4; leaves sessile. Stems decumbent to erect, to 50cm. Intrd-natd; open and rough ground, dunes; scattered in Br N to Angus
Large Speedwell - **V. austriaca** L.

9 Calyx-lobes 4; at least lower leaves petiolate. Stems procumbent to ascending, to 40cm. Native; banks, open woods, grassland and heathland on well-drained soils; common
Heath Speedwell - **V. officinalis** L.

Key B - Flowers in terminal racemes

1 Annuals with 1 root system, easily uprooted   2

1 Perennials, with non-flowering shoots and/or stems rooted more than just at base   8

   2 Plant glabrous. Stems erect, to 25cm. Intrd-surv; casual or persistent weed of gardens, nurseries and public flower-beds; scattered in BI N to C Sc   *American Speedwell* - **V. peregrina** L.

   2 Plant pubescent or minutely so, at least on capsules and inflorescence-axis   3

3 Bracts much longer than fruiting pedicels   4

3 Bracts shorter than to ± as long as fruiting pedicels   6

   4 At least some upper leaves lobed >1/2 way to midrib. Stems erect, to 15cm. Native; open places in poor grassland on dry sandy soils, often with *V. arvensis*; very local in W Suffolk
Spring Speedwell - **V. verna** L.

   4 All leaves entire to crenate-serrate, toothed <1/2 way to midrib   5

128. SCROPHULARIACEAE

5 All hairs glandular; leaves oblanceolate to narrowly oblong.
(see couplet 2) **V. peregrina**
5 Many hairs, at least below, non-glandular; leaves ovate. Stems erect to decumbent, to 30cm. Native; walls, banks, open acid or calcareous ground and cultivated land; common **Wall Speedwell - V. arvensis** L.

6 Lower bracts and upper leaves lobed much >1/2 way to base. Stems erect to ascending, to 20cm. Native; sandy arable fields; very local in W Norfolk and E & W Suffolk
Fingered Speedwell - **V. triphyllos** L.
6 Bracts and leaves toothed <1/2 way to midrib 7

7 Capsule notched to c.1/2 way; pedicels >2x as long as calyx; seeds flat. Stems erect, to 15cm. Intrd-surv; casual or persistent weed of gardens, nurseries and public flower-beds; scattered in S En
French Speedwell - **V. acinifolia** L.
7 Capsule notched to ≤1/4 way; pedicels <2x as long as calyx; seeds cup-shaped. Stems erect to ascending, to 20cm. Perhaps native; sandy arable fields; very local in W Suffolk and W Norfolk
Breckland Speedwell - **V. praecox** All.

8 Corolla-tube usually >2mm, longer than wide; racemes dense, long and many-flowered 9
8 Corolla-tube usually <2mm, wider than long; racemes lax, short and/or few-flowered 10

9 Leaves usually widest in middle 1/3, crenate to serrate with usually obtuse teeth, pubescent on both surfaces. Stems erect to ascending, woody near base, to 60(80)cm. Native; rocks and short grassland on limestone and other basic soils; very local in W Br from N Somerset to Westmorland, and in E Anglia *Spiked Speedwell* - **V. spicata** L.
9 Leaves usually widest in basal 1/3, serrate to biserrate with acute to subacuminate teeth, glabrous to sparsely and often minutely pubescent on both surfaces. Stems erect, woody near base, to 1.2m. Intrd-natd; waste and rough ground, banks, roadsides; scattered in Br, mainly S & C *Garden Speedwell* - **V. longifolia** L.

10 Corolla pink; style c.2x as long as capsule. Stems to 20cm. Intrd-natd; weed in lawns in very few places in Sc and N En (*V. reptans* D.H. Kent, illegitimate name) *Corsican Speedwell* - **V. repens** DC.
10 Corolla white to blue; style shorter than capsule 11

11 Capsule wider than long. Stems herbaceous, to 30cm, rooting at nodes. Native (*Thyme-leaved Speedwell* - **V. serpyllifolia** L.) 12
11 Capsule longer than wide 13

12 At least 1/2 of flowering stem upturned and erect; racemes ± glabrous or with eglandular hairs; pedicels c. as long as calyx; corolla 6-8mm across, whitish to pale blue with darker veins. Waste and cultivated ground, paths, lawns, open grassland, woodland rides and on mountains; common
V. serpyllifolia ssp. serpyllifolia
12 Most of flowering stem procumbent; racemes with glandular hairs; pedicels longer than calyx; corolla 7-10mm across, bright blue. Rock-ledges, flushes and wet gravel in mountains; Scottish

17. VERONICA

Highlands, less extreme plants in N En and Wa
V. serpyllifolia ssp. humifusa (Dicks.) Syme

13 Corolla >10mm across; stems woody at base; style >2mm; racemes with eglandular hairs. Stems woody below, to 20cm, erect to ascending. Native; alpine rocks above 500m; very local in C Sc
Rock Speedwell - **V. fruticans** Jacq.

13 Corolla <10mm across; stems herbaceous; style <2mm; racemes with glandular hairs. Stems to 15cm, herbaceous, erect to ascending from short rooting portion. Native; damp alpine rocks above 500m; local in mainland C Sc   *Alpine Speedwell* - **V. alpina** L.

Key C - Flowers solitary in leaf-axils
1   Calyx-lobes apparently 2, each bilobed at apex. Stems procumbent to decumbent, to 50cm. Intrd-natd; occasional casual in cultivated and rough ground and waste places; very scattered in En and Wa, natd in N Somerset and SW Ir *Crested Field-speedwell* - **V. crista-galli** Steven
1   Calyx-lobes 4, each acute to rounded at apex   2
   2   Perennial; stems rooting at nodes along length; pedicels >2x as long as leaves + petioles. Perennial with procumbent, minutely pubescent stems to 50cm. Intrd-natd; streamsides, lawns, grassy paths, banks and roadsides; common
Slender Speedwell - **V. filiformis** Sm.
   2   Annual; stems not rooting at nodes or doing so only near base; pedicels <2x as long as leaves + petioles   3
3   Calyx-lobes cordate at base; leaves with 3-7 shallow lobes or teeth. Stems procumbent to scrambling or ascending, to 60cm. Native; cultivated and waste ground, open woods, hedgerows, walls and banks; common (*Ivy-leaved Speedwell* - **V. hederifolia** L.)   4
3   Calyx-lobes cuneate to rounded at base; leaves crenate-serrate, most with >7 teeth   5
   4   Apical leaf-lobe usually wider than long; fruiting pedicels mostly 2-4x as long as calyx; calyx enlarging strongly after flowering, with marginal hairs mostly ≥0.9mm; corolla mostly ≥6mm across, whitish to blue; anthers blue, 0.7-1.2mm
V. hederifolia ssp. hederifolia
   4   Apical leaf-lobe usually longer than wide; fruiting pedicels mostly 3.5-7x as long as calyx; calyx enlarging slightly after flowering, with marginal hairs mostly ≤0.9mm; corolla mostly ≤6mm across, whitish to pale lilac-blue; anthers whitish to pale blue, 0.4-0.8mm   **V. hederifolia ssp. lucorum** (Klett & Richt.) Hartl
5   Lobes of capsule with apices diverging at c.90° from base; corolla mostly ≥8mm across. Stems procumbent to decumbent, to 50cm. Intrd-natd; cultivated and waste ground; throughout most of BI
Common Field-speedwell - **V. persica** Poir.
5   Lobes of capsule with apices ± parallel or diverging at narrow angle from base; corolla ≤8mm across   6
   6   Capsule with patent glandular hairs only. Stems procumbent to ascending, to 30cm. Native; cultivated ground; frequent in

most of BI
　　　　　　　　　　　　　　　Green Field-speedwell - **V. agrestis** L.
6　Capsule with many short eglandular arched hairs and variable numbers of patent glandular hairs. Stems procumbent to ascending, to 30cm. Native; cultivated ground; frequent in most of BI
　　　　　　　　　　　　　　　Grey Field-speedwell - **V. polita** Fr.

V. longifolia x *V. spicata* is probably the parentage of many garden and some natd plants; it is intermediate and fertile.

18. HEBE Juss. - *Hedge Veronicas*

Evergreen shrubs; leaves opposite, with pinnate venation but usually with obscured lateral veins; flowers in axillary racemes; corolla showy, as in *Veronica*; stamens 2. All the spp. are natd in rocky and scrubby places, almost always near the sea.

The terminal leaf-bud is composed of successively smaller developing leaves, without any modified bud-scales. The outermost 2 leaves enclose all the inner ones and their margins meet along 2 sides, but towards their base might (or might not) leave a gap (leaf-bud sinus) on either side due to the presence (or absence) of a distinct petiole.

1　Leaf-buds with distinct sinuses; leaves ± petiolate　　　　　　　2
1　Leaf-buds without sinuses; leaves ± sessile　　　　　　　　　　4
　2　Leaves ≤2.5 x 0.8cm; racemes mostly <3cm. Shrub to 2m. Intrd-natd; Scillies, Dorset　*Hooker's Hebe* - **H. brachysiphon** Summerh.
　2　Leaves ≥(2.5)3 x 0.7cm; racemes mostly >3cm　　　　　　　　3
3　Leaves linear-lanceolate, c.8-12x as long as wide, narrowly acute to acuminate at apex; racemes mostly >10cm; corolla usually white. Shrub to 2m. Intrd-natd; Devon, Cornwall and Man, marginally natd elsewhere in S & W Br, W Ir and CI
　　　　　　　　　Koromiko - **H. salicifolia** (G. Forst.) Pennell
3　Leaves oblanceolate to obovate or oblong-obovate, c.2-4x as long as wide, subacute to rounded at apex; racemes <10cm; corolla usually blue or pink to purple. Shrub to 1.5m. Intrd-natd; Devon, Cornwall, S & N Wa, S & W Ir and CI, less well natd elsewhere in Bl N to Shetland (by far the commonest natd taxon) (*H. elliptica* (G. Forst.) Pennell x *H. speciosa* (A. Cunn.) Cockayne & Allen)
　　　　　Hedge Veronica - **H. x franciscana** (Eastw.) Souster
　4　Leaves obovate to oblong-obovate, often shortly acuminate at apex, 2-3x as long as wide. Shrub to 1.5m. Intrd-natd; Man, Devon, Cornwall and Guernsey (*H. salicifolia* x *H. elliptica*)
　　　　　　Lewis's Hebe - **H. x lewisii** (J.B. Armstr.) A. Wall
　4　Leaves lanceolate to oblanceolate, subacute to obtuse at apex, 3-5x as long as wide　　　　　　　　　　　　　　　　　5
5　Leaves 4.5-8cm, mid-green; stem glabrous to finely and minutely pubescent, green. Spreading shrub to 1m. Intrd-natd; Devon and Cornwall
　　　Dieffenbach's Hebe - **H. dieffenbachii** (Benth.) Cockayne & Allan
5　Leaves 3-5cm, grey-green; stem glabrous, becoming purple. Erect

shrub to 2.5m. Intrd-natd; Devon and Cornwall
Barker's Hebe - H. barkeri (Cockayne) A. Wall

19. SIBTHORPIA L. - *Cornish Moneywort*
Stoloniferous perennials; leaves alternate, with palmate venation; flowers solitary in leaf-axils; corolla small and inconspicuous, ± actinomorphic, with (4-)5 subequal lobes slightly longer than tube; stamens (3)4(-5).

1 Stems procumbent, rooting at nodes, to 40cm; leaves reniform to orbicular, 5-9-lobed, pubescent, on long petioles; corolla 1-2.5mm across, whitish to yellowish or pinkish. Native; damp shady places; very local in SW En, S Wa, N Kerry, E Sussex and CI, natd in Outer Hebrides and rarely elsewhere   ***Cornish Moneywort* - S. europaea** L.

TRIBE 6 - PEDICULARIEAE (genera 20-26). Herbaceous annuals or sometimes perennials semi-parasitic on roots of many groups of angiosperms; leaves opposite or alternate; flowers strongly zygomorphic; corolla strongly 2-lipped, not spurred or pouched at base; stamens 4.

20. MELAMPYRUM L. - *Cow-wheats*
Annuals; leaves opposite, mostly entire; calyx not inflated, with 4 entire lobes; corolla mostly yellowish, with opening partially closed by boss-like swellings on lower lip, with 3 entire lobes on lower lip; capsules with 1-4 seeds; seeds smooth, with oil-body.

1 Bracts densely overlapping, concealing inflorescence axis at least in upper part, pink or purple at least near base, usually with >3 teeth on either side at base   2
1 Bracts not or scarcely overlapping, with inflorescence axis well exposed, green, usually with <3 teeth on either side at base   3
    2 Bracts cordate at base, strongly recurved, folded inwards along midrib proximally. Stems erect, to 50cm. Native; wood-borders and scrub; very local in E Anglia and adjacent C En
Crested Cow-wheat **- M. cristatum** L.
    2 Bracts rounded to cuneate at base, not recurved, not folded inwards along midrib. Stems erect, to 60cm. Possibly native; cornfields and grassy field margins; very local in Wight, N Essex and Beds   *Field Cow-wheat* **- M. arvense** L.
3 Lower lip of corolla strongly reflexed (turned down); lower 2 calyx-lobes patent, not upswept; fruit with two seeds. Stems erect, to 35cm. Native; upland woods and moorland; local in Sc N to E Ross, MW Yorks, Co Londonderry, Co Antrim
Small Cow-wheat **- M. sylvaticum** L.
3 Lower lip of corolla not reflexed, its underside forming a straight line with lower edge of tube; lower 2 calyx-lobes appressed to corolla and upswept; fruit with four seeds. Stems erect, to 60cm. Native; woods, scrub, heathland
(*Common Cow-wheat* **- M. pratense** L.)   4

4 Uppermost leaves (below bracts) (1)2-8(11)cm x (1)2-10(20)mm, mostly 7-15x as long as wide. Acid soils throughout most of Br and Ir **M. pratense ssp. pratense**

4 Uppermost leaves (below bracts) (3)4-7(10)cm x (4)8-20(27)mm, mostly 3-8x as long as wide. Calcareous soils in S En and SE Wa N to Herefs **M. pratense ssp. commutatum** (A. Kern.) C.E. Britton

21. EUPHRASIA L. - *Eyebrights*

Annuals; leaves mostly opposite, conspicuously toothed; calyx not inflated, with 4 entire lobes; corolla white to purple, usually with darker veins and yellow blotch on lower lip, rarely yellow all over, with open mouth, with lower lip with 3 emarginate lobes; capsules with many seeds; seeds furrowed longitudinally, without oil-body.

A highly critical genus. For a good chance of correct determination at least 5 or 6 well-grown (not stunted or spindly) and undamaged plants bearing some fruits as well as open flowers should be examined from a population. Ranges, rather than means, from these should be used. Single plants, or plants not agreeing with the above definition, are not allowed for in the key. The following key and accounts are based upon the classification of P.F. Yeo. Nodes are numbered from the base upwards, excluding the cotyledonary node. Corolla length is from base of tube to tip of upper lip in fresh state; dried specimens may have shrivelled or stretched (≤1mm) corollas. The name **E. officinalis** L. is often applied in an aggregate sense to the whole genus or to all spp. except *E. salisburgensis*. All spp. are known as *Eyebright*.

1 Middle and upper leaves with glandular hairs with stalk (6)10-12x as long as head 2
1 Middle and upper leaves without glandular hairs, or with glandular hairs with stalk ≤6x as long as head 11
    2 Capsule >2x as long as wide. Stems to 30(35)cm, with 0-5(6) pairs of branches sometimes again branched. Native; meadows and pastures, often replacing *E. nemorosa* as commonest sp. in N & W (**E. arctica** Rostrup) 3
    2 Capsule ≤2x as long as wide 4
3 Stem procumbent or flexuous at base, then erect; lower bracts suborbicular to broadly ovate; corolla 7-11(13)mm; capsule (5.5)6-7.5(8)mm. Orkney and Shetland **E. arctica ssp. arctica**
3 Stem erect from base; lower bracts narrowly to broadly ovate, rhombic or trullate; corolla 6-9(10)mm; capsule (4)4.5-6.5(7)mm. Ir, N & W Br except Shetland, very scattered in S & E Br
E. arctica ssp. borealis (F. Towns.) Yeo
    4 Corolla ≤7mm 5
    4 Corolla >7mm 6
5 Lowest flower at node 5-8; lower bracts 5-12mm, often longer than flowers; plant usually with 1-4 pairs of strong branches. Stems flexuous-erect, to 20(30)cm; branches (0)1-4(6) pairs, flexuous or arcuate, usually again branched. Native; in short turf on often damp

soils, heathland; local in C & S Br N to SW Sc, Ir **E. anglica** Pugsley
5 Lowest flower at node (2)3-5(6); lower bracts 3-6(7)mm, shorter than flowers; plant not branched or with 1-2 pairs of short branches. Stems flexuous-erect, to 15cm; branches 0-2 pairs, short; lowest flower usually at node (2)3-5(6). Native; damp mountain pastures and streamsides; very local in NW Wa, Lake District **E. rivularis** Pugsley
6 Lowest flower at node 2-5(6) 7
6 Lowest flower at node 5 or higher 9
7 Corolla ≤9mm; lower bracts 3-6(7)mm (see couplet 5) **E. rivularis**
7 Corolla 9-12.5mm; lower bracts 5-12(20)mm. Stems stout, erect, to 35cm; branches 0-5(more) pairs, often again branched. Native; grassland in hilly areas (**E. rostkoviana** Hayne) 8
   8 Branches 1-5(more) pairs, ascending, divergent or erect; internodes shorter than to 3(4)x as long as leaves; lowest flower usually at node 6-10; corolla (6.5)8-12mm. Often in damper places and by rivers, sometimes lowland; locally frequent in Ir, Wa, N En and C & S Sc **E. rostkoviana ssp. rostkoviana**
   8 Branches 0-3(4) pairs, erect; internodes 2-6(10)x as long as leaves; lowest flower usually at node 2-6; corolla (7)9-12.5mm. Usually in drier upland places; very local in W Wa, N En and S Sc
      E. rostkoviana ssp. montana (Jord.) Wettst.
9 Leaves dull greyish-green, often strongly suffused with violet or black; corolla usually lilac to purple. Stems erect, to 20(25)cm; branches 0-5(7) pairs, erect, often again branched. Native; *Ulex gallii/ Agrostis curtisii* heathland in Devon and Cornwall **E. vigursii** Davey
9 Leaves light or dark green, usually with little violet suffusion; corolla usually with at least lower lip white 10
   10 Stem usually erect, with erect or divergent branches; lower internodes of inflorescence mostly 1.5-3x as long as bracts; corolla 8-12mm (see couplet 7) **E. rostkoviana**
   10 Stem usually flexuous with flexuous or arched branches; lower internodes of inflorescence mostly <1.5x as long as bracts; corolla usually 6.5-8mm (see couplet 5) **E. anglica**
11 Capsule glabrous or with a few short hairs; at least 2 distal pairs of leaf-teeth (and sometimes all) not contiguous at base. Stems erect or flexuous, to 12cm; branches (0)1-7 pairs, slender, erect or patent, often again branched. Native; among limestone rocks and on dunes; W Ir from Co Limerick to E Donegal **E. salisburgensis** Funck.
11 Capsule with long ± numerous hairs in distal part; usually all leaf-teeth contiguous at base 12
   12 Corolla >7.5mm 13
   12 Corolla ≤7.5mm 18
13 Basal pair of teeth of lower bracts directed apically 14
13 Basal pair of teeth of lower bracts patent 15
   14 Stem and branches slender, flexuous; capsule usually c. as long as calyx; lower bracts mostly alternate. Stems to 20(45)cm; branches (0)2-8(10) pairs, usually ascending, usually branched again. Native; short, well-drained turf on moorland, heaths and

dunes; throughout most of BI **E. confusa** Pugsley
- 14 Stem stout, erect, with straight or evenly curved branches; capsule usually much shorter than calyx; lower bracts mostly opposite. ?Native; meadows and pastures; Guernsey **E. stricta** J.F. Lehm.
- 15 Lowest flower at node 8 or lower; capsule usually elliptic to obovate (see couplet 2) **E. arctica**
- 15 Lowest flower at node 9 or higher; capsule oblong to elliptic-oblong 16
  - 16 Stem and branches flexuous; leaves near base of branches usually very small (see couplet 14) **E. confusa**
  - 16 Stem and branches usually straight or gradually curved; leaves near base of branches not much smaller than others 17
- 17 Teeth of bracts acute to acuminate; capsule usually slightly shorter than calyx. Stems erect, to 35(40)cm; branches 1-9 pairs, ascending, often again branched. Native; pastures, scrub, woodland rides, marginal areas and heathland (dunes in Sc); throughout Bl, the commonest lowland sp. **E. nemorosa** (Pers.) Wallr.
- 17 Teeth of bracts mostly aristate; capsule much shorter than calyx. Stems erect or flexuous, to 20(30)cm; branches (0)3-8(10) pairs, ascending to patent, often again branched. Native; dry limestone (usually chalk) grassland, fens in E Anglia; S En, SW Wa, W Ir **E. pseudokerneri** Pugsley
  - 18 Calyx-tube whitish and membranous, with prominent green to blackish veins. Stems erect, to 10cm; branches 0-2 pairs, short, erect. Native; damp heathland near sea; Isle of Lewis (Outer Hebrides) **E. campbelliae** Pugsley
  - 18 Calyx-tube green, not membranous 19
- 19 Lowest flower at node 6 or higher 20
- 19 Lowest flower at node 5(-6) or lower 40
  - 20 Stem internodes mostly 2-6x as long as leaves 21
  - 20 Stem internodes mostly ≤2x as long as leaves 34
- 21 Basal pair of teeth of lower bracts directed apically 22
- 21 Basal pair of teeth of lower bracts patent 25
  - 22 Teeth of lower bracts obtuse to acute; corolla ≤6.5mm 23
  - 22 Teeth of lower bracts acute to aristate; corolla ≥7mm 24
- 23 Leaves strongly purple-tinged, not darker on lowerside; corolla usually lilac to purple; capsule shorter than calyx. Stems erect, slender, to 25cm; branches (0)2-7(10) pairs, slender, erect, usually again branched. Native; heathland, usually with *Calluna*, sometimes damp places; throughout most of BI except C & E En **E. micrantha** Rchb.
- 23 Leaves weakly or moderately purple-tinged, often darker on lowerside; corolla usually white; capsule at least as long as calyx. Stems erect, to 25cm; branches 0-4 pairs, long, erect to ascending. Native; wet moorland; Wa, NW En, Sc and Ir **E. scottica** Wettst.
  - 24 Capsule ≤3x as long as wide (see couplet 2) **E. arctica**
  - 24 Capsule ≥3x as long as wide (see couplet 14) **E. stricta**
- 25 Corolla ≥6.5mm 26
- 25 Corolla ≤6.5mm 27
  - 26 Lowest flower at node 9 or higher; leaves usually without

glandular hairs; lower bracts smaller than upper leaves
(see couplet 17) **E. nemorosa**
- 26 Lowest flower at node 8 or lower; leaves usually with glandular hairs; lower bracts larger than upper leaves (see couplet 2) **E. arctica**
- 27 Leaves subglabrous to sparsely pubescent 28
- 27 Leaves densely pubescent 30
  - 28 Stem and branches very slender, blackish; leaves strongly purple-tinged, not darker on lowerside; corolla usually lilac to purple
(see couplet 23) **E. micrantha**
  - 28 Stem and branches either stout or lightly pigmented; leaves weakly or moderately purple-tinged; corolla usually white 29
- 29 Lowest flower at node 8 or higher; stem stout; leaves not darker on lowerside; capsule usually shorter than calyx
(see couplet 17) **E. nemorosa**
- 29 Lowest flower at node 7 or lower; stem slender; leaves usually light green on upperside and purplish on lowerside; capsule usually longer than calyx (see couplet 23) **E. scottica**
  - 30 Lowest flower at node 9 or higher; stem ≤40cm; lower bracts often longer than wide (see couplet 17) **E. nemorosa**
  - 30 Lowest flower at node 8 or lower; stem ≤15cm; lower bracts c. as long as wide 31
- 31 Leaves pubescent mainly near apex, obovate to narrowly ovate to elliptic (see couplet 18) **E. campbelliae**
- 31 Leaves ± uniformly pubescent, usually suborbicular, ovate or ovate-oblong 32
  - 32 Teeth of lower bracts mostly wider than long; branches ≤3 pairs. Stems erect, to 10cm; branches 0-3 pairs, short, erect. Native; basic turf on sea-cliffs or dunes; extreme N Sc from N Ebudes and Outer Hebrides to Shetland **E. rotundifolia** Pugsley
  - 32 Teeth of lower bracts mostly as long as wide; branches ≤5 pairs 33
- 33 Corolla 5.5-7mm; capsule usually >2x as long as wide. Stems erect, to 12cm; branches (0)1-5 pairs, rather long, erect, sometimes again branched. Native; turf on sea-cliffs or dunes; extreme N Sc from N Ebudes and Outer Hebrides to Shetland **E. marshallii** Pugsley
- 33 Corolla 4.5-6mm; capsule ≤2x as long as wide. Stems erect or flexuous below, to 12(15)cm; branches 0-4(6) pairs, erect or ascending, sometimes again branched. Native; grassy, stony and sandy places, mostly near sea; N & NW Sc incl. Orkney and Shetland, Lake District, Caerns **E. ostenfeldii** (Pugsley) Yeo
  - 34 Basal pair of teeth of lower bracts directed apically 35
  - 34 Basal pair of teeth of lower bracts patent 36
- 35 Teeth of lower bracts not much longer than wide Return to 31
- 35 Teeth of lower bracts much longer than wide Return to 14
  - 36 Lowest flower at node 10 or higher 37
  - 36 Lowest flower at node 9 or lower 38
- 37 Stem erect, stout, with stout ascending branches; lower bracts mostly opposite (see couplet 17) **E. nemorosa**
- 37 Stem and branches slender and flexuous; lower bracts mostly

	alternate (see couplet 14) **E. confusa**
38	Leaves with numerous eglandular hairs Return to 31
38	Leaves with few eglandular hairs 39
39	Capsule 5.5-7mm, often slightly curved, as long as or longer than calyx. Stems erect from usually flexuous base, to 15cm; branches 0-4(5) pairs, erect or patent, sometimes again branched. Native; turf in salt-marshes or drier grassy places; extreme N, NW & CW Sc **E. heslop-harrisonii** Pugsley
39	Capsule usually ≤5.5mm, straight, usually shorter than calyx. Stems erect, stout, to 15(20)cm; branches 0-5(8) pairs, usually rather short and erect or ascending, but sometimes branched again, forming compact plant. Native; short turf on cliffs and dunes by sea, limestone pasture inland; coasts of BI except most of E En and N Sc, inland in parts of SW En **E. tetraquetra** (Bréb.) Arrond.
	40 Stem internodes mostly ≥2.5x as long as leaves 41
	40 Stem internodes mostly <2.5x as long as leaves 46
41	Capsule broadly elliptic to obovate-elliptic 42
41	Capsule oblong to narrowly elliptic 43
	42 Teeth of lower bracts mostly subacute, not longer than wide; corolla 4.5-7mm; lowest flower at node 2-4(5). Native; grassland on rock-ledges on mountains, mostly over 600m; Sc, Cheviot, Lake District, W Ir **E. frigida** Pugsley
	42 Teeth of lower bracts usually acute or acuminate, longer than wide; corolla ≥6.5mm; lowest flower usually at node 4 or higher (see couplet 2) **E. arctica**
43	Lower bracts deeply serrate, with basal pair of teeth directed apically (see couplet 14) **E. stricta**
43	Lower bracts crenate to shallowly serrate, with basal pair of teeth usually patent 44
	44 Upper leaves elliptic-ovate to narrowly obovate (see couplet 23) **E. scottica**
	44 Upper leaves suborbicular to broadly ovate or broadly obovate 45
45	Lowest flower at node 4 or lower; lower bracts often considerably larger than upper leaves (see couplet 42) **E. frigida**
45	Lowest flower at node 4 or higher; lower bracts scarcely larger than upper leaves Return to 31
	46 Corolla ≥6mm 47
	46 Corolla ≤6mm 49
47	Teeth of lower bracts usually very acute, all directed apically Return to 12
47	Teeth of lower bracts acute to subacute, the basal pair patent 48
	48 Capsule at least as long as calyx, usually emarginate (see couplet 42) **E. frigida**
	48 Capsule shorter than calyx, truncate to slightly emarginate (see couplet 39) **E. tetraquetra**
49	Lower bracts ovate to rhombic, with acute to aristate teeth, the basal pair directed forwards Return to 14
49	Lower bracts broadly ovate or rhombic to suborbicular, with obtuse

	to subacute teeth, the basal pair patent	50
50	Leaves with numerous hairs, all eglandular	51
50	Leaves with few eglandular hairs, sometimes with short glandular hairs	52
51	Lower bracts scarcely larger than upper leaves	Return to 31
51	Lower bracts considerably larger than upper leaves	(see couplet 42) **E. frigida**
	52 Capsule elliptic to obovate, emarginate. Stem flexuous, to 8cm; branches 0-2 pairs, flexuous. Native; mountain grassland and rock-ledges; Caerns and Merioneth	**E. cambrica** Pugsley
	52 Capsule oblong to elliptic-oblong, usually truncate	53
53	Capsule usually shorter than calyx; distal teeth of lower bracts not incurved	(see couplet 39) **E. tetraquetra**
53	Capsule as long as or longer than calyx; distal teeth of lower bracts ± incurved	54
	54 Capsule 4.5-5.5(7)mm, c.2x as long as wide, straight; upper leaves only obscurely petiolate, with margins of teeth not wavy. Stems erect, rather stout, to 6(9)cm; branches 1-3(4) pairs, short, ascending, occasionally again branched. Native; exposed short turf on sea-cliffs, sometimes behind saltmarshes; Outer Isles, N & W mainland Sc	**E. foulaensis** Wettst.
	54 Capsule (4.5)5.5-7mm, 2-3x as long as wide, often slightly curved; upper leaves ± distinctly petiolate, with margins of teeth wavy	(see couplet 39) **E. heslop-harrisonii**

All spp. have been reported to form hybrids; >60 combinations have been recorded. Of the 21 spp., 4 (*E. rostkoviana*, *E. rivularis*, *E. anglica* and *E. vigursii*) are diploids, the rest tetraploids.

Crosses between the tetraploids (excl. *E. salisburgensis*) are highly fertile and often common where 2 or more spp. occur together; 48 binary combinations have been reliably recorded in BI, as well as some triple hybrids. The following 11 hybrids are frequent and may occur in the absence of 1 or both parents, locally replacing them: *E. arctica* x *E. nemorosa*, x *E. confusa* and x *E. micrantha; E. tetraquetra* x *E. confusa; E. nemorosa* x *E. pseudokerneri*, x *E. confusa* and x *E. micrantha; E. confusa* x *E. micrantha* and x *E. scottica; E. frigida* x *E. scottica;* and *E. micrantha* x *E. scottica*.

Hybrids within the diploids are much less common, mainly because the spp. tend to be ± allopatric. 3 combinations are known, resulting in intergradation between the parent spp., mostly in Wa and W En.

Hybrids between the above 2 groups also occur; 10 combinations have been identified. First generation hybrids are probably highly but not totally sterile triploids, but most found in the field are ± fertile diploid introgressants.

E. arctica, *E. nemorosa* and *E. micrantha* have been found to hybridise with the tetraploid *E. salisburgensis* in Ir; these hybrids are highly but not totally sterile.

22. ODONTITES Ludw. - *Bartsias*

Annuals; leaves opposite, entire to toothed; calyx not inflated, with 4 entire lobes; corolla yellow or pinkish-purple, rarely white, with open mouth, with lower lip with 3 entire to slightly notched lobes; capsules with rather few seeds; seeds furrowed longitudinally, without oil-body. Intercalary leaves are those at nodes on the main stem between the topmost branches and the lowest bract.

1 Calyx 4-5mm; corolla 7-9mm, yellow, often tinged pinkish. Stems erect, to 50cm. Intrd-natd; gravelly rough ground near Aldermaston (Berks)   *French Bartsia* - **O. jaubertianus** (Boreau) Walp.
1 Calyx 5-8mm; corolla 8-10mm, pinkish-purple, rarely white. Stems erect, to 50cm. Native (*Red Bartsia* - **O. vernus** (Bellardi) Dumort.)   2
    2 Stems to 50cm, with (0)2-8 pairs of branches held at ≥50° to main stem; intercalary leaves 2-7 pairs; lowest flower at node 8-14. In habitats of both sspp. *vernus* and *litoralis*; frequent over most Bl except C & N Sc   **O. vernus ssp. serotinus** (Syme) Corb.
    2 Branches ≤4 pairs, held at ≤50° to main stem; intercalary leaves 0-1 pairs; lowest flower at node 4-9. Stems erect, to 25 cm   3
3 Calyx-teeth narrowly triangular, acute, as long as tube; style exserted from corolla at full anthesis. Stems with (0)1-4 pairs of branches. Grassy places, arable and waste ground, waysides; scattered in Ir and W & N Br, much over-recorded in S Br   **O. vernus ssp. vernus**
3 Calyx-teeth triangular, subacute to obtuse, shorter than tube; style included in upper lip of corolla of full anthesis. Stems with 0-3 pairs of branches. Gravelly and rocky sea-shores and salt-marshes; coasts of N & W Sc from Arran to Shetland
                                 O. vernus ssp. litoralis (Fr.) Nyman

23. BARTSIA L. - *Alpine Bartsia*

Perennials; leaves opposite, toothed; calyx not inflated, with 4 entire lobes; corolla dull purple, with open mouth, with lower lip with 3 entire lobes; capsules with few seeds; seeds with rather narrow longitudinal wings, without oil-body.

1 Stems erect, to 25cm; leaves and bracts glandular-pubescent; bracts all wholly or partly dark purple; corolla 15-20mm. Native; grassy places and rock-ledges on basic, often damp, soils in mountains; very local in hills of N En and C Sc   *Alpine Bartsia* - **B. alpina** L.

24. PARENTUCELLIA Viv. - *Yellow Bartsia*

Annuals; leaves opposite, toothed; calyx not inflated, with 4 entire lobes; corolla yellow, very rarely white, with open mouth, with lower lip with 3 entire lobes; capsules with numerous seeds; seeds ± smooth, without oil-body.

1 Stems erect, to 50cm; leaves and bracts glandular-pubescent; corolla 16-24mm. Native; damp grassy places mostly near coast; locally

frequent in S & W Br E to E Kent and N to Dunbarton, NW & SW Ir, CI *Yellow Bartsia* - **P. viscosa** (L.) Caruel

25. RHINANTHUS L. - *Yellow-rattles*
Annuals; leaves opposite, toothed; calyx inflated, especially at fruiting, with 4 entire lobes; corolla basically yellow to brownish-yellow, with semi-closed to ± open mouth, with upper lip with 1 subterminal white or violet tooth either side of tip, with lower lip with 3 ± entire lobes; capsule with numerous seeds; seeds discoid, usually with marginal wing, without oilbody.

Intercalary leaves are those at nodes on the main stem between the topmost branches and the lowest bract. Unbranched (starved) plants occur in most populations and are best ignored unless the prevalent sort; as for *Euphrasia*, a range of individuals should be used for determination. Corolla colour ignores the white or violet teeth on the upper lip.

1 Corolla (15)17-20mm, with lower lip held horizontal ± adjacent to upper lip, with teeth of upper lip mostly >1mm and longer than wide; dorsal line of corolla concavely curved proximally, merging into convexly curved upper lip to form overall sigmoid shape; seeds winged or not. Stems erect, to 60cm; intercalary leaves 0-2 pairs. Native; arable and grassy fields, rough ground, sandy open places on heathland or near sea; extremely local in Surrey, N Lincs and Angus *Greater Yellow-rattle* - **R. angustifolius** C.C. Gmel.
1 Corolla 12-15(17)mm, with lower lip turned down away from upper lip, with teeth of upper lip mostly <1mm and shorter than wide; dorsal line of corolla ± straight proximally, merging into convexly curved upper lip; seeds winged. Stems erect, to 50cm; intercalary leaves 0-6 pairs. Native (*Yellow-rattle* - **R. minor** L.) 2
  2 Calyx pubescent all over 3
  2 Calyx pubescent only on margins 4
3 Branches 0-2 pairs; intercalary leaves 0-3 pairs; lowest flower at node 7-10; leaves linear-lanceolate. Stems to 30cm. Grassy places on mountains; C & N Sc **R. minor ssp. lintonii** (Wilmott) P.D. Sell
3 Branches 0(-1) pairs; intercalary leaves 0; lowest flower at node 5-7(8); leaves linear-oblong. Stems to 20(28)cm. Grassy places on mountains; Sc, mostly C & N, Caerns, N & S Kerry

R. minor ssp. borealis (Sterneck) P.D. Sell
  4 Intercalary leaves mostly (2)3-6 pairs; lowest flower usually at node 14-19; leaves mostly linear. Stems to 50cm. Dry grassy places on chalk and limestone; from Dorset and W Gloucs to E Kent **R. minor ssp. calcareus** (Wilmott) E.F. Warb.
  4 Intercalary leaves mostly 0-2(4) pairs; lowest flower usually at node 6-13(15); leaves mostly linear-narrowly oblong to linearlanceolate 5
5 Intercalary leaves mostly 0(-1) pairs; lowest flower usually at node 6-9; leaves mostly parallel-sided for most of length. Stems to 40cm. Grassy places, especially on well-drained basic soils; ± throughout

BI, especially in lowlands, commoner in S **R. minor ssp. minor**
5 Intercalary leaves mostly (0)1-2(4) pairs; lowest flower usually at node (7)8-13(15); leaves mostly ± tapering from near base 6
  6 Stems ≤50cm, usually with several pairs of long flowering branches from basal and middle parts; leaves of main stems 1.5-4.5cm; corolla usually yellow. Stems to 50cm. Damp grassland and fens; throughout most of Br and Ir, commoner in N **R. minor ssp. stenophyllus** (Schur) O. Schwarz
  6 Stems ≤25cm, with 0-3 pairs of long flowering branches from near base; leaves of main stems 1-2.5cm; corolla usually dull- or brownish-yellow. Stems to 20(25)cm. Grassy places in hilly areas; local in Br N from MW Yorks, N Kerry and Co Londonderry
R. minor ssp. monticola (Sterneck) O. Schwarz

26. PEDICULARIS L. - *Louseworts*

Annuals to perennials; leaves alternate or mostly so, deeply pinnately lobed with crenate to lobed lobes; calyx becoming inflated, irregularly 2-5 lobed with toothed lobes; corolla pinkish-purple, rarely white, with open mouth, with lower lip with 3 ± entire lobes; capsules with rather few seeds; seeds ± smooth, without oil-body, winged or not.

1 Annual to biennial; stems usually single, usually erect, to 60cm; calyx pubescent, with 2 short, broad, variously dissected lobes; corolla with 2 lateral teeth on either side; capsule longer than calyx. Native; wet heaths and bogs; throughout Br and Ir, common in N & W, rare in C & E En *Marsh Lousewort* - **P. palustris** L.
1 (Biennial to) perennial; stems several, procumbent to ascending or some suberect, to 25cm; calyx with 4 short dissected lobes; corolla with 1 lateral tooth on either side; capsule shorter than or equalling calyx. Native; similar places to *P. palustris* and sometimes with it, but often in drier habitats (*Lousewort* - **P. sylvatica** L.) 2
  2 Calyx and pedicels glabrous. Throughout BI except parts of C En and where replaced by ssp. *hibernica* **P. sylvatica ssp. sylvatica**
  2 Calyx and pedicels pubescent. Prevalent in W Ir, very local in extreme W Sc, Lake District, E & S Ir and W Wa
P. sylvatica ssp. hibernica D.A. Webb

129. OROBANCHACEAE - *Broomrape family*

Brown to whitish, reddish or bluish, herbaceous, erect, perennial root-parasites; distinguished from Scrophulariaceae by the chlorophyll-less aerial parts and 1-celled ovary.

1 Plant rhizomatous; flowers pedicellate; calyx with 4 equal lobes
1. LATHRAEA
1 Plant not rhizomatous; flowers sessile except rarely some lower ones; calyx with 2-4(5) teeth arranged in 2 lateral lips **2. OROBANCHE**

1. LATHRAEA L. - *Toothworts*
Plants with rhizome with ± succulent scales; flowers pedicellate; calyx with 4 equal lobes; 2 lips of corolla held nearly parallel to one another.

1 Aerial stems whitish to cream or pale pink, to 30cm, pedicels shorter than calyx; corolla whitish-cream usually tinged with purple or pink, 14-20mm; capsule with numerous seeds. Native; in woods and hedgerows, usually on moist rich soils, on a range of woody plants especially *Ulmus* and *Corylus*; locally frequent in Br and Ir N to C Sc *Toothwort* - **L. squamaria** L.
1 Aerial stems 0; flowers arising from axils of scale-leaves near apex of rhizome; pedicels c. as long as to longer than calyx; corolla purple-violet, 40-50mm; capsule with 4-5 seeds. Intrd-natd; damp places on *Salix* and *Populus*; scattered in En, N Wa, S Sc and E Ir, Guernsey *Purple Toothwort* - **L. clandestina** L.

2. OROBANCHE L. - *Broomrapes*
Rhizomes 0; flowers sessile, rarely lower ones pedicellate; calyx with 2-4(5) teeth arranged in 2 lateral lips, the lips usually open to the base on upper or both sides; 2 lips of corolla held apart, the lower turned down.

Pressed plants are difficult to determine because the corolla shape and colour and the stigma colour are lost or obscured. At collection some corollas should be opened out by slitting up 1 side and pressing, the shape of the corolla in side view and of the lower lip in front view should be recorded, and the colour of the stem, corolla and stigmas noted. The host sp. is a useful character, but often difficult to ascertain with certainty. Corolla-lengths are from base of corolla-tube to tip of upper lip in a straight line.

1 Each flower with 2 bracteoles ± similar to the 4 calyx-teeth (1 bracteole and 2 calyx teeth each side of flower) in axil of each bract; stigmas white; capsule-valves free. Stems to 45cm, tinged bluish. Native; on *Achillea millefolium* and perhaps other Asteraceae; very local from Pembs and Dorset to N Lincs, CI

O. purpurea Jacq. - *Yarrow Broomrape*
1 Bracteoles 0; each flower with 2-4-toothed calyx (1-2 calyx-teeth on each side of flower); stigmas yellow, red or purplish, rarely white; capsule-valves coherent distally 2
  2 Lower lip of corolla with minute glandular hairs at margins 3
  2 Margins of lower lip of corolla glabrous or with very few glandular hairs, but latter often frequent elsewhere on corolla 5
3 Stigma-lobes yellow at flower opening; filaments glabrous in basal 1/3. Stems to 90cm, yellowish, tinged reddish-brown. Native; on various woody Fabaceae; local in Br N to S Sc, CI, S & SE Ir

Greater Broomrape - **O. rapum-genistae** Thuill.
3 Stigma-lobes red to purple at flower opening; filaments pubescent at least at base 4
  4 Stigma-lobes separate; corolla not suffused dark red, mostly >20mm; each calyx-lip with 1-2 teeth, shorter than corolla-tube.

Stems to 40cm, yellow tinged with purplish-brown. Native; on *Galium mollugo*; very local in E Kent

Bedstraw Broomrape - O. caryophyllacea Sm.

4 Stigma-lobes partly fused; corolla suffused dark red, mostly <20mm; each calyx-lip with 1 tooth, c. as long as corolla-tube. Stems to 25(35)cm, purplish-red. Native; on *Thymus*, perhaps other Lamiaceae; local in W & CE Sc, N & W Ir and N & SW En

Thyme Broomrape - O. alba Willd.

5 Calyx with 2 lateral lips partially fused on lowerside; stamens inserted (3)4-6mm above base of corolla-tube. Stems to 75cm, yellow to orange-brown. Native; on *Centaurea scabiosa*; on chalk and limestone in S & E En N to NE Yorks, Glam

Knapweed Broomrape - O. elatior Sutton

5 Calyx with 2 lateral lips free on both upperside and lowerside; stamens inserted 2-4(5)mm above base of corolla-tube   6

    6 Corolla with sparse dark glands mostly distally, with very strongly curved back so that mouth is nearly at right angles to base. Stems to 70cm, yellowish to purplish. Native; on *Carduus* and *Cirsium*; very local in Yorks

Thistle Broomrape - O. reticulata Wallr.

    6 Corolla without dark glands (often with pale ones), with nearly straight to slightly curved back   7

7 Corolla 20-30mm, with 2 upper and 3 lower lobes with conspicuous and patent margins; filaments sparsely pubescent along length. Stems to 80cm, yellowish to purplish. Intrd-natd; on herbaceous Fabaceae, often crop spp.; 1 part of S Essex

Bean Broomrape - O. crenata Forssk.

7 Corolla 10-22mm, with 0 or only 3 lower lobes with conspicuous and patent margins; filaments glabrous or pubescent only proximally   8

    8 Corolla-tube constricted just behind mouth; lower lip of corolla with acute to subacute lobes; stigmas usually yellow, rarely purplish. Stems to 60cm, brownish-purple, rarely yellowish. Native; on *Hedera*; S & W Br from Wight to Wigtowns, much of Ir, CI   **Ivy Broomrape - O. hederae** Duby

    8 Corolla-tube not constricted; lower lip of corolla with obtuse to rounded lobes; stigmas usually purplish, rarely yellow   9

9 Filaments with long white hairs at base, usually inserted ≥3mm above base of corolla-tube; bract equal to or longer than corolla; all 4 calyx-teeth long and filiform. Stems to 60cm, yellowish tinged with purple. Native; on *Picris* and *Crepis*; rare in Wight, W Sussex and E Kent   **Oxtongue Broomrape - O. artemisiae-campestris** Gaudin

9 Filaments glabrous to sparsely pubescent at base, usually inserted ≤3mm above base of corolla-tube; bract shorter than or equal to corolla; calyx-teeth various but not all 4 long and filiform. Stems to 60cm, yellowish, usually strongly tinged with red or purple. Native. On a very wide range of dicotyledons, including most specific for other spp.; throughout most of En, Wa, Ir (intrd) and CI   **Common Broomrape - O. minor** Sm.

130. GESNERIACEAE - *Pyrenean-violet family*

Herbaceous perennials; stem ± absent; leaves in basal rosette, simple, dentate, with long brown hairs; flowers 1-few on long peduncles arising from rosette, distinguished from those of Scrophulariaceae by the 1-celled ovary.

1. RAMONDA Rich. - *Pyrenean-violet*

1 Flowers 3-4cm across, violet with yellow centre and yellow exserted anthers. Intrd-surv; rock-face in Cwm Glas (Caerns) and on old estate in MW Yorks   *Pyrenean-violet* - **R. myconi** (L.) Rchb.

131. ACANTHACEAE - *Bear's-breech family*

Herbaceous perennials with most leaves in basal rosette; easily distinguished by the robust growth-habit, large pinnately lobed leaves, spiny bracts and unique flower structure (corolla 3.5-5cm, white with purple veins, with short corolla-tube, 3-lobed lower lip and 0 upper lip).

1. ACANTHUS L. - *Bear's-breeches*

1 Stems to 1m, glabrous at least above; leaves with acutely toothed but not spiny lobes, glabrous; bracts glabrous. Intrd-natd; waste places, roadside and railway banks, and scrub; scattered in S En, S Wa and CI, especially SW En   *Bear's-breech* - **A. mollis** L.
1 Stems to 80cm, often pubescent; leaves with spiny teeth, usually pubescent; bracts usually pubescent. Intrd-natd; similar places to *A. mollis* but much less common; very scattered in S En
     Spiny Bear's-breech - **A. spinosus** L.

132. LENTIBULARIACEAE - *Bladderwort family*

Rootless aquatic, insectivorous, perennials, or rooted, stemless, rosette-forming insectivorous perennials; easily recognized by the very different insectivorous habit of the 2 genera, both of which have a 2-lipped, spurred corolla, 2 stamens and free-central placentation.

1 Plant rooted, with basal rosette of simple entire leaves; corolla white to violet   **1. PINGUICULA**
1 Plant rootless, aquatic; leaves divided into linear to filiform segments; corolla yellow   **2. UTRICULARIA**

1. PINGUICULA L. - *Butterworts*
Roots present; leaves simple, entire, in basal rosette, covered in slime; flowers solitary on erect pedicels; calyx of 5 subequal lobes; corolla white to

violet, with open mouth; capsule opening by 2 valves.

1 Corolla white, with 1-2 yellow spots on lower lip. Overwintering as a bud; pedicels 5-11cm. Formerly native; boggy places in E Ross from 1831 to c.1900 *Alpine Butterwort* - **P. alpina** L.
1 Corolla pale lilac or pinkish to violet, very rarely white, sometimes with yellowish throat but never with yellow spots on lower lip 2
  2 Corolla 7-11mm incl. cylindrical spur 2-4mm, usually pale lilac. Overwintering as a rosette; pedicels 3-15cm. Native; bogs and wet heaths; locally frequent in Ir, W & N Sc, Man, SW Wa and SW & SC En *Pale Butterwort* - **P. lusitanica** L.
  2 Corolla 14-35mm incl. tapering spur 4-14mm, usually violet with whitish throat 3
3 Corolla 14-22(25)mm incl. spur 4-7(10)mm; lobes of lower lip separated laterally. Overwintering as a bud; pedicels 5-18cm. Native; bogs, wet heathland and limestone flushes; locally common over much of Br and Ir, especially N & W, absent from most of C & S En *Common Butterwort* - **P. vulgaris** L.
3 Corolla 25-35mm incl. spur 10-14mm; lobes of lower lip overlapping laterally. Overwintering as a bud, but leaves often persisting or precociously developing; pedicels 5-18cm. Native; bogs and damp moorland; locally common in SW Ir to Clare and E Cork, planted and persistent in scattered places in En and Wa
Large-flowered Butterwort - **P. grandiflora** Lam.

P. vulgaris x *P. grandiflora* = ***P. x scullyi*** Druce occurs rarely with the parents in S Kerry and Clare; it is intermediate and largely sterile.

2. UTRICULARIA L. - *Bladderworts*

Roots 0; plants free-floating or with lower stems in substratum; leaves divided into linear to filiform segments, some or all bearing tiny animal-catching bladders; flowers on erect racemes emerging from water; calyx of 2 obscurely-lobed lips; corolla yellow, with mouth ± closed by swollen upfolding of lower lip; capsule opening irregularly.

The leaf-segments usually have small marginal teeth (often extremely short) which bear 1 or more long bristles. The bladders are 1-4mm long and bear hairs on their inner and outer surfaces; those on the inner surface are 2- or 4-armed. Small circular or elliptic glands are present on the stems, leaves, outside of bladders and inside of corolla-spur. The presence of bristle-bearing teeth on the leaf-segment margins, the shape of the 4-armed bladder hairs ('quadrifids'), and the distribution of the glands on the inside of the corolla-spur are of diagnostic importance. The first can be seen with a strong lens, the second 2 need a microscope. Pressed, dried material is ideal for the first 2, but pressing often distorts the morphology of the quadrifids. At least 5-10 quadrifids should be examined and the range noted. Since the basal corolla-spur is forward-directed, its abaxial side is the side nearer the lower lip.

U. intermedia, *U. stygia* and *U. ochroleuca* have been much confused in the past and their distributions are very unclear. *U. ochroleuca* is probably the commonest at least in Sc, but *U. stygia* seems the commonest sp. in W Sc. All 3 flower very rarely but can be distinguished on microscopic vegetative characters. Distributions of *U. vulgaris* and *U. australis* are also unclear as distinction is uncertain without flowers, which are rarely produced in N Br.

1 Margins of leaf-segments without teeth with bristles; lower lip of corolla <8mm; spur 1-2mm. Stems usually of 2 sorts, sometimes all floating, to 40cm. Native; in boggy pools and fen-ditches; scattered in suitable places over Br and Ir   ***Lesser Bladderwort*** - **U. minor** L.   339
1 Margins of leaf-segments with teeth with bristles; lower lip of corolla ≥8mm; spur 3-10mm   2
   2 Stems of 1 sort, all bearing green leaves and bladders, free-floating; leaf-segments filiform; quadrifids with 2 long arms 1.8-2.8x as long as 2 short arms   3
   2 Stems of 2 sorts - free-floating ones bearing green leaves and 0 or few bladders, and ones often anchored in substratum and bearing very reduced non-green leaves and many bladders; leaf-segments linear; quadrifids with 2 long arms 1.2-2x as long as 2 short arms   4
3 Lower lip of corolla with reflexed margins (fresh material only); pedicel 8-15mm, recurved but not elongating after flowering; glands present on inside of only abaxial side of spur. Stems to 1m. Native; usually base-rich still or slow water; scattered in Br and Ir, commonest in E En   ***Greater Bladderwort*** - **U. vulgaris** L.   339
3 Lower lip of corolla with flat or slightly upturned margins; pedicel 8-15mm at flowering, becoming sinuous and 10-30mm after; glands present on inside of both abaxial and adaxial sides of spur. Stems to 60cm. Native; usually acidic still or slow water; scattered throughout Br and Ir, commonest in W & N Br   ***Bladderwort*** - **U. australis** R. Br.   339
   4 Green leaves totally without traps; apex of leaf-segments usually obtuse; spur 8-10mm, c. as long as lower lip; quadrifids with 2 shorter arms ± parallel or diverging at ≤21(37)°. Floating stems to 20(40)cm. Native; still, shallow water in peaty bogs and marshes; very scattered in Ir, Sc, N & SC En, E Anglia and Caerns
       Intermediate Bladderwort - **U. intermedia** Hayne   339
   4 Green leaves usually with some traps; apex of leaf-segments subulate; spur 3-5mm, c.1/2 as long as lower lip; quadrifids with 2 shorter arms diverging at >(30)52°   5
5 Margin of leaf-segments with 2-7 teeth with bristles; lower lip of corolla with flat or slightly upturned margins, 9-11 x 12-15mm; quadrifids with 2 shorter arms diverging at (30)52-97(140)°. Floating stems to 20cm. Native; similar places to *U. intermedia*; W Sc (Wigtowns to W Sutherland)   ***Nordic Bladderwort*** - **U. stygia** G. Thor   339
5 Margin of leaf-segments with 0-5 teeth with bristles; lower lip of corolla with flat margins at first, later with reflexed margins, c.8 x 9mm; quadrifids with 2 shorter arms diverging at (117)146-197(228)°. Floating stems to 20cm. Native; similar places to *U. intermedia*;

locally frequent in Sc, very scattered in Ir, N & SC En
Pale Bladderwort - **U. ochroleuca** R.W. Hartm.   3

133. CAMPANULACEAE - *Bellflower family*

Herbaceous annuals to perennials, often with white latex; differs from other families with fused petals and inferior ovaries by the 5 stamens borne on the receptacle (not on corolla), and numerous ovules on axile placentae.

1 Flowers densely packed into flattish heads or globose to elongated spikes; corolla divided nearly to base, with linear lobes   2
1 Flowers not all packed into dense heads, or if so then corolla divided <2/3 way to base   3
   2 Flowers in flattish heads with conspicuous region of flowerless bracts at base; each flower with 0 bract; flower buds straight; stigmas globose; stems pubescent   **6. JASIONE**
   2 Flowers in globose to elongated spikes, without flowerless bracts at base; each flower with 1 bract; flower buds curved; stigmas linear; stems ± glabrous   **5. PHYTEUMA**
3 Corolla distinctly zygomorphic; filaments fused laterally at least distally to form tube round style   4
3 Corolla actinomorphic; filaments free though often close together round style (anthers sometimes fused laterally)   6
   4 Stems procumbent to decumbent, rooting at nodes; leaves suborbicular; fruit a berry   **8. PRATIA**
   4 Stems erect to ascending, not rooting at nodes; leaves linear to obovate; fruit a capsule   5
5 Flowers and capsules pedicellate; ovary and capsules <1.5cm, widening distally, 2-celled   **7. LOBELIA**
5 Flowers and capsules sessile; ovary and capsules >2cm, cylindrical, 1-celled   **9. DOWNINGIA**
   6 Ovary and fruits >3x as long as wide; corolla shorter than calyx; annual   **2. LEGOUSIA**
   6 Ovary and fruits <2(3)x as long as wide; corolla longer than (rarely ± as long as) calyx; biennial or perennial   7
7 Corolla-tube <2mm wide; style >1.5x as long as corolla (tube + lobes)   **4. TRACHELIUM**
7 Corolla-tube >3mm wide; style not or scarcely longer than corolla   8
   8 Stems filiform, procumbent, with solitary axillary flowers on erect stalks much longer than corolla; all leaves petiolate; capsule opening apically (i.e. within calyx)   **3. WAHLENBERGIA**
   8 Usually at least flowering stems erect to ascending, if all procumbent then not all pedicels longer than corolla; usually at least uppermost leaves sessile or ± so; capsule opening laterally or basally (i.e. outside calyx)   **1. CAMPANULA**

1. CAMPANULA L. - *Bellflowers*

Biennials to perennials; flowers in racemes or panicles, sometimes in ± compact heads; corolla usually blue, actinomorphic, divided up to 1/2(2/3) way to base; filaments and anthers free; ovary 3-5-celled; style shorter than to slightly longer than corolla; stigmas 3-5, linear; capsule dehiscing by lateral or basal pores.

White-flowered variants of most spp. are not rare but have not been mentioned under each sp. Corolla-lengths given are those in the fresh state; considerable shrinking often occurs on drying.

1 Calyx with 5 sepal-like reflexed appendages alternating with 5 calyx-lobes   2
1 Calyx with 5 calyx-lobes but no extra appendages   3
   2 Biennial; all leaves cuneate at base; ovary and fruit 5-celled; stigmas 5. Stems hispid-pubescent, erect, to 60cm. Intrd-natd; waste and rough ground, grassy places and banks; scattered in Br, mainly C & S   *Canterbury-bells* - **C. medium** L.
   2 Perennial; basal and lower stem-leaves cordate at base; ovary and fruit 3-celled; stigmas 3. Stems pubescent, erect or ± so, to 70cm; corolla white. Intrd-natd; banks and rough ground; S En, especially by railways in CS & SW
                                Cornish Bellflower - **C. alliariifolia** Willd.
3 Capsule with pores in apical 1/2   4
3 Capsule with pores at or near base   7
   4 Calyx-lobes lanceolate to ovate, serrate; lower stem-leaves ovate to ovate-oblong. Stems sparsely scabrid-pubescent, erect, to 2m. Intrd-natd; waysides and in waste and rough ground, often in damp places; scattered in Br, especially N En and Sc
                                   Milky Bellflower - **C. lactiflora** M. Bieb.
   4 Calyx-lobes linear to lanceolate, entire or with 1-2 basal small teeth; stem-leaves linear to obovate   5
5 Perennial, with non-flowering rosettes arising from rhizomes; corollas mostly >3cm; stigmas >1/2 as long as styles. Stems glabrous, erect, to 80cm. Intrd-natd; waste and rough ground, grassy places and banks; scattered through most of Br
                                 Peach-leaved Bellflower - **C. persicifolia** L.
5 Usually biennial, without non-flowering rosettes; corollas mostly <3cm; stigmas <1/2 as long as style   6
   6 Tap-root thickened; inflorescence narrowly pyramidal; basal leaves abruptly narrowed to distinct petiole. Stems usually scabrid-pubescent, erect, to 80cm. Intrd-natd; rough grassy fields and banks; SE En   *Rampion Bellflower* - **C. rapunculus** L.
   6 Tap-root thin; inflorescence widely spreading; basal leaves gradually narrowed to indistinct petiole. Stems scabrid-pubescent, erect, to 60cm. Native; open woods, wood-borders, hedgebanks; very local in S Br N to Salop and Leics, also natd garden escape   *Spreading Bellflower* - **C. patula** L.
7 Flowers sessile. Stems pubescent, erect, to 80cm but often <20cm.

Native; chalk and limestone grassland, scrub and open woodland, cliffs and dunes by sea; mainly S & E Br N to CE Sc, scattered garden escape elsewhere ***Clustered Bellflower* - C. glomerata** L.
7 Flowers with distinct pedicels 8
    8 Calyx-teeth linear to filiform, <1mm wide at base 9
    8 Calyx-teeth lanceolate or narrowly triangular to ovate-oblong or -triangular, >1mm wide at base 10
9 Middle stem-leaves ovate-oblong or narrowly so, rounded at base, serrate. Stems sparsely pubescent, erect, to 60cm. Intrd-natd; shady river-bank in Dumfriess and roadside bank in Westmorland
***Broad-leaved Harebell* - C. rhomboidalis** L.
9 Middle stem-leaves linear to linear-elliptic, very gradually tapered to base, ± entire. Stems glabrous to sparsely pubescent, erect to decumbent, to 50cm. Native; grassy places, fixed dunes, rock-ledges, usually on acid often sandy soils; in suitable places throughout Br and Ir ***Harebell* - C. rotundifolia** L.
    10 Stems decumbent to ascending, ≤30(50)cm 11
    10 Stems erect, usually >50cm 13
11 Corolla funnel-shaped (diameter at apex much < length), lobed 1/4-2/5 way to base. Stems glabrous to sparsely pubescent, decumbent to ascending, to 30(50)cm. Intrd-natd; on walls and rocky banks; scattered in Br, mostly C & S, CI
***Adria Bellflower* - C. portenschlagiana** Schult.
11 Corolla broadly bell-shaped, ± star-shaped from above (diameter at apex c. equalling to wider than length), lobed 1/2-3/4 way to base 12
    12 Basal leaves 2-serrate; corolla-lobes c.5mm wide at base. Stems grey-pubescent, decumbent to ascending, to 30(50)cm. Intrd-natd; on walls and rocky banks; scattered in Br, mostly C & S
***Trailing Bellflower* - C. poscharskyana** Degen
    12 Basal leaves bluntly 1-serrate; corolla-lobes c.10mm wide at base. Stems glabrous to pubescent, decumbent to ascending, to 20(40)cm. Intrd-natd; on walls; Guernsey
***Italian Bellflower* - C. fragilis** Cirillo
13 Capsules erect; plant glabrous; inflorescence dense, pyramidal or cylindrical. Stems glabrous, erect, to 1m. Intrd-natd; on walls; Guernsey and W Kent, rare casual elsewhere
***Chimney Bellflower* - C. pyramidalis** L.
13 Capsules pendent; plant pubescent; inflorescence racemose 14
    14 Plant patch-forming, with shoots arising from rhizomes and/or root-buds; calyx-teeth patent to reflexed. Stems sparsely pubescent, erect, to 80cm. Intrd-natd; in fields, woods, banks and rough ground, very persistent; widely scattered in Br and Ir
***Creeping Bellflower* - C. rapunculoides** L.
    14 Plant tufted, without rhizomes or root-buds; calyx-teeth erect to erecto-patent 15
15 Middle and lower stem-leaves sessile, cuneate at base; stem bluntly ridged, softly pubescent to subglabrous. Stems erect, to 1.2m. Native; rich, often damp, mainly calcareous woods; most of Br but

very rare to absent in S En and N Sc, intrd in NE Ir
Giant Bellflower - **C. latifolia** L.
15 Middle and lower stem-leaves petiolate, cordate at base; stem sharply angled, sparsely hispid. Stems erect, to 80(100)cm. Native; mainly base-rich woods and hedgebanks; frequent in Br N to N Lincs and N Wa, SE Ir, garden escape elsewhere in Br and Ir
Nettle-leaved Bellflower - **C. trachelium** L.

2. LEGOUSIA Durande - *Venus's-looking-glasses*
Annuals; flowers in terminal cymes; corolla lilac to purple, actinomorphic, divided c.1/2 way to base; filaments and anthers free; ovary 3-celled; style shorter than corolla; stigmas 3, linear to ± globose; capsule dehiscing by subapical lateral pores.

1 Flowers mostly forming terminal corymbs; calyx-lobes oblong-lanceolate, acute to obtuse, c.2x as long as corolla, c.1/2 as long as ovary at anthesis; corolla 4-10mm across. Stems to 30cm. Native; arable fields; scattered in S, C & E En, mostly on calcareous soils
Venus's-looking-glass - **L. hybrida** (L.) Delarbre
1 Flowers forming terminal pyramidal inflorescence; calyx-lobes linear, acute to acuminate, slightly shorter than corolla, c. as long as ovary at anthesis; corolla 15-20mm across. Stems to 30cm. Probably intrd-natd; sporadic in cornfields in N Hants, casual elsewhere in
S En   *Large Venus's-looking-glass* - **L. speculum-veneris** (L.) Chaix

3. WAHLENBERGIA Roth - *Ivy-leaved Bellflower*
Perennials; flowers solitary, axillary; corolla blue, actinomorphic, divided 1/3-1/2 way to base; filaments and anthers free; ovary 3-celled; style shorter than corolla; stigmas 3, linear; capsule dehiscing by apical pores.

1 Stems filiform, procumbent, to 30cm; leaves broadly ovate to orbicular-reniform; flowers on long, filiform, erect stalks; corolla 6-10mm. Native; damp acid places on heaths and moors, in woods, by streams; W & S Br N to Argyll, S & SE Ir
Ivy-leaved Bellflower - **W. hederacea** (L.) Rchb.

4. TRACHELIUM L. - *Throatwort*
Perennials; flowers numerous in terminal corymbose compound cymes; corolla blue, actinomorphic, with narrow tube and small limb divided ≥1/2 way to base; filaments and anthers free; ovary 2-3-celled; style longer than corolla; stigmas 2-3, capitate; capsule dehiscing by 2-3 sub-basal pores.

1 Stems glabrous, erect, to 1m; lower leaves elliptic-ovate; flowers with tube 4-7 x c.0.5-1mm, with limb 2-3mm across. Intrd-natd; on walls; Guernsey, Jersey, Middlesex and W Kent *Throatwort* - **T. caeruleum** L.

5. PHYTEUMA L. - *Rampions*
Perennials; flowers numerous in terminal congested heads; corolla blue or

yellow, slightly zygomorphic, tubular and usually curved in bud but split nearly to base when open; filaments and anthers free but appressed around style; ovary 2-3-celled; style slightly shorter than corolla; stigmas 2-3, linear; capsule dehiscing by 2-3 lateral pores.

1 Corolla usually pale yellow (all native plants), rarely blue (most intrd plants); inflorescence oblong to cylindrical in flower. Stems glabrous, erect, to 80cm. Native; woods, scrub and hedgerows on acid soils; E Sussex, rare escape elsewhere in Br *Spiked Rampion* - **P. spicatum** L.
1 Corolla violet-blue; inflorescence globose to very shortly ovoid in flower 2
   2 Corolla nearly straight in bud; lowest bracts longer than inflorescence (though often reflexed); lowest leaves strongly cordate at base. Stems glabrous, erect to decumbent or pendent, to 40cm. Intrd-natd; limestone cracks at Inchnadamph, W Sutherland *Oxford Rampion* - **P. scheuchzeri** All.
   2 Corolla strongly curved in bud; bracts shorter than inflorescence; lowest leaves rounded to rarely subcordate at base. Stems glabrous to sparsely pubescent, erect, to 50cm (but often <15cm). Native; open chalk grassland; local in S En from N Wilts to E Sussex *Round-headed Rampion* - **P. orbiculare** L.

6. JASIONE L. - *Sheep's-bit*

Annuals to perennials; flowers numerous in terminal, congested heads; corolla blue, actinomorphic, tubular and straight in bud but split nearly to base when open; filaments free, anthers slightly laterally fused; ovary 2-celled; style longer than corolla; stigmas 2, subcapitate; capsule dehiscing by 2 apical short valves.

1 Stems pubescent, suberect to decumbent, to 50cm; inflorescence depressed-globose, 0.5-3.5cm across. Native; grassy or sandy or rocky places on acid soils, walls, cliffs, banks; locally common in BI, mainly in W *Sheep's-bit* - **J. montana** L.

7. LOBELIA L. - *Lobelias*

Annuals or perennials; flowers in terminal racemes; corolla pale lilac to blue, zygomorphic, with tube (with deep dorsal split) and expanded limb, the latter with 5 lobes arranged 2 in upper and 3 in lower lip; filaments and anthers fused laterally around style; ovary 2-celled; style shorter than corolla; stigma capitate, 2-lobed; capsule dehiscing by 2 apical valves.

1 Leaves all basal, linear, entire; plant submerged (except inflorescence) or at lakeside. Perennial; stems glabrous, erect, to 70(120)cm. Native; in stony, acid mainly montane lakes, rarely on wet ground adjacent; locally common in N & W Br S to S Wa, W, N & E Ir
Water Lobelia - **L. dortmanna** L.
1 Stem-leaves present, the lower ones obovate, serrate; plant terrestrial 2
   2 Corolla-lobes <2mm wide; pedicels <1cm. Perennial; stems

sparsely and shortly pubescent, erect, to 80cm. Native; acid heathy grassland, open woods, wood-borders; very local in S En from E Cornwall to W Kent **Heath Lobelia - L. urens** L.

2 Lower 3 corolla-lobes >2mm wide; at least lower pedicels >1cm. Annual; stems glabrous to sparsely pubescent, erect to decumbent, to 30cm. Intrd-natd; on tips and in pavement-cracks and rough ground; scattered in Br, CI **Garden Lobelia - L. erinus** L.

8. PRATIA Gaudich. - *Lawn Lobelia*

Perennials; flowers solitary in leaf-axils; corolla white; differs from *Lobelia* in fruit a berry.

1 Stems procumbent, rooting at nodes, glabrous, to 15cm; leaves ≤12mm, suborbicular; flowers 7-20mm. Intrd-natd; on damp lawns; scattered in Sc, SE En **Lawn Lobelia - P. angulata** (G. Forst.) Hook. f.

9. DOWNINGIA Torr. - *Californian Lobelia*

Annuals; corolla blue with white centre; differs from *Lobelia* in corolla with lower lip only shallowly lobed, flowers sessile with long, pedicel-like ovary, ovary 1-celled, and capsule dehiscing by 3-5 longitudinal slits.

1 Stems erect, glabrous, to 25cm; flowers 8-18mm. Intrd-casual; in grassy places; sporadic in SE En
Californian Lobelia - **D. elegans** (Lindl.) Torr.

134. RUBIACEAE - *Bedstraw family*

Annual to perennial herbs, evergreen climbers or rarely shrubs; easily recognized by the small flowers with 4-5 petals fused into an (often very short) tube, 0 or minute calyx and inferior 2-celled ovary with 1 ovule per cell. Most spp. have apparently whorled leaves, 4 corolla-lobes and distinctive paired nutlets.

1	Leaves opposite, usually with stipules or smaller leaves also at same node	2
1	Leaves in whorls of ≥4, ± all same size in 1 whorl	4
	2 Evergreen shrub	**1. COPROSMA**
	2 Procumbent to ascending herb	3
3	Leaves linear or nearly so; fruit of 2 nutlets	**5. ASPERULA**
3	Leaves ovate to suborbicular; fruit succulent	**2. NERTERA**
	4 Most or all flowers with 5 corolla-lobes	5
	4 Most or all flowers with 4 corolla-lobes	6
5	Procumbent to ascending annual; leaves ≥6 in a whorl; corolla pink; fruit dry	**4. PHUOPSIS**
5	Evergreen climber or scrambler; leaves 4-6 in a whorl; corolla yellowish-green; fruit succulent	**8. RUBIA**

		6	Calyx distinct, c.0.5-1mm at first, slightly enlarging in fruit	
				3. SHERARDIA
		6	Calyx absent or vestigial	7
7			Corolla-tube >1mm	8
7			Corolla-tube <1mm	9
		8	Ovary and fruit smooth to papillose	**5. ASPERULA**
		8	Ovary and fruit covered with hooked bristles	**6. GALIUM**
9			At least some whorls with >4 leaves	**6. GALIUM**
9			All whorls with 4 leaves	10
		10	Flowers in dense axillary whorls; ovary and fruit smooth	
				7. CRUCIATA
		10	Flowers in terminal panicles; ovary and fruit covered with hooked bristles	**6. GALIUM**

1. COPROSMA J.R. & G. Forst. - *Tree Bedstraw*

Evergreen, usually dioecious shrubs; leaves opposite, with small stipules, petiolate; flowers inconspicuous, <1cm, in axillary clusters with 2 partly fused bracts below; calyx minute; corolla greenish, 4-5-lobed, with long tube; fruit succulent, with 2 nuts.

1 Shrub to 3m; leaves broadly ovate-oblong, (2)5-8cm, very shiny on upperside, with recurved margins. Intrd-natd; planted as windbreak in Scillies, sometimes self-sown   *Tree Bedstraw* - **C. repens** A. Rich.

2. NERTERA Gaertn. - *Beadplant*

Herbaceous perennials; leaves opposite, with small stipules, shortly petiolate; flowers solitary, axillary and terminal, <5mm; calyx minute; corolla greenish, 4-lobed; fruit succulent, with 2 nuts.

1 Stems procumbent, to 15cm, glabrous; leaves ovate to suborbicular, with recurved margins; fruit globose, c.4mm, bright reddish-orange. Intrd-natd; on damp lawns; few places in WC Sc

Beadplant - **N. granadensis** (L.f.) Druce

3. SHERARDIA L. - *Field Madder*

Annuals; leaves in whorls of 4-6, sessile; flowers 4-10 in dense terminal and axillary clusters with whorl of 8-10 leaf-like bracts at base, the clusters stalked; calyx 0.5-1mm at first, slightly enlarging in fruit, with 4-6 deeply toothed lobes; corolla pale to deep mauvish-pink, 4-lobed; fruit a pair of scabrid nutlets with persistent calyx on top.

1 Stems procumbent to ascending, to 40cm, glabrous to pubescent; corolla 4-5mm, with tube longer than lobes. Native; arable fields, waste places, thin grassland and lawns; frequent almost throughout BI but local in Sc.   *Field Madder* - **S. arvensis** L.

4. PHUOPSIS (Griseb.) Hook. f. - *Caucasian Crosswort*

Annuals to perennials; leaves in whorls of 6-9, sessile; flowers many in dense terminal clusters with whorl of many leaf-like bracts at base; calyx minute; corolla deep pink, 5-lobed; fruit a pair of glabrous, papillose nutlets.

1 Stems procumbent to sprawling, to 30(70)cm, sparsely scabrid-pubescent; corolla 12-15mm, with narrow tube much longer than lobes. Intrd-natd; on waste and rough ground and tips; scattered in C & S Br **Caucasian Crosswort - P. stylosa** (Trin.) B.D. Jacks.

5. ASPERULA L. - *Woodruffs*

Annuals to herbaceous perennials; leaves in whorls of 4-8 and all equal, or in whorls of 4 with 2 long and 2 short, sessile; flowers in dense terminal clusters with whorl of leaf-like bracts at base or in loose terminal panicles; calyx minute; corolla various colours, 4-lobed; fruit a pair of smooth to papillose nutlets.

1 Leaves in whorls of 6-8; corolla blue. Subglabrous annual with erect stems to 50cm. Intrd-casual; tips and waste places; scattered in most of Br and CI, mainly S **Blue Woodruff - A. arvensis** L.
1 Leaves in whorls of 4, equal or 2 short and 2 long; corolla white to pink 2
   2 Leaves equal in all whorls, narrowly ovate to narrowly elliptic; inflorescence a compact cluster with whorl of leaf-like bracts below. Pubescent perennial with erect stems to 50cm. Intrd-natd; in damp woods; local in C Sc, rare and impermanent further S
                                              Pink Woodruff - A. taurina L.
   2 Leaves at upper whorls 2 long and 2 short, linear-oblanceolate; inflorescence a diffuse panicle. Subglabrous perennial with procumbent to ascending stems to 50cm. Native; limestone and chalk grassland, calcareous dunes
(*Squinancywort* - **A. cynanchica** L.) 3
3 Rhizomes 0 or brown; leaves mostly linear, mostly 10-20 x 0.5-1mm; pedicels 0-1mm; corolla-lobes usually distinctly shorter than -tube. Locally common in S Br and W Ir, scattered N to Westmorland and SE Yorks **A. cynanchica ssp. cynanchica**
3 Rhizomes orange; leaves linear to oblanceolate, often <10mm and >1mm wide; pedicels usually 0; corolla-lobes usually c. as long as -tube. Calcareous dunes in S Wa, W Ir
                            A. cynanchica ssp. occidentalis (Rouy) Stace

6. GALIUM L. - *Bedstraws*

Annuals to herbaceous perennials; leaves in whorls of 4-12 and all equal, sessile; flowers in terminal panicles or axillary cymes; calyx minute; corolla white to yellow, 4-lobed; fruit a pair of smooth to bristly nutlets, the bristles sometimes hooked.

1 Ovaries and fruits with hooked bristles 2
1 Ovaries and fruits smooth to rugose or papillose 5
   2 All whorls with 4 leaves. Erect perennial to 45cm. Native; damp

grassy, rocky and gravelly places, also on sand-dunes; locally frequent in Br N from MW Yorks, scattered in W, C & N Ir, very local in Wa ***Northern Bedstraw* - G. boreale** L.

2 Most or all whorls with ≥5 leaves   3

3 Rhizomatous perennial; flowers in terminal panicles; corolla-tube >1mm. Erect perennial to 45cm. Native; damp, base-rich woods and hedgerows; frequent throughout BI except CI and Outer Isles
***Woodruff* - G. odoratum** (L.) Scop.

3 Annual; flowers in axillary cymes; corolla-tube <1mm   4

  4 Fruit >3mm (excl. bristles), its bristles with bulbous base; corolla >1.4mm across. Procumbent to scrambling-erect annual to 3m, with strongly recurved prickles on stems. Native; cultivated and arable land, hedgerows and scrub, other open ground; common   ***Cleavers* - G. aparine** L.

  4 Fruit <3mm (excl. bristles), its bristles wider but not bulbous at base; corolla <1.4mm across. Like *G. aparine* but slightly more slender; stems to 1m. Probably intrd-natd; natd arable weed around Saffron Waldron (N Essex), casual or temporarily natd in Br N to S Sc   ***False Cleavers* - G. spurium** L.

5 Corolla bright yellow. Procumbent to erect, smooth perennial to 1m. Native; dry grassy places especially on calcareous soils, often by sea; common   ***Lady's Bedstraw* - G. verum** L.

5 Corolla white to pale cream (N.B. *G. x pomeranicum*), sometimes tinged pink   6

  6 Annuals of open ground, usually easily uprooted, with sparse rooting system   7

  6 Perennials, firmly rooted, often in grassland or wet ground   9

7 Leaves with forward-directed marginal prickles. Slender, procumbent to ascending annual to 30cm, with small recurved prickles on stems. Native; walls and sandy banks; scattered in E Anglia and SE En   ***Wall Bedstraw* - G. parisiense** L.

7 Leaves with backward-directed marginal prickles   8

  8 Fruit smooth, <3mm; peduncles and pedicels divaricate at various angles at fruiting, but straight   (see couplet 4) **G. spurium**

  8 Fruit papillose, >3mm; peduncles and/or pedicels strongly recurved. Habit of *G. aparine* but stems to 60cm. Probably intrd-natd; arable and waste places; rare and sporadic in C & SE En
***Corn Cleavers* - G. tricornutum** Dandy

9 Leaves obtuse to acute, never apiculate or mucronate   10

9 Leaves apiculate to mucronate at apex   12

  10 Leaves linear; pedicels scarcely divaricate at fruiting; inflorescence obconical, widest near top. Decumbent to ascending or scrambling, smooth or slightly scabrid perennial to 40cm. Native; marshy places, ditches and pondsides; very local in S En, SE Yorks and CI   ***Slender Marsh-bedstraw* - G. constrictum** Chaub.

  10 Leaves linear-oblong to oblanceolate or narrowly elliptic; pedicels strongly divaricate at fruiting; inflorescence usually conical to cylindrical, widest well away from apex. Decumbent to ascending

or scrambling, smooth or more often scabrid perennial to 1m.
Native; damp meadows, pondsides, ditches, marshes and fens;
common (*Common Marsh-bedstraw* - **G. palustre** L.) 11

11 Most leaves <20mm; inflorescence ± cylindrical; pedicels mostly
<4mm at flowering; corolla mostly 2-3.5mm across; fruit c.1.6mm
G. palustre ssp. palustre

11 Most leaves >20mm; inflorescence ± conical; pedicels mostly >4mm
at flowering; corolla mostly 3-4.5mm across; fruit c.1.9mm. Probably
the commoner ssp. **G. palustre ssp. elongatum** (C. Presl) Arcang.

12 Stems to 60cm, rough, with projecting minute papillae or
pricklets. Native; fens and base-rich marshy places, scattered in
BI except N Sc, N Ir and CI   *Fen Bedstraw* - **G. uliginosum** L.

12 Stems perfectly smooth 13

13 Corolla-lobes apiculate to strongly mucronate at apex, with point
≥0.2mm; fruit minutely wrinkled. Decumbent to erect, smooth
perennial to 1.5m. Native; all sorts of grassy places, hedgerows,
mainly on well-drained base-rich soils; throughout most of BI but
rarer in N & W (*Hedge Bedstraw* - **G. mollugo** L.) 14

13 Corolla-lobes acute to minutely apiculate at apex, with point
<0.2mm; fruit minutely tuberculate 15

14 Leaves mostly oblanceolate to narrowly obovate; inflorescence
broad, with branches mostly at ≥45°, rather lax; corolla 2-3mm
across; pedicels strongly divaricate at fruiting. Relative
distributions of sspp. unknown **G. mollugo ssp. mollugo**

14 Leaves mostly linear-oblanceolate to oblanceolate; inflorescence
rather narrow, with branches mostly at <45°, rather dense; corolla
2.5-5mm across; pedicels weakly divaricate at fruiting. Mostly on
drier, more calcareous soils **G. mollugo ssp. erectum** Syme

15 Leaf-margins with forward-directed prickles; leaves on flowering
shoots oblanceolate. Decumbent to ascending, smooth perennial to
30cm. Native; dry grassland, rocky places and open woods on acid
soils; common *Heath Bedstraw* - **G. saxatile** L.

15 Leaf-margins with at least some backward-directed prickles; leaves
on flowering shoots linear to linear-elliptic or -oblanceolate 16

16 Fruit with minute high-domed subacute tubercles. Decumbent
to ascending, smooth perennial to 30cm. Native; limestone or
other base-rich grassland or rocks; local in BI NW of line from
R. Severn to R. Humber *Limestone Bedstraw* - **G. sterneri** Ehrend.

16 Fruit with minute low-domed to rounded tubercles. Decumbent
to erect, smooth perennial to 40cm. Native; dry chalk and
limestone grassland; scattered and local in En SE of line from
R. Severn to R. Humber *Slender Bedstraw* - **G. pumilum** Murray

G. verum x *G. mollugo* = ***G. x pomeranicum*** Retz. is frequent with the parents in Br N to N Aberdeen and in CI, and rare in S & E Ir; it is intermediate in all characters, notably petal- and leaf-shape and flower-colour, and somewhat variable, suggesting backcrossing. *G. sterneri* x *G. saxatile* has been cytologically confirmed in Caerns, M Perth, Easterness and W

Sutherland; it is intermediate and highly sterile.

7. CRUCIATA Mill. - *Crosswort*
Herbaceous perennials; leaves in whorls of 4, all equal, ± sessile; flowers in dense axillary whorls of cymes, the terminal flower bisexual and the laterals male in each cyme; calyx minute; corolla yellow, 4-lobed, with very short tube; fruit 1 or a pair of smooth nutlets.

1 Stems erect, to 60cm, conspicuously pubescent; leaves elliptic to oblong- or ovate-elliptic, 10-20mm, yellowish-green. Native; grassy places, hedgerows, scrub and rough ground, mostly on calcareous soils; common in Br N to C Sc, intrd in Ir   *Crosswort* - **C. laevipes** Opiz

8. RUBIA L. - *Madders*
Evergreen scrambling perennials; leaves 4-6 per whorl, all equal, narrowed to ± sessile base; flowers in diffuse axillary and terminal panicles; calyx minute; corolla pale yellowish-green, 5-lobed, with very short tube; fruit succulent, with 1 seed.

1 Stems trailing to scrambling, to 1.5m, glabrous, with strong recurved prickles; leaves 1-5cm, elliptic to narrowly so, leathery, 1-veined, glabrous, with strong recurved prickles on margins. Native; hedges, scrub, rocky places; locally common in S & W Br from E Kent to N Wa, S, E & C Ir, CI, mainly coastal   ***Wild Madder* - R. peregrina** L.

135. CAPRIFOLIACEAE - *Honeysuckle family*

Deciduous or evergreen shrubs (small and procumbent in *Linnaea*), small trees or woody climbers, rarely (*Sambucus ebulus*) herbaceous perennials; the only woody plants with fused petals, inferior ovary and stamens only 4 or 5.

1 Leaves pinnate **1. SAMBUCUS**
1 Leaves simple (sometimes deeply lobed) 2
  2 Flowers numerous in corymbose compound cymes; style ± 0
2. VIBURNUM
  2 Flowers 2-few, not corymbose; style conspicuous 3
3 Main stems procumbent, the flowers in pairs terminal on erect lateral stems **4. LINNAEA**
3 Main stems ± erect or climbing 4
  4 Bracts ≥15mm, leaf-like, purple, or green strongly tinged purple
5. LEYCESTERIA
  4 Bracts <15mm, usually not purple 5
5 Deciduous shrub; ovary 4-celled, with 2 fertile and 2 sterile cells, the former each with 1 ovule; corolla actinomorphic, <10mm
3. SYMPHORICARPOS
5 Ovary 2-3-celled, all cells fertile and with >1 ovule; corolla strongly zygomorphic to ± actinomorphic, if <10mm then plant an evergreen

shrub or corolla strongly zygomorphic or both 6
- 6 Fruit a capsule; corolla >20mm, weakly zygomorphic, scarcely 2-lipped **6. WEIGELA**
- 6 Fruit a berry; if corolla >20mm then strongly zygomorphic and 2-lipped **7. LONICERA**

1. SAMBUCUS L. - *Elders*

Deciduous shrubs or herbaceous perennials; leaves pinnate; flowers numerous in corymbose or paniculate compound cymes, actinomorphic; stamens 5; ovary 3-5-celled, each cell with 1 ovule; style 0; stigmas as many as carpels; fruit a drupe with 3-5 seeds.

- 1 Inflorescence an ovoid to ± globose panicle; ripe fruits red; stipules represented by stalked glands. Shrub to 4m. Intrd-natd; in hedges, woods and shrubberies; frequent in Br N from Derbys and Cheshire, rare further S *Red-berried Elder* - **S. racemosa** L.
- 1 Inflorescence a slightly convex to slightly concave corymb; ripe fruits black to purplish-black, rarely red or greenish-yellow or -white; stipules 0 or subulate to ovate 2
  - 2 Rhizomatous herbaceous perennial; stipules conspicuous, ovate or narrowly so; anthers purple. Stems erect, to 1.5m. Possibly native in En; waysides, rough and waste ground; scattered over most of BI *Dwarf Elder* - **S. ebulus** L.
  - 2 Erect shrub; stipules small and subulate; anthers cream 3
- 3 Fruits black, rarely greenish-yellow or -white; 2nd year twigs with numerous lenticels; leaflets (3)5(-7); not rhizomatous. Shrub or small tree to 10m. Native; hedges, woods, shrubberies, waste and rough ground, especially on manured soils; common *Elder* - **S. nigra** L.
- 3 Fruits purplish-black, rarely red; 2nd year twigs with few lenticels; leaflets (5)7(-11); rhizomatous. Suckering shrub to 4m. Intrd-natd; in scrub, rough ground and on railway banks; very scattered in Sc, N En and Surrey *American Elder* - **S. canadensis** L.

2. VIBURNUM L. - *Viburnums*

Deciduous or evergreen shrubs; leaves simple, sometimes lobed; flowers numerous in corymbose compound cymes, actinomorphic, sometimes some sterile; stamens 5; ovary 3-celled, but appearing 1-celled due to abortion of 2 cells, with 1 ovule; style 0; stigmas 3; fruit a drupe with 1 seed.

- 1 Leaves deciduous, lobed or serrate 2
- 1 Leaves evergreen, entire or obscurely denticulate 3
  - 2 Leaves lobed; outer flowers sterile, much larger than inner; fruits red, subglobose. Deciduous shrub to 4m. Native; woods, scrub and hedges; frequent throughout Br and Ir except N Sc
  Guelder-rose - **V. opulus** L.
  - 2 Leaves serrate; all flowers fertile, uniform in size; fruits red, then black, compressed. Deciduous shrub to 6m. Native; woods, scrub and hedges, especially on base-rich soils; common in Br

SE of line from Glam to S Lincs, scattered elsewhere and
probably intrd *Wayfaring-tree* - **V. lantana** L.
3 Leaves smooth; first-year twigs glabrous or sparsely pubescent.
Evergreen shrub to 6m. Intrd-natd; cliffs, banks and rough ground;
widespread in S En and S & N Wa, Man *Laurustinus* - **V. tinus** L.
3 Leaves strongly wrinkled; first-year twigs tomentose. Evergreen
shrub to 6m. Intrd-natd; a relic, rarely natd, in old woodland or
parkland; very scattered in En N to Leics
Wrinkled Viburnum - **V. rhytidophyllum** Hemsl.

V. lantana x *V. rhytidophyllum* = ***V. x rhytidophylloides*** J.V. Suringar is grown in gardens and plants have been found in the wild in W Kent and Surrey.

3. SYMPHORICARPOS Duhamel - *Snowberries*
Deciduous shrubs; leaves simple, entire, sometimes deeply lobed; flowers in dense terminal spikes, actinomorphic; stamens (4-)5; ovary 4-celled, with 2 fertile cells each with 1 ovule and 2 sterile cells; style present; stigma 1, capitate; fruit a drupe with 2 seeds.

1 Suckering from rhizomes; fruits pure white; style glabrous. Erect then
arching shrub to 2m. Intrd-natd; in woods, scrub, rough ground;
frequent throughout BI *Snowberry* - **S. albus** (L.) S.F. Blake
1 Rooting from stoloniferous stem-tips; fruits pink or pink-flushed;
style pubescent 2
  2 Fruits ± uniformly pink, 4-6mm. Habit as for *S. x chenaultii*.
Intrd-natd; open scrub; W Kent *Coralberry* - **S. orbiculatus** Moench
  2 Fruits white, flushed pink on exposed side, 6-10mm. Arching
shrub to 1.5m, with procumbent or arching then procumbent
non-flowering stems rooting at tips. Intrd-natd; woods, scrub,
rough ground; scattered in Br N to C Sc (*S. microphyllus* Kunth x
S. orbiculatus Moench) *Hybrid Coralberry* - **S. x chenaultii** Rehder

4. LINNAEA L. - *Twinflower*
Procumbent, evergreen, dwarf shrubs; leaves simple, crenate; flowers 2, each with 1 bract at base and 2 bracteoles at apex of pedicel, actinomorphic; stamens 4; ovary with 1 fertile cell with 1 ovule and 2 sterile cells; style present; stigma 1, bilobed; fruit an achene.

1 Stems procumbent, to 40cm; flowers in pairs borne on erect leafless
stems to 8cm; corolla 5-10mm, pink. Native; on barish ground under
shade of rocks or trees, mostly in woods, especially of *Pinus*; very
local in E Sc N to Caithness *Twinflower* - **L. borealis** L.

5. LEYCESTERIA Wall. - *Himalayan Honeysuckle*
Deciduous shrub, often semi-herbaceous; leaves simple, entire to serrate; flowers in crowded terminal spikes with large purple or purple-green bracts, ± actinomorphic; stamens 5; ovary 5-celled, each cell with several

ovules; style present; stigma 1, capitate; fruit a several-seeded berry.

1 Stems erect, to 2m; corolla 10-20mm, pinkish-purple; fruit purple, subglobose. Intrd-natd; woods, shrubberies and rough ground; scattered ± throughout BI   *Himalayan Honeysuckle* - **L. formosa** Wall.

6. WEIGELA Thunb. - *Weigelia*
Deciduous shrubs; leaves simple, serrate; flowers in small axillary cymes; corolla weakly zygomorphic, with 5 lobes scarcely arranged in 2 lips; stamens 5; ovary 2-celled, with many ovules per cell; style present; stigma 1, capitate or 2-lobed; fruit a capsule.

1 Erect arching shrub to 2(3)m; corolla 25-35mm, pink to red. Intrd-natd; in rough and marginal ground; very scattered in En and N to C Sc, Man   *Weigelia* - **W. florida** (Bunge) A. DC.

7. LONICERA L. - *Honeysuckles*
Deciduous or evergreen shrubs or climbers; leaves simple, sometimes lobed, entire; flowers sessile, in pedunculate axillary pairs or in terminal heads; corolla zygomorphic with 4-lobed upper and 1-lobed lower lip, or ± actinomorphic with 5 lobes; stamens 5; ovary 2-3-celled with several ovules per cell; style long; stigma capitate or slightly lobed; fruit a several-seeded berry.

1 Flowers and fruit sessile in terminal and subterminal whorls; climbers 2
1 Flowers and fruit in pairs, sessile at apex of common axillary stalk, sometimes crowded near branch ends   4
   2 All leaves separate, not fused in pairs; berry red. Deciduous climber to 6(10)m. Native; woods, scrub and hedges; common
Honeysuckle - **L. periclymenum** L.
   2 At least most apical pair of leaves on each branch fused around stem at base; berry orange   3
3 Bracteoles at base of each flower 0 or minute. Deciduous climber to 6(10)m. Intrd-natd; hedges and rough ground; scattered in Br N to C Sc   *Perfoliate Honeysuckle* - **L. caprifolium** L.
3 Bractoles c.1mm, obscuring base of ovary. Deciduous climber to 6(10)m. Intrd-natd; marginal and rough places; SE En and E Anglia (*L. caprifolium* x *L. etrusca* Santi)
Garden Honeysuckle - **L. x italica** Tausch
   4 Stems twining   5
   4 Stems not twining   6
5 Corolla 15-25mm, glabrous on outside; 2 bracts at base of each flower-pair subulate. Evergreen climber to 5(10)m. Intrd-natd; hedges, rough and marginal ground; Surrey and Herts
Henry's Honeysuckle - **L. henryi** Hemsl.
5 Corolla 30-50mm, pubescent on outside; 2 bracts at base of each flower-pair leaf-like. Semi-evergreen climber to 5(10)m. Intrd-natd;

hedges, scrub, banks and rough ground; scattered in C & S Br
Japanese Honeysuckle - **L. japonica** Thunb.
6   Two bracts at base of each flower-pair (and bracteoles within) ovate, obscuring base of flower, purple and enlarging in fruit. Deciduous shrub to 2m, with spreading or arching branches. Intrd-natd; rough and marginal ground; scattered in En (mostly N), Man, SW Sc and Ir
Californian Honeysuckle - **L. involucrata** (Richardson) Spreng.
6   Two bracts at base of each flower-pair subulate to linear-lanceolate, not obscuring ovaries, scarcely enlarging in fruit   7
7   Corolla distinctly 2-lipped; leaves deciduous, (20)30-80mm; berry red   8
7   Corolla ± actinomorphic; leaves evergreen, (4)6-32mm; berry violet   9
8   Corolla 8-15mm, pale yellow to cream, sometimes tinged pink; young stems and leaves pubescent. Deciduous shrub to 2m. Possibly native; woods and scrub on chalk in W Sussex, also widely natd in hedges, woods and scrub through much of Br and Ir *Fly Honeysuckle* - **L. xylosteum** L.
8   Corolla 15-25mm, pink to red, sometimes white; young stems and leaves glabrous. Deciduous shrub to 4m. Intrd-natd; bird-sown in hedges and rough ground; very scattered in En N to Cheshire *Tartarian Honeysuckle* - **L. tatarica** L.
9   Often >1m; leaves (4)6-16mm, mostly ovate, rounded to subcordate at base. Evergreen shrub to 1.8m, with erect to arching branches. Intrd-natd; scrub, hedges, woodland, banks and rough ground; scattered ± throughout Bl *Wilson's Honeysuckle* - **L. nitida** E.H. Wilson
9   Rarely >1m; leaves (6)12-32mm, mostly oblong-elliptic to narrowly so, cuneate at base. Evergreen shrub to 1m, with spreading branches. Intrd-natd; much grown in shrubberies and road-borders, sometimes self-sown; very scattered in Br N to C Sc, Man
Box-leaved Honeysuckle - **L. pileata** Oliv.

136. ADOXACEAE - *Moschatel family*

Perennial, rhizomatous herbs; instantly recognizable by the small yellowish-green 'townhall clock' flower-head.

1. ADOXA L. - *Moschatel*

1   Long-petioled 2-3-ternate leaves, and erect flowering stems to 15cm with 2 1-ternate opposite leaves, arising from short, white, scaly rhizome. Native; woods, hedges, shady rocky places; frequent throughout Br except N Sc, Co Antrim, intrd in Co Dublin
Moschatel - **A. moschatellina** L.

137. VALERIANACEAE - *Valerian family*

Annual to perennial herbs; distinguished by the inferior ovary with 1 ovule but 2 other sterile cells (these often obscure), 0 or minute calyx at flowering, 5-lobed tubular corolla often pouched or spurred at base, and 1 or 3 stamens.

1 Stems forked into 2 at each node; calyx remaining minute at fruiting
   1. VALERIANELLA
1 Main stem simple or with lateral branches; calyx developing long
   feathery projections at fruiting   2
   2 Stamen 1   **3. CENTRANTHUS**
   2 Stamens 3   **2. VALERIANA**

1. VALERIANELLA Mill. - *Cornsalads*

Annuals with stems repeatedly forked; leaves simple, entire to serrate or sparsely lobed; flowers in rather lax to dense compound cymes, bisexual; calyx ± 0 or small, persistent but remaining small on top of fruit, usually unequal; corolla-tube not pouched or spurred; stamens 3; stigmas 3; sterile cells of ovary small to large.

Ripe fruits are essential for determination; fruit lengths exclude the calyx.

1 Calyx in fruit absent or vestigial, <1/10 as long as rest of fruit   2
1 Calyx in fruit distinct, c.1/4 to nearly as long as rest of fruit   3
   2 Fruit c. as wide as thick, much longer than wide or thick, with a
     very deep groove on abaxial face. Stems erect, to 15(40)cm.
     Native; arable and rough ground, bare places in grassland, on
     banks, walls and rocky outcrops; scattered in Br N to Berwicks,
     CI and (intrd) S & E Ir *Keeled-fruited Cornsalad* -**V. carinata** Loisel.
   2 Fruit c.2x as thick as wide, scarcely longer than thick, shallowly
     grooved on abaxial face. Stems erect, to 15(40)cm. Native; arable
     and rough ground, bare places in grassland, on banks, walls,
     rocky outcrops and dunes; frequent throughout BI
     Common Cornsalad - **V. locusta** (L.) Laterr.
3 Calyx in fruit with short tube, with usually 6 teeth, >2/3 as long as
   rest of fruit, nearly as wide as fruit. Stems erect, to 15(40)cm. Native;
   banks, walls and rough ground; very scattered in extreme S En
   and CI   *Hairy-fruited Cornsalad* - **V. eriocarpa** Desv.
3 Calyx in fruit with very short or 0 tube, with <6 (often 1) teeth,
   <1/2(2/3) as long as rest of fruit, <1/2 as wide as fruit   4
   4 Main tooth of calyx in fruit scarcely or not toothed; fruit ±
     smooth on all faces, with 2-6 fine grooves and/or longitudinal
     ridges, with easily broken walls. Stems erect, to 15(40)cm.
     Native; cornfields and rough ground; very local in S En and
     (intrd) Ir   *Broad-fruited Cornsalad* - **V. rimosa** Bastard
   4 Main tooth of calyx in fruit usually with 2 or more distinct teeth;
     fruit with 2 distinct ribs on abaxial face delimiting ovate ± flat
     area, with hard walls. Stems erect, to 15(40)cm. Native; cornfields

and rough ground; scattered in Br and (intrd) Ir N to Cheviot
Narrow-fruited Cornsalad - **V. dentata** (L.) Pollich

2. VALERIANA L. - *Valerians*

Perennials with main and lateral stems; leaves pinnate and/or simple on each plant; flowers in rather dense compound cymes, bisexual or dioecious; calyx developing long feathery projections at fruiting; corolla-tube not or slightly pouched at base; stamens 3; stigmas 3; sterile cells of ovary scarcely discernible.

1 Stem-leaves and basal leaves pinnate, with several lateral leaflets as large as terminal one. Stems erect, to 2m, sometimes with short stolons. Native; dry or damp grassy places and rough ground; frequent ± throughout Br and Ir *Common Valerian* - **V. officinalis** L.
1 At least basal leaves simple   2
   2 Basal leaves entire, cuneate to rounded at base; stem-leaves with several lateral leaflets as large as terminal one. Stems erect, to 40cm; stolons well developed. Native; marshes, fens and bogs; frequent in Br N to S Sc *Marsh Valerian* - **V. dioica** L.
   2 Basal leaves dentate, cordate at base; stem-leaves simple or with 1 large terminal and 1-2 much smaller pairs of lateral leaflets. Stems erect, to 1.2m; stolons 0. Intrd-natd; damp woods and shady hedgebanks; N & W Br from E Cornwall to N Aberdeen, mostly Sc, rare in W & NE Ir *Pyrenean Valerian* - **V. pyrenaica** L.

3. CENTRANTHUS DC. - *Valerians*

Annuals or perennials with main and usually lateral stems; leaves simple and entire to deeply pinnately lobed or ± pinnate; flowers in dense compound cymes, bisexual; calyx developing long feathery projections at fruiting; corolla-tube with backward-directed spur; stamen 1; stigma 1; sterile cells of ovary scarcely discernible.

1 Erect perennial to 80cm; leaves simple, somewhat glaucous; flowers red, pink or white; corolla-spur (2)5-12mm. Intrd-natd; on walls, dry rocky or sandy places, cliffs and banks; common in CI and S & C Br, extending to N & E Ir and C Sc *Red Valerian* - **C. ruber** (L.) DC.
1 Erect annual to 30cm; leaves deeply pinnately lobed to ± pinnate; flowers pink; corolla-spur <2mm. Intrd-natd; on open ground in churchyard; Surrey *Annual Valerian* - **C. calcitrapae** (L.) Dufr.

138. DIPSACACEAE - *Teasel family*

Biennial to perennial herbs; the flowers with 4 free stamens and borne in a capitulum, and the ovary and fruit enclosed in a tubular epicalyx, are unique.

The calyx, which arises from the top of the ovary outside the corolla, is often less conspicuous than the *epicalyx*, which arises at the base of the ovary but encloses the latter in a tubular structure and often expands into lobes around the calyx. The ripe fruit remains enclosed in the epicalyx. The receptacle bears bracts at the base of the capitulum and usually a bract associated with each flower within the capitulum. Before using the key a capitulum should be dissected carefully to distinguish between calyx and epicalyx.

1 Stems, and usually midribs on lowerside of leaves, prickly
 1. DIPSACUS
1 Stems and leaves glabrous to pubescent, not prickly 2
 2 Corolla 5-lobed; epicalyx expanded at apex into membranous, veined funnel **5. SCABIOSA**
 2 Corolla 4-lobed; epicalyx variously expanded at apex but not membranous 3
3 Corolla cream; calyx without teeth or bristles **2. CEPHALARIA**
3 Corolla blue to purple or violet, rarely white or pinkish; calyx with 4-8 teeth or bristles 4
 4 Flowers ± all equal-sized; calyx with 4-5 bristles; receptacle bearing bracts, 1 subtending each flower as well as some at base **4. SUCCISA**
 4 Outer flowers much longer than inner ones; calyx with 8 bristles; receptacle bearing bracts at base but not subtending each flower
 3. KNAUTIA

1. DIPSACUS L. - *Teasels*
Biennials; stems prickly; leaves simple or pinnate; receptacle with spine-tipped bracts subtending each flower and very long spiny ones at base; flowers all ± 1 size; epicalyx 4-angled, scarcely toothed at apex; calyx cup-shaped, scarcely toothed; corolla 4-lobed.

1 Capitula ovoid-cylindrical; upper stem-leaves sessile, fused in pairs round stem at base 2
1 Capitula globose; upper stem-leaves petiolate, not fused in opposite pairs 4
 2 Leaves pinnately dissected at least 1/2 way to midrib. Stems to 3(4)m. Intrd-natd; rough and marginal ground; Middlesex and Oxon *Cut-leaved Teasel* - **D. laciniatus** L.
 2 Leaves entire to dentate or serrate 3
3 Bracts on receptacle with stiff but flexible, straight apical spine. Stems to 2(3)m. Native; marginal habitats and rough ground by roads, railways, streams, woods and fields; frequent in CI and Br N to C Sc, local in Ir *Wild Teasel* - **D. fullonum** L.
3 Bracts on receptacle with stiff, rigid, recurved apical spine. Stems to 2(3)m. Intrd-natd; casual from birdseed and sometimes natd on tips and waste ground; scattered in CI and Br, Co Dublin
 Fuller's Teasel - **D. sativus** (L.) Honck.
 4 Capitula 15-28mm across (incl. bracts 7-13mm) in full flower or fruit. Stems erect, to 1.5m. Native; damp places in open woods,

hedgerows and by streams; rather scattered in En and Wa
Small Teasel - **D. pilosus** L.

4 Capitula 30-40mm across (incl. bracts 14-20mm) in full flower or fruit. Stems erect, to 1.5m. Intrd-natd; rough and waste places; Cambs, rare casual casual elsewhere in En
Yellow-flowered Teasel - **D. strigosus** Willd.

D. fullonum x *D. sativus* occurs rarely on waste ground where the 2 parents occurred earlier. *D. fullonum* x *D. laciniatus* = *D.* x *pseudosilvester* Schur occurs with both parents in Oxon.

2. CEPHALARIA Schrad. - *Giant Scabious*

Bisexual perennials; stems sparsely pubescent; leaves simple and deeply pinnately lobed to pinnate; receptacle with rather leathery bracts subtending each flower and similar ones at base; outer flowers longer than inner; epicalyx 8-ridged, with 8 apical teeth; calyx cup-shaped, scarcely toothed; corolla 4-lobed.

1 Stems erect, to 2m, with long branches; leaves all deeply pinnately lobed or pinnate; capitula 4-10cm across; corolla pale yellow. Intrd-natd; rough grassy places and waste ground; scattered in C & S En, rarely further N   *Giant Scabious* - **C. gigantea** (Ledeb.) Bobrov

3. KNAUTIA L. - *Field Scabious*

Gynodioecious perennials; stems pubescent; leaves simple and crenate to pinnate; receptacle with herbaceous bracts at base but none subtending each flower; outer flowers longer than inner; epicalyx 4-ridged, scarcely toothed but with dense hairs at apex; calyx with 8 long bristles; corolla 4-lobed.

1 Stems erect to ascending, to 1m; lowest leaves simple and crenate, upper ones pinnate; bisexual capitula 2.5-4cm and female ones 1.5-3cm across; corolla bluish-lilac. Native; dry grassy places on light soils; frequent over most of BI but very local in N Ir and N & W Sc.   *Field Scabious* - **K. arvensis** (L.) Coult.

4. SUCCISA Haller - *Devil's-bit Scabious*

Gynodioecious perennials; stems sparsely pubescent; leaves all simple, entire or distantly toothed; receptacle with herbaceous bracts at base and smaller ones subtending each flower; flowers all ± same size; epicalyx 4-angled, with 4 teeth; calyx with 4-5 bristle-tipped teeth; corolla 4-lobed.

1 Stems erect to ascending, to 1m; bisexual capitula 2-3cm and female ones 1.5-2.5cm across; corolla usually bluish-violet. Native; many sorts of grassy places, wet or dry, acid or calcareous, in open or shade; common   *Devil's-bit Scabious* - **S. pratensis** Moench

5. SCABIOSA L. - *Scabiouses*

Bisexual perennials; stems sparsely pubescent; leaves simple and serrate or lobed, to pinnate; receptacle with herbaceous bracts at base and narrower ones subtending each flower; outer flowers longer than inner; epicalyx 8-ridged, expanded at apex into membranous funnel; calyx with 5 long bristles at top of stalk (*hypanthium*); corolla 5-lobed.

1 Corolla bluish-lilac; fruiting capitula up to 1.5(2)cm long; epicalyx-tube 2-3mm, with 16-24-veined membranous funnel 0.8-1.5mm; hypanthium shorter than fruit. Stems erect, to 70cm. Native; dry calcareous grassland and rocky places; locally common in Br N to S Sc *Small Scabious* - **S. columbaria** L.
1 Corolla dark purple to pale lilac; fruiting capitula up to 2.5(3)cm long; epicalyx-tube 1.5-2.5mm, with 8-ribbed funnel c. as long and inrolled at apex; hypanthium longer than fruit. Stems erect, to 70cm. Intrd-natd; rough ground by sea; W Cornwall and E Kent
Sweet Scabious - **S. atropurpurea** L.

139. ASTERACEAE - *Daisy family*

Annual to perennial herbs, often woody near base, rarely shrubs; flowers usually numerous, small, borne on common receptacle in dense terminal heads (*capitula*), zygomorphic or actinomorphic (often both in same capitulum), bisexual to variously monoecious or dioecious, epigynous; sepals 0 or represented by *pappus* of scales, teeth, bristles, hairs or a membranous ring, often enlarging in fruit; petals 5 (rarely 0 or 4), fused into tube with distal limb of either (a) 5 (rarely 4) actinomorphic (or nearly so) lobes or teeth (*tubular* flowers), or (b) unilateral strap-like *ligule* often with 3 or 5 apical teeth (*ligulate* flowers), the tube often extremely short in latter type; stamens 5, borne on corolla-tube, the anthers fused laterally into cylinder around style (not in *Xanthium*); ovary 1-celled with 1 ovule; style 1, usually branched, each branch with linear stigmatic surface, fruit an achene, often with persistent pappus.

The flowers borne in capitula, with 5 stamens with laterally fused anthers (except in *Xanthium*), and the fruit an achene, are a unique combination.

The capitula bear around the outside of the flower-bearing area a series of often sepal-like bracts or *phyllaries*. They may also bear mixed in with the flowers (often 1 per flower) small *receptacular scales* or *bristles*. Each achene is often inserted on the receptacle into a minute *achene-pit*. Each capitulum may be *discoid*, with tubular flowers only, *ligulate*, with ligulate flowers (usually with 5-toothed ligules) only, or *radiate*, with a central region of tubular flowers (*disc flowers*) and an outer region of ligulate flowers (*ray flowers*) (usually with 3-toothed ligules). Sometimes each capitulum has very few flowers (very rarely only 1, e.g. *Olearia paniculata*). In *Echinops* the capitula have only 1 flower but are aggregated into large spherical compound heads. In *Ambrosia* and *Xanthium* the male and female capitula are separate on the same plant; the females contain only 1 or 2 flowers respectively. Otherwise, except in dioecious spp., the disc flowers are bisexual, but ray flowers may be bisexual, female or sterile.

Before using the keys a capitulum should be dissected to identify all its component parts. Rare variants without ray flowers, or with all the flowers ligulate (*flore pleno*), are not covered in the key, though more frequent variations in the presence or absence of ray flowers (e.g. in *Aster*, *Bidens*, *Senecio*) are.

General key
1. Shrubs, with new growth arising each year from older woody stems
 Key A
1. Herbs; stems sometimes woody at base 2
    2. Flowers 1 per capitulum, the capitula aggregated into tight globose heads **1. ECHINOPS**
    2. Flowers >1 per capitulum at least in some capitula, the capitula not in tight globose heads 3
3. Male and female capitula separate on same plant, the male more apical and with several flowers, the female lower down and with 1-2 flowers 4
3. If male and female capitula separate then on different plants, both with several-many flowers 5
    4. Leaves alternate; fruiting heads with 2 prominent terminal processes, covered with stiff hooked bristles **84. XANTHIUM**
    4. Leaves mostly opposite; fruiting heads without terminal processes, with short straight spines or with tubercles **82. AMBROSIA**
5. Flowers all ligulate, the ligules usually 5-toothed at ends; milky latex usually present **Key B**
5. Flowers all tubular or tubular and ligulate in same head, the ligules usually 3-toothed at ends; milky latex 0 6
    6. Style with ring of minute hairs, or with glabrous thickened ring, just below the branches; anthers with long 'tails' at base; leaves and/or phyllaries often spiny; flowers mostly pink or blue to purple, all tubular **Key C**
    6. Style without ring of hairs or thickened zone below branches; anthers mostly without basal 'tails'; plant rarely with spines; flowers various colours, often yellow, the marginal ones often ligulate 7
7. At least the lower leaves opposite **Key D**
7. All leaves basal or alternate 8
    8. Capitula discoid, with ligules 0 or inconspicuous and not exceeding inner phyllaries **Key E**
    8. Capitula radiate, with ≥1 obvious ligules exceeding inner phyllaries 9
9. Pappus of at least inner flowers of hairs (sometimes also of scales)
 Key F
9. Pappus 0, or of scales and/or few bristles 10
    10. Receptacular scales or abundant bristles present **Key G**
    10. Receptacular scales and bristles 0, but sometimes short fringes round achene-pits present **Key H**

139. ASTERACEAE

Key A - Shrubs (all aliens)
1 Marginal flowers conspicuously ligulate 2
1 Flowers 1, or few to many and all tubular, the outer sometimes with very short ligule-like lobes not or scarcely exceeding phyllaries (but beware ligule-like inner phyllaries in *Helichrysum*) 5
   2 Leaves pinnate or divided nearly to midrib, or if not then ligules purple **70. SENECIO**
   2 Leaves entire to toothed <1/2 way to midrib; ligules white or yellow 3
3 Leaves dying in winter, very sticky; glandular hairs abundant **42. DITTRICHIA**
3 Leaves evergreen, not sticky; glandular hairs 0 4
   4 Ligules white **53. OLEARIA**
   4 Ligules yellow **74. BRACHYGLOTTIS**
5 Leaves divided >1/2 way to midrib, very fragrant when bruised 6
5 Leaves entire to shallowly lobed or toothed, usually not fragrant 7
   6 Leaves all <5mm wide, with crowded incurved lobes; capitula >5mm wide, solitary on long stalks **59. SANTOLINA**
   6 At least lower leaves >1cm wide, with spreading lobes; capitula <5mm wide, clustered **58. ARTEMISIA**
7 Stems procumbent, rooting, <60cm; inner phyllaries patent in flower, white, ligule-like **40. HELICHRYSUM**
7 Stems erect to scrambling, often >60cm; inner phyllaries not patent and ligule-like 8
   8 Young stems and leaf lowersides densely white- to buff-tomentose 9
   8 Stems and leaves glabrous or nearly so 11
9 At least larger leaves remotely sinuate-lobed **74. BRACHYGLOTTIS**
9 Leaves entire 10
   10 Leaves green on upperside **53. OLEARIA**
   10 Leaves white-tomentose on upperside **40. HELICHRYSUM**
11 Leaves linear, <5mm wide **49. CHRYSOCOMA**
11 Larger leaves lanceolate to orbicular, >1cm wide 12
   12 Erect shrub; leaves rhombic to obovate, cuneate at base; flowers white, dioecious **54. BACCHARIS**
   12 Scrambler; leaves ± ivy-shaped, cordate to hastate at base; flowers yellow, bisexual **72. DELAIREA**

Key B - Flowers all ligulate (tribe Lactuceae)
1 Pappus 0 or of scales 2
1 Pappus, at least in central flowers, of hairs 7
   2 Leaves spiny; plant thistle-like **14. SCOLYMUS**
   2 Leaves not spiny; plant not thistle-like 3
3 Pappus of scales 4
3 Pappus 0; achene often terminating in minute collar 6
   4 Capitula <2cm across; ligules yellow **19. HEDYPNOIS**
   4 Capitula >2cm across; ligules blue, rarely white 5
5 Receptacle without scales; capitula sessile or short-stalked, several forming terminal raceme; most or all leaves at flowering time on

	stems	**15. CICHORIUM**
5	Receptacle with scales; capitula solitary on long stalks; most or all leaves at flowering time basal	**16. CATANANCHE**
	6 Leaves all basal; peduncles strongly dilated below capitula	**17. ARNOSERIS**
	6 Stems leafy; peduncles not or scarcely dilated below capitula	**18. LAPSANA**
7	Pappus-hairs feathery, with slender lateral branches visible to naked eye, at least on some achenes	8
7	Pappus-hairs all simple, smooth or shortly toothed (lens)	12
	8 Aerial stems with 0 or extremely reduced leaves	9
	8 Aerial stems with well-developed leaves	10
9	Receptacular scales present among flowers	**20. HYPOCHAERIS**
9	Receptacular scales 0	**21. LEONTODON**
	10 Plant hispid, with harsh, hooked bristles	**22. PICRIS**
	10 Plant glabrous or with very soft hairs	11
11	Phyllaries c.8, all in 1 row and of 1 length	**24. TRAGOPOGON**
11	Phyllaries >10, the outermost c.1/2 as long as innermost	**23. SCORZONERA**
	12 Leaves with strong spines; receptacular scales present, wrapped round achenes; pappus-hairs 2-4	**14. SCOLYMUS**
	12 Leaves not or weakly spiny; receptacular scales usually 0; pappus-hairs numerous	13
13	Achenes distinctly flattened	14
13	Achenes not or scarcely flattened	17
	14 Achenes with distinct narrow beak at apex, or at least markedly narrowed distally	15
	14 Achenes without beak and scarcely narrowed distally	16
15	Pappus-hairs in 2 equal rows; phyllaries in several rows	**27. LACTUCA**
15	Pappus-hairs in 2 unequal rows; phyllaries in 2 distinct unequal rows	**29. MYCELIS**
	16 Ligules yellow	**26. SONCHUS**
	16 Ligules blue to mauve	**28. CICERBITA**
17	Capitulum 1 per stem; stems without leaves or scales; rhizomes and stolons 0	**30. TARAXACUM**
17	Capitula >1 per stem; *or* stems with leaves or scales; *or* rhizomes and/or stolons present	18
	18 Pappus-hairs pure white	19
	18 Pappus-hairs yellowish-white to pale brown	20
19	Capitulum 1 per stem; long thin rhizomes present	**25. AETHEORHIZA**
19	Capitula normally >1 per stem; rhizomes 0	**31. CREPIS**
	20 Stolons usually present; achenes ≤2(2.5)mm, each rib ending in a small point at apex of achene	**32. PILOSELLA**
	20 Stolons 0; achenes >(1.5)2.5mm, each rib ending in a smooth ring at apex of achene	21
21	Plant glabrous except for phyllaries; phyllaries in distinct inner and outer rows	**31. CREPIS**
21	Plant with hairs (often dense) on some (often most) parts; phyllaries	

graduated between innermost and outermost **33. HIERACIUM**

Key C - Style with ring of minute hairs or with glabrous thickened ring just below branches (tribe Cardueae)
1 Phyllaries strongly hooked at tip, stiff, subulate **3. ARCTIUM**
1 Phyllaries not hooked at tip 2
   2 Corolla pale yellow to reddish-orange (beware inner phyllaries of *Carlina*) 3
   2 Corolla pink or blue to purple or mauve, rarely white 5
3 Pappus-hairs feathery, with slender lateral branches visible to naked eye or weak lens **6. CIRSIUM**
3 Pappus 0, or of narrow scales, or of simple to toothed hairs 4
   4 Outer phyllaries large and leaf-like **13. CARTHAMUS**
   4 Outer (and inner) phyllaries scale-like, with spiny apical portion **12. CENTAUREA**
5 Pappus-hairs feathery, with slender lateral branches visible to naked eye or weak lens 6
5 Pappus 0, of narrow scales, or of simple to toothed hairs 9
   6 Inner phyllaries with distinct apical portion (sometimes only a stout spine but always abruptly delimited from basal portion) **8. CYNARA**
   6 Phyllaries without distinct apical and basal portions, if with apical spine then gradually narrowed into it 7
7 Phyllaries all obtuse to rounded at apex; leaves ± entire to distantly toothed, the teeth not spinose or bristle-like **4. SAUSSUREA**
7 At least outer phyllaries spinose, mucronate or acuminate at apex; leaves spiny or at least with fine bristle-like teeth 8
   8 Outer phyllaries leaf-like; inner ones scarious, patent in dry weather, pale yellow on upperside, appearing like ligules **2. CARLINA**
   8 All phyllaries scale-like to ± subulate **6. CIRSIUM**
9 Leaves with sharp spines 10
9 Leaves without spines 12
   10 Receptacle glabrous, but achene-pits fringed with teeth **7. ONOPORDUM**
   10 Receptacle densely pubescent or bristly 11
11 Stem-leaves not decurrent down stem; stems not spiny; outer phyllaries with spine-tipped lateral lobes or teeth **9. SILYBUM**
11 Stem-leaves decurrent down stem in a spiny wing; phyllaries all entire, with terminal spine **5. CARDUUS**
   12 Phyllaries simple, entire **10. SERRATULA**
   12 At least inner phyllaries with distinct apical portion which is scarious, toothed, or spiny 13
13 All flowers bisexual and of same size; apical portion of phyllaries scarious, not separated from main part by constriction **11. ACROPTILON**
13 Marginal flowers functionally female though sometimes with sterile stamens, often longer than inner flowers; apical portion of phyllaries

usually toothed or spiny, if merely scarious then separated from main part by constriction **12. CENTAUREA**

Key D - Plant herbaceous; at least lower leaves opposite
1 Capitula discoid, with only tubular flowers 2
1 Capitula radiate, with ≥1 marginal ligulate flower 6
   2 Terminal capitula male only, in elongated bractless racemes **82. AMBROSIA**
   2 All capitula bisexual, variously arranged 3
3 Pappus of hairs **98. EUPATORIUM**
3 Pappus 0 or of scales or stout bristles 4
   4 Leaves pinnate, or simple with narrowly cuneate base; pappus of barbed bristles **90. BIDENS**
   4 Leaves simple, broadly cuneate to cordate at base; pappus 0 or of scales 5
5 Flowers blue; receptacular scales 0; pappus of scales **99. AGERATUM**
5 Flowers greenish-white; receptacular scales present; pappus 0 **83. IVA**
   6 Capitula with 1 ligulate flower **95. SCHKUHRIA**
   6 Capitula with 3-numerous ligulate flowers 7
7 Plant with large underground tubers 8
7 Plant without underground tubers 9
   8 Leaves (bi-)pinnate or (bi-)ternate; ligules often red to pink or purple, or white **93. DAHLIA**
   8 Leaves simple; ligules yellow **88. HELIANTHUS**
9 Ligules pink or white 10
9 Ligules yellow to greenish- or brownish-yellow 12
   10 Leaf-lobes linear to filiform; ligules usually pink, rarely white **92. COSMOS**
   10 Leaf-lobes lanceolate to ovate; ligules white, rarely purplish 11
11 Capitula <7mm across excl. ligules; pappus of scales **89. GALINSOGA**
11 Capitula >7mm across excl. ligules; pappus of barbed strong bristles **90. BIDENS**
   12 Pappus of barbed, strong, persistent bristles **90. BIDENS**
   12 Pappus 0 or of weak deciduous bristles and/or of scales 13
13 Most leaves divided ± to base or to midrib 14
13 Leaves simple, subentire to toothed or shallowly lobed 15
   14 Pappus of conspicuous scales; receptacular scales 0 **94. TAGETES**
   14 Pappus 0, or minute, or of 2 small scales; receptacular scales present **91. COREOPSIS**
15 Ligules <1cm; outer phyllaries linear, with dense glandular hairs **86. SIGESBECKIA**
15 Ligules >1cm; outer phyllaries lanceolate to ovate, not or scarcely glandular 16
   16 Annual or perennial; leaves petiolate, or sessile but not clasping stem; achenes flattened in radial plane **88. HELIANTHUS**
   16 Annual; leaves sessile, clasping stem at base; achenes flattened in tangential plane **85. GUIZOTIA**

Key E - Plant herbaceous; leaves alternate or all basal; capitula discoid or with very inconspicuous ligules (excl. Cardueae)
1 Pappus 0, or of small scales or a few bristles 2
1 Pappus of hairs (somewhat expanded apically in male *Antennaria* and *Anaphalis*) 15
   2 Leaves divided <1/2 way to midrib 3
   2 Leaves divided >1/2 way to midrib 7
3 Receptacular scales present; plant densely white-woolly **60. OTANTHUS**
3 Receptacular scales 0; plant not densely woolly 4
   4 Pappus of 1-8 stiff barbed bristles and sometimes also some small scales **46. CALOTIS**
   4 Pappus 0 or a small membranous ring 5
5 Capitula very numerous, in racemes or panicles **58. ARTEMISIA**
5 Capitula 1-many, if many then in corymbs 6
   6 Achenes ± compressed; stems procumbent to weakly erect, <30cm **69. COTULA**
   6 Achenes scarcely compressed; stems strong, erect, usually >50cm **56. TANACETUM**
7 Capitula very numerous, in racemes or panicles; corolla brownish-, reddish- or greenish-yellow 8
7 Capitula 1-few, or if ± numerous then in corymbs and corolla bright yellow 9
   8 Flowers all identical, with functional male and female parts **57. SERIPHIDIUM**
   8 Outer flowers functionally female, with filiform corolla; inner ones bisexual and tubular **58. ARTEMISIA**
9 Corolla pubescent; stem and leaves densely pubescent **58. ARTEMISIA**
9 Corolla glabrous; stem and leaves glabrous to pubescent 10
   10 Receptacular scales present 11
   10 Receptacular scales 0 12
11 Corolla with small pouch at base, obscuring top of ovary in 1 plane **62. CHAMAEMELUM**
11 Corolla not pouched at base, not obscuring top of ovary **63. ANTHEMIS**
   12 Achenes compressed; stems procumbent to weakly erect, <30cm **69. COTULA**
   12 Achenes not or scarcely compressed; stems usually stiffly erect to ascending 13
13 Capitula flat, in corymbs; perennial **56. TANACETUM**
13 Capitula conical to convex, not in corymbs; annual 14
   14 Achenes strongly 3-ribbed, with usually 2 resin-glands near apex **68. TRIPLEUROSPERMUM**
   14 Achenes with 4-5 ribs, without resin-glands **67. MATRICARIA**
15 Leaves broadly ovate to orbicular, cordate at base, petiolate 16
15 Leaves linear to ovate, cuneate at base, petiolate or sessile 18
   16 Plant with scrambling stems bearing ivy-shaped leaves **72. DELAIREA**

	16 Plants with aerial stems bearing only reduced, bract-like leaves, the main leaves all basal	17
17	Capitula solitary	**80. HOMOGYNE**
17	Capitula in inflorescences	**79. PETASITES**
	18 Plant densely glandular, very sticky	**42. DITTRICHIA**
	18 Plant glabrous to woolly, not obviously glandular, not sticky	19
19	Plant glabrous to pubescent, not woolly	20
19	Plant woolly at least in part, especially near tops of stems	25
	20 Leaves conspicuously serrate to deeply lobed or ± pinnate; phyllaries in 1 main row with much shorter supplementary ones near base of capitulum	**70. SENECIO**
	20 Leaves entire to slightly or remotely serrate; phyllaries in several rows	21
21	Phyllaries totally scarious, coloured or white	**40. HELICHRYSUM**
21	Phyllaries at least partly herbaceous and green	22
	22 Stems glabrous, sometimes weakly scabrid	23
	22 Stems pubescent	24
23	Pappus straw-coloured to pale reddish or brownish; at least lower leaves >2.5cm, either succulent or flat	**48. ASTER**
23	Pappus pure white; all leaves <2.5cm, with margins rolled under	**49. CHRYSOCOMA**
	24 Flowers yellow; outer phyllaries with patent to recurved tips	**41. INULA**
	24 Flowers white to cream or pinkish; phyllaries appressed to flowers	**51. CONYZA**
25	Annual with simple root system	26
25	Perennial with rhizomes or stolons (often short)	27
	26 Marginal florets with receptacular scales; outer phyllaries herbaceous, woolly beyond 1/2 way to apex	**36. FILAGO**
	26 Receptacular scales 0; all phyllaries scarious, glabrous or woolly in lower 1/2	**39. GNAPHALIUM**
27	Capitula in elongate panicles; capitula bisexual	**39. GNAPHALIUM**
27	Capitula in ± crowded terminal subcorymbose clusters; plants dioecious or subdioecious	28
	28 Largest leaves forming basal rosettes, the stem-leaves much narrower; stems rarely >20cm; stoloniferous	**37. ANTENNARIA**
	28 Leaves not forming basal rosette, the largest ones being on the stems; stems rarely <25cm; rhizomatous	**38. ANAPHALIS**

Key F - Plant herbaceous; leaves alternate or all basal; capitula radiate; pappus of hairs

1	Ligules white, or pink or blue to purple or mauve	2
1	Ligules yellow to orange	9
	2 Main leaves all basal, broadly ovate to orbicular, cordate at base; flowering stems with only reduced bract-like leaves	**79. PETASITES**
	2 Leaves on flowering stems, or if ± all basal then not cordate at base	3
3	Outer phyllaries broad, green, ± leafy	**52. CALLISTEPHUS**

3	Outer phyllaries similar to or smaller than inner ones	4
	4 At least the larger leaves truncate to cordate at base	5
	4 Leaves all cuneate at base	7
5	Phyllaries in a graded series of rows	**48. ASTER**
5	Phyllaries in 1 main row, with smaller supplementary ones near base of capitulum	6
	6 Leaves pinnately veined; phyllaries in 1 main row with smaller supplementary ones at base of capitulum; ligules white	**70. SENECIO**
	6 Leaves palmately veined; capitula without supplementary phyllaries; ligules white or coloured	**71. PERICALLIS**
7	Ligules linear to narrowly elliptic, >1mm wide; phyllaries all green or more green in apical than basal 1/2	**48. ASTER**
7	Ligules filiform to linear, usually <0.6mm but sometimes up to 1.5mm wide; phyllaries more green in basal than in apical 1/2	8
	8 Ligules ≤1mm; central tubular flowers fewer than peripheral filiform ones	**51. CONYZA**
	8 Ligules >1mm; central tubular flowers more numerous than peripheral filiform ones (or the latter 0)	**50. ERIGERON**
9	Main leaves all basal; flowering stems with 1 capitulum and only reduced bract-like leaves	**78. TUSSILAGO**
9	Leaves on flowering stems, or if ± all basal then capitula >1 per stem	10
	10 Phyllaries in 1 main row, often with smaller supplementary ones near base of capitulum	11
	10 Phyllaries in 2 or more (often indistinct) rows	15
11	Capitula with only 2-4 disc and 2-4 ray flowers, numerous in cylindrical or pyramidal panicles	12
11	Capitula with >5 disc and ≥4 ray flowers, not in dense pyramidal panicles	13
	12 Lower leaves deeply palmately lobed; petioles sheathing stem; ray flowers usually 2	**76. LIGULARIA**
	12 Lower leaves deeply pinnately lobed; petioles not sheathing stem; ray flowers usually 3	**75. SINACALIA**
13	Stem-leaves and basal leaves with petioles with broad sheathing bases	**76. LIGULARIA**
13	Stem-leaves and basal leaves without broad sheathing petioles	14
	14 Phyllaries in 1 main row with supplementary smaller ones near base of capitulum	**70. SENECIO**
	14 Phyllaries in 1 row, with no supplementary small ones near base of capitulum	**73. TEPHROSERIS**
15	Phyllaries in 2 distinct rows of ± equal length	**77. DORONICUM**
15	Phyllaries in 2 or more indistinct rows, progressively longer towards the inside	16
	16 Plant densely glandular, very sticky	**42. DITTRICHIA**
	16 Plant not or slightly glandular, not sticky	17
17	Pappus of inner row of hairs and outer row of small (often laterally fused) scales	**43. PULICARIA**
17	Pappus entirely of hairs	18

18	Capitula >(1.5)2cm across (incl. ligules); ligules >10mm; anthers with long filiform basal appendages	**41. INULA**
18	Capitula <1.5(2)cm across (incl. ligules); ligules <10mm; anthers without basal appendages	**47. SOLIDAGO**

Key G - Plant herbaceous; leaves alternate or all basal; capitula radiate; pappus 0, or of scales or bristles; receptacular scales present

1	Receptacle with abundant bristles only; pappus of 5-10 scales with long apical bristles	**96. GAILLARDIA**
1	Receptacular scales present; pappus 0, or of small scales not bristle-tipped	2
	2 Achenes strongly compressed, >2x as wide as thick, with lateral strong ribs or narrow wings	3
	2 Achenes angular, or not or slightly compressed, <2x as wide as thick, without lateral ribs or wings	4
3	Capitula <2cm across (incl. ligules); ligules <1cm	**61. ACHILLEA**
3	Capitula >2cm across (incl. ligules); ligules >1cm	**91. COREOPSIS**
	4 Ligules white	5
	4 Ligules yellow to orange	6
5	Corolla of tubular flowers with small pouch at base, obscuring top of ovary in 1 plane	**62. CHAMAEMELUM**
5	Corolla of tubular flowers not pouched, not obscuring top of ovary	**63. ANTHEMIS**
	6 Phyllaries with broad scarious margins and tips	**63. ANTHEMIS**
	6 Phyllaries entirely herbaceous	7
7	Ligules <2mm wide; receptacle slightly convex; lower leaves cordate at base	**44. TELEKIA**
7	Ligules >2mm wide; receptacle conical; leaves all cuneate at base	**87. RUDBECKIA**

Key H - Plant herbaceous; leaves alternate or all basal; capitula radiate; pappus 0, or of scales or bristles; receptacular scales 0

1	Capitula with 1 ligulate flower	**95. SCHKUHRIA**
1	Capitula with outer row of ligulate flowers	2
	2 Ligules yellow to orange or brownish-red, at least in part; pappus often of distinct scales or bristles	3
	2 Ligules white to pink; pappus 0 or a minute rim	9
3	At least some achenes very strongly curved, very warty on outer face	**81. CALENDULA**
3	Achenes not or slightly curved, if warty then not just on outer face	4
	4 Pappus of 1-8 bristles, sometimes with minute scales as well	5
	4 Pappus 0, or a minute rim, or of scales	6
5	Pappus bristles barbed, persistent; fruiting capitula <1cm across, not resinous	**46. CALOTIS**
5	Pappus bristles smooth or forwardly serrated, deciduous; fruiting capitula >1cm across, very resinous	**45. GRINDELIA**
	6 Pappus 0	**64. CHRYSANTHEMUM**
	6 Pappus of distinct scales	7

7	Leaves subglabrous to sparsely pubescent, decurrent on stems as wings; receptacle markedly convex	**97. HELENIUM**
7	Leaves densely white-tomentose on lowerside, not decurrent; receptacle ± flat to slightly convex	8
	8 Ligules pale yellow on upperside, purplish on lowerside; phyllaries free, glabrous to pubescent	**34. ARCTOTHECA**
	8 Ligules orange-yellow, with basal black blotch bearing central white spot, or rarely plain yellow; outer phyllaries fused into cup-like structure, white-tomentose	**35. GAZANIA**
9	Rosette-plant; capitula solitary on leafless stems	**55. BELLIS**
9	Flowering stems bearing leaves	10
	10 Stem-leaves simple, shallowly to deeply toothed but not to midrib, the teeth simple	11
	10 Stem-leaves pinnately lobed to midrib or nearly so, the lobes further lobed	13
11	Ligules <10mm	**56. TANACETUM**
11	Ligules >10mm	12
	12 Lowest (tubular) part of corolla of ray flowers with 2 narrow transparent wings	**66. LEUCANTHEMUM**
	12 Lowest (tubular) part of corolla of ray flowers not winged	**65. LEUCANTHEMELLA**
13	Ultimate leaf-segments lanceolate to ovate, flat	**56. TANACETUM**
13	Ultimate leaf-segments linear to filiform, not or scarcely flattened	14
	14 Achenes strongly 3-ribbed, with usually 2 resin-glands near apex	**68. TRIPLEUROSPERMUM**
	14 Achenes with 4-5 ribs, without resin-glands	**67. MATRICARIA**

SUBFAMILY 1 - LACTUCOIDEAE (tribes 1-3; genera 1-35). Plants often producing white latex; stem-leaves usually spiral ('alternate'), sometimes 0; capitula ligulate, discoid or rarely (Arctotideae) radiate; tubular flowers (if present) usually with long narrow lobes, usually blue to red, rarely yellow; filaments joining anthers on back; 2 style branches usually each with 1 broad stigmatic surface on inner face.

TRIBE 1 - CARDUEAE (genera 1-13). Plant not producing white latex, often spiny; capitula discoid, with the outermost florets often longer and with larger lobes and hence pseudo-radiate, usually red to blue (or white), rarely yellow.

1. ECHINOPS L. - *Globe-thistles*
Biennials to perennials; leaves deeply pinnately lobed, white- to grey-tomentose on lowerside, with teeth ending in rather weak spines; capitula each with 1 flower, many arranged in compact, globose heads ≥2.5cm; phyllaries in c.3 rows, herbaceous, long-pointed and with long fine teeth, with a zone of bristles outside them; corolla white to blue; pappus of partially fused scale-like bristles.

1 Phyllaries strongly recurved at tip; glandular hairs 0. Stems erect, to

2.5m; corolla white to greyish. Intrd-natd; occurrence as for *E. sphaerocephalus*, but commoner **Globe-thistle - E. exaltatus** Schrad.
1 Phyllaries erect or very slightly curved at tip; glandular hairs present at least on leaf upperside 2
  2 Phyllaries without glandular hairs; corolla bluish. Stems erect, to 1.25m, with eglandular hairs. Intrd-natd; occurrence as for *E. sphaerocephalus*, the commonest sp. in Br
  Blue Globe-thistle - E. bannaticus Schrad.
  2 Phyllaries with abundant glandular hairs; corolla greyish. Stems erect, to 2.5m. Intrd-natd; waste places and rough ground, on road- and railway-banks; scattered throughout much of Br
  Glandular Globe-thistle - E. sphaerocephalus L.

2. CARLINA L. - *Carline Thistle*

Biennials; leaves pinnately lobed, very spiny; outer phyllaries ± leaf-like, innermost linear, entire, scarious, patent and ligule-like in dry weather, straw-yellow on upperside; corolla purple; pappus of feathery hairs often variously united proximally.

1 Stems erect, to 60cm, not spiny; leaves very spiny; capitula 1-6 in terminal corymb. Native; open grassland, on usually calcareous but sometimes sandy soils, fixed dunes and cliff-tops; frequent throughout most of BI, only coastal in N Ir and Sc
  Carline Thistle - C. vulgaris L.

3. ARCTIUM L. - *Burdocks*

Biennials; leaves simple, entire to remotely denticulate, the lower ovate-cordate and petiolate, not spiny; phyllaries very numerous, stiff, subulate, strongly hooked at apex, spreading to form subglobose head; corolla purple, rarely white; pappus of rough yellowish free hairs.

1 Petiole of basal leaves solid at least at base; capitula forming ± corymbose clusters on each main branch; capitula on distal part of each main branch with peduncles ≥2.5cm. Stems erect, well branched, to 2m; phyllaries exceeding corolla by 1-5mm; middle phyllaries 0.9-1.7mm wide. Native; waysides, field-borders, wood-clearings, waste places; rather scattered in S & C Br, very scattered in Ir **Greater Burdock - A. lappa** L.
1 Petiole of basal leaves hollow at least near base; capitula on distal part of each main branch sessile or with peduncles usually ≤2cm, if >2cm then capitula forming clearly racemose clusters on each main branch 2
  2 Capitula on distal part of each main branch sessile; middle phyllaries (1.6)1.7-2.5mm wide; phyllaries exceeding corolla by 1.2-6mm; corolla glabrous. Stems erect, well branched, to 2m. Native; mostly in open woods and semi-shaded disturbed ground, especially on calcareous soils; distribution uncertain, but less common than *A. minus* (*A. minus* ssp. *nemorosum* (Lej.)

3. ARCTIUM 465

Syme) *Wood Burdock* - **A. nemorosum** Lej.
2 Capitula on distal part of each main branch sessile or pedunculate; middle phyllaries ≤1.6mm wide, or if ≤1.8mm wide then corolla equalling or exceeding phyllaries or corolla pubescent or both. Stems erect, well branched, to 2m. Native; similar places to *A. lappa*; common (*A. minus* ssp. *minus* & ssp. *pubens* (Bab.) P. Fourn.) *Lesser Burdock* - **A. minus** (Hill) Bernh.

A. nemorosum and *A. minus* are much confused but hybrids between them are unknown. Plants with pedunculate distal capitula, or with middle phyllaries <1.6mm wide, or with a glandular-pubescent corolla, or with corolla equalling or exceeding phyllaries, or with capitula ≤19 x 27mm are *A. minus*, but the reverse is not necessarily true.

A. lappa x *A. minus* = ***A. x nothum*** (Ruhmer) J. Weiss occurs infrequently in S Br with the parents; it has reduced achene fertility.

4. SAUSSUREA DC. - *Alpine Saw-wort*

Perennials; leaves simple, subentire to denticulate, not spiny; phyllaries in many rows, simple, entire, rounded to obtuse at apex; corolla purple; pappus of feathery hairs in 1 row, often ± united proximally, with an outer row of simple shorter hairs, the former often deciduous.

1 Stems erect, to 45cm; leaves ovate or elliptic to narrowly so, densely white-pubescent on lowerside; capitula 15-20mm, 1-several in ± dense terminal cluster. Native; mountain cliffs and scree; local in N En, N Wa, Sc and Ir *Alpine Saw-wort* - **S. alpina** (L.) DC.

5. CARDUUS L. - *Thistles*

Annuals or biennials; stems with spiny wings at least in part; leaves variously lobed, sharply and densely spiny; phyllaries simple, linear-subulate to narrowly ovate, spine-tipped, in many rows; corolla purple (or white); pappus of many rows of simple hairs united proximally.

1 Capitula subcylindrical, <14mm across (excl. flowers); corolla with 5 ± equal lobes 2
1 Capitula globose to bell-shaped, >14mm across (excl. flowers); corolla with 1 lobe distinctly more deeply delimited than other 4 3
   2 Capitula in clusters of 3-10; stems with spiny wings right up to base of capitula; phyllaries thin and transparent on margins, without strongly thickened midrib except sometimes near apex. Stems erect, to 60(80)cm. Native; waysides, rough and open ground; locally frequent by coasts in BI except most of N & W Sc, very scattered inland *Slender Thistle* - **C. tenuiflorus** Curtis
   2 Capitula in clusters of 1-3; stems with discontinuous spiny wings, with at least some peduncles unwinged distally; phyllaries with strongly thickened margins and midrib for at least distal 1/2. Stems erect, to 60(80)cm. Intrd-natd; on open limestone cliff

at Plymouth (S Devon), infrequent casual elsewhere
Plymouth Thistle - **C. pycnocephalus** L.
3 Capitula 15-25(30)mm across (excl. flowers), in clusters of (1)2-4(5), ± erect; phyllaries linear-subulate, not narrowed just above base; corolla 12-15mm. Stems erect, to 1.5m. Native; hedgerows, ditchsides and streamsides, rough ground; frequent in En and Wa, scattered in Ir and Sc *Welted Thistle* - **C. crispus** L.
3 Capitula (20)30-60mm across (excl. flowers), usually solitary, pendent; phyllaries lanceolate, narrowed just above base; corolla 15-25mm. Stems erect, to 1m. Native; grassy and bare places, waysides and rough ground, mostly on calcareous soils; locally frequent in CI and Br N to S Sc, rare and intrd elsewhere
Musk Thistle - **C. nutans** L.

C. crispus x *C. nutans* = **C. x stangii** Nyman (?*C. x dubius* Balb.) is found scattered in Br N to Yorks; it is partially fertile.

6. CIRSIUM Mill. - *Thistles*

Biennials to perennials; stems with or without spiny wings; leaves denticulate to deeply lobed, spiny or at least with bristle-pointed teeth; phyllaries simple, ovate to linear-subulate, spine-tipped to mucronate or acuminate at apex, in many rows; corolla purple (or white), rarely yellow; pappus of many rows of feathery hairs united proximally.

1 Leaves with rigid bristles on upperside 2
1 Leaves glabrous or with soft hairs on upperside 3
    2 Stem with discontinuous spiny wings. Erect biennial to 1.5m. Native; grassland, waysides, cultivated, rough and waste ground; common *Spear Thistle* - **C. vulgare** (Savi) Ten.
    2 Stem not winged. Erect biennial to 1.5m. Native; dry grassland, scrub and banks on calcareous soil; locally frequent in Br N to Co Durham *Woolly Thistle* - **C. eriophorum** (L.) Scop.
3 Flowers yellow 4
3 Flowers purple, rarely white 5
    4 Upper part of stem with only very reduced, distant leaves; capitula pendent. Erect perennial to 1.5m. Intrd-natd; in disused quarry; N Somerset *Yellow Thistle* - **C. erisithales** (Jacq.) Scop.
    4 Upper part of stem with large yellowish-green leaves exceeding the erect capitula. Erect perennial to 1.5m. Intrd-natd; in marshes and by streams; E Perth, S Lancs and Fermanagh, rare and impermanent elsewhere *Cabbage Thistle* - **C. oleraceum** (L.) Scop.
5 Stems continuously spiny-winged; biennial with tap-root. Erect biennial to 2m. Native; marshes, damp grassland and open woods, ditchsides; common *Marsh Thistle* - **C. palustre** (L.) Scop.
5 Stems not winged or with very short wings below each leaf; perennial with at least short rhizomes 6
    6 Distal broad part of corolla c.1/2 as long as proximal narrow part, lobed ≥3/4 way to base; stem usually well branched. Perennial

with long rhizomes and erect stems to 1.2m. Native; grassland, hedgerows, arable, waste and rough ground; common
Creeping Thistle - **C. arvense** (L.) Scop.
6 Distal broad part of corolla c. as long as proximal narrow part, lobed c.1/2 way to base; stem usually unbranched 7
7 Stem 0-10cm, or if up to 30cm then with ≥1 well-developed leaf near top. Perennial with basal leaf-rosette and 1-few capitula sessile in centre or on wingless leafy stems to 10(30)cm. Native; short, base-rich grassland; locally frequent in Br N to NE Yorks, Alderney
Dwarf Thistle - **C. acaule** (L.) Scop.
7 Stem >10cm, with only distant much reduced leaves in distal 1/4 8
8 Stem-leaves widened proximally to broad base clasping stem; capitula mostly >20mm to tip of uppermost phyllaries. Erect perennial to 1.2m. Native; grassland, scrub, open woodland and streamsides in hilly country; locally common in Br N from Derbys and Rads, rare in NC Ir
Melancholy Thistle - **C. heterophyllum** (L.) Hill
8 Stem-leaves narrowed to base, not or scarcely clasping stem; capitula mostly <20mm to tip of uppermost phyllaries 9
9 Lower stem-leaves deeply lobed, the lobes deeply lobed, green on lowerside; some roots swollen into tubers. Erect perennial to 60cm. Native; dry calcareous grassland; very local in N & S Wilts, Glam and Cambs *Tuberous Thistle* - **C. tuberosum** (L.) All.
9 Lower stem-leaves not or rather shallowly lobed, the lobes scarcely or not lobed, white on lowerside; roots not swollen. Erect perennial to 80cm. Native; fens, bogs, wet fields on peaty soil; local in En, Wa and SW Sc, throughout Ir *Meadow Thistle* - **C. dissectum** (L.) Hill

Hybrids occur rarely but 11 combinations have been found; they are highly sterile to partially fertile. Probably least rare are *C. dissectum* x *C. palustre* = **C. x forsteri** (Sm.) Loudon and *C. heterophyllum* x *C. palustre* = **C. x wankelii** Reichardt. *C. palustre* x *C. arvense* = **C. x celakovskianum** Knaf is scattered in Br and Ir but surprisingly rare.

7. ONOPORDUM L. - *Cotton Thistles*
Biennials; stems with spiny wings; leaves dentate to shallowly lobed, strongly spiny; phyllaries simple, linear-lanceolate, strongly spine-tipped, in many rows; corolla purple (rarely white); pappus of many rows of simple hairs united proximally.

1 All parts of stems and leaves greyish-white with cottony hairs; outer phyllaries 1.5-2.5mm wide at widest point; corolla 15-25mm. Stems erect, to 2.5m. Probably intrd-natd; fields, marginal habitats, waste and rough ground; locally frequent in En (? native in E Anglia), very scattered in Wa, Sc and CI *Cotton Thistle* - **O. acanthium** L.
1 Upper parts of stem and upper leaves green, with sparse pubescence and strongly contrasting raised white veins on leaf lowerside; outer phyllaries 4-6mm wide at widest point; corolla

30-35mm. Stems erect, to 2.5m. Intrd-natd; waste and rough ground; few places in S En *Reticulate Thistle* - **O. nervosum** Boiss.

8. CYNARA L. - *Globe Artichoke*
Perennials; stems not winged, spiny or not; leaves deeply lobed, spiny or not; phyllaries in many rows, the outer with or without apical spine, the inner with distinct apical portion with or without apical spine; corolla mauve or violet to blue; pappus of many rows of feathery hairs united proximally.

1   Stems erect, to 1.8m; leaves spiny or not, tomentose on lowerside; capitula 3.5-7.5 x 3.5-9.5cm (excl. flowers). Intrd-surv; derelict or rough ground; scattered in CI and En. Var. *cardunculus* (*Cardoon*) has strongly spiny leaves and phyllaries; var. *scolymus* (L.) Fiori (*Globe Artichoke*) is spineless *Globe Artichoke* - **C. cardunculus** L.

9. SILYBUM Adans. - *Milk Thistle*
Annuals to biennials; stems not spiny; leaves shallowly to rather deeply lobed, strongly spiny; phyllaries in many rows, the outer with spine-tipped lateral lobes or teeth and strong apical spine; corolla purple (rarely white); pappus of many rows of simple hairs united proximally.

1   Stems erect, to 1m; leaves bright green, usually veined or marbled with white; capitula 2.5-4 x 5-14cm (excl. flowers). Intrd-natd; waste and rough ground; scattered throughout BI
Milk Thistle - **S. marianum** (L.) Gaertn.

10. SERRATULA L. - *Saw-wort*
Perennials; stems and leaves not spiny; lower leaves usually lobed almost to midrib; phyllaries simple, acute to acuminate, not spiny, in many rows; corolla purple (rarely white); pappus of many rows of free, simple hairs, the outermost much shorter than the inner ones.

1   Stems erect, to 70(100)cm, but often very short (<10cm); leaves subglabrous, with bristle-tipped teeth. Native; grassland, scrub, open woodland, cliff-tops and rocky streamsides on well-drained soils; local in Br N to SW Sc, Jersey *Saw-wort* - **S. tinctoria** L.

11. ACROPTILON Cass. - *Russian Knapweed*
Perennials; stems and leaves not spiny; leaves simple, entire to narrowly lobed; phyllaries with conspicuous broad apical scarious border, in many rows; corolla pink, the outermost no longer than the inner; pappus of hairs, soon falling.

1   Stems erect, well branched, to 70cm; leaves densely grey-pubescent; phyllaries pale brown with paler, minutely pubescent, scarious border at apex, becoming jagged later. Intrd-natd; on waste ground; Hereford (Herefs) *Russian Knapweed* - **A. repens** (L.) DC.

12. CENTAUREA L. - *Knapweeds*

Annuals to perennials; stems and leaves not spiny; leaves simple and entire to ± pinnate; phyllaries with distinct apical portion which is scarious, toothed or spiny, in many rows; corolla purple to pink or blue, white or yellow, that of outermost flowers often much longer than that of inner flowers (pseudo-radiate); pappus 0, or of many rows of simple to toothed free hairs, sometimes also with some scales.

1	Apical portion of phyllaries with ≥1 sharp spines	2
1	Apical portion of phyllaries merely scarious or variously toothed, not spiny	6
	2 Flowers yellow; stems strongly winged	3
	2 Flowers purple, rarely white; stems not winged	4
3	Apical phyllary spines <10mm, with lateral spines arranged pinnately along its proximal 1/2; corolla with minute sessile glands. Erect to spreading annual to biennial to 60cm. Intrd-casual; tips and waste ground; scattered mainly in S & C Br	
	Maltese Star-thistle - **C. melitensis** L.	
3	At least some apical phyllary spines >10mm, with lateral spines arranged palmately at its base; corolla not glandular. Erect to spreading annual to biennial to 60cm. Intrd-casual; waste ground; scattered mainly in C & S Br *Yellow Star-thistle* - **C. solstitialis** L.	
	4 Apical portion of phyllaries scarious, variously toothed, with 1(-few) terminal spines. Erect perennial (often behaving as annual) to 80cm. Intrd-casual; on waste ground and tips; scattered mainly in S & C Br *Lesser Star-thistle* - **C. diluta** Aiton	
	4 Whole of apical portion of phyllaries modified into spines	5
5	Apical phyllary spine >10mm, >3x as long as longest laterals. Erect to spreading biennial to 60cm. Intrd-natd; waste and rough ground, waysides, on well-drained soils; scattered places in S En, casual elsewhere *Red Star-thistle* - **C. calcitrapa** L.	
5	Apical phyllary spine <5mm, <1.5x as long as longest laterals. Erect to spreading perennial to 80cm. Intrd-natd; on maritime dunes in Jersey and Guernsey, casual or ± natd in few places in S Br	
	Rough Star-thistle - **C. aspera** L.	
	6 Apical portion of phyllaries with 1(-few) distinct terminal subspinose teeth distinctly different from lateral teeth (see couplet 4) **C. diluta**	
6	Apical portion of phyllaries similarly toothed at apex and sides	7
7	Flowers yellow; apical portion of outer phyllaries >1cm wide. Erect perennial to 1m. Intrd-natd; waste and rough ground; E Suffolk	
	Giant Knapweed - **C. macrocephala** Willd.	
7	Flowers purple to blue, rarely white; apical portion of outer phyllaries <1cm wide	8
	8 Apical portion of phyllaries strongly delimited from basal portion, with slight constriction between	9
	8 Apical portion of phyllaries ill-delimited from basal portion, the former decurrent down sides of latter	10

9 Apical portion of outer phyllaries dark brown to black, deeply and very regularly toothed; pappus present or absent. Erect perennial to 1m. Native; grassy places, rough ground, waysides; common
Common Knapweed - **C. nigra** L.
9 Apical portion of outer phyllaries pale to dark brown, irregularly toothed or deeply jagged; pappus absent or ± so. Erect perennial to 60cm. Native, from crosses between *C. nigra* and formerly natd *C. jacea*; grassy places; very scattered in S En and CI
(*C. jacea* x *C. nigra*)   *Hybrid Knapweed* - **C. x moncktonii** C.E. Britton
    10 Easily uprooted annual; basal leaves 0 at flowering time. Erect annual to 80cm. Native; traditionally natd in cornfields, now mostly casual in waste places; scattered throughout BI
Cornflower - **C. cyanus** L.
    10 Deeply rooted perennial; basal leaves or non-flowering shoots present at flowering time   11
11 Leaves simple, entire, strongly decurrent and forming distinct wings on stem; strongly rhizomatous; flowers usually blue. Erect perennial to 80cm. Intrd-natd; grassy places and rough ground; most of Br, especially C & N   *Perennial Cornflower* - **C. montana** L.
11 Leaves usually deeply lobed, rarely simple and entire, not decurrent on stem; scarcely rhizomatous; flowers usually reddish-purple. Erect perennial to 1.2m. Native; grassland, rough ground, cliffs and waysides mainly on calcareous soils; locally common in C & N BI, very local in Sc   *Greater Knapweed* - **C. scabiosa** L.

13. CARTHAMUS L. - *Safflowers*

Annuals; leaves entire or with acute lobes and apex tipped with spines or bristles; phyllaries in many rows, the outer leaf-like, the inner usually spine-tipped; corolla yellow to orange; pappus 0 or of narrow pointed scales.

1 Leaves and outer phyllaries entire to shallowly lobed with bristly or softly spine-tipped lobes; pappus 0. Stems erect, glabrous, to 60cm. Intrd-casual; tips, waste and rough ground; rather frequent in S Br, very scattered elsewhere   *Safflower* - **C. tinctorius** L.
1 Leaves and outer phyllaries with apex and usually deep lobes with strong terminal spines; pappus c. as long as achene. Stems erect, densely pubescent, to 60cm. Intrd-casual; similar habitats and distribution to *C. tinctorius*   *Downy Safflower* - **C. lanatus** L.

TRIBE 2 - LACTUCEAE (genera 14-33). Plants producing white latex, rarely spiny; capitula ligulate, the flowers all bisexual and ligulate, usually with 5-toothed ligules, usually yellow.

14. SCOLYMUS L. - *Golden Thistle*

Thistle-like (annuals,) biennials or perennials; rhizomes and stolons 0; stems leafy, with several capitula; phyllaries in several rows; receptacular scales present, wrapped round achenes; pappus (0 or) of a few rigid hairs; ligules yellow; achenes flattened, not beaked.

1 Stems erect, to 80cm, pubescent; leaves linear to ovate, lobed, with
strong rigid spines; capitula 2.5-4cm across. Intrd-casual; waste
places; very scattered in S En and Wa   *Golden Thistle* - **S. hispanicus** L.

15. CICHORIUM L. - *Chicory*
Perennials; rhizomes and stolons 0; stems leafy, with many capitula; phyllaries in 2 rows; receptacular scales present, small; pappus of short scales; ligules blue, rarely white; achenes not flattened, somewhat angular, not beaked.

1 Stems stiff, procumbent to erect, glabrous to pubescent, to 1m; lower
leaves deeply lobed to toothed; capitula 2.5-4cm across. Possibly
native; roadsides, rough grassland, waste places, especially on
calcareous soils; locally common in C & S Br, scattered elsewhere
Chicory - **C. intybus** L.

16. CATANANCHE L. - *Blue Cupidone*
Perennials; rhizomes and stolons 0; stems not or very sparsely leafy, branched sparingly, with 1 capitulum at end of each long branch; phyllaries in several rows, translucent-scarious except for dark midrib; receptacular scales present; pappus of scales; ligules blue, rarely white; achenes not flattened, 5-10-ribbed, not beaked.

1 Stems erect, to 80cm, pubescent; leaves mostly or all basal, to 30cm,
linear to elliptic-linear, entire or with few narrow lateral lobes;
capitulum 2.5-4cm across, with silvery papery phyllaries. Intrd-
natd; on roadside; N Somerset   *Blue Cupidone* - **C. caerulea** L.

17. ARNOSERIS Gaertn. - *Lamb's Succory*
Annuals; stems leafless, with small scale-like bracts, branched or not; phyllaries in 2 rows, the outer small and incomplete; receptacular scales 0; pappus 0; ligules yellow; achenes somewhat flattened, c.5-ribbed, not beaked.

1 Stems erect, to 30cm, ± glabrous, conspicuously dilated some way
below capitula; leaves all basal, lobed to toothed; capitula 7-12mm
across. Probably native; sandy arable fields; formerly very local in
E En and E Sc, extinct since 1971
Lamb's Succory - **A. minima** (L.) Schweigg. & Körte

18. LAPSANA L. - *Nipplewort*
Annuals to perennials; rhizomes and stolons 0; stems leafy, with many capitula; phyllaries in 2 rows, the outer small and incomplete; receptacular scales 0; pappus 0; ligules yellow; achenes slightly flattened, c.20-ribbed, not beaked.

1 Stems erect, to 1m, pubescent at least below; lower leaves ± pinnate

with lateral leaflets much smaller than terminal one
(***Nipplewort*** - **L. communis** L.) 2
- 2 Annual; upper stem-leaves lanceolate to ovate or rhombic, usually well toothed; capitula 1.5-2cm across; ligules c.1.5x as long as inner phyllaries; ripe achenes c.1/2 as long as phyllaries. Native; open woods, hedgerows, waste and rough ground; common **L. communis ssp. communis**
- 2 Usually perennial; upper stem-leaves linear to linear-lanceolate, entire or slightly toothed; capitula 2.5-3cm across; ligules 2-2.5x as long as inner phyllaries; ripe achenes <1/2 as long as phyllaries. Intrd-natd; calcareous grassland and banks; Middlesex, Beds, Flints and Caerns **L. communis ssp. intermedia** (M. Bieb.) Hayek

19. HEDYPNOIS Mill. - *Scaly Hawkbit*

Annuals; stems leafy, with several capitula; phyllaries in 2 rows, the outer very small; receptacular scales 0; pappus of scales, those of inner achenes long and narrow, those of outer achenes short and ± fused laterally to form ring; ligules yellow; achenes not flattened, ribbed, not beaked.

1 Stems erect to decumbent, pubescent, to 45cm, conspicuously dilated below capitula at fruiting; leaves entire to lobed; capitula 5-15mm across. Intrd-casual; casual in waste places; occasional in En *Scaly Hawkbit* - **H. cretica** (L.) Dum. Cours

20. HYPOCHAERIS L. - *Cat's-ears*

Annuals to perennials; rhizomes and stolons 0; stems usually leafless, usually with small scale-like bracts, mostly with >1 capitula; phyllaries in several rows; receptacular scales present, reaching at least base of ligule; pappus of 1-2 rows of dirty-white to pale brown hairs, the single or inner row feathery; ligules yellow; achenes not flattened, finely ribbed, beaked or not.

- 1 Pappus-hairs all feathery, in 1 row; leaves usually spotted or streaked with purple; outer phyllaries usually uniformly pubescent. Stems to 60cm, erect, usually pubescent ± throughout. Native; grassy or open ground, mostly calcareous or sandy, maritime cliffs; very local in Br N to Westmorland, Jersey *Spotted Cat's-ear* - **H. maculata** L.
- 1 Pappus-hairs in 2 rows, outer usually simple, inner feathery; leaves rarely with purple markings; outer phyllaries glabrous to very sparsely or patchily pubescent 2
  - 2 Central achenes 8-17mm (incl. beak), beaked; marginal achenes usually beaked; capitula 2-4cm across, opening every day; ligules c.4x as long as wide. Stems to 60cm, erect or ascending, glabrous or pubescent near base. Native; grassy places in many situations; very common *Cat's-ear* - **H. radicata** L.
  - 2 Central achenes 6-9(13.5)mm (incl. beak), beaked or not; marginal achenes not beaked; capitula 1-1.5cm across, opening only in bright sunshine; ligules c.2x as long as wide. Stems to

40cm, erect or ascending, glabrous. Native; grassy or open ground on sandy soils; frequent in E Anglia and CI, very scattered elsewhere in Br **Smooth Cat's-ear** - **H. glabra** L.

H. radicata x *H. glabra* = **H. x intermedia** Richt. occurs rarely in SE En, Merioneth and Fife, but may be overlooked; it is intermediate in capitulum characters and <5% fertile. The capitula open in dull and sunny weather.

21. LEONTODON L. - *Hawkbits*

Perennials; rhizomes and stolons 0; stems leafless, usually with small scale-like bracts, simple or branched; phyllaries in several rows; receptacular scales 0; pappus of 1-2 rows of pale brown to dirty-white hairs, the single or inner row feathery, the outer row simple, or outer achenes with pappus a scaly ring; ligules yellow; achenes not flattened, finely ribbed, not or indistinctly beaked.

1 Leaves glabrous or with simple hairs; pappus of 1 row of feathery hairs; stems usually branched. Stems to 60cm, glabrous, or sparsely pubescent below. Native; grassy places in very many situations; very common **Autumn Hawkbit** - **L. autumnalis** L.
1 Leaves with at least some hairs forked; pappus of 2 rows of hairs, the inner feathery, the outer simple; stems simple 2
   2 All achenes with pappus of hairs; phyllaries (9)11-13(15)mm; whole of stem and phyllaries usually conspicuously pubescent. Stems to 60cm, usually conspicuously pubescent. Native; basic, often calcareous, grassland; common in Br N to S Sc, locally frequent in Ir, rare in N & C Sc **Rough Hawkbit** - **L. hispidus** L.
   2 Outer achenes with pappus a scaly ring; phyllaries 7-11mm; upper part of stem and phyllaries usually glabrous to sparsely pubescent. Stems to 40cm, glabrous or sparsely pubescent. Native; similar places and distribution to *L. hispidus* but also in CI and Man **Lesser Hawkbit** - **L. saxatilis** Lam.

L. hispidus x *L. saxatilis* occurs rarely in En but may be overlooked; it is intermediate in habit and pubescence, but resembles *L. saxatilis* in pappus-type and is <1% fertile.

22. PICRIS L. - *Oxtongues*

Annuals to perennials; rhizomes and stolons 0; stems leafy, with numerous stiff bristles hooked at apex, with several capitula; phyllaries in several rows; receptacular scales 0; pappus of 2 rows of white to off-white hairs, the inner or both rows feathery; ligules yellow; achenes somewhat flattened, weakly ribbed, transversely wrinkled, beaked or not.

1 Leaves entire to dentate; inner phyllaries 12-20mm, the outer 3-5 ovate-cordate, >2x as wide as inner; achenes with beak ± as long as body; pappus white. Probably intrd-natd; marginal, disturbed and rough ground and waste places; frequent in S & E En, scattered N

to N En, CI, very local in S & E Ir   *Bristly Oxtongue* - **P. echioides** L.
1   Leaves sinuate-toothed to -lobed; inner phyllaries 11-13mm, the outer ones similar to inner but shorter and often patent; achenes with beak 0 or <1/2 as long as body; pappus off-white. Native; grassland and open or rough mostly calcareous ground; similar distribution to *P. echioides*   *Hawkweed Oxtongue* - **P. hieracioides** L.

23. SCORZONERA L. - *Viper's-grass*

Perennials; rhizomes and stolons 0; stems leafy, usually simple, sometimes with 1 or 2 branches; phyllaries in several rows; receptacular scales 0; pappus of several rows of dirty-white feathery hairs; ligules yellow, the outermost flushed crimson on lowerside; achenes slightly flattened, ribbed, not beaked.

1   Stems erect, to 50cm, woolly when young, becoming glabrous; leaves entire, narrowly elliptic to linear-lanceolate; capitula 2-3cm across. Native; marshy fields; Glam and Dorset   *Viper's-grass* - **S. humilis** L.

24. TRAGOPOGON L. - *Goat's-beards*

Annuals to perennials; rhizomes and stolons 0; stems leafy, simple or branched; phyllaries in 1 row; receptacular scales 0; pappus of 1 row of mainly feathery but some simple hairs, or the outer achenes with entirely simple hairs, dirty-white to pale brown; ligules yellow or purple; achenes not flattened, ribbed, beaked. The capitula open in the morning but close regularly about noon.

1   Ligules yellow; outer achenes often <3cm (incl. beak). Stems erect, to 75cm (*Goat's-beard* - **T. pratensis** L.)   2
1   Ligules pink to purple; outer achenes ≥3cm (incl. beak)   3
    2   Ligules as long as or longer than phyllaries. Intrd-natd; grassy places and open or rough ground; very scattered in S & C Br
        T. pratensis ssp. pratensis
    2   Ligules c.1/2-3/4 as long as phyllaries. Native; grassy places, rough and cultivated ground, roadsides; common in Br N to C Sc, scattered in N Sc and Ir, Alderney
        T. pratensis ssp. minor (Mill.) Wahlenb.
3   Marginal achenes with pappus of simple hairs; achenes very gradually narrowed into beak. Stems erect, to 50cm. Intrd-casual; gardens, parks, tips and waste ground; scattered in S Br
    Slender Salsify - **T. hybridus** L.
3   All achenes with pappus with some feathery hairs; achenes abruptly narrowed into beak. Stems erect, to 1m. Intrd-natd; waste and rough ground, waysides; very scattered in BI, mainly S
    Salsify - **T. porrifolius** L.

T. pratensis x *T. porrifolius* = ***T. x mirabilis*** Rouy occurs rarely in C & S En near *T. porrifolius* but rarely persists; it has yellow ligules suffused purple distally, so that the capitulum appears purple with a yellow centre, and has a low level of fertility.

25. AETHEORHIZA Cass. - *Tuberous Hawk's-beard*
Perennials with long thin rhizomes bearing large tubers; stems simple, with 1 capitulum and 0(-2) leaves; phyllaries in several rows, often weakly 2-rowed; receptacular scales 0; pappus of several rows of white, simple hairs; ligules yellow; achenes not flattened, with 4 deep grooves, not beaked.

1 Stems to 30cm, erect, glabrous; leaves sinuate-dentate; phyllaries (13)14-15(16)mm; achenes 3-4.5mm. Intrd-surv; very persistent in old gardens; Co Wexford and Co Armagh
Tuberous Hawk's-beard - **A. bulbosa** (L.) Cass.

26. SONCHUS L. - *Sowthistles*
Annuals or perennials, sometimes with rhizomes; stems leafy, usually branched; phyllaries in several rows; receptacular scales 0; pappus of 2 or more rows of white, simple, hairs; ligules yellow; achenes distinctly flattened, distinctly ribbed, not beaked.

1 Plant annual or biennial, with main root and laterals 2
1 Plant perennial, rhizomatous or with thick ± erect underground portion 3
  2 Auricles of stem-leaves pointed; achenes transversely rugose. Stems erect, to 1.5m. Native; waste and cultivated ground, roadsides; abundant *Smooth Sowthistle* - **S. oleraceus** L.
2 Auricles of stem-leaves rounded; achenes not rugose. Stems erect, to 1.5m. Native; similar places and distribution to *S. oleraceus*, often with it *Prickly Sowthistle* - **S. asper** (L.) Hill
3 Auricles of stem-leaves pointed; stems arising from thick underground root-like organ; achenes straw-coloured. Stems erect, to 2.5m. Native; marshes, fens and riversides; SE En from S Hants to E Norfolk, intrd in parts of C En *Marsh Sowthistle* - **S. palustris** L.
3 Auricles of stem-leaves rounded; plant strongly rhizomatous; achenes bright brown. Stems erect, to 1.5m. Native; arable and waste land, waysides, dunes and shingle by sea, ditches and river-banks; common *Perennial Sowthistle* - **S. arvensis** L.

S. oleraceus x *S. asper* has occurred very rarely in C & S En; it has leaf-auricles rounded and dentate as in *S. asper* but with one long pointed tooth, and is sterile.

27. LACTUCA L. - *Lettuces*
Annuals or biennials, or perennials with rhizomes; stems leafy, branched at least above, with many capitula; phyllaries in several rows; receptacular scales 0; pappus of 2 rows of white, simple hairs; ligules yellow or blue; achenes distinctly flattened, ribbed, beaked.

1 Achenes with beak <1/2 as long as body and of same colour; ligules

blue; perennial. Stems erect, to 80cm. Intrd-natd; rough and waste
ground mostly by sea; very scattered in En, Wa, Man, Guernsey,
NE Galway *Blue Lettuce* - **L. tatarica** (L.) C.A. Mey.
1 Achenes with beak ≥1/2 as long as body and much lighter in colour;
ligules yellow; annual to biennial 2
   2 Plant <1m; leaf midrib on lowerside glabrous to sparsely hispid;
achenes without bristles 3
   2 Plant often 1-2(2.5)m; leaf midrib on lowerside with strong
prickles; achenes with minute bristles just below beak 4
3 Stem-leaves oblong-ovate, cordate and clasping stem at base;
inflorescence subcorymbose. Stems erect, to 75cm. Intrd-casual; on
tips, waste ground and abandoned arable land; scattered in Br,
mainly S *Garden Lettuce* - **L. sativa** L.
3 Middle and upper stem-leaves linear-oblong, sagittate and clasping
stem at base; inflorescence very narrow. Stems erect, to 75(100)cm.
Native; salt-marshes, shingle, waste places and sea-walls near sea;
very local in S Essex, W Kent and E Sussex *Least Lettuce* - **L. saligna** L.
   4 Ripe achenes (excl. beak) (4)4.2-4.8(5.2)mm, maroon to blackish;
stems and midribs strongly tinged maroon. Stems erect, to 2(2.5)m.
Probably native; similar distribution to *L. serriola* but scattered
N to C Sc and less common in E & SE En *Great Lettuce* - **L. virosa** L.
   4 Ripe achenes (excl. beak) (2.8)3-4(4.2)mm, olive-grey; stems and
midribs greenish-white. Stems erect, to 1.5(2)m. Probably native;
waysides and waste and rough ground; frequent in En SE of
line from Severn to Humber, very scattered elsewhere in En,
Wa and CI *Prickly Lettuce* - **L. serriola** L.

28. CICERBITA Wallr. - *Blue-sowthistles*

Perennials, often with rhizomes; stems leafy, branched above, with many
capitula; phyllaries in several rows; receptacular scales 0; pappus of 2 rows
of simple, dirty- to yellowish-white hairs; ligules blue to mauve; achenes
flattened, ribbed, not beaked.

1 Plant glabrous. Not rhizomatous; stems erect, to 1.3m. Intrd-natd;
rough and waste ground and roadsides; very scattered in En and Sc
 Hairless Blue-sowthistle - **C. plumieri** (L.) Kirschl.
1 Peduncles and/or upper parts of stems with simple or glandular hairs 2
   2 Upper parts of stems and sometimes peduncles with simple
hairs; glandular hairs 0. Not rhizomatous; stems erect, to 2m.
Intrd-natd; rough and waste ground and roadsides; very
scattered in En, Sc and Man
 Pontic Blue-sowthistle - **C. bourgaei** (Boiss.) Beauverd
   2 Peduncles and usually upper parts of stems with glandular hairs 3
3 Inflorescence narrowly pyramidal; lower leaves glabrous, with
sharply triangular apical lobe. Not rhizomatous; stems erect, to 1.3m.
Native; moist mountain rock-ledges; extremely local in 4 sites in
Angus and S Aberdeen *Alpine Blue-sowthistle* - **C. alpina** (L.) Wallr.
3 Inflorescence subcorymbose; lower leaves pubescent on veins on

lowerside, with ovate-subcordate apical lobe. Strongly rhizomatous; stems erect, to 2m. Intrd-natd; rough and waste ground and roadsides; frequent throughout most of Br, scattered in Ir
Common Blue-sowthistle - **C. macrophylla** (Willd.) Wallr.

29. MYCELIS Cass. - *Wall Lettuce*

Perennials, often short-lived; stems leafy, branched above, with many capitula; phyllaries in 2 very unequal rows; receptacular scales 0; pappus of 2 unequal rows of white simple hairs; ligules yellow; achenes distinctly flattened, ribbed, very shortly beaked.

1 Usually suffused maroon; stems erect, to 1m, glabrous; leaves deeply pinnately lobed, sharply dentate; capitula 1-1.5cm across, with c.5 flowers. Native; shady places in woods, on walls and rocks and in hedgerows; locally common in En and Wa, scattered and probably intrd in Sc, Ir and Man *Wall Lettuce* - **M. muralis** (L.) Dumort.

30. TARAXACUM F.H. Wigg. - *Dandelions*

Perennials with tap-roots; stems usually leafless, with 1 capitulum; phyllaries in 2 often very different rows; receptacular scales 0; pappus of several rows of white, simple hairs; ligules yellow, usually with coloured stripe(s) on lowerside; achenes not flattened, finely ribbed, usually spinulose near apex, beaked.

A very critical genus in which apomixis is the rule; 229 microspp. are currently recognized in BI, of which >40 are probably endemic but <1/2 of the rest native. In this work the microspp. are not treated in full but are aggregated into 9 rather ill-defined sections, determination of which is often not easy even after much experience.

In most spp. the achene is spinulose near its apex, but between that region and the beak there is a short, usually pyramidal region known as the *cone*. Descriptions apply only to fully ripe achenes; *achene length* excludes cone and beak. Leaves produced in summer do not maintain all diagnostic characters, so determination should be attempted only with specimens collected during the first main flush of flowering (usually Apr to early May in the lowlands). Plants from shaded, heavily trodden or grazed, or mown areas should be avoided.

1 Plants delicate, usually with strongly dissected (often nearly pinnate) leaves; outer row of phyllaries mostly <8mm, with small outgrowth near apex on lowerside; capitula rarely >3cm across 2
1 Plants usually medium to robust, rarely with nearly pinnate leaves; outer row of phyllaries mostly >8mm, without subapical outgrowth; capitulum usually >3cm across 3
  2 Achenes greyish-brown, with pyramidal cone c.0.4mm; leaves often with ≥6 pairs of lateral lobes. 2 microspp. currently placed here. Native; open sandy turf by sea; local on coasts of BI, commonest in Sc, but S to CI **T. sect. Obliqua** (Dahlst.) Dahlst.
  2 Achenes usually purplish-violet, reddish or yellowish-brown,

with cylindrical cone mostly 0.6-1mm; leaves rarely with >6 pairs of lateral lobes. 30 microspp. currently placed here. Mostly native; dry exposed places, usually on well-drained soils, short grassland, heathland, dunes; throughout BI, commonest lowland section after *Hamata* and *Ruderalia*, mostly maritime in N **T. sect. Erythrosperma** (H. Lindb.) Dahlst.

3 Outer row of phyllaries appressed, ovate, with broad scarious border; leaves very narrow, usually scarcely lobed. 4 microspp. currently placed here. Native; wet usually base-rich meadows and fen grassland; local, scattered in BI **T. sect. Palustria** (H. Lindb.) Dahlst.

3 Outer row of phyllaries appressed to recurved, linear to narrowly ovate, with no or with narrow to very narrow scarious border; leaves broader, usually distinctly lobed 4

    4 Leaves and petioles green; rare plants of a few mountain cliffs in Sc. 6 microspp. currently placed here. Native; mostly base-rich mountain rock-ledges and flushes; very local in highlands of Sc
    T. sect. Taraxacum

    4 Lowland plants, or if on mountain cliffs then leaves usually dark or blotched or spotted with purple and petiole usually purple 5

5 Achenes (excl. cone and beak) ≥4mm, nearly cylindrical; outer row of phyllaries erect to appressed; ligules usually with dark red stripes on lowerside; pollen usually 0. 3 microspp. currently placed here. Native; damp or wet acidic grassy places, often in upland areas, also roadsides etc.; throughout BI **T. sect. Spectabilia** (Dahlst.) Dahlst.

5 Achenes very rarely >4mm, narrowly top-shaped; outer row of phyllaries rarely appressed; ligule stripes rarely dark red; pollen present or 0 6

    6 Leaves with large dark spots covering >10% of surface. 12 microspp. currently placed here. Mostly native; habitat and distribution as for sect. *Spectabilia* but uncommon in SE En
    T. sect. Naevosa M.P. Christ.

    6 Leaves unspotted or with spots covering <10% of blade (beware leaves damaged or attacked by pathogens) 7

7 Petiole and midrib uppersides green or solid red or purple; outer row of phyllaries mostly 9-16mm, usually recurved, not dark on lowerside; leaves often complexly lobed and folded in 3 dimensions. 120 microspp. currently placed here. Native (c.26 microspp.) and intrd-natd and -casual; habitat and distribution as for sect. *Hamata*. By far the commonest section, especially as weeds in lowland areas **T. sect. Ruderalia** Kirschner, H. Øllg. & Stepánek

7 Petiole and midrib uppersides usually minutely (lens) striped red or purple; outer row of phyllaries mostly 7-12mm, often (often not) patent to erect and dark on lowerside; leaves ± flat, relatively simply lobed 8

    8 Lateral leaf-lobes broad-based, with convex front and concave rear edge, commonly 4 pairs; outer row of phyllaries usually arched to varying degrees, often subobtuse. 18 microspp. currently placed here. Native (c. 7 microspp.) and intrd-natd;

damp and dry grassland, roadsides and rough ground; throughout BI, usually weedy **T. sect. Hamata** H. Øllg.
8 Lateral leaf-lobes rarely as above, often 5-6 pairs; outer row of phyllaries erect to recurved all ± to same degree, often acute. 34 microspp. currently placed here. Mostly native; mostly wet places in lowland grassland and in upland flushes and on rock-ledges; throughout BI **T. sect. Celtica** A.J. Richards

31. CREPIS L. - *Hawk's-beards*
Annuals to perennials, sometimes shortly rhizomatous; stems branched, leafy or (*C. praemorsa*) all basal; phyllaries in 2 rows; receptacular scales 0 but receptacle often pubescent, sometimes each achene-pit with membranous fringe; pappus of several rows of white or (*C. paludosa*) yellowish-white, simple hairs; ligules yellow; achenes not flattened, ribbed, beaked or not (if not, usually slightly tapered distally).

1 Flowering stems leafless. Erect, sparsely pubescent perennial to 60cm. Possibly native; on natural calcareous grassy bank; Westmorland
Leafless Hawk's-beard - **C. praemorsa** (L.) Walther
1 Flowering stems bearing leaves 2
   2 Outer achenes with short or 0 beak, distinctly different from inner slender-beaked ones. Erect, pubescent annual or biennial to 60cm, stinking when fresh. Native; on maritime shingle; Dungeness (E Kent) *Stinking Hawk's-beard* - **C. foetida** L.
   2 Achenes all the same, or inner and outer slightly different but grading into one another 3
3 Achenes distinctly beaked, the beak usually ≥1/2 as long as body of achene 4
3 Achenes not beaked, but often narrowed at apex 6
   4 Basal lobes of upper stem-leaves not clasping stem; achenes with beak scarcely 1/2 as long as body. Erect to ascending subglabrous annual to 75cm. Intrd-casual; roadsides, disturbed soil and re-seeded verges; scattered in En, Ir and Sc
Narrow-leaved Hawk's-beard - **C. tectorum** L.
   4 Basal lobes of upper stem-leaves clasping stem; achenes with beak c. as long as body 5
5 Upper parts of plant nearly always with many patent stiff bristles; achenes (incl. beak) 3-5.5mm. Erect, usually hispid (very rarely subglabrous) annual or biennial to 75cm. Intrd-casual; with crops or grass or in rough ground; scattered in Br, mostly C & S
Bristly Hawk's-beard - **C. setosa** Haller f.
5 Upper parts of plant without patent stiff bristles; achenes (incl. beak) (5)6-8(9)mm. Erect, pubescent perennial to 80cm. Intrd-natd; grassy places, waysides, walls and rough ground; common in S & CE En, scattered in N En, CI and Ir *Beaked Hawk's-beard* - **C. vesicaria** L.
   6 Phyllaries pubescent on their inner faces 7
   6 Phyllaries glabrous on their inner faces 8

7 Achenes (2.5)3-4(4.5)mm, with 10 ribs distinctly rough towards apex
(see couplet 4) **C. tectorum**
7 Achenes 4-7.5mm, with 13-20 ± smooth ribs. Erect pubescent biennial
to 1.2m. Probably native; rough grassy places; waysides; scattered in
Br and Ir N to C Sc *Rough Hawk's-beard* - **C. biennis** L.
  8 Pappus-hairs yellowish-white, brittle. Erect subglabrous
  perennial to 80cm. Native; wet places in open woodland,
  grassland, fens; Bl N from S Ir, S Wa and C En
  Marsh Hawk's-beard - **C. paludosa** (L.) Moench
  8 Pappus-hairs pure white, ± flexible   9
9 Achenes c.20-ribbed; perennial arising from short rhizome. Erect
subglabrous to sparsely pubescent perennial to 60cm. Native;
grassy, often damp slopes or hills; very local from MW Yorks to
E Perth *Northern Hawk's-beard* - **C. mollis** (Jacq.) Asch.
9 Achenes 10-ribbed; annual or biennial with tap-root   10
  10 Achenes 2.5-3.8mm; outer phyllaries patent to erecto-patent;
  receptacle with laciniate membranous fringes around achene-pits.
  Erect to ascending rather sparsely pubescent annual or biennial
  to 80cm. Intrd-casual; roadsides, disturbed soil and re-seeded
  verges; scattered in S En *French Hawk's-beard* - **C. nicaeensis** Balb.
  10 Achenes 1.4-2.5mm; outer phyllaries appressed to inner;
  receptacle with a few hairs around achene-pits. Erect to
  decumbent glabrous to sparsely pubescent annual or biennial to
  75cm. Native; grassy places, rough and waste ground; common
  Smooth Hawk's-beard - **C. capillaris** (L.) Wallr.

32. PILOSELLA Hill - *Mouse-ear-hawkweeds*
Perennials, usually stoloniferous; stems usually leafless, sometimes with
few leaves, with 1-many capitula; basal leaves oblanceolate to narrowly
obovate or narrowly elliptic, pubescent, subentire; phyllaries in several
rows; receptacular scales 0 but achene-pits often variously fringed; pappus
of 1 row of dirty-white to pale brown, simple hairs; ligules yellow to orange;
achenes not flattened, 10-ribbed, not beaked, scarcely tapered towards apex

1 All flowering stems with only 1 capitulum   2
1 At least some flowering stems with >1 capitulum   5
  2 Stolons elongated, slender, with spaced out small leaves, 0 or few
  ending in leaf-rosette. Scapes to 30(50)cm. Native; short grassland
  on well-drained soils, banks, rocky places; locally common
  throughout BI except Shetland
  ***Mouse-ear-hawkweed* - P. officinarum** F.W. Schultz & Sch. Bip.
  2 Stolons few or 0, short (<5cm), stout, with full-sized ± crowded
  leaves, often ending in leaf-rosette. Scapes to 30cm. Native; short
  grassland on well-drained soils, dunes; very local, but commoner
  in CI than *P. officinarum* (*Shaggy Mouse-ear-hawkweed* -
  P. peleteriana (Mérat) F.W. Schultz & Sch. Bip.)   3
3 Scapes up to 12(18)cm; rosette-leaves 9-20mm wide, not or scarcely
tapered at base. Phyllaries 11-15mm, lanceolate; capitula 12-20mm

across excl. ligules. CI (all islands), Dorset and Wight
P. peleteriana ssp. **peleteriana**
3 Scapes (6)10-30cm; rosette-leaves 4-12(18)mm wide, long-tapered at
base 4
    4 Phyllaries 11-15mm, lanceolate; capitula 12-17mm across excl.
ligules. Craig Breidden (Monts)
P. peleteriana ssp. **subpeleteriana** (Nägeli & Peter) P.D. Sell
    4 Phyllaries 10-12(13)mm, linear-lanceolate; capitula (9)10-12(14)mm
across excl. ligules. Jersey, S Devon, Staffs, Derbys and MW Yorks
P. peleteriana ssp. **tenuiscapa** (Pugsley) P.D. Sell & C. West
5 Ligules orange-brown to brick-red, often turning purplish when dried.
Scapes to 40(65)cm. Intrd-natd; garden escape (setting abundant seed)
on rough ground, walls, roadsides and railway banks
(***Fox-and-cubs*** - **P. aurantiaca** (L.) F.W. Schultz & Sch. Bip.) 6
5 Ligules yellow, sometimes red-striped on lowerside 7
    6 Spreading mostly by rhizomes; basal leaves mostly 10-20 x 2-6cm;
phyllaries 8-11mm. Scattered in Br **P. aurantiaca** ssp. **aurantiaca**
    6 Spreading mostly by stolons; basal leaves mostly 6-10(16) x
1.2-2(3)cm; phyllaries 5-8mm. Frequent throughout BI
P. aurantiaca ssp. **carpathicola** (Nägeli & Peter) Soják
7 Capitula (1)2-4(7) per stem, not crowded; phyllaries (8)9-12mm
(**P. flagellaris** (Willd.) P.D. Sell & C. West) 8
7 Capitula (3)6-50 per stem, many closely crowded; phyllaries 5-9mm 9
    8 Scapes to 40cm; capitula 2-4(7); peduncles with simple eglandular
hairs 2-3mm; phyllaries with few to numerous simple eglandular
hairs ≤1.5mm. Intrd-natd; grassy roadsides and railway banks;
scattered in C & CS En and CE Sc
Spreading Mouse-ear-hawkweed - **P. flagellaris** ssp. **flagellaris**
    8 Scapes to 18cm; capitula (1)2(-4); peduncles with simple
eglandular hairs ≤7.5mm; phyllaries with dense simple
eglandular hairs ≤2.5mm. Native; dry rocky pastures, rocky
slopes and outcrops; 3 localities in Shetland
Shetland Mouse-ear-hawkweed -
P. flagellaris ssp. **bicapitata** P.D. Sell & C. West
9 Largest leaves ≤12(20)mm wide; phyllaries mostly <1mm wide, acute.
Scapes to 65cm. Intrd-natd; grassy roadsides, walls and railway
banks (***Tall Mouse-ear-hawkweed*** - **P. praealta** (Gochnat)
F.W. Schultz & Sch. Bip.) 10
9 Largest leaves ≥(12)15mm wide; phyllaries mostly >1mm wide,
obtuse 11
    10 Stolons 0 or very short; phyllaries with numerous glandular and
0 or few eglandular hairs. Scattered localities in Br N to Ayrs
P. praealta ssp. **praealta**
    10 Stolons long and slender; phyllaries with numerous glandular
and 0 to numerous eglandular hairs. Scattered localities in SC En
and W Lothian **P. praealta** ssp. **thaumasia** (Peter) P.D. Sell
11 Leaves scarcely glaucous; capitula >10 on well-developed
inflorescences. Scapes to 50(80)cm. Intrd-natd; rough ground, walls

and railway banks; scattered in Br and S Ir

Yellow Fox-and-cubs - **P. caespitosa** (Dumort.) P.D. Sell & C. West

11 Leaves distinctly glaucous; capitula <10 per inflorescence. Scapes to 35(45)cm. Intrd-natd; in *Calluna/Erica* heathland; S Hants
(*P. lactucella* x *P. caespitosa*)

Irish Fox-and-cubs - **P. x floribunda** (Wimm. & Grab.) Arv.-Touv.

Hybridization is frequent wherever 2 spp. occur together. *P. peleteriana* x *P. officinarum* = ***P. x longisquama*** (Peter) Holub has occurred in Jersey, Guernsey, Staffs and E Kent; *P. officinarum* x *P. aurantiaca* = ***P. x stoloniflora*** (Waldst. & Kit.) F.W. Schultz & Sch. Bip. occurs in scattered places from Guernsey to N Sc.

33. HIERACIUM L. - *Hawkweeds*

Perennials, without stolons or rhizomes; stems leafy or sometimes not, with (1)few-several capitula, with or without basal rosette of leaves at flowering; phyllaries in several rows; receptacular scales 0 but achene-pits often variously fringed; pappus of 1 row of dirty-white to pale brown, simple hairs; ligules yellow; achenes not flattened, 10-ribbed, not beaked, scarcely tapered towards apex.

All the taxa are obligate apomicts except the single sp. of section *Hieracioides* (*H. umbellatum* L.), which exists as both sexual and apomictic plants. 261 microspp. are currently recognized in BI, of which many are endemic and probably a considerable number are aliens. In this work they are not treated in full, but are aggregated into 15 sections that are recognizable after a little practice.

Plants often exhibit a second phase of flowering on new growth, either naturally or if the first growth is damaged. Only the first growth provides reliable diagnostic characters. As a rule of thumb, identification should not be attempted on plants with 0-1 stem-leaves after mid-Jun, on plants with 2-8 stem-leaves after mid-Jul, and on others after mid-Aug.

1 Stem-leaves 8-many except in dwarfed plants; rosette of leaves usually 0 at flowering 2
1 Stem-leaves 0-8(15); rosette of leaves usually present at flowering 7
  2 Middle stem-leaves not or scarcely clasping stem at base 3
  2 Middle stem-leaves distinctly clasping stem at base, though often very narrowly so 5
3 Leaves all sessile, often linear-lanceolate, ± all of similar shape, with recurved margins; phyllaries (except innermost) with recurved tips; styles yellow when fresh. Plants normally >30cm; stem-leaves normally >15. 1 microsp. placed here (**H. umbellatum** L.). Native; sandy heathland, dunes and dry rocky places, often near the coast; scattered throughout BI, mostly W & S   **H. sect. Hieracioides** Dumort.
3 Lower leaves petiolate, usually broader, middle and upper ones sessile or nearly so, not with recurved margins; phyllaries very rarely with recurved tips; styles usually dark when fresh 4
  4 Stem-leaves rarely <15, often crowded, upper ones with broad

rounded bases. Plants normally >30cm. 5 microspp. currently placed here. Native; rough ground, grassy and marginal places and on roadside and railway banks; En and Wa, rather local in Sc, very local in E Ir **H. sect. Sabauda** (Fr.) Arv.-Touv.

4 Stem-leaves usually <15, rarely crowded, upper ones narrowed to base. Stems normally >30cm. 21 microspp. currently placed here. Native; grassy, rocky and marginal habitats; frequent in Br, rare in W & N Ir **H. sect. Tridentata** (Fr.) Arv.-Touv.

5 Middle stem-leaves slightly constricted just above the broad clasping base; peduncles with dense glandular hairs; achenes pale brown. Stems normally >30cm. 2 microspp. placed here. Native; grassy and rocky places, often on limestone; local in N Br S to S Wa and Peak District, Co Antrim **H. sect. Prenanthoidea** W.D.J. Koch

5 Middle stem-leaves not constricted, with narrow clasping base; peduncles with 0-few glandular hairs; achenes purplish- or blackish-brown 6

 6 Stem-leaves c.10-30, the lower ones clasping stem at base to merely sessile; phyllaries sparsely pubescent and glandular; ligules glabrous at tip. Plants normally >30cm. 10 microspp. currently placed here. Native; grassy and rocky places; locally common in N Br S to S Wa and Peak District, very local in E, N & W Ir **H. sect. Foliosa** (Fr.) Arv.-Touv.

 6 Stem-leaves c.2-10(15), the lower ones subpetiolate; phyllaries moderately pubescent and glandular; ligules glabrous or pubescent at tip. Plants normally >20cm. 19 microspp. currently placed here. Native; rocky places, cliffs, hillsides; 14 microspp. endemic to Shetland, others very local in Sc and N En **H. sect. Alpestria** (Fr.) Arv.-Touv.

7 Stem-leaves 1-7(12), clasping stem at base 8
7 Stem-leaves 0-8(12), not clasping stem at base 9

 8 Stem-leaves yellowish-green; plant sticky-glandular. Plants 10-60cm. 3 microspp. placed here. Intrd-natd; walls and rough ground; very scattered in En and Sc

H. sect. Amplexicaulia (Griseb.) Scheele

 8 Stem-leaves glaucous-green; plant not sticky-glandular. Plants 10-60cm. 10 microspp. currently placed here. Native; cliffs and rocky streamsides; coastal and upland areas of Sc, N En, N & W Ir, 1 site in Wa **H. sect. Cerinthoidea** Monnier

9 Stems, leaves and phyllaries with dense, white, patent hairs. Plants 10-50cm. 1 microsp. (**H. lanatum** Vill.) placed here. Intrd-natd; on coastal dunes; E Norfolk **H. sect. Andryaloidea** Monnier

9 White patent hairs not dense on stems, leaves and phyllaries 10

 10 Leaves with small glandular hairs on margins and sometimes on surface; phyllaries usually with shaggy hairs 11

 10 Leaves without stalked glands; phyllaries without shaggy hairs; widespread 12

11 Stem-leaves 0-4, narrow and bract-like; capitula 1(-5). Plants 5-15(25)cm. 30 microspp. currently placed here. Native; rock-ledges,

barish slopes and scree, grassy banks, usually above 650m; 27
microspp. endemic to mainland Sc, 3 others (2 endemic) in Sc,
Lake District and Snowdonia   **H. sect. Alpina** (Griseb.) Gremli
11 Stem-leaves (0)1-4, usually at least one leaf-like; capitula (1)2-5.
Plants 20-50cm. 31 microspp. currently placed here. Native; rock-
ledges and rocky streamsides usually above 450m; local in mainland
Sc, 2 microspp. extend to N En, 1 to Co Antrim, 1 is endemic to Lake
District   **H. sect. Subalpina** Pugsley
  12 Leaves usually bristly at least along margins; phyllaries erect
  in bud, without dense white stellate hairs   13
  12 Leaves variously pubescent but not bristly; phyllaries incurved
  in bud, with dense white stellate hairs at least on margins   14
13 Stem-leaves 2-10(12); basal leaves few (mostly ≤4). Plants mostly
10-60cm. 15 microspp. currently placed here. Native; cliffs, rocky
and grassy banks, often on limestone; scattered in Ir, Wa, Sc and
W, C & N En   **H. sect. Oreadea** (Fr.) Arv.-Touv.
13 Stem-leaves 0-1(2); basal leaves numerous. Plants mostly 10-60cm.
30 microspp. currently placed here. Native; cliffs, rocky and grassy
banks, often on limestone; scattered in Ir, Wa, Sc and W, C & N En
                                        H. sect. Stelligera Zahn
  14 Stem-leaves 0-2(3); basal leaves numerous. Plants mostly
  >(10)20cm. 53 microspp. currently placed here. Native and
  intrd-natd; rough ground, woodland, marginal habitats, cliffs
  and rocky places; throughout BI   **H. sect. Hieracium**
  14 Stem-leaves 2-8(15); basal leaves few (commonly 2-4), often
  withering at flowering. Plants mostly >(10)20cm. 30 microspp.
  currently placed here. Native and intrd-natd; rough ground,
  woodland, marginal habitats, cliffs and rocky places;
  throughout BI   **H. sect. Vulgata** (Griseb.) Willk. & Lange

TRIBE 3 - ARCTOTIDEAE (genera 34-35). Plants not producing white latex, not spiny; capitula radiate, with sterile ligulate flowers with yellow to orange, usually 3-toothed ligules.

34. ARCTOTHECA J.C. Wendl. - *Plain Treasureflower*
Annuals; lower leaves deeply pinnately lobed, white-tomentose on lowerside; phyllaries in several rows, free, glabrous to sparsely pubescent, with conspicuous scarious, rounded to obtuse tips; receptacular scales 0; achenes densely pubescent; pappus of distinct scales.

1 Stems decumbent, leafy only near base, to 40cm, white-pubescent;
all leaves usually deeply lobed; ligules pale yellow on upperside,
purplish on lowerside. Intrd-casual; arable fields and waste places;
scattered in Br   *Plain Treasureflower* - **A. calendula** (L.) Levyns

35. GAZANIA Gaertn. - *Treasureflower*
Perennials differing from *Arctotheca* in outer phyllaries fused into cup-like structure, white-tomentose on lowerside, with acute to acuminate scarious

tips; leaves sometimes all simple.

1 Stems decumbent to ascending, to 50cm, often ± woody at base, white-tomentose below; lower leaves deeply pinnately lobed, upper ones ± entire; ligules orange-yellow with basal black blotch bearing central white spot, or rarely plain yellow. Intrd-natd; walls, rocks and cliffs near sea; Scillies and CI *Treasureflower* - **G. rigens** (L.) Gaertn.

SUBFAMILY 2 - ASTEROIDEAE (tribes 4-10; genera 36-99). Plant not producing white latex; stem-leaves usually spiral ('alternate'), sometimes opposite or 0; capitula mostly radiate, sometimes discoid; tubular flowers most commonly yellow to orange, usually with short lobes or teeth; filaments joining anthers at base; 2 style branches each with 2 stigmatic zones, 1 near each margin of inner face.

TRIBE 4 - INULEAE (genera 36-44). Annual to perennial herbs, rarely shrubs; leaves alternate, simple; capitula discoid or radiate; phyllaries in several rows, herbaceous to scarious; receptacular scales present or (usually) 0; pappus usually of hairs, sometimes of short scales or of scales and hairs; flowers usually yellow or brown to whitish.

36. FILAGO L. - *Cudweeds*
Annuals with stems and leaves ± covered with woolly hairs; capitula small, brownish, borne in clusters of 2-c.40; phyllaries in few ill-defined rows, the outer herbaceous, the inner scarious; receptacle conical, with scales associated with outer (female) florets only; flowers all tubular, the inner bisexual, with wider corollas than the outer female; pappus of bisexual flowers of simple hairs, of female flowers of simple hairs or 0.

1 Capitula 2-7(14) in each cluster; outer phyllaries obtuse to subacute, patent in fruit 2
1 Capitula (5)10-c.40 in each cluster; outer phyllaries acuminate, erect in fruit 3
  2 Leaves 4-10mm, the most apical ones not overtopping clusters of capitula. Stems erect, to 25cm, irregularly branching. Native; similar habitats and distribution to *F. vulgaris*
              Small Cudweed - **F. minima** (Sm.) Pers.
  2 Leaves (8)12-20(25)mm, the most apical ones overtopping clusters of capitula. Stems erect, to 25cm, irregularly branching. Intrd-natd; in sandy and gravelly ground; Sark
              Narrow-leaved Cudweed - **F. gallica** L.
3 Leaves widest in basal 1/2, the most apical ones not overtopping clusters of capitula; capitula in clusters of (15)20-40. Stems erect, to 40cm, branching below each cluster of capitula. Native; barish places on sandy soils, e.g. heaths, waysides, sand-pits; throughout most of BI except most of N & W Sc *Common Cudweed* - **F. vulgaris** Lam.
3 Leaves widest in apical 1/2, the most apical ones usually overtopping clusters of capitula; capitula in clusters of (5)10-20(25) 4

4 Clusters of capitula each overtopped by (0)1-2 leaves; outer phyllaries with erect red-tinged points; plant usually yellowish-woolly. Stems erect, to 40cm, ± irregularly branched. Native; barish places on sandy soils, e.g. heaths, waysides, sand-pits; very local in S & E En from S Hants to SE Yorks
Red-tipped Cudweed - **F. lutescens** Jord.

4 Clusters of capitula each overtopped by 2-4(5) leaves; outer phyllaries with recurved yellowish points; plant white-woolly. Stems erect, to 40cm, branching below each cluster of capitula. Native; barish places on sandy soils, e.g. heaths, waysides, sand-pits; very local in S En   *Broad-leaved Cudweed* - **F. pyramidata** L.

37. ANTENNARIA Gaertn. - *Mountain Everlasting*

Dioecious whitish-woolly perennials; capitula pale to deep pink, sometimes whitish, borne in terminal umbel-like clusters of 2-8; phyllaries in several rows, scarious, the outer ones of male capitula patent and perianth-like in flower, white to pink; receptacle flat, without scales; flowers all tubular, males with wider corollas than females; pappus of female flowers of simple hairs, of male flowers of simple hairs widened distally.

1 Stems erect, to 20cm, with basal leaf-rosette; surface-creeping leafy stolons present; leaves green on upperside, white-woolly on lowerside. Native; heaths, moors, mountain slopes; common in much of N 1/2 of BI, scattered S to S Ir, W Cornwall and Northants
Mountain Everlasting - **A. dioica** (L.) Gaertn.

38. ANAPHALIS DC. - *Pearly Everlasting*

Dioecious to variably sexed white-woolly perennials; capitula white, with yellow flowers, borne in large terminal corymbose inflorescences; phyllaries in several rows, scarious, pearly white; receptacle convex, without scales; flowers all tubular, male and female variously arranged, the males with wider corollas; pappus of 1 row of hairs.

1 Stems erect to ascending, to 1m, without basal leaf-rosette; rhizomes present; leaves green on upperside, white-woolly on lowerside. Intrd-natd; relic or throwout by rivers and in grassland, marginal and rough ground; scattered in BI, especially W
Pearly Everlasting - **A. margaritacea** (L.) Benth.

39. GNAPHALIUM L. - *Cudweeds*

Annuals or perennials ± covered with whitish woolly hairs; capitula small, yellowish to brown, variously arranged; phyllaries whitish to yellowish or brown, subherbaceous to scarious, in several rows; receptacle flat, without scales; flowers all tubular, the inner bisexual, with wider corollas than the outer female; pappus of simple hairs.

1 Capitula in terminal, subglobose to subcorymbose clusters; annual   2
1 Capitula in elongated, racemose clusters, sometimes few or rarely 1;

39. GNAPHALIUM

annual to perennial　　　　　　　　　　　　　　　　　　　　　　　　4
- 2 Phyllaries brown; clusters of capitula conspicuously leafy. Stems decumbent to erect, to 25cm. Native; damp places in fields and arable land and by ponds and paths; common

　　　　　　　　　　　　　　　　Marsh Cudweed - **G. uliginosum** L.
- 2 Phyllaries uniformly white to yellowish: clusters of capitula not leafy　　　　　　　　　　　　　　　　　　　　　　　　　　3

3 Leaves white-woolly on both sides, not decurrent down stem. Stems erect, to 50cm. Native; sandy fields, waste places and sand-dunes; very local in CI and W Norfolk　　*Jersey Cudweed* - **G. luteoalbum** L.

3 Leaves green on upperside, white-woolly on lowerside, decurrent down stem. Stems erect, to 80cm. Intrd-natd; rough ground, cliffs, marginal habitats; CI (all main islands), E Cornwall. Resembles a small *Anaphalis*　　　　　　　　*Cape Cudweed* - **G. undulatum** L.
- 4 Annual to biennial without rhizome; achenes <1mm, glabrous; phyllaries acute to acuminate. Stems decumbent to erect, to 40cm. Intrd-natd; in churchyard; Surrey

　　　　　　　　　　　　　　American Cudweed - **G. purpureum** L.
- 4 Perennial with short ± surface rhizome; achenes >1mm, pubescent; phyllaries obtuse to rounded or retuse　　　　　5

5 Capitula <10 per stem; pappus-hairs free, falling separately. Stems erect, to 12(20)cm. Native; mountain rocks and gravel; local in C, N & W mainland Sc and Skye (N Ebudes)　　*Dwarf Cudweed* - **G. supinum** L.

5 Capitula normally >10 per stem; pappus-hairs united at base, falling as a unit　　　　　　　　　　　　　　　　　　　　　　　　　6
- 6 Stem-leaves 1-(or indistinctly 3-)veined, steadily diminishing in size up the stem. Stems erect, to 60cm. Native; rather open ground on heaths, banks, woodland rides; locally frequent in much of Br and Ir　　　　　　*Heath Cudweed* - **G. sylvaticum** L.
- 6 Stem-leaves 3(-5)-veined, scarcely diminishing in size until above 1/2 way up stem. Stems erect, to 30cm. Native; mountain rocks and gravel; very local in C Sc

　　　　　　　　　　Highland Cudweed - **G. norvegicum** Gunnerus

40. HELICHRYSUM Mill. - *Everlastingflowers*

Woody perennials; capitula conspicuous, terminal, solitary; phyllaries in several rows, scarious, white; receptacle slightly convex, without scales; flowers all tubular, the outer female, the inner bisexual and with wider corollas; pappus of 1 row of hairs.

1 Mat-forming dwarf evergreen shrub with decumbent stems to 60cm; leaves *Thymus*-like, silvery on lowerside; capitula on erect stems up to 10cm, c.2-3cm across with patent white inner phyllaries up to 1cm. Intrd-natd; in rocky turf by stream; 1 place in Shetland

　　New Zealand Everlastingflower - **H. bellidioides** (G. Forst.) Willd.

41. INULA L. - *Fleabanes*

Perennial herbs, sometimes woody at base; capitula 1-many, terminal, usually subcorymbose, usually showy, yellow, usually radiate, less often discoid; phyllaries in several rows, herbaceous; receptacle flat or slightly convex, without scales; pappus of 1 row of hairs.

1 Stem and leaves succulent, glabrous. Stems erect to decumbent, to 1m. Native; salt-marshes, shingle, cliffs, rocks and ditchsides by sea; local on coasts of BI N to S Sc *Golden-samphire* - **I. crithmoides** L.
1 Stem and leaves not succulent, very sparsely to densely pubescent 2
   2 Ligules 0 or <1mm; capitula numerous on each stem. Stems erect, to 1.25m. Native; scrub, grassland and barish places on calcareous soils; locally common in En, Wa and CI
Ploughman's-spikenard - **I. conyzae** (Griess.) Meikle
   2 Ligules conspicuous, >1cm; capitula 1-c.5 on each stem 3
3 Outer phyllaries ovate; capitula >5cm across (incl. ligules); stems rarely <1m. Stems erect, to 2.5m. Intrd-natd; in fields, waysides, marginal habitats, rough ground; scattered throughout BI
Elecampane - **I. helenium** L.
3 Outer phyllaries lanceolate; capitula <5cm across (incl. ligules); stems <1m 4
   4 Leaves subglabrous to sparsely pubescent on lowerside, prominently reticulate-veined on upperside; achenes glabrous. Stems erect, to 70cm. Native; stony limestone shores; Lough Derg (N Tipperary) *Irish Fleabane* - **I. salicina** L.
   4 Leaves densely pubescent on lowerside, obscurely veined on upperside; achenes pubescent. Stems erect, to 60cm. Intrd-natd; derelict gardens and rough ground; 2 sites in Man
Hairy Fleabane - **I. oculus-christi** L.

42. DITTRICHIA Greuter - *Fleabanes*

Low shrubs or annual herbs with glandular-sticky stems and leaves; capitula several in racemose inflorescences, rather showy, yellow, radiate but ligules often very short; phyllaries in several rows, herbaceous; receptacle flat or slightly convex, without scales; pappus of 1 row of hairs fused at base.

1 Perennial with resinous smell when crushed; stems ascending to erect, woody, to 1m; ligules ≤10mm, much longer than phyllaries. Intrd-natd; rough ground in E Suffolk and by harbour in E Sussex, casual elsewhere in S En *Woody Fleabane* - **D. viscosa** (L.) Greuter
1 Annual with strong camphorous smell when crushed; stems erect, to 50cm; ligules ≤3mm, not or scarcely exceeding phyllaries. Intrd-casual; wool-alien in fields, etc.; scattered in En
Stinking Fleabane - **D. graveolens** (L.) Greuter

43. PULICARIA Gaertn. - *Fleabanes*

Annuals or perennials; capitula several to many, terminal, usually subcorymbose, usually showy, yellow, radiate; phyllaries in several rows, herbaceous; receptacle flat, without scales; pappus of 1 row of hairs plus an outer row of free or fused scales.

1 Densely pubescent perennial with extensive rhizomes; stems erect, to 1m; stem-leaves cordate at base; capitula 1.5-3cm across. Native; marshes, ditches, wet fields, hedgebanks; common in lowland BI N to N En, rare elsewhere   *Common Fleabane* - **P. dysenterica** (L.) Bernh.
1 Pubescent annual; stems erect, to 45cm; stem-leaves rounded to cuneate at base; capitula 0.6-1.2cm across. Native; sandy places flooded in winter, often by ponds; very local in Surrey and
S Hants   *Small Fleabane* - **P. vulgaris** Gaertn.

44. TELEKIA Baumg. - *Yellow Oxeye*
Herbaceous perennials; capitula 1 to several, terminal, subcorymbose, showy, yellow, radiate; phyllaries in several rows, herbaceous; receptacle convex, with scales; pappus of fused scales.

1 Stems erect, pubescent, to 2m; lower stem-leaves petiolate and cordate at base, upper ones sessile and rounded to broadly cuneate at base; capitula 5-8cm across. Intrd-natd; in rough ground and by lakes and rivers; scattered throughout most of Br, mostly N
Yellow Oxeye - **T. speciosa** (Schreb.) Baumg.

TRIBE 5 - ASTEREAE (genera 45-55)
Annual to (usually) perennial herbs, rarely shrubs; leaves alternate or all basal, simple; capitula discoid or radiate; phyllaries in 2-several rows, usually herbaceous; receptacular scales 0; pappus usually of hairs, sometimes 0 or of strong bristles; flowers various colours.

45. GRINDELIA Willd. - *Gumplants*
Herbaceous perennials; stem-leaves sessile, clasping stem, serrate; capitula radiate, with yellow ray and disc flowers; phyllaries sticky, in several rows, herbaceous, with recurved tips; pappus of 2-8 stiff bristles.

1 Stems erect, to 75cm, sparsely shaggy-pubescent; capitula 3-5cm across; ligules 8-15mm. Intrd-natd; on sea-cliffs; Whitby (NE Yorks)
Coastal Gumplant - **G. stricta** DC.

46. CALOTIS R. Br. - *Bur Daisy*
Perennials; stem-leaves various; capitula radiate or ± discoid, with yellow disc and yellow, white or mauve ray flowers; phyllaries in ± 2 rows, herbaceous; pappus of (1)3(-6) rigid barbed bristles and usually some extra shorter bristles or scales; fruiting capitula forming a globose bur 5-9mm across.

1 Perennial with branching, erect to procumbent stems to 30(60)cm; leaves obtriangular, narrowed to petiole, dentate at distal end; capitula 1-2cm across in flower, with white or mauve ligules.

Intrd-casual; wool-alien in fields, etc.; scattered in En
Bur Daisy - **C. cuneifolia** R. Br.

47. SOLIDAGO L. - *Goldenrods*
Perennials; stem-leaves narrowly elliptic to oblanceolate or obovate, serrate, narrowed to base; capitula small, numerous, ± crowded, radiate, yellow; phyllaries in many rows, herbaceous; pappus of 1-2 rows of hairs.

The N American spp. are numerous and very difficult; they have possibly given rise in cultivation in Br to new taxa that add to the problems of identification.

1 Capitula sessile, in small clusters forming corymbose inflorescence; leaves gland-spotted, ligules 1-1.5mm. Stems erect, to 1.5m. Intrd-natd; waste land, banks, waysides and rough grassland; very scattered in En   *Grass-leaved Goldenrod* - **S. graminifolia** (L.) Salisb.
1 At least most capitula stalked, forming pyramidal to ± cylindrical inflorescence; leaves not gland-spotted; ligules ≥1.5mm   2
  2 Leaves with many pairs of short lateral veins (though often inconspicuous)   3
  2 Leaves with 1(-2) pairs of main lateral veins from near base, running parallel with midrib for most of length   4
3 Most stem-leaves rounded to acute at apex; inner phyllaries >4.5mm; disc flowers ≥10; ligules 4-9mm. Stems erect, to 70(100)cm, but often much less. Native; open woodland, grassland, hedgerows, rocky places, cliffs; frequent over most of BI   *Goldenrod* - **S. virgaurea** L.
3 Most stem-leaves acute to acuminate at apex; inner phyllaries ≤4.5mm; disc flowers ≤8; ligules 1.5-4mm. Stems erect, to 1.5m. Intrd-natd; on waste land, banks, waysides and rough grassland; Renfrews and Dunbarton   *Rough-stemmed Goldenrod* - **S. rugosa** Mill.
  4 Leaves scabrid-pubescent on surfaces, stems pubescent at least in top 1/2. Stems erect, to 2.5m. Intrd-natd; waste land, waysides, banks and rough grassland; frequent throughout C & S Br and CI, scattered in N Br and Ir   *Canadian Goldenrod* - **S. canadensis** L.
  4 Leaves glabrous on surfaces or pubescent only on lowerside veins; stems ± glabrous. Stems erect, to 2.5m. Intrd-natd; occurrence as for *S. canadensis*, often with it but less common
Early Goldenrod - **S. gigantea** Aiton

S. virgaurea x *S. canadensis* = **S. x niederederi** Khek has been found in W Kent, W Gloucs and Cheshire; it is closer to *S. canadensis* in inflorescence shape and capitulum size and to *S. virgaurea* in leaf venation, and is sterile.

48. ASTER L. - *Michaelmas-daisies*
Perennials; stem-leaves ovate to linear, entire, with various bases; capitula conspicuous, radiate or discoid, with yellow disc flowers and white to blue, pink or purple ligules; phyllaries in 2-several rows, herbaceous or partly membranous; pappus of 1-2 rows of hairs.

The cultivated *Michaelmas-daisies* that are found in the wild are difficult to determine due to hybridization between *A. novi-belgii* and 2 other spp. These 2 hybrids and *A. lanceolatus* appear to be the commonest taxa, and show every grade of variation from 1 parent to the other.

1 Leaves all 1-veined, linear to very narrowly elliptic, gland-spotted, not succulent; maritime cliffs. Stems erect to decumbent, to 50cm, glabrous. Native; limestone sea-cliffs; very local in W Br from S Devon to Westmorland   *Goldilocks Aster* - **A. linosyris** (L.) Bernh.
1 Leaves mostly with well-developed lateral veins, if all 1-veined then succulent, not gland-spotted but sometimes with stalked glands; widespread   2
   2 Leaves succulent, with 0-few lateral veins mostly running parallel with midrib; mostly maritime. Biennial or sometimes annual; stems erect, to 1m, glabrous. Native; salt-marshes, less often cliffs and rocks on coasts around whole BI, rare in inland saline areas   *Sea Aster* - **A. tripolium** L.
   2 Leaves not succulent, usually with normally developed lateral veins, widespread   3
3 Upper part of plant with abundant long patent hairs and shorter stalked glands. Stems erect, to 2m. Intrd-natd; on waste and rough ground; very scattered in Br, mainly S & C
Hairy Michaelmas-daisy - **A. novae-angliae** L.
3 Plant with 0 or rather sparse long patent hairs; stalked glands 0   4
   4 Upper leaves tapering to base, not clasping stem; leaves rarely >1cm wide; phyllaries ≤5mm; ligules usually white (see also *A. x salignus*). Stems erect, to 1.2m. Intrd-natd; on waste and rough ground; frequent throughout much of BI
Narrow-leaved Michaelmas-daisy - **A. lanceolatus** Willd.
   4 Upper leaves tapering to base or not, but distinctly (though often narrowly) clasping stem at base; some leaves >1cm wide; phyllaries >5mm; ligules usually coloured   5
5 Phyllaries with wide white borders in basal 1/2 and narrow ones in apical 1/2, leaving elliptic to trullate green patch in centre near apex; outer phyllaries reaching ≤1/2 as high as inner ones   6
5 Phyllaries wholly or mainly green in apical 1/2, hence appearing leafy near apex; outer phyllaries usually reaching ≥1/2 as high as inner ones   7
   6 Leaves distinctly glaucous on upperside; outer phyllaries usually reaching distinctly <1/2 as high as inner ones; plant usually ≤1m. Stems erect, to 1m. Intrd-natd; on waste and rough ground; rather rare and scattered in En, Tyrone
Glaucous Michaelmas-daisy - **A. laevis** L.
   6 Leaves not glaucous; outer phyllaries often reaching c.1/2 as high as inner ones; plant usually 1-2m. Stems erect, to 2m. Intrd-natd; on waste and rough ground; scattered in Br (*A. laevis* x *A. novi-belgii*)   *Late Michaelmas-daisy* - **A. x versicolor** Willd.
7 Middle stem-leaves mostly 2.5-5x as long as wide, conspicuously

clasping stem; outer phyllaries usually c.1/2-3/4 as high as inner ones
(see couplet 6) **A. x versicolor**
7 Middle stem-leaves mostly 4-10x as long as wide, usually very
narrowly clasping stem; outer phyllaries nearly as long as inner ones 8
    8 Outer phyllaries widest below middle, rather neatly appressed
to capitulum. Stems erect, to 1.5m. Intrd-natd; on waste and
rough ground; by far the commonest taxon in Br (*A. novi-belgii* x
A. lanceolatus)   *Common Michaelmas-daisy* - **A. x salignus** Willd.
    8 Outer phyllaries widest at or just above middle, with conspicuous
leafy apical 1/2 loosely or unevenly appressed to capitulum.
Stems erect, to 1.5m. Intrd-natd; on waste and rough ground;
scattered over Br, greatly over-recorded for *A. x salignus*
Confused Michaelmas-daisy - **A. novi-belgii** L.

49. CHRYSOCOMA L. - *Shrub Goldilocks*

Glabrous shrublets; stem-leaves linear to filiform, the edges rolled under,
entire, sessile; capitula yellow, discoid, 1 on end of each branch; phyllaries
in several rows, narrow, herbaceous with membranous margins; pappus of
1 row of hairs.

1 Leaves linear, absent from apical 1-3cm of stem below capitula;
capitula c.10-15mm across. Stems erect to ascending, to 60cm.
Intrd-natd; walls, dunes and open ground; Scillies
Shrub Goldilocks - **C. coma-aurea** L.
1 Leaves filiform, ascending to within <1cm of capitula; capitula
c.5-10mm across. Habit as *C. coma-aurea*. Intrd-casual; fairly frequent
wool-alien; scattered in En
Fine-leaved Goldilocks - **C. tenuifolia** P.J. Bergius

50. ERIGERON L. - *Fleabanes*

Annuals to perennials; stem-leaves linear to obovate, entire or toothed,
sessile or shortly petiolate; capitula 1-many per stem, radiate, with whitish
to yellow disc flowers and whitish to pink or mauve ligules; central flowers
tubular, bisexual, more numerous than peripheral female filiform flowers or
the latter 0, the outermost female flowers with obvious ligules at least as
long as tubular part; phyllaries in several rows, narrow, herbaceous with ±
membranous margins; pappus of 1 row of hairs, sometimes with an outer
row of very short hairs, the ray flowers sometimes with only very short
hairs or narrow scales.

1 Ligules ≤4mm, not or scarcely exceeding pappus. Stems erect, to
60cm. Native; barish sandy or calcareous soils, banks, walls and
dunes; locally frequent in En, Wa and CI, rare in Ir, casual in Sc
Blue Fleabane - **E. acer** L.
1 Ligules ≥4mm, exceeding pappus by ≥2mm   2
    2 Capitula very showy, 3-5cm across, with bluish-mauve or pale
mauve ligules 9-20mm; leaves succulent. Stems procumbent to
ascending, to 50cm. Intrd-natd; in rocky places and on cliffs;

S En (especially Wight), Denbs, Ayrs and CI
Seaside Daisy - **E. glaucus** Ker Gawl.
2 Capitula less showy, 1.5-3cm across, with white to pinkish (mauve in 1 alpine sp.) ligules 4-10mm; leaves not succulent 3
3 Stem procumbent to ascending; lower leaves with 1 pair of lateral lobes or teeth. Stems procumbent to ascending, to 50cm. Intrd-natd; walls, banks and stony ground; scattered in BI N to Co Armagh, Denbs and E Norfolk, especially SW En and CI
Mexican Fleabane - **E. karvinskianus** DC.
3 Stem erect; lower leaves entire or toothed, but not regularly toothed as in last 4
4 Stems rarely >20cm, with 1(-3) capitula; ligules mauve. Stems erect, to 20cm. Native; mountain rock-ledges above 800m; very rare in Scottish Highlands
Alpine Fleabane - **E. borealis** (Vierh.) Simmons
4 Stems rarely <20cm, with numerous capitula; ligules white or pale pink or blue; lowland aliens; leaves often serrate 5
5 Leaves clasping stem; pappus of ray and of disc flowers of long hairs only. Stems erect, to 75cm. Intrd-natd; on walls and in rough ground; very scattered in Br *Robin's-plantain* - **E. philadelphicus** L.
5 Leaves not clasping stem; pappus of disc flowers of long hairs and outer short scales, of ray flowers of short scales only. Stems erect, to 70(100)cm. Intrd-natd; in sandy places and rough ground; very scattered in SW En, rare casual elsewhere
Tall Fleabane - **E. annuus** (L.) Pers.

50 x 51. ERIGERON x CONYZA = X CONYZIGERON Rauschert

X C. huelsenii (Vatke) Rauschert (*E. acer* x *C. canadensis*) occurs sporadically with the parents in disturbed sandy places in S En; it is intermediate in pubescence and capitulum size (ligules pale mauve, 1-2mm), and sterile.

51. CONYZA Less. - *Fleabanes*

Annuals; stem-leaves linear to narrowly elliptic or oblanceolate, entire or toothed, sessile or shortly petiolate; capitula numerous, discoid or very inconspicuously radiate, with white to cream or pinkish flowers; central flowers tubular, bisexual; peripheral flowers female, more numerous, the outermost often with very short ligules (shorter than tubular part); phyllaries as in *Erigeron*; pappus of 1 row of hairs.

1 Phyllaries yellowish-green, glabrous to sparsely pubescent; disc flowers with 4-lobed corolla; inflorescence ± cylindrical (see also under *C. bilbaoana* below). Stems erect, to 1m. Intrd-natd; in waste and rough ground, walls, waysides and dunes on well-drained soils; common in SE En and CI, much sparser to N & W
Canadian Fleabane - **C. canadensis** (L.) Cronquist
1 Phyllaries greyish-green, pubescent to densely so; disc flowers with 5-lobed corolla; inflorescence pyramidal to corymbose 2

2   Inflorescence pyramidal; pappus yellowish-white or cream; phyllaries not or minutely red-tipped. Stems erect, to 1(2)m. Intrd-natd; in waste and rough ground in protected sunny spots; frequent in parts of London area and CI, scattered elsewhere in S En and E Ir   *Guernsey Fleabane* - **C. sumatrensis** (Retz.) E. Walker

2   Inflorescence with long lateral branches, usually subcorymbose; pappus greyish- or off-white; phyllaries usually conspicuously red-tipped. Stems erect, to 60cm. Intrd-natd; casual in waste and cultivated ground, natd in Middlesex; scattered in CI, En and Sc
Argentine Fleabane - **C. bonariensis** (L.) Cronquist

Plants similar to *C. canadensis* but more robust and hispid and with larger capitula, 5-lobed corolla in disc flowers and a subcorymbose inflorescence, found since 1992 in S En and E Ir, have been called **C. bilbaoana** J. Rémy; their true identity, and distinction from and relationship to the other spp., need further study.

C. canadensis x *C. bonariensis* was found as a single sterile intermediate plant in Middlesex in 1993.

52. CALLISTEPHUS Cass. - *China Aster*
Annuals; stem-leaves ovate, deeply toothed or lobed, the lower petiolate; capitula very conspicuous, radiate but often *flore pleno*, with yellow disc flowers when present, with white or blue to pink or purple ligules; phyllaries in several rows, the outer herbaceous and very leafy, the inner membranous-bordered; pappus of 2 rows of hairs.

1   Stems erect, to 75cm, stiffly pubescent; capitula few, up to 12cm across, very showy. Intrd-casual; garden throwout on tips and waste ground; scattered in En   *China Aster* - **C. chinensis** (L.) Nees

53. OLEARIA Moench - *Daisy-bushes*
Strong shrubs to small trees, evergreen; leaves simple, alternate or opposite, white-tomentose on lowerside, entire to sharply toothed, petiolate; capitula numerous, ± crowded in lateral or terminal corymbose panicles, radiate or discoid, with yellow to reddish disc flowers and (if present) white ligules; phyllaries in several rows, rather scarious; pappus of 1 row of hairs. See *Pittosporum* (Pittosporaceae) for differences.

1   Leaves with conspicuous acute teeth on margin. Shrub to 3(6)m. Intrd-natd; hedges, scrub, banks and rough ground; scattered in Ir, Man and W Br N to Wigtowns, Guernsey
New Zealand Holly - **O. macrodonta** Baker
1   Leaves entire   2
    2   Leaves opposite. Shrub or tree to 10m. Intrd-natd; grown as hedging in very mild areas, frequent relic, self-sown in W Cornwall and Man   *Ake-ake* - **O. traversii** (F. Muell.) Hook. f.
    2   Leaves alternate   3
3   Leaves undulate at margin; inflorescences axillary, flowering

Oct-Nov; capitula with only 1 floret. Shrub to 3(6)m. Intrd-natd;
grown in very mild areas for hedging, frequent relic, ± natd in
W Cornwall   *Akiraho* - **O. paniculata** (J.R. & G. Forst.) Druce
3 Leaves flat; inflorescences terminal, flowering Jul-Aug; capitula
  with ≥2 florets   4
  4 Leaves 1-3cm; capitula with 3-5 ray flowers. Shrub to 2(3)m.
    Intrd-natd; the hardiest and most grown sp., frequent relic,
    rarely self-sown on walls and in open ground; very scattered
    in SW and WC Br   *Daisy-bush* - **O. x haastii** Hook. f.
  4 Leaves 4-10cm; capitula with 0-2 ray flowers. Shrub to 6m.
    Intrd-natd; on dunes; Scillies
       Mangrove-leaved Daisy-bush - **O. avicenniifolia** (Raoul) Hook. f.

54. BACCHARIS L. - *Tree Groundsel*
Dioecious, deciduous shrubs; leaves simple, alternate, roughly toothed in distal 1/2, tapered to petiole; capitula ± numerous, in terminal, loose leafy panicles, small, discoid, whitish; phyllaries in several rows, herbaceous with scarious borders; pappus of 1 row of hairs, shorter in male plants.

1 Erect, ± sticky shrub to 4m; leaves obovate, glabrous; capitula in
  wide, terminal panicles, c.2mm across, white, produced in Oct.
  Intrd-surv; grown by sea in S due to salt-tolerance, persistent in
  S Hants   *Tree Groundsel* - **B. halimiifolia** L

55. BELLIS L. - *Daisy*
Herbaceous perennials; leaves all basal, in rosette, simple, toothed, petiolate; capitula single on leafless stalks, radiate, with white to pink or red ligules and yellow disc flowers, or *flore pleno*; phyllaries in 2 rows, herbaceous; pappus 0.

1 Leaves obovate, irregularly serrate; stems procumbent to erect, to
  12(20)cm, leafless, with 1 capitulum; capitula 12-25mm across, up
  to 80mm across and often *flore pleno* in cultivars. Native; mostly in
  short grassland; abundant   *Daisy* - **B. perennis** L.

TRIBE 6 - ANTHEMIDEAE (genera 56-69). Annual to perennial herbs, rarely shrubs; leaves alternate, simple to pinnate, often finely and deeply divided; capitula discoid or radiate; phyllaries in 2-several rows, herbaceous with scarious margins and apex; receptacular scales 0 or present; pappus usually 0, sometimes a short rim; usually with yellow disc flowers and white ligules but exceptions not rare.

56. TANACETUM L. - *Tansies*
Strongly aromatic perennial herbs; leaves simple and toothed to deeply pinnately lobed or pinnate; capitula radiate or discoid, rarely *flore pleno*; disc flowers yellow; ligules white or 0; receptacular scales 0; pappus a very short rim.

1 Leaves toothed, divided much <1/2 way to midrib. Stems erect, to
  1.2m; capitula discoid. Intrd-natd; garden outcast; very scattered in
  C & S Br   *Costmary* - **T. balsamita** L.
1 All or most leaves pinnate, or pinnately lobed much >1/2 way to
  midrib   2
  2 Rhizomes 0; ultimate leaf-lobes obtuse to subacute, sometimes
    apiculate. Stems erect, to 70cm; capitula usually radiate. Intrd-
    natd; on walls, waste ground and waysides; frequent
                                    Feverfew - **T. parthenium** (L.) Sch. Bip.
  2 Rhizomatous; ultimate leaf-lobes acute to acuminate   3
3 Capitula discoid, >5mm across. Stems erect, to 1.2m; capitula
  discoid. Possibly native; grassy places, waysides, rough ground;
  frequent   *Tansy* - **T. vulgare** L.
3 Capitula radiate, ≤5mm across excl. ligules; ligules white. Stems
  erect, to 1.2m; capitula radiate. Intrd-natd; grassy places and
  waysides; very scattered in En and Sc
            Rayed Tansy - **T. macrophyllum** (Waldst. & Kit.) Sch. Bip.

57. SERIPHIDIUM (Hook.) Fourr. - *Sea Wormwood*
Aromatic perennials; differ from *Artemisia* in flowers all similar and
functionally bisexual.

1 Stems decumbent to erect, woody below; leaves white-woolly,
  1-2-pinnate with linear ultimate segments. Native; dry parts of
  saltmarshes, sea-walls and rough ground by sea; local on coasts of
  Br N to C Sc, E & W Ir   *Sea Wormwood* - **S. maritimum** (L.) Polj.

58. ARTEMISIA L. - *Mugworts*
Annual to perennial herbs or small shrubs, often aromatic; leaves entire to
finely divided; capitula discoid, small, brownish overall; flowers usually
yellowish, the outer female, with filiform corolla, the inner bisexual, with
tubular corolla; receptacular scales 0; pappus 0.

1 Leaves most or all entire. Aromatic perennial to 1.2m. Intrd-surv;
  persistent on tips and waste ground; very scattered in S En
                                         Tarragon - **A. dracunculus** L.
1 Most or all leaves deeply divided   2
  2 Stems woody ± to top. Very aromatic shrub to 1.2m. Intrd-surv;
    persistent on tips and waste ground; sporadic in S & C Br.
    Rarely flowers   *Southernwood* - **A. abrotanum** L.
  2 Stems herbaceous, or woody only near base   3
3 Capitula 1-2(5); stems <10cm. Aromatic, rosette-perennial to 8cm.
  Native; at 3 sites at c.800m on barish mountain-tops in E & W Ross
                                  Norwegian Mugwort - **A. norvegica** Fr.
3 Capitula normally >10; stems >10cm   4
  4 Annual or biennial with simple root system and 0 non-flowering
    shoots   5
  4 Perennial with strong underground portion and non-flowering

58. ARTEMISIA

shoots ... 6

5 Leaves in inflorescence projecting laterally well beyond capitula, with many primary divisions >(1.5)2cm x c.1-3mm. Scarcely aromatic, erect annual (to biennial) to 1.5m. Intrd-natd; casual on waste ground and reservoir mud, ± natd in few sites in S Br
Slender Mugwort - **A. biennis** Willd.

5 Leaves in inflorescence extending laterally less far than capitula, with primary divisions <1(1.5)cm x c.0.5-1mm. Aromatic, erect annual to 1.5m. Intrd-casual; habitat as *A. biennis* but rarer and not natd; very scattered in S Br   *Annual Mugwort* - **A. annua** L.

   6 Mature leaves densely (often whitish-)pubescent on upperside ... 7

   6 Mature leaves glabrous or subglabrous on upperside (beware mildew) ... 8

7 Plant not aromatic, rhizomatous; receptacle glabrous; capitula 6-10 x 5-9mm excl. flowers. Stems to 60cm. Intrd-natd; on maritime dunes; Kirkcudbrights, Clyde Is and Ayrs
Hoary Mugwort - **A. stelleriana** Besser

7 Plant aromatic, at least when fresh, not rhizomatous; receptacle pubescent; capitula 1.5-3.5 x 3-5mm excl. flowers. Tufted perennial to 1m. Native; similar places to *A. vulgaris*; frequent in En, Wa and CI, very scattered and intrd in Sc and Ir   *Wormwood* - **A. absinthium** L.

   8 All leaf-lobes <2mm wide; plant not aromatic; achenes usually produced only by marginal flowers. Stems to 75cm. Native; grassy places by roads and on heathland; very local in W Suffolk and W Norfolk, natd in Glam   *Field Wormwood* - **A. campestris** L.

   8 Most or all leaf-lobes >2mm wide; plant aromatic, at least when fresh; achenes produced by all flowers ... 9

9 Plant not or scarcely rhizomatous; terminal untoothed portion of middle stem-leaves usually <3cm; stem with central (white) pith region occupying c.4/5 of total (white + green) pith diameter; flowers Jul-Sep. Aromatic, tufted perennial to 1.5(2)m. Native; rough ground, waste places, waysides; common throughout lowland BI   *Mugwort* - **A. vulgaris** L.

9 Plant strongly rhizomatous; terminal untoothed portion of middle stem-leaves usually >3cm; stem with central (white) pith region occupying c.1/3 of total (white + green) pith diameter; flowers Oct-Dec. Aromatic perennial to 1.5m. Intrd-natd; in similar places to *A. vulgaris*; frequent in London area, especially near R Thames, very scattered elsewhere in S & C En, Caerns, Moray, Easterness, Guernsey   *Chinese Mugwort* - **A. verlotiorum** Lamotte

A. vulgaris x *A. verlotiorum* occurs in Middlesex, S Essex and Surrey; it is intermediate in all characters (white part of pith c.3/5 total pith width) and completely sterile (flowers appear Oct-Dec but have abortive stamens).

59. SANTOLINA L. - *Lavender-cotton*

Evergreen shrubs; leaves neatly and closely pinnately lobed; capitula discoid, yellow; receptacular scales present; pappus 0.

1 Stems decumbent to suberect, to 60cm; whole plant white- to grey-tomentose; capitula 6-10mm across, solitary on erect stems. Intrd-natd; persistent on tips, rough ground, old gardens and rockeries; scattered in S & C Br *Lavender-cotton* - **S. chamaecyparissus** L.

60. OTANTHUS Hoffmanns. & Link - *Cottonweed*
Perennial, densely white-woolly herbs; leaves simple, crenate; capitula discoid with yellow flowers; phyllaries obscured by dense hairs; receptacular scales present; pappus 0.

1 Stems erect to ascending, to 30cm; leaves oblong-obovate; capitula few, subcorymbose, 6-9mm across. Native; maritime fixed sand and shingle; now restricted to 1 place in Co Wexford
Cottonweed - **O. maritimus** (L.) Hoffmanns. & Link

61. ACHILLEA L. - *Yarrows*
Perennial herbs; leaves simple and very shallowly toothed to deeply and finely dissected; capitula radiate, rarely *flore pleno*; disc flowers and ligules white to deep pink, rarely yellow; receptacular scales present; pappus 0.

1 Leaves simple, toothed much <1/2 way to midrib; capitula >1cm across. Stems erect, to 60cm. Native; damp grassy places and marshy fields; frequent in Br and most of Ir, casual in CI
Sneezewort - **A. ptarmica** L.
1 Leaves compound, or simple and divided much >1/2 way to midrib; capitula ≤1cm across 2
    2 Middle stem-leaves <3x as long as wide; with <10 pairs of primary lateral lobes. Stems erect, to 50cm. Intrd-natd; waste ground; Newport Docks (Mons) *Southern Yarrow* - **A. ligustica** All.
    2 Middle stem-leaves >3x as long as wide, with >15 pairs of primary lateral lobes 3
3 Leaves ± flat in fresh state, the primary lateral lobes ± contiguous on the rhachis; inner phyllaries >3.5mm. Stems erect, to 1.3m. Intrd-natd; grassy places; Derbys and MW Yorks
Tall Yarrow - **A. distans** Waldst. & Kit. ex Willd.
3 Leaves with lobes spreading in 3 dimensions in fresh state, the primary lateral lobes separated by a length of winged rhachis; inner phyllaries ≤3.5mm. Stems erect, to 80cm. Native; grassland (usually short), banks and waysides; very common *Yarrow* - **A. millefolium** L.

62. CHAMAEMELUM Mill. - *Chamomile*
Annual or perennial herbs; leaves deeply and finely dissected; capitula radiate, rarely discoid; disc flowers yellow, with short pouch at base of tube; ligules white; receptacular scales present; pappus 0.

1 Strongly aromatic; stems procumbent to ascending, to 30cm; receptacular scales oblong to narrowly obovate, acuminate. Native; short grassy places on sandy soils; locally frequent in CI, S Br and

SW Ir, intrd N to C Sc   *Chamomile* - **C. nobile** (L.) All.

63. ANTHEMIS L. - *Chamomiles*
Aromatic annual to perennial herbs; leaves deeply and finely dissected; capitula radiate, rarely discoid; disc flowers yellow; ligules white or yellow; receptacular scales present; pappus 0 or a short rim.

1 Ligules yellow, occasionally 0; achenes distinctly compressed. Biennial or perennial to 50cm. Intrd-natd; waste places, rough and marginal land; rather frequent in S & C Br, rare in N
Yellow Chamomile - **A. tinctoria** L.
1 Ligules white, very rarely 0; achenes not or scarcely compressed   2
  2 Receptacular scales only on inner (upper) part of receptacle, linear-subulate; achenes tuberculate on ribs; fresh plant with unpleasant scent. Annual to 50cm. Native; similar habitats to *A. arvensis* but often on heavier soils; locally frequent in S & C Br, rare and mainly casual in N Br and CI, natd in S & E Ir
Stinking Chamomile - **A. cotula** L.
  2 Receptacular scales all over receptacle, at least the inner ones lanceolate to oblanceolate; achenes ribbed or scarcely so, but not tuberculate; fresh plant with sweet scent   3
3 Perennial, often woody near base and with non-flowering shoots; at least outer receptacular scales 3-toothed; achenes not or slightly ribbed. Stems to 60cm. Intrd-natd; rough and marginal ground and on cliffs, mostly near sea; very scattered in S Br
Sicilian Chamomile - **A. punctata** Vahl
3 Annual or biennial, not woody at base and usually without non-flowering shoots; receptacular scales with single slender apex; achenes strongly ribbed. Stems to 50cm. Native; arable land, waste places and rough ground, usually on calcareous soils, also a grass-seed alien; locally frequent in S & C Br, rare and mainly casual in N Br and CI   *Corn Chamomile* - **A. arvensis** L.

63 x 68. ANTHEMIS x TRIPLEUROSPERMUM = X TRIPLEUROTHEMIS Stace

X T. maleolens (P. Fourn.) Stace (*A. cotula* x *T. inodorum*) has been found in Berks and Salop; it is intermediate in the irregular presence of receptacular scales that are intermediate between those of the *Anthemis* parent and the phyllaries, and in the sterile achenes with intermediate rib development and traces of subapical oil-glands.

64. CHRYSANTHEMUM L. - *Crown Daisies*
Annual herbs; leaves simple, shallowly to deeply lobed; capitula radiate; disc flowers yellow; ligules yellow, cream or yellow and cream; receptacular scales 0; pappus 0.

1 Leaves glaucous; ligules yellow; achenes 2.5-3mm, not winged.

Stems decumbent to erect, to 60cm. Intrd-natd; weed of arable fields, waste places and waysides; locally frequent

Corn Marigold - **C. segetum** L.

1 Leaves green; ligules cream, yellow, or cream and yellow; achenes 3-3.5mm, the inner with adaxial wing, the marginal with 2 lateral and 1 adaxial wings. Stems ascending to erect, to 80cm. Intrd-casual; similar places to *C. segetum* but much rarer; very scattered in En and Wa *Crown Daisy* - **C. coronarium** L.

65. LEUCANTHEMELLA Tzvelev - *Autumn Oxeye*

Perennial herbs; leaves simple, sharply serrate; capitula radiate; disc flowers yellow; ligules white; receptacular scales 0; pappus ± 0.

1 Stems erect, to 1.5m; resembles *Leucanthemum x superbum* but leaves paler green and more sharply and deeply serrate, flowers later (Sep-Oct), and tubular part of corolla of ray flowers unwinged. Intrd-natd; rough ground and by ditches and ponds; scattered in S En

Autumn Oxeye - **L. serotina** (L.) Tzvelev

66. LEUCANTHEMUM Mill. - *Oxeye Daisies*

Differ from *Leucanthemella* in tubular part of ray flowers with 2 narrow translucent wings; and achenes with translucent secretory canals.

1 Basal and lower stem-leaves obovate-spathulate, abruptly contracted to broadly cuneate base; upper stem-leaves usually deeply serrate; capitula 2.5-6(7.5)cm across. Stems erect to ascending, to 75cm. Native; grassy places, especially on rich soils; common

Oxeye Daisy - **L. vulgare** Lam.

1 Basal and lower stem-leaves elliptic-oblong, gradually contracted to narrowly cuneate base; upper stem-leaves usually shallowly serrate to subentire; capitula (5)6-10cm across, often *flore pleno*. Stems erect to ascending, to 1.2(1.5)m. Intrd-natd; waste and rough ground and grassy waysides; scattered throughout Br and CI (*L. lacustre* (Brot.) Samp. x *L. maximum* (Ramond) DC.)

Shasta Daisy - **L. x superbum** (J.W. Ingram) D.H. Kent

67. MATRICARIA L. - *Mayweeds*

Annual herbs, differing from *Tripleurospermum* in much more conical, hollow (not solid) receptacle; ligules often 0; and achenes with 4-5 weak (not 3 strong) ribs and without (not with) oil-glands.

1 Superficially much like *Tripleurospermum inodorum* but usually more strongly and sweetly scented when fresh; phyllaries with very pale brown (not deep brown) scarious margins; ligules soon very strongly reflexed. Native; in similar places to and often with *T. inodorum* but less common and more restricted to arable ground on light soils; locally common in CI, En and Wa, rare in Sc and Ir

Scented Mayweed - **M. recutita** L.

67. MATRICARIA

1 Plant erect, to 35cm; ligules 0; differs from rare ligule-less plants of
M. recutita in sweet pineapple-like scent, much wider, white scarious
margins to phyllaries, and disc flowers with 4-lobed (not 5-lobed)
corolla. Intrd-natd; weed of barish places by paths and waste
places; common   *Pineappleweed* - **M. discoidea** DC.

68. TRIPLEUROSPERMUM Sch. Bip. - *Mayweeds*
Annual to perennial herbs; leaves deeply and finely dissected; capitula
radiate, rarely discoid; disc flowers yellow; ligules white; receptacular scales
0; pappus a very short rim.

1 Erect to procumbent (biennial to) perennial to 60cm; achenes 1.8-
3.5mm, with 3 strong ribs ± touching laterally on 1 face, with 2
subapical distinctly elongated oil-glands on opposite face. Native;
sand, shingle, rocks, walls, cliffs and waste ground near sea; locally
common round most coasts of BI
Sea Mayweed - **T. maritimum** (L.) W.D.J. Koch
1 Erect to ascending annual to 60cm; achenes 1.3-2.2mm, with 3 strong
ribs on 1 face separated by 2 distinct granular areas, with 2 subapical
orbicular to angular oil-glands on opposite face. Native; waste,
rough and cultivated land; common
Scentless Mayweed - **T. inodorum** (L.) Sch. Bip.

T. maritimum x *T. inodorum* is intermediate in leaf and achene characters and
is ≥80 per cent fertile (with backcrossing occurring); it is not infrequent in
coastal areas.

69. COTULA L. - *Buttonweeds*
Annual to perennial herbs: leaves entire to deeply pinnately divided;
capitula discoid, bisexual or dioecious, yellow or white, with pedicellate
flowers; in bisexual capitula outer flowers are female with 0 or minute
corolla, inner ones bisexual with 4-lobed corolla; in dioecious capitula males
and females both with minutely 4-lobed corolla; receptacular scales 0;
pappus 0.

1 Leaves entire to very irregularly pinnately lobed with usually <6
lobes, ± succulent; capitula 8-12mm across, bright yellow. Glabrous
annual to perennial with procumbent to ascending, often rooting
stems to 30cm. Intrd-natd; wet, usually saline places; Cheshire,
MW & SW Yorks, S Hants and W Cork, rare casual elsewhere
Buttonweed - **C. coronopifolia** L.
1 Leaves regularly pinnately (to 2-pinnately) lobed with usually ≥6
lobes, not succulent; capitula 3-10mm across, white or dull yellow   2
  2 Annual; capitula bisexual, white, the female (outer) flowers
with 0 corolla; phyllaries not purple-tinged. Stems to 15cm,
suberect to decumbent, pubescent. Intrd-natd; rough or arable

ground; very scattered in En, natd in S Devon

Annual Buttonweed - **C. australis** (Spreng.) Hook. f.
2 Procumbent perennial with rooting stems; capitula dioecious, dull yellow, the female flowers with corolla; phyllaries strongly purple-tinged 3
3 Leaves with oblong-triangular abruptly apiculate teeth or shallow lobes. Stems to 20cm, rooting along length. Intrd-surv; established in mown lawns, often not flowering; very scattered in BI

Hairless Leptinella - **C. dioica** (Hook. f.) Hook. f.
3 Leaves lobed nearly to midrib, the lobes with lanceolate, acute to acuminate teeth. Stems to 20cm, rooting along length. Intrd-natd; lawns and roadside grassland; scattered in Br and Ir, especially Sc

Leptinella - **C. squalida** (Hook. f.) Hook. f.

TRIBE 7 - SENECIONEAE (genera 70-80). Annual to perennial herbs, sometimes shrubs, rarely weak climbers; leaves alternate or all basal; capitula discoid or radiate; phyllaries usually in 1 or 2 rows, often in 1 main row and 1 much shorter row, herbaceous; receptacular scales 0; pappus of 1-many rows of simple hairs; corolla most often yellow in both ray and disc flowers.

70. SENECIO L. - *Ragworts*
Annual to perennial herbs, rarely shrubby; leaves alternate, pinnately veined; capitula discoid or radiate; phyllaries in 1 main row with short supplementary ones at base of capitulum; corolla of disc flowers yellow, of ray flowers usually yellow (rarely white or purple).

1 Stems woody at least towards base 2
1 Stems entirely herbaceous 6
2 Ligules purple. Erect perennial to 1m. Intrd-natd; in open woodland and on rubbish tip; Scillies

Woad-leaved Ragwort - **S. glastifolius** L. f.
2 Ligules yellow 3
3 Stems and leaves ± glabrous 4
3 At least leaf lowersides grey- or white-pubescent 5
4 Leaves linear, ± entire. Spreading perennial to 80cm. Intrd-natd; natd on sandy beach in E Kent, rare casual in waste places elsewhere *Narrow-leaved Ragwort* - **S. inaequidens** DC.
4 Leaves conspicuously toothed to deeply lobed. Erect to ascending annual to perennial to 50cm. Intrd-natd; waste ground, walls and waysides; common in En and Wa, local elsewhere *Oxford Ragwort* - **S. squalidus** L.
5 Leaves lobed ± to midrib, with obtuse to rounded lobes; phyllaries tomentose. Spreading woody perennial to 1m. Intrd-natd; on cliffs and rough ground mostly near sea; S & SW En, Wa, CI, Co Dublin

Silver Ragwort - **S. cineraria** DC.
5 Leaves serrate, acuminate at apex; phyllaries glabrous. Erect perennial to 1.5m (usually much less). Intrd-casual; occasional

wool-alien in fields and waste land; very scattered in En
Shoddy Ragwort - **S. pterophorus** DC.
- 6 Ligules white or purple 7
- 6 Ligules yellow to orange, or 0 8
- 7 Leaves with linear lobes, divided >1/2 way to midrib; ligules usually purple. Erect perennial to 1.5m. Intrd-natd; rough ground; Guernsey
Purple Ragwort - **S. grandiflorus** P.J. Bergius
- 7 Leaves very shallowly and irregularly toothed; ligules white. Erect perennial to 1m. Intrd-natd; grassy places and by streams; extreme N Caithness, Orkney, Shetland *Magellan Ragwort* - **S. smithii** DC.
  - 8 Leaves simple, entire to shallowly toothed 9
  - 8 At least some leaves lobed ≥1/2 way to midrib 15
- 9 Ligules 4-8 10
- 9 Ligules >8 12
  - 10 Middle and upper stem-leaves shortly but distinctly petiolate. Erect perennial to 1.5m. Intrd-natd; in damp shady places; MW Yorks and S & W Lancs
Wood Ragwort - **S. ovatus** (P. Gaertn., B. Mey. & Scherb.) Willd.
  - 10 Middle and upper stem-leaves sessile 11
- 11 Leaf-teeth divergent, often obtuse; phyllaries and peduncles usually glabrous. Erect perennial to 1.5m. Intrd-natd; by streams and in wet meadows; Kirkcudbrights, S Ebudes, Offaly, Man
Golden Ragwort - **S. doria** L.
- 11 Leaf-teeth with acute, ± incurved apex; phyllaries and peduncles pubescent. Intrd-natd; by streams and ponds and in fens and swampy ground; scattered in Br and Ir N to C Sc
Broad-leaved Ragwort - **S. fluviatilis** Wallr.
  - 12 Leaves linear, <5mm wide (see couplet 4) **S. inaequidens**
  - 12 Leaves ovate or lanceolate to oblanceolate, most >5mm wide 13
- 13 Capitula 1-3(4); phyllaries 10-15mm. Erect perennial to 60cm. Intrd-natd; on river banks; M Perth *Chamois Ragwort* - **S. doronicum** (L.) L.
- 13 Capitula numerous; phyllaries 6-10mm 14
  - 14 Phyllaries conspicuously black-tipped; leaves ± glabrous
(see couplet 4) **S. squalidus**
  - 14 Phyllaries not black-tipped; leaves pubescent on lowerside. Erect perennial to 1.5(2)m. Native; fenland ditches; 1 place in Cambs *Fen Ragwort* - **S. paludosus** L.
- 15 Glandular hairs present at least on peduncles, often also on leaves and stems 16
- 15 Glandular hairs 0 17
  - 16 Achenes glabrous; supplementary phyllaries at base of capitulum 1/3-1/2 as long as main ones; plant sticky. Erect annual to 60cm. Possibly native; waste and rough ground, railway tracks, roadsides, walls; frequent in most of Br, intrd and local in CI, N & C Sc and Ir *Sticky Groundsel* - **S. viscosus** L.
  - 16 Achenes minutely pubescent; supplementary phyllaries ≤1/4(1/3) as long as main ones; plant not or scarcely sticky. Erect annual to 70cm. Native; open ground on heaths, banks and sandy

places; locally common *Heath Groundsel* - **S. sylvaticus** L.
17 Biennial to perennial, firmly rooted, usually with very short thick rhizome; phyllaries without black tips 18
17 Annual to perennial, usually easily uprooted, without rhizome; at least supplementary phyllaries with black tips 20
    18 Supplementary phyllaries at base of capitulum c.1/2 as long as main ones; all achenes shortly pubescent; leaves grey-pubescent on lowerside. Erect perennial to 1.2m. Native; grassy places, banks, waysides and field-borders, usually on base-rich soils; common in most of En and Wa, very local in E Ir, rare alien in Sc *Hoary Ragwort* - **S. erucifolius** L.
    18 Supplementary phyllaries c.1/4-2/5 as long as main ones; achenes of ray flowers glabrous; leaves glabrous to sparsely pubescent on lowerside 19
19 Achenes of disc flowers pubescent; stem-leaves with several pairs of lateral lobes and terminal lobe not much larger; corymbs dense. Erect perennial to 1.5m (often much shorter). Native; grassland, waysides, waste ground, sand-dunes; common
Common Ragwort - **S. jacobaea** L.
19 Achenes of disc flowers glabrous to sparsely pubescent; stem-leaves with 1-few pairs of lateral lobes and terminal lobe much larger; corymbs lax. Erect biennial to perennial to 80cm. Native; marshes, damp meadows and streamsides; common in much of W BI, scattered further E *Marsh Ragwort* - **S. aquaticus** Hill
    20 Ligules <8mm or 0; capitula (excl. ligules) cylindrical in flower, c.2x as long as wide 21
    20 Ligules usually ≥8mm, rarely shorter or 0; capitula (excl. ligules) bell-shaped in flower, <1.5x as long as wide 22
21 Ligules usually 0; achenes ≤2.5mm; pollen grains 20-25 microns across, 3-pored. Usually erect annual to 30(45)cm. Native; open and rough ground; very common *Groundsel* - **S. vulgaris** L.
21 Ligules usually present; achenes >3mm; pollen grains 30-36 microns across, mostly 4-pored. Erect annual to 30(50)cm. Native; waste ground and waysides; Flints, Denbs, Salop and Midlothian
Welsh Groundsel - **S. cambrensis** Rosser
    22 Leaves ± flat, usually with lateral lobes much longer than width of central undivided portion, usually glabrous or nearly so
(see couplet 4) **S. squalidus**
    22 Leaves usually undulate, with lateral lobes c. as long as width of central undivided portion, usually conspicuously pubescent. Erect annual to 50cm. Intrd-casual; road-verges and newly landscaped areas; sporadic in Br and Man
Eastern Groundsel - **S. vernalis** Waldst. & Kit.

Hybrids occur in 7 combinations, but only *S. jacobaea* x *S. aquaticus* = ***S. x ostenfeldii*** Druce, which is partially fertile and backcrosses, and may be commoner than either parent in parts of Ir and W Br, is more than sporadic. *S. cineraria* x *S. jacobaea* = ***S. x albescens*** Burb. & Colgan is also fertile and backcrosses. *S. squalidus* x *S. vulgaris* = **S. x baxteri** Druce is rare and highly sterile but it gave rise to *S. cambrensis* and apparently produces radiate *S. vulgaris* by introgression.

71. PERICALLIS D. Don - *Cineraria*
Annual to perennial herbs; leaves alternate, palmately veined; capitula radiate; phyllaries all in 1 main row; colour of disc flowers and ligules usually contrasting, the former darker, blue, red or pink to purple, never yellow.

1 Stems erect, pubescent, to 80cm; leaves petiolate, palmately lobed; capitula in ± dense corymbose masses, 1.5-4cm across (incl. ligules). Intrd-natd; open ground, walls and waysides; Scillies, rarely on mainland W Cornwall *Cineraria* - **P. hybrida** B. Nord.

72. DELAIREA Lem. - *German-ivy*
Trailing or climbing, ± glabrous woody perennial; leaves alternate, palmately veined; capitula discoid; phyllaries in 1 main row with short supplementary ones at base of capitulum; disc flowers yellow.

1 Stems 3m; leaves succulent, palmately lobed; capitula numerous in dense axillary and terminal panicles, produced in Nov. Intrd-natd; clambering over hedges and walls; CI and Scillies, rarely mainland E & W Cornwall *German-ivy* - **D. odorata** Lem.

73. TEPHROSERIS (Rchb.) Rchb. - *Fleaworts*
(Biennial to) perennial herbs; leaves alternate, pinnately veined; capitula radiate; phyllaries all in 1 main row; disc flowers and ligules yellow.

1 Basal leaves usually withered before flowering; stem-leaves very numerous, lanceolate, sessile, dentate, ± clasping stem at base; capitula often >12, 2-3cm across; ligules c.21. Erect, densely pubescent perennial to 1m. Native; fen ditches; formerly local in E En, extinct since 1899 *Marsh Fleawort* - **T. palustris** (L.) Fourr.
1 Basal leaves oblong-ovate, petiolate, entire to coarsely dentate; stem-leaves much smaller, lanceolate, sessile, rarely >10; capitula ≤12(15), 1.5-2.5cm across; ligules 12-15. Erect, densely pubescent perennial. Native; short natural grassland
(*Field Fleawort* - **T. integrifolia** (L.) Holub) 2
  2 Stems to 30(40)cm; leaves entire to denticulate; stem-leaves usually ≤6; capitula rarely >6; phyllaries 6-8.5mm. On chalk and limestone; local in S En N to Cambs and E Gloucs
T. integrifolia ssp. integrifolia
  2 Stems to 60(90)cm; leaves usually dentate; stem-leaves often >6; capitula often >6; phyllaries 8-12mm. On glacial drift on sea-cliffs; extremely local on Holyhead Island (Anglesey)
T. integrifolia ssp. maritima (Syme) B. Nord.

74. BRACHYGLOTTIS J.R. & G. Forst. - *Ragworts*
Evergreen shrubs; leaves alternate, pinnately veined, densely white-felted on lowerside; capitula discoid or radiate; phyllaries in 1 main row with short supplementary ones at base of capitulum; disc flowers cream or yellow; ligules yellow.

1 Many leaves >8cm, distantly sinuate-lobed; capitula <5mm across, cream; ligules 0; phyllaries with woolly hairs only at base. Shrub or small tree to 6m. Intrd-surv; used as hedging in Scillies, often long persistent after neglect   *Hedge Ragwort* - **B. repanda** J.R. & G. Forst.
1 Leaves <8cm, entire to denticulate or tightly undulate; capitula >1cm across, yellow; ligules conspicuous; phyllaries with woolly hairs along ± whole length   2
  2 Leaves <4cm, tightly crenate-undulate. Spreading shrub to 1m. Intrd-natd; persistent and ± natd in Man and on dunes near Llandudno (Caerns)
Monro's Ragwort - **B. monroi** (Hook. f.) B. Nord.
  2 Many leaves >4cm, entire to remotely denticulate. Spreading shrub to 1(2)m. Intrd-natd; persistent on rough ground; scattered in Br and Man N to C Sc   *Shrub Ragwort* - **B. 'Sunshine'**

75. SINACALIA H. Rob. & Brettell - *Chinese Ragwort*
Rhizomatous, ± glabrous herbaceous perennials; leaves alternate, pinnately veined; capitula radiate, with 3 or 4 disc and 3 or 4 ray flowers; phyllaries all in 1 main row, but with small bracts some way below base of capitulum; disc flowers and ligules yellow.

1 Stems erect, to 2m, unbranched except in inflorescence; leaves up to 20cm, ovate, deeply pinnately lobed; capitula numerous in large terminal panicles. Intrd-natd; damp shady places; scattered in N Wa, Man, N En and C Sc   *Chinese Ragwort* - **S. tangutica** (Maxim.) B. Nord.

76. LIGULARIA Cass. - *Leopardplants*
Herbaceous perennials; leaves mostly basal, those on stems alternate, cordate at base, palmately veined, with sheathing petiole bases; capitula radiate; phyllaries all in 1 row; disc flowers brownish-yellow; ligules yellow to orange.

1 Basal leaves reniform, dentate; capitula 4-10cm across, several in subcorymbose terminal cluster, with numerous disc flowers and 10-15 orange ligules 15-40mm. Stems erect, to 1.2m. Intrd-natd; damp or shady places; very scattered in En and Sc, especially N
Leopardplant - **L. dentata** (A. Gray) H. Hara
1 Basal leaves deeply palmately lobed, the lobes lobed or toothed; capitula 1.5-3cm across, numerous in long narrow terminal raceme-like panicle, with usually only 3 disc flowers and 2 yellow ligules 6-15mm. Stems erect, to 1.8m. Intrd-natd; persistent by R. Tyne (S Northumb)   *Przewalski's Leopardplant* - **L. przewalskii** (Maxim.) Diels

77. DORONICUM L. - *Leopard's-banes*
Rhizomatous herbaceous perennials; leaves alternate, ± palmately veined, the basal ones ± withered by flowering time; capitula radiate; phyllaries in 2 rows of equal length; disc flowers and ligules yellow.

1 Basal leaves all cuneate at base. Stems erect, to 1m, with only short glandular hairs. Intrd-natd; woods and shady places; scattered in Br **Plantain-leaved Leopard's-bane - D. plantagineum** L.
1 Most or all basal leaves cordate to rounded or truncate at base 2
   2 Petioles of basal leaves with many long (≥1mm) flexuous or patent hairs; capitula usually 3-8 per stem. Stems erect, to 80cm, with long eglandular and short glandular hairs. Intrd-natd; woods and shady places; frequent throughout Br
Leopard's-bane - **D. pardalianches** L.
   2 Petioles of basal leaves with 0-very few long hairs; capitula 1-2(3) per stem 3
3 Basal leaves deeply cordate at base; all hairs on stems short (<1mm) and glandular. Stems erect, to 60cm. Intrd-natd; by path and on bank of reservoir; Surrey *Eastern Leopard's-bane* - **D. columnae** Ten.
3 Basal leaves shallowly cordate to truncate or rounded at base; stems usually with a few long (≥1mm) eglandular as well as short glandular hairs 4
   4 Basal leaves acute, mainly shallowly cordate at base, with prominent teeth >2mm. Stems erect, to 1m. Intrd-natd; in woods and shady places; scattered in Br
Harpur-Crewe's Leopard's-bane - **D. x excelsum** (N.E. Br.) Stace
   4 Basal leaves obtuse, mainly rounded to truncate at base, with less prominent teeth <2mm. Stems erect, to 1m. Intrd-natd; in woods and shady places; scattered in Br
Willdenow's Leopard's-bane - **D. x willdenowii** (Rouy) A.W. Hill

78. TUSSILAGO L. - *Colt's-foot*
Rhizomatous, herbaceous perennials; leaves all basal, cordate at base, ± palmately veined, cottony-pubescent on lowerside; flowering stems bearing many bracts and 1 terminal capitulum, cottony-pubescent; capitula radiate; phyllaries all in 1 row; disc flowers and ligules yellow.

1 Stems erect, to 15cm, appearing before leaves; leaves broadly ovate, 20-30cm across, shallowly palmately lobed, the lobes dentate to denticulate; capitula 1.5-3.5cm across. Native: open or semi-open or disturbed ground in many habitats, including arable land and maritime sand and shingle; common *Colt's-foot* - **T. farfara** L.

79. PETASITES Mill. - *Butterburs*
Dioecious, rhizomatous, herbaceous perennials; leaves all basal, cordate at base, ± palmately veined, cottony-pubescent on lowerside; flowering stems bearing few to many bracts and a terminal raceme or panicle of capitula, cottony-pubescent; male capitula composed of male flowers with clavate

sterile stigmas and sometimes some female-like sterile discoid or radiate flowers; female capitula composed of discoid female flowers and 1 or few central male-like sterile flowers; flowers white to purple or cream.

1 Marginal flowers ligulate (ligules <1cm); inflorescences appearing Nov-Feb, with basal leaves present, always male; leaves with small teeth all of 1 size. Leaves up to 20cm across, not lobed; petioles up to 30cm; flowering stems erect, to 30cm. Intrd-natd; waste and rough ground and waysides; throughout BI, common in S, local in C & N
Winter Heliotrope - **P. fragrans** (Vill.) C. Presl
1 Marginal flowers tubular; inflorescences appearing Feb-Apr, before basal leaves, male or female; leaves unevenly dentate, with large teeth or short lobes dispersed among small teeth 2
   2 Basal leaf sinus with parallel or divergent sides, bordered by 0-1 veins on each side; flowers pure white; leaves ≤30cm across. Leaves with well-developed acute lobes; petioles up to 30cm; flowering stems erect, to 30cm (to 70cm in fruit). Intrd-natd; rough ground, waysides and woods; throughout Br, rare in S, frequent in N, N Ir   *White Butterbur* - **P. albus** (L.) Gaertn.
   2 Basal leaf sinus with convergent sides, bordered by ≥2 veins on each side; flowers usually cream or with purplish tinge; leaves often >30cm across 3
3 Leaves distinctly but very shallowly lobed; mature inflorescences ± cylindrical; upper part of stem below inflorescence with bracts <1cm wide; phyllaries and/or florets usually with anthocyanin; corollas white to purple-tinged. Leaves ≤90cm across; petioles up to 1.5m; flowering stems erect, to 30cm (to 1m in fruit). Native; by rivers and ditches, in damp fields and waysides; male plant frequent throughout most of BI; female plant frequent in N & C En, sporadic elsewhere   *Butterbur* - **P. hybridus** (L.) P. Gaertn., B. Mey. & Scherb.
3 Leaves scarcely or not lobed, unevenly dentate; mature inflorescences ± hemisperical; upper part of stem below inflorescence with bracts >1cm wide; anthocyanin absent; corollas cream in male capitula, whitish in female ones. Leaf-size as in *P. hybridus*; flowering stems erect, to 30cm. Intrd-natd; by rivers and in damp places; scattered throughout Br. Female plants rarely or ? never natd here
Giant Butterbur - **P. japonicus** (Siebold & Zucc.) Maxim.

80. HOMOGYNE Cass. - *Purple Colt's-foot*

Rhizomatous, herbaceous perennials; leaves all basal, cordate at base, palmately veined, rather sparsely pubescent on lowerside; flowering stems bearing few bracts and 1 terminal capitulum, cottony-pubescent; capitula discoid, the central flowers bisexual with 5-lobed corolla, the outermost row female with obliquely truncate tubular corolla; phyllaries ± in 1 row; flowers purple.

1 Stems erect, to 35cm, appearing with leaves; leaves reniform-orbicular, up to 4cm across, shallowly crenate-dentate; capitula

80. HOMOGYNE

10-15mm across. Probably intrd-natd; 1 locality in Angus at c.600m
Purple Colt's-foot - **H. alpina** (L.) Cass.

TRIBE 8 - CALENDULEAE (genus 81). Annual to perennial herbs with distinctive scent; leaves alternate, simple, ± sessile; capitula radiate; phyllaries in 1-2 rows, herbaceous with scarious margin; receptacular scales 0; achenes varying in one capitulum in size, degree of curving and wartiness, all with pappus 0; flowers yellow to orange, often *flore pleno*.

81. CALENDULA L. - *Marigolds*

1 Capitula 4-7cm across; ligules c.2x as long as phyllaries. Perennial, often behaving as annual; stems erect to procumbent, to 80cm. Intrd-natd; escape or throwout on tips and waste ground; scattered in BI, rarely natd in S   *Pot Marigold* - **C. officinalis** L.
1 Capitula 1-2.5cm across; ligules <2x as long as phyllaries. Erect to procumbent annual to 30cm. Intrd-natd; weed of cultivated ground in Guernsey and Scillies, rare casual elsewhere in S Br
Field Marigold - **C. arvensis** L.

TRIBE 9 - HELIANTHEAE (genera 82-97). Annual to perennial herbs; leaves often opposite, sometimes alternate; capitula discoid or radiate; phyllaries in 1-several rows, all herbaceous or the inner scarious; receptacular scales usually present, sometimes 0; pappus 0 or minute, sometimes of barbed bristles; corolla yellow to brown in disc flowers, yellow to brown, orange, red, purple or white in ray flowers. *Ambrosia* and *Xanthium* have unusual aberrant floral structure (q.v.).

82. AMBROSIA L. - *Ragweeds*
Annuals or perennials; leaves all opposite or the upper ones alternate, variously deeply lobed, often nearly pinnate; phyllaries 5-12, ± herbaceous, fused proximally; capitula monoecious, <5mm across, discoid, the male in dense elongated terminal racemes without subtending bracts, the female just below in axils of leaf-like bracts and with only 1 flower, greenish; receptacle flat, with scales; pappus 0.

1 Leaves palmately 3-5-lobed or the upper ones not lobed, all opposite; larger leaf-lobes >1cm wide. Stems erect, to 2.5m. Intrd-casual; tips and waste ground; scattered in Br   *Giant Ragweed* - **A. trifida** L.
1 Leaves pinnately divided nearly to base, the upper ones alternate; leaf-lobes <1cm wide   2
   2 Annual; female phyllaries in fruit with small erect spines near or above the middle. Stems erect, to 1m. Intrd-casual; waste ground and tips; fairly frequent in En, Wa and CI
Ragweed - **A. artemisiifolia** L.
   2 Perennial, with stems arising from creeping roots; female phyllaries in fruit tuberculate, without spines. Stems erect, to 1m. Intrd-natd; waste places, rough ground and dunes, natd

locally (especially Ayrs and S & W Lancs); scattered in Br
Perennial Ragweed - **A. psilostachya** DC.

83. IVA L. - *Marsh-elder*

Annuals; leaves opposite below, alternate above, simple, ovate, petiolate; phyllaries 5, ± herbaceous, with 5 similar but membranous receptacular scales immediately within; capitula <5mm across, discoid, with 5 outer female and 8-20 inner male flowers, greenish; receptacle flat, with scales; pappus 0.

1 Stems erect, to 1(2)m; leaves and inflorescences ± resembling those of a *Chenopodium*, but ultimate clusters are ± sessile capitula. Intrd-casual; waste places and rough ground, sometimes semi-natd; very scattered in S Br, mainly SE *Marsh-elder* - **I. xanthiifolia** Nutt.

84. XANTHIUM L. - *Cockleburs*

Annuals; leaves all alternate, scarcely to very deeply pinnately or palmately lobed; capitula monoecious, <1cm across, discoid, the male more apical with numerous, free subherbaceous phyllaries in 1(-3) rows, many flowers, and a cylindrical receptacle with scales, the female with phyllaries fused into an ellipsoid bur around 2 flowers, the bur developing conspicuous hooked spines in fruit; pappus 0.

1 Spines 0 at base of each leaf; leaf-blades with undivided portions >3cm wide; fruiting burs with 2 straight or curved apical beaks. Often strong-smelling; stems erect to decumbent, to 1m, scabrid. Intrd-casual; tips and waste ground, sometimes ± natd; scattered in En and Wa *Rough Cocklebur* - **X. strumarium** L.
1 Strong 3-pronged spines present at base of each leaf; leaf-blades with undivided portions <2cm wide; fruiting burs with 1 straight apical beak; stems smooth but often pubescent 2
  2 Leaves sessile or with petioles <1cm, simple or with 1-2 pairs of lobes. Stems erect to decumbent, to 1m. Intrd-casual; tips and waste ground, sometimes ± natd; scattered in En and Wa
  Spiny Cocklebur - **X. spinosum** L.
  2 Leaves with petioles >2cm, pinnately divided nearly to midrib, the larger lobes toothed or lobed. Stems erect to decumbent, to 1m. Intrd-casual; tips and waste ground, sometimes ± natd; scattered in En *Argentine Cocklebur* - **X. ambrosioides** Hook. & Arn.

85. GUIZOTIA Cass. - *Niger*

Annuals; leaves mostly opposite, alternate above, simple, sessile; phyllaries in 2 dissimilar rows, the outer longer and wider, herbaceous, the inner scarious; capitula radiate; receptacle convex, with scales; pappus 0; ligules c.8, yellow.

1 Stems erect, to 1(2)m; capitula 2-4cm across, incl. ligules 1-2cm.

Intrd-casual; tips and waste ground; scattered in Br and CI
Niger - **G. abyssinica** (L. f.) Cass.

86. SIGESBECKIA L. - *St Paul's-worts*
Annuals; leaves all opposite, simple, with winged petioles; phyllaries in 2 dissimilar rows, the outer much longer and narrower, with many stalked glands, herbaceous; capitula ≤1cm across, radiate; receptacle flat or slightly convex, with scales; pappus 0; ligules numerous, yellow.

1 Leaves triangular-hastate, irregularly dentate or lobed; petioles winged distally, the wings tapering proximally and ± absent at base. Stems to 1.2m, erect. Intrd-casual; occasional wool-alien; scattered in C & S En *Eastern St Paul's-wort* - **S. orientalis** L.
1 Leaves ovate, cordate to broadly cuneate at base, shallowly crenate or serrate; petioles winged to base, the wings tapering proximally but widened at base and clasping stem. Stems to 1.2m, erect. Intrd-natd; waste and cultivated ground; natd in S Lancs, occasional casual elsewhere in En *Western St Paul's-wort* - **S. serrata** DC.

87. RUDBECKIA L. - *Coneflowers*
Perennials; leaves alternate, simple to deeply lobed; phyllaries in 2 or more rows, herbaceous; capitula radiate; receptacle conical, with scales partly enclosing achenes; pappus 0 or a short rim; ligules numerous, yellow to orange.

1 Stems erect, to 80cm; leaves simple, roughly pubescent, entire or nearly so; capitula 5-10cm across; disc flowers brownish-purple. Intrd-casual; persistent in rough ground and waste places; very scattered in S Br *Black-eyed-Susan* - **R. hirta** L.
1 Stems erect, to 3m; leaves deeply divided, the lowest ± pinnate, glabrous or nearly so; capitula 7-14cm across; disc flowers greenish-yellow. Intrd-natd; waste and rough ground; scattered in Br, mainly S & C *Coneflower* - **R. laciniata** L.

88. HELIANTHUS L. - *Sunflowers*
Annuals to perennials; leaves opposite below, alternate above, simple; phyllaries in 2 or more rows, herbaceous; capitula radiate; receptacle flat or slightly convex, with scales partly enclosing achenes; pappus of 2 narrow scales soon falling off, sometimes with some shorter extra scales; ligules numerous, yellow (often *flore pleno*).

1 Plant annual, with simple tap-root 2
1 Plant perennial, clump-forming, with (often very short) rhizomes 3
  2 Phyllaries ovate, abruptly contracted to acuminate tip; central receptacular scales inconspicuously pubescent. Stems erect, to 3m, usually unbranched. Intrd-casual; on tips and in waste places; throughout most of urban BI *Sunflower* - **H. annuus** L.
  2 Phyllaries lanceolate to narrowly ovate, gradually tapered to apex;

central receptacular scales with conspicuous long white hairs at apex. Stems usually ≤1m. Intrd-casual; tips and waste places; frequent in London area *Lesser Sunflower* - **H. petiolaris** Nutt.

3 Stems ± glabrous in lower half; at least some phyllaries much exceeding edge of receptacle. Stems erect, to 1.5(2)m. Intrd-natd; garden escape or throwout; scattered in Br, mainly S & C (*H. annuus* x *H. decapetalus* L.) *Thin-leaved Sunflower* - **H. x multiflorus** L.

3 Stems roughly pubescent almost to base; phyllaries not or scarcely exceeding edge of receptacle 4

4 Rhizomes with swollen tubers; stems often >2m (to 3m); phyllaries not or loosely appressed to receptacle. Intrd-natd; very persistent in waste places and old gardens; scattered through Br and CI *Jerusalem Artichoke* - **H. tuberosus** L.

4 Rhizomes without swollen tubers; stems rarely >2m; phyllaries closely appressed to receptacle. Intrd-natd; garden escape or throwout; throughout most of BI (*H. pauciflorus* Nutt. x *H. tuberosus*) *Perennial Sunflower* - **H. x laetiflorus** Pers.

89. GALINSOGA Ruiz & Pav. - *Gallant-soldiers*

Annuals; leaves all opposite, simple, ovate, petiolate; phyllaries in 2 rows, largely herbaceous, the outer much shorter, the inner with membranous margins; capitula <1cm across, radiate; receptacle conical, with scales; pappus of scales; ligules few (usually 5), white or pinkish; disc flowers yellow.

1 Peduncles rather sparsely pubescent with glandular and eglandular hairs c.0.2mm; receptacular scales mostly distinctly 3-lobed, the central lobe the largest; pappus-scales fringed with hairs, without a terminal projection. Stems erect to ascending, to 80cm, glabrous or sparsely pubescent. Intrd-natd; weed of cultivated and waste ground; locally frequent in CI and Br N to C Sc, especially in cities
Gallant-soldier - **G. parviflora** Cav.

1 Peduncles with many glandular and eglandular hairs c.0.5mm; receptacular scales mostly simple, some with 1 or 2 weak lateral lobes; pappus-scales fringed with hairs and with a fine terminal projection. Stems erect to ascending, to 80cm, conspicuously pubescent. Intrd-natd; similar habitats and distribution to *G. parviflora*, often with it, also rare in Ir
Shaggy-soldier - **G. quadriradiata** Ruiz & Pav.

90. BIDENS L. - *Bur-marigolds*

Annuals; leaves all opposite, simple and toothed to pinnate; phyllaries in 2 dissimilar rows, the outer herbaceous, the inner ± membranous with a usually scarious border; capitula usually discoid, rarely radiate; receptacle flat or slightly convex, with scales; pappus of 2-5 barbed (forwardly or backwardly), strong bristles; ligules 0, rarely yellow, very rarely white.

The achenes provide important diagnostic characters, but some of those traditionally used, e.g. bristle number and direction of barbs on bristles, are

90. BIDENS

sometimes unreliable. Throughout the account 'achenes' refers to the central ones in the capitulum; the outer ones may differ considerably. All spp. are usually eligulate, but *B. cernua* sometimes has conspicuous yellow rays and *B. frondosa* and *B. pilosa* white ones.

1 Leaves not lobed, or lobed <1/2 way to midrib 2
1 At least lower leaves pinnate, or lobed nearly to midrib 4
   2 Achenes scarcely flattened (<2x as wide as thick), the faces between the 4 ridges warty. Stems erect, to 75cm. Intrd-natd; by canals; scattered in En, especially London area
London Bur-marigold - **B. connata** Willd.
   2 Achenes strongly flattened (>2x as wide as thick), the faces between the 2-4 ridges smooth 3
3 At least lower leaves distinctly petiolate, with 1(-2) pairs of distinct lobes. Stems erect, to 75cm. Native; similar habitat and distribution to *B. cernua*, and often with it   *Trifid Bur-marigold* - **B. tripartita** L.
3 All leaves tapered to base but sessile, unlobed (but strongly serrate). Stems erect, to 75cm. Native; by ponds and streams and canals and in ditches and marshy fields; locally common in C & S Br and Ir, rare and very scattered in N   *Nodding Bur-marigold* - **B. cernua** L.
   4 Leaflets lobed again to midrib or ± so. Stems erect, to 1m. Intrd-casual; a characteristic wool-alien; scattered in En and Sc
Spanish-needles - **B. bipinnata** L.
   4 Leaflets unlobed 5
5 Petioles winged to base; apical (main) lobe of leaf scarcely stalked or with winged stalk; barbs on edge of achenes (?always) backward-directed   (see couplet 3) **B. tripartita**
5 At least lower leaves with ± unwinged petioles and with apical (main) lobe with distinct ± wingless stalk; barbs on edge of achenes (but not on apical bristles) (?always) forward-directed 6
   6 Leaflet-teeth mostly wider than long; achenes ± parallel-sided, slightly tapered at each end. Stems erect, to 1m. Intrd-casual; a characteristic wool-alien; scattered in En   *Black-jack* - **B. pilosa** L.
   6 Leaflet-teeth mostly longer than wide; achenes tapered ± from apex to base. Stems erect, to 1m. Intrd-natd; by canals and rivers and on damp ground and waste places; scattered in En and Wa, frequent near Birmingham and London *Beggarticks* - **B. frondosa** L.

91. COREOPSIS L. - *Tickseeds*

Annuals to perennials; leaves (usually all) opposite, all or most pinnately or ternately divided to midrib or ± so, the primary divisions often divided again; phyllaries in 2 dissimilar rows, the outer narrower and shorter and ± herbaceous, the inner partially membranous; capitula radiate; receptacle flat or slightly convex, with scales; pappus 0 or of few very short teeth or bristles; ligules c.8, yellow, sometimes with dark basal blotch.

1 Tufted perennial; stems erect, to 1m, well-branched above; capitula on long slender peduncles, uniformly yellow, 3-5cm across. Intrd-

natd; garden escape or throwout; SE En
Large-flowered Tickseed - **C. grandiflora** Sweet

92. COSMOS Cav. - *Mexican Aster*
Annuals; leaves all opposite, 2-3-pinnate with linear to filiform segments; phyllaries in 2 dissimilar rows, the outer narrower, herbaceous with membranous border, the inner membranous; capitula radiate; receptacle flat, with scales; pappus of (0)2(-3) bristles with usually backward-directed barbs; ligules numerous, pinkish-purple, rarely white; disc flowers yellow.

1 Stems erect, to 2m; capitula 4-9cm across, incl. ligules 1.5-4cm. Intrd-natd; persistent on tips and in waste places; scattered in Br, mostly C & S *Mexican Aster* - **C. bipinnatus** Cav.

93. DAHLIA Cav. - *Dahlia*
Perennials (but killed by first frosts); leaves all opposite, petiolate, (bi-)pinnate or (bi-)ternate, the uppermost simple, phyllaries in 2 dissimilar rows, the inner membranous, the outer ± herbaceous; capitula radiate (but most often *flore pleno*); receptacle flat or slightly convex, with scales; pappus 0 or of 2 obscure teeth; ligules numerous, yellow, white, pink or purple; disc flowers yellow.

1 Stems erect, to 2m, rather succulent; capitula extremely variable in size (up to 30cm across but often <10cm), colour, shape of ligules, and degree to which they are *flore pleno*. Intrd-casual; tips and waste ground; scattered in En, mainly SE and SW *Dahlia* - **D. pinnata** Cav.

94. TAGETES L. - *Marigolds*
Aromatic annuals; leaves opposite below, alternate above, pinnate; phyllaries in 1 row, fused for most of length to form sheath round capitulum; capitula radiate; receptacle flat, without scales; pappus of unequal scales; ligules c.3-8, yellow to orange or brownish, often *flore pleno*.

1 Ligules <3mm; phyllary-sheath <4mm wide. Stems erect, to 1.2m. Intrd-casual; rather characteristic wool-alien; scattered in En
 Southern Marigold - **T. minuta** L.
1 Ligules >(5)10mm; phyllary-sheath >5mm wide 2
  2 Peduncles conspicuously swollen below capitula; phyllaries mostly >1.5cm; ligules mostly 1-3cm. Stems erect, to 50(100)cm. Intrd-casual; garden escape or throwout; scattered in En
 African Marigold - **T. erecta** L.
  2 Peduncles not or scarcely swollen below capitula; phyllaries mostly <1.5cm; ligules mostly (0.5)1-1.5cm. Stems erect, to 40cm. Intrd-casual; garden escape or throwout; scattered in En
 French Marigold - **T. patula** L.

95. SCHKUHRIA Roth - *Dwarf Marigold*
Annuals; leaves usually alternate, rarely some opposite, pinnate with linear

to filiform leaflets with many minute sunken glands; phyllaries few in 1 overlapping row, herbaceous with membranous tips; capitula <1cm across, radiate with only 1 short yellow ligule; receptacle concave, without scales; pappus of scales.

1 Stems erect, to 60cm; capitula numerous in subcorymbose panicle, obconical, 6-10 x 2-6mm (excl. ligule). Intrd-casual; characteristic wool-alien; scattered in En   *Dwarf Marigold* - **S. pinnata** (Lam.) Kuntze

96. GAILLARDIA Foug. - *Blanketflower*
Annuals to perennials; leaves all alternate, simple, coarsely dentate, tapered to base but ± sessile; phyllaries in 2-3 rows, ± herbaceous, becoming reflexed before fruiting; capitula radiate; receptacle strongly convex to subglobose, with bristle-like scales; pappus of scales with apical bristle; ligules numerous, yellow or more often purple in proximal 1/4-3/4.

1 Annual or short-lived perennial; stems erect, to 70cm; capitula 3.5-10cm across, very showy. Intrd-natd; tips and rough ground, sometimes ± natd especially on sand and shingle by sea; local in S En (*G. aristata* Pursh x *G. pulchella* Foug.)
Blanketflower - **G. x grandiflora** Van Houtte

97. HELENIUM L. - *Sneezeweed*
Tufted perennials; leaves all alternate, simple, subentire to shallowly dentate, tapered to base, decurrent on stem; phyllaries in 2-3 rows, herbaceous, becoming reflexed before fruiting; capitula radiate; receptacle convex, without scales; pappus of scales with apical bristle; ligules numerous, yellow to brownish-purple.

1 Stems erect, to 1m; capitula 4-6.5cm across, the ligules soon becoming reflexed; disc and ray flowers yellow to brownish-purple in various combinations. Intrd-natd; garden relic or throwout; sporadic in SE En   *Sneezeweed* - **H. autumnale** L.

TRIBE 10 - EUPATORIEAE (genera 98-99). Annual to perennial herbs; leaves opposite; capitula discoid; phyllaries in several rows, herbaceous; receptacular scales 0; pappus of 1 row of hairs or pointed scales; corolla blue or pinkish-purple, sometimes white.

98. EUPATORIUM L. - *Hemp-agrimony*
Perennials; at least some leaves very deeply 3-(5-)lobed or palmate, cuneate at base; pappus of hairs; corolla pinkish-purple, rarely white.

1 Stems erect, to 1.5m; capitula 2-5mm across, very numerous in compound subcorymbose panicles Native; all sorts of damp places and by water, in shade or open, sometimes in dry grassland or rough ground; common in most of En and Wa, frequent in Ir and CI, local and mostly coastal in Sc   *Hemp-agrimony* - **E. cannabinum** L.

99. AGERATUM L. - *Flossflower*
Annuals; leaves simple, ovate, cordate; pappus of narrow scales; corolla usually blue, rarely pink or white.

1 Stems erect, to 60cm; capitula 6-10mm across, rather numerous in compound subcorymbose panicles. Intrd-casual; in waste ground or on tips; occasional in SE En, ± natd in hedgerow in Scillies
Flossflower - **A. houstonianum** Mill.

L I L I I D A E — MONOCOTYLEDONS

Very rarely trees, rarely shrubs; rarely with secondary thickening and never from a permanent vascular cambium; vascular bundles usually scattered through stem; primary root usually short-lived; leaves usually with parallel major venation, and minor venation scarcely or not reticulate; flower parts mostly in threes; pollen grains mostly bilaterally symmetrical, commonly with 1 pore and/or furrow; cotyledon normally 1; endosperm typically helobial. Numerous exceptions to all the above occur.

140. BUTOMACEAE - *Flowering-rush family*

Glabrous, aquatic perennials rooted in mud, emergent through water; distinctive among petaloid monocots in its 6 ± free, dehiscent follicles, and similar sepals and petals.

1. BUTOMUS L. - *Flowering-rush*

1 Stems erect, to 1.5m, bearing simple terminal umbel; petals and sepals 1-1.5cm. Native; in ponds, canals, ditches and river-edges; rather scattered in Br N to C Sc and in Ir, but often intrd
Flowering-rush - **B. umbellatus** L.

141. ALISMATACEAE - *Water-plantain family*

Glabrous, aquatic annuals or perennials rooted in mud, often emergent through water; distinguished from Butomaceae in the very different petals and sepals, and 1-2(few)-seeded indehiscent fruits.

Submerged leaves are often ribbon-like, often very different from the diagnostically-shaped aerial leaves, and should be ignored.

1 Flowers monoecious, male and female in same inflorescence; stamens >6 **1. SAGITTARIA**
1 Flowers bisexual; stamens 6   2
  2 Stems procumbent or floating, rooting and producing tufts of leaves and inflorescences   3
  2 Stems erect, leafless, all leaves basal   4
3 Floating and aerial leaves obtuse; carpels in an irregular whorl or flattish mass **3. LURONIUM**
3 Floating and aerial leaves acute; carpels spiral in a ± globose

	(*Ranunculus*-like) head	**2. BALDELLIA**
	4 Carpels spiral in ± globose (*Ranunculus*-like) head	**2. BALDELLIA**
	4 Carpels in a single (often irregular) whorl	5
5	Fruits curved inwards, ± unbeaked, 1-seeded	**4. ALISMA**
5	Fruits divergent outwards, beaked, usually 2(-few)-seeded	
		5. DAMASONIUM

1. SAGITTARIA L. - *Arrowheads*

Leaves all basal, linear and/or long-petiolate and sagittate; flowers conspicuous, usually in whorls, monoecious, the male flowers in upper and the female in lower whorls; petals white; stamens 7-numerous; carpels spiral in ± globose head, each with 1 ovule. In autumn stolons tipped by small bud-like propagules are formed.

Leaf-shape is notoriously variable in this genus and needs to be used with great caution; emergent or floating leaves are required for identification.

1 Most or all emergent leaves strongly sagittate, with two long basal lobes 2
1 Most or all emergent or floating leaves linear to elliptic, rarely a few with short basal lobes 3
    2 Achenes 4-6mm, with apical beak <1mm; petals usually with purple blotch at base; anthers purple. Stems emergent, to 1m. Native; in ponds, canals and slow rivers; frequent in En, very scattered in Wa and Ir *Arrowhead* - **S. sagittifolia** L.
    2 Achenes 2.5-4mm, with subapical beak >1mm; petals without basal purple blotch; anthers yellow. Stems emergent, to 1m. Intrd-natd; in ponds and by streams; Surrey and Jersey
Duck-potato - **S. latifolia** Willd.
3 Flowers and leaves floating; many leaves linear, usually some or all floating ones elliptic; filaments glabrous. Stems to 30cm, producing flowers on water surface. Intrd-natd; in acid pond; N Hants
Narrow-leaved Arrowhead - **S. subulata** (L.) Buchenau
3 Flowers and leaves emergent; emergent leaves elliptic or rarely some with short basal lobes; filaments with scale-like hairs. Stems emergent, to 75cm. Intrd-natd; in canals; 2 places in S Devon
Canadian Arrowhead - **S. rigida** Pursh

2. BALDELLIA Parl. - *Lesser Water-plantain*

Varying in vegetative and inflorescence habit from *Alisma*-like to *Luronium*-like; petals pale mauve to ± white with yellow basal blotch; stamens 6; carpels spiral in ± globose head, each with 1 ovule.

1 Rosette-plant with erect stem to 20cm bearing inflorescence in simple whorl or with 1(-2) whorls below, or plant with trailing stems producing tufts of leaves and reduced (often 1-flowered) inflorescences. Native; wet places or shallow water in ditches, streamsides and pondsides; scattered over most of BI
Lesser Water-plantain - **B. ranunculoides** (L.) Parl.

3. LURONIUM Raf. - *Floating Water-plantain*
Stems procumbent or floating, rooting at intervals and producing tufts of leaves and inflorescences; submerged leaves linear, floating ones petiolate, elliptic, obtuse; flowers solitary or 2-5 in simple umbels, bisexual; petals pale mauve to ± white with yellow basal blotch; stamens 6; carpels in irregular whorl or flattish mass, each with 1 ovule.

1 Stems to 75cm, often much less; floating leaves 1-2.5(4)cm. Native; in acid ponds and canals; local in Wa and N & C En
Floating Water-plantain - **L. natans** (L.) Raf.

4. ALISMA L. - *Water-plantains*
Leaves all basal, linear and/or long-petiolate and narrowly elliptic to elliptic-ovate; flowers in whorled panicles, or (in small plants) in simple whorls or umbels, bisexual; petals pale mauve to ± white with basal yellow blotch; stamens 6; carpels in single whorl with lateral or subterminal style, each with 1 ovule.

1 Leaves elliptic to ovate-elliptic, rounded to subcordate at base; style arising c.1/2 way up fruit. Stems erect, to 1m. Native; in or by ponds, ditches, canals and slow rivers; common
Water-plantain - **A. plantago-aquatica** L.
1 Leaves linear to narrowly elliptic or lanceolate-elliptic, cuneate at base; style arising in upper 1/2 of fruit 2
  2 Fruits widest near middle, with ± straight, erect style. Stems erect, to 1m. Native; similar places to *A. plantago-aquatica*; scattered in Ir, in Br N to Yorks, and in Man
Narrow-leaved Water-plantain - **A. lanceolatum** With.
  2 Fruits widest in upper 1/2, with strongly recurved style. Stems erect, to 30(50)cm. Possibly native, more likely sporadically intrd; in shallow ponds; Worcs since 1920, for short periods in S Lincs, W Norfolk and Cambs
Ribbon-leaved Water-plantain - **A. gramineum** Lej.

A. plantago-aquatica x *A. lanceolatum* = ***A. x rhicnocarpum*** Schotsman is intermediate in all characters and sterile; there are records from Ir, Man, C Sc, London area and E Anglia, but all need confirming.

5. DAMASONIUM Mill. - *Starfruit*
Leaves all basal, long-petiolate, submerged, floating or sometimes emergent, ovate-oblong with cordate base; flowers in whorls, bisexual; petals white, with basal yellow blotch; stamens 6; carpels 6-10 in 1 whorl, with terminal style, each with 2-several ovules.

1 Stems erect, to 30(60)cm; leaves 3-6(8)cm. Native; muddy margins of acid ponds; sporadic and scattered in S En   *Starfruit* - **D. alisma** Mill.

142. HYDROCHARITACEAE - *Frogbit family*

Glabrous, aquatic perennials, free floating or rooted in mud; distinctive among monocots in having an inferior ovary combined with 3 sepals, usually 3 petals and dioecious flowers.

1 Leaves with long petioles and suborbicular cordate blades
1. HYDROCHARIS
1 Leaves sessile, tapering apically or at both ends 2
  2 Leaves all in basal rosette 3
  2 Leaves borne along stems 4
3 Leaves sharply serrate all along margins, rigid; petals conspicuous, white **2. STRATIOTES**
3 Leaves denticulate only near apex, flaccid; petals vestigial
7. VALLISNERIA
  4 Leaves variously whorled to spiral **6. LAGAROSIPHON**
  4 Leaves all whorled or opposite 5
5 Middle and upper leaves in whorls of 3-6(8), with 2 minute (c.0.5mm) fringed scales at base **5. HYDRILLA**
5 Middle and upper leaves in whorls of 3-4(5), with or without 2 entire scales at base <0.5mm 6
  6 Leaves in whorls of (3)4-5; petals >5mm, white; in industrially warmed water only **3. EGERIA**
  6 Leaves in whorls of (2)3-4(5); petals <5mm, inconspicuous; widespread **4. ELODEA**

1. HYDROCHARIS L. - *Frogbit*

Plants usually floating, the roots hanging in water; leaves with long petioles and floating suborbicular cordate blades, all in basal rosette, entire, with large stipules; flowers mostly dioecious, c.5-10% monoecious but all of 1 sex from each rosette, arising on pedicels from a stalked spathe (1 in female, 1-3(4) in male), conspicuous, with petals much larger than sepals; stamens 9-12, some usually as sterile staminodes; female flowers with 6 staminodes only; styles 6, bifid.

1 Main stems are floating stolons to 50(100)cm forming over-wintering terminal buds in autumn; leaves 1.6-5cm across. Native; in ponds, canals and ditches; locally frequent in En, very scattered in Wa and Ir *Frogbit* - **H. morsus-ranae** L.

2. STRATIOTES L. - *Water-soldier*

Plant usually floating, mostly submerged, rising to surface at flowering; leaves sessile, linear-lanceolate, tapering from base, all in basal rosette, spinous-serrate, without stipules; flowers mostly dioecious, arising from a stalked spathe (1 ± sessile in female, several pedicelled in male), conspicuous, with petals much larger than sepals; stamens 12, with sterile staminodes surrounding them; female flowers with staminodes only; styles 6, bifid.

2. STRATIOTES

1 Rosettes large, sturdy; leaves up to 50 x 2cm. Native; ponds, dykes and canals, usually calcareous; very local in E Anglia, N & S Lincs, SE Yorks and Denbs, intrd in scattered places in BI N to C Sc
Water-soldier - **S. aloides** L.

3. EGERIA Planch. - *Large-flowered Waterweed*

Stems long, branched, rooted in mud, submerged; leaves sessile, in whorls of (3)4-5, narrowly oblong-linear, minutely serrate, without stipules; flowers dioecious, pedicellate, arising from sessile axillary spathe (1 in female, 2-4 in male), conspicuous; petals white, much larger than sepals; stamens 9; styles 3, bifid.

1 Stems to 2m (?more); leaves 10-30 x 1.5-4mm, 0.5-1mm wide 0.5mm behind apex, with acute apex. Intrd-natd; in warmed water of canals and mill-lodges; very local in S Lancs, Glam and Middlesex
Large-flowered Waterweed - **E. densa** Planch.

4. ELODEA Michx. - *Waterweeds*

Stems long, branched, rooted in mud, submerged; leaves sessile, the lower opposite, the upper in whorls of 3-4(5), minutely serrate, with 2 minute entire basal scales; flowers dioecious, solitary from sessile axillary spathe, inconspicuous; petals whitish to reddish, c. as large as sepals; stamens 9; styles 3, bifid. Only female plants occur in BI.

1 Leaf apices obtuse to subacute; leaves (0.7)0.8-2.3mm wide 0.5mm behind apex. Stems to 3m; leaves in whorls of (2)3(-4), 4.5-17 x 1.4-5.6mm. Intrd-natd; ponds, lakes, canals, slow rivers; common
Canadian Waterweed - **E. canadensis** Michx.
1 Leaf apices acute to acuminate; leaves 0.2-0.7(0.8)mm wide 0.5mm behind apex   2
   2 Usually some leaves strongly recurved and/or twisted, with marginal teeth 0.05-0.1mm; root-tips white to greyish-green when fresh; sepals 1.6-2.5mm. Stems to 3m; leaves in whorls of (2)3-4(5), 5.5-35 x 0.8-3mm. Intrd-natd; habitat as for *E. canadensis*; locally common in En, very scattered in Wa, Sc, Ir and Jersey, spreading   *Nuttall's Waterweed* - **E. nuttallii** (Planch.) H. St. John
   2 Usually no leaves strongly recurved or twisted, with marginal teeth usually 0.1-0.15mm; root-tips red when fresh; sepals 3-4.3mm. Stems to 3m; leaves in whorls of 3, 9-25 x 0.7-2.2mm. Intrd-natd; habitat as for *E. canadensis* but possibly not fully natd; very local in S En and S Wa
South American Waterweed - **E. callitrichoides** (Rich.) Casp.

5. HYDRILLA Rich. - *Esthwaite Waterweed*

Stems long, branched, rooted in mud, submerged; leaves sessile, the lower opposite, the upper in whorls of 3-6(8), minutely serrate, with 2 minute fringed scales at base; flowers dioecious, solitary from sessile axillary spathe, inconspicuous; petals transparent with red streaks, c. as large as sepals; stamens 3; styles 3(-5), simple.

1　Stems to 1m (?more); leaves 5-20 x 0.7-2mm, 0.2-0.7mm wide 0.5mm behind apex, with narrowly acute to acuminate apex. Native; lakes; Rusheenduff Lough (W Galway), Esthwaite Water (Westmorland) 1914-c.1945　　　　*Esthwaite Waterweed* - **H. verticillata** (L. f.) Royle

6. LAGAROSIPHON Harv. - *Curly Waterweed*
Stems long, branched, rooted in mud, submerged; leaves variously whorled to spiral, the lowest always spiral (not opposite), subentire to minutely denticulate, with 2 minute entire basal scales; flowers dioecious, inconspicuous, arising from sessile axillary spathe (1 in female, several in male); petals reddish, c. as large as sepals; stamens 3; styles 3, bifid.

1　Stems to 3m; leaves 6-30 x 1-3mm, usually strongly recurved, 0.2-0.5mm wide 0.5mm behind apex, with narrowly acute to acuminate apex. Intrd-natd; ponds, lakes, canals, slow rivers; locally frequent in CI and Br N to C Sc, M Cork
　　　　　　　　　　　　　　　Curly Waterweed - **L. major** (Ridl.) Wager

7. VALLISNERIA L. - *Tapegrass*
Plant rooted in mud, submerged, with stems as stolons; leaves sessile, all in basal rosette, linear, denticulate near apex, without stipules; flowers dioecious (or monoecious?), solitary (female) or many (male) in stalked spathes, inconspicuous; petals 0 or vestigial; stamens (1)2(-3); styles 3, bifid.

1　Leaves 2-80cm x 1-10mm, ribbon-like; stalks of female spathes very long, spiralling after flowering. Intrd-natd; slow rivers and canals (often not permanent), usually where water is heated; very local in S En, S Lancs, SW Yorks　　　　　　　*Tapegrass* - **V. spiralis** L.

143. APONOGETONACEAE - *Cape-pondweed family*

Glabrous, aquatic perennials with elongated stems rooted in mud at tuberous base; the forked, white inflorescence borne just above the water surface is unique.

1. APONOGETON L. f. - *Cape-pondweed*

1　Leaves oblong-elliptic, up to 25 x 7cm; spikes up to 6cm, each with up to 10 flowers; tepals 10-20mm. Intrd-natd; persistent or ± natd where planted in ponds; scattered in Br　*Cape-pondweed* - **A. distachyos** L. f.

144. SCHEUCHZERIACEAE - *Rannoch-rush family*

Glabrous, *Juncus*-like perennials with rhizomes clothed with old leaf-bases

144. SCHEUCHZERIACEAE

and erect leafy flowering stems; leaves alternate, with distinctive large apical pore; carpels free (usually 3), each with usually only 2 seeds.

1. SCHEUCHZERIA L. - *Rannoch-rush*

1 Stems to 25(40)cm; upper stem-leaves usually overtopping inflorescence. Native; in pools or wet *Sphagnum* bogs; very rare in 2 places in M Perth *Rannoch-rush* - **S. palustris** L.

145. JUNCAGINACEAE - *Arrowgrass family*

Glabrous perennials with rhizomes and erect, ± leafless, flowering stems; leaves ± all basal, in rosette or a few alternate near stem base; *Juncus*-like or *Plantago*-like superficially, but the 3 or 6 1-seeded fruit segments are diagnostic.

1. TRIGLOCHIN L. - *Arrowgrasses*

1 Leaves usually deeply furrowed on upperside near base; fruit with 3 fertile cells, 7-10mm; stigmas long-fringed. Stems erect, to 60cm. Native; marshy places and wet fields, sometimes at back of salt-marshes; throughout BI, mostly N *Marsh Arrowgrass* - **T. palustre** L.
1 Leaves usually flat on upperside near base; fruit with 6 fertile cells, 3-5mm; stigmas papillate. Stems erect, to 60cm. Native; in salt-marshes and salt-sprayed grassland; round coasts of whole BI, rare in inland salty areas *Sea Arrowgrass* - **T. maritimum** L.

146. POTAMOGETONACEAE - *Pondweed family*

Glabrous aquatic (or ± so) perennials, often with rhizomes, with leafy submerged flowering stems; distinguished from other pondweeds by the 4 tepals and 4 stamens.

1 All or most leaves alternate, all with membranous sheath or stipule
 1. POTAMOGETON
1 All leaves opposite (or some in whorls of 3), only the uppermost with membranous stipules **2. GROENLANDIA**

1. POTAMOGETON L. - *Pondweeds*
Leaves all alternate, or just those subtending inflorescences opposite, all with membranous sheath or stipules; fruits with thick pericarp, soft on outside but with bony inner layer.

For accurate identification it is essential first to examine thoroughly the leaf morphology, including range of leaf-shape, leaf-blade venation, and morphology of basal sheath/stipules. The key covers only mature plants that have reached flowering or are about to flower; beware recent flooding or drying out of habitat when distinguishing floating and submerged leaves. In taxa that can have floating leaves (whether actually present or not) the upper submerged leaves often approach the former in certain respects and may be quite different from the middle and lower submerged leaves, which are the diagnostic ones. In the keys and descriptions 'veins' refers to the midrib plus its laterals that run ± parallel to it for nearly the whole leaf length. Fruit lengths include the beak. Fresh material is best for determination; the stipule characters are diffcult to see in dried material, but are sometimes necessary for certain identification.

General key
1. Floating leaves present — *Key A*
1. All leaves submerged — 2
   2. Leaves slightly to strongly convex-sided, >6mm (usually >10mm) wide, often with ≥7 main veins — *Key B*
   2. Leaves grass-like, parallel-sided for almost whole length, <5(7)mm wide, with 3-5(7) main veins — *Key C*

Key A - Floating leaves present
1. Floating leaves with distinct hinge-like joint at junction with petiole; submerged leaves (if present) <3.5mm wide. Floating leaves opaque, elliptic to ovate-elliptic, up to 10(14) x 4.5(8)cm, very rarely 0. Native; lakes, ponds, rivers, ditches, usually over rich soils; frequent (see also Key C, couplet 1) **Broad-leaved Pondweed - P. natans** L.
1. Floating leaves merging gradually or abruptly into petiole, but without distinct joint; at least some submerged leaves (if present) >3.5mm wide — 2
   2. Submerged leaves always present, sessile or with petiole <1cm — 3
   2. Submerged leaves usually present, if so then some or all distinctly petiolate with petiole >1cm — 8
3. All submerged leaves strictly parallel-sided ± throughout length, mostly ≤8mm wide. Floating leaves opaque, narrowly elliptic-oblong, up to 8 x 2.2cm. Native; 3 lakes in S Uist (Outer Hebrides), natd in canals in S Lancs and SW Yorks*American Pondweed* - **P. epihydrus** Raf.
3. At least some submerged leaves convex-sided for at least part of length, often >8mm wide — 4
   4. At least lower submerged leaves broadly rounded at base, ± clasping stem. Floating leaves ± opaque, elliptic or narrowly so, up to 6.5 x 2.3cm, often 0. Native; lakes, canals, ponds, streams; scattered throughout most of Br and Ir, rare in S (*P. gramineus* x *P. perfoliatus*) (see also Key B, couplet 4) **Bright-leaved Pondweed - P. x nitens** Weber
   4. All submerged leaves narrowed to base, not clasping stem — 5
5. Submerged leaves obtuse to rounded at apex — 6
5. Submerged leaves acute, cuspidate or acuminate at apex — 7
   6. Stem terete; at least some submerged leaves >1.5cm wide, with >7 veins; fertile. Floating leaves ± translucent, elliptic to rather narrowly so, up to 9 x 2.5cm, sometimes 0. Native; lakes, canals,

streams, especially on peaty soil; fairly frequent ± throughout BI except SW En

(see also Key B, couplet 10) *Red Pondweed* - **P. alpinus** Balb.

6 Stem compressed; submerged leaves ≤1.5cm wide, all or most 5-7-veined; sterile. Floating leaves reported sometimes, not seen by me. Native; rivers; very local in S Wa, E Sc and NE En (*P. alpinus* x *P. crispus*) (see also Key B, couplet 11)

Graceful Pondweed - **P. x olivaceus** G. Fisch.

7 Submerged leaves ≤1.2cm wide; stipules mostly 1-2.5cm on main stems. Floating leaves opaque, elliptic, up to 9.5 x 3.4cm, sometimes 0. Native; lakes, ponds, canals, streams, usually on acid soils; scattered throughout most of Br and Ir, rare in S

(see also Key B, couplet 5) *Various-leaved Pondweed* - **P. gramineus** L.

7 Submerged leaves ≥1cm wide; stipules >2cm on main stems. Floating leaves ± opaque, elliptic or narrowly so, up to 10.5 x 4cm, sometimes 0. Native; lakes, ponds, streams, canals; scattered throughout C & N Br and Ir (*P. lucens* x *P. gramineus*)

(see also Key B, couplet 6) *Long-leaved Pondweed* - **P. x zizii** Roth

8 Floating leaves translucent, the vein network clearly visible (fresh or dried); fruits 1.5-1.9mm. Floating leaves ovate, up to 8.5 x 5.5cm. Native; ponds and pools, on base-rich peat; local over most of BI except N & E Sc   *Fen Pondweed* - **P. coloratus** Hornem.

8 Floating leaves opaque, the vein network difficult to see; fruits ≥1.9mm or not developing   9

9 Margins of young submerged leaves minutely denticulate; fruits 2.7-4.1mm. Floating leaves elliptic to narrowly so, translucent, up to 13 x 5cm. Native; slow-flowing base-rich rivers; very local in S En

Loddon Pondweed - **P. nodosus** Poir.

9 Margins of young submerged leaves ± entire; fruits <2.7mm or not formed   10

10 Most leaves floating (or ± aerial); submerged leaves with midrib not prominent; fruits 1.9-2.6mm. Floating (or emergent) leaves opaque, similar in shape to those of *P. natans*, up to 10.5 x 7cm. Native; shallow ponds, bogs, ditches, small streams, on acid soil; common   *Bog Pondweed* - **P. polygonifolius** Pourr.

10 Most leaves submerged, with prominent midrib; fruits not formed   11

11 Petioles of submerged leaves (6)10-35cm. Floating leaves opaque, narrowly elliptic to elliptic, up to 14 x 4.5cm. Native; in R. Stour in Dorset (*P. natans* x *P. nodosus*)

Schreber's Pondweed - **P. x schreberi** G. Fisch.

11 Petioles of submerged leaves ≤5.5cm. Floating leaves opaque, elliptic to narrowly so, up to 11.5 x 3.5cm. Native; lakes, ponds, canals, streams; scattered in Br and Ir (*P. natans* x *P. gramineus*)

Ribbon-leaved Pondweed - **P. x sparganiifolius** Fr.

FIG 526 - Floating leaves of *Potamogeton*.
1, *P. natans*. 2, *P. x sparganiifolius*.
3, *P. polygonifolius*. 4, *P. coloratus*. 5, *P. nodosus*. 6, *P. alpinus*.
7, *P. gramineus*. 8, *P. x nitens*. 9, *P. x zizii*. 10, *P. epihydrus*.

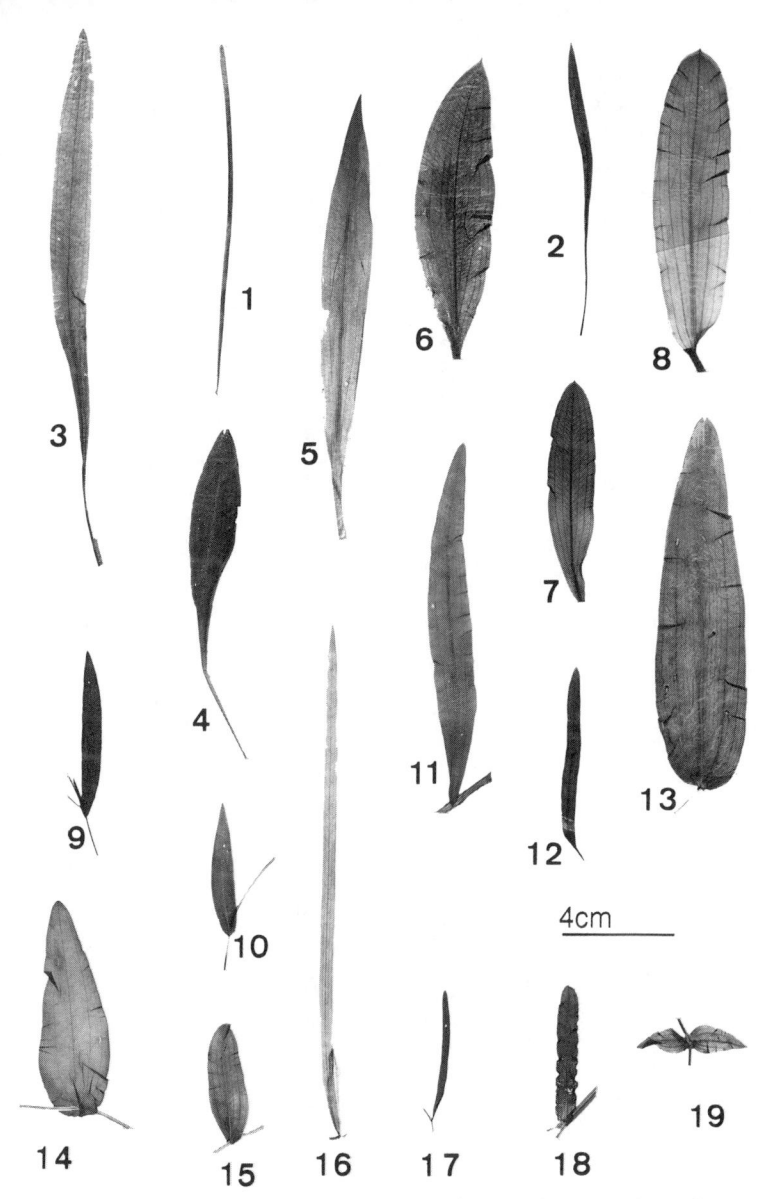

FIG 527 - Submerged leaves of **Potamogetonaceae**. 1, *Potamogeton natans*. 2, *P. x sparganiifolius*. 3, *P. polygonifolius*. 4, *P. coloratus*. 5, *P. nodosus*. 6, *P. lucens*. 7, *P. x zizii*. 8, *P. x salicifolius*. 9, *P. gramineus*. 10, *P. x nitens*. 11, *P. alpinus*. 12, *P. x olivaceus*. 13, *P. praelongus*. 14, *P. perfoliatus*. 15, *P. x cooperi*. 16, *P. epihydrus*. 17, *P. x lintonii*. 18, *P. crispus*. 19, *Groenlandia densa*.

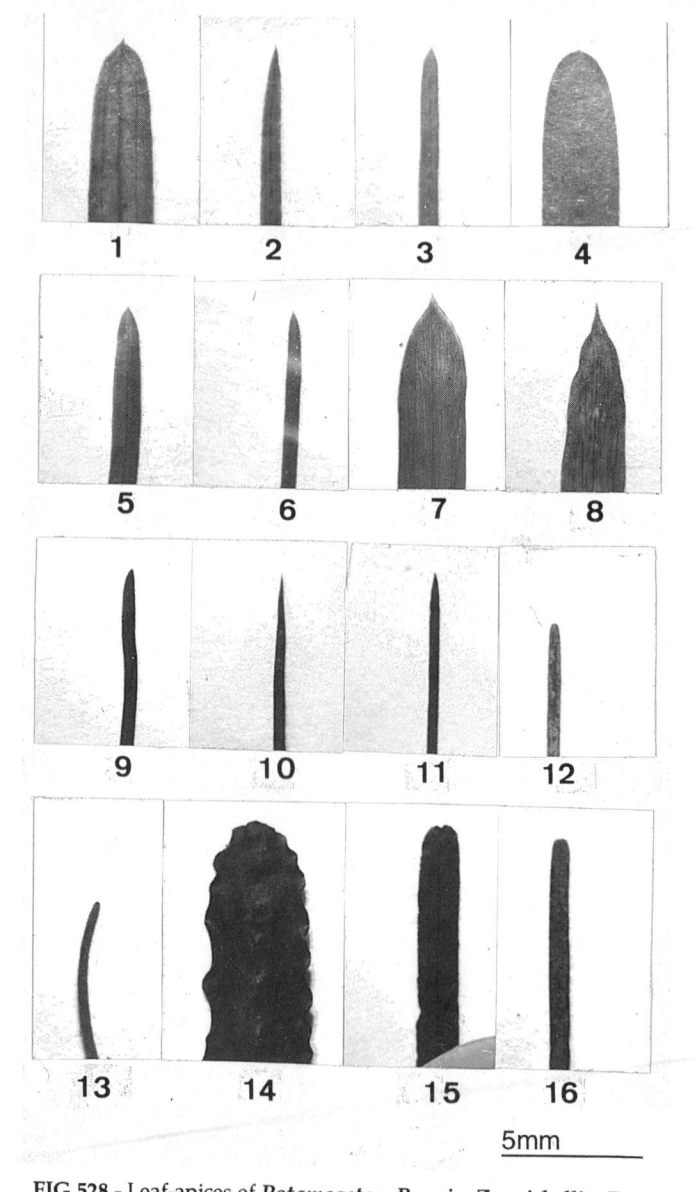

FIG 528 - Leaf-apices of *Potamogeton, Ruppia, Zannichellia, Zostera*.
1-10, *Potamogeton*. 1, *P. friesii*. 2, *P. rutilus*. 3, *P. pusillus*.
4, *P. obtusifolius*. 5, *P. berchtoldii*. 6, *P. trichoides*. 7, *P. compressus*.
8, *P. acutifolius*. 9, *P. filiformis*. 10, *P. pectinatus*.
11-12, *Ruppia*. 11, *R. maritima*. 12, *R. cirrhosa*. 13, *Zannichellia palustris*.
14-16, *Zostera*. 14, *Z. marina*. 15, *Z. angustifolia*. 16, *Z. noltei*.

1. POTAMOGETON

Key B - All leaves submerged, convex-sided, >6mm wide, often with ≥7 veins

1 Leaves minutely serrate along ± whole margin to naked eye, usually regularly undulate; fruits with beak ≥1/2 as long as body. Leaves (3)5-12(18)mm wide, acute to rounded at apex, 3-5(7)-veined. Native; lakes, ponds, canals, rivers, streams; frequent to common
                               ***Curled Pondweed* - P. crispus** L.   **527**

1 Leaves entire even with x10 lens, or obscurely serrate just near apex, not regularly undulate; fruits with beak <1/2 as long as body   2

  2 Leaves with acute, acuminate or cuspidate apex, or if obtuse or rounded then also mucronate   3

  2 Leaves with obtuse to rounded, sometimes hooded, not mucronate apex   7

3 At least some leaves rounded at base and ± clasping stem; sterile hybrids   4

3 All leaves tapered to base, not clasping stem   5

  4 Stipules 2-7cm; leaves mostly >2cm wide. Leaves 14-40mm wide, acute to cuspidate at apex; 5-17-veined. Native; ponds, canals, rivers; scattered in C & S En, very local elsewhere (*P. lucens* x *P. perfoliatus*)   ***Willow-leaved Pondweed* - P. x salicifolius** Wolfg.   **527**

  4 Stipules 1-3cm; leaves mostly <2cm wide. Submerged leaves 9-23mm wide, acute to apiculate or obtuse at apex, 7-17-veined
                            (see Key A, couplet 4) **P. x nitens**

5 Leaves 5-12mm wide; stipules 1-2.5(3.5)cm; fruits 2.4-3.1mm. Submerged leaves 5-12mm wide, acute to obtuse and mucronate at apex, 7-9-veined   (see Key A, couplet 7) **P. gramineus**

5 Leaves 10-65mm wide; stipules ≥2cm; fruits 2.7-4.5mm   6

  6 Leaves all or mostly <12 x 2.5cm; only some leaves petiolate, the petiole usually narrowly winged; fruits 2.7-3.4mm. Submerged leaves (8.5)10-25(30)mm wide, acute to obtuse and mucronate at apex, 9-13-veined   (see Key A, couplet 7) **P. x zizii**

  6 Leaves all or mostly >12 x 2.5cm; all leaves petiolate, the petiole unwinged near base; fruits 3.2-4.5mm. Leaves(17)25-65mm wide, acuminate to rounded and mucronate at apex, 9-13-veined. Native; lakes, ponds, canals, slow rivers on base-rich soil; locally common in En and Ir, rare elsewhere   ***Shining Pondweed* - P. lucens** L.   **527**

7 At least some leaves rounded at base, usually ± clasping stem   8

7 Leaves narrowed to base, not clasping stem   10

  8 Stipules conspicuous, >1cm (often much more so), persistent; leaves mostly >10cm, with ≥3 faint lateral strands between midrib and nearest strong lateral vein. Leaves 14-40mm wide, blunt and hooded at apex (bifid when pressed), 11-19-veined. Native; lakes, rivers, canals, streams; scattered throughout Br and Ir except S En   ***Long-stalked Pondweed* - P. praelongus** Wulfen   **527**

  8 Stipules inconspicuous, at least some <1cm, soon disappearing; leaves all or most <10cm, with <3 faint lateral strands between midrib and nearest strong lateral vein   9

9 Stem compressed; leaves oblong-lanceolate, most or all <2cm wide;

146. POTAMOGETONACEAE

sterile. Leaves 8-25mm wide, rounded at apex, 7-13-veined. Native; lakes, ponds, canals, streams; very scattered in Br and Ir (*P. perfoliatus* x *P. crispus*) **Cooper's Pondweed - P. x cooperi** (Fryer) Fryer
9 Stem terete; leaves oblong-ovate, most or all >2cm wide; fertile. Leaves 7-42mm wide, obtuse to rounded and often ± hooded at apex, 11-25-veined. Native; lakes, ponds, canals, rivers, streams; frequent **Perfoliate Pondweed - P. perfoliatus** L.
10 Stem terete; leaves with >7 veins; fertile. Submerged leaves (6.5)10-25(33)mm wide, obtuse and often hooded at apex, 9-15-veined (see Key A, couplet 6) **P. alpinus**
10 Stem compressed; leaves 3-7-veined; sterile 11
11 Leaves ≤5mm wide, with (3-)5 veins. Leaves 1.7-5mm wide, rounded at apex. Native; rivers, canals, streams; very local in En, Monts, Kirkcudbrights, E Ir (*P. friesii* x *P. crispus*)
 Linton's Pondweed - P. x lintonii Fryer
11 Leaves >5mm wide, with 5-7 veins. Submerged leaves 6-15mm wide, obtuse to subacute at apex (see Key A, couplet 6) **P. x olivaceus**

Key C - All leaves submerged, parallel-sided for almost whole length, <5(7)mm wide, with 3-5(7) main veins
1 Stipules mostly >5cm; leaves actually blade-less petioles, opaque, without obvious midrib, 0.4-3.5mm wide, obtuse to acuminate at apex
 (see Key A, couplet 1) **P. natans**
1 Stipules <5cm; leaves translucent, with obvious midrib 2
 2 Stipules fused to base of leaf, forming sheathing leaf-base, free distally to form ligule 3
 2 Stipules free from leaf, forming stipule-like outgrowth from node 6
3 Sheathing leaf-base not fused in tube round stem, but with margins overlapping 4
3 Some or all leaves with sheathing leaf-base fused in tube round stem proximally when young 5
 4 Leaves acute to acuminate at apex; style c.0.2mm; fruits 3.3-4.7mm. Leaves 0.2-4mm wide, 3-5-veined but laterals very faint. Native; similar habitats to *P. filiformis*; frequent over most of BI **Fennel Pondweed - P. pectinatus** L.
 4 Leaves truncate to obtuse at apex; style 0; fruits not developing. Leaves otherwise ± as in *P. pectinatus* (*P. pectinatus* x *P. vaginatus* Turcz.) **Bothnian Pondweed - P. x bottnicus** Hagstr.
5 Fruits 2.2-3.2mm; style never present; pollen well-formed. Leaves 0.3-1.2mm wide, obtuse to rounded at apex, 3-veined but 2 laterals very faint and submarginal. Native; lakes, rivers, streams, dykes, sometimes brackish; scattered in Sc and N & W Ir, S Northumb
 Slender-leaved Pondweed - P. filiformis Pers.
5 Fruits not developing; style sometimes present; pollen misshapen. Leaves ± as in *P. pectinatus*. Native; local in MW & NW Yorks (outside range of *P. filiformis*), coasts of Sc and N & SW Ir (*P. filiformis* x *P. pectinatus*) **Swedish Pondweed - P. x suecicus** K. Richt.
 6 Most leaves <2mm wide, with 3(-5) veins 7

1. POTAMOGETON

6 Most leaves >2mm wide, with 3-5(many) veins 10
7 Stipules fused in tube round stem proximally when young; fruits 1.8-2.3mm 8
7 Stipules not fused in tube round stem, but with margins overlapping; fruits 1.8-3.2mm 9
   8 Leaves very gradually and finely acute to long-acuminate, rigid; C & N Sc only. Leaves 0.5-1.1mm wide, 3(-5)-veined. Native; lakes; local in N & NW Sc   **Shetland Pondweed - P. rutilus** Wolfg.   528
   8 Leaves acute to obtuse and abruptly mucronate, not rigid; widespread. Leaves 0.5-1.4(1.9)mm wide, 3(-5)-veined. Native; lakes, ponds, canals, streams, mostly in base-rich water; frequent ± throughout BI, but less so than *P. berchtoldii*
                                                 Lesser Pondweed - P. pusillus L.   528
9 Leaves acute to finely acute, mostly <1mm wide; carpels 1(-2) per flower; fruits usually warty or ± toothed near base. Leaves 0.3-1(1.8)mm wide, 3-veined. Native; ponds, canals, streams; scattered in C & S Br, very local in C Sc and Guernsey
                           Hairlike Pondweed - P. trichoides Cham. & Schltdl.   528
9 Leaves obtuse to subacute, often mucronate, mostly >1mm wide; carpels (3)4-5(7) per flower; fruits not warty. Leaves 0.5-2.3mm wide, 3(-5)-veined. Native; lakes, ponds, canals, rivers, streams; fairly common throughout most of BI   **Small Pondweed - P. berchtoldii** Fieber   528
   10 Leaves with 3 or 5 main veins and many finer strands between them 11
   10 Leaves with 3 or 5(-7) main veins only 12
11 Leaves with 3 main veins, acute to acuminate (mostly acuminate); fruit usually with basal wart or tooth, usually with erect symmetric beak. Leaves 1.5-5.5mm wide, most or all <10cm long. Native; ponds, canals, streams, mostly on calcareous soils; local in En from Dorset to E Norfolk   **Sharp-leaved Pondweed - P. acutifolius** Link   528
11 Leaves with 5 main veins (2 submarginal), acute to acuminate but mostly mucronate; fruit not warted, with asymmetric, curved beak. Leaves 3-6mm wide, most >10cm long. Native; lakes, ponds, canals, streams; locally frequent in C En and E Wa, S Aberdeen
                                      Grass-wrack Pondweed - P. compressus L.   528
   12 Leaves usually minutely serrate along at least part of margin; fruits tapered to beak ≥1/2 as long as body
                                    (see Key B, couplet 1) **P. crispus**
   12 Leaves entire or minutely serrate just near apex; fruits, if formed, with beak <1/2 as long as body 13
13 Stems with many lateral branches closely placed, forming fan-like sprays; stipules not fused in tube round stem. Leaves (1)2.5-3.5mm wide, usually <10cm long, obtuse or rounded and apiculate at apex, 3-5-veined. Native; lakes, ponds, canals, streams; locally frequent ± throughout mainland Br and Ir
              Blunt-leaved Pondweed - P. obtusifolius Mert. & W.D.J. Koch   528
13 Stems without many closely placed lateral branches; stipules fused in tube round stem proximally when young 14

14 Leaves ≤5mm wide, usually serrulate distally, rounded to acute at apex; sterile (see Key B, couplet 11) **P. x lintonii**
14 Leaves ≤3.5(4)mm wide, entire, mucronate at apex; fruits 2.4-3mm. Leaves 1.5-3.5(4)mm wide, (3)5(-7)-veined. Native; lakes, ponds, canals; frequent in En, scattered in Wa, Sc and Ir
Flat-stalked Pondweed - **P. friesii** Rupr.

Although 27 hybrids have been found in BI, hybridization is not common and most hybrids are very rare. 10 that sometimes occur in the absence of both parents are treated above. All hybrids except *P. x zizii* are sterile, with no fruit developed; the fruits of *P. x zizii* are of unproven viability. Hybrids are variously intermediate between the parents.

2. GROENLANDIA J. Gay - *Opposite-leaved Pondweed*

Leaves all opposite (or rarely some in whorls of 3), only the uppermost with 2 membranous stipules fused to edges of leaf-base; fruits with thin, papery pericarp.

1 Leaves all submerged, ovate to lanceolate, up to 4.2 x 1.3cm, acute to obtuse at apex, clasping stem. Native; ponds, ditches, streams (often fast-flowing); locally frequent in En, scattered in S Wa and Ir, very rare in Sc
Opposite-leaved Pondweed - **G. densa** (L.) Fourr.

147. RUPPIACEAE - *Tasselweed family*

Glabrous aquatic perennials rooted in substratum, with leafy submerged stems; leaves mostly alternate but the upper opposite, linear, <1.5mm wide, sessile, with sheathing base; distinguished from *Potamogeton* in its terminal inflorescence, lack of perianth, fruits becoming long-stalked and leaves with only a midrib.

1. RUPPIA L. - *Tasselweeds*

1 Leaves usually acute, with slightly inflated sheath; peduncles ≤2.6cm, straight to curved or flexuous in fruit; anthers ≤1mm; drupelets 2-2.8mm, incl. beak. Native; in brackish ditches and pools; local round most coasts of Br and Ir
Beaked Tasselweed - **R. maritima** L.
1 Leaves usually obtuse to rounded, with strongly inflated sheath; peduncles ≥4cm, usually spiral in fruit; anthers >1mm; drupelets 2.7-3.4mm, incl. beak. Native; similar places to *R. maritima*, sometimes with it but more local, frequent only in E & SE En and Shetland
Spiral Tasselweed - **R. cirrhosa** (Petagna) Grande

148. NAJADACEAE - *Naiad family*

Glabrous, aquatic annuals or perennials rooted in substratum, with leafy

148. NAJADACEAE

submerged stems; distinguished from *Potamogetonaceae* in its linear opposite (or whorled) leaves with only a midrib, and unisexual sessile axillary flowers with 0 perianth and 1 carpel.

1. NAJAS L. - *Naiads*

1 Leaves 1-2.5(4)cm, <1mm wide incl. teeth, ± entire or minutely denticulate; fruit 2.5-3.5 x 1-1.5mm, plus long filiform style. Native; clean lakes; local in W Br from Westmorland to Outer Hebrides, W Ir   *Slender Naiad* - **N. flexilis** (Willd.) Rostk. & W.L.E. Schmidt
1 Leaves 1-4.5cm, 1-6mm wide incl. teeth, conspicuously spinose-dentate; fruit (3)4-6(8) x 1.5-3mm, plus short thick style. Native; slightly brackish waterways; extremely local in NE E Norfolk
Holly-leaved Naiad - **N. marina** L.

149. ZANNICHELLIACEAE - *Horned Pondweed family*

Glabrous, aquatic rhizomatous annuals or perennials rooted in substratum, with leafy submerged stems; distinguished from *Potamogetonaceae* in its linear mostly opposite leaves and unisexual axillary flowers with 0 perianth; and from *Najas* in its several carpels and entire leaves.

1. ZANNICHELLIA L. - *Horned Pondweed*

1 Leaves 2-10cm x 0.4-1(2)mm, entire, acute to obtuse at apex, 1(-3)-veined; fruits 3-6mm incl. style, variably stalked, variably winged and toothed on dorsal and ventral edges. Native; rivers, streams, ditches and ponds, fresh or brackish; frequent throughout most of BI   *Horned Pondweed* - **Z. palustris** L.

150. ZOSTERACEAE - *Eelgrass family*

Glabrous, marine perennials rooted in substratum, with leafy submerged stems often exposed at low tide; leaves alternate, linear, sessile, with sheathing base with ligule at top, entire; distinguished from other marine or brackish pondweeds in the complex, congested inflorescence ± enclosed in leaf-sheath.

1. ZOSTERA L. - *Eelgrasses*

1 Flowering stems lateral, unbranched or with few branches near base; leaves of sterile shoots 0.5-1.5mm wide, with sheaths clasping stems but not fused into tube. Native; similar habitats and distribution to *Z. angustifolia* and often with it, but ± never below low-water mark
Dwarf Eelgrass - **Z. noltei** Hornem.
1 Flowering stems terminal, branched; leaves of sterile shoots 1-10mm

wide, with sheaths fused into tube round stem 2
2 Leaves (2)4-10mm wide; stigmas c.2x as long as style. Native; sea-coasts, c.0-4(9)m below low-water mark; scattered round coasts of BI *Eelgrass* - **Z. marina** L. 52
2 Leaves 1-2(3)mm wide; stigmas c. as long as style. Native; sea-coasts and estuaries, from half-tide to low-tide mark or rarely down to 4m below; scattered round coasts of BI
Narrow-leaved Eelgrass - **Z. angustifolia** (Hornem.) Rchb. 52

151. ARACEAE - *Lords-and-Ladies family*

Glabrous, herbaceous perennials with rhizomes or underground tubers giving rise to aerial leaves and flowering stems with 0-few leaves; the minute numerous flowers packed on to an axis (*spadix*), which often extends distally as a succulent *appendix*, and which is subtended or partially enclosed by a leaf-like but often coloured *spathe* except in the very distinctive *Acorus*, are diagnostic.

1 Leaves linear, *Iris*-like; spadix apparently lateral and lacking spathe
1. ACORUS
1 Leaves lanceolate to ovate, narrowed at base and usually petiolate; spadix terminal, with spathe 2
  2 Flowers covering spadix to its apex 3
  2 Spadix with succulent sterile appendix distal to flowers 5
3 Spathe ± flat, not enclosing spadix even at extreme base **3. CALLA**
3 Spathe wrapped round basal part of spadix 4
  4 Leaves truncate to cuneate at base, with petioles shorter than blade; tepals 4 **2. LYSICHITON**
  4 Leaves cordate at base, with petiole longer than blade; tepals 0
4. ZANTEDESCHIA
5 Leaves palmately divided **6. DRACUNCULUS**
5 Leaves simple 6
  6 Spathe fused into tube proximally, with distal filiform projection ≥5cm **7. ARISARUM**
  6 Spathe overlapping at base, not fused into tube, no more than acuminate at apex **5. ARUM**

1. ACORUS L. - *Sweet-flags*

Rhizomatous; leaves linear, *Iris*-like, entire, sessile; spadix apparently lateral, without appendix; spathe apparently 0; flowers bisexual; tepals 6; stamens 6; ovary 2-3-celled; fruit not forming. Fresh leaves have strong spicy scent when bruised.

1 Leaves 50-125 x 0.7-2.5cm, with well-defined midrib; spadix 5-9 x 0.6-1.2cm. Intrd-natd; in shallow water at edges of lakes, ponds, rivers and canals; scattered over most of BI *Sweet-flag* - **A. calamus** L.
1 Leaves 8-50 x 0.2-0.8cm, without obvious midrib; spadix 5-10 x

1. ACORUS

0.3-0.5cm. Intrd-natd; by lake; Surrey
Slender Sweet-flag - **A. gramineus** Aiton

2. LYSICHITON Schott - *Skunk-cabbages*
Rhizomatous; leaves ovate-oblong, entire, truncate to cuneate at base, shortly petiolate; spadix terminal, without appendix; spathe wrapped round and concealing spadix at base, falling off after flowering; flowers bisexual; tepals 4; stamens 4; ovary (1-)2-celled; fruit a green berry with 2 seeds.

1 Flowers foul-smelling; spathe 10-35cm, yellow; spadix 3.5-12cm; tepals 3-4mm; anthers 0.9-2mm. Intrd-natd; swampy ground; scattered throughout Ir and S & W Br but see next sp.
American Skunk-cabbage - **L. americanus** Hultén & H. St. John
1 Flowers ± scentless; spathe and spadix slightly smaller; spathe white; tepals 2-3mm; anthers 0.6-0.8mm. Intrd-natd; similar places to *L. americanus*; distribution uncertain due to confusion with latter but probably rarer   *Asian Skunk-cabbage* - **L. camtschatcensis** (L.) Schott

3. CALLA L. - *Bog Arum*
Rhizomatous; leaves ovate to broadly so, cordate, entire, with long petiole; spadix terminal, without appendix; spathe open, ± flat, not concealing spadix; flowers mostly bisexual but uppermost usually male; tepals 0; stamens 6; ovary 1-celled; fruit a red berry with several seeds.

1 Leaf-blades 5-12 x 4-10cm; spathe 3-8 x 3-6cm, white or greenish-white; spadix 1-3 x 0.7-2cm. Intrd-natd; marshy ground and shallow ponds, often in shade; very scattered in Br from SE En to C Sc   *Bog Arum* - **C. palustris** L.

4. ZANTEDESCHIA Spreng. - *Altar-lily*
Rhizomes short, tuberous; leaves ovate or broadly so, cordate, entire, with long petiole; spadix terminal, without appendix; spathe wrapped round and concealing spadix at base; flowers unisexual; tepals 0; stamens 2-3; ovary with (1-)3 cells; fruit a yellow berry with several seeds; but very seldom (?never) produced.

1 Leaf-blades 10-45 x 10-25cm; petiole up to 50(75)cm; spathe 10-25cm, pure white; spadix up to 15cm, bright yellow. Intrd-natd; ditches, damp hedgerows and scrub, and neglected fields; CI, SW En, S Wa and W Ir   *Altar-lily* - **Z. aethiopica** (L.) Spreng.

5. ARUM L. - *Lords-and-Ladies*
Rhizomes short, tuberous; leaves triangular-ovate, hastate to sagittate, entire, with long petiole; spadix terminal, with long appendix; spathe wrapped round and concealing spadix at base, pale greenish-yellow; flowers unisexual; tepals 0; stamens 3-4; ovary 1-celled; fruit a red berry with 1-several seeds.

1 Leaf-blades appearing in early spring, often blackish-purple-spotted, with concolorous midrib; spadix appendage purple or yellow, usually reaching c.1/2 way up expanded part of spathe. Native; woods and hedgerows, usually on base-rich soils; frequent in BI N to C Sc, natd further N **Lords-and-Ladies - A. maculatum** L.

1 Leaf-blades appearing in early winter, with pale midrib; spadix appendage yellow, usually reaching c.1/3 way up expanded part of spathe (*Italian Lords-and-Ladies* - **A. italicum** Mill.) 2

  2 Leaves sometimes dark-spotted, with veins slightly paler than rest of leaf, with basal lobes somewhat convergent, sometimes overlapping; fruits with 1-2 seeds. Native; hedgerows, scrub and stony field-borders; extreme S & SW En and CI, very rarely natd elsewhere **A. italicum ssp. neglectum** (F. Towns.) Prime

  2 Leaves never dark-spotted, with whitish veins, with basal lobes divergent; fruits with 2-4 seeds. Intrd-natd; in similar places to ssp. *neglectum*; scattered in CI, Br N to Man, Dunbarton and S Lincs, E Ir **A. italicum ssp. italicum**

A. maculatum x *A. italicum* occurs rarely in S & SW En, S Wa and CI; it is intermediate and probably sterile, but has leaves appearing in early winter and often spotted.

6. DRACUNCULUS Mill. - *Dragon Arum*

Rhizomes short, tuberous; leaves deeply ± palmately lobed, cordate at base, with entire lobes and long petiole; spadix terminal, with long appendix; spathe wrapped round and concealing spadix at base; flowers unisexual; tepals 0; stamens 2-4; ovary 1-celled; fruit a red berry with several seeds.

1 Leaf-lobes up to 20cm; spadix appendage dark purple, nearly as long as spathe; spathe 25-40cm, dark purple. Intrd-natd; hedges, rough ground and old gardens; scattered in S & SE En and CI

Dragon Arum - **D. vulgaris** Schott

7. ARISARUM Mill. - *Mousetailplant*

Rhizomatous; leaves triangular-ovate, sagittate, entire, with long petiole; spadix terminal, with long appendix; spathe fused in tube round spadix and concealing most of it, extended apically into filiform projection; flowers unisexual; tepals 0; stamen 1; ovary 1-celled; fruit green, with several seeds.

1 Leaf-blades 6-15cm; spadix appendage whitish, concealed within spathe; spathe 2-4cm excl. filiform projection 5-15cm, dark or greenish-brown. Intrd-natd; hedges, rough ground and old gardens; very scattered in S En *Mousetailplant* - **A. proboscideum** (L.) Savi

152. LEMNACEAE - *Duckweed family*

Aquatic perennial plants reduced to ± undifferentiated pad-like *frond* to

152. LEMNACEAE

15mm (but often much less) floating on or under water surface (sometimes stranded on mud), not or variously adhering together, with 0-21 roots per frond; the floating pad-like plants are unique. Usually not flowering.

Only well-grown spring or summer fronds should be used; poorly grown ones or those produced in autumn and over-wintering are often atypical, being smaller and often with fewer veins and fewer or 0 roots.

1 Fronds rootless and veinless, spherical to ellipsoid **3. WOLFFIA**
1 Each frond with (0)1-16(21) roots and 1-16(21) veins, ± flattened at least on upperside 2
   2 Each frond with (0-)1 root and 1-5(7) veins **2. LEMNA**
   2 Each frond with 7-16(21) roots and veins **1. SPIRODELA**

1. SPIRODELA Schleid. - *Greater Duckweed*
Fronds with 7-16(21) roots, with 7-16(21) veins, floating on water surface.

1 Fronds 1.5-10 x 1.5-8mm, ± flattened on both surfaces. Native; canals, ditches and ponds; rather local in C & S Br, very scattered in N Br and Ir *Greater Duckweed* - **S. polyrhiza** (L.) Schleid.

2. LEMNA L. - *Duckweeds*
Fronds with (0-)1 root, with 1-5(7) veins, floating on or below water surface.

1 Fronds narrowed to a stalk-like portion at 1 end, usually submerged, usually cohering in branched chains of 3-50. Fronds 3-15 (plus stalk 2-20) x 1-5mm, with (1-)3 veins, with 0-1 root. Native; ponds, ditches and canals; frequent in most of BI *Ivy-leaved Duckweed* - **L. trisulca** L.
1 Fronds orbicular to ellipsoid, without stalk-like portion, usually on water surface, cohering in small groups (not chains) 2
   2 Fronds usually strongly swollen on lowerside, ± hemispherical, with (3)4-5(7) veins originating from 1 point. Fronds 1-8 x 0.8-6mm, with larger air-spaces (visible as reticulum on frond upperside) >0.3mm across. Native; ponds, ditches and canals, usually in rich, often brackish water; frequent in C & S Br, very scattered in CI and Ir *Fat Duckweed* - **L. gibba** L.
   2 Fronds ± flattened on both surfaces, with 1-3(5) veins, if 4 or 5 then 3 originating from 1 point and the outermost 1 or 2 extras branching from near base of inner laterals 3
3 Fronds 0.8-3(4) x 0.5-2.5mm long, usually elliptic, with 1 vein, with larger air-spaces ≤0.25mm across. Intrd-natd; same habitats as *L. minor*; scattered in CI and Br N to E Norfolk and Flints
Least Duckweed - **L. minuta** Kunth
3 Fronds (1)2-5(8) x 0.6-5mm long, usually ovate, with 3(-5) veins, with larger air-spaces ≤0.3mm across. Native; ponds, ditches, canals and slow parts of rivers and streams; common throughout BI except rare in N Sc *Common Duckweed* - **L. minor** L.

3. WOLFFIA Schleid. - *Rootless Duckweed*
Fronds with 0 roots, with 0 veins, usually floating on water surface.

1 Fronds 0.5-1.5 x 0.4-1.2mm, strongly swollen on both sides (thicker than wide). Native; ponds and ditches; very local in S En and Mons
Rootless Duckweed - **W. arrhiza** (L.) Wimm.

153. COMMELINACEAE - *Spiderwort family*

Herbaceous perennials; easily recognized by the flowers with 3 green sepals, 3 white or coloured petals, and 6 stamens with conspicuous long hairs on filaments.

1. TRADESCANTIA L. - *Spiderworts*

1 Tufted, shortly rhizomatous perennial; stems erect, to 60cm; leaves linear, 15-35cm; flowers 2.5-3.5cm across; petals usually violet, sometimes white, pink or purple. Intrd-surv; on tips and waste ground where thrown out; rare in SE En   *Spiderwort* - **T. virginiana** L.
1 Stems trailing, rooting at nodes, to 1m or more; leaves ovate-oblong, 1.5-5cm; flowers 1-1.5cm across; petals white. Intrd-surv; in shrubberies and frost-free rough ground where thrown out; rare in S En and CI   *Wandering-jew* - **T. fluminensis** Vell.

154. ERIOCAULACEAE - *Pipewort family*

Aquatic (or ± so) herbaceous perennials rooted in substratum; unique in the subulate basal leaf-rosette and erect leafless stems bearing small capitate inflorescences of many unisexual flowers.

1. ERIOCAULON L. - *Pipewort*

1 Leaves clearly transversely septate, very finely pointed, up to 10cm; stems erect, usually emergent and varying in height according to water level, up to 20(150)cm; inflorescence 5-12(20)mm across. Native; in shallow lakes and pools or in bare wet peaty ground; very local in W Sc and W Ir   *Pipewort* - **E. aquaticum** (Hill) Druce

155. JUNCACEAE - *Rush family*

Annuals or herbaceous perennials, often ± aquatic; leaves alternate or all basal, grass-like (*bifacial*) to rush-like (cylindrical to flattened but *unifacial*), with sheathing base with membranous ligule at top of sheath, the blade simple, linear and entire or 0; distinguished from other rush-, sedge- or grass-like plants by the flowers with 6 tepals, (3-)6 stamens and a single 1-3-

155. JUNCACEAE

celled ovary with 3-many ovules.

1 Leaves bifacial to unifacial, glabrous; ovary with many ovules; capsule with many seeds **1. JUNCUS**
1 Leaves bifacial, usually pubescent at least near base when young; ovary with 3 ovules; capsule with 3 seeds **2. LUZULA**

1. JUNCUS L. - *Rushes*
Annuals to perennials; leaves various, bifacial to unifacial, glabrous; ovary 1-3-celled, with many ovules; capsule with many seeds.

'Leaves' refers to stem-leaves and/or basal leaves, but excludes leaf-like bracts immediately below or within the inflorescence. In subg. *Genuini* (Key B, couplets 2-6) the lowest inflorescence bract is cylindrical and stem-like, making the inflorescence appear lateral. A sharp scalpel or razor-blade is needed to cut longitudinal and transverse sections of stem and leaves to see the internal structure; dried material usually needs resuscitation by boiling in water. Seed length includes terminal appendages.

General key
1 Leaves distinctly bifacial, flat and ± grass-like with 2 opposite surfaces but sometimes inrolled, or subcylindrical and ± rush-like but with a distinct deep channel on upperside for most or all of length **Key A**
1 Leaves unifacial or ± so, or apparently absent, cylindrical to flattened-cylindrical and rush-like, not deeply channelled or with deep channel only near ligule and extending <1/2 way to leaf apex, sometimes with shallow grooves **Key B**

Key A - Leaves bifacial, flat, or subcylindrical but with a deep channel on upperside

1 Easily uprooted annual, with simple fibrous root-system 2
1 Rhizomatous perennial, usually firmly rooted (rhizomes often very short and plant densely tufted) 5
   2 Stems unbranched, with basal leaves and leaf-like bracts at top but bare between. Stems erect, to 5cm and often much less. Native; barish ground on heaths, usually where water stands in winter; extremely local in W Cornwall, Anglesey and CI
Dwarf Rush - **J. capitatus** Weigel
   2 Stems branched, leafy (but leaves often short and very narrow) 3
3 Leaves usually >1.5mm wide; tepals usually with dark line either side of midrib; anthers 1.2-5x as long as filaments; seeds with longitudinal ridges (10-15 in side view) clearly visible with x20 lens. Stems to 35(50)cm, erect to ascending. Native; muddy margins of areas of fresh water, wet fields, marshes and ditches; scattered in W & S BI *Leafy Rush* - **J. foliosus** Desf.
3 Leaves rarely >1.5mm wide; tepals rarely with dark line on either side of midrib; anthers usually 0.3-1.1x as long as filaments; seeds without longitudinal ridges visible with x20 lens (beware shrivelled seeds) 4
   4 Inner tepals rounded to emarginate and mucronate at apex;

capsule truncate at apex, at least as long as inner tepals. Stems erect to procumbent, to 17cm. Native; damp brackish habitats near coast and inland, and on damp lime-waste; scattered throughout BI   *Frog Rush* - **J. ambiguus** Guss.

4   Inner tepals acute to subacute at apex; capsule acute to obtuse at apex, rarely ± truncate, usually shorter than inner tepals. Stems erect to procumbent, to 35(50)cm, often much less. Native; all kinds of damp habitats, fresh-water and brackish, natural and artificial; common   *Toad Rush* - **J. bufonius** L.

5 Outer 3 tepals obtuse to rounded at apex   6
5 Outer 3 tepals acute to acuminate at apex   7

    6 Anthers 0.5-1mm, 1-2x as long as filaments; style 0.1-0.3mm, <1/2 as long as stigmas; seeds 0.35-0.5mm. Loosely tufted to extensively rhizomatous; stems erect, to 50cm. Native; marshes and wet meadows, often near sea; scattered in En and Wa, rare in C Ir   *Round-fruited Rush* - **J. compressus** Jacq.

    6 Anthers 1-2mm, 2-3x as long as filaments; style 0.5-0.8mm, c. as long as stigmas; seeds 0.5-0.7mm. Usually more extensively rhizomatous than *J. compressus*; stems erect, to 50cm. Native; salt-marshes and inland saline areas; abundant round coasts of BI, very scattered inland   *Saltmarsh Rush* - **J. gerardii** Loisel.

7 Lowest 2 bracts of inflorescence leaf-like, usually far exceeding inflorescence; stem with (1)2-4(5) well-developed leaves (usually near base)   8
7 All bracts of inflorescence mostly scarious, usually much shorter than inflorescence; stem with 0(-1) well-developed leaves   9

    8 Inflorescence of 1-3(4) flowers in tight cluster; anthers longer than filaments; seeds 0.9-1.6mm, with long appendage at each end. Densely tufted; stems erect, to 30cm. Native; barish places on mountains; locally frequent in C & W Sc, Shetland
Three-leaved Rush - **J. trifidus** L.

    8 Inflorescence of 5-40 flowers, usually ± diffuse; anthers shorter than filaments; seeds 0.3-0.4mm, with short appendages. Densely tufted; stems erect, to 80cm. Intrd-natd; damp barish ground on roadsides, tracks and paths; locally frequent throughout BI
Slender Rush - **J. tenuis** Willd.

9 Leaves flat or ± inrolled; stamens 3. Loosely tufted; stems erect, to 30cm. Intrd-natd; damp pathsides, lake shores and wet meadows; local in W Galway   *Broad-leaved Rush* - **J. planifolius** R. Br.

9 Leaves rounded on lowerside, with deep channel on upperside; stamens 6. Densely tufted; stems erect, to 50cm. Native; bogs, wet moors and heaths, on acid soil; common throughout Br and Ir where acid soils exist   *Heath Rush* - **J. squarrosus** L.

Key B - Leaves unifacial or ± so, cylindrical to flattened-cylindrical

1 Leaves on stems represented only by blade-less scarious sheaths near base   2
1 At least 1 stem-leaf with well-developed green blade   7

1. JUNCUS

2 Stems strongly glaucous, with pith conspicuously and regularly interrupted at least in region just below inflorescence. Densely tufted; stems erect, to 1.2m. Native; marshes, dune-slacks, wet meadows, ditches, by lakes and rivers, usually on neutral or base-rich soils; common N to C Sc   ***Hard Rush* - J. inflexus** L.

2 Stems not glaucous, with pith well formed and conspicuous or ill-formed and irregular   3

3 Rhizomes extended, forming straight lines or diffuse patches of aerial stems; inflorescence usually <20-flowered; stems rarely >50cm   4

3 Rhizomes short, forming dense clumps of aerial stems (or large dense patches in very old plants); inflorescence usually >20-flowered; stems usually >50cm   5

    4 Inflorescence in lower 2/3(-3/4) of apparent stem, ± globose; stem with subepidermal sclerenchyma girders, with fine longitudinal ridges when dry. Rather weakly rhizomatous; stems erect, to 45cm. Native; on stony, silty edges of lakes and reservoirs; local in Br from Leics to Easterness   ***Thread Rush* - J. filiformis** L.

    4 Inflorescence in upper 1/4 of apparent stem, usually elongated; stem without subepidermal sclerenchyma girders, not ridged when dry. Strongly rhizomatous; stems erect, to 75cm. Native; maritime dune-slacks, rarely on upland river terraces, on bare or grassy ground; local in N Sc S to M Ebudes and Fife, S Lancs
                                                                                ***Baltic Rush* - J. balticus** Willd.

5 Fresh stems dull, ridged, with usually <35 ridges; main (stem-like) bract opened out and ± flat at base adjacent to inflorescence, causing it to hinge over backwards at end of season. Densely tufted; stems erect, to 1m. Native; similar places to *J. effusus*; common throughout BI   ***Compact Rush* - J. conglomeratus** L.

5 Fresh stems glossy, smooth, becoming finely ridged with usually >35 ridges when dry; main bract scarcely opened out at base adjacent to inflorescence, not hingeing over backwards at end of season   6

    6 Capsule as long as or longer than tepals, 2.5-3.5mm when fully fertile; stamens 6. Densely tufted; stems erect, to 2(3)m. Intrd-casual; waste ground and fields; formerly natd, now sporadic in En   ***Great Soft-rush* - J. pallidus** R. Br.

    6 Capsule shorter than tepals, 2-2.5mm when fully fertile; stamens 3(-6). Densely tufted, sometimes patch-forming; stems erect, to 1.2m. Native; marshes, ditches, bogs, wet meadows, by rivers and lakes, damp woods, mostly on acid soils; abundant
                                                                                        ***Soft-rush* - J. effusus** L.

7 Leaves and main bract with very sharp apex, with subepidermal sclerenchyma girders, with vascular bundles scattered through pith   8

7 Leaves and main bract with soft apex, without subepidermal sclerenchyma girders, without vascular bundles in pith   9

    8 Capsule 2.5-3.5mm, not or slightly longer than tepals; inner tepals obtuse to subacute, the extreme apex not exceeded by membranous margins. Densely tufted to clearly rhizomatous; stems erect, very stiff, to 1m. Native; saltmarshes; common on

coasts of BI except extreme N Sc *Sea Rush* - **J. maritimus** Lam.
8 Capsule 4-6mm, much longer than tepals; inner tepals retuse, with membranous margins extended into lobes on each side of extreme apex and exceeding it. Densely tufted; stems erect, extremely stiff, to 1.5m. Native; sandy sea-shores and drier parts of saltmarshes; very local in BI N to W Norfolk, Caerns and Co Dublin
Sharp Rush - **J. acutus** L.
9 Leaves cylindrical, with continuous pith within vascular cylinder; each flower with 2 small bracteoles immediately beneath tepals. Rhizomatous, patch-forming; stems rather weak, ± erect, to 1m. Intrd-natd; saltmarsh in N Somerset, and wet reclaimed land by docks in Stirlings *Somerset Rush* - **J. subulatus** Forssk.
9 Leaves cylindrical to flattened-cylindrical, with pith usually interrupted by transverse septa of stronger tissue and often with large cavities; flowers without bracteoles at base 10
  10 Easily uprooted annual, with simple fibrous root system. Stems erect to ascending, to 8cm. Native; damp hollows and rutted tracks on heathland; Lizard area (W Cornwall)
Pigmy Rush - **J. pygmaeus** Thuill.
  10 Perennial, usually firmly rooted, either rhizomatous or with ± swollen stem-bases (widespread) 11
11 Anthers <1/3 as long as filaments; seeds with conspicuous whitish appendages at each end each c. as long as actual seed (alpine) 12
11 Anthers >1/3 (up to 2x) as long as filaments; seeds with at most minute points at each end (lowland or alpine) 14
  12 Outer tepals acute; capsule ≥6mm; stems usually solitary from rhizome system; anthers >1mm. Stems to 30cm. Native; boggy places and flushes on mountains; very local in C, W & NW Sc
Chestnut Rush - **J. castaneus** Sm.
  12 Outer tepals obtuse to rounded; capsule <6mm; stems usually in small tufts; anthers <1mm 13
13 Capsule 3.2-4.5mm (excl. style), retuse at apex; lowest bract usually exceeding inflorescence; flowers mostly 2 per inflorescence. Stems tufted, to 12cm, erect. Native; barish rocky places on mountains; very local in C, W & NW Sc *Two-flowered Rush* - **J. biglumis** L.
13 Capsule 4-5.5mm (excl. style), obtuse at apex; lowest bract usually shorter than inflorescence; flowers mostly 3 per inflorescence. Stems tufted, to 20cm, erect. Native; boggy and rocky places on mountains; local in C, W & NW Sc, very local in N En and N Wa
Three-flowered Rush - **J. triglumis** L.
  14 Leaves with >2 empty or loosely pith-filled longitudinal cavities separated by thin walls bearing a few vascular bundles 15
  14 Leaves with 1 empty or loosely pith-filled longitudinal cavity 16
15 Rhizomatous; outer tepals obtuse, incurved at apex; flowers very rarely vegetatively proliferating. Stems erect, to 1.2m. Native; fens, marshes and dune-slacks on peaty base-rich soil; locally frequent in En, Wa and Ir, very local elsewhere
Blunt-flowered Rush - **J. subnodulosus** Schrank

1. JUNCUS

15 Not rhizomatous; stem-base usually swollen; outer tepals acute, not incurved; flowers commonly vegetatively proliferating. Stems loosely tufted, erect to procumbent or floating, to 30cm, often rooting at nodes. Native; in all kinds of wet and damp places, often submerged; common **Bulbous Rush - J. bulbosus** L.

16 Tepals acuminate, the outer with recurved apical points. Rhizomatous; stems erect, to 1.1m. Native; marshes, bogs, damp grassland, margins of rivers and ponds; common
Sharp-flowered Rush - J. acutiflorus Hoffm.

16 Outer tepals obtuse to shortly acuminate with erect apical points; inner tepals acute to rounded 17

17 Outer tepals subacute to obtuse; inner tepals obtuse to rounded; capsule obtuse (ignore beak). Rhizomatous; stems erect to ascending, to 40cm. Native; marshes, flushes and streamsides on mountains; local in mainland Br N from MW Yorks
Alpine Rush - J. alpinoarticulatus Chaix

17 Outer tepals acute; inner tepals acute to subacute; capsule acute (ignore beak). Rhizomatous; stems erect to decumbent, to 80cm but often much less. Native; damp grassland, heaths, moors, marshes, dune-slacks, margins of rivers and ponds; common
Jointed Rush - J. articulatus L.

8 hybrids, varying in fertility and abundance, have been recorded. *J. articulatus* x *J. acutiflorus* = **J. x surrejanus** Stace & Lambinon occurs with the parents throughout BI and is commoner than either in some places; it is intermediate in tepal shape and size and has low fertility. *J. inflexus* x *J. effusus* = **J. x diffusus** Hoppe occurs sporadically with the parents, usually as isolated plants; the stems are not glaucous and have continuous pith, the inflorescence shape is ± as in *J. inflexus*, and fertility is low. *J. effusus* x *J. conglomeratus* = **J. x kern-reichgeltii** Reichg. occurs sporadically with the parents in Br (mainly N & W) and E & W Cork; it is intermediate in diagnostic characters and highly fertile, forming back-crosses.

2. LUZULA DC. - *Wood-rushes*

Perennials, vegetatively grass-like; leaves bifacial, variously pubescent but rarely without hairs near base of leaf on margins; ovary 1-celled, with 3 ovules; capsules with 3 seeds.

1 Flowers all or most borne singly in inflorescence each on distinct pedicels >3mm, rarely some in pairs 2
1 Flowers mostly borne in groups of 2 or more, each one in a group sessile or with pedicels <2mm, often a few solitary 3
  2 Basal leaves rarely >4mm wide; inflorescence branches erect to widely erecto-patent in fruit; seeds with terminal appendage ≤1/2 as long as rest of seed, ± straight. Tufted, with very short rhizomes; stems ± erect, to 35cm. Native; woods and hedgerows; local in CI and S Br **Southern Wood-rush - L. forsteri** (Sm.) DC.
  2 Some basal leaves usually >4mm wide; lower inflorescence

branches reflexed in fruit; seeds with terminal appendage >1/2 as long as (often longer than) rest of seed, often curved or hooked. Tufted, with very short rhizomes; stems erect, to 35cm. Native; woods, hedgerows, among heather on moors; common in most of Br, scattered in Ir **Hairy Wood-rush - L. pilosa** (L.) Willd.

3 Tepals white to pale straw-coloured. Tufted, with short rhizomes; stems ± erect, to 70cm. Intrd-natd; in woods and by shady streams; scattered through most of Br
White Wood-rush - L. luzuloides (Lam.) Dandy & Wilmott

3 Tepals yellowish- to dark-brown 4

    4 All or most basal leaves >8mm wide. Densely tufted and with long rhizomes; stems erect, to 80cm. Native; woods, moorland, shady streamsides; locally common in BI except parts of E En
Great Wood-rush - L. sylvatica (Huds.) Gaudin

    4 All leaves <8mm wide 5

5 Inflorescence drooping, spike-like, with the flower groups subsessile along main axis, or the lower themselves forming lateral spikes. Tufted, with short rhizomes; stems erect but pendent at apex, to 30cm. Native; open stony ground on mountains; local in C & N Sc
Spiked Wood-rush - **L. spicata** (L.) DC.

5 Inflorescence without single main axis; either all flower clusters congested in dense head or some or all with distinct stalks arising from short main axis near base of inflorescence 6

    6 Leaves deeply channelled, glabrous or sparsely pubescent just near base; seeds with inconspicuous appendage ≤1/10 as long as rest of seed. Tufted, with short rhizomes; stems erect, to 10cm. Native; open stony ground on high mountains; very local in C & N mainland Sc **Curved Wood-rush - L. arcuata** (Wahlenb.) Sw.

    6 Leaves ± flat, conspicuously pubescent; seeds with conspicuous whitish terminal appendage c.1/4-1/2 as long as rest of seed; widespread 7

7 Rhizomatous or stoloniferous; anthers (2.5)3-4x as long as filaments; style (excl. stigmas) (0.9)1.1-1.6mm. Tufted, with short rhizomes; stems erect, to 15(25)cm. Native; short grassland and similar places; very common **Field Wood-rush - L. campestris** (L.) DC.

7 Rhizomes and stolons 0; anthers 0.8-2.2(2.5)x as long as filaments; style (excl. stigmas) 0.2-0.8(0.9)mm 8

    8 Outer tepals 2-2.6(2.8)mm; peduncles densely minutely papillose; seeds 0.5-0.6mm wide; style (excl. stigmas) 0.2-0.3mm; stigmas 0.5-0.6mm. Tufted, usually with 0 rhizomes; stems erect, to 30cm. Native; open grassy places in dry parts of fens; Hunts, Co Antrim and Offaly, perhaps extinct **Fen Wood-rush - L. pallidula** Kirschner

    8 Outer tepals 2.6-3.3(3.5)mm; peduncles smooth or distally sparsely papillose; seeds 0.7-0.9mm wide; style (excl. stigmas) 0.4-0.9mm; stigmas 1.2-2.4(3.1)mm. Tufted, usually with 0 rhizomes; stems erect, to 60cm. Native; grassland, heaths, moors, woods on acid soil (**Heath Wood-rush - L. multiflora** (Ehrh.) Lej.) 9

9 Seeds 1.2-1.5 x 0.9-1mm (excl. appendage); all or most flower clusters

2. LUZULA

subsessile in compact lobed head. Common throughout BI except parts of C & E En **L. multiflora ssp. congesta** (Thuill.) Arcang.
9 Seeds 0.8-1.2 x 0.7-0.9mm (excl. appendage); flowers in several mostly stalked corymbose clusters 10
   10 Flower-clusters all on erect peduncles; seed appendages 0.4-0.5mm; most basal leaves >3mm wide; capsules 2-2.8mm. Common throughout BI except parts of C & E En and C & W Ir
L. multiflora ssp. multiflora
   10 Some flower-clusters on recurved peduncles; seed appendages 0.2-0.3mm; most basal leaves <3mm wide; capsules 1.9-2.2mm. Widespread in W & C Ir, largely replacing ssp. *multiflora*
L. multiflora ssp. hibernica Kirschner & T.C.G. Rich

L. forsteri x *L. pilosa* = **L. x borreri** Bab. occurs frequently within the range of *L. forsteri* in Br, and formerly outside it in Co Wicklow; it is intermediate in inflorescence shape and leaf width, with very low fertility.

156. CYPERACEAE - *Sedge family*

Herbaceous, usually rhizomatous, perennials, rarely annuals, with usually solid, often 3-angled stems, mostly aquatic or in wet places; flowers much reduced, 1 each in axil of bract-like *glume*, 1-many in discrete units (*spikelets*), the spikelets terminal and solitary or in terminal spikes, racemes or panicles, often with extra sterile glumes; easily told from other grass- or rush-like plants except Poaceae by the (often unisexual) very reduced flowers with perianth 0 or in the form of bristles, and the 1-celled, 1-ovuled ovary, and from Poaceae by the absence of a bract (*palea*) *above* each flower and usually solid, often 3-angled stems. Some Poaceae lack a palea but all these have hollow stems; the only member of the Cyperaceae with this combination is *Cladium*.

1 Stems hollow; leaves usually with fierce saw-edged margins and lowerside of midrib, easily cutting the skin **14. CLADIUM**
1 Stems solid (centre often occupied by very soft pith); leaves not saw-edged or very mildly so 2
   2 Perianth represented by bristles which elongate and greatly exceed glumes at fruiting, forming a whitish cottony head 3
   2 Perianth 0 or represented by inconspicuous bristles shorter than glumes 4
3 Perianth-bristles 4-6 per flower; spikelet 1, terminal, <1cm excl. bristles (extinct) **2. TRICHOPHORUM**
3 Perianth-bristles numerous per flower; spikelets 1-several, >1cm excl. bristles **1. ERIOPHORUM**
   4 Flowers all unisexual, the male and female in different spikes or different parts of the same spike, or rarely on different plants; ovary and fruit enclosed or closely enfolded in membranous innermost glume 5

4	Flowers all bisexual; ovary and fruit not enclosed or closely enfolded in innermost glume	6
5	Ovary and fruit entirely enclosed in fused membranous glume usually ending in a short or long beak; male and female flowers in same or different spikes or on different plants, the spikes variously crowded or distant, and stalked or sessile; stigmas and stamens 2 or 3	**16. CAREX**
5	Ovary and fruit closely enfolded in innermost glume which is not fused, leaving fruit exposed at top; male and female flowers in same spikes, the spikes crowded and ± sessile; stigmas and stamens 3	**15. KOBRESIA**
6	Inflorescence of 1 terminal spikelet; lowest bract not leaf-like or stem-like, shorter than spikelet	7
6	Inflorescence of ≥2 spikelets, or sometimes of 1 but then lowest bract leaf-like or stem-like and exceeding spikelet	9
7	Most or all leaf-sheaths on stems with leafy blades	**9. ELEOGITON**
7	Most or all leaf-sheaths on stems without blades	8
8	Uppermost leaf-sheath on stem with short blade	**2. TRICHOPHORUM**
8	Uppermost leaf-sheath on stem (and all or most below it) without a blade	**3. ELEOCHARIS**
9	Inflorescence with ≥2 bifacial leaf-like bracts very close together at base	10
9	Inflorescence with basal bracts stem-like or leaf-like, if leaf-like then either 1 or ≥2 and well spaced out	12
10	Spikelets flattened, with glumes on 2 opposite sides of axis	**11. CYPERUS**
10	Spikelets ± terete, with glumes spirally arranged	11
11	Inflorescence dense; spikelets >8mm	**4. BOLBOSCHOENUS**
11	Inflorescence diffuse; spikelets <5mm	**5. SCIRPUS**
12	Inflorescence a flattened compact terminal head, with spikelets only on 2 opposite sides of main axis	**10. BLYSMUS**
12	Inflorescence various, if a compact terminal head then spikelets not only on 2 opposite sides of axis	13
13	Spikelets flattened, with glumes on 2 opposite sides of axis	**12. SCHOENUS**
13	Spikelets terete, with glumes spirally arranged	14
14	Inflorescence obviously terminal with leaf-like main bract; stems with several well-developed leaf-blades	**13. RHYNCHOSPORA**
14	Inflorescence usually apparently lateral, with main bract ± stem-like and continuing stem apically; stems with 0-1(2) reduced leaf-blades	15
15	Stems very slender, <1mm wide, rarely >20cm	**8. ISOLEPIS**
15	Stems stouter, >1.5mm wide, rarely <30cm	16
16	Inflorescence composed of (1-)several sessile to stalked globose apparent spikelets	**6. SCIRPOIDES**
16	Inflorescence composed of (1-)several sessile to stalked ovoid spikelets	**7. SCHOENOPLECTUS**

1. ERIOPHORUM L. - *Cottongrasses*

Perennials with long or short rhizomes; stems terete to triangular in section, leafy; leaves variously shaped in section; inflorescence of 1-several large spikelets in a terminal umbel; lowest bract leaf-like or glume-like; flowers bisexual; perianth of numerous (>6) bristles elongating to form conspicuous white, cottony head in fruit; stamens 3; ovary not enfolded or enclosed by glume; stigmas 3.

1 Spikelet 1, erect, without leaf-like bract at base; leaf-blades ± triangular in section, 0 or very reduced on uppermost stem leaf-sheath. Rhizomes very short; stems densely tufted, often tussock-forming, erect, to 50cm. Native; wet peaty places, especially on moorland bogs; common in Ir and W, C & N Br, very local in C, E & S En
Hare's-tail Cottongrass - **E. vaginatum** L.

1 Spikelets ≥(1)2, ± pendent in fruit, with 1-3 ± leaf-like bracts at base; leaf-blades flat to V-shaped in section, well developed on uppermost stem leaf-sheath   2

   2 Stalks of spikelets smooth; stems ± terete to very bluntly 3-angled; anthers >2mm. Rhizomes long; stems scattered, erect, to 60cm. Native; wet usually acid bogs; common in suitable places
Common Cottongrass - **E. angustifolium** Honck.

   2 Stalks of spikelets with numerous minute forward-pointed bristles; stems distinctly 3-angled; anthers ≤2mm   3

3 Leaf-blades 0.5-2mm wide; glumes with midrib plus several shorter parallel veins on either side; plant with long rhizomes, with solitary stems. Habit like a slender *E. angustifolium*. Native; similar places to *E. angustifolium*; very local in S Br and C & W Ir
Slender Cottongrass - **E. gracile** Roth

3 Leaf-blades 3-8mm wide; glumes with only midrib; plant loosely tufted. Rhizomes long; stems erect, to 60cm. Native; wet base-rich marshes and flushes; scattered throughout Br and Ir
Broad-leaved Cottongrass - **E. latifolium** Hoppe

2. TRICHOPHORUM Pers. - *Deergrasses*

Tufted perennials; stems with 3 rounded angles to terete, with only the uppermost leaf-sheath with a blade; leaves thick-crescent-shaped in section, very narrow; inflorescence of 1 terminal spikelet; lowest bract glume-like; flowers bisexual; perianth of 4-6 bristles, elongating or not in fruit; stamens 3; ovary not enfolded or enclosed by glume; stigmas 3.

1 Rather diffusely tufted; stems with 3 rounded angles, slightly scabrid near apex; perianth-bristles elongating to 10-25mm, forming white cottony head, at fruiting. Stems erect, to 20(30)cm. Native; bog in Angus from 1791 to c.1813   *Cotton Deergrass* - **T. alpinum** (L.) Pers.

1 Densely tufted; stems ± terete, smooth; perianth-bristles remaining shorter than glumes, pale brown. Stems erect, to 35cm. Native; bogs, wet moors and heaths (*Deergrass* - **T. cespitosum** (L.) Hartm.)   2

   2 Spikelets with 8-20 flowers; uppermost leaf-sheath with oblique,

elliptic opening c.2-3 x 1mm, with leaf-blade c. 2x as long as opening. Common throughout Br and Ir in suitables places
T. cespitosum ssp. germanicum (Palla) Hegi
2 Spikelets with 3-10 flowers; uppermost leaf-sheath (pull out terminal stem!) with transverse ± orbicular opening c.1 x 1mm, with leaf-blade c. 5-10x as long as opening. Forming smaller tufts with weaker stems and smaller spikelets than ssp. *germanicum*. Usually in wetter places than ssp. *germanicum*; S Northumb and Cheviot **T. cespitosum ssp. cespitosum**

3. ELEOCHARIS R. Br. - *Spike-rushes*

Perennials with long or short stout and/or slender rhizomes; stems terete to ridged, with blade-less leaf-sheaths; leaf-blades 0 (some spikelet-less stems may resemble basal leaves); inflorescence of 1 terminal spikelet; lowest bract glume-like; flowers bisexual; perianth of 0-6 bristles, not elongating in fruit; stamens 3; ovary not enfolded or enclosed by glume; stigmas 2 or 3.

1 Lowest glume >(2/5-)1/2 as long as spikelet; spikelets 3-12-flowered 2
1 Lowest glume <2/5(-1/2) as long as spikelet; spikelets 10-many-flowered 4
   2 Glumes greenish; uppermost stem leaf-sheath delicate, inconspicuous; very slender whitish rhizomes ending in small whitish tubers (c.2-5mm) present. Stems sparsely tufted, to 8cm. Native; wet muddy places by sea and in estuaries; very local in SW Br from S Hants to Caerns *Dwarf Spike-rush* -
E. parvula (Roem. & Schult.) Bluff, Nees & Schauer
   2 Glumes brown (often with green midrib); uppermost stem leaf-sheath conspicuous, brownish; rhizomes brownish, not bearing tubers 3
3 Stems <0.5mm wide, usually 4-ridged; spikelets 2-5mm; lowest glume 1.5-2.5mm. Stems sparsely tufted, to 10cm (more if submerged). Native; in and by pond and lake margins; scattered in Br and Ir
Needle Spike-rush - **E. acicularis** (L.) Roem. & Schult.
3 Stems ≥0.5mm wide, ± terete; spikelets 4-10mm; lowest glume 2.5-5mm. Stems loosely to rather densely tufted, to 30cm. Native; wet places in fens, dune-slacks and moorland; scattered throughout BI
Few-flowered Spike-rush - **E. quinqueflora** (Hartmann) O. Schwarz
   4 Lowest glume ± completely encircling spikelet at base 5
   4 Lowest glume ≤1/2(-3/4)-encircling spikelet at base 6
5 Uppermost stem leaf-sheath oblique (c.45°) at apex; stigmas 3; nuts 3-angled. Stems densely tufted, to 40cm. Native; bogs and wet peaty places, usually on acid soils; throughout BI, common in W, sparse in E *Many-stalked Spike-rush* - **E. multicaulis** (Sm.) Desv.
5 Uppermost stem leaf-sheath ± truncate at apex; stigmas 2; nuts biconvex. Stems loosely to rather densely tufted, to 50cm. Native; marshes and dune-slacks; scattered throughout Br and Ir, mostly coastal *Slender Spike-rush* - **E. uniglumis** (Link) Schult.
   6 Stems with 10-16 vascular bundles, showing up as fine ridges in

3. ELEOCHARIS

dried state; bristles (4)5(-6); spikelets conical; swollen style-base on top of fruit c.1-1.5x as long as wide, slightly constricted at base. Stems loosely to rather densely tufted, to 50cm. Native; wet, marshy and flushed areas in or by rivers; local in N En and S Sc **Northern Spike-rush - E. austriaca** Hayek

6 Stems with ≥20 vascular bundles, showing up as fine ridges in dried state; bristles (0-)4; spikelets cylindrical- to ellipsoid-conical; swollen style-base on top of fruit wider than long, strongly constricted at base. Stems loosely to rather densely tufted, to 75cm, often much less. Native; in or by ponds, marshes, ditches, rivers (*Common Spike-rush* - **E. palustris** (L.) Roem. & Schult.) 7

7 Spikelets usually with <40 flowers; glumes from middle of spikelet 3.5-4.5mm; nut (1.3)1.5-2mm. Frequent throughout BI
E. palustris ssp. vulgaris Walters

7 Spikelets usually with >40 flowers; glumes from middle of spikelet 2.7-3.5mm; nut 1.2-1.4(1.5)mm. S & C En, much rarer than ssp. *vulgaris* **E. palustris ssp. palustris**

E. palustris x *E. uniglumis* occurs within populations of *E. uniglumis* near to those of *E. palustris* on the W coasts of Br; it is intermediate and fertile.

4. BOLBOSCHOENUS (Asch.) Palla - *Sea Club-rush*

Strongly rhizomatous perennials; stems with 3 acute angles, leafy; leaves flattened, widely V-shaped in section; inflorescence of (1)3-many spikelets either sessile or variously clustered on 1-several stalks; lowest bract leaf-like; flowers bisexual; perianth-bristles 1-6, not elongating in fruit; stamens 3; ovary not enfolded or enclosed by glume; stigmas 2-3.

1 Stems strong, erect, to 1m; leaves long, 2-10mm wide; spikelets 10-30mm, dark brown. Native; wet muddy places in estuaries or by sea; common round most coasts of BI, rarely inland
Sea Club-rush - B. maritimus (L.) Palla

5. SCIRPUS L. - *Wood Club-rush*

Strongly rhizomatous perennials; stems with 3 rounded angles, leafy; leaves flat; inflorescence of very numerous spikelets 1-several on ends of diffusely branching panicle; lowest bract leaf-like; flowers bisexual; perianth-bristles 6, not elongating in fruit; stamens 3; ovary not enfolded or enclosed by glume; stigmas 3.

1 Stems strong, erect, to 1.2m; leaves long, 5-20mm wide; spikelets 3-4mm, greenish-brown. Native; by streams and in marshes and damp woods or shady places; locally frequent over Br and Ir N to C Sc **Wood Club-rush - S. sylvaticus** L.

6. SCIRPOIDES Ség. - *Round-headed Club-rush*

Strongly rhizomatous perennials; stems terete; leaf-sheaths mostly blade-less but uppermost 1(-2) with well-developed blade; leaves semi-circular in section; inflorescence of (1)5-many variously stalked or sessile globular heads (apparent spikelets) each actually of numerous tightly packed spikelets; lowest bract ± stem-like, making inflorescence appear lateral; flowers bisexual; perianth-bristles 0; stamens 3; ovary not enfolded or enclosed by glume; stigmas 3.

1 Stems erect, to 1.5m; leaves semicircular in section; heads 3-10mm across, with spikelets 2.5-4mm, brown. Native; damp sandy places near sea; very rare in N Devon and N Somerset, rarely intrd elsewhere in S Br **Round-headed Club-rush - S. holoschoenus** (L.) Soják

7. SCHOENOPLECTUS (Rchb.) Palla - *Club-rushes*

Strongly rhizomatous perennials; stems terete or triangular in section with acute angles; leaf-sheaths mostly blade-less but uppermost 1(-3) with rather short blade; leaves crescent-shaped in section; inflorescence of (1-)few to numerous variously stalked or sessile ovoid spikelets 5-8mm; lowest bract ± stem-like, making inflorescence appear lateral; flowers bisexual; perianth-bristles 0-6, not elongating in fruit; stamens 3; ovary not enfolded or enclosed by glume; stigmas 2-3.

1 Stems terete 2
1 Stems triangular in section with acute angles 3
    2 Glumes (except apical projection) smooth; stigmas mostly 3; nut 2.5-3mm, mostly 3 angled. Stems erect, to 3m. Native; in shallow water of lakes, ponds, slow rivers, canals and dykes; frequent over most of Br and Ir *Common Club-rush* - **S. lacustris** (L.) Palla
    2 Glumes minutely (x20 lens) but densely papillose at least near midrib and apex; stigmas 2; nut 2-2.5mm, biconvex or planoconvex. Stems erect, to 1.5m. Native; in similar places to *S. lacustris* but also in marshes, dune-slacks and wet peaty places; frequent near sea, very scattered inland
        Grey Club-rush - S. tabernaemontani (C.C. Gmel.) Palla
3 Glumes with rounded to obtuse lobe on either side of apical projection; stems with uppermost 1(-2) leaf-sheaths with blades; perianth-bristles 6, >1/2 as long as nut. Stems erect, to 1(1.5)m. Native; in tidal mud of rivers; very local in S En and W Ir
        Triangular Club-rush - S. triqueter (L.) Palla
3 Glumes with acute to subacute lobe on either side of apical projection; stems with uppermost 2-3 leaf-sheaths with well-developed blades; perianth-bristles 0-6, <1/2 as long as nut. Stems erect, to 60cm. Native; pond-margin in Jersey (extinct), wet dune-slacks in S Lancs (extinct, re-intrd from same stock) **Sharp Club-rush - S. pungens** (Vahl) Palla

Hybrids of *S. triqueter* with both *S. lacustris* (= **S. x carinatus** (Sm.) Palla) and *S. tabernaemontani* (=**S. x kuekenthalianus** (Junge) D.H. Kent) occur with the parents in S En, the former perhaps extinct.

8. ISOLEPIS R. Br. - *Club-rushes*
Densely tufted annuals (to perennials); stems terete; leaf-sheaths confined to near base of stem, the upper 1(-2) with short blades; leaves crescent-shaped in section; inflorescence of 1-4 sessile spikelets 2-5mm; lowest bract usually ± stem-like, making inflorescence appear lateral, sometimes very short and ± glume-like; flowers bisexual; perianth-bristles 0; stamens 1-2; ovary not enfolded or enclosed by glume; stigmas (2-)3.

1 Main bract usually distinctly longer than inflorescence, ± stem-like; spikelets 1-4; glumes reddish-brown with green midrib; nut shiny, with longitudinal ridges. Stems erect to ascending, to 15(30)cm but usually <10cm. Native; wet ± open ground in ditches, fens, marshes and dune-slacks, on heaths, and by ponds and lakes; frequent throughout BI   *Bristle Club-rush* - **I. setacea** (L.) R. Br.
1 Main bract at most only slightly longer than inflorescence, often ± glume-like; spikelets 1(-3); glumes with brown area usually only a blotch either side of midrib; nut matt, smooth. Native; similar places to *I. setacea* but mostly near sea; frequent in Ir and extreme W Br E to S Hants, E Norfolk *Slender Club-rush* - **I. cernua** (Vahl) Roem. & Schult.

9. ELEOGITON Link - *Floating Club-rush*
Stoloniferous perennials, usually in water; stems terete, leafy; leaves ± flat; inflorescence of 1 terminal spikelet 2-5mm; lowest bract ± glume-like; flowers bisexual; perianth-bristles 0; stamens 3; ovary not enfolded or enclosed by glume; stigmas 2-3.

1 Stems usually floating, rooting at nodes, to 50cm, sometimes on mud or gravel and much shorter; spikelets green to pale brown. Native; in or by peaty ponds, lakes and ditches; fairly frequent throughout BI, commoner in W   *Floating Club-rush* - **E. fluitans** (L.) Link

10. BLYSMUS Schult. - *Flat-sedges*
Rhizomatous perennials; stems subterete, leafy; leaves flat to strongly inrolled; inflorescence a flattened ± compact terminal head with spikelets on 2 opposite sides of axis, the spikelets 4-10mm; lowest bract leaf-like to ± glume-like; flowers bisexual; perianth-bristles 0-6; stamens 3; ovary not enfolded or enclosed by glume; stigmas 2.

1 Leaves flat to slightly keeled, rough, grass-like; spikelets usually 10-20, reddish-brown; nut 1.5-2mm. Stems erect, to 40cm. Native; marshy, rather open ground; locally frequent in En, very local in Sc   *Flat-sedge* - **B. compressus** (L.) Link
1 Leaves strongly inrolled, smooth, rush-like; spikelets usually 3-8, dark brown; nut 3-4mm. Native; saltmarshes and dune-slacks in turf; locally frequent on coasts of Br and Ir S to N Lincs and Pembs
               Saltmarsh Flat-sedge - **B. rufus** (Huds.) Link

11. CYPERUS L. - *Galingales*

Rhizomatous perennials or tufted annuals or perennials; stems triangular in section with acute to rounded angles, leafy at base; leaves flat to keeled, grass-like; inflorescence a simple or more often compound umbel or umbel-like raceme, with grass-like many-flowered spikelets usually clustered on ultimate branches or all clustered in ± dense head; lowest 2-10 bracts leaf-like, often much exceeding inflorescence; flowers bisexual; perianth-bristles 0; stamens 1-3; ovary not enfolded or enclosed by glume; stigmas (2-)3.

1 Tufted annual; leaves <3(-5)mm wide; glumes <1.5mm. Stems erect, to 20cm. Native; damp barish ground by ponds and in ditches; very rare in S En and Jersey **Brown Galingale - C. fuscus** L.
1 Tufted or rhizomatous perennial; widest leaves ≥(2-)4mm wide; glumes >1.5mm 2
   2 Spikelets reddish-brown, ≤2mm wide; inflorescence ± diffuse; rhizomes long; stamens 3. Stems erect, to 1m. Native; marshes, pondsides and ditches; very local near coast in CI and SW Br, intrd in few places elsewhere in S & C En **Galingale - C. longus** L.
   2 Spikelets greenish- to yellowish-brown, ≥2mm wide; inflorescence ± compact; rhizomes very short; stamen 1. Stems erect, to 60cm. Intrd-natd; roadsides, rough and waste ground and by water; scattered in CI and S Br **Pale Galingale - C. eragrostis** Lam.

12. SCHOENUS L. - *Bog-rushes*

Densely tufted perennials; stems terete, with leaf-sheaths only at or near base and bearing short or long blades; leaves very thickly crescent-shaped in section to subterete; inflorescence a compact head of 1-4-flowered flattened spikelets; lowest bract leaf-like to ± glume-like; flowers bisexual; perianth-bristles 0 or up to 6; stamens 3; ovary not enfolded or enclosed by glume; stigmas 3.

1 Inflorescence of (2)5-10 spikelets, with lowest bract usually conspicuously exceeding it; glumes minutely rough (x20 lens) on keel. Stems erect, to 75cm. Native; damp peaty places, serpentine heathland, bogs, saltmarshes, fens, flushes; locally frequent in BI, mostly near W coasts and in E Anglia **Black Bog-rush - S. nigricans** L.
1 Inflorescence of 1-3 spikelets, with lowest bract shorter than to c. as long as it; glumes smooth on keel. Stems erect, to 40cm. Native; semi-open ground in base-rich flushes; very rare in M & E Perth
                                          Brown Bog-rush - S. ferrugineus L.

13. RHYNCHOSPORA Vahl - *Beak-sedges*

Tufted to creeping rhizomatous perennials; stems with 3 rounded angles or terete, leafy; leaves channelled; inflorescence of 1-few rather compact heads each of several 1-3-flowered spikelets; lowest bract leaf-like to ± glume-like; flowers bisexual; perianth-bristles 5-13; stamens 2-3; ovary not enfolded or enclosed by glume; stigmas 2, the common style-base persistent and forming beak to fruit.

1 Inflorescence whitish at flowering; lowest bract of terminal head not or sometimes slightly longer than head. Stems ± tufted, erect, to 40cm. Native; wet acid peaty places, locally common in Br and Ir, mainly W **White Beak-sedge - R. alba** (L.) Vahl
1 Inflorescence brown at flowering; lowest bract of terminal head >1cm longer than head. Stems ± scattered, erect, to 30cm. Native; similar places to *R. alba* and usually with it; very local in W & C Ir and S & W Br **Brown Beak-sedge - R. fusca** (L.) W.T. Aiton

14. CLADIUM P. Browne - *Great Fen-sedge*

Rhizomatous vigorous perennials; stems with 3 rounded angles or terete, leafy; leaves channelled, usually with fiercely serrate edges and keel; inflorescence much-branched, with many rather compact heads each of several 1-3-flowered spikelets; lowest bract of each head leaf-like to glume-like, but inflorescence branches with leaf-like, long bracts at base; flowers bisexual; perianth-bristles 0; stamens 2(-3); ovary not enfolded or enclosed by glume; stigmas (2-)3.

1 Stems erect, to 3m; leaves up to 2m x 2cm, grey-green; inflorescence up to 70 x 10cm. Native; wet, base-rich areas in fens and by streams and ponds; locally common but very scattered in BI

Great Fen-sedge - C. mariscus (L.) Pohl

15. KOBRESIA Willd. - *False Sedge*

Rather densely tufted perennials; stems with 3 rounded angles, leafy at extreme base; leaves channelled; inflorescence of 1-flowered spikelets arranged in terminal cluster of 3-10 spikes; lowest bract a sheath with short, usually brown blade; flowers unisexual, the upper spikelets male and lower female in each spike; perianth-bristles 0; stamens 3; female flowers with an extra inner glume folded (but not fused) around ovary; stigmas 3.

1 Stems erect, to 20cm; leaves 0.5-1.5mm wide; inflorescence 1-2.5cm x 2-6mm. Native; flushed grassy or barish areas on mountains; very local in Upper Teesdale and M Perth

False Sedge - K. simpliciuscula (Wahlenb.) Mack.

16. CAREX L. - *Sedges*

Extensively rhizomatous to densely tufted perennials; stems erect, usually leafy but often only at extreme base, triangular in section with acute to rounded angles or terete; leaves flat to channelled or inrolled; inflorescence of 1-flowered spikelets grouped in variously arranged spikes, all except the terminal subtended by a bract; lowest bract leaf-like to glume-like; flowers unisexual, each in axil of 1 glume, the sexes variously arranged from mixed in 1 spike to dioecious, but commonly the upper spikes entirely male and the lower entirely female; perianth-bristles 0; stamens 2 or 3; female flowers with an extra inner glume (*utricle*) completely fused around ovary, forming a false fruit enclosing nut and usually with long or short distal beak; stigmas 2 or 3.

Ripe fruits are essential for keying down *Carex* spp.; the length of the utricle includes any beak. Two important diagnostic characters might present some difficulties. Genuinely 1-spiked inflorescences should not be confused with those with several congested spikes forming a single ± lobed head. In the former case there is 1 simple axis bearing flowers or fruits directly upon it; in the latter case the axis has lateral (often very short) branches, usually with a bract at the base of each. Depauperate stems of several spp. may rarely possess only 1 spike, but more normal stems should also be available. The number of stigmas (2 or 3) is important. In material with ripe fruits the stigmas might have disappeared, but the shape of the nut and often that of the utricle then provides the clue (see General Key, couplet 4). The beak of the utricle is often bifid, and these projections must not be mistaken for the stigmas. The glumes subtending the male and female flowers are called 'male glumes' and 'female glumes' respectively. The length of the sheath of the lowest bract refers to only the portion fused round the stem.

General key

1 Spike 1, terminal   **Key A**
1 Spikes >1 (sometimes very close together)   2
  2 Spikes all ± similar in appearance, often very close together and forming single lobed head, the terminal spike usually at least partly female   **Key B**
  2 Spikes dissimilar in appearance, the upper ≥1 all or mostly male, the lower ≥1 all or mostly female, usually clearly separate and sometimes remote   3
3 Utricles pubescent on part or whole of main body (excl. beak and edges) (x10 lens)   **Key C**
3 Utricles glabrous on main body (sometimes pubescent on beak or along edges, sometimes papillose on main body)   4
  4 Stigmas 2; utricles usually biconvex or plano-convex; nuts biconvex   **Key D**
  4 Stigmas 3; utricles usually 3-angled to terete; nuts 3-angled   5
5 At least lowest spike pendent, stalked   **Key E**
5 All spikes erect to patent, often sessile   **Key F**

Key A - Spike 1, terminal

1 Spike all male   2
1 Spike female at least at base   3
  2 Plant densely tufted; stems usually scabrid and with 3 rounded angles above. Stems to 40cm. Possibly native; calcareous fen; N Somerset, extinct since c.1845   *Davall's Sedge* - **C. davalliana** Sm.   56●
  2 Plant rhizomatous; stems usually smooth and terete. ;Stems loosely tufted, to (20)30cm. Native; base-rich bogs and flushes; throughout most of Br, common in N, very scattered in C and absent from most of S, scattered in Ir   *Dioecious Sedge* - **C. dioica** L.   56●
3 Stigmas 2; utricles usually biconvex or plano-convex; nuts biconvex   4
3 Stigmas 3; utricles usually 3-angled to terete; nuts 3-angled   6

4 Utricles 4-6mm, not ribbed, strongly reflexed at maturity; plants monoecious, with spikes male at apex, female at base. Densely tufted to shortly rhizomatous; stems to 30cm, terete, smooth. Native; bogs, fens and flushes, usually base-rich; frequent in most of BI except much of C & E En *Flea Sedge* - **C. pulicaris** L. 563

4 Utricles 2.5-4.5mm, distinctly ribbed, not or weakly reflexed at maturity; plants usually dioecious, sometimes monoecious 5

5 Utricles 2.5-3.5mm, abruptly contracted to scabrid beak
 (see couplet 2) **C. dioica**

5 Utricles 3.5-4.5mm, gradually contracted to smooth beak
 (see couplet 2) **C. davalliana**

6 Utricles erecto-patent to erect when ripe, obovoid, usually ≤3.5mm; leaves curved or curly. Shortly rhizomatous; stems loosely tufted, to 20cm, with acute angles, ± smooth. Native; on rock ledges or stony ground on limestone or calcareous-flushed sandstone mostly above 600m; locally frequent in N & C Sc
 Rock Sedge - **C. rupestris** All. 563

6 Utricles patent to reflexed when ripe, narrowly ovoid to narrowly ellipsoid, 3.5-7.5mm; leaves ± straight 7

7 Utricles 3.5-5(6)mm, with fine bristle (as well as style base) protruding 1-2mm from beak; female glumes c.2mm. Rhizomatous; stems usually single, to 12cm, with rounded angles or terete, smooth. Native; base-rich flushes on open stony slopes at 600-900m, 1 locality in M Perth *Bristle Sedge* - **C. microglochin** Wahlenb. 563

7 Utricles 5-7.5mm, with only style-base protruding from beak; female glumes c.4mm. Shortly rhizomatous; stems not or very loosely tufted, to 25cm, with rounded angles, smooth. Native; acid blanket bogs; frequent in C & N Sc, scattered S to NE Yorks and Caerns, Co Antrim *Few-flowered Sedge* - **C. pauciflora** Lightf. 563

Key B - Spikes >1, all ± similar in appearance

1 Stigmas 3; utricles usually 3-angled to terete; nuts 3-angled 2
1 Stigmas 2; utricles usually biconvex or plano-convex; nuts biconvex 5
2 Lowest spike erect on short rigid stalk 3
2 Lowest spike pendent to patent on distinct flexible stalk 4

3 All spikes clustered and greatly overlapping; utricles 1.8-2.5mm, greenish-brown, minutely papillose, longer than acute female glumes. Densely tufted; stems to 30cm, with rounded angles, smooth or rough distally. Native; N-facing damp rock-ledges and rocky slopes at 690-990m, where snow lies late; 5 places in C Sc
 Close-headed Alpine-sedge - **C. norvegica** Retz. 563

3 Spikes not or scarcely overlapping, the lowest arising ≥1cm below next; utricles 3-4.5mm, pale green, smooth, shorter than acuminate female glumes. Shortly rhizomatous; stems not or loosely tufted, to 70cm, with acute angles, smooth. Native; wet fens; very rare in Argyll, Westerness and Easterness *Club Sedge* - **C. buxbaumii** Wahlenb. 563

4 Lowest bract with sheath 0-3mm; terminal spike female at top, male below. Shortly rhizomatous; stems loosely tufted, to 55cm,

with acute angles, smooth or rough distally. Native; wet rock-ledges above 720m; locally frequent in C & NW Sc, very rare in S Sc, Lake District and Caerns *Black Alpine-sedge* - **C. atrata** L. 56

4 Lowest bract with sheath ≥5mm; terminal spike male at top, female below. Shortly rhizomatous; stems loosely tufted, to 35cm, with acute to rounded angles, smooth. Native; mountain flushes 540-1050m; very rare in M Perth, Westerness and Argyll
Scorched Alpine-sedge - **C. atrofusca** Schkuhr 56

5 Rhizomes long; stems very loosely tufted or scattered 6
5 Rhizomes short; stems densely tufted 13
    6 Terminal spike (not necessarily that extending highest) female at least at apex 7
    6 Terminal spike male at least at apex 8
7 Utricles 4-5.5(7)mm, reddish-brown, narrowly winged. Rhizomatous, with stems borne singly or in pairs, to 1m, with acute angles, rough. Native; marshes, fens and wet meadows; frequent throughout most of BI *Brown Sedge* - **C. disticha** Huds. 56
7 Utricles 2-3mm, yellowish-green, not winged. Shortly rhizomatous, with stems loosely tufted, to 50cm, with acute angles, rough distally. Native; wet acid places on heaths, in bogs and boggy woods and on mountainsides; throughout most of Br and Ir, common in N
White Sedge - **C. curta** Gooden. 56
    8 Utricles narrowly winged on body 9
    8 Utricles not winged on body 10
9 Terminal spike all male; glumes (male and female) ≥5mm; leaf-sheaths hyaline on side opposite blade. Very extensively rhizomatous, with stems borne singly, to 90cm (often much less), with acute angles, slightly rough. Native; bare or grassy dunes or sandy places; round coasts of whole BI, very local inland
Sand Sedge - **C. arenaria** L. 56
9 Terminal spike male only at apex; glumes (male and female) ≤5mm; leaf-sheaths herbaceous on side opposite blade, except for apical hyaline rim (see couplet 7) **C. disticha**
    10 Stems ± terete, smooth 11
    10 Stems triangular in section with acute to rounded angles, rough on angles near top 12
11 Stems rarely >15cm; leaves usually curved, reaching or nearly reaching inflorescence, crescent-shaped in section when fresh. Long-rhizomatous, with stems single or loosely grouped, to 18cm, terete, smooth. Native; sand-dunes and damp sandy places; local on coasts of N Br S to Cheviot and Westmorland, mainly N Sc
Curved Sedge - **C. maritima** Gunnerus 56
11 Stems rarely <15cm; leaves usually straight, falling well short of inflorescence, flat or V-shaped in section when fresh. Rhizomatous, with stems borne singly, to 40cm, ± terete, smooth. Native; very wet acid bogs; very rare in W Sutherland and Easterness
String Sedge - **C. chordorrhiza** L. f. 56
    12 Lowest bract leaf-like, mostly at least as long as whole

inflorescence; beak of utricle <1/2 as long as body; utricles pale brown. Rhizomatous but stems often clustered, to 80cm, with rounded angles, rough distally. Native; damp, usually brackish grassy places, in marshes, pastures and ditches; locally frequent around coasts of S & C Br, scattered N to Cheviot and Pembs, Jersey, Co Wexford *Divided Sedge* - **C. divisa** Huds. 560

12 Lowest bract not leaf-like, much shorter than inflorescence; beak of utricle >1/2 as long as body; utricles blackish-brown. Usually with rhizomes extended, sometimes tufted but rarely forming tussocks; stems to 60cm, with acute angles, rough. Native; wet peaty and acid places in ditches, meadows, marshes and scrub; scattered in Br and Ir *Lesser Tussock-sedge* - **C. diandra** Schrank 560

13 All spikes with female flowers in apical part 14
13 At least some spikes (sometimes only uppermost or lowermost) with male flowers in apical part 19

    14 Lowest bract easily exceeding inflorescence, leaf-like. Densely tufted; stems to 75cm, with acute angles (or compressed distally), rough distally. Native; woods, hedgerows, shady banks and ditchsides; frequent to common *Remote Sedge* - **C. remota** L. 560

    14 Lowest bract shorter than inflorescence, usually not leaf-like 15

15 Spikes not longer than wide, each with <10 utricles; utricles patent. Densely tufted; stems to 40cm, with rounded angles or subterete, rough distally. Native; wide range of acid to basic bogs and marshes; throughout BI, common in W & N *Star Sedge* - **C. echinata** Murray 560

15 Spikes longer than wide, each usually with >10 utricles; utricles erect to erecto-patent 16

    16 Utricles winged in upper 1/2. Densely tufted; stems to (40)90cm, with acute angles, rough distally. Native; damp or dry grassy places; common *Oval Sedge* - **C. ovalis** Gooden. 560

    16 Utricles not winged 17

17 Utricles pale green to yellowish- or pale brownish-green
(see couplet 7) **C. curta**
17 Utricles reddish- to dark-brown 18

    18 Spikes (5)8-12(18); utricles divaricate, without slit in beak; lowland wet places. Densely tufted; stems to 80cm, with acute angles, rough. Native; damp places in wet meadows and boggy woods, by ditches and streams; scattered in Br N to Dunbarton, N Ir *Elongated Sedge* - **C. elongata** L. 560

    18 Spikes (2)3-4(6); utricles appressed, with slit down back of beak; high mountains. Shortly rhizomatous, with stems tufted, to 20(30)cm, with rounded angles, rough distally. Native; wet acid places on mountains ≥750m, usually where snow lies late; very local in C Sc *Hare's-foot Sedge* - **C. lachenalii** Schkuhr 560

19 Utricles 2-2.6mm. Densely tufted; stems to 1m, with acute angles, slightly rough. Intrd-natd; rough ground; scattered in Br N to C Sc
American Fox-sedge - **C. vulpinoidea** Michx. 560

19 Utricles ≥2.7mm 20

    20 Utricles biconvex (weakly to strongly convex on adaxial and

strongly convex on abaxial side) 21
- 20 Utricles plano-convex (flat on adaxial and weakly convex on abaxial side) 23
21 Utricles conspicuously winged in upper 1/2. Densely tufted, often forming large tussocks; stems to 1.5m, with acute angles, rough. Native; by lakes and streams, in marshes, fens and wet woods, on usually base-rich soils; frequent
Greater Tussock-sedge - **C. paniculata** L. 56
- 21 Utricles not or scarcely winged 22
  - 22 Lowest leaf-sheaths remaining whole; all spikes usually sessile (see couplet 12) **C. diandra**
  - 22 Lowest leaf-sheaths decaying into fibres; lowest spikes usually stalked. Densely tufted; stems to 1m, with acute angles, rough. Native; similar places to *C. paniculata*; very local in E Anglia, N En, S Sc and C Ir *Fibrous Tussock-sedge* - **C. appropinquata** Schumach. 56
23 Stems >2mm wide; leaves mostly ≥4mm wide; utricles with distinct, often prominent veins 24
- 23 Stems <2mm wide; leaves mostly ≤4mm wide; utricles with obscure veins 25
  - 24 Ligule truncate; leaf-sheaths transversely wrinkled on side opposite blade; utricles matt, papillose, with slit down back of beak. Densely tufted; stems to 1m, stout, with acute ± winged rough angles. Native; wet places on heavy soils, in ditches and marshes and by streams; local in En N to SW Yorks, mainly in SE *True Fox-sedge* - **C. vulpina** L. 56
  - 24 Ligule acute; leaf-sheaths not wrinkled on side opposite blade; utricles shiny, smooth, without slit in beak. Densely tufted; stems to 1m, stout, with acute ± winged rough angles. Native; wet places on heavy soils in a range of habitats; frequent in most of BI but ± entirely coastal in Sc *False Fox-sedge* - **C. otrubae** Podp. 56
25 Roots and often base of plant purple-tinged; ligule acute to obtuse, distinctly longer than wide; utricles thickened and corky at base. Densely tufted; stems to 80cm, with acute angles, rough. Native; damp grassy places in fields, on banks and waysides and by rivers and ponds; frequent throughout most of En, very scattered in Wa, C & S Sc, Ir and Guernsey *Spiked Sedge* - **C. spicata** Huds. 56
- 25 Plant not purple-tinged; ligule rounded at apex, c. as long as wide; utricles not thickened at base 26
  - 26 Lowest 2-4 spikes or clusters of spikes separated by a gap >2x their own length; ripe utricles appressed to axis. Densely tufted; stems with rounded angles, rough. Native; hedgerows, wood borders, grassy rough ground; frequent in S Br, scattered N to N En and in Ir and CI *Grey Sedge* - **C. divulsa** Stokes **ssp. divulsa** 56
  - 26 Lowest spikes separated by a gap ≤2x their own length; ripe utricles divaricate from axis 27
27 Utricles 4.5-5mm, cuneate at base; inflorescence 3-5(8)cm. Habit as ssp. *divulsa*. Native; similar places to ssp. *divulsa* but mostly on chalk

and limestone; local in Br N to Midlothian
Grey Sedge - **C. divulsa** ssp. **leersii** (Kneuck.) W. Koch 560
27 Utricles 2.6-4.5mm, truncate to rounded at base; inflorescence
(1)2-3(4)cm. Densely tufted; stems with rounded angles, rough
(*Prickly Sedge* - **C. muricata** L.) 28
   28 Utricles (3.5)4-4.5mm; female glumes shorter and darker than the
utricles. Native; steep, dry, limestone slopes; 5 places in Br from
W Gloucs to Berwicks   **C. muricata** ssp. **muricata** 560
   28 Utricles 2.6-3.5(4)mm; female glumes nearly as long as and similar
in colour to or paler than the utricles. Native; open grassy places
usually on dry acid soils; frequent throughout Br N to C Sc, CI,
S & E Ir   **C. muricata** ssp. **lamprocarpa** Celak. 560

Key C - Spikes >1, the upper (male) different in appearance from the lower
(female); utricles pubescent
1 Utricles with conspicuously bifid beak ≥0.5mm   2
1 Utricles with truncate to notched beak 0-0.5mm   3
   2 Lower leaf-sheaths usually pubescent; utricles 4.5-7mm, with beak
1.5-2.5mm. Rhizomatous or shortly so, with stems loosely tufted,
to 70cm, with rounded angles, ± smooth. Native; damp grassy
places in many habitats; common   *Hairy Sedge* - **C. hirta** L. 561
   2 Leaf-sheaths glabrous; utricles 3.5-5mm, with beak 0.5-1mm.
Rhizomatous; stems to 1.2m, with rounded angles, smooth.
Native; bogs and fens; scattered in Br and Ir except most of
S & C En   *Slender Sedge* - **C. lasiocarpa** Ehrh. 561
3 Rhizomes extended; stems not or loosely tufted, often borne singly   4
3 Rhizomes very short; stems densely tufted   7
   4 Male spikes (1)2-3; lowest female spike clearly stalked, pendent
to erecto-patent. Glaucous, rhizomatous; stems loosely tufted, to
60cm, with rounded angles or subterete, smooth. Native; wet or
dry grassland on chalk and limestone, sand-dunes and base-rich
clay, mountain flushes; common   *Glaucous Sedge* - **C. flacca** Schreb. 561
   4 Male spike 1; lowest female spike sessile or with concealed stalk,
erect   5
5 Leaves ± glaucous, erect; lowest living leaf-sheaths reddish-brown;
stems usually >20cm. Rhizomatous; stems not or loosely tufted, to
50cm, with rounded angles, smooth to slightly rough distally. Native;
damp grassy places in fields and on waysides and in woodland
rides; very local in S En   *Downy-fruited Sedge* - **C. filiformis** L. 562
5 Leaves not glaucous, usually ± recurved; lowest living leaf-sheaths
mid- to dark-brown; stems usually <20cm   6
   6 Lowest bract with sheath 3-5mm; female glumes acute, green to
brown, with 0 or narrow scarious border. Shortly rhizomatous;
stems loosely tufted, to 30cm, with acute to rounded angles,
smooth. Native; acid or basic dry to damp short grassland;
frequent   *Spring-sedge* - **C. caryophyllea** Latourr. 562
   6 Lowest bract with sheath 0-2mm; female glumes obtuse to
rounded, purplish-black, with wide scarious border. Shortly

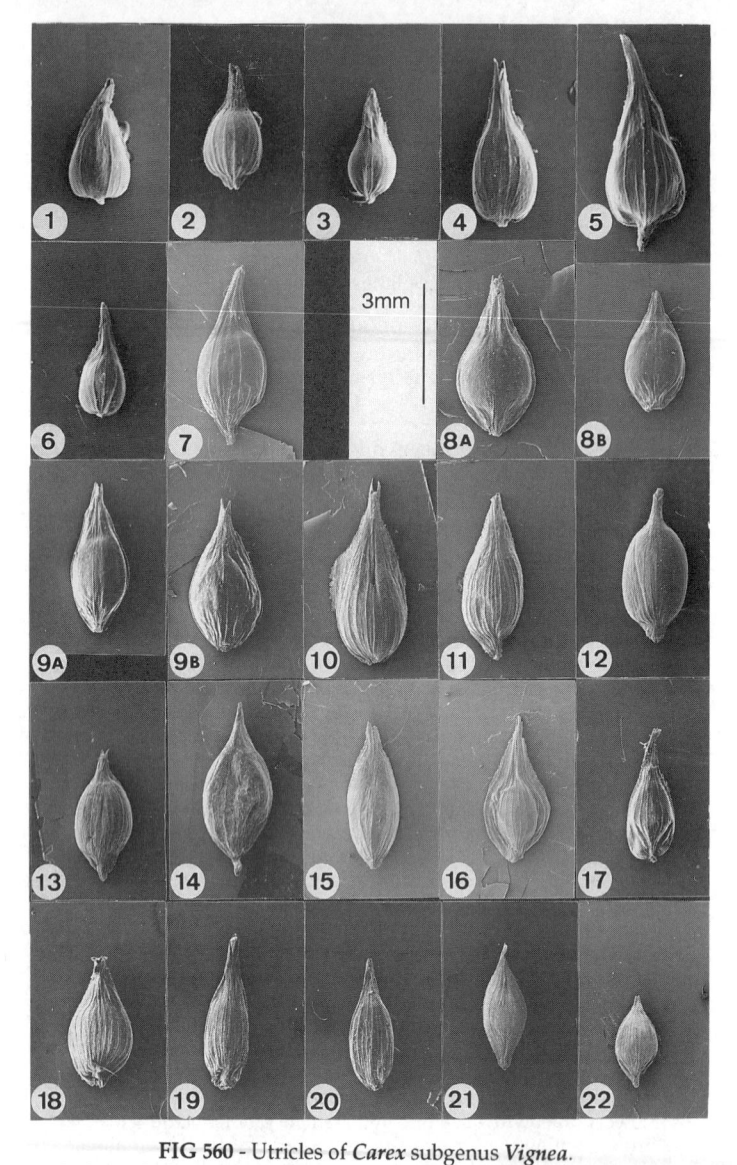

FIG 560 - Utricles of *Carex* subgenus *Vignea*.
1, *C. paniculata*. 2, *C. approprinquata*. 3, *C. diandra*. 4, *C. vulpina*.
5, *C. otrubae*. 6, *C. vulpinoidea*. 7. *C. spicata*.
8A, *C. muricata* ssp. *muricata*. 8B, *C. muricata* ssp. *lamprocarpa*.
9A, *C. divulsa* ssp. *divulsa*. 9B, *C. divulsa* ssp. *leersii*.
10, *C. arenaria*. 11, *C. disticha*. 12, *C. chordorrhiza*. 13, *C. divisa*.
14, *C. maritima*. 15, *C. remota*. 16, *C. ovalis*. 17, *C. echinata*. 18, *C. dioica*.
19, *C. davalliana*. 20, *C. elongata*. 21, *C. lachenalii*. 22, *C. curta*.

FIG 561 - Utricles of *Carex* subgenus *Carex*. 23, *C. hirta*. 24, *C. lasiocarpa*. 25, *C. acutiformis*. 26, *C. riparia*. 27, *C. pseudocyperus*. 28, *C. rostrata*. 29, *C. vesicaria*. 30, *C. x grahamii*. 31, *C. saxatilis*. 32, *C. pendula*. 33, *C. sylvatica*. 34, *C. capillaris*. 35, *C. strigosa*. 36, *C. flacca*. 37, *C. panicea*. 38, *C. vaginata*. 39, *C. depauperata*.

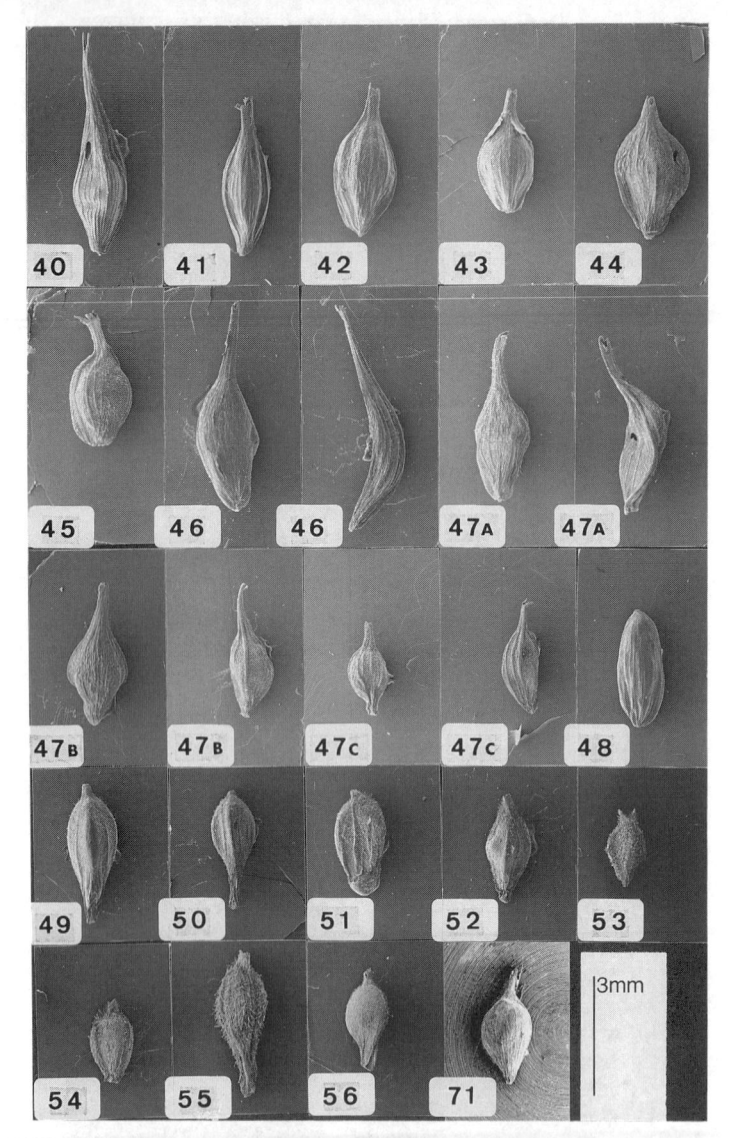

FIG 562 - Utricles of *Carex* subgenus *Carex*. 40, *C. laevigata*. 41, *C. binervis*. 42, *C. distans*. 43, *C. punctata*. 44, *C. extensa*. 45, *C. hostiana*. 46, *C. flava* (abaxial and lateral views). 47A, *C. viridula* ssp. *brachyrrhyncha* (abaxial and lateral views). 47B, *C. viridula* ssp. *oedocarpa* (abaxial and lateral views). 47C, *C. viridula* ssp. *viridula* (abaxial and lateral views). 48, *C. pallescens*. 49, *C. digitata*. 50, *C. ornithopoda*. 51, *C. humilis*. 52, *C. caryophyllea*. 53, *C. filiformis*. 54, *C. ericetorum*. 55, *C. montana*. 56, *C. pilulifera*. 71, *C. buchananii*.

FIG 563 - Utricles of *Carex* subgenera *Carex* and *Primocarex*.
57, *C. atrofusca*. 58, *C. limosa*. 59, *C. rariflora*. 60, *C. magellanica*.
61, *C. atrata*. 62, *C. buxbaumii*. 63, *C. norvegica*. 64, *C. recta*.
65, *C. aquatilis*. 66, *C. acuta*. 67, *C. trinervis*. 68, *C. nigra*. 69, *C. elata*.
70, *C. bigelowii*. 72, *C. microglochin*.
73, *C. pauciflora*. 74, *C. rupestris*. 75, *C. pulicaris*.

rhizomatous; stems loosely mat-forming, to 20cm, with rounded angles, smooth. Native; dry short calcareous grassland; very locally common in E & N En *Rare Spring-sedge* - **C. ericetorum** Pollich 56

7 Inflorescence occupying >1/2 of stem length; stems much shorter than leaves; female spikes with 2-4 flowers. Densely tufted; stems to 10(15)cm, subterete, smooth. Native; short limestone grassland; very locally common in SW En *Dwarf Sedge* - **C. humilis** Leyss. 56

7 Inflorescence occupying <1/4 of stem length; stems usually longer than leaves; female spikes usually with >4 flowers 8

    8 Flowering stems arising laterally, from leaf axils, leafless; female spikes ≤3mm wide, overtopping male 9

    8 Flowering stems terminal, leafy at base; female spikes ≥4mm wide, falling short of top of male 10

9 Utricles 3-4mm; female glumes purplish-brown, c. as long as utricles; female spikes arising 1 above the other; basal leaf-sheaths crimson. Densely tufted; stems to 25cm, subterete, smooth. Native; open woodland, scrub and grassy rocky slopes on chalk and limestone; very local in En, Mons *Fingered Sedge* - **C. digitata** L. 56

9 Utricles 2-3mm; female glumes pale brown, much shorter than utricles; female spikes all arising at ± same point; basal leaf-sheaths brown. Densely tufted; stems to 15(20)cm, subterete, smooth. Native; short limestone grassland; very local in N En
Bird's-foot Sedge - **C. ornithopoda** Willd. 56

    10 Lowest bract with sheath 3-5mm (see couplet 6) **C. caryophyllea**

    10 Lowest bract with sheath 0-2mm 11

11 Lowest bract usually green, leaf-like; female glumes brown or reddish-brown; beak of utricle 0.3-0.5mm. Densely tufted; stems to 40cm, with acute angles, rough distally. Native; grassy or barish places in open or woodland on base-poor sandy or peaty soils; frequent *Pill Sedge* - **C. pilulifera** L. 56

11 Lowest bract brown, glume-like or bristle-like; female glumes purplish-black; beak of utricle ≤0.3mm 12

    12 Female glumes obtuse, with scarious minutely pubescent margin; utricles 2-3mm; leaves mostly >2mm wide, rigid, recurved
(see couplet 6) **C. ericetorum**

    12 Female glumes subacute (to obtuse) and mucronate, with hyaline (but not scarious) glabrous margin; utricles 3-4.5mm; leaves mostly <2mm wide, soft, ± erect. Rhizomatous; stems loosely tufted, to 40cm, with acute to rounded angles, rough distally. Native; wet or dry grassy places in open or light shade on acid or basic soils; very local in S Br *Soft-leaved Sedge* - **C. montana** L. 56

Key D - Spikes >1, the upper (male) different in appearance from the lower (female); utricles glabrous; stigmas 2

1 Utricles with distinct forked or notched beak >0.3mm 2
1 Utricles with 0 or indistinct truncate or minutely notched beak ≤0.3mm 5

    2 Female glumes almost entirely hyaline so that female spikes are

silvery-white. Densely tufted, with stems and leaves strongly red-coloured; stems to 60cm, subterete, smooth. Intrd-natd; on rough ground; 1 site in Glasgow (Lanarks)
Silver-spiked Sedge - **C. buchananii** Berggr. **562**

2 Female glumes hyaline only at edges; female spikes not silvery-white 3

3 Utricles not inflated; female glumes 3-4mm, acute to acuminate; female spikes up to 5cm. Rhizomatous; stems tufted, to 1.5m, with acute angles, rough. Native; marshes, wet meadows and swamps, by ponds and streams; common in lowland Br and Ir, rare in upland areas *Lesser Pond-sedge* - **C. acutiformis** Ehrh. **561**

3 Utricles inflated; female glumes 2-3mm, subacute; female spikes up to 3cm 4

4 Utricles 3-3.5mm, containing nut, ± not ribbed. Rhizomatous, with stems in small tufts, to 40cm, with rounded angles, ± rough distally. Native; wet places, especially where snow lies late on mountains >750m; local in N & W Sc *Russet Sedge* - **C. saxatilis** L. **561**

4 Utricles 4-5mm, empty, distinctly ribbed. Rhizomatous, with stems in small tufts, to 50cm, with rounded angles, ± rough distally. Native; mountain flushes >750m; very local in C Sc (*C. saxatilis* x ?*C. vesicaria*)
Mountain Bladder-sedge - **C. x grahamii** Boott **561**

5 At least some female glumes with apical points >1/2 as long as rest of glume; glumes ≥2x as long as utricles. Rhizomatous; stems tufted, to 1.1m, with acute to rounded angles, smooth. Native; forming extensive patches in wet estuarine areas; very local in NE Sc
Estuarine Sedge - **C. recta** Boott **563**

5 Female glumes rounded or obtuse to acuminate with 0 or short apical point (except sometimes the lowest 1(-3)); glumes ≤1.5x as long as utricles 6

6 Leaf-sheaths breaking into conspicuous ladder-like fibres on side opposite blade; stems densely tufted, often forming large tussocks. Stems to 1m, with acute angles, rough distally. Native; bogs, fens, reedswamps and by rivers and lakes; locally frequent in Ir and C Br, very scattered elsewhere in Br *Tufted-sedge* - **C. elata** All. **563**

6 Leaf-sheaths not breaking into fibres; stems usually scattered, sometimes tufted but not tussock-forming 7

7 Utricles 3.5-5mm, prominently veined; female glumes 3-veined. Rhizomatous; stems scarcely tufted, to 40cm, with rounded angles, smooth. Possibly native, now extinct; E Norfolk in 1869
Three-nerved Sedge - **C. trinervis** Degl. **563**

7 Utricles 2-3.5mm; female glumes 1-veined, or 3-veined and then utricles veinless 8

8 Utricles without visible veins 9

8 Utricles with distinct faint to ± prominent veins 10

9 Lowest bract exceeding inflorescence; stems usually >25cm, with 3 rounded angles, brittle; female glumes often 3-veined. Rhizomatous; stems tufted, to 1.1m, with acute to rounded angles, smooth. Native;

swampy areas by lakes and rivers and in marshes; locally frequent
in N Br, scattered in Ir *Water Sedge* - **C. aquatilis** Wahlenb. 56
9 Lowest bract shorter than inflorescence; stems usually <25cm, with
3 acute angles, not brittle; female glumes 1-veined. Shortly
rhizomatous; stems scarcely tufted but often close, to 30cm, with
acute angles, rough. Native; stony and heathy areas, often where
snow lies late, and in flushed gullies, above 600m; common in
highlands of Sc, scattered in N En, N Wa and N & W Ir
Stiff Sedge - **C. bigelowii** Schwein. 56
10 Leaves 1-3(5)mm wide, the margins rolling inwards on drying;
lowest bract rarely as long as inflorescence; male spike usually 1.
Rhizomatous to very shortly so; stems tufted to single, to 70cm,
with rounded angles, smooth. Native; wide range of wet acid to
basic places, especially marshes and flushes; common
Common Sedge - **C. nigra** (L.) Reichard 56
10 Leaves 3-10mm wide, the margins rolling outwards on drying;
lowest bract usually exceeding inflorescence; male spikes
usually 2-4. Rhizomatous; stems tufted, to 1.1m, with acute
angles, rough distally. Native; ponds, ditches, canals, rivers
and marshes; locally frequent *Slender Tufted-sedge* - **C. acuta** L. 56

Key E - Spikes >1, the upper (male) different in appearance from the lower
(female); utricles glabrous; stigmas 3; lowest spike pendent
1 Lower leaf-sheaths and lowerside of blades pubescent. Densely
tufted; stems to 60cm, with acute angles, rough distally. Native;
damp grassland, woodland clearings and stream-banks; frequent
in most of Br, scattered in Ir *Pale Sedge* - **C. pallescens** L. 56
1 Leaf-sheaths and blades glabrous 2
   2 Utricles with distinct forked or notched beak usually ≥1mm
(<1mm in *C. atrofusca* and *C. acutiformis*) 3
   2 Utricles with beak 0 or ≤1mm and with truncate, oblique or very
slightly notched apex 12
3 Male spikes ≥2 4
3 Male spike 1 8
   4 Rhizomes extended; stems not or loosely tufted, often borne singly 5
   4 Rhizomes very short; stems densely tufted 7
5 Female glumes 6-10mm, exceeding utricles. Rhizomatous; stems
tufted, to 1.5m, with acute angles, rough. Native; marshes, wet
meadows and swamps, by ponds and streams; common in C & S Br,
scattered elsewhere *Greater Pond-sedge* - **C. riparia** Curtis 56
5 Female glumes 4-6mm, mostly shorter than utricles 6
   6 Utricles 3.5-5mm, with beak <1mm
(see Key D, couplet 3) **C. acutiformis**
   6 Utricles (4)5-8mm, with beak 1.5-2.5mm. Shortly rhizomatous;
stems slightly tufted, to 1.2m, with rounded angles, rough
distally. Native; swamps, marshes, lake margins; frequent
Bladder-sedge - **C. vesicaria** L. 56
7 Female spikes 6-8mm wide, on peduncles with exposed portion

usually shorter than spike. Densely tufted; stems to 1.2m, with rounded angles, smooth. Native; damp shady places, especially woods on heavy soils; scattered throughout most of BI, except most of C & E En and N Sc   *Smooth-stalked Sedge* - **C. laevigata** Sm.   562

7 Female spikes 3-5mm wide, on peduncles with exposed portion usually much longer than spike. Densely tufted; stems to 70cm, with rounded angles, smooth. Native; woods, hedgerows and scrub on damp usually heavy soils; frequent   *Wood-sedge* - **C. sylvatica** Huds.   561

   8 Female glumes (except midrib) and utricles both purplish-black
(see Key B, couplet 4) **C. atrofusca**

   8 Female glumes and/or utricles brownish or greenish   9

9 Utricles with smooth beaks; lower female spikes with peduncles >1/2 exposed   10

9 Utricles with scabrid beaks; lower female spikes with peduncles >1/2 ensheathed   11

   10 Female spikes 3-5mm wide; female glumes 3-5mm; ligules <5mm
(see couplet 7) **C. sylvatica**

   10 Female spikes 6-10mm wide; female glumes 5-10mm; ligules >5mm. Shortly rhizomatous, with stems loosely tufted, to 90cm, with acute angles, rough. Native; in marshes and swamps, by ponds, rivers and canals; locally common in C & S Br, very scattered elsewhere   *Cyperus Sedge* - **C. pseudocyperus** L.   561

11 Leaves 5-12mm wide; female glumes acuminate; ligules 7-15mm
(see couplet 7) **C. laevigata**

11 Leaves 2-5(7)mm wide; female glumes obtuse and mucronate; ligules 1-2mm. Densely tufted; stems to 1.2m, often much less, subterete, smooth. Native; damp heaths, moors, rocky places and mountainsides; frequent   *Green-ribbed Sedge* - **C. binervis** Sm.   562

   12 Male spikes (1)2-3; utricles papillose   (see Key C, couplet 4) **C. flacca**

   12 Male spike usually 1; utricles not papillose   13

13 Rhizomes very short; stems densely tufted   14

13 Rhizomes extended; stems not or loosely tufted, often borne singly   16

   14 Female spikes ≤2.5cm, arising very close together; all leaves <3mm wide; plant rarely >30cm. Densely tufted; stems to 20(40)cm, with rounded angles, smooth. Native; base-rich or calcareous flushes, mineral-rich bogs; local in N En and Sc, Caerns   *Hair Sedge* - **C. capillaris** L.   561

   14 Female spikes ≥2.5cm, well spaced out along stem; largest leaves >4mm wide; plant rarely <30cm   15

15 Female spikes <3mm wide, with peduncle c.1/2 exposed. Densely tufted; stems to 75cm, with rounded angles or subterete, smooth. Native; damp base-rich soils in woods or woodland clearings; local in Br N to NE Yorks, scattered in N & C Ir
Thin-spiked Wood-sedge - **C. strigosa** Huds.   561

15 Female spikes >3mm wide, with peduncle ± entirely ensheathed. Densely tufted; stems to 1.8m, with rounded angles, smooth. Native; rich heavy soils in woods and damp copses; common in S Br, scattered elsewhere   *Pendulous Sedge* - **C. pendula** Huds.   561

16 Female glumes distinctly narrower than utricles, acuminate at apex, 5-6.5mm, >1.5x as long as utricles; lowest spike with 1-2 male flowers at base. Shortly rhizomatous; stems to 40cm, with rounded angles, smooth. Native; wet bogs; scattered in Br from Lake District to NW Sc, N & WC Wa, Co Antrim
Tall Bog-sedge - **C. magellanica** Lam. 56
16 Female glumes at least as wide as utricles, acute to obtuse (sometimes mucronate) at apex, 3-4.5mm, <1.5x as long as utricles; lowest spike entirely female 17
17 Female spikes 3-4mm wide, with 5-8 flowers; utricles ± beakless; stems usually smooth. Shortly rhizomatous; stems loosely tufted or carpet-forming, to 20cm, with rounded angles. Native; wet peaty mountain slopes or flushes at 750-1050m, where snow lies late; very local in CE Sc *Mountain Bog-sedge* - **C. rariflora** (Wahlenb.) Sm. 56
17 Female spikes 5-7mm wide, with 7-20 flowers; utricles with distinct beak 0.1-0.5mm; stems usually rough distally. Rhizomatous; stems not or loosely tufted, to 40cm, with rounded angles. Native; very wet blanket- or valley-bogs; local in N & C Ir and N & W Br, very scattered in S Ir and En *Bog-sedge* - **C. limosa** L. 56

Key F - Spikes >1, the upper (male) different in appearance from the lower (female); utricles glabrous; stigmas 3; lowest spike erect to patent
1 Lower leaf-sheaths and lowerside of blades pubescent
(see Key E, couplet 1) **C. pallescens**
1 Leaf-sheaths and blades glabrous 2
   2 Utricles papillose (see Key C, couplet 4) **C. flacca**
   2 Utricles not papillose 3
3 Lowest bract not sheathing at base, or with sheath ≤2mm 4
3 Lowest bract with distinct sheathing base >3mm 9
   4 Male glumes 7-9mm, acuminate; female glumes longer than utricles (see Key E, couplet 5) **C. riparia**
   4 Male glumes 3-7mm, obtuse to acute; female glumes shorter than utricles 5
5 Beak >1mm 6
5 Beak <1mm 7
   6 Utricles 3.5-6.5mm, usually patent, abruptly contracted into beak 1-1.5mm; female glumes acute. Rhizomatous; stems slightly tufted or not, to 1m, with rounded angles, rough distally. Native; acid swamps, lake-margins and reed-beds; throughout Br and Ir, common in N & W *Bottle Sedge* - **C. rostrata** Stokes 56
   6 Utricles (4)5-8mm, usually erecto-patent, gradually contracted into beak 1.5-2.5mm; female glumes acuminate
(see Key E, couplet 6) **C. vesicaria**
7 Stems solid, triangular in section with concave faces and acute angles; all or most leaves >5mm wide; female spikes 2-5cm
(see Key D, couplet 3) **C. acutiformis**
7 Stems hollow, triangular in section with flat to convex faces and rounded angles; all or most leaves <5mm wide; female spikes 1-3cm 8

8	Utricles 3-3.5mm, containing nut, ± not ribbed. (see Key D, couplet 4) **C. saxatilis**	
8	Utricles 4-5mm, empty, distinctly ribbed (see Key D, couplet 4) **C. x grahamii**	
9	Rhizomes extended; stems not or loosely tufted	10
9	Rhizomes very short; stems densely tufted	14
	10 Utricles ≥5mm, with beak ≥1.5mm	11
	10 Utricles ≤5mm, with beak <1.5mm	12
11	Female spikes <10-flowered; utricles with beak ≥2.5mm; male spike 1. Shortly rhizomatous; stems to 1m, loosely tufted, with rounded angles or subterete, smooth. Native; dry woods and hedgebanks on chalk or limestone; very rare in N Somerset, Surrey and M Cork *Starved Wood-sedge* - **C. depauperata** With.	**561**
11	Female spikes >10-flowered; utricles with beak <2.5mm; male spikes 2-4 (see Key E, couplet 6) **C. vesicaria**	
	12 Utricles with scabrid, clearly bifid beak. Shortly rhizomatous; stems loosely tufted, to 65cm, with rounded angles, smooth. Native; marshes, flushes and fens; common in W & N Br, locally frequent in Ir and rest of Br *Tawny Sedge* - **C. hostiana** DC.	**562**
	12 Utricles with smooth, truncate to shallowly notched, often very short beak	13
13	Sheaths of lowest bract inflated, loose from stem; utricles with beak 0.5-1mm; leaves green or yellowish-green. Rhizomatous; stems tufted, to 40cm, with rounded angles or subterete, smooth. Native; wet rocky places, damp slopes and flushes above 600m; local in S & C mainland Sc *Sheathed Sedge* - **C. vaginata** Tausch	**561**
13	Sheaths of lowest bract not inflated, close to stem; utricles with beak <0.5mm; leaves glaucous. Shortly rhizomatous; stems loosely tufted, to 60cm, with rounded angles or subterete, smooth. Native; wet, usually acid, heaths and moors, bogs, mountain flushes; throughout BI *Carnation Sedge* - **C. panicea** L.	**561**
	14 At least 1/2 of female spikes close-set to terminal male spike; 1-several bracts far exceeding inflorescence	15
	14 At most 1 female spike close-set to terminal male spike; bracts usually shorter than inflorescence, sometimes just exceeding it	19
15	Utricles erecto-patent, greyish-green often purple-blotched; leaves glaucous or dark- or greyish-green, deeply channelled and/or with inrolled margins. Stems densely tufted, to 40cm, with rounded angles, smooth. Native; muddy or sandy brackish places in estuaries and by sea; frequent round coasts of most of BI *Long-bracted Sedge* - **C. extensa** Gooden.	**562**
15	At least lower utricles in each spike patent to reflexed, bright- or yellowish-green; leaves bright- or yellowish-green, flat or V-shaped	16
	16 Beaks of utricles curved or bent, usually ≥1/2 as long as the usually ± curved body	17
	16 Beaks of utricles straight, usually <1/2 as long as and continuing line of ± straight body at least when fresh. Densely tufted; stems	

to 40(75)cm, with rounded angles, smooth
(*Yellow-sedge* - **C. viridula** Michx.) 18
17 Utricles 5.5-6.5mm; male spike usually subsessile; leaves ≤7mm wide, ≥2/3 as long as stems. Densely tufted; stems to 70cm, with rounded angles, smooth. Native; base-rich fen by lake; Roudsea Wood (Westmorland), as hybrids with *C. viridula* in MW Yorks, Hants and NE Galway *Large Yellow-sedge* - **C. flava** L.
17 Utricles 3.5-5mm; male spike usually clearly stalked; leaves ≤4mm wide, <2/3 as long as stems. Native; base-rich fens, flushes, and lakesides; frequent throughout most of Br and Ir except in acid areas and in S Br **C. viridula ssp. brachyrrhyncha** (Celak.) B. Schmid
   18 Utricles 3-4mm, with beak 0.8-1.3mm; male spike usually clearly stalked; lowest female spike usually distant from others. Native; acid or medium base-rich fens, bogs, flushes, wet fields and by lakes; common over most of BI
   C. viridula ssp. oedocarpa (Andersson) B. Schmid
   18 Utricles 1.75-3.5mm, with beak 0.25-1mm; male spike usually sessile; lowest female spike usually bunched with others. Native; acid (mostly) or basic wet places in bogs, marshes and dune-slacks and by lakes; scattered through most of BI **C. viridula ssp. viridula**
19 Female spikes 2-3mm wide, lax-flowered; apex of sheath of uppermost stem-leaves (not bracts) truncate on side opposite blade
(see Key E, couplet 15) **C. strigosa**
19 Female spikes ≥4mm wide, dense-flowered; apex of sheath of uppermost stem-leaves (not bracts) either concave on side opposite blade or with an apical projection 20
   20 Utricles patent 21
   20 Utricles erecto-patent to erect 22
21 Utricles yellowish-green, not shiny, contrasting with female glumes (dark brown with broad scarious margins) (see couplet 12) **C. hostiana**
21 Utricles pale green, often minutely dark-dotted, shiny, scarcely contrasting with female glumes (pale brown with narrow scarious margins). Densely tufted; stems to 1m, with rounded angles, smooth. Native; marshes and wet rocky places near sea; local in S, W & N Ir, CI and S & W Br *Dotted Sedge* - **C. punctata** Gaudin
   22 Leaves 5-12mm wide; female glumes acuminate; ligules 7-15mm
   (see Key E, couplet 7) **C. laevigata**
   22 Leaves 2-5(7)mm wide; female glumes acute to obtuse, often mucronate or apiculate; ligules 1-3mm 23
23 Female spikes 1.5-4.5cm; female glumes dark reddish- or blackish-brown; utricles with 2 conspicuous green lateral ribs distinct from others (see Key E, couplet 11) **C. binervis**
23 Female spikes 1-2cm; female glumes either pale reddish-brown or dark brown with broad scarious margins; utricles with several ± equally prominent ribs 24
   24 Leaf-blades and lowest bract rather abruptly contracted to narrow parallel-sided point; utricles with beak 0.8-1.2mm; female

glumes dark brown with broad scarious margins
(see couplet 12) **C. hostiana**

24 Leaf-blades and lowest bract gradually contracted to apex; utricles with beak 0.7-1mm; female glumes pale- to mid-brown with narrow scarious margins. Densely tufted; stems to 1m, subterete, smooth. Native; brackish and fresh-water marshes, wet rocky places, mostly near sea; round most coasts of BI, inland in S Br, rarely so elsewhere   *Distant Sedge* - **C. distans** L.   **562**

Nearly 40 hybrids have been recorded but none except *C. hostiana* x *C. viridula* = ***C. x fulva*** Gooden. is common. All are variously intermediate and highly sterile, with empty though often well-developed utricles, except for partial fertility exhibited by *C. flava* x *C. viridula* = ***C. x alsatica*** Zahn and by hybrids involving *C. nigra* and its relatives (spp. in Key D, couplets 5-10).

157. POACEAE - *Grass family*

Annuals or herbaceous perennials (rarely woody perennials - *bamboos*), often with rhizomes or stolons, with usually hollow, cylindrical (rarely flattened or other shapes but not 3-angled) stems; leaves alternate, with long, usually linear, entire, thin (but often rolled up or folded along long axis) blade (*leaf*), with long, stem-sheathing, often cylindrical lower part (*sheath*), usually with *ligule* (a membrane, a fringe of hairs, or a membrane with a distal fringe of hairs) at top of sheath on adaxial side, sometimes with small wing-like extension (*auricle*) on either side at top of sheath. Flowers much reduced, 1-many in discrete units (*spikelets*) very variously arranged in terminal inflorescences, mostly bisexual but often unisexual and bisexual mixed in same spikelet, rarely male and female in different spikelets or parts of plants (dioecious in *Cortaderia*), hypogynous; perianth represented by 2 minute scales (*lodicules*) at base of ovary (rarely fused or 0, 1 or 3); stamens usually 3, rarely 2, 4 or 6; ovary 1-celled, with 1 ovule; styles 2, rarely 1 or 3; stigmas elongated, feathery; fruit a typical *caryopsis*, rarely the wall not fused to the seed inside. Spikelets consisting of a series of bracts; usually 2 (sterile) *glumes* (*lower* and *upper*) at base, rarely 1 or 0, with empty axils; 1-many *florets* above consisting of the bisexual or unisexual flower proper plus 2 (fertile) glumes on either side - *lemma* on abaxial and *palea* on adaxial side (latter sometimes 0); the florets borne on slender axis (*rhachilla*), often 1 or more sterile or even reduced to vestigial scales; lemmas often with horny region (*callus*) at base, this often vestigial but sometimes well developed (often pointed and bristly); lemmas and/or glumes often with short to long dorsal to terminal bristles (*awns*).

Easily distinguished from all other grass-like plants by the distinctive inflorescence and flower structure, and usually from the Cyperaceae by the hollow stems.

Before attempting use of the keys it is essential to dissect a spikelet and to understand thoroughly its structure, the detailed characteristics of all its parts, and the arrangement of spikelets in the inflorescence. The growth-

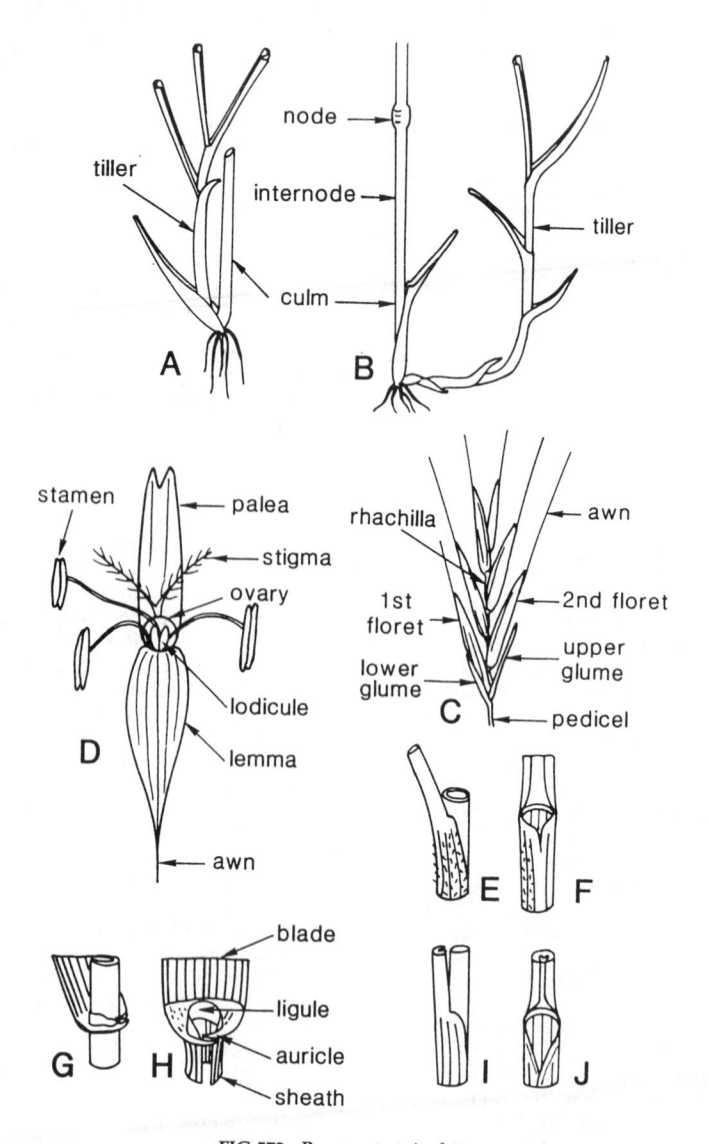

FIG 572 - Poaceae terminology.
A, intravaginal innovation shoot. B, extravaginal innovation shoot.
C, spikelet. D, floret with lemma pulled back.
E-F, innovation leaf-sheath of *Festuca rubra*. G-H, innovation leaf sheath
of *F. pratensis*. I-J, innovation leaf-sheath of *F. ovina*.
Drawings by S. Ogden.

157. POACEAE

form of the plant is also important; perennials can be distinguished by the presence of sterile leafy shoots (*tillers*) as well as flowering stems (*culms*), and they might have rhizomes and/or stolons as well. Spikelet, glume and lemma lengths exclude awns unless otherwise stated.

General key

1. Ligule a dense fringe of hairs, or membranous but breaking into dense fringe of hairs distally **Key A**
1. Ligule membranous, sometimes jagged or pubescent but not densely fringed with hairs distally, sometimes 0   2
    2. Bamboos - stems woody; leaves with distinct short petiole between blade and sheath **Key B**
    2. Stems not woody; leaves without petiole between blade and sheath   3
3. *Maize* - female spikelets in simple raceme (*cob*) low down on plant; male spikelets in terminal panicle or umbel of racemes (*tassel*) **94. ZEA**
3. Male and female spikelets not in separate inflorescences   4
    4. Spikelets arising in groups of 2-7, one fatter and bisexual, the other 1-6 thinner and sterile or male **Key C**
    4. Spikelets all bisexual and similar   5
5. Inflorescence a simple spike, or a simple raceme whose spikelets have pedicels ≤2mm **Key D**
5. Inflorescence more complex than a simple spike or raceme of spikelets (but often very condensed)   6
    6. Inflorescence an umbel or raceme of spikes, or of racemes whose spikelets have pedicels <2mm **Key E**
    6. Inflorescence a panicle, or a raceme whose spikelets have pedicels >3mm   7
7. Spikelets regularly proliferating to form small leafy plantlets (sexual spikelets present or not)   8
7. Spikelets not or only irregularly proliferating (if so, diseased, or because of sterility, or very late in season)   10
    8. Lemmas (whether proliferating or not) distinctly keeled on back along midrib **22. POA**
    8. Lemmas (whether proliferating or not) rounded on back   9
9. Lemmas with awn arising from dorsal surface, with tuft of hairs arising just below base **38. DESCHAMPSIA**
9. Lemmas with awn 0 or terminal, without tuft of hairs arising just below base **15. FESTUCA**
    10. Spikelets with only 1 floret (bisexual, not accompanied by vestigial florets or scales) **Key F**
    10. Spikelets with ≥2 bisexual florets, or with only 1 fertile floret but also ≥1 male or sterile florets or scales   11
11. Spikelets with 1 bisexual terminal floret and 1 or more male or sterile florets or scales below it **Key G**
11. Spikelets with ≥2 bisexual florets, or with only 1 fertile floret but also ≥1 male or sterile florets or scales above it   12
    12. Ovary with pubescent terminal appendage extending beyond

	base of styles	**Key H**
12	Ovary glabrous or pubescent, but style bases on apex of ovary (sometimes wide apart) and not exceeded by ovary appendage	13
13	Lemmas with dorsal or subterminal, usually bent awn, often bifid at apex, sometimes awnless and then clearly bifid at apex	**Key I**
13	Lemmas awnless and entire at apex, or with terminal straight or curved awn and then sometimes bifid or several-toothed at apex	**Key J**

Key A - Ligule, at least distally, a dense fringe of hairs

1	*Maize* - female spikelets in simple raceme (*cob*) low down on plant; male spikelets in terminal panicle or umbel of racemes (*tassel*)	**94. ZEA**
1	Male and female spikelets not in separate inflorescences	2
	2 Spikelets arising in pairs, one fatter and bisexual (sessile, often awned), the other thinner and male or sterile (awnless, often stalked)	**93. SORGHUM**
	2 Spikelets all bisexual, all similar or in pairs with the two in each pair differing in pedicel length only	3
3	Spikelets with stout hooked spines on dorsal surface of glume	**83. TRAGUS**
3	Spikelets without hooked spines (but sometimes with barbed awns)	4
	4 Inflorescence a spike or contracted spike-like panicle; spikelets or groups of spikelets with 1-several (sometimes proximally fused) barbed bristles at base	5
	4 Inflorescence of several spikes or racemes or obviously a panicle; spikelets without barbed bristles at base (though often with soft hairs)	6
5	Bristles fused proximally to form small cupule around spikelets, falling with spikelets to form a bur	**92. CENCHRUS**
5	Bristles remaining on axis when spikelets or florets fall, not fused	**90. SETARIA**
	6 Inflorescence an umbel or raceme of spikes, or of racemes whose spikelets have pedicels ≤2mm	7
	6 Inflorescence a panicle	17
7	Spikes all arising from same point at tip of stem	8
7	At least some spikes arising at different (though often very close) points along apical part of stem	11
	8 Spikes terminating in bare prolongation of axis	**78. DACTYLOCTENIUM**
	8 Spikes terminating in spikelet	9
9	Strongly stoloniferous; spikelets with 1 (bisexual) floret only	**81. CYNODON**
9	Plant tufted, not stoloniferous; spikelets with 1-several bisexual florets, if only 1 then with sterile florets or scales distal to it	10
	10 Spikelets with ≥3 bisexual florets; lemmas not awned	**77. ELEUSINE**
	10 Spikelets with 1 bisexual floret plus 1-few sterile ones more distal; lemmas awned	**80. CHLORIS**
11	Spikelets in pairs, the two in each a pair differing ± only in pedicel	

157. POACEAE

    length (scarcely natd aliens not treated here) **(MISCANTHUS)**
11 Spikelets not in pairs 12
  12 Spikelets with ≥3 florets **77. ELEUSINE**
  12 Spikelets with 1-2 florets 13
13 Spikelets ≥8mm **82. SPARTINA**
13 Spikelets ≤6mm 14
  14 Spikelets with small globose swelling at base, immediately below glumes **88. ERIOCHLOA**
  14 Spikelets without globose swelling at base 15
15 Spikelets with 2 scales (upper and lower glume) at base <2/3 as long as spikelet, with 1 (bisexual) floret **79. SPOROBOLUS**
15 Spikelets with 0 or 1 scale (lower glume) at base <2/3 as long as spikelet, with 2 florets (the upper bisexual, the lower male or sterile) 16
  16 Upper lemma with awn 0.3-1mm (hidden between upper glume and lower lemma); lower lemma and upper glume shortly acuminate **87. UROCHLOA**
  16 Upper lemma awnless; lower lemma and upper glume obtuse **86. BRACHIARIA**
17 Spikelets in pairs, the two in each a pair differing ± only in pedicel length (scarcely natd aliens not treated here) **(MISCANTHUS)**
17 Spikelets not in pairs 18
  18 Most leaves >1m long, with very rough, cutting edges; plants dioecious or gynodioecious **72. CORTADERIA**
  18 Leaves <1m, without cutting edges; plants bisexual 19
19 Spikelets consisting of 3 florets, 2 lower empty distinctly larger than 1 upper fertile **10. EHRHARTA**
19 Spikelets rarely consistently with 3 florets, if so then the lowest bisexual 20
  20 Spikelets with basal tuft of long silky hairs becoming very conspicuous in fruit **74. PHRAGMITES**
  20 Spikelets glabrous to pubescent, but not with basal tuft of long silky hairs 21
21 Spikelets with 1 floret only 22
21 Spikelets with ≥2 florets, sometimes only 1 bisexual and ≥1 male or sterile 25
  22 Lemmas conspicuously awned 23
  22 Lemmas awnless 24
23 Lemma with single bent awn **12. STIPA**
23 Lemma with trifid straight awn (scarcely natd aliens not treated here) **(ARISTIDA)**
  24 Lemmas with 1 vein; spikelets all with 1 floret **79. SPOROBOLUS**
  24 Lemmas with 3-5 veins; usually many spikelets with >1 floret **73. MOLINIA**
25 Spikelets with 2 florets, the distal bisexual, the proximal male or sterile **84. PANICUM**
25 Spikelets with ≥2 florets, at least the lowest bisexual 26
  26 Lemma entire at apex, not awned, (1)3(-5)-veined; glumes much shorter than rest of spikelet 27

26 Lemma notched or 2-3-lobed at apex, sometimes awned,
5-9-veined; glumes as long as spikelet (excl. awns) or nearly so 28
27 Perennial of wet peaty areas; lemmas ≥3mm **73. MOLINIA**
27 Alien annuals or sometimes perennials; lemmas <3mm
76. ERAGROSTIS
    28 Lemmas with bent awn >5mm **70. RYTIDOSPERMA**
    28 Lemmas awnless or with straight awn <1mm 29
29 Lemmas deeply 2-lobed; spikelets falling whole; lemmas <3mm, with very broad hyaline margins; annual **71. SCHISMUS**
29 Lemmas minutely 2-3-toothed; florets falling separately from glumes; lemmas >4mm, with very narrow hyaline margins; perennial
69. DANTHONIA

Key B - Bamboos (stems woody; leaf-blades with short petiole)
1 Main stems ± square in section **8. CHIMONOBAMBUSA**
1 Main stems cylindrical, or flattened or grooved on one side 2
    2 Main stems flattened or grooved on 1 side at least at upper internodes 3
    2 Main stems ± cylindrical throughout except sometimes just above each node 4
3 Nodes of mid-region of main stems mostly with 2 unequal branches and often a very small 3rd one; stems flattened or grooved on 1 side throughout (scarcely natd aliens not treated here) **(PHYLLOSTACHYS)**
3 Nodes of mid-region of main stems mostly with 3-5 branches; stems often flattened or grooved only at upper internodes
1. SEMIARUNDINARIA
    4 Nodes of mid-region of main stems mostly with 1(-2) lateral branches 5
    4 Nodes of mid-region of main stems mostly with ≥3 lateral branches 8
5 Leaves pubescent on lowerside **6. SASAELLA**
5 Leaves glabrous (except sometimes on margin) 6
    6 Leaves 4-5x as long as wide, with 5-14 veins on either side of midrib **5. SASA**
    6 Leaves >6x as long as wide, with 2-9 veins on either side of midrib 7
7 Stems 2.5-5m; leaves 15-30cm, 2-4cm wide, with 5-9 veins on either side of midrib **7. PSEUDOSASA**
7 Stems 0.2-2m; leaves 2.5-20cm, 0.3-2.5cm wide, with 2-7 veins on either side of midrib **4. PLEIOBLASTUS**
    8 Stems ≤2.5m **4. PLEIOBLASTUS**
    8 Stems >2.5m 9
9 Leaves 15-25mm wide, mostly with 4-7 veins on either side of midrib
4. PLEIOBLASTUS
9 Leaves ≤13(20)mm wide, mostly with 2-4 veins on either side of midrib 10
    10 Stems bearing sheaths with distinct falcate auricles and many stiff bristles at apex **2. YUSHANIA**
    10 Stems bearing sheaths without auricles and with no or few

157. POACEAE

bristles at apex   11
11 Sheaths very shortly pubescent when young, without apical bristles; leaf-sheaths with ligules ≤2mm; inflorescence an open panicle with narrow sheaths at base   **3. FARGESIA**
11 Sheaths glabrous, with few apical bristles; leaf-sheaths with ligules >2mm; inflorescence short, partly enclosed by large, wide sheaths (scarcely natd aliens not treated here)   **(THAMNOCALAMUS)**

Key C - Spikelets arising in groups of 2-7, one fatter and bisexual, the other 1-6 thinner and male or sterile (N.B. *Tragus* (Keys A & D) may have 2-4 fertile and 1-3 sterile spikelets in groups)

1 Inflorescence a spike with 3 spikelets (each with 1 floret) per node, the central one bisexual, the 2 laterals male or sterile   **65. HORDEUM**
1 Inflorescence a panicle (sometimes strongly contracted and spike-like), not with regularly 3 spikelets per node and not all spikelets with 1 floret   2
   2 Sterile or male spikelets with ≤3 florets, with glumes at least nearly as long as spikelet   3
   2 Sterile or male spikelets clearly with >5 florets, with glumes much shorter than spikelet   4
3 Spikelets in pairs, 1 bisexual, 1 male or sterile, not falling as a unit   **93. SORGHUM**
3 Spikelets in groups of 3-7, 1 bisexual, the rest male or sterile, all falling as a unit   **44. PHALARIS**
   4 Bisexual spikelets with 1 floret plus a sterile vestige, accompanied by 2-4 sterile spikelets, all falling as a unit   **19. LAMARCKIA**
   4 Bisexual spikelets with (1)2-5 florets, accompanied by 1-few sterile spikelets, the latter and the glumes of fertile spikelets not falling   **18. CYNOSURUS**

Key D - Spikelets all bisexual and similar; inflorescence a simple spike or raceme with pedicels ≤2mm

1 Spikelets with stout hooked spines on dorsal surface of glume   **83. TRAGUS**
1 Spikelets without hooked spines (but sometimes with barbed awns)   2
   2 Spikelets or groups of spikelets with 1-several (sometimes proximally fused) stiff bristles at base (NB do not confuse with bristle-like glumes)   3
   2 Spikelets without stiff bristles at base   4
3 Bristles fused proximally to form small cupule around spikelets, falling with spikelets to form a bur   **92. CENCHRUS**
3 Bristles remaining on axis when spikelets or grains fall, not fused   **90. SETARIA**
   4 One 1-floreted spikelet at each node   5
   4 Each node with >1 spikelet or with 1 spikelet with >1 floret   8
5 Densely tufted perennial; lemma awned; glumes 1-2, much shorter than floret   **11. NARDUS**
5 Annual; lemma not awned; glumes 1-2, at least as long as floret   6

6	Spikelets not sunk in hollows in axis; glumes obtuse; lemma pubescent	**51. MIBORA**
6	Spikelets sunk in hollows in axis; glumes acute to acuminate; lemma glabrous	7
7	Glumes 2 on all spikelets	**27. PAROPHOLIS**
7	Glumes 2 on terminal spikelet, 1 on lateral spikelets	**28. HAINARDIA**
	8 Spikelets 2-3 at each node	9
	8 Spikelets 1 at each node	13
9	Spikelets with 1-2 florets; lemmas long-awned	10
9	Spikelets with (2)3-6 florets; lemmas awned or not	12
	10 Spikelets 2 per node, with mostly 2 florets (upper reduced and sterile) (scarcely natd aliens not treated here)	**(TAENIATHERUM)**
	10 Spikelets all or mostly 3 per node, with mostly 1 floret	11
11	Two glumes of each spikelet free to base	**65. HORDEUM**
11	Two glumes of each spikelet fused at base	**64. HORDELYMUS**
	12 Lowest lemma >14mm, awnless; leaves very glaucous; all or ± all nodes with 2 spikelets	**63. LEYMUS**
	12 Lowest lemma <14mm, awned; leaves not glaucous; upper nodes with 1 spikelet	**62 x 65. X ELYTRORDEUM**
13	Lower glume (except in terminal spikelet) 0 or ≤3/4 as long as upper	14
13	Lower glume ≥3/4 as long as upper	17
	14 Lemma with a bent awn from dorsal surface	**34. GAUDINIA**
	14 Lemma awnless or with straight to curved terminal or subterminal awn	15
15	Upper glume with 1-3 veins; lemma very gradually narrowed to long terminal awn	**17. VULPIA**
15	Upper glume with 5(-9) veins; lemma awnless or rather abruptly narrowed to long or short terminal or subterminal awn	16
	16 Lower glume 0 except in terminal spikelet	**16. LOLIUM**
	16 Lower glume present on all or most spikelets	**15 x 16. X FESTULOLIUM**
17	Perennial; sterile shoots and often rhizomes present	18
17	Annual; sterile shoots and rhizomes 0	22
	18 Spikelets scarcely flattened, on distinct pedicels 0.5-2mm	**60. BRACHYPODIUM**
	18 Spikelets flattened, sessile or with vestigial pedicel <0.5mm	19
19	Inflorescence axis not breaking up at maturity	20
19	Inflorescence axis breaking up at maturity, 1 segment falling off with each spikelet	21
	20 Plant densely tufted, without rhizomes; lemmas usually with awns >7mm; spikelets breaking up below each lemma at maturity, leaving glumes on rhachis; anthers ≤3mm	**61. ELYMUS**
	20 Plant with long rhizomes, not densely tufted; lemmas rarely with awns >7mm; spikelets eventually falling whole, not leaving 2 glumes alone on rhachis; anthers ≥3.5mm	**62. ELYTRIGIA**
21	Lemmas awned	**62 x 65. X ELYTRORDEUM**
21	Lemmas not awned	**62. ELYTRIGIA**
	22 Glumes truncate at apex (awned or not)	23

157. POACEAE

22 Glumes acuminate to obtuse, sometimes shouldered or notched, at apex (awned or not) — 24
23 Glumes rounded on back (except near apex) (scarcely natd aliens not treated here) **(AEGILOPS)**
23 Glumes 1-2-keeled **68. TRITICUM**
   24 Glumes and lemmas <4mm, not awned **25. CATAPODIUM**
   24 Glumes and lemmas >4mm, the lemmas long-awned — 25
25 Glumes linear-lanceolate; spikelets with 2(-3) florets (all bisexual) **66. SECALE**
25 Glumes lanceolate to ovate; spikelets with 2-3 bisexual florets plus 1-few distal sterile ones, or with >4 bisexual florets — 26
   26 Spikelets with >4 bisexual florets, usually flattened narrow-side to inflorescence axis **60. BRACHYPODIUM**
   26 Spikelets with 2-3 bisexual plus 1-few distal sterile florets, flattened broad-side to inflorescence axis **67. X TRITICOSECALE**

Key E - Spikelets all bisexual and similar; inflorescence an umbel or raceme of ≥2 spikes or racemes with pedicels ≤2mm

1 Spikes all arising from same point at tip of stem — 2
1 At least some spikes arising at different (though often very close) points along apical part of stem — 7
   2 Lemmas conspicuously awned **80. CHLORIS**
   2 Lemmas not awned — 3
3 Spikelets with 2-c.10 florets, if with 2 then lower one bisexual — 4
3 Spikelets with 1-2 florets, if with 2 then lower one male or sterile — 5
   4 Spikes terminating in bare prolongation of axis **78. DACTYLOCTENIUM**
   4 Spikes terminating in a spikelet **77. ELEUSINE**
5 Fertile floret with 3-4 scales (1-2 glumes, lemma and palea of lower sterile floret) below it; annual **91. DIGITARIA**
5 Fertile floret with 2 scales (2 glumes, or upper glume and lemma representing lower sterile floret) below it; stoloniferous perennial — 6
   6 Spikelet with 2 lower scales (glumes) 1-veined, shorter than fertile lemma **81. CYNODON**
   6 Spikelet with 2 lower scales (glume and sterile lemma) 3-veined, longer than fertile lemma **89. PASPALUM**
7 Spikelets with 2 florets, the upper bisexual, the lower male or sterile — 8
7 Spikelets with 1-many florets, at least the lowest bisexual — 12
   8 Ligule 0 **85. ECHINOCHLOA**
   8 Ligule present (membranous or a fringe of hairs) — 9
9 Ligule membranous Return to 5
9 Ligule a fringe of hairs or membranous with a fringe of hairs distally — 10
   10 Spikelets with small bead-like swelling at base; lower glume ± 0 **88. ERIOCHLOA**
   10 Spikelets without bead-like swelling at base; lower glume present — 11
11 Upper lemma with awn 0.3-1mm (hidden between upper glume and lower lemma) **87. UROCHLOA**
11 Upper (and lower) lemma awnless **86. BRACHIARIA**

12	Spikelets with 1-2 florets	13
12	Spikelets with ≥3 florets	17
13	Spikelets ≥8mm	**82. SPARTINA**
13	Spikelets ≤6mm	14
14	Ligule membranous	15
14	Ligule a fringe of hairs	16
15	Spikes 2(-4), 2-7cm; spikelets appressed-pubescent on 1 side, glabrous on other	**89. PASPALUM**
15	Spikes usually >4, 0.5-2cm; spikelets glabrous on both sides	**54. BECKMANNIA**
16	Two basal scales of spikelet much shorter than rest of spikelet, glabrous	**79. SPOROBOLUS**
16	Two basal scales of spikelet longer than rest of spikelet, densely silky-pubescent	**88. ERIOCHLOA**
17	All except terminal spikelet in each branch with 1 glume	**16. LOLIUM**
17	All spikelets with 2 glumes	18
18	Lemmas 5-veined, >5.5mm	**15 x 16. X FESTULOLIUM**
18	Lemmas 1-3-veined (sometimes with 1-3 extra veins very close to midrib, forming thickened keel), <5.5mm, keeled	19
19	Axis of inflorescence much longer than longest spike; tip of lemma bifid (often awned from notch)	**75. LEPTOCHLOA**
19	Axis of inflorescence usually shorter (rarely slightly longer) than longest spike; tip of lemma acute to obtuse (often awned from tip)	**77. ELEUSINE**

Key F - Spikelets all bisexual and similar; inflorescence a panicle; spikelets with 1 (bisexual) floret and no other scales or sterile florets

1	Spikelets with small bead-like swelling at base	**88. ERIOCHLOA**
1	Spikelets without bead-like swelling at base	2
2	Glumes ± 0 (reduced to rims below floret)	**9. LEERSIA**
2	Both glumes well developed	3
3	Panicle a soft woolly ovoid dense head; lemmas tapered to 2 apical bristles, with a longer awn	**49. LAGURUS**
3	Panicle rarely a soft woolly ovoid dense head, if so then lemmas blunt and awnless; lemmas awned or not but without 2 long apical bristles	4
4	Both glumes notched at apex and with awn from sinus; lemma with dorsal awn	5
4	Glumes usually not awned, often tapered to fine apex but not notched at apex, if awned then lemma awnless	6
5	Awns of glumes (not lemma) >3mm; lemma awned from apex; fertile annual with deciduous spikelets	**52. POLYPOGON**
5	Awns of glumes <3mm; lemma awned from below apex; sterile perennial with ± persistent spikelets	**45 x 52. X AGROPOGON**
6	Lemmas with long terminal awn with much twisted, stout proximal part	**12. STIPA**
6	Lemmas with terminal, dorsal or 0 awn but without twisted proximal part	7
7	Floret with tuft of white hairs at base, the hairs >1/4 as long as lemma	8

157. POACEAE

7	Floret variously pubescent or glabrous, but without basal tuft of white hairs >1/4 as long as lemma	12
	8 Spikelets <9(10)mm; anthers <3mm	9
	8 Spikelets >(9)10mm; anthers >3mm	11
9	Lemma keeled, acute to obtuse, awnless; palea almost as long as lemma	**22. POA**
9	Lemma rounded on back, truncate or variously toothed at apex, often awned; palea c.2/3-3/4 as long as lemma	10
	10 Tuft of hairs at base of lemma c.0.3-0.6mm	**45. AGROSTIS**
	10 Tuft of hairs at base of lemma >1mm	**46. CALAMAGROSTIS**
11	Panicle very pale, spike-like; spikelets 10-16mm; lemmas with basal hairs <1/2 as long as lemma, with awn 0-1mm; anthers 4-7mm, shedding pollen	**47. AMMOPHILA**
11	Panicle usually purplish-green, usually ± lobed; spikelets 9-12mm; lemmas with basal hairs ≥1/2 as long as lemma, with awn 1-2mm; anthers 3-4.5mm, not opening	**46 x 47. X CALAMMOPHILA**
	12 Both glumes ≤3/4 as long as spikelet, usually 1 or both c.1/2 as long	13
	12 Both glumes >3/4 as long as spikelet, usually 1 or both as long or longer	16
13	Lemmas with awns >3mm; annual	**50. APERA**
13	Lemmas awnless; normally perennial	14
	14 Lemmas ± truncate; ligule membranous	**24. CATABROSA**
	14 Lemmas obtuse to acute; ligule a fringe of hairs	15
15	Lemmas with 1 vein, <3mm; spikelets all with 1 floret	**79. SPOROBOLUS**
15	Lemmas with 3-5 veins, >3mm; usually some spikelets with >1 floret	**73. MOLINIA**
	16 Glumes with swollen ± hemispherical base and long very tapering distal part	**48. GASTRIDIUM**
	16 Glumes without swollen base	17
17	Lemma shiny and much harder and tougher than glumes when mature	18
17	Lemma about same texture as glumes or more flimsy and delicate	19
	18 Lemmas awnless; ligules 2-10mm	**14. MILIUM**
	18 Lemmas with deciduous awn 2-5mm; ligules c.1mm	**13. ORYZOPSIS**
19	Lemmas with subterminal awn 4-10mm	**50. APERA**
19	Lemmas with awn 0 or <4mm, or with awn ≥4mm and arising from lower 1/2 of back of lemma	20
	20 Panicle very compact, ovoid or oblong to cylindrical, spike-like, with very short branches	21
	20 Panicle diffuse or contracted but with obvious branches	22
21	Lemmas awnless; palea present; florets falling at maturity leaving glumes on panicle	**55. PHLEUM**
21	Lemmas usually awned (often shortly); palea absent; spikelets falling as a whole at maturity	**53. ALOPECURUS**
	22 Lemmas acute to obtuse, strongly keeled	**22. POA**

22 Lemmas truncate or variously toothed at apex, rounded on back 23
23 Glumes with short pricklets or at least rough all over back; palea nearly as long as lemma; lemma c.1/2 as long as glumes
52. POLYPOGON
23 Glumes often with short pricklets on midrib but ± smooth otherwise; palea ≤3/4 as long as lemma; lemma c.2/3-3/4 as long as glumes 24
    24 Panicle branches bearing spikelets ± to base, disarticulating (if at all) near base of pedicels; very rare **45 x 52. X AGROPOGON**
    24 Panicle branches with clear bare region at base, disarticulating (sometimes not) at base of lemmas; very common **45. AGROSTIS**

Key G - Spikelets all bisexual and similar; inflorescence a panicle; spikelets with 1 bisexual floret plus 1 or more male or sterile florets or scales below it
1 Spikelets or groups of spikelets with 1-several (proximally fused or free) stiff bristles at base 2
1 Spikelets without stiff bristles at base 3
    2 Bristles fused proximally to form small cupule around spikelets, falling with spikelets to form bur **92. CENCHRUS**
    2 Bristles not fused, retained on axis when spikelets or florets fall
90. SETARIA
3 Ligule a fringe of hairs, or membranous with a distal fringe of hairs 4
3 Ligule membranous or 0 5
    4 Upper glume as long as spikelet; spikelet of 2 florets, the lower sterile **84. PANICUM**
    4 Upper glume much shorter than spikelet; spikelet of 3 florets, the lower 2 sterile **10. EHRHARTA**
5 Bisexual floret with 1 male or sterile floret below it 6
5 Bisexual floret with 2 male or sterile florets below it 8
    6 Lower lemma awned from proximal 1/2 of back; perennial
32. ARRHENATHERUM
    6 Lower lemma awnless or with terminal awn; annual 7
7 Ligule membranous; sterile floret a minute scale; glumes equal, keeled **44. PHALARIS**
7 Ligule 0; sterile floret as large as fertile one; glumes very unequal, not keeled **85. ECHINOCHLOA**
    8 Lower glume c.1/2 as long as upper; lemmas of lower 2 florets with awns ≥1.5mm **43. ANTHOXANTHUM**
    8 Glumes equal or subequal; lemmas of lower 2 florets awnless or with awn <1mm 9
9 Panicle ± diffuse; lower 2 florets male, longer than bisexual floret
42. HIEROCHLOE
9 Panicle compact; lower 2 florets sterile, shorter than bisexual floret
44. PHALARIS

Key H - Spikelets all bisexual and similar; inflorescence a panicle or a raceme with pedicels >3mm; spikelets with ≥2 bisexual florets; ovary with pubescent terminal appendage extending beyond base

157. POACEAE

of styles (NB *Brachypodium* and *Gaudinia* have similar ovaries, but the inflorescence is a spike or raceme, not a panicle - Key D).

1 At least lowest lemma with conspicuously bent awn arising dorsally from lower 2/3 of its length, or awnless with glumes longer than rest of spikelet   2
1 Lemmas with straight or slightly bent awns arising from apex or subapically, or awnless with glumes much shorter than rest of spikelet   4
    2 Easily uprooted annual without non-flowering shoots; upper glume >(15)20mm   **33. AVENA**
    2 Firmly rooted perennial with non-flowering shoots; upper glume <15(20)mm   3
3 Spikelets usually <11mm (excl. awns); awn of lowest lemma arising c.1/3 from base; upper glume 7-11mm; lowest floret usually male (sometimes bisexual)   **32. ARRHENATHERUM**
3 Spikelets usually >11mm (excl. awns); all lemmas with awn arising near middle or above it; upper glume 10-20mm; lowest floret bisexual   **31. HELICTOTRICHON**
    4 Lemmas strongly keeled on back   **59. CERATOCHLOA**
    4 Lemmas rounded on back or keeled near apex   5
5 Perennial, with sterile shoots at flowering time, often with rhizomes   6
5 Annual (or biennial), without sterile shoots at flowering time, without rhizomes   7
    6 Spikelets <15mm (excl. awns), narrowed to apex; lemmas ≤8mm (excl. awns)   **15. FESTUCA**
    6 Spikelets ≥15mm (excl. awns), ± parallel-sided almost to apex; lemmas ≥8mm (excl. awns)   **57. BROMOPSIS**
7 Spikelets ovate to lanceolate, slightly compressed, markedly narrowed towards top; lemmas 5-11mm, with awn c. as long or shorter; lower glume 3-7-veined; upper glume 5-9-veined   **56. BROMUS**
7 Spikelets ± straight-sided, widening distally; lemmas 9-36mm, with awn c. as long or longer, lower glume 1(-3)-veined; upper glume 3(-5)-veined   **58. ANISANTHA**

Key I - Spikelets all bisexual and similar; inflorescence a panicle or a raceme with pedicels >3mm; spikelets with ≥2 bisexual florets, or with only 1 and that basal and with male or sterile florets distal to it; ovary glabrous or pubescent but without pubescent terminal appendage extending beyond base of styles; lemmas with subterminal to dorsal, usually bent awn, often bifid at apex, sometimes unawned and then always bifid at apex

1 Lemmas >6mm, usually with awn ≥10mm   2
1 Lemmas <6mm, never with awn >10mm   4
    2 Easily uprooted annual without non-flowering shoots; upper glume >(15)20mm   **33. AVENA**
    2 Firmly rooted perennial with non-flowering shoots; upper glume <15(20)mm   3
3 Spikelets usually ≤11mm (excl. awns); awn of lowest lemma arising

c.1/3 from base; upper glume 7-11mm; upper lemma(s) often
± unawned **32. ARRHENATHERUM**
3 Spikelets usually ≥11mm (excl. awns); all lemmas with awns arising
near middle or above it; upper glume 10-20mm
31. HELICTOTRICHON
4 Easily uprooted annual without non-flowering shoots 5
4 Firmly rooted (unless in sand) perennial with non-flowering
shoots 6
5 Spikelets with 2 florets; lemmas with bent awn arising from below
1/2 way **41. AIRA**
5 Spikelets with 3-5 florets; lemmas with usually ± straight subterminal
awn **37. ROSTRARIA**
6 Awns slightly but distinctly widened towards apex (club-shaped)
40. CORYNEPHORUS
6 Awns parallel-sided or tapered to apex 7
7 Glumes pubescent at least along midrib, falling with florets at
maturity; lower lemma unawned; upper floret usually male
39. HOLCUS
7 Glumes glabrous, remaining on plant when florets fall; lowest lemma
awned; all florets usually bisexual 8
8 Spikelets with 2 florets; upper glume as long as spikelet or almost
so; lemmas truncate to very blunt and usually jagged at apex
38. DESCHAMPSIA
8 Spikelets with 2-4 florets; upper glume distinctly shorter than
spikelet; lemmas acute and finely bifid at apex **35. TRISETUM**

Key J - Spikelets all bisexual and similar; inflorescence a panicle or a raceme
with pedicels >3mm; spikelets with ≥2 bisexual florets, or with
only 1 and that basal and with male or sterile florets distal to it;
ovary glabrous or pubescent but without pubescent terminal
appendage extending beyond base of styles; lemmas unawned and
entire at apex, or with terminal straight to curved awn and then
sometimes bifid or several-toothed at apex
1 Sheaths fused almost to apex to form tube round stem 2
1 Sheaths with free, usually overlapping margins 4
2 Lemmas finely pointed or awned, 5-veined **15. FESTUCA**
2 Lemmas acute to rounded or ± 3-lobed at apex, 7(-9)-veined 3
3 Fertile florets 1-3, with a club-shaped group of sterile florets beyond
30. MELICA
3 Fertile florets 4-c.16, with 0 or 1 reduced sterile floret beyond
29. GLYCERIA
4 Lemmas with 3-5 short very pointed or shortly awned teeth at
± truncate apex 5
4 Lemmas 1-pointed to rounded or 2-lobed at apex, awned or not 6
5 Ligule a fringe of hairs; glumes 3-5-veined (lateral veins often short);
lemmas 7-9-veined, the veins ending short of the apical points
69. DANTHONIA
5 Ligule membranous; glumes 1-veined; lemmas 3-5-veined, the veins

157. POACEAE

	running into the apical points	**26. SESLERIA**
	6 Lemmas 2-lobed or -toothed at apex, awned or not from the sinus	7
	6 Lemmas 1-pointed (sometimes very minutely notched) to rounded at apex, awned or not from tip	9
7	Lemmas with bent awn >5mm	**70. RYTIDOSPERMA**
7	Lemmas awnless or with straight awn <1mm	8
	8 Lemmas deeply 2-lobed; spikelets falling whole; annual	**71. SCHISMUS**
	8 Lemmas minutely 2-toothed; florets falling separately from glumes; perennial	**69. DANTHONIA**
9	Ligule a fringe of hairs	10
9	Ligule membranous	11
	10 Perennial of wet peaty areas; lemmas ≥3mm	**73. MOLINIA**
	10 Alien annuals or sometimes perennials; lemmas <3mm	**76. ERAGROSTIS**
11	Lemma strongly keeled throughout length	12
11	Lemma rounded on back or keeled only distally	14
	12 Lemmas 3-veined, with wide membranous ± shiny margins, rendering panicle silvery	**36. KOELERIA**
	12 Lemmas 5-veined, often with membranous margins but scarcely shiny and panicle not silvery	13
13	Spikelets borne in dense 1-sided clusters; lemmas long-acuminate to shortly awned	**23. DACTYLIS**
13	Spikelets not aggregated into dense 1-sided clusters; lemmas obtuse to acute	**22. POA**
	14 Lemmas awned	15
	14 Lemmas not awned	18
15	Easily uprooted annual without non-flowering shoots	**17. VULPIA**
15	Firmly rooted perennial with non-flowering shoots	16
	16 Lower glume mostly >3/4 as long as upper; anthers dehiscent	**15. FESTUCA**
	16 Lower glume mostly <3/4 as long as upper; anthers indehiscent	17
17	Pointed auricles present at junction of sheath and leaf; lemmas rather abruptly narrowed to awn; upper glume with 5-9 veins	**15 x 16. X FESTULOLIUM**
17	Auricles 0; lemmas very gradually narrowed to awn; upper glume with 1-3 veins	**15 x 17. X FESTULPIA**
	18 Lemmas acute to acuminate at apex	19
	18 Lemmas subacute to rounded at apex	20
19	Both glumes >1/2 as long as spikelet; lemmas 3-veined	**36. KOELERIA**
19	At least lower glume, and usually both glumes, <1/2 as long as spikelet; lemmas 5-veined	**15. FESTUCA**
	20 Lemmas strongly cordate at base, wider than long; spikelets pendent at maturity	**21. BRIZA**
	20 Lemmas not or scarcely cordate at base, longer than wide; spikelets not pendent	21
21	Lemmas 3-veined; spikelets with 2-3 florets; sheaths compressed, keeled on the back	**24. CATABROSA**

21 Lemmas 5-veined; spikelets usually with >4 florets; sheaths
 rounded on the back 22
  22 Pointed auricles present at junction of sheath and leaf; lower
 glume mostly <3/5 as long as upper **15 x 16. X FESTULOLIUM**
  22 Auricles 0; lower glume ≥3/5 as long as upper 23
23 Lemmas minutely pubescent at base, with conspicuous membranous
 tips and margins; usually perennial **20. PUCCINELLIA**
23 Lemmas glabrous, with very narrow membranous margins; annual
 25. CATAPODIUM

SUBFAMILY 1 - BAMBUSOIDEAE (tribes 1-3; genera 1-10). Woody bamboos or herbaceous perennials; leaves often with short false-petiole separating them from sheath, often with conspicuous cross-veins as well as longitudinal veins; ligule membranous, rarely a membrane fringed with hairs; sheaths not fused, the lower ones often without or with very reduced leaves; inflorescence a panicle; spikelets with ≥3 florets, with all florets (except often the distal ones) bisexual, or with 1 bisexual floret with or without 2 large sterile ones below it; glumes (0)2(-3); lemmas firm, 5-many-veined, awnless or with terminal awn; paleas with 1-many veins, 1 on mid-line; stamens 3-6; stigmas 2 or 3; lodicules 2 or 3.

TRIBE 1 - BAMBUSEAE (genera 1-8). Woody bamboos; leaves with short false-petiole and with cross-veins as well as longitudinal veins; ligule membranous; lower sheaths usually without or with very reduced leaves; spikelets with ≥3 florets, with all (except often the distal) florets bisexual; glumes (0-)2(3); lemmas 5-many-veined, awnless; paleas with ≥4 veins; stamens 3-6; stigmas 2 or 3; lodicules 2 or 3.

Sheaths refers to those on the main stems and not bearing leaf-blades, as opposed to *leaf-sheaths*. Bamboos become natd in neglected woods, parks, gardens, shrubberies and estates.

1. SEMIARUNDINARIA Nakai - *Narihira Bamboo*
Stems >3m, densely clump-forming, flattened or grooved on 1 side (alternating in successive internodes, at least at upper internodes); nodes with 3 lateral branches or 3 larger ones plus some smaller ones; leaves 1.5-4cm wide, glabrous, mostly with 4-6 veins on either side of midrib; stamens 3; stigmas 3.

1 Stems 3-6(8)m, 1.5-4(8)cm thick, green suffused with purple; leaves
 12-24 x 1.5-2.5(4)cm, often purplish. Intrd-natd; bank of R. Thames
 in Surrey *Narihira Bamboo* - **S. fastuosa** (Mitford) Nakai

2. YUSHANIA Keng f. - *Fountain-bamboos*
Stems mostly >3m, terete except just above nodes; nodes mostly with ≥3 lateral branches; leaves 5-12mm wide, ± glabrous, mostly with 2-4 veins on either side of midrib; stamens 3; stigmas 2-3.

1 Stems 3-6m, 0.7-2cm thick, purplish-green; leaves 6-12(16) x 0.5-

1.2(2)cm. Intrd-natd; in scattered places in W & S Br, MW Yorks, Ir and CI (*Sinarundinaria anceps* (Mitford) C.S. Chao & Renvoize)
Indian Fountain-bamboo - **Y. anceps** (Mitford) W.C. Lin

3. FARGESIA Franch. - *Fountain-bamboos*
Stems mostly >3m, terete except just above nodes; nodes with ≥3 lateral branches; leaves 5-15mm wide, glabrous, mostly with 3-4 veins on either side of midrib; stamens 3; stigmas 2-3.

1 Stems 4-6m, 1-1.5cm thick, purplish-green; leaves 3.5-7(10) x 0.5-1.3cm. Intrd-natd; scattered over En, Cards (*Sinarundinaria nitida* (Stapf) Nakai, *Thamnocalamus spathaceus* (Franch.) Soderstr.)
Chinese Fountain-bamboo - **F. spathacea** Franch.

4. PLEIOBLASTUS Nakai - *Bamboos*
Stems 0.2-5(8)m, terete except just above nodes; nodes with 1-many lateral branches; leaves 0.3-3cm wide, glabrous, mostly with 2-7 veins on either side of midrib; stamens 3; stigmas 3.

1 Stems (0.2)0.4-0.75(1.2)m, 1-3mm thick; branches 1-2 per node; leaves 2.5-8 x 0.3-1.5cm. Intrd-natd; scattered in BI, especially Guernsey
Dwarf Bamboo - **P. pygmaeus** (Miq.) Nakai
1 Stems 1.2-5(8)m, >5mm thick; branches 2-many per node; most leaves >10cm 2
  2 Stems 1.2-2(3)m, 1-1.5cm thick; leaves (8)12-20(25) x 0.8-3cm, uniformly green on lowerside. Intrd-natd; scattered in Br and Ir
Maximowicz's Bamboo - **P. chino** (Franch. & Sav.) Makino
  2 Stems (2.5)3-5(8)m, 2-3cm thick; leaves 15-25 x 1-2.5(3.5)cm, on lowerside green on 1 side and greyish-green on other side. Intrd-natd; few sites in SW En, SW Ir and Guernsey
Simon's Bamboo - **P. simonii** (Carrière) Nakai

5. SASA Makino & Shibata - *Bamboos*
Stems 0.5-3m, terete except just above nodes; nodes mostly with 1 lateral branch; leaves 2.5-9cm wide, glabrous except often with marginal hairs or bristles, mostly with 5-13 veins on either side of midrib; stamens 6; stigmas 3.

1 Stems 2-3m, 7-10mm thick; leaves 12-30(40) x 3.5-9cm, with 8-14 veins on either side of midrib; false-petioles usually green; sheaths glabrous. Intrd-natd; widely natd in BI
Broad-leaved Bamboo - **S. palmata** (Burb.) E.G. Camus
1 Stems 0.5-1.5m, 5-7mm thick; leaves 10-25 x 2.5-6cm, with 5-9 veins on either side of midrib; false-petioles often purplish; sheaths pubescent especially near base when young. Intrd-natd; widely but rarely natd in BI, mostly S En
Veitch's Bamboo - **S. veitchii** (Carrière) Rehder

6. SASAELLA Makino - *Hairy Bamboo*
Stems 0.5-1.5m, terete except just above nodes; nodes mostly with 1 lateral branch; leaves 1-3cm wide, sparsely pubescent on upperside, pubescent to densely so on lowerside, mostly with 3-5 veins either side of midrib; stamens 6; stigmas 3.

1   Stems 0.5-1.5m, 3-8mm thick; leaves 8-20 x 1-3cm. Intrd-natd; scattered in Br and Ir, especially in woods
Hairy Bamboo - **S. ramosa** (Makino) Makino

7. PSEUDOSASA Nakai - *Arrow Bamboo*
Stems to 5m, terete except just above nodes; nodes mostly with 1 lateral branch; leaves 2-5cm wide, glabrous, mostly with 5-9 veins either side of midrib; stamens 3; stigmas 3.

1   Stems 2.5-5m, 1-2cm thick; leaves 15-30 x 2-4cm. Intrd-natd; frequent throughout much of BI, easily the commonest grown and natd bamboo
Arrow Bamboo - **P. japonica** (Steud.) Nakai

8. CHIMONOBAMBUSA Makino - *Square-stemmed Bamboo*
Stems to 8m, square in section with rough edges and faces; nodes mostly with 3 lateral branches; leaves 1-3cm wide, minutely pubescent when young, mostly with 8-14 veins on either side of midrib; stamens 3; stigmas 2.

1   Stems 5-8m, 2-4cm thick; leaves 10-20 x 1-3cm. Intrd-natd; SW En and SW Ir
Square-stemmed Bamboo - **C. quadrangularis** (Franceschi) Makino

TRIBE 2 - ORYZEAE (genus 9). Herbaceous rhizomatous perennials; leaves without false-petiole, without cross-veins; ligule membranous; spikelets with 1 floret; glumes 0 or vestigial; lemmas 5-veined, awnless; palea 3-veined; stamens 3; lodicules 2.

9. LEERSIA Sw. - *Cut-grass*

1   Culms to 1.2m, the sheaths and leaves rough with downward-pointed bristles; panicle diffuse or remaining 1/2-enclosed in sheaths. Native; wet meadows, ditches, canal-banks and riversides; very locally frequent from S Somerset to Surrey and W Sussex
Cut-grass - **L. oryzoides** (L.) Sw.

TRIBE 3 - EHRHARTEAE (genus 10). Herbaceous annuals or rhizomatous perennials; leaves without false-petiole, without cross-veins; ligule a membrane fringed with hairs; spikelets with 1 small bisexual floret with awnless 3-7-veined lemma and 1-veined palea, and two larger sterile florets below each consisting of 1 awned 5-7-veined lemma only, the former obscured by the latter two; glumes 2, often very small; stamens 4-6; lodicules 2.

10. EHRHARTA Thunb. - *Weeping-grass*

1 Erect to procumbent perennial to 70cm, the leaves rough with upward-directed bristles; panicle sparsely branched, narrow. Intrd-casual; rather infrequent wool-alien; scattered over En
Weeping-grass - **E. stipoides** Labill.

SUBFAMILY 2 - POOIDEAE (tribes 4-14, genera 11-68). Annual to perennial herbs; leaves without false-petiole or cross-veins; ligule membranous, sometimes a membrane fringed with hairs; sheaths usually not fused, sometimes fused; inflorescence a spike, raceme or panicle; spikelets with 1-many florets, with all florets (except often the distal ones) bisexual or some variously male or sterile; glumes (0-)2; lemmas firm to membranous, 3-9-veined, awnless or with terminal or dorsal awn; palea normally 2(-4)-veined without vein on mid-line, rarely with midrib or 0; stamens 1-3; stigmas 1-2; lodicules 0 or 2(-3), or 2 fused laterally.

TRIBE 4 - NARDEAE (genus 11). Perennials; ligule membranous; sheaths not fused; inflorescence a ± 1-sided spike with 1 spikelet at each node; spikelets with 1 bisexual floret; glumes very short, upper often 0; lemma with 2-3 keels, 3-veined, with terminal awn; stamens 3; stigma 1; lodicules 0; ovary glabrous.

11. NARDUS L. - *Mat-grass*

1 Densely tufted; culms to 40(60)cm, wiry; leaves tightly inrolled. Native; acid heaths, moors and mountain slopes; throughout BI in suitable places
Mat-grass - **N. stricta** L.

TRIBE 5 - STIPEAE (genera 12-14). Annuals or perennials; ligule membranous or sometimes a membrane with a fringe of hairs; sheaths not fused; inflorescence a ± diffuse panicle; spikelets with 1 bisexual floret; glumes equal or ± so, longer than body of lemma; lemma rounded on back, 3-5-veined, awnless or with long terminal awn, becoming hard and tightly wrapped round caryopsis; stamens 3; stigmas 2; lodicules 2-3; ovary glabrous.

12. STIPA L. - *Needle-grasses*

Annuals or perennials; ligule a membrane with a fringe of hairs; lemma with long basal callus pointed and with apically directed bristles, with long terminal awn with stouter spirally twisted proximal part separated from narrow distal part by conspicuous bend; lodicules 2-3.

1 Perennial; glumes with long fine 3-pronged apex, hyaline at margins and apex; lemma c.10mm incl. callus, with membranous cupule at apex. Culms to 60cm. Intrd-natd; wool-alien; scattered in En, sometimes ± natd in SE
American Needle-grass - **S. neesiana** Trin. & Rupr.

1 Annual; glumes with long fine 1-pointed apex, wholly hyaline; lemma 6-8mm incl. callus, with cupule of hairs at apex. Culms to 60cm. Intrd-casual; wool-alien; scattered in En
Mediterranean Needle-grass - **S. capensis** Thunb.

13. ORYZOPSIS Michx. - *Smilo-grass*
Perennials; ligule membranous; lemma shiny, with short smooth basal callus, with long straight deciduous terminal awn; lodicules 2.

1 Stems to 1.5m, densely tufted; panicle with long central axis bearing at each node many slender branches with spikelets clustered at ends. Intrd-natd; casual wool-alien and ornamental; scattered in En and S Wa, natd in Jersey and W Kent
Smilo-grass - **O. miliacea** (L.) Asch. & Schweinf.

14. MILIUM L. - *Millets*
Annuals or perennials; ligule membranous; lemma shiny, with minute callus, awnless; lodicules 2.

1 Tufted perennial; stems erect, to 1.5m; leaves 5-15mm wide; panicle 10-40cm, with patent to reflexed slender whorled branches. Native; moist shady woods; locally frequent throughout En, scattered in Wa, Ir and lowland Sc   *Wood Millet* - **M. effusum** L.
1 Annual; stems procumbent to decumbent, to 10(15)cm; leaves 1-3mm wide; panicle 1-4cm, with appressed branches. Native; short turf on sand-dunes and cliffs by sea; 2 places in Guernsey
Early Millet - **M. vernale** M. Bieb.

TRIBE 6 - POEAE (genera 15-26). Annuals or perennials; ligule membranous; sheaths not fused or sometimes fused; inflorescence a panicle, or a raceme or spike with normally 1 spikelet at each node; spikelets with (1)3-many florets all (except the most apical) bisexual or with a group of apical male or sterile florets, sometimes spikelets in pairs, one of each pair entirely sterile; glumes equal to very unequal, sometimes the lower vestigial or 0; lemma rounded or keeled on back, 3-9-veined, awnless or with terminal awn; stamens 1-3; stigmas 2; lodicules 2; ovary glabrous or pubescent at apex.

15. FESTUCA L. - *Fescues*
Perennials with or without rhizomes, without stolons; sheaths not fused or fused; inflorescence a panicle; spikelets with (2)3-many florets all (except the most apical) bisexual, sometimes proliferating; glumes subequal; lemmas rounded on back, 3-5-veined, acuminate to subacute at apex, awned or not; stamens 3.

The *F. rubra* (couplets 8-15) and *F. ovina* (couplets16-24) aggs are extremely critical, and accurate identification requires considerable experience and often the aid of leaf-sections, leaf epidermis characters and chromosome numbers. Shaded, droughted, crowded-out or otherwise starved plants should be avoided.

Leaf, sheath and ligule characters refer to leaves on tillers unless otherwise stated. Spikelet lengths are not total spikelet lengths, but the length from the base of the lower glume to the apex (excl. awn) of the fourth lemma; lengths of spikelets with only 3 florets are obtained by addition to the total length of an increment equal to the distance between the tips of the second and third lemmas. Lemma lengths exclude awns, and refer to only the lowest 2 per spikelet. Awn lengths are calculated by averaging the lengths of all the awns of one spikelet, and then finding the mean of this value for 5-10 spikelets. All measurements given are a range of means of 5-10 measurements per plant (not a range of extremes). The distinction between *extravaginal* and *intravaginal* tillers is fundamental. The latter arise parallel to the parent shoot and remain enclosed within the parental sheath for some distance. The former arise ± at right angles to the parent shoot and break through the parental sheath at its base. Rhizomes always start off extravaginally; hence the presence of rhizomes indicates the existence of extravaginal shoots, but the absence of rhizomes does not necessarily mean that all tillers are intravaginal. Sheath fusion should be observed on the next to most apical sheath of a tiller by stripping off all the more mature sheaths below it.

1 Base of leaf on each side extended into pointed auricles clasping stem at level of and on opposite side to ligule   2
1 Leaves without auricles, or with short rounded auricles not clasping stem   4
   2 Lemmas with awns longer than body; exposed nodes of culms dark violet-purple. Culms to 100(150)cm; rhizomes 0. Native; woods, hedgerows and other shady places; common in Br and Ir except N & NW Sc   *Giant Fescue* - **F. gigantea** (L.) Vill.
   2 Lemmas with awns 0 or much shorter than body; exposed nodes of culms green, sometimes tinged purplish   3
3 Auricles glabrous; lowest 2 panicle nodes with two unequal branches, the shorter with 1-2(3) spikelets. Culms to 80(120)cm; rhizomes 0. Native; meadows, hedgerows, waysides, by ditches and rivers, usually on rich moist soil; frequent *Meadow Fescue* - **F. pratensis** Huds.
3 Auricles usually fringed with minute hairs (often few, wearing off with age); lowest 2 panicle nodes with two subequal branches, the shorter with (3)4-many spikelets. Culms to 120(200)cm; rhizomes 0. Native; grassy places, rough and marginal ground on wide range of soils; common   *Tall Fescue* - **F. arundinacea** Schreb.
   4 Leaves of culms and tillers flat (or folded longitudinally when dry), >4mm wide (or >2mm from midrib to edge)   5
   4 Leaves of tillers and usually culms folded longitudinally or the edges also inrolled, ≤4mm wide (or ≤2mm from midrib to edge)   6
5 Leaves 4-14mm wide; ligules >1mm; lemmas 3-veined, awnless; rhizomes 0; ovary with pubescent apex. Culms to 120(150)cm; rhizomes 0. Native; moist stony slopes and ravines in woods and copses; scattered in Br from S Wa to N mainland Sc, E Sussex, very

scattered in Ir   **Wood Fescue** - **F. altissima** All.
5 Leaves ≤5mm wide; ligules <1mm; lemmas 5-veined, awned; rhizomes present; ovary with glabrous apex

(see couplet 12) **F. rubra ssp. megastachys**
- 6 Ovary obovoid, free from palea, with pubescent apex; most of lemma width translucent, not green; leaves all folded, with sharp apices; ligules 0.5-1mm. Culms to 45cm, arising from mat-forming vegetative shoots; rhizomes 0. Intrd-natd; in limestone quarry; NW Yorks   **Spiky Fescue** - **F. gautieri** (Hack.) K. Richt.
- 6 Ovary ellipsoid to oblong, adherent to palea, usually glabrous, if with pubescent apex then culm leaves flat; most of lemma width green, not translucent; leaves without sharp apices; ligules ≤0.5mm   7

7 Young leaves on tillers with sheaths fused almost up to top; some or all tillers extravaginal (*F. rubra* agg.)   8   57

7 Young leaves on tillers with sheaths not fused near apex but with overlapping margins; all tillers intravaginal (*F. ovina* agg.)   16   57
- 8 Ovary (and caryopsis) with pubescent apex; leaves with 3(-5) veins; leaves of culms and tillers markedly different, the former flat and 2-4mm wide, the latter folded and ≤0.6mm from midrib to edge. Culms to 100(120)cm, densely tufted; rhizomes 0. Intrd-natd; woods and wood-borders on light soils; scattered in Br, mainly S En, Co Limerick

    Various-leaved Fescue - **F. heterophylla** Lam.
- 8 Ovary (and caryopsis) glabrous; leaves with 5-9(11) veins; leaves of culms and tillers similar to obviously different   9

9 Leaves with densely pubescent adaxial ribs, rounded on midrib abaxially, with abaxial sclerenchyma usually continuous or semi-continuous, with distinct sclerenchyma bundles in adaxial ribs; always coastal. Culms to 75(90)cm, scattered; rhizomes very long. Native; mobile sand-dunes and sandy shingle by sea; frequent on coasts of Br N to E Ross and S Wa, rare to W Ross and in E Ir, CI

    Rush-leaved Fescue - **F. arenaria** Osbeck
9 Leaves with scabrid or sparsely pubescent adaxial ribs, obtuse to keeled on midrib abaxially, with abaxial sclerenchyma in discrete bundles, often without or with very sparse sclerenchyma in adaxial ribs; coastal or inland (**Red Fescue** - **F. rubra** L.)   10   57
- 10 Rhizomes 0 or very few and very short. Culms to 75cm, densely tufted. Native; grassy places and rough ground, usually in well-drained soils; probably throughout Br but under-recorded

    F. rubra ssp. commutata Gaudin
- 10 Rhizomes well developed (plants densely tufted or not)   11

11 Leaves 0.8-1.4(2.5)mm from midrib to edge, with distinct islets of sclerenchyma in adaxial ribs; lemmas usually 6-8mm   12

11 Leaves 0.5-1.2mm from midrib to edge, usually without or with very sparse sclerenchyma in adaxial ribs (except in ssp. *juncea*); lemmas 4.2-6.5mm   13
- 12 Leaves of culms and often tillers flat; culms up to 1m; panicle

diffuse. Rhizomes well developed, forming diffuse patches; culms to 100cm. Probably intrd-natd; grassy places, especially on waysides where it is much planted; scattered throughout Br
F. rubra ssp. megastachys Gaudin

12 Leaves of culms and tillers folded; culms up to 70cm; panicle compact with ± appressed branches. Rhizomes well developed, forming loose patches. Native; grassy and rocky places from near sea-level to >800m; Sc from Argyll to Shetland and Outer Hebrides, Cumberland **F. rubra ssp. scotica** Al-Bermani

13 Lemmas 4.2-6mm, with awns 0.1-1.1mm, glaucous, usually with dense white hairs (sometimes glabrous but usually some pubescent plants nearby). Rhizomes well developed, forming loose patches; culms to 50cm. Native; wet mountain slopes and gulleys, rock-crevices and flushes down to sea-level, often on serpentine; scattered from C Sc to Shetland, Caerns, Lake District **F. rubra ssp. arctica** Govor.

13 Lemmas 4.4-8mm, with awns 0.5-2.8mm, glaucous or not, the hairs (if present) not dense and white 14

14 Spikelets 8.7-11.2mm; lemmas 5.7-8 x ≥2mm, with awns 1.1-2.8mm; saline sand or mud, often forming dense mats. Rhizomes rather short; culms to 55cm. Native; salt-marshes and other muddy or sandy saline areas; probably round coasts of BI
F. rubra ssp. litoralis (G. Mey.) Auquier

14 Spikelets 6.8-10.2mm; lemmas 4.4-6.7 x 1.5-2.4mm, with awns 0.5-2.2mm; rarely in saline sand or mud but often maritime 15

15 Rhizomes short, forming dense tufts, often some or all plants in population with very glaucous leaves. Culms to 75cm. Native; maritime cliffs and inland grassy rocky places; round coasts of BI and in hilly areas of N Br **F. rubra ssp. juncea** (Hack.) K. Richt.

15 Rhizomes medium to long, forming loose patches; plants rarely glaucous. Rhizomes well developed, usually forming loose patches; Culms to 75cm. Native; all kinds of grassy places; common
F. rubra ssp. rubra

16 Leaves with 5-9 veins, with 4(-6) adaxial grooves; lemmas with awns usually >1.2mm and often >1.6mm 17

16 Leaves with 5-7 veins, with 2(-4) adaxial grooves; lemmas with awns <1.6mm and often <1.2mm 18

17 Pedicels 1.2-2.8mm; spikelets 6.1-8.5mm; lemmas with awns 1.2-2.6mm; sheaths often sparsely pubescent; leaves not or slightly glaucous, with abaxial sclerenchyma in 3 main islets (at midrib and edges). Densely tufted, without rhizomes; culms to 70cm. Intrd-natd; roadsides, commons and rough ground especially on acid well-drained soils; frequent in SE En and E Anglia, scattered elsewhere ***Hard Fescue* - F. brevipila** R. Tracey

17 Pedicels 0.5-1.8mm; spikelets 5.4-7mm; lemmas with awns 0.5-1.5mm; sheaths glabrous; leaves usually very glaucous, with abaxial sclerenchyma in ≥5 main islets (at midrib, edges and variously in between). Densely tufted, without rhizomes; culms to 40cm. Native; very dry acid heaths and maritime cliff-tops; W Suffolk, N Lincs,

S Devon and CI   *Blue Fescue* - **F. longifolia** Thuill.

18 Spikelets 4.7-7.5mm; lemmas with awns usually 0-1mm; panicles mostly <8cm, with ≤26 spikelets and lowest 2 nodes <2cm apart; leaves usually <0.57mm midrib to edge   19

18 Spikelets often >7mm, rarely <6mm; lemmas with awns usually >1mm; panicles ≤13cm, with ≤40 spikelets and lowest 2 nodes ≤3.7cm apart; leaves >0.57mm midrib to edge   23

19 Spikelets all or mostly proliferating; sexual florets (if present) with lemmas 3.4-4.2mm and awns 0-0.2mm. Densely tufted, without rhizomes; culms to 44(50)cm. Native; grassy places in hilly districts, usually on rocky ground; common in C & N Sc, local in S Sc, N En, Wa and Ir   *Viviparous Sheep's-fescue* - **F. vivipara** (L.) Sm.

19 Spikelets not or rarely some proliferating; sexual florets with lemmas 2.5-4.9mm and awns 0-1mm   20

   20 Spikelets ≤5.5mm; lemmas ≤3.5mm, with awns 0-0.6mm; leaves glabrous, 0.3-0.45(0.53)mm from midrib to edge. Densely tufted, without rhizomes; culms to 35cm. Native; grassy places on usually acid sandy soils; frequent
                  Fine-leaved Sheep's-fescue - **F. filiformis** Pourr.

   20 Spikelets ≥5.3mm; lemmas ≥3.3mm, with awns 0-1.6mm; leaves often pubescent at base, 0.3-0.75mm from midrib to edge. Densely tufted, without rhizomes. Native (*Sheep's-fescue* - **F. ovina** L.)   21

21 Spikelets 5.5-7.5mm; lemmas 3.6-4.9mm; leaves with (5-)7 veins. Culms 20-50cm. Native; grassy places on well-drained, often calcareous or serpentine soils; locally common
                  F. ovina ssp. **ophioliticola** (Kerguélen) M.J. Wilk.

21 Spikelets 5.3-6.3mm; lemmas 3.1-4.2mm; leaves with 5-7 veins   22

   22 Awns 0-0.8mm; leaves usually pubescent at base; lemmas usually pubescent; stomata mostly >31.5 microns. Culms 6-45cm. Native; grassy places on well-drained, usually acid soils; common
                  F. ovina ssp. **hirtula** (Travis) M.J. Wilk.

   22 Awns 0-1.2mm; leaves and lemmas usually scabrid; stomata mostly <31.5 microns. Culms 10-35cm. Native; grassy places on well-drained, usually acid soils; common in N, C & SW Br, very sparse in SC & SE En, ?Ir   **F. ovina** ssp. **ovina**

23 Culms 19-66cm; panicles 3.7-8.6cm; pedicels 0.6-2.5mm; widespread. Densely tufted, without rhizomes. Culms to 66cm. Probably native; grassy places on well-drained, acid or calcareous soils, often with *F. ovina*; very scattered in Br   *Confused Fescue* - **F. lemanii** Bastard

23 Culms 11-35cm; panicles 2.3-5.9cm; pedicels 0.8-1.5mm; coastal in CI only   24

   24 Panicles well exserted from sheath at anthesis; leaves usually not glaucous; culms to 40cm, erect. Densely tufted, without rhizomes. Native; fixed dunes; W coast of Jersey
                  Breton Fescue - **F. armoricana** Kerguélen

   24 Panicles not completely or only just exserted from sheath at anthesis; leaves often slightly glaucous; culms to 25cm, erect to procumbent. Densely tufted, without rhizomes. Native; grassy

cliff-tops and -bases; all islands in CI
Huon's Fescue - **F. huonii** Auquier

Hybrids occur in all 3 combinations between *F. pratensis*, *F. arundinacea* and *F. gigantea*; all are sterile and rare.

15 x 16. FESTUCA x LOLIUM = X FESTULOLIUM Asch. & Graebn.

Inflorescences variously intermediate, at 1 extreme a simple raceme (rarely a spike as in *Lolium*), at the other a branching panicle, often a raceme with a few racemose branches near base; most spikelets with 2 glumes, but lower much shorter than upper; anthers ± indehiscent with ± empty pollen grains, but some degree of fertility exists and some backcrossing may occur.

L. perenne and *L. multiflorum* hybridise with *F. pratensis*, *F. arundinacea* and *F. gigantea*; of the 6 possible combinations only *F. gigantea* x *L. multiflorum* has not been found.

The only common combination is *F. pratensis* x *L. perenne* = **X F. loliaceum** (Huds.) P. Fourn. (*Hybrid Fescue*); it has glabrous auricles and awnless lemmas and the inflorescence is usually a simple or little-branched raceme. Native; pastures, meadows, riversides and roadsides, often on damp rich soils; throughout most of BI, commonest in S En.

15 x 17. FESTUCA X VULPIA = X FESTULPIA Stace & R. Cotton

Plants perennial and vegetatively close to the *Festuca* parent, but with fewer and shorter (or 0) rhizomes and some overlapping sheaths; panicles narrower and less branched than in *Festuca*, with markedly longer awns; lower glume c.1/2 as long as upper; anthers indehiscent with ± empty pollen grains, but a very small degree of fertility may exist.

F. rubra hybridises with *V. fasciculata*, *V. bromoides* and *V. myuros*, and *F. arenaria* with *V. fasciculata*. Only *F. rubra* x *V. fasciculata* = **X F. hubbardii** Stace & R. Cotton is not merely sporadic; it has lower glume 2.4-4.4mm, upper glume (incl. awn) 3.5-7.2mm, lemmas 6-9.5mm plus awn 2-5.5mm, and anthers 3, 1.5-2mm. Native; on open sand-dunes with parents; frequent in CI and Br N to Westmorland, possibly wherever *V. fasciculata* occurs.

16. LOLIUM L. - *Rye-grasses*

Annuals to perennials without rhizomes or stolons; sheaths not fused; inflorescence normally a spike with spikelets compressed and orientated with backs of lemmas adjacent to spike axes; spikelets partly lying in concavities in spike axis, with 2-many florets all (except the most apical) bisexual; glumes 2 in terminal spikelet, 1 (the upper, abaxial) in lateral spikelets; lemmas rounded on back, 5-9-veined, obtuse to subacute or minutely bifid at apex, with or without subterminal awn; stamens 3.

Abnormal plants with branched inflorescences are not rare, and all spp. may have awned or awnless lemmas.

1	Lemmas ovate to elliptic, ≤3x as long as wide, becoming thick and hard at base in fruit; caryopsis ≤3x as long as wide	2
1	Lemmas narrowly oblong-ovate, >3x as long as wide, not becoming	

thick or hard; caryopsis >3x as long as wide 3
2 Lowest 2 lemmas (4.6)5-8.5mm; glume (7)10-30mm. Annual to 75(100)cm. Intrd-casual; tips and waste places; scattered in BI
***Darnel* - L. temulentum** L.
2 Lowest 2 lemmas 3.5-5(5.5)mm; glume 5-12(15)mm. Annual to 75cm. Intrd-casual; tips and waste places; scattered and sporadic in En   ***Flaxfield Rye-grass* - L. remotum** Schrank
3 Perennial, with tillers at flowering and fruiting time; leaves folded along midrib when young; lemmas usually unawned. Culms to 50(90)cm. Native; grassy places, waste and rough ground; abundant
***Perennial Rye-grass* - L. perenne** L.
3 Annual or biennial, without tillers at flowering and fruiting time; leaves rolled along long axis when young; lemmas awned or not 4
   4 Lemmas nearly always awned; spikelets usually with ≥11 florets. Annual or biennial to 100(120)cm. Intrd-natd; rough and waste ground, field borders, waysides; scattered throughout BI, common in lowlands   ***Italian Rye-grass* - L. multiflorum** Lam.
   4 Lemmas usually unawned; spikelets usually with ≤11 florets. Annual to 70cm. Intrd-casual; waste ground and tips; scattered in Br   ***Mediterranean Rye-grass* - L. rigidum** Gaudin

L. perenne x *L. multiflorum* = ***L. x boucheanum*** Kunth occurs occasionally in Br and Ir as a natural product and commonly in lowland Br and CI as an escape from its cultivation as an important pasture or meadow grass; it is a fertile annual to perennial, and has rolled young leaves and intermediate spikelet structure (always shortly awned).

17. VULPIA C.C. Gmel. - *Fescues*

Annuals; sheaths not fused; inflorescence a sparsely branched narrow panicle or a raceme; spikelets with 2-many florets all (except the most apical) bisexual or with a group of sterile florets at apex; glumes 2, very unequal, the lower at most 3/4 as long as upper; lemmas rounded on back, 3-5-veined, acute or acuminate with long terminal awn; stamens 1-3.

Glume lengths and ratios must be measured in spikelets that are not apical on the inflorescence or its branches; in apical spikelets the lower glume is often much longer than normal.

1 Lemmas with basal pointed minutely scabrid callus; ovary and caryopsis with minute apical hairy appendage; lemmas 8-18mm excl. awn; upper glume 10-30mm incl. awn. Culms to 50cm. Native; open parts of sand-dunes; locally frequent on coasts of BI N to Cheviot, Man and Co Louth, commoner in W
***Dune Fescue* - V. fasciculata** (Forssk.) Fritsch
1 Lemma with basal rounded glabrous callus; ovary and caryopsis glabrous; lemmas 3-7.5mm excl. awn; upper glume 1.5-9mm incl. awn if present 2
   2 Anthers 3, 0.7-1.3(1.9)mm, well exserted at anthesis; lemmas 3-5mm excl. awn; inflorescence ± always a raceme. Culms to

17. VULPIA

    40cm. Native; open grassy places on chalk, also in waste places and waysides; very scattered in Br N to E Gloucs and W Norfolk
Mat-grass Fescue - **V. unilateralis** (L.) Stace
2 Anthers 1(-3), 0.4-0.8(1.8)mm, usually not exserted at anthesis; lemmas 4-7.5mm; inflorescence a panicle except in starved plants   3
3 Spikelets with 1-3 bisexual and 3-7 distal sterile (but scarcely smaller) florets; lemma of fertile florets 3(-5)-veined. Culms to 45cm. Native; on maritime or submaritime sand or shingle; local in S Br N to N Lincs and Merioneth, CI   *Bearded Fescue* - **V. ciliata** Dumort.
3 Spikelets with 2-5 bisexual and 1-2 distal much reduced sterile florets; lemma of fertile florets 5-veined   4
    4 Lower glume 2.5-5mm, 1/2-3/4 as long as upper; lemmas usually >1.3mm wide when flattened; inflorescence normally well exserted from uppermost sheath at maturity. Culms to 50cm. Native; open grassy places on well drained soils, rough and waste ground; frequent over most of Bl, common in S Br
Squirreltail Fescue - **V. bromoides** (L.) Gray
    4 Lower glume 0.4-2.5mm, 1/10-2/5 as long as upper; lemmas usually <1.3mm wide when flattened; inflorescence normally not fully exserted from uppermost sheath at maturity. Culms to 65cm. Probably native; open ground, walls, rough or waste ground, roadsides and by railways; throughout most of Bl, common in S Br   *Rat's-tail Fescue* - **V. myuros** (L.) C.C. Gmel.

18. CYNOSURUS L. - *Dog's-tails*

Annuals or perennials without rhizomes or stolons; sheaths not fused; inflorescence a compact spike-like panicle; spikelets of 2 kinds - fertile with (1)2-5 florets all (except the most apical) bisexual, and sterile consisting of numerous very narrow acuminate lemmas in herring-bone arrangement, normally 1 fertile and 1 sterile together; glumes 2, subequal; fertile lemmas rounded on back, 5-veined, acute to obtuse or minutely bifid, awnless or with long terminal or subterminal awn; stamens 3.

1 Panicle linear-oblong, 1-10(14) x 0.4-1cm; fertile lemmas 3-4mm plus awn 0-1mm; leaves 1-4mm wide. Tufted perennial to 75cm. Native; grassy places on a great range of soils; common
Crested Dog's-tail - **C. cristatus** L.
1 Panicle asymmetrically ovoid, 1-4(8) x 0.7-2cm; fertile lemmas 4.5-7mm plus awn 6-16mm; leaves 2-10mm wide. Tufted annual to 75(100)cm. Intrd-natd; casual on waste and rough open ground; scattered in Bl N to C Sc, natd in sunny places on sandy or rocky ground on coasts of S En and CI   *Rough Dog's-tail* - **C. echinatus** L.

19. LAMARCKIA Moench - *Golden Dog's-tail*

Annuals; sheaths not fused; inflorescence a rather compact panicle; spikelets of 2 kinds - fertile with 1 bisexual and 1 vestigial floret, and sterile consisting of numerous flat, overlapping obtuse lemmas, normally 3 sterile and 2 fertile together, but 1 of the 2 fertile often reduced and not producing a caryopsis; glumes 2, subequal; lemmas rounded on back, 5-veined, minutely bifid, with long awn from sinus; stamens 3.

1 Culms tufted, to 20(30)cm; leaves 2-6mm wide; panicle 3-9 x 2-3cm, oblong. Intrd-casual; casual occasionally persisting for few years in waste or rough ground and on tips; very scattered in Br, mainly S & C En and S Wa   *Golden Dog's-tail* - **L. aurea** (L.) Moench

20. PUCCINELLIA Parl. - *Saltmarsh-grasses*

Annuals to perennials without rhizomes, with or without stolons; sheaths not fused; inflorescence a panicle; spikelets with 2-many florets all (except the most apical) bisexual; glumes 2, slightly unequal; lemmas rounded on back, 5-veined, subacute to rounded at apex, unawned; stamens 3.

1 Lemmas 2.8-4.6mm   2
1 Lemmas 1.8-2.5(2.8)mm   3
  2 Perennial with many tillers and usually rooting stolons at flowering; anthers 1.3-2.5mm. Culms to 80cm. Native; bare or semi-bare mud in salt-marshes and estuaries, rarely saline areas and by salted roads inland; common round coasts of BI
Common Saltmarsh-grass - **P. maritima** (Huds.) Parl.
  2 Annual or biennial with 0 or few tillers and 0 stolons at flowering; anthers 0.75-1mm. Culms to 40cm. Native; in bare places on mud and clay and among rocks and stones; coasts of S & E Br from Pembs to Yorks
Stiff Saltmarsh-grass - **P. rupestris** (With.) Fernald & Weath.
3 At least some of panicle branches at lower nodes bearing spikelets ± to base; lemmas subacute to obtuse, with midrib reaching apex; anthers mostly <0.75mm. Tufted perennial; culms to 50cm. Native; in barish places, on sea-walls and banks and by dykes; locally frequent on coasts of S Br from Carms to W Norfolk, rarely inland by salted roads   *Borrer's Saltmarsh-grass* - **P. fasciculata** (Torr.) E.P. Bicknell
3 Panicle branches at lower nodes ± all with conspicuous basal region bare of spikelets; lemmas broadly obtuse to rounded, with midrib falling short of apex; anthers mostly >0.75mm. Tufted perennial
(*Reflexed Saltmarsh-grass* - **P. distans** (Jacq.) Parl.)   4
  4 Leaves usually flat, 1.5-4mm wide; lower panicle branches strongly reflexed at maturity; lemmas 2-2.5mm. Culms to 60cm. Native; semi-bare mud, rough and waste ground in estuaries, upper edges of salt-marshes, inland saline areas and by salted main roads; round coasts of BI N to CE Sc, common inland in C & E En   **P. distans ssp. distans**
  4 Leaves usually folded along midrib, 1-2mm wide; lower panicle branches patent (sometimes weakly reflexed) to suberect at maturity; lemmas (1.8)2.2-2.8mm. Culms to 40cm. Native; stony or rocky, sometimes sandy places and on sea-walls; coasts of N & E Sc S to Midlothian, Outer Isles
P. distans ssp. borealis (Holmb.) W.E. Hughes

Sterile hybrids occur rarely in 4 combinations; *P. distans* hybridizes with the other 3 spp. and *P. maritima* with *P. rupestris*.

21. BRIZA L. - *Quaking-grasses*
Annuals, or perennials with short rhizomes; sheaths overlapping; inflorescence a panicle or sometimes a raceme with pedicels >5mm; spikelets with 4-many florets all (except most apical) bisexual, characteristically flattened, broadly ovate and pendent; glumes 2, subequal; lemmas rounded on back, cordate at base, rounded to very obtuse at apex, 7-9-veined, the veins not reaching apex, unawned; stamens 3.

1 Perennial, with tillers at flowering; ligule 0.5-1.5mm; leaves 2-4mm wide. Culms to 75cm. Native; grassland on light to heavy, acid to calcareous, very dry to damp, but usually base-rich soils; locally common **Quaking-grass - B. media** L.
1 Annual, without tillers at flowering; ligule 2-6mm; leaves 2-10mm wide 2
   2 Spikelets 2.5-5mm, >20 per panicle. Culms to 60cm. Possibly native; arable fields, bulb-fields, waste places; locally frequent in SW & CS En and CI, rare casual elsewhere
Lesser Quaking-grass - **B. minor** L.
   2 Spikelets 8-25mm, <15 per panicle. Culms to 75cm. Intrd-natd; dry open places and on banks and field-margins; distribution as for *B. minor*, usually as garden escape
Greater Quaking-grass - **B. maxima** L.

22. POA L. - *Meadow-grasses*
Annuals or perennials with or without stolons or rhizomes; sheaths overlapping; inflorescence a panicle; spikelets with 1-many florets all (or all except the most apical) bisexual, sometimes proliferating; glumes 2, subequal; lemmas keeled on back, 5-veined, awnless or rarely with short terminal awn, its callus often with tuft of cottony hairs; stamens 3.

1 At least some spikelets proliferating 2
1 Spikelets not proliferating 4
   2 Base of culms swollen, bulb-like; plant green. Tufted perennial; culms erect to decumbent, to 40cm. Native; barish places in short grassland and open ground on sandy soil, shingle or limestone near sea, very rare inland; coasts of S Br from Glam to N Lincs, CI, Co Wexford **Bulbous Meadow-grass - P. bulbosa** L.
   2 Base of culms not swollen; plant usually glaucous 3
3 Leaves 2-4.5mm wide when flattened, parallel-sided and ± abruptly narrowed to apex, the uppermost usually arising below 1/2 way up culm. Tufted perennial; culms erect to ± pendent, to 40cm. Native; damp mountain rock-ledges and -crevices and rocky slopes at 300-1200m; local in C & N Sc, NW En, Caerns, S Kerry, Co Sligo
Alpine Meadow-grass - **P. alpina** L.
3 Leaves 1-2mm wide when flattened, gradually tapered to apex, the

uppermost usually arising above 1/2 way up culm. Tufted perennial; culms erect, to 25cm. Native; similar habitat and range to *P. flexuosa*, often (?always) with it and *P. alpina* (*P. alpina* x *P. flexuosa*)
Swedish Meadow-grass - **P. x jemtlandica** (Almq.) K. Richt.
4 Plant with distinct, often far-creeping rhizomes 5
4 Plant without rhizomes, sometimes with stolons 8
5 Culms strongly compressed, 4-6(9)-noded, usually slightly bent at each node. Rhizomatous perennial; culms erect to ascending, to 60cm. Native; walls, paths, waysides, stony ground and banks on well-drained soils; rather scattered throughout BI except N Sc but intrd in Ir *Flattened Meadow-grass* - **P. compressa** L.
5 Culms terete to somewhat compressed, 2-4-noded, usually straight except near base 6
  6 Glumes subequal, both usually 3-veined and distinctly acuminate; culms usually all solitary; sheaths usually with some hairs at junction with blade. Perennial with extensive rhizomes; culms erect, to 30(40)cm. Native; grassland, roadsides, on old walls, usually on sandy soil but often near water; throughout BI
Spreading Meadow-grass - **P. humilis** Hoffm.
  6 Glumes distinctly unequal, the lower often 1-veined, acute; culms usually in small or dense clusters; sheaths glabrous at junction with blade 7
7 Tiller leaves 0.5-2mm wide; lemmas 2-3mm; culms usually in dense clusters. Perennial with rhizomes; culms erect, to 70cm. Native; grassy places, rough ground, on and by walls, banks, on well-drained soil; probably frequent throughout BI
Narrow-leaved Meadow-grass - **P. angustifolia** L.
7 Tiller leaves 2-4(5)mm wide; lowest lemmas 3-4mm; culms usually in small clusters. Perennial with strong but short rhizomes; culms erect, to 75(100)cm. Native; meadows, pastures, waysides, rough and waste ground; very common *Smooth Meadow-grass* - **P. pratensis** L.
8 All or some leaves >5mm wide 9
8 Leaves ≤5mm wide 11
9 Panicle-branches erect; lowest lemma usually >4.5mm, with awn 1.2-2.4mm. Very densely tufted perennial forming large tussocks up to 1m x 1m (excl. culms); culms erect, to 80cm. Intrd-surv; very persistent where planted in yards and on walls in Shetland
Tussac-grass - **P. flabellata** (Lam.) Raspail
9 Panicle-branches patent; lowest lemma ≤4.5mm, awnless 10
  10 Leaves ≤6mm wide; lowest lemma ≤3mm; ligules often >2mm; uppermost culm-leaf usually as long or longer than its sheath. Tufted perennial; culms erect, to 1(1.5)m. Intrd-natd; marshes, fens, ditches and damp grassland; sporadic in BI N to C Sc (mainly S & C En) *Swamp Meadow-grass* - **P. palustris** L.
  10 Leaves often >6mm wide; lowest lemma >3mm; ligules <2mm; uppermost culm-leaf much shorter than its sheath. Densely tufted perennial; culms erect, to 1.3m. Intrd-natd; woods and

copses; scattered throughout Br, Co Down

Broad-leaved Meadow-grass - **P. chaixii** Vill.

11 Base of culms swollen, bulb-like (see couplet 2) **P. bulbosa**
11 Base of culms not swollen 12
   12 Base of culms surrounded by dense mass of dead sheaths; mountains 13
   12 Base of culms with few or no persistent dead sheaths; widespread 14
13 Leaves 2-4.5mm wide when flattened, parallel-sided and ± abruptly narrowed to apex, the uppermost usually arising below 1/2 way up culm (see couplet 3) **P. alpina**
13 Leaves 1-2mm wide when flattened, gradually tapered to apex, the uppermost usually arising above 1/2 way up culm. Tufted perennial; culms erect, to 25cm. Native; mountain screes and ledges at 800-1100m; very local in C Sc   *Wavy Meadow-grass* - **P. flexuosa** Sm.
   14 Annual, or perennial due to procumbent stems rooting; culms usually procumbent to ascending; anthers ≤1(1.3)mm; usually some leaves transversely wrinkled 15
   14 Perennial; culms usually erect; anthers ≥1.3mm; leaves not transversely wrinkled 16
15 Anthers 0.6-0.8(1.3)mm, 2-3x as long as wide; panicle-branches usually patent to reflexed at fruiting. Annual, or perennial due to procumbent stems rooting; culms erect to procumbent, to 20(30)cm. Native; rough, waste and cultivated ground, waysides, on paths and in close-cut turf; abundant   *Annual Meadow-grass* - **P. annua** L.
15 Anthers 0.2-0.5mm, 1-1.5x as long as wide; panicle-branches usually erect to suberect at fruiting. Annual; culms erect to procumbent, to 10(25)cm. Native; rough ground, waysides and on paths, usually near sea; CI, Scillies, SW En E to S Hants

Early Meadow-grass - **P. infirma** Kunth

   16 Ligule of uppermost culm-leaf 4-10mm, acute; sheaths rough. Perennial with many, often procumbent tillers, some becoming stolons; culms erect, to 70(90)cm. Native; open woods, marshes, ditches, riversides, damp grassland, by ponds and lakes, cultivated and rough ground; very common

Rough Meadow-grass - **P. trivialis** L.

   16 Ligule of uppermost culm-leaf ≤5mm, obtuse to rounded; sheaths smooth 17
17 Ligule of uppermost culm-leaf 0.2-0.5mm, usually truncate. Tufted perennial; culms erect, to 75(90)cm. Native; woods, hedgebanks, walls and other shady places; frequent to common in most of BI, but intrd in much of Ir and NW Br   *Wood Meadow-grass* - **P. nemoralis** L.
17 Ligule of uppermost culm-leaf 0.8-4(5)mm, usually obtuse to rounded 18
   18 Lowland plant c.30-150cm; lowest panicle node with (3)4-6(8) branches; ligule of uppermost culm-leaf 2-4(5)mm

(see couplet 10) **P. palustris**

   18 Mountain plant 10-40cm; lowest panicle node with 1-2(4)

branches; ligule of uppermost culm-leaf 1-2.5(3)mm. Tufted perennial; culms erect, to 40cm. Native; damp mountain rock-ledges and -crevices and rocky slopes at 610-910m; very local in C & N Sc, Lake District, Caerns

Glaucous Meadow-grass - **P. glauca** Vahl

23. DACTYLIS L. - *Cock's-foot*

Perennials without stolons or rhizomes, with strongly compressed tillers; sheaths overlapping; inflorescence a ± 1-sided panicle simply lobed or formed of stalked dense clusters of spikelets; spikelets with 2-5 florets all (except the most apical) bisexual; glumes 2, unequal; lemma keeled on back, 5-veined, without or with short terminal awn; stamens 3.

1 Leaves and sheaths very rough, ± glaucous; lemmas with hairs or prickles on keel, with awn 1.5-2mm. Culms densely tufted, to 1.4m. Native; grassland, open woodland and rough, waste and cultivated ground; very common *Cock's-foot* - **D. glomerata** L.
1 Leaves and sheaths not or only slightly rough, green; lemmas glabrous to obscurely prickly on keel, with awn 0 or <0.5mm. Slenderer than *D. glomerata*. Intrd-natd; in woods; scattered in S En (Bucks and Surrey to Dorset) *Slender Cock's-foot* - **D. polygama** Horv.

24. CATABROSA P. Beauv. - *Whorl-grass*

Perennials with stolons, without rhizomes; sheaths overlapping; inflorescence a diffuse panicle; spikelets with 1-3 florets all or all except the most apical bisexual; glumes 2, unequal; lemma rounded on back, 3-veined, truncate, awnless; stamens 3.

1 Culms erect to decumbent, to 75cm, often purple-tinged, glabrous or ± so. Native; wet meadows, marshes, ditches, by ponds and streams, often on barish mud; scattered throughout most of lowland BI, rare in S, often coastal in NW *Whorl-grass* - **C. aquatica** (L.) P. Beauv.

25. CATAPODIUM Link - *Fern-grasses*

Annuals; sheaths not fused, inflorescence a stiff raceme or little-branched panicle; spikelets with (3)5-14 florets all (except the most apical) bisexual; glumes 2, subequal; lemma rounded on back or keeled distally, 5-veined, acute to obtuse or emarginate at apex, awnless; stamens 3.

1 Lower glume (1)1.3-2mm; upper glume (1.4)1.5-2.3(2.5)mm; lemmas 2-2.6(3)mm. Culms erect to procumbent, to 15(60)cm. Native; dry, barish places on banks, walls, sand, shingle, chalk and stony ground, especially near sea; locally common in BI N to C Sc
Fern-grass - **C. rigidum** (L.) C.E. Hubb.
1 Lower glume 2-3mm; upper glume 2.3-3.3mm; lemmas 2.2-3mm. Culms erect to procumbent, to 25cm. Native; dry barish places by sea on walls, banks, sand and shingle, often with *C. rigidum*; locally common round coasts of BI *Sea Fern-grass* - **C. marinum** (L.) C.E. Hubb.

1 plant of *C. rigidum* x *C. marinum* was found in 1960 in Merioneth.

26. SESLERIA Scop. - *Blue Moor-grass*
Tufted perennials with short rhizomes, without stolons; sheaths not fused; inflorescence a small, very compact panicle; spikelets with 2(-3) florets all bisexual; glumes 2, subequal; lemma rounded on back, 3-5-veined, 3-5-toothed at apex, each tooth with awn 0-1.5mm; stamens 3.

1 Culms to 45cm; panicle 1-3cm, ovoid, usually bluish-violet tinged. Native; barish grassland, rock-crevices and -ledges, screes and 'pavement' on limestone and (in Sc) calcareous mica-schists; locally common in N En S to MW Yorks and in W Ir, very local in C Ir and C Sc, 1 site in Derbys **Blue Moor-grass - S. caerulea** (L.) Ard.

TRIBE 7 - HAINARDIEAE (genera 27-28). Annuals; ligule membranous; sheaths not fused; inflorescence a very slender cylindrical spike with alternating spikelets partly sunk in cavities in rhachis which breaks up into 1-spikeleted segments at fruiting; spikelets with 1 (bisexual) floret; glumes 1-2, strongly veined and ± horny; lemma delicate, 3-veined with very short lateral veins, acute, unawned; stamens 3; stigmas 2; ovary with rounded glabrous appendage beyond style-bases; lodicules 2; ovary glabrous.

27. PAPHOLIS C.E. Hubb. - *Hard-grasses*
Glumes 2, inserted side-by-side and together covering rhachis-cavity except at anthesis; lemma with its side towards rhachis.

1 Anthers 1.5-3(3.5)mm. Culms usually erect, sometimes ascending or curved, rarely procumbent, to 25(40)cm; spike fully exserted from uppermost sheath or not, 2-20cm. Native; sparsely-grassed ground on salty soil; frequent on coasts of BI N to C Sc
Hard-grass - P. strigosa (Dumort.) C.E. Hubb.

1 Anthers 0.5-1(1.5)mm. Culms decumbent to ascending, to 10(20)cm; spike usually not fully exserted from uppermost sheath, 1-8(15)cm. Native; similar places to *P. strigosa* but also drier spots on cliff-tops and banks; local on coasts of Br N to Caerns and S Northumb, CI, SE Ir **Curved Hard-grass - P. incurva** (L.) C.E. Hubb.

28. HAINARDIA Greuter - *One-glumed Hard-grass*
Glumes 2 in terminal spikelet, 1 (the upper) in all others and inserted so as to cover the rhachis-cavity except at anthesis; lemma with its back towards rhachis.

1 Culms erect to ascending, straight or curved, to 30(45)cm; spike fully exserted from uppermost sheath or not, 2-25cm; anthers 1.5-3.5mm. Intrd-casual; on tips and waste ground; scattered in S En
One-glumed Hard-grass - H. cylindrica (Willd.) Greuter

TRIBE 8 - MELICEAE (genera 29-30). Perennials with rhizomes or stolons;

ligule membranous; sheaths fused into tube (often splitting later); inflorescence a little- or much-branched panicle or a raceme with pedicels >3mm; spikelets with 1-many bisexual florets, if with <4 florets then with group of sterile ones beyond; glumes 2, subequal or equal; lemmas rounded on back, 7-9-veined, subacute to rounded, awnless; stamens 3; stigmas 2; lodicules fused laterally into single scale shorter than wide; ovary glabrous.

29. GLYCERIA R. Br. - *Sweet-grasses*

Aquatic or marsh grasses with rhizomes and/or stolons; panicles much- to rather little-branched; ligules acute or acuminate to rounded, 2-10(15)mm; spikelets with 4-16 florets all (except the most apical) bisexual and each falling separately when fruit ripe; glumes 1(-3)-veined; lemmas 7-veined.

1 Spikelets 5-12mm, with 4-10 florets; paleas not winged on keels; culms erect and self-supporting, to 2.5m. Native; in and by rivers, canals, ponds and lakes, usually in deeper water than other spp.; common in most of En except N, scattered elsewhere
Reed Sweet-grass - **G. maxima** (Hartm.) Holmb.
1 Spikelets 10-35mm, with 6-17 florets; paleas winged on keels distally; culms decumbent to ascending, if erect not self-supporting, rarely >1m   2
  2 Anthers remaining indehiscent; pollen grains all or mostly empty and shrunken; spikelets remaining intact after flowering, not forming fruits. Culms to 1m. Native; with 1, both or neither parents; scattered over most of BI (*G. fluitans* x *G. notata*)
Hybrid Sweet-grass - **G. x pedicellata** F. Towns.
  2 Anthers dehiscent; pollen grains full and turgid; spikelets breaking up between florets when fruit ripe   3
3 Lemmas 5.5-6.5(7.5)mm; anthers 1.5-2.5(3)mm. Culms to 1m. Native; on mud or in shallow water by ponds, rivers and canals and in marshes, ditches and wet meadows; common
Floating Sweet-grass - **G. fluitans** (L.) R. Br.
3 Lemmas 3.5-5mm; anthers 0.6-1.3mm   4
  4 Lemmas distinctly 3(-5)-toothed at apex, exceeded by 2 sharply pointed apical teeth of palea. Culms to 60cm. Native; similar places to *G. fluitans*; scattered throughout BI
Small Sweet-grass - **G. declinata** Bréb.
  4 Lemmas not or scarcely toothed at apex, not exceeded by 2 (very short) apical teeth of palea. Culms to 1m. Native; similar places to *G. fluitans*; frequent throughout most of BI
Plicate Sweet-grass - **G. notata** Cheval.

G. fluitans x *G. declinata* is a sterile hybrid differing from *G.* x *pedicellata* in its obscurely 3-toothed lemmas; very scattered in En and E Cork.

30. MELICA L. - *Melicks*

Woodland or mountain grasses with short rhizomes; inflorescence a raceme or a sparsely branched panicle; ligules truncate, <2mm; spikelets with 1-3

30. MELICA

bisexual florets plus distal ± club-shaped cluster of sterile vestiges, all the florets falling as a unit when fruit ripe; glumes 3-5-veined; lemmas 7-9-veined.

1 Sheaths without apical bristle; spikelets usually in a simple raceme, pendent, with 2-3 fertile florets. Culms to 60cm. Native; woods, scrub and shady rock-crevices on limestone; scattered in N & W Br S to Mons and Northants *Mountain Melick* - **M. nutans** L.
1 Sheaths with long bristle at apex on side opposite ligule; spikelets in a sparsely branched panicle, erect, with 1 fertile floret. Culms to 60cm. Native; woods and shady hedgebanks; locally common in BI except in N Sc and CI *Wood Melick* - **M. uniflora** Retz.

TRIBE 9 - AVENEAE (genera 31-41). Annuals or perennials with or without rhizomes, without stolons; ligule membranous; sheaths not fused or rarely fused to form tube; inflorescence a panicle, sometimes densely contracted, rarely a spike; spikelets with 2-6(11) florets, sometimes 1 or more floret (basal or apical) male, sometimes proliferating; glumes 2, equal to unequal, often nearly as long to longer than rest of spikelet, often with wide ± shiny hyaline margins; lemmas 5-9-veined, rounded on back, acute or obtuse to bifid or variously toothed at apex, awnless or with short terminal awn or more often with dorsal awn with conspicuous bend; stamens 3; stigmas 2; lodicules 2; ovary glabrous or pubescent at apex or all over.

31. HELICTOTRICHON Schult. & Schult. f. - *Oat-grasses*
Tufted perennials with short or 0 rhizomes; inflorescence a rather sparsely branched, ± diffuse panicle; spikelets with 2-7 florets all except the most apical bisexual; glumes clearly unequal, the lower 1-3-veined, the upper 3-5-veined; lemmas 5-7-veined, variously shortly toothed at apex, with long, bent, dorsal awn; rhachilla-segments pubescent, the hairs in a longer tuft at apex of each segment around lemma-base; ovary glabrous or pubescent at apex.

1 Ovary and caryopsis glabrous; lower glume 4-5mm; upper glume 5-7mm; lemmas 5-8mm (excl. awn), papillose. Culms erect, to 1m. Intrd-casual; tips, waste ground and fields; scattered in En
 Swamp Wallaby-grass - **H. neesii** (Steud.) Stace
1 Ovary and caryopsis with pubescent apex; lower glume 7-15mm; upper glume 10-20mm; lemmas 9-17mm (excl. awn), smooth to scabrid 2
  2 Lower culm-sheaths softly pubescent; spikelets with 2-3(4) florets; rhachilla hair-tuft 3-6(7)mm; palea with smooth keels. Culms erect, to 1m. Native; grassland usually on base-rich soils; ± throughout Bl *Downy Oat-grass* - **H. pubescens** (Huds.) Pilg.
  2 Culm-sheaths glabrous; spikelets with 3-6(8) florets; rhachilla hair-tuft 1-3mm; palea with scabrid keels. Culms usually erect, to 80cm. Native; similar places to *H. pubescens* and often with it, but usually in shorter turf and commoner in mountains; ±

throughout Br   *Meadow Oat-grass* - **H. pratense** (L.) Besser

32. ARRHENATHERUM P. Beauv. - *False Oat-grass*

Loosely tufted perennials; inflorescence a fairly well branched, ± diffuse panicle; spikelets with 2(-5) florets, the lower (lowest) male, the upper bisexual, rarely both bisexual; glumes unequal, the lower 1-veined, the upper 3-veined; lemmas 7-veined, bifid at apex; lemma of male floret with long bent dorsal awn, that of bisexual floret(s) awnless or with short terminal awn or rarely with dorsal bent awn; rhachilla-segments with apical hair-tuft 1-2mm; ovary pubescent at apex.

1 Culms usually erect, to 1.8m; spikelets 7-11mm; lowest lemma 7-10mm with awn up to 20mm. Native; coarse grassy places, waysides, hedgerows, maritime sand and shingle, rough and waste ground; abundant   *False Oat-grass* - **A. elatius** (L.) J. & C. Presl

33. AVENA L. - *Oats*

Annuals; inflorescence a diffuse panicle; spikelets with 2-3 florets all bisexual or the distal 1 or 2 reduced and male or sterile; glumes subequal, 7-11-veined; lemmas 7-9-veined, bifid or with 2 bristles at apex, with or without long, bent, dorsal awn; rhachilla-segments with or without hair-tuft; ovary pubescent at apex or all over.

A difficult genus, in which general appearance and size of parts are often of little value. For accurate determination spikelets with fully ripe fruits are needed.

1 Lemmas bifid, the 2 apical points (1)3-9mm and each with 1 or more veins entering from main body of lemma and reaching apex   2
1 Lemmas bifid, the 2 apical points 0.5-2mm and without veins or with vein(s) not reaching apex   3
   2 Rhachilla disarticulating between florets at maturity, releasing 1-fruited disseminules each with elliptic basal scar; lemmas with dense long hairs on lower 1/2. Stems to 1m. Intrd-natd; rare grain-alien, natd in Guernsey   *Slender Oat* - **A. barbata** Link
   2 Rhachilla not disarticulating at maturity, whole spikelets acting as disseminules, or the florets breaking away irregularly without basal scar; lemmas glabrous or sparsely pubescent on lower 1/2. Stems to 1.2m. Intrd-natd; casual in waste places and very local cornfield weed sometimes natd, especially in Wa, Sc and Ir
                Bristle Oat - **A. strigosa** Schreb.
3 Rhachilla not disarticulating at maturity, whole spikelets acting as disseminules, or the florets breaking away irregularly without basal scar; lemmas usually unawned, if awned then awn nearly straight, usually glabrous. Stems to 1(1.5)m. Intrd-natd; locally common crop, frequent on tips and waysides and as grain-alien; throughout BI   *Oat* - **A. sativa** L.
3 Rhachilla disarticulating at maturity at least above glumes, often also between florets, hence at least lowest floret with basal scar; lemmas

33. AVENA

with long, strongly bent awns, usually pubescent 4
4 Rhachilla disarticulating at maturity above glumes only, releasing 2-3-fruited disseminules, hence only lowest floret with (ovate) basal scar; longer glume mostly 25-30mm. Stems to 1.5m. Intrd-natd; similar places to *A. fatua* but usually on heavy soils and replacing it there; scattered in CS & SE En
Winter Wild-oat - A. sterilis L.
4 Rhachilla disarticulating at maturity between florets, releasing 1-fruited disseminules each with ovate basal scar; longer glume mostly 18-25mm. Stems to 1.5m. Intrd-natd; weed of arable, waste and rough ground; common in most of En, very scattered elsewhere **Wild-oat - A. fatua** L.

A. fatua x *A. sativa* occurs rarely in Br in and around fields of *A. sativa* infested by *A. fatua*; it has low fertility.

34. GAUDINIA P. Beauv. - *French Oat-grass*
Annuals sometimes lasting a few years; inflorescence a spike whose axis breaks giving 1-spikeleted segments at fruiting; spikelets with 4-11 florets all (except most apical) bisexual; lower glume c.1/2 as long as upper, 3-5- and 5-11-veined respectively; lemmas 5-9-veined, minutely bifid at apex, with long, bent, dorsal awn; rhachilla segments ± glabrous; ovary with distinct hairy apex remaining conspicuous as projection on fruit.

1 Stems to 45(100)cm; lowest lemma 7-11mm with awn 5-13mm. Intrd-natd; grassy fields, rough ground and waysides; natd locally in SC En, SW Ir and CI **French Oat-grass - G. fragilis** (L.) P. Beauv.

35. TRISETUM Pers. - *Yellow Oat-grass*
Perennials without rhizomes or stolons; inflorescence a well-branched panicle; spikelets with 2-4 florets all (except the most apical) bisexual; glumes unequal, the lower 1-, the upper 3-veined; lemmas 5-veined, bifid with 2 short bristle-points at apex, with long, bent dorsal awn; rhachilla-segments pubescent, the hairs <1mm at apex; ovary glabrous.

1 Stems loosely tufted, to 80cm; lowest lemma 4-5.5mm with awn 4.5-9mm. Native; meadows, pastures, grassy waysides, especially on base-rich soil, throughout most of lowland BI, common in En and SE Sc **Yellow Oat-grass - T. flavescens** (L.) P. Beauv.

36. KOELERIA Pers. - *Hair-grasses*
Tufted perennials without rhizomes or stolons; inflorescence a spike-like panicle with very short branches; spikelets with 2-3(5) florets all (except most apical) bisexual; glumes unequal, the lower 1-, the upper 3-veined; lemmas 3-veined, acute or with extremely short apical awn, otherwise awnless; rhachilla-segments shortly pubescent; ovary glabrous.

1 Lower sheaths very persistent and rotting to form reticulated network

of fibres round swollen culm-bases. Stems erect to procumbent, to 40cm. Native; short limestone grassland; Mendip Hills (N Somerset)
Somerset Hair-grass - **K. vallesiana** (Honck.) Gaudin
1   Lower sheaths not very persistent, not forming reticulated fibres and culm-base not swollen. Stems erect to ascending, to 60cm. Native; short limestone or sandy base-rich grassland, dunes, less often on inland sandy soils; throughout most of BI
Crested Hair-grass - **K. macrantha** (Ledeb.) Schult.

K. vallesiana x *K. macrantha* occurs in most of the *K. vallesiana* populations; it is sterile and has intermediate sheath characters.

37. ROSTRARIA Trin. - *Mediterranean Hair-grass*
Annuals; inflorescence a spike-like panicle with very short branches; spikelets with 3-5(11) florets all (except apical 1-few) bisexual; glumes unequal, the lower 1-, the upper 3-veined; lemmas 5-veined, shortly bifid, with short subterminal awn; rhachilla-segments pubescent; ovary glabrous.

1   Stems erect, to 20(60)cm; panicles 1-10 x 0.4-1cm; spikelets 3-8mm; awns 1-3mm. Intrd-casual; tips and waste places; scattered in Br
Mediterranean Hair-grass - **R. cristata** (L.) Tzvelev

38. DESCHAMPSIA P. Beauv. - *Hair-grasses*
Densely tufted perennials usually without rhizomes or stolons; inflorescence a very diffuse panicle with fine branches; spikelets with 2 florets both bisexual or sometimes proliferating; glumes unequal, the lower 1-, the upper (1-)3-veined; lemmas 4-5-veined, rounded, obtuse or jagged-toothed at apex, with dorsal (rarely subterminal) straight or bent awn; rhachilla-segments pubescent with longer hair-tuft at base of each; ovary glabrous.

1   At least some spikelets proliferating. Stems to 1.5(2)m. Native
    (*Tufted Hair-grass* - **D. cespitosa** (L.) P. Beauv.)   2
1   Spikelets not proliferating   3
    2   Leaves scarcely or not hooded at apex; panicle-branches and
        pedicels with minute (sometimes very sparse) pricklets, the main
        branches rarely reflexed; awn arising from lower 1/2 of lemma.
        Damp meadows, waysides and ditches; common throughout BI
D. cespitosa ssp. cespitosa
    2   Leaves distinctly hooded at apex; panicle-branches and pedicels
        smooth, the main branches usually reflexed; awn arising from
        middle or upper 1/2 of lemma. Damp grassy places on mountains;
        frequent in W & C Highlands of Sc, very local in Caerns, Lake
        District and W Ir   **D. cespitosa ssp. alpina** (L.) Hook. f.
3   Leaves >1mm even if rolled up; awns of lemmas not or scarcely
    exceeding glumes (**D. cespitosa** - see couplet 1)   4
3   Leaves <1mm even if opened out; awns of lemmas conspicuously
    exceeding glumes   5

38. DESCHAMPSIA

4 Spikelets 2-3(3.5)mm; hair-tuft at base of lower lemma not reaching apex of rhachilla-segment above. Woods and shady hedgerows; common in lowland Br, ?Ir
D. cespitosa ssp. parviflora (Thuill.) Dumort.

4 Spikelets (3)3.5-5(6)mm; hair-tuft at base of lower lemma reaching apex of rhachilla-segment above
(see couplet 2) **D. cespitosa ssp. cespitosa**

5 Lemmas 2-3mm, toothed at apex with marginal teeth the longest; ligule 2-8mm, very acute; palea bifid. Stems to 70cm. Native; bogs and boggy pools and ditches; very local and scattered in Br and Ir
Bog Hair-grass - **D. setacea** (Huds.) Hack.

5 Lemmas 3-5.5mm, subacute to minutely toothed at apex with marginal teeth not longer than inner ones; ligule 0.5-3mm, obtuse; palea entire at apex. Stems to 60(100)cm. Native; acid heaths, moors and open woods, drier parts of bogs; on suitable soils throughout BI
Wavy Hair-grass - **D. flexuosa** (L.) Trin.

39. HOLCUS L. - *Soft-grasses*

Densely tufted or rhizomatous perennials; inflorescence a rather compact panicle; spikelets with 2 florets, the lower bisexual, the upper male; glumes subequal in length but unequal in width, the lower 1-, the upper 3-veined; lemmas 5-veined, ± rounded at apex, the lower awnless, the upper with dorsal awn arising from upper 1/2; rhachilla-segments ± glabrous but lemmas with basal tuft of hairs; ovary glabrous.

1 Culms softly and (sparsely to) densely patent-pubescent; glumes obtuse to subacute at apex; awn of upper lemma recurved to backwardly-hooked, included in glumes. Stems densely tufted, to 1m; rhizomes 0. Native; rough grassland, lawns, arable, rough and waste ground, open woods; common *Yorkshire-fog* - **H. lanatus** L.

1 Uppermost culm-sheath glabrous with conspicuously patent-pubescent node below it; glumes acute to acuminate at apex; awn of upper lemma slightly bent, well exserted from glumes. Stems loosely tufted, to 1m; rhizomes present. Native; woods, hedgerows, less often open grassland, mostly on acid soils; common
Creeping Soft-grass - **H. mollis** L.

H. lanatus x *H. mollis* = **H. x hybridus** Wein is very scattered in Br and Ir; it is sterile and more closely resembles *H. mollis*.

40. CORYNEPHORUS P. Beauv. - *Grey Hair-grass*

Densely tufted perennial without rhizomes; inflorescence a rather compact panicle; spikelets with 2 florets, both bisexual; glumes subequal, 1-veined or the upper also with 2 very short laterals; lemmas with 1 central and often 2 pairs of very short lateral veins; obtuse to very shortly bifid at apex, with dorsal bent awn twisted proximally and with club-shaped apex; rhachilla-segments pubescent and with tuft of hairs at base of each lemma; ovary glabrous.

1 Stems erect to decumbent, to 35cm; leaves very glaucous, <1mm wide and tightly inrolled; ligules 2-4mm, very acute. Native; open sand on leached fixed dunes and inland sandy heathland on acid soils; very local in Jersey and E Anglia, scattered and probably intrd elsewhere in En and Sc **Grey Hair-grass - C. canescens** (L.) P. Beauv.

41. AIRA L. - *Hair-grasses*

Annuals; inflorescence a compact to diffuse panicle; spikelets with 2 florets, both bisexual; glumes subequal, 1-3-veined; lemmas with 5 short veins, shortly bifid, with dorsal, slightly bent awn; rhachilla-segments extremely short; lemma with short hair-tuft at base; ovary glabrous.

1 Sheaths slightly rough; panicle diffuse, with conspicuous suberect to erecto-patent branches. Stems erect to decumbent, to 25(50)cm. Native; dry sandy, gravelly or rocky ground, on walls, heaths and dunes; frequent **Silver Hair-grass - A. caryophyllea** L.
1 Sheaths smooth; panicle compact, with short erect branches largely obscured. Stems erect to procumbent, to 10(15)cm. Native; similar places to *A. caryophyllea*; common **Early Hair-grass - A. praecox** L.

TRIBE 10 - PHALARIDEAE (genera 42-44). Annuals or perennials with or without rhizomes, without stolons; ligule membranous; sheaths not fused; inflorescence a panicle, usually contracted; spikelets with (2-)3 florets, the lower (1-)2 male or sterile and often much reduced, the upper 1 bisexual, rarely the spikelets in groups consisting of 1 central bisexual and 4-6 surrounding sterile (or male) spikelets; glumes 2, equal or unequal, 1 or both nearly as long as to longer than rest of spikelet; lemmas of bisexual florets 3-7-veined, keeled or rounded on back, awnless; palea with 1 vein; lower 2 lemmas minute to longer than upper 1, 0- or 4-5-veined, awnless or with dorsal, bent awn; stamens 2 or 3; stigmas 2; lodicules 0 or 2; ovary glabrous.

42. HIEROCHLOE R. Br. - *Holy-grass*

Rhizomatous perennials; inflorescence a diffuse panicle; lower 2 florets with 3 stamens, with 5-veined lemma slightly longer than lemma of bisexual floret; terminal floret with 2 stamens, with 5-veined lemma, with 2 lodicules; glumes subequal, keeled, slightly shorter than rest of spikelet, 1-3-veined. Crushed or dried plant smells strongly of new-mown hay.

1 Stems to 60cm; spikelets 3.5-5mm, greenish-purple, becoming characteristic golden-brown. Native; banks of rivers and lakes, wet meadows, flushed cliff-bases by sea; very local on and near coast in Sc, Co Antrim **Holy-grass - H. odorata** (L.) P. Beauv.

43. ANTHOXANTHUM L. - *Vernal-grasses*

Annuals or tufted perennials; inflorescence a contracted panicle; lower 2 florets sterile, with 4-5-veined lemma slightly longer than lemma of bisexual floret and with long dorsal awn; terminal floret with 2 stamens, 5-veined awnless lemma and 1-veined palea; lodicules 0; glumes very unequal, lower 1-veined, upper 3-veined and longer than rest of spikelet. Crushed or dried plant smells strongly of new-mown hay.

1 Tufted perennial; ligules 1-5mm; glumes usually pubescent; awns not or only slightly exceeding glumes. Culms unbranched, to 50(100)cm. Native; in all kinds of grassy places; abundant
Sweet Vernal-grass - **A. odoratum** L.
1 Tufted annual; ligules 0.6-2mm; glumes glabrous; awns much exceeding glumes. Culms usually well branched, to 40cm. Intrd-casual; waste ground on well-drained soil; scattered in En, S Wa and CI *Annual Vernal-grass* - **A. aristatum** Boiss.

44. PHALARIS L. - *Canary-grasses*

Annuals or rhizomatous perennials; inflorescence a contracted (often spike-like) panicle; lower 2 florets reduced to scales or rarely only 1 present; terminal floret with 3 stamens, with 5-veined awnless lemma, with 2 lodicules; glumes equal, sharply keeled, longer than rest of spikelet.

1 Perennial with very short to long rhizomes and tillers at flowering time 2
1 Annual without rhizomes or tillers 3
   2 Panicle at least distinctly lobed, usually with conspicuous branches; glumes strongly keeled but not winged. Perennial to 2m, with long rhizomes. Native; by lakes and rivers, in ditches, wet meadows and marshes, also rough and waste ground; common *Reed Canary-grass* - **P. arundinacea** L.
   2 Panicle not lobed, oblong to lanceolate in outline, without visible branches; glumes with distinct wing on keel. Perennial to 1.5m, with short rhizomes. Intrd-natd; in fields and rough ground; frequent in C & S En, more widely scattered as casual
Bulbous Canary-grass - **P. aquatica** L.
3 Spikelets in groups of 3-7, one bisexual the rest sterile, falling as a group when fruits ripe. Annual to 1.6m. Intrd-natd; weed of arable fields; frequent in C & S En and S Wa, more widespread in Br as casual *Awned Canary-grass* - **P. paradoxa** L.
3 Spikelets all bisexual, the (2-)3 florets of each falling at maturity leaving glumes on panicle 4
   4 At least 1 glume with its keel-wing minutely toothed on at least some spikelets. Annual to 60cm. Intrd-natd; sandy places; Guernsey and Scillies, more widespread in Br as casual
Lesser Canary-grass - **P. minor** Retz.
   4 Wings of glume-keels entire 5
5 Sterile florets 2, ≥1/2 as long as fertile floret. Annual to 1.2m. Intrd-natd; casual alien on tips and waste ground, sometimes ± natd; frequent *Canary-grass* - **P. canariensis** L.
5 Sterile florets 1-2, <1/3 as long as fertile floret. Annual to 60cm. Intrd-

casual; tips and waste ground; scattered in En
>><< *Confused Canary-grass* - **P. brachystachys** Link

TRIBE 11 - AGROSTIDEAE (genera 45-55). Annuals or perennials with or without rhizomes and/or stolons; ligule membranous; sheaths not fused; inflorescence a panicle, contracted or diffuse, often spike-like, rarely a raceme or spike; spikelets with 1 (bisexual) floret (in *Beckmannia* with 2 bisexual or 1 bisexual plus 1 male or sterile florets); glumes 2, usually equal or nearly equal, usually both at least as long as rest of spikelet; lemmas 3-7-veined, rounded on back, awnless or with terminal, subterminal or dorsal awn; stamens 3; stigmas 2; lodicules 0 or 2; ovary glabrous.

45. AGROSTIS L. - *Bents*

Annuals or perennials with or without rhizomes and/or stolons; inflorescence a slightly contracted to very diffuse panicle with obvious branches; glumes 2, equal or nearly so, 1-3-veined, longer than rest of spikelet; lemmas 3-5-veined, awnless or with subterminal or dorsal awn, with or without hair-tuft on callus; palea vestigial to nearly as long as lemma, (usually weakly) 2-veined or veinless; disarticulation at maturity at base of lemma.

A difficult genus due to plasticity and genetic variation, hybridization, and over-use in the past of unreliable characters, notably presence of awns. The palea length and presence of hair-tufts on the lemma-callus are important (though minute) characters (x≥10 lens essential). Presence of rhizomes (as opposed to stolons) is also very reliable but their absence is not, as they are often not developed until late on (in fruit) and in some habitats never.

1 Palea minute, <2/5 as long as lemma 2
1 Palea >2/5 as long as lemma (c.1/2 as long to nearly as long) 6
    2 Anthers 0.2-0.6mm; main panicle-branches bare of spikelets for proximal ≥2/3 of length; alien (often casual) of waste ground 3
    2 Anthers 1-2mm; main panicle-branches bare of spikelets for proximal ≤1/2 of length; natives 4
3 Spikelets ≥2mm; lemma 1.5-1.7mm, distinctly exceeding caryopsis; leaves 1-3mm wide. Tufted perennial (often annual with us) to 60cm. Intrd-natd; waste and rough ground and by roads and railways; scattered in Br *Rough Bent* - **A. scabra** Willd.
3 Spikelets <2mm; lemma 1-1.2mm, not or scarcely exceeding caryopsis; leaves <1mm wide. Tufted perennial (often annual with us) to 40cm. Intrd-casual; wool- and grain-alien; scattered in Br
>><< *Small Bent* - **A. hyemalis** (Walter) Britton, Sterns & Poggenb.
    4 Tiller leaves ≤0.3mm wide, bristle-like; panicle always with ± erect branches; rhizomes and stolons 0. Densely tufted perennial to 60cm. Native; dry sandy or peaty heaths; locally common in SW En, extending to S Wa and Surrey
>><< *Bristle Bent* - **A. curtisii** Kerguélen
    4 Tiller leaves flat or inrolled, >(0.6)1mm wide even when inrolled;

45. AGROSTIS

  panicle often with patent to erecto-patent branches at or after
  flowering; rhizomes or stolons usually present   **5**

5 Stolons 0; rhizomes usually present; ligule on 2nd culm-leaf usually ≤1.5x as long as wide, acute to obtuse. Perennial to 60cm. Native; dry sandy or peaty heaths, moors and hillsides; frequent to common
            Brown Bent - **A. vinealis** Schreb.

5 Rhizomes 0; stolons usually present, bearing tufts of leaves or shoots at nodes; ligule on 2nd culm-leaf usually ≥1.5x as long as wide, acute to acuminate. Perennial to 75cm. Native; damp or wet meadows, marshes, ditches, pondsides, on acid soils; frequent to common
              Velvet Bent - **A. canina** L.

 6 Anthers 0.2-0.8mm; lemma-callus with tuft of hairs ≥0.3mm   **7**
 6 Anthers 1-1.5mm; lemma-callus glabrous or pubescent   **8**

7 Lemmas with awn ≥2mm, well exserted from glumes; main panicle-branches bare of spikelets for proximal ≥2/3 of length, patent to erecto-patent after flowering; rhachilla extended above lemma-base, reaching ≥1/2 way up lemma. Tufted annual (at least with us) to 60cm. Intrd-casual; waste and rough ground and by roads and railways; scattered in Br   *Blown-grass* - **A. avenacea** J.F. Gmel.

7 Lemmas with awn 0-0.5mm, not exserted from glumes; main panicle-branches bare of spikelets for proximal ≤1/2 of length, erect to erecto-patent after flowering; rhachilla not extended above lemma-base. Loosely tufted annual (with us) or perennial. Intrd-casual; tips and waste ground; scattered in En   *African Bent* - **A. lachnantha** Nees

 8 Lemma-callus with tuft of hairs 0.2-0.3mm; awn 0 or, if present, arising from basal 1/3 of lemma, often exceeding glumes. Rhizomatous perennial to 60cm. Intrd-natd; lawns, roadsides, amenity and sports areas where sown and escaped; throughout Br   *Highland Bent* - **A. castellana** Boiss. & Reut.

 8 Lemma-callus glabrous or with hairs <0.2mm; awn 0 or, if present, arising from apical 1/2 of lemma, rarely exceeding glumes   **9**

9 Panicle contracted after flowering; rhizomes 0 or short and with ≤3 scale-leaves; ligules of culm-leaves subacute to rounded; stolons usually well developed. Perennial to 75cm. Native; damp meadows, ditches, marshes, by lakes, ponds, canals and rivers, damp arable and rough ground, dune-slacks; abundant
            Creeping Bent - **A. stolonifera** L.

9 Panicle-branches patent or nearly so after flowering; rhizomes usually present, with >3 scale-leaves; ligules of culm-leaves truncate; stolons 0 or poorly developed   **10**

 10 Ligules of tillers shorter than wide; fruiting panicle-branches with spikelets all well separated; leaves rarely >5mm wide. Rhizomatous perennial to 75cm. Native; all kinds of grassy places and rough ground, especially on acid soils; abundant
            Common Bent - **A. capillaris** L.

 10 Ligules of tillers longer than wide; fruiting panicle-branches bearing spikelets in small ± dense clusters at tips; leaves often >5mm wide. Rhizomatous perennial to 1(1.2)m. Native; grassy places, rough, cultivated and waste ground, mostly on disturbed sandy soils; throughout BI, common in S & C En

Black Bent - **A. gigantea** Roth

6 hybrid combinations have been found; none is common but probably they are under-recorded. The commonest is *A. capillaris* x *A. stolonifera* = **A. x murbeckii** P. Fourn. All are highly sterile except *A. capillaris* x *A. castellana* = **A. x fouilladei** P. Fourn., which is partially fertile.

45 x 52. AGROSTIS x POLYPOGON = X AGROPOGON P. Fourn.

Sterile hybrids between *A. stolonifera* and the 2 spp. of *Polypogon*. Variously intermediate, differing from *A. stolonifera* in more compact panicle with shorter pedicels and disarticulation (if any) at maturity near base of pedicel. *A. stolonifera* x *P. monspeliensis* = **X A. littoralis** (Sm.) C.E. Hubb. (*Perennial Beard-grass*) is sporadic with the parents on maritime damp sand or mud on coasts of En from Dorset to W Norfolk, casual elsewhere in Br. *A. stolonifera* x *P. viridis* = **X A. robinsonii** (Druce) Melderis & D.C. McClint. has occurred with the parents in Guernsey.

46. CALAMAGROSTIS Adans. - *Small-reeds*

Rhizomatous perennials; inflorescence a ± diffuse to slightly contracted panicle; glumes 2, equal or nearly so, 1-3-veined, longer than rest of spikelet; lemmas 3-5-veined, with apical or dorsal awn, with conspicuous basal hair-tuft ≥1/2 as long as lemma; palea c.2/3 as long as lemma, 2-veined; disarticulation at maturity at base of lemma.

1 Hairs at base of lemma not reaching lemma-apex; lemma minutely rough, with awn arising 1/4-1/3 way up 2
1 Hairs at base of lemma reaching at least to apex of lemma; lemma smooth, with awn arising from middle, apex or upper 1/2 3
   2 Spikelets 3-4(4.5)mm; lower glume acute; culms usually rough just below panicle. Culms to 1m, with 2-3 nodes. Native; marshes, fens and lakesides; very scattered in N BI S to W Suffolk and Co Antrim *Narrow Small-reed* - **C. stricta** (Timm) Koeler
   2 Spikelets (4)4.5-6mm; lower glume acuminate; culms smooth throughout. Culms to 1m, with 2-3 nodes. Native; marshes and fens; 1 locality in Caithness
Scottish Small-reed - **C. scotica** (Druce) Druce
3 Culms mostly with 5-8 nodes; ligules 7-10(14)mm; pollen 0; anthers indehiscent. Culms to 1.5m. Native; fens, marshes, ditches and lakesides; few places from Westmorland to S Aberdeen
Scandinavian Small-reed - **C. purpurea** (Trin.) Trin.
3 Culms mostly with 2-5 nodes; ligules (1)2-9(12)mm; pollen present; anthers dehiscent 4
   4 Leaf uppersides (often sparsely) pubescent; ligules (1)2-4(6)mm; lemmas 3-5-veined, with basal hairs 1.1-1.5x as long as lemma. Culms to 1.2m, with 3-5 nodes. Native; fens, marshes and open

46. CALAMAGROSTIS

wet woods; scattered in En and S Sc
Purple Small-reed - **C. canescens** (F.H. Wigg.) Roth
4 Leaf uppersides scabrid, not pubescent; ligules 4-9(12)mm; lemmas 3-veined, with basal hairs >1.5x as long as lemma. Culms to 2m, with 2-4 nodes. Native; damp woods and wood-margins, ditches, fens, dune-slacks; scattered throughout much of Br
Wood Small-reed - **C. epigejos** (L.) Roth

C. canescens x *C. stricta* = *C. x gracilescens* (Blytt) Blytt occurs for certain only by a canal in SE Yorks, but other populations of *C. stricta* in En and Sc probably have been introgressed by *C. canescens* in the past.

46 x 47. CALAMAGROSTIS x AMMOPHILA = X CALAMMOPHILA
Brand - *Purple Marram*
X C. baltica (Schrad.) Brand (*C. epigejos* x *A. arenaria*) (*Purple Marram*) occurs naturally on maritime dunes with *A. arenaria* and near *C. epigejos*, and is also planted for sand-binding, in E En, S Hants and W Sutherland. It is vigorous, sterile and more closely resembles *A. arenaria*.

47. AMMOPHILA Host - *Marram*
Strongly rhizomatous perennials; inflorescence a compact linear-ellipsoid panicle; glumes 2, subequal, the lower 1-3-, the upper 3-veined, slightly longer than rest of spikelet; lemmas 5-7-veined, awnless or with minute subapical awn <1mm, with basal hair-tuft <1/2 as long as lemma; palea nearly as long as lemma, 2-4-veined; disarticulation at maturity at base of lemma.

1 Culms to 1.2m; ligules 10-30mm. Native; on mobile sand-dunes; common round coasts of BI, rare casual inland
Marram - **A. arenaria** (L.) Link

48. GASTRIDIUM P. Beauv. - *Nit-grasses*
Annuals; inflorescence a compact linear-ellipsoid panicle; glumes 2, unequal, 1-veined, much longer than rest of spikelet, linear-lanceolate, with swollen ± hemispherical base; lemmas 5-veined, with dorsal usually bent awn, without basal hair-tuft; palea ± as long as lemma, 2-veined; disarticulation at maturity at base of lemma.

1 Stems procumbent to erect; spikelets (2)3-5mm; lemma subglabrous to sparsely pubescent, with awn 0-4mm, rarely exceeding glumes. Stems to 50(90)cm. Native; barish or sparsely grassed, well-drained, calcareous ground; very local mostly near coast in SW Br from S Devon and Wight to Glam, also casual scattered in En and S Wa
Nit-grass - **G. ventricosum** (Gouan) Schinz & Thell.
1 Stems usually erect; spikelets (4)5-8mm; lemma pubescent to densely so, with awn 4-7(8)mm, often exceeding glumes. Stems to 50(90)cm. Intrd-casual; rather infrequent wool-alien; scattered in En
Eastern Nit-grass - **G. phleoides** (Nees & Meyen) C.E. Hubb.

49. LAGURUS L. - *Hare's-tail*
Annuals; inflorescence a very compact, ovoid, densely silky-pubescent panicle; glumes 2, equal, 1-veined, linear-lanceolate and tapered to apical awn, longer than rest of spikelet, lemmas 5-veined, with 2 apical bristles reaching c. as far as glume-awns, with dorsal bent awn well exserted from glumes, pubescent but without basal hair-tuft; palea shorter than body of lemma, 2-veined; disarticulation at maturity at base of lemma.

1 Stems to 60cm, erect; panicle 1-7 x 0.5-2cm; spikelets 7-10mm; awns 8-20mm. Intrd-natd; abundant on mobile sand-dunes in Jersey and Guernsey, rare casual or locally natd elsewhere in S BI
Hare's-tail - **L. ovatus** L.

50. APERA Adans. - *Silky-bents*
Annuals; inflorescence a diffuse to ± contracted panicle; glumes 2, unequal, the lower 1-, the upper 3-veined, ± as long as rest of spikelet; lemmas 5-veined, with subterminal ± straight awn much longer than body, minutely pubescent at base; palea shorter than to ± as long as body of lemma, 2-veined; disarticulation at maturity at base of lemma.

1 Panicles up to 25 x 15cm, very diffuse, the longer branches bare for proximal 1/2; spikelets 2.4-3mm; anthers 1-2mm. Stems to 1m. Possibly native; frequent in dry sandy arable fields and marginal habitats in E Anglia, scattered to SE En and Humber basin, casual elsewhere
Loose Silky-bent - **A. spica-venti** (L.) P. Beauv.
1 Panicles up to 20 x 1.5cm, loosely contracted, the branches with spikelets nearly to base; spikelets 1.8-2.5mm; anthers 0.3-0.4mm. Stems to 40(70)cm. Possibly native; dry sandy fields and rough ground; locally frequent in E Anglia, scattered to SE Yorks and Berks
Dense Silky-bent - **A. interrupta** (L.) P. Beauv.

51. MIBORA Adans. - *Early Sand-grass*
Annuals; inflorescence a slender 1-sided raceme with pedicels <0.5mm; glumes 2, equal, 1-veined, longer than rest of spikelet; lemmas 5-veined, awnless, pubescent but without basal hair-tuft; palea as long as lemma, 2-veined; disarticulation at maturity at base of lemma.

1 Culms very slender, to 8(15)cm; racemes 0.5-2cm; spikelets 1.8-3mm. Native; on loose sand near sea; Anglesey and CI, natd in few places in Br N to E Lothian
Early Sand-grass - **M. minima** (L.) Desv.

52. POLYPOGON Desf. - *Beard-grasses*
Annuals or perennials with stolons; inflorescence a contracted or semi-diffuse panicle; glumes 2, equal, 1-veined or the upper 3-veined, much longer than rest of spikelet, awned from apex or unawned; lemmas 5-veined, truncate and finely toothed at apex, without basal hair-tuft, awnless or with short terminal awn; palea nearly as long as lemma, 2-veined; disarticulation at maturity near base or apex of pedicel.

52. POLYPOGON

1 Annual; inflorescence densely contracted; glumes notched at apex, each with apical awn 3.5-7mm; lemma with awn (0)1-2mm. Culms to 80cm. Native; drier parts of salt-marshes and damp places near sea; S & SE En from Dorset to E Norfolk, casual elsewhere
Annual Beard-grass - **P. monspeliensis** (L.) Desf.
1 Stoloniferous perennial; inflorescence semi-diffuse; glumes obtuse, unawned; lemma unawned. Culms to 60(100)cm. Intrd-natd; roadsides and rough ground and by pools; Guernsey and Jersey, casual elsewhere *Water Bent* - **P. viridis** (Gouan) Breistr.

53. ALOPECURUS L. - *Foxtails*

Annuals or perennials without rhizomes, sometimes with stolons; inflorescence a very contracted spike-like panicle; glumes 2, equal, 3-veined, sometimes with their margins fused proximally round spikelet, very slightly shorter to very slightly longer than rest of spikelet, keeled, rounded to very acute or apiculate at apex; lemmas 4-veined, obtuse to truncate or notched, without basal hair-tuft, sometimes with margins fused proximally round carpel and stamens, with dorsal long or short awn from lower 1/2; palea 0; disarticulation at maturity near base of pedicel.

1 Lemmas unawned or with awn shorter than body and not exserted from glumes or exserted by ≤0.5mm 2
1 Lemmas with awn longer than body and exserted from glumes by ≥1mm 3
    2 Panicles >3x as long as wide; glumes with hairs <0.5mm; lemma with margins fused proximally for 1/3-1/2 their length. Annual to perennial; culms usually decumbent to ascending, sometimes rooting at lower nodes, to 40cm. Native; wet meadows, marshes, ditches, pondsides; scattered in C & S Br N to NW Yorks, Man, E Cork *Orange Foxtail* - **A. aequalis** Sobol.
    2 Panicles <3x as long as wide; glumes with hairs >(0.5)1mm; lemma with margins fused proximally for <1/4 their length. Perennial; culms ± erect, to 50cm. Native; mountain springs and flushes at 600-1200m; very local in N Br from Westmorland to E Ross *Alpine Foxtail* - **A. borealis** Trin.
3 Margins of lemma free or fused proximally for <1/4 their length; 2 glumes fused only at extreme base 4
3 Margins of lemma fused proximally for c.1/3-1/2 their length; 2 glumes fused proximally for c.1/4-1/2 their length 5
    4 Glumes acute; basal culm internode swollen, (1)2-4.5(6)mm wider than normal culm width. Perennial; culms usually erect to ascending, not rooting at nodes, to 30(40)cm. Native; wet grassy places near sea or in estuaries; local on coasts of S & E Br from Carms and E Cornwall to N Lincs, Guernsey
Bulbous Foxtail - **A. bulbosus** Gouan
    4 Glumes obtuse; basal culm internode 0-1(1.5)mm wider than normal culm width. Perennial; culms usually decumbent to ascending, often rooting at lower nodes, to 40(50)cm. Native;

wet meadows, marshes, ditches, pondsides; frequent to common *Marsh Foxtail* - **A. geniculatus** L.
5 Annual; glumes fused proximally for 1/3-1/2 their length, subglabrous or with hairs <0.5mm on keel, margins and at base, with winged keel. Annual; culms erect, to 80cm. Native; weed of arable fields and waste ground; frequent in S, C & E En, very scattered elsewhere *Black-grass* - **A. myosuroides** Huds.
5 Perennial; glumes fused proximally for c.1/4 their length, conspicuously pubescent with hairs >0.5mm, with keel unwinged. Perennial; culms usually erect, to 1.2m. Native; grassy places, mostly on damp rich soils; abundant *Meadow Foxtail* - **A. pratensis** L.

A. geniculatus hybridises with *A. pratensis*, *A. bulbosus* and *A. aequalis* in mixed populations; the hybrids are uncommon and highly sterile.

54. BECKMANNIA Host - *Slough-grasses*
Annuals or perennials without stolons, rarely with rhizomes; inflorescence a long raceme of closely packed appressed spikes forming a long narrow inflorescence; glumes 2, equal, 3-veined with connecting veins between, enclosing rest of spikelet except for apiculate tip of lemma(s), strongly keeled and hooded; florets 1 or 2, the 1st bisexual, the second bisexual, male or sterile; lemmas 5-veined, apiculate but not otherwise awned, without basal hair-tuft; palea nearly as long as lemma, 2-veined; disarticulation at maturity below glumes.

1 Inflorescence up to 15(25)cm; spikes 5-20mm; spikelets 2.2-3.2mm. Culms to 50(100)cm. Intrd-casual; waste ground and tips; scattered in S En and S Wa *American Slough-grass* - **B. syzigachne** (Steud.) Fernald

55. PHLEUM L. - *Cat's-tails*
Annuals or perennials, sometimes with rhizomes or stolons; inflorescence a very contracted spike-like panicle; glumes 2, equal, 3-veined, strongly keeled, usually with stiff hairs on keel, apiculate to shortly awned at apex; lemmas 3-7-veined, irregularly truncate to rounded at apex, without basal hair-tuft, unawned; palea as long or nearly as long as lemma, 2-veined; disarticulation at maturity below the lemma.

1 Annual without tillers at flowering; glumes acute to subacute at apex, gradually narrowed to awn; anthers ≤1mm. Culms erect, to 20(30)cm. Native; maritime sand-dunes and inland on sandy heaths; frequent on coasts of BI, inland in E Anglia *Sand Cat's-tail* - **P. arenarium** L.
1 Perennial with tillers at flowering; glumes obtuse or truncate at apex, abruptly or very abruptly narrowed to awn; anthers ≥1mm 2
  2 Glumes 5-8.5mm incl. awns 2-3mm; panicles (1)2-3(5)x as long as wide. Culms erect, to 50cm. Native; grassy, rocky or mossy wet places on mountains at 600-1200m; very local in N Br from Westmorland to E Ross *Alpine Cat's-tail* - **P. alpinum** L.
  2 Glumes 2-5.5mm incl. awns ≤2mm; panicles (2)3-20(30)x as long

55. PHLEUM

    as wide   3
3 Glumes obtuse (and shortly awned) at apex; culms not swollen at base; ligules 0.5-2mm. Culms erect, to 60cm, the lower sheaths often purplish-tinged. Native; dry sandy and chalky pastures and adjacent rough ground; very local in CE En from Beds to E Suffolk
Purple-stem Cat's-tail - **P. phleoides** (L.) H. Karst.
3 Glumes truncate (and shortly awned) at apex; culms usually swollen at base; ligules 1-9mm   4
   4 Spikelets (3.5)4-5.5mm incl. awns (0.8)1-2mm; panicle 6-10mm wide; leaves 3-9mm wide; ligule usually obtuse. Culms erect, to 1.5m. Native; grassy places and rough ground; common
Timothy - **P. pratense** L.
   4 Spikelets 2-3.5mm incl. awns 0.2-1(1.2)mm; panicle 3-6mm wide; leaves 2-6mm wide; ligule usually acute. Culms to 50(100)cm. Native; grassy places; probably throughout BI
Smaller Cat's-tail - **P. bertolonii** DC.

TRIBE 12 - BROMEAE (genera 56-59). Annuals or perennials without stolons, with or without rhizomes; ligule membranous; sheaths usually fused when young but soon splitting; inflorescence a panicle, rarely slightly contracted; spikelets with several to many bisexual florets, the apical 1 or 2 (sometimes more) often reduced and male or sterile; glumes 2, unequal, 1-9-veined, much shorter than rest of spikelet, unawned; lemmas 5-11-veined, rounded or keeled on back, usually minutely bifid at apex, usually with long subterminal awn; stamens 2-3; stigmas 2; lodicules 2; ovary with pubescent terminal appendage (the styles arising below it).

56. BROMUS L. - *Bromes*
Annuals; spikelets ovoid to narrowly so, terete or slightly compressed; lower glume 3-5(7)-veined; upper glume 5-7(9)-veined; lemmas 7-9(11)-veined, rounded on back, subacute to obtuse or rounded-obtuse, and minutely bifid at apex; stamens 3.

  All the spp. are very variable in habit, becoming very small (often with only 1 spikelet) in dry conditions, and in extreme cases spikelet and lemma measurements also vary outside the normal range. Dwarf, starved plants should be avoided; more normal plants are usually nearby. Lemma measurements should be made from the middle or low part of spikelets.

1 Caryopsis thick, with inrolled margins; lemma with margins wrapped around caryopsis when mature, hence lemma margins not overlapping next higher lemma, but rhachilla ± revealed between florets; rhachilla disarticulating tardily   2
1 Caryopsis thin, flat or with weakly inrolled margins; lemma margins not wrapped round caryopsis, overlapping next higher lemma and obscuring rhachilla; rhachilla disarticulating readily   3
   2 Spikelets 12-20mm, glabrous or pubescent; lemmas 6.5-9(10)mm; palea equalling lemma; caryopsis 6-9mm; sheaths usually glabrous or sparsely pubescent. Culms erect, to 1.2m. Intrd-

casual; weed of cereals, marginal and waste ground; very
scattered in BI   *Rye Brome* - **B. secalinus** L.
2 Spikelets 8-12mm, glabrous; lemmas 5-6mm; palea shorter than
lemma; caryopsis 4-4.5mm; sheaths pubescent. Culms erect, to
60cm. Intrd-natd; grassy fields and waysides; very scattered in
Br and Ir   *Smith's Brome* - **B. pseudosecalinus** P.M. Sm.
3 Palea divided nearly to base; panicle with mostly subsessile spikelets
densely clustered in groups of 3. Culms erect, to 1m. Native;
arable and waste land, especially as weed in *Onobrychis, Lolium* or
Trifolium crops; formerly scattered in S En, last seen in 1972 in
Cambs   *Interrupted Brome* - **B. interruptus** (Hack.) Druce
3 Palea entire to shortly bifid; panicle various but spikelets not
subsessile in groups of 3   4
  4 Anthers 3.5-5mm, ≥1/2 as long as lemmas; panicle branches long,
forming very open panicle. Culms erect, to 1m. Intrd-natd; casual
weed of arable and grass fields and waste ground; very scattered
in C & S Br, natd in E Gloucs   *Field Brome* - **B. arvensis** L.
  4 Anthers 0.2-3mm, <1/2 as long as lemmas; panicles various, often
not very open   5
5 Awns curved or bent outwards at maturity, their apices widely
diverging   6
5 Awns ± straight to slightly flexuous or curved at maturity, the apices
of those of the more apical lemmas ± parallel or even convergent (if
± curved outwards the culms procumbent to ascending)   9
  6 Panicle-branches and pedicels much shorter than spikelets;
pedicels ≤10mm; spikelets ≤18mm; awn ± cylindrical from base
(*Soft-brome* - **B. hordeaceus** L.)   7
  6 At least some panicle-branches and pedicels on well grown plants
longer than spikelets; pedicels often >15mm; spikelets ≥18mm;
awn strongly flattened at base   8
7 Lemmas 6.5-8.5mm; culms usually <15cm, with usually <10 spikelets.
Stems to 15(20)cm, erect to ascending. Native; grassy cliff-tops and
sandy or shingly ground by sea; locally frequent on coasts of CI and
Br scattered N to Angus **B. hordeaceus ssp. ferronii** (Mabille) P.M. Sm.
7 Lemmas 8-11mm; culms usually >15cm, with usually >10 spikelets.
Stems to 60cm, erect. Intrd-casual; waste places and waysides;
scattered in En
        B. hordeaceus ssp. divaricatus (Bonnier & Layens) Kerguélen
  8 Lemmas 11-18mm; panicles usually rather stiffly erect. Culms
erect, to 70cm. Intrd-casual; waste places; scattered in C & S Br
        Large-headed Brome - **B. lanceolatus** Roth
  8 Lemmas 8-10mm; panicles usually lax with patent to pendent
branches. Culms erect, to 80cm. Intrd-casual; waste places;
scattered in En   *Thunberg's Brome* - **B. japonicus** Thunb.
9 Lemmas (4.5)5.5-6.5mm; caryopsis longer than palea. Culms erect, to
80cm. Probably intrd-natd; grassland, waysides and rough ground;
scattered throughout BI   *Slender Soft-brome* - **B. lepidus** Holmb.
9 Lemmas 6.5-11mm; caryopsis shorter than to as long as palea   10

10 Panicle ± lax with at least some pedicels longer than spikelets; lemmas rather coriaceous, with rather obscure veins; anthers 1-3mm ... 11
10 Panicle ± dense, usually with all pedicels shorter than spikelets; lemmas papery, with prominent veins; anthers 0.5-1.5(2)mm ... 12
11 Lemmas 8-11mm; anthers mostly 1-1.5mm; spikelets 15-28mm; lowest rhachilla-segment mostly 1.3-1.7mm. Culms erect, to 1m. Native; grassy places, waysides and rough ground, especially in damp rich meadows; local in C & S Br *Meadow Brome* - **B. commutatus** Schrad.
11 Lemmas 6.5-8mm; anthers mostly 1.5-3mm; spikelets 10-16mm; lowest rhachilla-segment mostly 0.7-1mm. Culms erect, to 1m. Native; similar places and distribution to *B. commutatus*
Smooth Brome - **B. racemosus** L.
12 Lemmas 8-11mm, usually pubescent. Stems to 80cm (often much less), erect. Native; grassy places, waysides, rough ground; frequent throughout lowland BI **B. hordeaceus ssp. hordeaceus** ... 13
12 Lemmas (6)6.5-8mm, usually glabrous
13 Culms to 8(12)cm, procumbent to ascending; caryopsis shorter than palea. Native; sandy places by sea; coasts of CI and Br probably N to Sc **B. hordeaceus ssp. thominei** (Hardouin) Braun-Blanq.
13 Culms usually >10cm (to 60cm), usually erect; caryopsis c. as long as palea. Probably native; grassland, waysides, rough ground; scattered throughout BI, frequent in S & C Br (*B. hordeaceus* x *B. lepidus*) *Lesser Soft-brome* - **B. x pseudothominei** P.M. Sm.

B. commutatus x *B. racemosus* occurs rather frequently with the parents in C & S Br; it is fertile and backcrosses.

57. BROMOPSIS (Dumort.) Fourr. - *Bromes*

Perennials with long to very short rhizomes; spikelets narrowly oblong then tapered to apex, terete or slightly compressed; lower glume 1(-3)-veined; upper glume 3(-5)-veined; lemmas 5-7-veined, rounded or slightly keeled on back, acute to shortly acuminate and minutely bifid at apex; stamens 3.

1 Inflorescence very lax, the branches pendent or all swept to 1 side; sheaths with distinct pointed auricles at apex ... 2
1 Inflorescence dense to fairly lax, the branches erect to erecto-patent; sheaths without or with short rounded auricles at apex ... 3
2 Lowest panicle-node with usually >2 branches, some with 1 or very few spikelets, and with small ± glabrous scale; panicle branches swept to 1 side. Rhizomes very short; culms to 1.2m. Native; similar places to *B. ramosa*, often with it; very scattered in mainland Br *Lesser Hairy-brome* - **B. benekenii** (Lange) Holub
2 Lowest panicle-node usually with 2 branches, both long and with >3 spikelets, and with small pubescent scale; panicle branches pendent. Rhizomes very short; culms to 2m. Native; woods, wood-margins and hedgerows; frequent throughout lowland BI
Hairy-brome - **B. ramosa** (Huds.) Holub

3 Plant densely tufted, with short rhizomes; lemmas with awns (2)3-8mm; leaves of tillers usually folded or inrolled along long axis. Rhizomes very short; culms to 1.2m. Native; dry grassland and grassy slopes; common on base-rich soils in C, S & E En, scattered elsewhere in BI   **Upright Brome - B. erecta** (Huds.) Fourr.
3 Plant not densely tufted, with long rhizomes; lemmas awnless or less often with awns up to 3(6)mm; leaves of tillers usually flat. Intrd-natd; rough grassy places, waysides and field-margins (***Hungarian Brome* - B. inermis** (Leyss.) Holub)   4
   4 Sheaths usually glabrous; culm-nodes glabrous or with short hairs just below; lemmas glabrous to scabrid or with sparse hairs on margins, with awn 0(-3)mm. Scattered in Br
B. inermis ssp. inermis
   4 Sheaths usually pubescent; culm-nodes pubescent; lemmas appressed-pubescent on margins, with awn 0-6mm. S Essex
B. inermis ssp. pumpelliana (Scribn.) W.A. Weber

58. ANISANTHA K. Koch - *Bromes*

Annuals; spikelets ± parallel-sided or widening distally, slightly compressed; lower glume 1(-3)-veined; upper glume 3(-5)-veined; lemmas 7-veined, rounded on back, acute to acuminate and minutely bifid at apex; stamens 2 or 3.

Lemma-lengths should be measured on only the two basal florets.

1 Lemmas 20-36mm   2
1 Lemmas 9-20mm   3
   2 Panicle lax, with branches spreading laterally or pendent; callus-scar at base of lemma ovate, rounded at end. Stems to 80cm. Intrd-natd; rough and waste ground, waysides and open grassland on warm sandy soils; frequent in CI and E Anglia, very scattered casual elsewhere in Br   ***Great Brome* - A. diandra** (Roth) Tzvelev
   2 Panicle dense, with erect branches; callus-scar at base of lemma elliptic, pointed at end. Stems to 60cm. Intrd-natd; similar places to *A. diandra*; rather infrequent in CI and S Br, rare casual elsewhere   ***Ripgut Brome* - A. rigida** (Roth) Hyl.
3 Panicle lax, with branches spreading laterally or pendent   4
3 Panicle dense, with stiffly erect branches   5
   4 Lemmas 9-13mm; inflorescence compound, the larger branches with 3-8 spikelets (except in depauperate plants); spikelets with >3 apical sterile florets. Stems to 60cm. Intrd-natd; similar places to *A. diandra*; W Suffolk and W Norfolk, infrequent casual elsewhere   ***Drooping Brome* - A. tectorum** (L.) Nevski
   4 Lemmas 13-20mm; inflorescence simple or the larger branches slightly branched and with up to 3(5) spikelets; spikelets with 1-2 sterile apical florets. Stems to 80cm. Native; rough and waste ground, waysides, open grassland, weed of arable land and gardens; throughout lowland BI, common in C & S Br
***Barren Brome* - A. sterilis** (L.) Nevski

58. ANISANTHA

5 Lemmas mostly 12-20mm; spikelets with 1-2(3) sterile apical florets. Stems to 60cm. Intrd-natd; similar places to *A. diandra*; local in SW En, S Wa and CI, occasional casual elsewhere
Compact Brome - **A. madritensis** (L.) Nevski

5 Lemmas 9-13(15)mm; spikelets with >(2)3 apical sterile florets. Stems to 40cm. Intrd-casual; waste places; very scattered in Br
Foxtail Brome - **A. rubens** (L.) Nevski

59. CERATOCHLOA DC. & P. Beauv. - *Bromes*

Perennials, often short-lived (?sometimes annual or biennial), usually without rhizomes; spikelets ovoid to narrowly so, compressed; lower glume 3-5-veined; upper glume 5-7-veined; lemmas 7-11(13)-veined, strongly keeled on back, acute to shortly acuminate and minutely bifid at apex; stamens 3.

1 Leaves <3(4)mm wide; lemmas 8-13mm; leaves and sheaths densely pubescent with long patent hairs 2
1 Leaves 4-10mm wide; lemmas (10)12-18mm; leaves and sheaths glabrous to conspicuously long-patent-pubescent 3
    2 Lemmas glabrous or sparsely pubescent, awnless or with awn up to 1(2)mm. Culms to 1m. Intrd-casual; tips and waste places; very scattered in Br   *Patagonian Brome* - **C. brevis** (Steud.) B.D. Jacks.
    2 Lemmas sparsely to densely pubescent, with awn (3)4-8(12)mm. Culms to 1m. Intrd-casual; rough ground and waysides; very scattered in S En   *Southern Brome* - **C. staminea** (Desv.) Stace
3 Lemmas awnless or with awn up to 3(5)mm, with 9-11(13) veins; palea 1/2-3/4 as long as body of lemma. Culms to 1m. Intrd-natd; rough ground, roadsides and field-borders; scattered in C & S Br and CI   *Rescue Brome* - **C. cathartica** (Vahl) Herter
3 Lemmas with awn (3)4-10mm, with 7-9 veins; palea 3/4-1x as long as body of lemma 4
    4 Lemmas glabrous to sparsely and shortly pubescent, with awns (4)6-10(12)mm; leaves and sheaths glabrous to sparsely pubescent. Culms to 80cm. Intrd-natd; rough ground, field borders, waysides and on river-banks; scattered in Br, mostly S, Guernsey, Co Dublin   *California Brome* - **C. carinata** (Hook. & Arn.) Tutin
    4 Lemmas conspicuously pubescent, with awns (3)4-6(7)mm; leaves and sheaths sparsely pubescent to pubescent. Culms to 1m. Intrd-natd; rough and waste ground; very scattered in SE En, mostly casual   *Western Brome* - **C. marginata** (Steud.) B.D. Jacks.

TRIBE 13 - BRACHYPODIEAE (genus 60). Annuals or rhizomatous perennials; ligule membranous; sheaths not fused; inflorescence a raceme with usually 1 spikelet at each node with pedicels ≤2(2.8)mm; spikelets with many bisexual florets, the apical 1 or 2 reduced and male or sterile; glumes 2, unequal, 3-9-veined, much shorter than rest of spikelet, sometimes shortly awned; lemmas mostly 7-veined, rounded on back, acute to acuminate and usually with short to long awn; stamens 3; stigmas 2; lodicules 2; ovary with pubescent terminal appendage (the styles arising from below it).

60. BRACHYPODIUM P. Beauv. - *False Bromes*

1 Annual; anthers 0.3-1mm; spikelets distinctly compressed. Culms stiffly erect, to 15cm. Intrd-casual; waste places; very scattered in En
Stiff Brome - **B. distachyon** (L.) P. Beauv.
1 Perennials with rhizomes and tillers; anthers 3-6(8)mm; spikelets subterete 2
   2 Plant weakly rhizomatous, usually densely tufted; culms with 4-8 internodes; raceme usually pendent at apex; leaves usually >6mm wide; lemmas with awns 7-15mm. Culms to 1m. Native; woods, scrub and shady wood-borders and hedgerows, in open grassland mainly in the N; common through most of BI
False Brome - **B. sylvaticum** (Huds.) P. Beauv.
   2 Plant strongly rhizomatous, usually scarcely tufted; culms with 3-4(5) internodes; raceme usually erect; leaves usually <6mm wide; lemmas with awns 1-5mm. Culms to 1.2m. Native; grassland, mainly on chalk and limestone; common in suitable places in C, S & E En, very scattered in NW & SW En, Wa and Ir, sporadic in Sc and CI *Tor-grass* - **B. pinnatum** (L.) P. Beauv.

B. pinnatum x *B. sylvaticum* = **B. x cugnacii** A. Camus is probably the identity of some variably sterile intermediates found very scattered in En and Ir.

TRIBE 14 - TRITICEAE (genera 61-68). Annuals or perennials with or without rhizomes, without stolons; ligule membranous; sheaths not fused; inflorescence a spike with 1-3 spikelets at each node; spikelets with 1-many florets, often some florets (or some spikelets if spikelets >1 per node) male or sterile; glumes 2, equal or slightly unequal, 1-11-veined, often with long terminal awn; lemmas mostly 5-7-veined, rounded on back, very acute to obtuse, awnless or with terminal awn; stamens 3; stigmas 2; lodicules 2; ovary usually pubescent or pubescent at apex, sometimes with pubescent terminal appendage (the styles arising from below it).

61. ELYMUS L. - *Couches*
Perennials without rhizomes; spikelets 1 per node, with several to many florets with all but the apical 1 or 2 bisexual, flattened broadside on to rhachis; glumes 2-5-veined, acute or narrowly acute, often awned; lemmas 5-veined, usually long-awned, sometimes awnless or short-awned; spikelets breaking up below each lemma at maturity, leaving glumes on rhachis.

1 Spikelets ± contiguous; glumes very acute, finely pointed with awn ≤4mm, reaching >1/2 way up body of adjacent lemma. Culms to 1.2m. Native; woods, hedgerows, shady river-banks and mountain gullies and cliff-ledges; scattered throughout Br and Ir
Bearded Couch - **E. caninus** (L.) L.
1 Spikelets ± distant, each (excl. awns) not reaching as far as next

spikelet on opposite side of rhachis; glumes acute, unawned, reaching ≤1/2 way up body of adjacent lemma. Culms to 1.2m. Intrd-casual; fields and waste ground; scattered in En

Australian Couch - **E. scabrus** (Labill.) Á. Löve

62. ELYTRIGIA Desv. - *Couches*

Perennials with long rhizomes; spikelets 1 per node, with several to many florets with all but the apical 1-2 bisexual, flattened broadside on to rhachis; glumes 3-11-veined, acute to very obtuse, rarely awned; lemmas 5-veined, unawned or with short to rarely long awns; spikelets not breaking up easily at maturity, usually falling whole or rhachis breaking up.

1 Rhachis breaking up between each spikelet at maturity, smooth; ribs on leaf upperside densely minutely pubescent. Culms to 60(80)cm. Native; maritime sand-dunes; common round coasts of BI

Sand Couch - **E. juncea** (L.) Nevski

1 Rhachis not breaking up between each spikelet at maturity, scabrid; ribs on leaf upperside glabrous to scabrid, sometimes with sparse long hairs   2

  2 At least middle and lower sheaths with minute (often sparse) fringe of hairs on exposed free margin; leaves usually inrolled even when fresh, their upperside ribs ± flat-topped. Culms to 1.2m. Native; wet sandy, gravelly or muddy places by sea, often at margins of dunes, creeks or salt-marshes; frequent round coasts of BI N to SW Sc *Sea Couch* - **E. atherica** (Link) Carreras Mart.

  2 Sheaths with glabrous margin; leaves usually flat when fresh and turgid, their upperside ribs with rounded tops. Culms to 1.5m. Native (*Common Couch* - **E. repens** (L.) Nevski)   3

3 Leaves mostly flat when fresh, green or glaucous, basal ones 3-10mm wide, upper ones 2-8.5mm wide, with fine well-spaced ribs on upperside; spikes up to 20(30)cm; spikelets 10-20mm, with 3-8 florets; glumes 7-12mm, with 3-7 veins, usually awnless; lemmas 8-13mm, usually awnless but sometimes with awn up to 15mm. Cultivated, waste and rough ground; abundant   **E. repens ssp. repens**

3 Leaves mostly inrolled, ± glaucous, basal ones 3-5mm wide, upper ones 1-3.4mm wide, with thick (0.15-0.2mm wide), close ribs on upperside; spikes up to 9cm; spikelets (7)9-14mm, with (2)3-6 florets; glumes (4.4)6-9(10)mm, with 3(-7) veins, usually with awn (0.2)1-2.3mm; lemmas (5.5)7-10(12)mm, with awn (0.3)1.8-2.8(4.6)mm. Maritime sand, often on dunes; S & E coasts of En from S Somerset to W Norfolk, Guernsey   **E. repens ssp. arenosa** (Spenn.) Á. Löve

Hybrids occur fairly frequently on the coasts in all 3 combinations; all are sterile, with indehiscent anthers, empty pollen and no seed-set.

62 x 65. ELYTRIGIA x HORDEUM = X ELYTRORDEUM Hyl.

X E. langei (K. Richt.) Hyl. (*E. repens* x *H. secalinum*) has been found in wet meadows, fields and roadsides in a few scattered localities in En from

Scillies to Northumb. It exists in two variants: 1 clearly intermediate; the other resembling an awned *E. repens*.

63. LEYMUS Hochst. - *Lyme-grass*
Perennials with long rhizomes; spikelets 2(-3) per node, with 3-6 florets, with all but the most apical bisexual, flattened broadside on to rhachis; glumes 3-5-veined, finely pointed but not awned; lemmas mostly 7-veined, acute, unawned; spikelets breaking at maturity below each lemma.

1 Culms to 1.5(2)m, very glaucous; leaves flat, becoming inrolled, 8-20mm wide; spike up to 35cm, dense; spikelets 20-32mm. Native; mobile sand on maritime dunes, rarely intrd inland; round coasts of most of BI *Lyme-grass* - **L. arenarius** (L.) Hochst.

64. HORDELYMUS (Jess.) Harz - *Wood Barley*
Perennials with very short rhizomes; spikelets (2-)3 per node, with 1(-2) florets, all bisexual; glumes 1-3-veined, finely pointed and with long awn, each pair fused at base; lemmas 5-veined, with very long awn; spikelets breaking at maturity above the glumes but often not until late autumn.

1 Culms to 1.2m; leaves flat, 5-14mm wide; spike 5-10cm, dense; lemmas 8-10mm with awns 15-25mm. Native; woods and copses; local in Br N to S Northumb *Wood Barley* - **H. europaeus** (L.) Harz

65. HORDEUM L. - *Barleys*
Annuals or less often perennials without rhizomes; spikelets 3 per node, each with 1 floret, the central spikelet bisexual, the laterals bisexual, male or sterile; glumes very narrow, 1-3-veined, with long awn; lemmas of bisexual florets 5-veined, with very long awn; rhachis breaking up at each node at maturity, or (in cultivated taxa) below each bisexual lemma.

Most of the spp. are superficially very similar. For accurate identification a *triplet* of spikelets (the 3 spikelets at 1 rhachis node) from near the middle of the spike should be isolated, and the 6 glumes and 3 florets identified. The spikelets, florets and lemmas are here referred to as central or lateral according to their position in the triplet.

1 Rhachis not breaking up at maturity, the caryopsis-containing florets breaking away from the rest of the spikelet which remains on the rhachis; awns of central lemmas usually >10cm (rarely very short) 2
1 Rhachis breaking up at maturity, the triplet of spikelets forming the dispersal unit; all awns ≤10cm 3
  2 All 3 florets in each triplet producing a caryopsis and with a long-awned lemma. Annual to 1m. Intrd-casual; waste places, waysides and field-borders; fairly frequent throughout BI
Six-rowed Barley - **H. vulgare** L.
  2 Only central floret of each triplet producing a caryopsis; lateral lemmas awnless or ± so. Annual to 75cm. Intrd-natd; relic in

waste places, fields and waysides; throughout BI
Two-rowed Barley - **H. distichon** L.

3 Glumes of lateral spikelets >3cm, awn-like from base to apex; lateral florets extremely reduced, usually simply an awn-like outgrowth; awn of central lemma usually >5cm. Tufted perennial to 60cm. Intrd-natd; waste places and along salted main roads; locally frequent in E Br from E Kent to C Sc *Foxtail Barley* - **H. jubatum** L.

3 Glumes of lateral spikelets <3cm, if >2cm then at least 1 of each pair distinctly widened at base; lateral florets male or sterile but with obvious floret construction; awns of central lemma <5cm 4
    4 Perennial, with tillers at flowering 5
    4 Annual, without tillers 6

5 Proximal part of glumes with very short soft hairs >0.1mm; anthers 0.8-2mm; awns strongly divergent at maturity; upper leaf-blades 1.5-2(3)mm wide. Tufted perennial to 40cm. Intrd-casual; tips, waste ground and fields; scattered in En

Antarctic Barley - **H. pubiflorum** Hook. f.

5 Proximal part of glumes with minute rough prickles <0.1mm; anthers 3-4mm; awns stiffly erect at maturity; upper leaf-blades 2-6mm wide. Tufted perennial to 80m. Native; meadows and pastures, mostly on heavy soils; common in C, S & E En, very scattered W & N to Scillies and Durham, rare in CI and Ir *Meadow Barley* - **H. secalinum** Schreb.

    6 Glumes of central spikelet with conspicuous marginal hairs >0.5mm; leaves usually with well-developed pointed auricles. Annual to 60cm (*Wall Barley* - **H. murinum** L.) 7
    6 Glumes of central spikelets with only pricklets <0.1mm; leaves usually without or with small rounded auricles 9

7 Lemma-body and palea of central floret longer than those of lateral florets; central floret with stalk (above glumes) <0.6mm. Native; weed of waste and rough ground and barish patches in rough grassland; common in C, S & E En and CI, scattered N & W to NE & SW Sc and N Wa and in Ir **H. murinum ssp. murinum**

7 Lemma-body and palea of central floret shorter than those of lateral florets; central floret with stalk (above glumes) 0.6-1.5mm 8
    8 Anthers of central florets usually blackish, <0.6mm, ≤1/3 as long as those of lateral florets; leaves usually glaucous. Intrd-casual; habitat and distribution as for ssp. *leporinum*, but rarer

H. murinum ssp. glaucum (Steud.) Tzvelev
    8 Anthers of central florets usually yellowish, >0.6mm long, 1/2-1x as long as those of lateral florets; leaves not glaucous. Intrd-casual; waste ground; scattered in En and Sc

H. murinum ssp. leporinum (Link) Arcang.

9 Lateral florets distinctly stalked, the stalk (above glumes) c.1mm and c. as long as stalk (below glumes) of lateral spikelets; longest awns of triplet usually <1cm 10

9 Lateral florets sessile or nearly so, the stalk (above glumes) 0-0.5mm and much shorter than 1-1.5mm stalk (below glumes) of lateral spikelets; longest awns of triplet usually >1cm 11

10 Lateral lemmas obtuse to acute, 1.7-3.3mm. Annual to 45cm. Intrd-casual; habitat and distribution as for *H. pusillum*
***Argentine Barley* - H. euclaston** Steud.

10 Lateral lemmas strongly acuminate or with awn ≤2mm, 2.8-6mm incl. awn. Annual to 45cm. Intrd-casual; tips and waste ground and in fields; scattered in En ***Little Barley* - H. pusillum** Nutt.

11 Glumes of lateral spikelets slightly heteromorphic, the inner with flattened basal part 0.3-0.7mm wide (c.2x as wide as basal part of outer glume); lower sheaths pubescent with hairs ≥0.5mm. Annual to 40cm. Intrd-casual; tips, waste ground and in fields; scattered in Br ***Mediterranean Barley* - H. geniculatum** All.

11 Glumes of lateral spikelets strongly heteromorphic, the inner with ± winged basal part 0.7-1.2mm wide (c.3-4x as wide as basal part of outer glume); lower sheaths glabrous to pubescent with hairs ≤0.25mm. Annual to 40cm. Native; barish or sparsely grassed often salty ground near sea, by saltmarshes, on banks and walls and in rough or waste ground; locally common in S Br N to S Lincs and Flints ***Sea Barley* - H. marinum** Huds.

66. SECALE L. - *Rye*

Annuals; spikelets 1 per node, each with 2(-3) bisexual florets; glumes very narrow, acute, 1-veined, awnless or shortly awned; lemmas 5-veined, keeled, acuminate, usually very long-awned; spikelets disarticulating at maturity below each caryopsis, leaving glumes, lemma and palea on rhachis.

1 Culms to 1.5m; spikes 5-15cm, usually pendent at maturity; lemmas with awn c.2-5cm. Intrd-casual; relic and from grain on tips and in waste places; frequent throughout most of BI ***Rye* - S. cereale** L.

67. X TRITICOSECALE A. Camus (*SECALE* x *TRITICUM*) - *Triticale*

A new grain-crop, derived by artificial hybridization of *Triticum* and *Secale*, now being grown on a field scale and increasingly found as a relic. It has a pendent spike and long awns and varies greatly in height, sometimes up to 1.8m. It differs from *Triticum* in its obtuse (not truncate) glumes and from *Secale* in its broader glumes and distal sterile florets. There is no valid specific epithet.

68. TRITICUM L. - *Wheats*

Annuals; spikelets 1 per node, each with 3-7(9) florets, the apical ≥2 sterile and reduced; glumes keeled, truncate to bifid, apiculate to shortly awned; lemmas 5-veined, keeled, truncate to bifid, apiculate to very long-awned; spikelets disarticulating at maturity below each caryopsis, leaving glumes, lemma and palea on rhachis.

1 Rhachis glabrous; glumes strongly keeled in upper 1/2, scarcely so in lower 1/2; lemmas awnless or with awn up to 16cm. Culms to 1.5m. Intrd-natd; relic in fields and waste ground and on roadsides and

tips; common throughout BI *Bread Wheat* - **T. aestivum** L.
1 Rhachis with hair-tufts at each node; glumes strongly keeled throughout; lemmas with awn 8-16cm. Culms to 1.5m. Intrd-casual; occasional casual as relic or grain-alien; scattered in Br and Ir
Rivet Wheat - **T. turgidum** L.

SUBFAMILY 3 - ARUNDINOIDEAE (tribe 15, genera 69-74). Annual or perennial herbs; leaves without false-petiole or cross-veins; ligule a fringe of hairs; sheaths not fused; inflorescence a panicle; spikelets with (1)2-many florets, with all florets (except often the distal ones) bisexual or the lowest male or sterile; glumes 2; lemmas firm, 1-9-veined, sometimes bifid at apex, awnless or with terminal awn; stamens 3; stigmas 2; lodicules 2.

TRIBE 15 - ARUNDINEAE (genera 69-74).

69. DANTHONIA DC. - *Heath-grass*
Densely tufted perennials; inflorescence a small panicle with rarely >12 spikelets; spikelets with 4-6 florets, all or all except most apical bisexual; glumes ovate, c. or nearly as long as rest of spikelet, 3-7-veined; lemmas 7-9-veined, minutely 3-toothed at apex, awnless, with tuft of short hairs at base and fringe up each side to c.1/2 way.

1 Culms decumbent to erect, to 40(60)cm; panicle 2-7cm; spikelets 6-12mm. Native; sandy or peaty often damp soil, usually acid but also mountain limestones, mostly on heaths, moors and mountains; throughout BI *Heath-grass* - **D. decumbens** (L.) DC.

70. RYTIDOSPERMA Steud. - *Wallaby-grass*
Densely tufted perennials; inflorescence a rather compact to elongated panicle; spikelets with 6-10 florets, lower ones bisexual, upper ones male or sterile and reduced; glumes lanceolate to narrowly ovate, with wide hyaline margins, slightly longer to slightly shorter than rest of spikelet (excl. awns), 5-7-veined; lemmas 7-veined, with 2 long acuminate lobes at apex tipped with straight awns, with long, bent terminal awn from sinus, with dense white silky hairs at base and middle reaching or nearly reaching apex of body of lemma.

1 Culms erect, to 60cm; panicles 3-5cm; spikelets 7-16mm excl. awns; lemmas with bent terminal awn 5-15mm, with lateral awns 2-8mm. Intrd-casual; tips, fields and waste places; scattered in En
Wallaby-grass - **R. racemosum** (R. Br.) Connor & Edgar

71. SCHISMUS P. Beauv. - *Kelch-grass*
Tufted annuals; inflorescence a rather compact panicle; spikelets with 4-10 florets, lower ones bisexual, upper ones male or sterile and reduced; glumes lanceolate, with wide hyaline margins, slightly longer to slightly shorter than rest of spikelet, 5-7-veined; lemmas 9-veined, deeply and acutely 2-lobed at apex, awnless or ± so, with short hairs at base and long silky hairs

on back not reaching apex.

1 Culms erect, to 25cm; panicle 1-4cm; spikelets 5-6mm; lemmas c.2mm. Intrd-casual; tips, fields and waste land; scattered in En
Kelch-grass - **S. barbatus** (L.) Thell.

72. CORTADERIA Stapf - *Pampas-grasses*
Densely tufted perennials with fiercely serrated leaves; inflorescence a very large spreading panicle; dioecious or gynodioecious; spikelets with 2-7 florets; glumes lanceolate, hyaline, slightly unequal, 1-veined, at least upper ± as long to longer than rest of spikelet and with long terminal awn; lemmas 3-5-veined, long-awned at apex, with tuft of long fine hairs at base reaching ± to apex of lemma body.

1 Leaves 0.9-2.7m; panicles 40-120cm, silvery-white or red-tinged; lemmas bifid, with bristle-like points, pubescent all over; flowers Aug-Nov. Culms to 3m. Intrd-natd; rough ground, waysides, old gardens, maritime cliffs and dunes; scattered in BI, mostly S En, S Wa and CI *Pampas-grass* - **C. selloana** (Schult. & Schult. f.) Asch. & Graebn.
1 Leaves 0.6-1.2m; panicles 30-60cm, silvery- to yellowish-white; lemmas acuminate, glabrous distally; flowers Jul-Aug. Culms to 3m. Intrd-natd; similar places to *C. selloana*; several sites in S En and Wa
Early Pampas-grass - **C. richardii** (Endl.) Zotov

73. MOLINIA Schrank - *Purple Moor-grass*
Densely tufted perennials; inflorescence a ± diffuse to ± contracted panicle; spikelets with 1-4 florets, all except most apical bisexual; glumes ovate, slightly unequal, 1-3-veined, much shorter than rest of spikelet; lemmas 3-5-veined, acute to obtuse, awnless, glabrous.

1 Plant often forming tussocks. Native
(*Purple Moor-grass* - **M. caerulea** (L.) Moench) 2
  2 Culms usually <65cm; panicle usually <30cm, narrow, with branches mostly <5cm; spikelets 3-5.5mm; lemmas 3-4mm. Heaths, moors, bogs, fens, mountain grassland and cliffs and lake-shores, always on at least seasonally wet ground; common in suitable places throughout BI **M. caerulea ssp. caerulea**
  2 Culms mostly 65-125(160)cm; panicles mostly 30-60cm, with very uneven-lengthed branches often >10cm, usually spreading at least during flowering; spikelets (3)4-7.5mm; lemmas (3.2)3.5-5.4(5.7)mm. Fens, fen-scrub, fen-type vegetation by rivers and canals; scattered in suitable places in C & S Br, very scattered elsewhere **M. caerulea ssp. arundinacea** (Schrank) K. Richt.

74. PHRAGMITES Adans. - *Common Reed*
Extensively rhizomatous perennials; inflorescence a very large spreading panicle; spikelets with 2-6(more) florets, the lowest male or sterile, the rest bisexual; glumes unequal, narrowly elliptic-ovate, 3-5-veined, much shorter

than rest of spikelet; lemmas lanceolate, acute to acuminate at apex, with 1-3 veins, awnless; rhachilla-segments with long, white, silky hairs becoming very conspicuous in fruit.

1 Culms to 3.5m but sometimes <1m; panicles 20-60cm, usually purple; spikelets 8-16mm, with rhachilla-hairs up to 10mm. Native; on mud or in shallow water by lakes, rivers, canals, marshes, fens, bog-margins and edges of salt-marshes and estuaries; common throughout BI ***Common Reed* - P. australis** (Cav.) Steud.

SUBFAMILY 4 - CHLORIDOIDEAE (tribes 16-17, genera 75-83). Annual to perennial herbs; leaves without false-petiole or cross-veins; ligules usually a fringe of hairs, sometimes membranous or a membrane fringed with hairs; sheaths not fused; inflorescence usually an umbel or a raceme of spikes or racemes, sometimes a panicle; spikelets with 1-many florets with all florets (except often the distal ones) bisexual; glumes 2; lemmas membranous to firm, 1-3(9)-veined, awnless or less often awned; stamens 3; stigmas 2; lodicules 0 or 2.

TRIBE 16 - ERAGROSTIDEAE (genera 75-79). Annuals or perennials with or without rhizomes; ligules membranous or a fringe of hairs; inflorescence a diffuse to contracted panicle, an umbel of spikes, or a raceme of racemes; spikelets with 3-many florets (or with 1 floret in *Sporobolus*), all (except 1-2 most apical) bisexual; glumes 2, much shorter than rest of spikelet; lemmas (1-)3-veined (with extra veins close to midrib in *Eleusine*), usually keeled, awnless (very shortly awned in *Leptochloa*).

75. **LEPTOCHLOA** P. Beauv. - *Beetle-grasses*
Rhizomatous perennials (but annual with us); ligule membranous; inflorescence a loose, long raceme of racemes, the spikelets well spaced out, the rhachis ending in a spikelet; spikelets with 6-10(14) florets; glumes unequal, 1-veined; lemmas 3-veined, slightly keeled with prominent midrib and submarginal laterals, with long silky hairs at base and on margins and base of dorsal midline, with 1 tooth on either side at apex, with very short apical awn; spikelets disarticulating between florets; pericarp adherent to seed.

1 Culms to 1m; inflorescences up to 40cm, with long, straight erecto-patent unbranched racemes from main axis; spikelets 8-15mm. Intrd-casual; tips, fields and waste places; scattered in En
***Brown Beetle-grass* - L. fusca** (L.) Kunth

76. **ERAGROSTIS** Wolf - *Love-grasses*
Annuals or tufted perennials; ligule a fringe of hairs; inflorescence a usually diffuse panicle; spikelets with 3-many florets, often very narrow and parallel-sided; glumes subequal to unequal, (0)1(-3)-veined; lemmas 3-veined, keeled, acute to obtuse, rounded or emarginate; awnless (sometimes apiculate); spikelets disarticulating between florets or the caryopsis falling

free leaving persistent rhachilla; pericarp adherent to seed.

Superficially often similar to *Poa*, but the 3-veined lemmas and ligule a ring of hairs distinguish it.

1 Anthers (0.8)1-1.3mm; plant potentially perennial. Culms tufted, erect, to 1.2m. Intrd-natd; tips and waste ground; casual scattered in En and Wa, ± natd in Southampton (S Hants)

African Love-grass - **E. curvula** (Schrad.) Nees
1 Anthers 0.2-0.6mm; plant annual (at least with us) 2
  2 Plant with minute, sessile, wart-like glands (x≥10 lens) on leaf-margins, sheath-midrib, lemma- and glume-veins, panicle-branches and/or pedicels (if 0 on leaves then always ≥1 on pedicels) 3
  2 Plant without minute sessile glands 4
3 Leaves often >5mm wide; spikelets ≥2mm wide; lemmas (1.7)2-2.8mm; sessile glands prominent on lemma veins, usually not so on pedicels. Culms erect or ascending, to 75cm. Intrd-casual; tips and waste ground; scattered in En and Wa *Stink-grass* - **E. cilianensis** (All.) Janch.
3 Leaves <5mm wide; spikelets ≤2mm wide; lemmas 1.5-2mm; 1-2 sessile glands prominent near pedicel-apex, usually none on lemmas. Culms erect or ascending, to 50cm. Intrd-casual; tips and waste ground; scattered in C & S Br *Small Love-grass* - **E. minor** Host
  4 Caryopsis 1-1.3mm; upper glume 1.7-3mm; lemmas 2-2.7mm. Culms erect, to 1m. Intrd-casual; tips and waste ground; scattered in En *Teff* - **E. tef** (Zucc.) Trotter
  4 Caryopsis 0.5-0.8mm; upper glume 0.7-1.4mm; lemmas 1.2-1.8mm 5
5 Panicle-branches erecto-patent at maturity, with spikelets borne nearly to base; lemmas greenish-grey, purple-tinged at apex. Culms erect or ascending, to 70cm. Intrd-natd; wool-, birdseed- and grain-alien, natd in Jersey *Jersey Love-grass* - **E. pilosa** (L.) P. Beauv.
5 Panicle-branches patent at maturity, bare of spikelets for lowest c.1/4-1/3; lemmas dark grey, not purple-tinged. Culms to 60cm. Intrd-casual; tips and waste ground; scattered in En

Weeping Love-grass - **E. parviflora** (R.Br.) Trin.

77. ELEUSINE Gaertn. - *Yard-grasses*

Annuals or tufted perennials; ligule membranous with a sparse or dense fringe of hairs; inflorescence an umbel or very short raceme of spikes with very crowded spikelets, the rhachis ending in a spikelet; spikelets with 3-6 florets; glumes subequal to unequal, 1-7-veined; lemmas with 3 main veins and 2-4 extra veins close to midrib, keeled, acute to apiculate, not awned; spikelets disarticulating between florets; pericarp not adherent to seed, which eventually falls out separately.

1 Lemmas and glumes obtuse, with hooded tip; perennial. Culms to 40cm. Intrd-casual; tips and waste ground; scattered in En

American Yard-grass - **E. tristachya** (Lam.) Lam.
1 Lemmas and glumes pointed (either acute, or subacute to obtuse and

apiculate), without hooded tip; annual   2
- 2   Spikes 1-3cm, in very short terminal raceme; lemmas c.1.5mm wide from keel to edge. Culms to 45cm. Intrd-casual; tips and waste ground; scattered in En

    Fat-spiked Yard-grass - **E. multiflora** A. Rich.
- 2   Spikes (3.5)5-15cm, all or most in terminal umbel; lemmas <1mm wide from keel to edge. Culms to 90cm. Intrd-casual; birdseed-, grain-, wool-, cotton- or pulse-alien, on tips and waste ground; scattered in En (*Yard-grass* - **E. indica** (L.) Gaertn.)   3
- 3   Ligule sparsely and minutely pubescent at apex; lower glume 1-veined, 1.1-2.3mm; upper glume 1.8-3mm; lemmas 2.4-4mm; seeds 1-1.3mm, with very fine close striations between and at right-angles to main ridges (x≥20 lens)   **E. indica ssp. indica**
- 3   Ligule with strong pubescent fringe at apex; lower glume (1)2-3-veined, 2-3.5mm; upper glume 3-4.7mm; lemmas 3.7-5mm; seeds 1.2-1.6mm, with granulations between main ridges (x≥20 lens)

    E. indica ssp. africana (Kenn.-O'Byrne) S.M. Phillips

78. DACTYLOCTENIUM Willd. - *Button-grass*

Annuals but sometimes rooting at lower nodes; ligule membranous, sometimes slightly fringed at apex; inflorescence an umbel of spikes with very crowded spikelets, the rhachis ending in a short projection; spikelets with 3-5 florets; glumes unequal, 1-veined, the upper with a long awn-like point; lemmas 3-veined, keeled, acuminate to acute-apiculate, not awned; spikelets disarticulating above glumes (not between florets); pericarp not adherent to seed, which eventually falls out separately.

1   Stems erect to decumbent, to 40cm; inflorescence of 4-10 crowded umbellate spikes each (0.5)1-2cm; spikelets 3-5mm. Intrd-casual; tips and waste ground; scattered in En

Button-grass - **D. radulans** (R. Br.) P. Beauv.

79. SPOROBOLUS R. Br. - *Dropseeds*

Tufted perennials; ligule a fringe of hairs; inflorescence a narrow panicle with short to long, closely appressed, erect branches (the ultimate branches with closely borne small spikelets resembling a spikelet with many florets at a casual glance); spikelets with 1 floret; glumes unequal, 0-1-veined, much shorter than rest of spikelet; lemma 1-3-veined, rounded on back, acute to acuminate, awnless, inrolled and ± cylindrical but tapering at apex; spikelet disarticulating below lemma; pericarp not adherent to seed, which eventually falls out separately.

1   Culms erect to ascending, to 1m; panicle very narrow (<1cm), up to 35cm, with short (≤2cm) erect branches; spikelets (lemma) 2.1-2.5mm. Intrd-casual; tips and rough ground; scattered in En

African Dropseed - **S. africanus** (Poir.) A. Robyns & Tournay

TRIBE 17 - **CYNODONTEAE** (genera 80-83). Annuals or perennials, with or without rhizomes or stolons; ligules membranous or a fringe of hairs; inflorescence an umbel or raceme of spikes, or a spike-like panicle; spikelets with 1 bisexual floret, sometimes with 1-2(3) extra sterile or male florets distal to it; glumes (1-)2, shorter to longer than rest of spikelet; lemmas 1-3(9)-veined, usually keeled, awned or not.

80. CHLORIS Sw. - *Rhodes-grasses*

Annuals or perennials, sometimes with stolons; ligule membranous with a well-marked fringe of hairs; inflorescence an umbel of (4)6-many slender long spikes; spikelets with 2-3(4) florets, the lowest bisexual, the others reduced and male or sterile; glumes unequal, 1-veined, narrowly acute, shorter to slightly longer than rest of spikelet; lowest (bisexual) lemma 3-veined, keeled, minutely to deeply bifid at apex with long terminal or subterminal straight awn; upper lemma(s) variously reduced, but 2nd of similar shape and only that long-awned (hence spikelets 2-awned); spikelets disarticulating above glumes.

1 Lemma with low rounded to transversely or obliquely truncate lobes either side of awn, forming very shallow notch. Perennial, often stoloniferous; culms to 45cm. Intrd-casual; tips and waste ground and in fields; scattered in En   ***Windmill-grass* - C. truncata** R.Br.
1 Lemma with sharply acute lobes or teeth either side of awn, forming deep notch   2
   2 Fertile lemma with dense tuft of silky hairs at apex, producing feathery spikes. Annual; culms to 60(100)cm. Intrd-casual; habitat and distribution as for *C. truncata*
***Feathery Rhodes-grass* - C. virgata** Sw.
   2 Fertile lemma without apical tuft of hairs, producing ± glabrous spikes. Tufted perennial (?stoloniferous); culms to 60cm. Intrd-casual; habitat and distribution as for *C. truncata*
***Australian Rhodes-grass* - C. divaricata** R. Br.

81. CYNODON Rich. - *Bermuda-grasses*

Perennials with rhizomes and/or stolons; ligule membranous or a fringe of hairs; inflorescence an umbel of 3-6 slender long spikes; spikelets with 1 (bisexual) floret; glumes subequal, shorter than rest of spikelet, 1-veined, narrowly acute; lemma 3-veined, keeled, subacute, unawned; spikelets disarticulating above glumes.

1 Ligule a fringe of short hairs <0.5mm with longer tuft at each edge; rhachilla of spikelet continued beyond base of floret as fine projection between upper glume and floret >1/2 as long as floret. Culms to 30cm. Probably intrd-natd; rough sandy ground, waysides and short grassland near sea; local in SW En, S Wa and CI, casual elsewhere in En and Wa   ***Bermuda-grass* - C. dactylon** (L.) Pers.
1 Ligule membranous, 0.4-1mm, with sparse hairs at apex and longer tuft at base of each edge; rhachilla of spikelet not extended beyond base of floret. Culms to 30cm. Intrd-casual; tips and rough ground;

81. CYNODON

scattered in En *African Bermuda-grass* - **C. incompletus** Nees

82. SPARTINA Schreb. - *Cord-grasses*
Strongly rhizomatous perennials; ligule a dense fringe of hairs; inflorescence a raceme of (1)2-12(30) long, ± erect spikes; spikelets with 1 (bisexual) floret; glumes unequal, the upper as long as or longer than rest of spikelet, 1-9-veined, narrowly acute, awned or not; lemma 1-3-veined, keeled, acute or minutely notched, unawned; spikelets falling (tardily) entire at maturity.

1 Upper glume very scabrid on keel with rigid pricklets ≥0.3mm, with awn 3-8mm; spikes with 2 rows of spikelets each crowded 4-10 per cm. Culms to 1.8m. Intrd-natd; by fresh-water lakes; W Galway, S Northumb and N Hants *Prairie Cord-grass* - **S. pectinata** Link
1 Upper glume glabrous or with soft hairs <0.3mm on keel, awnless; spikes with 2 rows of spikelets each spaced out 1-3 per cm; coastal   2
   2 Glumes glabrous or with hairs on keel only, sometimes very sparse on body also. Culms to 1.2m. Intrd-natd; tidal mud-flats; S Hants, E Ross, S Essex and Dorset
Smooth Cord-grass - **S. alterniflora** Loisel.
   2 Glumes softly pubescent on keel and body   3
3 Ligules 1.8-3mm at longest point (beware damaged ones); anthers (5)7-10(13)mm. Culms to 1.3m. Native; tidal mud-flats; common on coasts of Bl N to C Sc *Common Cord-grass* - **S. anglica** C.E. Hubb.
3 Ligules 0.2-1.8mm at longest point; anthers 4-8(10)mm, if >7mm then indehiscent   4
   4 Ligules 0.2-0.6mm; anthers 4-6.5mm, dehiscent. Culms to 50(80)cm. Native; tidal sandy or muddy bare places by sea or in estuaries; local in S & E En from Wight to N Lincs, intrd in Co Dublin *Small Cord-grass* - **S. maritima** (Curtis) Fernald
   4 Ligules 1-1.8mm; anthers 5-7(10)mm, indehiscent. Culms to 1.3m. Native; tidal mud-flats; scattered in Br and E Ir (*S. maritima* x *S. alterniflora*) *Townsend's Cord-grass* - **S. x townsendii** H. & J. Groves

83. TRAGUS Haller - *Bur-grasses*
Annuals; ligule a dense fringe of hairs; inflorescence a spike or spike-like, with 2-5 spikelets on extremely short branch at each node; spikelets with 1 (bisexual) floret; glumes very unequal, the lower 0 or vestigial, the upper at least as long as floret, 5-7-veined, each vein with line of strong hooked spines, acute, unawned; lemma 3-veined, not keeled, acute; each nodal group of spikelets falling as a bur at maturity, the spikelets facing the centre of the bur with the glume-hooks outermost.

Distinctive vegetatively in the strong curved spines on proximal part of leaf-margins.

1 Upper glume 2-3mm. Culms to 60cm. Intrd-casual; tips and rough ground and in fields; scattered in En
African Bur-grass - **T. berteronianus** Schult.

1 Upper glume 3.5-4.5mm 2
  2 Spikelets 2 at all or almost all nodes; upper glume with 5 veins
    (and rows of spines). Culms to 40cm. Intrd-casual; tips and
    rough ground and in fields; scattered in En
                    Australian Bur-grass - **T. australianus** S.T. Blake
  2 Spikelets 3-5 at all or almost all nodes; upper glume with 7 veins
    (and rows of spines) (one pair of veins sometimes thinner and
    with smaller spines). Culms to 40cm. Intrd-casual; tips and
    rough ground and in fields; scattered in En
                    European Bur-grass - **T. racemosus** (L.) All.

SUBFAMILY 5 - PANICOIDEAE (tribes 18-19, genera 84-94). Annual to perennial herbs; leaves without false-petiole or cross-veins; ligule 0 or a fringe of hairs or a membrane fringed with hairs, rarely membranous; sheaths not fused; inflorescence a panicle, a spike, or an umbel or raceme of racemes or spikes; spikelets with 2 florets, the upper bisexual, the lower male or sterile (*Zea* is monoecious; *Sorghum* has paired spikelets, 1 bisexual the other male or sterile); glumes 2; lemmas firm to thick, 5-11-veined, awnless or with terminal awn; stamens 3; stigmas 2; lodicules 2 (0 in female *Zea*).

TRIBE 18 - PANICEAE (genera 84-92). Annuals or less often perennials with or without rhizomes or stolons; ligule 0, membranous, a fringe of hairs or a membrane fringed with hairs; inflorescence a panicle, a spike, or an umbel or raceme of racemes or spikes; spikelets all the same, all bisexual.

84. PANICUM L. - *Millets*
Annuals; ligule a dense fringe of hairs or membranous with distal fringe of hairs; inflorescence a diffuse panicle; spikelets with 2 florets, the lower male or sterile with lemma ± as long as spikelet, the upper bisexual, smaller, concealed between upper glume and lower lemma; glumes unequal, the lower much shorter than, the upper ± as long as the spikelet, the upper closely resembling the lower lemma; lower lemma 5-11-veined, awnless; upper lemma awnless; spikelets falling whole at maturity.

1 Sheaths with long patent hairs; lower glume >1/3 as long as spikelet 2
1 Sheaths glabrous; lower glume ≤1/3 as long as spikelet 3
  2 Spikelets (4)4.5-5.5(6.5)mm. Culms to 1m. Intrd-casual; tips and
    waste ground; scattered in BI *Common Millet* - **P. miliaceum** L.
  2 Spikelets 2-3.5mm. Culms to 1m. Intrd-casual; tips and waste
    ground; scattered in Br and CI *Witch-grass* - **P. capillare** L.
3 Spikelets 2-2.8mm, subacute to obtuse at apex; lower floret usually
  male, with well developed palea >1/2 as long as lemma. Culms to
  1m. Intrd-casual; tips and waste ground; scattered in En
                    Transvaal Millet -**P. schinzii** Hack.
3 Spikelets 2.7-3.5mm, acute to acuminate at apex; lower floret sterile,
  with 0 or much reduced palea. Culms to 1m. Intrd-casual; tips and

waste ground; scattered in S En

Autumn Millet - **P. dichotomiflorum** Michx.

85. ECHINOCHLOA P. Beauv. - *Cockspurs*

Annuals; ligule 0; inflorescence a raceme of ± dense spikes or racemes, or the secondary racemes again racemosely branched, often with long stiff hairs especially in tufts at branch-points, the spikelets usually in >2 rows; spikelets with 2 florets, the lower male or sterile with lemma ± as long as spikelet, the upper bisexual, smaller, concealed between upper glume and lower lemma; glumes unequal, the lower much shorter than, the upper ± as long as the spikelet, the upper closely resembling the lower lemma; lower lemma 5-7-veined, awned or awnless; upper lemma awnless; spikelets falling whole at maturity.

1 Inflorescence with lateral spikes or racemes all or mostly clearly separate, obviously branched; lower floret male or sterile, its lemma awned or not   2
1 Inflorescence with fat lateral spikes or racemes close together and forming entirely or for most part a single, lobed, elongate head; lower floret sterile, its lemma not awned   3
    2 Primary branches of inflorescence simple, ≤3cm; spikelets 1.5-3mm; lower floret usually male; lower lemma acute to apiculate, with awn 0-2mm; leaf-blades ≤8mm wide. Culms to 1.2m. Intrd-casual; tips and waste ground; occasional in S Br

Shama Millet - **E. colona** (L.) Link
    2 Lower primary branches of inflorescence usually branched again; spikelets 3-4mm; lower floret usually sterile; lower lemma often awned (awn ≤5cm); leaf-blades mostly >1cm wide. Culms to 1.2m. Intrd-natd; casual on tips, waysides and waste ground, weed of cultivated ground sometimes natd; scattered throughout most of BI, especially S   *Cockspur* - **E. crus-galli** (L.) P. Beauv.
3 Glumes and lower lemma yellowish-green to straw-coloured; spikelets 2.5-3.5mm; lower lemma acute to subacute, sometimes minutely apiculate. Culms to 1.2m. Intrd-casual; tips and waste ground; scattered in Br, mainly S   *White Millet* - **E. frumentacea** Link
3 Glumes and lower lemma bright green usually strongly tinged purplish, sometimes completely purplish; spikelets 3-4mm; lower lemma acuminate. Culms to 1.2m. Intrd-casual; tips and waste ground; scattered in Br, mainly S

Japanese Millet - **E. esculenta** (A. Braun) H. Scholz

86. BRACHIARIA (Trin.) Griseb. - *Signal-grasses*

Annuals; ligule a dense fringe of hairs; inflorescence a raceme of racemes with spikelets in 2 rows on 1 side of rhachis; spikelets with 2 florets, the lower male or sterile with lemma ± as long as spikelet, the upper bisexual, smaller, concealed between upper glume and lower lemma; glumes unequal, the lower <1/2 as long as upper, the upper closely resembling the lower lemma; lower lemma 5-7-veined, obtuse, awnless; upper lemma obtuse to rounded, awnless; spikelets falling whole at maturity.

1 Culms decumbent to erect, to 50cm; racemes 2-6, 3-8cm; spikelets
3.5-4.5mm, glabrous. Intrd-casual; tips and waste ground; scattered
in S En   **Broad-leaved Signal-grass - B. platyphylla** (Griseb.) Nash

87. UROCHLOA P. Beauv. - *Signal-grasses*

Differs from *Brachiaria* in upper glume and lower lemma shortly acuminate, 7-veined; and upper lemma with distinct terminal awn.

1 Culms decumbent to erect, to 50cm; racemes 2-6, 3-8cm; spikelets
3.5-5mm, glabrous to pubescent. Intrd-casual; tips and waste
ground; scattered in C & S En
          Sharp-flowered Signal-grass - **U. panicoides** P. Beauv.

88. ERIOCHLOA Kunth - *Cup-grasses*

Annuals or perennials; ligule a dense fringe of hairs; inflorescence a rather irregular raceme of racemes, the main branches ± appressed to main axis and often slightly branched again, with spikelets scarcely in recognisable rows; spikelets with 2 florets, the lower sterile with lemma almost as long as spikelet and palea 0, the upper bisexual, smaller, concealed between upper glume and lower lemma and with palea, with small bead-like swelling at apex of pedicel; lower glume ± 0; upper glume as long as spikelet (slightly longer than lower lemma but otherwise very similar); lower lemma 5-veined, acuminate to awned to 2mm; upper lemma obtuse to rounded, with awn 0.3-1mm (as in *Urochloa*); spikelets falling whole, disarticulating immediately below bead-like swelling.

The ± absence of a lower glume, and the lower floret being represented by only a glume-like lemma, produces an apparently 2-glumed spikelet with 1 floret.

1 Perennials; culms to 60(100)cm; inflorescence up to 15 x 1cm; spikelets
3.6-6mm incl. awn ≤2mm, with dense white silky hairs. Intrd-casual;
tips and in fields and waste places; scattered in En
          Perennial Cup-grass - **E. pseudoacrotricha** (Thell.) S.T. Blake

89. PASPALUM L. - *Finger-grasses*

Perennials with stolons; ligule membranous; inflorescence a raceme of 2(-4) racemes with spikelets in 2 rows on 1 side of rhachis; spikelets with 2 florets, the lower sterile with lemma as long as spikelet and palea very small, the upper bisexual, smaller, concealed between upper glume and lower lemma and with palea; lower glume ± 0; upper glume very similar to lower lemma; lower lemma 3-5-veined, acute or slightly apiculate; upper lemma subacute, apiculate; spikelets falling whole.

The ± absence of a lower glume, and the lower floret being represented by little more than a glume-like lemma, produces an apparently 2-glumed spikelet with 1 floret.

1   Decumbent stoloniferous perennial; culms to 50cm but usually <20cm high, subglabrous; racemes 2-7cm; spikelets 2.5-3.5mm. Intrd-natd; damp ground by sea at Mousehole (W Cornwall), by canal in E London (Middlesex)   *Water Finger-grass* - **P. distichum** L.

90. SETARIA P. Beauv. - *Bristle-grasses*

Annuals, or perennials with rhizomes; ligule a dense fringe of hairs; inflorescence a dense spike-like panicle with very short or ± vestigial crowded branches, sometimes interrupted in lower part; spikelets with 2 florets, the lower male or sterile with lemma as long as spikelet, the upper bisexual, slightly smaller, concealed between or protruding from within upper glume and lower lemma, with 1-c.12 strong bristles borne on pedicel and usually awn-like and exceeding spikelets; glumes unequal, the lower ± 0 to c.2/3 as long as spikelet, the upper c.2/3 to as long as spikelet; lower lemma 5-veined, obtuse, awnless; upper lemma obtuse to rounded, awnless; spikelets falling whole at maturity, or (in *S. italica*), the upper floret falling leaving the lower floret and glumes on panicle, the bristles always remaining on panicle.

The number of bristles (1-3, or more) borne on the pedicel below each spikelet is a valuable character but easily over-estimated where spikelets have aborted; in such cases the bristles of aborted spikelets can be wrongly counted in with those of an adjacent well-developed spikelet, but close inspection shows the presence of spikelet-less pedicel(s).

1   Bristles (4)6-8(12) below each spikelet; upper glume scarcely longer than lower glume, c.1/2-2/3 as long as spikelet; lower (sterile) floret with palea almost as long as lemma   2
1   Bristles 1-3 below each spikelet (see note above); upper glume much longer than lower glume, c.2/3-1x as long as spikelet; lower (sterile) floret with palea ≤1/2 as long as lemma   3
    2   Spikes ≥6mm wide when mature (excl. bristles); spikelets (2.5)3-3.3mm. Annual; culms to 75cm. Intrd-casual; weed of cultivated and waste ground and on tips; occasional in S & C Br and CI, very scattered in N Br and E Ir
                                Yellow Bristle-grass - **S. pumila** (Poir.) Schult.
    2   Spikes <5mm wide when mature (excl. bristles); spikelets 2-2.5(3)mm. Shortly rhizomatous perennial; culms to 75cm. Intrd-casual; tips and waste ground; scattered in S En and S Wa
                              Knotroot Bristle-grass - **S. parviflora** (Poir.) Kerguélen
3   Main rhachis (often rather sparsely) hispid, with pricklets <0.2mm; bristles usually with backward-directed (rarely with forward-directed) barbs   4
3   Main rhachis densely pubescent with hairs >(0.2)0.5mm; bristles always with forward-directed barbs   5
    4   Spikelets 1.7-2mm; sheaths glabrous. Annual; culms to 60cm. Intrd-casual; tips and waste and cultivated ground; scattered in S En and CI   *Adherent Bristle-grass* - **S. adhaerens** (Forssk.) Chiov.
    4   Spikelets 2-2.3mm; sheaths pubescent on margin. Annual; culms

to 60cm. Intrd-casual; weed of cultivated and waste ground and on tips; occasional in S & C Br and CI, very scattered in N Br and E Ir *Rough Bristle-grass* - **S. verticillata** (L.) P. Beauv.
5 Spikelets disarticulating below upper lemma, leaving glumes and lower lemma on rhachis; panicle often >15 x 1.5cm, the bristles often not or scarcely longer than spikelet-clusters; upper lemma smooth. Annual; culms to 1.5m. Intrd-casual; tips and waste ground; scattered in Br and CI, mostly S *Foxtail Bristle-grass* - **S. italica** (L.) P. Beauv.
5 Spikelets falling whole, leaving only pedicels and bristles on rhachis; panicle rarely as much as 15 x 1.5cm, the bristles always much longer than spikelets; upper lemma finely transversely rugose 6
 6 Upper glume c.3/4 as long as spikelet; spikelets (2.5)2.7-3mm; leaves pubescent, often sparsely so. Annual; culms to 1m. Intrd-casual; characteristic alien from soyabean waste, also from grain; scattered in S En *Nodding Bristle-grass* - **S. faberi** Herrm.
 6 Upper glume as long or almost as long as spikelet; spikelets (1.8)2-2.5(2.7)mm; leaves glabrous. Annual; culms to 1m. Intrd-natd; weed of cultivated and waste ground, also casual on tips, etc.; frequent in S & C Br and CI, sporadic elsewhere
Green Bristle-grass - **S. viridis** (L.) P. Beauv.

91. DIGITARIA Haller - *Finger-grasses*

Annuals sometimes rooting at lower nodes or rarely perennials; ligule membranous; inflorescence an umbel of 2-many long narrow racemes or with some racemes borne just below terminal cluster; spikelets with 2 florets, the lower sterile with lemma as long as spikelet, the upper bisexual, as long or slightly shorter, concealed between or protruding from within upper glume and lower lemma; glumes very unequal, the lower very short to ± 0, the upper 1/3-1x as long as spikelet; lower lemma 5-7-veined, obtuse, awnless; upper lemma acute to subacute, awnless; spikelet falling whole.

1 Upper glume and lower lemma ± same length; spikelets 2-2.3(2.5)mm, conspicuously minutely pubescent; sheaths glabrous except at mouth; pedicel-apex slightly cup-shaped. Culms decumbent to erect, to 35cm. Intrd-natd; weed of cultivated ground, waste places and tips, etc., sometimes natd; very scattered in S En
Smooth Finger-grass - **D. ischaemum** (Schweigg.) Muhl.
1 Upper glume ≤3/4 as long as lower lemma; spikelets 2.5-3.5(3.8)mm, sparsely appressed-pubescent; sheaths usually pubescent, sometimes not; pedicels often slightly thicker near apex but simply truncate 2
 2 Lower lemma with smooth veins, often pubescent at edges; upper glume (1/2)2/3-3/4 as long as spikelet, tapering-acute. Culms decumbent to erect, to 50cm. Intrd-casual; similar places to *D. sanguinalis*; scattered in S Br
Tropical Finger-grass - **D. ciliaris** (Retz.) Koeler
 2 Lower lemma with minutely scabrid veins (x20 lens), rarely pubescent at edges; upper glume 1/3-1/2(2/3) as long as spikelet, rather abruptly acute. Culms decumbent to erect, to 50cm. Intrd-

natd; weed of similar habitats to *D. ischaemum*, but commoner on tips and less common in cultivated ground; scattered in S Br and CI **Hairy Finger-grass - D. sanguinalis** (L.) Scop.

92. CENCHRUS L. - *Sandburs*

Annuals to tufted perennials; ligule a fringe of hairs; inflorescence spike-like, with rhachis bearing groups of 1-few spikelets on very short stalk, the group surrounded and enclosed by spiny bur composed of fused spines and bristles; spikelets with 2 florets, the lower sterile, the upper bisexual; glumes very unequal, the lower often ± 0; lower lemma 5-veined, awnless; spikelets falling as group within bur.

1 Culms to 60cm; panicles to 10cm; burs ± globose, 8-15mm across incl. spines. Intrd-casual; tips and waste ground; scattered in S En
Spiny Sandbur - C. echinatus L.

TRIBE 19 - ANDROPOGONEAE (genera 93-94). Annuals, or perennials with rhizomes; ligule membranous, breaking into distal fringe of hairs; inflorescence a compact to diffuse panicle or umbel of racemes; spikelets in pairs, 1 bisexual and the other male or sterile in *Sorghum*, separated into male and female panicles in *Zea* and only the male paired.

93. SORGHUM Moench - *Millets*

Annuals or perennials; inflorescence a large panicle with spikelets in pairs, 1 bisexual and sessile, the other much thinner, sterile or male and stalked; bisexual spikelets with 2 florets, the upper bisexual, the lower reduced to a lemma; glumes 2, both long, the lower becoming hardened and ± enclosing florets; spikelets falling whole or persistent.

1 Rhizomatous perennial to 1.5m; leaves usually <2cm wide; panicle ± diffuse at anthesis; bisexual spikelets usually with bent twisted awn to 16mm. Intrd-casual; tips and waste ground; scattered in S Br and CI **Johnson-grass - S. halepense** (L.) Pers.
1 Annual to 2m; leaves usually >2cm wide; panicle compact at anthesis; bisexual spikelets usually awnless. Intrd-casual; tips and waste ground; scattered in S Br and CI **Great Millet - S. bicolor** (L.) Moench

94. ZEA L. - *Maize*

Annuals; male and female inflorescences separate, the male a large terminal panicle of spike-like racemes, the female axillary (the familiar 'cob') forming a compact elongated mass seated on a spongy axis; male spikelets in pairs, 1 subsessile the other stalked, with 2 florets and equal glumes, awnless; female spikelets with 2 florets, the lower sterile and much reduced, the upper female, with equal glumes, awnless; spikelets persistent.

1 Culms to 3(-more)m; leaves 3-c.12cm wide; male panicle up to 20cm; female inflorescence up to c.20cm. Intrd-casual; tips and waste ground; scattered in S & C Br and CI **Maize - Z. mays** L.

158. SPARGANIACEAE - *Bur-reed family*

Glabrous, aquatic or semi-aquatic, rhizomatous, herbaceous perennials rooted in mud, with unmistakable globose unisexual heads of flowers and fruits.

1. SPARGANIUM L. - *Bur-reeds*

1 Inflorescence branched, the male heads borne at apex of branches as well as of main axis; tepals dark-tipped. Stems erect, to 1.5m. Native; by ponds, lakes, slow rivers and canals, in marshy fields and ditches; common throughout most of Bl (*Branched Bur-reed* - **S. erectum** L.)   2
1 Inflorescence not branched, the male heads all at apex of main axis; tepals ± translucent, not dark-tipped   5
    2 Fruits distinctly shouldered below beak   3
    2 Fruits gradually rounded below beak   4
3 Fruit with flat top (excl. beak) (3)4-6(7)mm across. Mainly C & S Br, Man, rare in Ir   **S. erectum ssp. erectum**
3 Fruit with rounded (domed) top (excl. beak) 2.5-4.5mm across. Over most of Br, Ir and Man   **S. erectum ssp. microcarpum** (Neuman) Domin
    4 Fruit ellipsoid, gradually tapered to beak, 2-4.5mm across. Throughout Br and Ir   **S. erectum ssp. neglectum** (Beeby) K. Richt.
    4 Fruit subglobose, abruptly contracted to beak, 4-7mm across. Ir, S & C Br, CI   **S. erectum ssp. oocarpum** (Celak.) Domin
5 Stem-leaves not inflated but strongly keeled at base; male heads 3-10, clearly separated. Stems erect, sometimes ± floating, to 60cm. Native; similar places to *S. erectum* but rarely not in water; frequent throughout BI   *Unbranched Bur-reed* - **S. emersum** Rehmann
5 Stem-leaves often inflated but not keeled at base; male heads 1-2(3), if >1 then close together and appearing ± as 1 elongated head   6
    6 Bract of lowest female head ≥10cm, ≥2x as long as inflorescence; lowest female peduncle arising from bract axil but fused to main axis for short distance above axil; male heads mostly 2. Stems usually floating, to 1m. Native; in peaty, acid lakes or pools; local in NW Br and Ir   *Floating Bur-reed* - **S. angustifolium** Michx.
    6 Bract of lowest female head <10cm, barely longer than inflorescence; lowest female peduncle arising directly from bract axil; male head usually 1. Stems usually floating, to 50cm. Native; in acid or alkaline lakes, pools or ditches with high organic content; scattered over most of Br and Ir except most of C & S En and S Wa   *Least Bur-reed* - **S. natans** L.

S. emersum x *S. angustifolium* = ***S. x diversifolium*** Graebn. has been found in a few places in W Sc from Wigtowns to W Sutherland; it is fertile.

159. TYPHACEAE - *Bulrush family*

Glabrous, aquatic or semi-aquatic, rhizomatous, herbaceous perennials rooted in mud, with unmistakable cylindrical inflorescence, male distally, female proximally.

1. TYPHA L. - *Bulrushes*

1 Leaves 8-24mm wide; male and female parts of spike usually contiguous (rarely ≤2.5cm apart). Stems erect, to 3m. Native; reed-swamps, lakes, ponds, slow rivers, ditches; frequent except in N & W Sc **Bulrush - T. latifolia** L.
1 Leaves 3-6(10)mm wide; male and female parts of inflorescence separated by (0.5)3-8(12)cm. Stems erect, to 3m. Native; similar places to *T. latifolia* but often on more organic soils; scattered throughout most of BI but absent from C & N Sc, Man and most of Ir **Lesser Bulrush - T. angustifolia** L.

T. latifolia x *T. angustifolia* = **T. x glauca** Godr. occurs in scattered places throughout En and Wa and in Stirlings; it is highly (?completely) sterile.

160. BROMELIACEAE - *Rhodostachys family*

Glabrous, glaucous, dome-shaped, pineapple-like, evergreen, almost woody plants with spiny linear leaves and globose inflorescences.

1 Inflorescences ± sessile in terminal leaf-rosettes; petals blue, with basal scale-like nectaries; leaves flat distally, with marginal spines ± all apically directed **1. FASCICULARIA**
1 Inflorescences arising from terminal leaf-rosettes on distinct stalks ≥10cm; petals pink, without scale-like nectaries; leaves concave throughout, with the lower marginal spines patent to recurved **2. OCHAGAVIA**

1. FASCICULARIA Mez - *Rhodostachys*
Inflorescences sessile; petals blue, with basal scale-like nectaries.

1 Plant to 75cm high and 2m across; leaves up to 35cm. Intrd-surv; on maritime dunes or shingle in Scillies and Guernsey (*F. pitcairniifolia* auct.) **Rhodostachys - F. bicolor** (Ruiz & Pav.) Mez

2. OCHAGAVIA Phil. - *Tresco Rhodostachys*
Inflorescences stalked; petals pink, without scale-like nectaries.

1 Plant to 75cm high and 2m across; leaves up to 25cm. Intrd-surv; on dunes in Tresco (Scillies) **Tresco Rhodostachys - O. carnea** (Beer) L.B. Sm. & Looser

161. PONTEDERIACEAE - *Pickerelweed family*

Aquatic, glabrous, rhizomatous, herbaceous perennials rooted in mud; the only aquatic with ovate-cordate leaves and a spike of blue flowers.

1. PONTEDERIA L. - *Pickerelweed*

1 Semi-submerged, with creeping or floating stems to 1m; leaves emergent, 5-25cm; spike 3-15cm, dense. Intrd-natd; edges of ponds; scattered in S En, S Lancs **Pickerelweed - P. cordata** L.

162. LILIACEAE - *Lily family*

Usually erect, mostly glabrous, herbaceous perennials or occasionally small evergreen shrubs, rhizomatous or with a corm or bulb or tuberous roots. A very variable family, usually recognized by conspicuous flowers with 6 petaloid tepals, 6 stamens and 3-celled ovary; exceptions are *Maianthemum*, *Ruscus* and *Paris*. All but *Ruscus* are herbaceous perennials arising from a corm, bulb, rhizome or tuberous roots. See Pontederiaceae, Iridaceae and Agavaceae for differences.

General key
1 Leaves *Iris*-like, i.e. vertical, with 2 identical surfaces, borne on 2 opposite sides of stem, each with leaf-base sheathing that of next higher leaf 2
1 Leaves not *Iris*-like 3
   2 Styles 3; tepals creamy- or greenish-white; filaments glabrous; leaf- margin with dense minute hairs near leaf-apex **1. TOFIELDIA**
   2 Style 1; tepals yellow; filaments densely pubescent; leaf-margin glabrous and smooth throughout **2. NARTHECIUM**
3 Leaves (3)4(-8), in 1 whorl on stem below single flower **18. PARIS**
3 Leaves not all in 1 whorl on stem below flower; flowers often >1 4
   4 Flowers *Crocus*-like, appearing from soil in autumn without leaves or stems; ovary subterranean, emerging with leaves to fruit in spring **7. COLCHICUM**
   4 Flowers with stems and leaves; if flower ± *Crocus*-like then ovary above ground at flowering 5
5 Leaves all reduced to small scales, with green ± leaf-like stems (cladodes) arising from main stems in their axils 6
5 At least some leaves green and photosynthetic 7
   6 Evergreen shrub; cladodes elliptic, very rigid, borne singly **38. RUSCUS**
   6 Herb; cladodes linear, soft, borne in clusters of ≥4 **37. ASPARAGUS**
7 Flowers in an umbel with 1-few spathe-like bracts at base, or solitary with spathe-like bract(s) at base, sometimes flowers replaced by bulbils *Key A*
7 Flowers in a cyme or raceme with or without bracts but without basal

162. LILIACEAE

spathe-like bract(s), or rarely solitary or in umbel without spathe-like bract(s) at base **Key B**

Key A - Flowers in an umbel or solitary, with 1-few spathe-like bracts at base (see also *Colchicum*)

1. Flowers entirely replaced by bulbils 2
1. At least some flowers present 3
    2. Stems <4cm; leaves <10cm **9. GAGEA**
    2. Stems >5cm; at least some leaves >15cm **25. ALLIUM**
3. Funnel- or collar-like corona present inside perianth 4
3. Corona absent 5
    4. Stamens arising from below corona and not fused to it; flowers winter to spring, white to yellow **35. NARCISSUS**
    4. Stamens fused to corona at base; flowers summer, white **36. PANCRATIUM**
5. Ovary inferior or semi-inferior 6
5. Ovary superior 11
    6. Perianth yellow **32. STERNBERGIA**
    6. Perianth white, pink or red, often tinged green 7
7. Flowers pink or red, often tinged green, rarely white but then without green or yellow patches 8
7. Flowers white with green or yellow patches 10
    8. Ovary semi-inferior; perianth <2cm, greenish-red, with free tepals **26. NECTAROSCORDUM**
    8. Ovary inferior; perianth >4cm, bright pink, with a proximal tube 9
9. Perianth-tube <2cm, funnel-shaped, gradually widened into lobed part of perianth; spathe herbaceous; flowers appearing before leaves; leaves flat **30. AMARYLLIS**
9. Perianth-tube >3cm, narrowly tubular for most part, abruptly widened into lobed part of perianth; spathe scarious; flowers appearing after leaves; leaves channelled **31. CRINUM**
    10. All 6 tepals similar **33. LEUCOJUM**
    10. 3 inner tepals much shorter and blunter than 3 outer **34. GALANTHUS**
11. Perianth yellow 12
11. Perianth white to various shades of red or blue 13
    12. Leaves >12mm wide; bracts (spathes) at base of inflorescence ovate, <15mm; flowers >5 **25. ALLIUM**
    12. Leaves ≤12mm wide; bracts (spathes) at base of inflorescence linear, >15mm; flowers 1-5 **9. GAGEA**
13. Perianth with tube >10mm, usually some shade of blue 14
13. Perianth with tepals free or fused for <5mm, very rarely blue or bluish 15
    14. Flowers in umbel, held horizontally, slightly zygomorphic, 3-5cm **28. AGAPANTHUS**
    14. Flowers solitary, erect, actinomorphic, 2.5-3.5cm **29. TRISTAGMA**
15. Tepals free; style arising from base of ovary; plant with onion-like smell when fresh **25. ALLIUM**

15 Tepals fused at base; style arising from top of ovary; plant without
 onion-like smell **27. NOTHOSCORDUM**

Key B - Flowers in a cyme or raceme, rarely in an umbel or solitary, without
 spathe-like bract at base (see also *Tofieldia, Narthecium, Paris*)
1 Ovary inferior **39. ALSTROEMERIA**
1 Ovary superior 2
  2 Tepals and stamens 4; leaves strongly cordate at base
   16. MAIANTHEMUM
  2 Tepals and stamens 6(-8); leaves cuneate to rounded at base 3
3 Tepals united into proximal tube >1/5 of their length 4
3 Tepals free or united just at extreme base 11
  4 Perianth yellow to orange or red, >3.5cm 5
  4 Perianth white to blue, pink or purple, very rarely pale yellow,
   ≤3.5cm 6
5 Flowers very numerous; perianth <5cm, tubular to narrowly bell-
 shaped **6. KNIPHOFIA**
5 Flowers <c.20; perianth >5cm, funnel-shaped **5. HEMEROCALLIS**
  6 Flowers borne in groups of 1-c.5 in axils of main foliage leaves
   15. POLYGONATUM
  6 Flowers terminal, or in terminal inflorescences, with bracts 0 or
   much reduced from leaves 7
7 Leaves linear to narrowly elliptic, narrowed at base; plant
 rhizomatous 8
7 Leaves linear, not narrowed at base; plant with bulb 9
  8 Flowers pendent, stalked, usually white, with perianth-tube
   longer than -lobes; leaves with distinct petiole, up to 30 x 10cm
   incl. petiole **14. CONVALLARIA**
  8 Flowers erect to patent, sessile, pink, with perianth-tube shorter
   than -lobes; leaves only slightly narrowed at base, up to 40 x 2cm
   17. REINECKEA
9 Perianth-tube >2x as long as lobes; corolla contracted at mouth
 24. MUSCARI
9 Perianth-tube <2x as long as lobes; corolla spread open at mouth 10
  10 Perianth-tube much shorter than -lobes, the lobes bent outwards
   at junction with tube **23. CHIONODOXA**
  10 Perianth-tube c. as long as -lobes, the lobes gradually curved
   outwards **22. HYACINTHUS**
11 Inflorescence with few-many flowers, each bractless or with bracts
 much reduced from leaves; leaves all basal 12
11 Inflorescence with 1-several flowers; stems bearing at least 1 leaf, or
 if leaves all basal at least lowest bract ± leaf-like or flower 1 16
  12 Inflorescence a panicle; filaments densely pubescent **4. SIMETHIS**
  12 Inflorescence a raceme; filaments glabrous 13
13 Bracts 2 per flower; tepals fused at extreme base **21. HYACINTHOIDES**
13 Bracts 0 or 1 per flower; tepals free 14
  14 Tepals usually blue, sometimes pink or pure white **20. SCILLA**
  14 Tepals white with green to reddish-brown stripe on abaxial

	surface	15
15	Plant with bulb; bracts whitish	**19. ORNITHOGALUM**
15	Plant without bulb, with swollen roots; bracts brown	**3. ASPHODELUS**
16	Tepals <2cm	17
16	Tepals ≥2cm	18
17	Tepals white with purplish veins	**8. LLOYDIA**
17	Tepals yellow with green stripes or tinge on abaxial side	**9. GAGEA**
18	Stigmas sessile	**11. TULIPA**
18	Stigmas on obvious style	19
19	Stigmas not or scarcely longer than wide; filaments loosely fixed to middle of anther; flowers several	**13. LILIUM**
19	Stigmas linear; filaments ± rigidly fixed to base of anthers; flowers mostly 1	20
20	Stem-leaves 2, near stem-base, reddish-blotched; tepals strongly reflexed	**10. ERYTHRONIUM**
20	Stem-leaves 3-6(8), most not near stem-base, not blotched; tepals not reflexed	**12. FRITILLARIA**

SUBFAMILY 1 - MELANTHIOIDEAE (genera 1-2). Plant rhizomatous; leaves all or mostly basal, *Iris*-like (vertical, flat with 2 identical faces); inflorescence a terminal raceme; tepals yellow to greenish-white, free; styles 1 or 3; ovary superior; fruit a capsule.

1. TOFIELDIA Huds. - *Scottish Asphodel*
Tepals creamy- or greenish-white; filaments glabrous; anthers c. as long as wide, dehiscing inwards; styles 3; capsule splitting where ovary-cells meet; seeds ovoid-curved.

1 Stems to 20cm, with 5-10 flowers near apex; leaves up to 8cm x 3mm; tepals 1.5-2.5mm. Native; by streams and in flushes on mountains; very local in N En (Upper Teesdale), locally frequent in C & N Sc
Scottish Asphodel - **T. pusilla** (Michx.) Pers.

2. NARTHECIUM Huds. - *Bog Asphodel*
Tepals yellow; filaments densely pubescent; anthers >2x as long as wide, dehiscing outwards; style 1; capsule splitting along centre of ovary-cells; seeds with long fine projections at each end. Vegetatively distinct from *Tofieldia* in glabrous leaf-margin (see General generic key, couplet 2).

1 Stems to 45cm, with 6-20 flowers near apex; leaves up to 30cm x 5mm; tepals 6-9mm. Native; bogs and other wet peaty acid places on heaths, moors and mountains; common in Ir and W & N Br, absent from most of C & E En *Bog Asphodel* - **N. ossifragum** (L.) Huds.

SUBFAMILY 2 - ASPHODELOIDEAE (genera 3-6). Plant rhizomatous or with swollen roots; leaves all or nearly all basal, linear; inflorescence a raceme or terminal compound cyme; tepals various colours (not blue); ovary superior; style 1; fruit a capsule splitting along centre of ovary-cells.

3. ASPHODELUS L. - *White Asphodel*

Plant with swollen roots; inflorescence racemose; tepals white with greenish to reddish-purple stripe on outside, erecto-patent, free; filaments glabrous; ovules 2 per cell; flowers actinomorphic.

1 Stems to 1m, with dense ± unbranched terminal raceme; leaves up to 60 x 3cm, flat and strongly keeled; tepals 15-20mm. Intrd-natd; on grassy bank; Jersey **White Asphodel - A. albus** Mill.

4. SIMETHIS Kunth - *Kerry Lily*

Rhizomatous; inflorescence cymose; tepals purplish outside, white inside, ± patent, free; filaments densely pubescent; ovules 2 per cell; flowers actinomorphic.

1 Stems to 40cm, with terminal ± lax panicle; leaves up to 50cm x 7.5mm; tepals 8-11mm. Native; rocky heathland near sea with *Ulex*; near Derrynane (S Kerry) **Kerry Lily - S. planifolia** (L.) Gren.

5. HEMEROCALLIS L. - *Day-lilies*

Rhizomatous; inflorescence cymose; tepals yellow to orange, fused to form proximal tube, funnel- to trumpet-shaped; filaments glabrous; ovules many per cell; flowers slightly zygomorphic, at least by upward curvature of stamens and style in laterally-directed flowers.

1 Flowers dull orange, 7-10cm, ± scentless; stems to c.1m, with up to c.20 flowers; leaves up to 90 x 2.5cm. Intrd-natd; rough ground, banks and grassy places; scattered ± throughout Br and CI
Orange Day-lily - H. fulva (L.) L.
1 Flowers yellow, 7-8cm, sweetly scented; stems to c.80cm, with up to c.12 flowers; leaves up to 65 x 1.5cm. Intrd-natd; occurrence as for *H. fulva*; scattered in Br, less common than *H. fulva* in En and Wa but more common in Sc **Yellow Day-lily - H. lilioasphodelus** L.

6. KNIPHOFIA Moench - *Red-hot-pokers*

Densely tufted, with short rhizomes; inflorescence a dense raceme; tepals red at first, becoming yellow, fused to form proximal tube; filaments glabrous; ovules many per cell; flowers cylindrical to narrowly bell-shaped, very slightly zygomorphic due to curvature of perianth and stamens.

1 Stems to 1.2m; leaves up to 80 x 1.8cm; bracts 3-9mm, ovate to oblong-ovate, rounded to subacute at apex; raceme up to 12 x 6cm; perianth 2.8-4cm; stamens included or just exserted. Intrd-natd; dunes or waste ground usually near sea; scattered in S & W Br, extending to E Kent and E Suffolk, CI **Red-hot-poker - K. uvaria** (L.) Oken
1 Stems to 2m; leaves up to 200 x 4cm; bracts 8-12mm, lanceolate to linear-oblong, acute to acuminate at apex or rounded at extreme apex; raceme 12-30 x 6-7cm; perianth 2.4-3.5cm; stamens exserted 4-15mm ± as soon as corolla opens. Intrd-natd; occurrence as for

6. KNIPHOFIA

K. uvaria and much confused with it; E & W Cornwall and CI and
N to Caerns and Flints   **Greater Red-hot-poker - K. x praecox** Baker

SUBFAMILY 3 - WURMBAEOIDEAE (genus 7). Plant with a corm; leaves ± all on stem, appearing in spring with fruits, linear-oblong; flowers appearing in autumn without leaves, 1-few each arising from ground; tepals pinkish to pale purple, united into long tube proximally; ovary superior, but subterranean at flowering, emerging above ground at stem apex at fruiting; styles 3; fruit a capsule splitting where ovary-cells meet.

Crocus-like, but with 6 (not 3) stamens and 3 simple stigmas.

7. COLCHICUM L. - *Meadow Saffron*

1   Tepals with narrow erect tube 5-20cm; capsule(s) produced on stem with sheathing leaves; leaves up to 35 x 5cm. Native; damp meadows and open woods on rich soils; local in C & S Br, very local in S Ir, natd elsewhere   **Meadow Saffron - C. autumnale** L.

SUBFAMILY 4 - LILIOIDEAE (genera 8-18). Plant rhizomatous or with a bulb; leaves all basal, all on stem, or both, linear to ovate or elliptic; inflorescence a single flower or a spike or raceme; tepals various colours but not blue, free or fused; style 1 or 0; ovary superior; fruit a berry or a capsule splitting along centre of ovary-cells.

8. LLOYDIA Rchb. - *Snowdon Lily*
Plant with a bulb; leaves mostly basal, some on stem, linear or ± so; flowers 1-2(3) at stem-apex; tepals free, erecto-patent; style 1; fruit a capsule.

1   Stems to 15cm, with 2-4 leaves; tepals 9-12mm, white with purple veins. Native; cracks in basic mountain rocks; very local in Caerns
    Snowdon Lily - L. serotina (L.) Rchb.

9. GAGEA Salisb. - *Star-of-Bethlehems*
Plant with 1-2 bulbs; leaves basal (1-4) and on stem (1-4), linear or ± so; flowers 1-5 at stem-apex; tepals yellow, free, patent; style 1; fruit a capsule.

1   Stems to 25cm, with 1 bulb at base, with 2-3 leaf-like bracts; basal leaf usually 1, 15-45cm x 7-15mm. Native; damp base-rich woods, hedgerows and rough fields; scattered in Br N to C Sc
    Yellow Star-of-Bethlehem - G. lutea (L.) Ker Gawl.
1   Stems to 4cm, with 2 bulbs at base, with usually 4-6 leaf-like bracts; basal leaves usually 2 per bulb, 4-9cm x c.1mm. Native; cracks and ledges of basic rocks; 1 site in Rads
    Early Star-of-Bethlehem - G. bohemica (Zauschn.) Schult. & Schult. f.

10. ERYTHRONIUM L. - *Dog's-tooth-violet*
Plant with a bulb; leaves 2, at base of stem, elliptic-oblong, reddish-blotched; flower 1, pendent; tepals free, sharply reflexed; style 1; fruit a

capsule.

1 Stems to 30cm; leaves up to 9 x 4cm; tepals 18-30mm, bright pink or sometimes white, reflexed back to expose long-exserted stamens and style. Intrd-natd; woodland, old estates and parks; Midlothian and Fife **Dog's-tooth-violet - E. dens-canis** L.

11. TULIPA L. - *Tulips*
Plant with a bulb; leaves basal and on stem, elliptic to linear-elliptic; flowers 1-2 at stem apex, erect at maturity; tepals free, forming cup- or bowl-shaped flower; style 0; fruit a capsule.

1 Flower rounded at base; filaments completely glabrous; buds erect. Stems to 60cm. Intrd-surv; the common garden tulip, persistent on tips, waysides and rough ground; scattered in C & S En and CI
***Garden Tulip* - T. gesneriana** L.
1 Flowers cuneate at extreme base due to narrowing of perianth; filaments pubescent near base; buds pendent 2
   2 Tepals ± uniform yellow. Stems to 50cm. Intrd-natd; woods, meadows and neglected estates; scattered in En and S & C Sc
***Wild Tulip* - T. sylvestris** L.
   2 Tepals pink to purple with yellow blotch on inside at base. Stems to 50cm. Intrd-natd; stony rough ground; Tresco (Scillies)
***Cretan Tulip* - T. saxatilis** Sieber ex Spreng.

12. FRITILLARIA L. - *Fritillary*
Plant with a bulb; leaves all or most on stem, linear; flowers 1(-few) at stem apex, pendent even at maturity; tepals free, forming cup-, bowl- or bell-shaped flower; style 1, with 3 linear stigmas; fruit a capsule.

1 Stems to 30(50)cm; perianth cup- to bowl-shaped, conspicuously chequered light and dark purple and cream, sometimes white, 3-5cm. Doubtfully native; damp meadows and pastures; local in En N to Staffs and W Norfolk, natd elsewhere ***Fritillary* - F. meleagris** L.

13. LILIUM L. - *Lilies*
Plant with a bulb; leaves all or most on stem, linear to elliptic; flowers few to many in terminal raceme, with free tepals; flowers pendent with tepals diverging and very strongly rolled back forming 'Turk's-cap' shape in our 2 spp.; style 1, with 3-lobed stigma; fruit a capsule.

1 Leaves oblanceolate to elliptic, at least some in whorls; flowers c.4cm across, purple with darker spots, sometimes white. Stems to 1.5. Intrd-natd; woods; scattered in Br N to C Sc, Man
***Martagon Lily* - L. martagon** L.
1 Leaves linear, spiral; flowers c.5cm across, greenish- to orange-yellow with darker spots. Stems to 1m. Intrd-natd; woods, hedgerows and

field margins; scattered in Br, mostly W & N, Man, Jersey
Pyrenean Lily - **L. pyrenaicum** Gouan

14. CONVALLARIA L. - *Lily-of-the-valley*
Plant rhizomatous; leaves all basal or ± at base of stem, elliptic or narrowly so; flowers 6-20 in long terminal raceme, pendent; tepals fused into tube longer than lobes, forming bell-shaped flower, white; style 1; fruit a red berry.

1 Stems to 25cm; leaves up to 30 x 10cm; flowers 5-10 x 5-10mm. Native; dry woods, scrub and hedgebanks usually on base-rich soil; scattered through most of Br, but often natd *Lily-of-the-valley* - **C. majalis** L.

15. POLYGONATUM Mill. - *Solomon's-seals*
Plant rhizomatous; leaves all on stem, linear-elliptic to elliptic; flowers 1-6 in axillary, stalked, pendent clusters; tepals fused into ± cylindrical tube longer than lobes, white or cream with green markings; style 1; fruit a purple or bluish-black berry.

1 Leaves linear, most in whorls of 3-8. Stems erect, to 80cm. Native; mountain woods; very rare in M & E Perth, Angus and S Northumb
Whorled Solomon's-seal - **P. verticillatum** (L.) All.
1 Leaves elliptic, all alternate 2
   2 Perianths not contracted in middle; filaments glabrous; flowers 1 or 2 per leaf-axil. Stems erect or erect then arching, to 40cm. Native; in woods on limestone; very local in NW En, Peak District, and around Severn estuary, very scattered in S Wa
Angular Solomon's-seal - **P. odoratum** (Mill.) Druce
   2 Perianths slightly contracted in middle; filaments sparsely pubescent; flowers 1-6 per leaf-axil 3
3 Perianths 9-15(20) x 2-4mm; stems terete. Stems erect then arching, to 80cm. Native; woods, mostly on basic soils; locally frequent in S Br, scattered N to N En *Solomon's-seal* - **P. multiflorum** (L.) All.
3 Perianths 15-22(25) x 3-6mm; stems ridged to slightly angled. Stems erect, then arching, to 1m. Intrd-natd; woods, scrub and rough ground; throughout CI and Br, very scattered in Ir (*P. multiflorum* x *P. odoratum*) *Garden Solomon's-seal* - **P. x hybridum** Brügger

16. MAIANTHEMUM Weber - *May Lilies*
Plant rhizomatous; leaves few from rhizome, 2(-3) on stem, ovate, cordate; flowers numerous in terminal raceme; tepals 4, free, patent, white; stamens 4; style 1; fruit a red berry.

1 Stems to 20cm, pubescent distally; leaves 3-6 cm; raceme 1-4(5)cm. Native; woods on acid soils; extremely local in Durham, NE Yorks and N Lincs, probably intrd in W Norfolk and W Lothian
May Lily - **M. bifolium** (L.) F.W. Schmidt
1 Stems to 35(45)cm, glabrous; leaves 5-11(20)cm; raceme 2.5-7.5cm.

Intrd-natd; woodland; S Somerset and MW Yorks
False Lily-of-the-valley - **M. kamtschaticum** (Cham.) Nakai

17. REINECKEA Kunth - *Reineckea*
Plant rhizomatous; leaves all basal, linear; flowers numerous in short terminal spike, erect to patent; tepals fused into tube slightly shorter than patent to reflexed lobes, pink; style 1; fruit a red berry.

1 Leaves in rosette, up to 40 x 2cm; spike 4-9cm on stalk 3-5cm; flowers rather crowded, with tube c.5mm and lobes c.7mm. Intrd-natd; in woodland; near Lizard (W Cornwall)
Reineckea - **R. carnea** (Haw.) Kunth

18. PARIS L. - *Herb-Paris*
Plant rhizomatous; leaves (3)4(-8), all in 1 whorl at top of stem, elliptic to obovate; inflorescence a single terminal, erect, long-stalked flower; tepals 8(-12), 4(-6) outer lanceolate, 4(-6) inner linear, all green, free, patent; stamens 8(-12); ovary superior, 4(-5)-celled; style 1, short; stigmas 4(-5), linear; fruit a dehiscent black berry.

1 Stems to 40cm; leaves up to 15 x 8cm; tepals 2-3.5cm. Native; in moist woods on calcareous soils; rather local in Br, absent from most of Wa, SW En and N & W Sc *Herb-Paris* - **P. quadrifolia** L.

SUBFAMILY 5 - SCILLOIDEAE (genera 19-24). Plant with a bulb; leaves all basal, linear or nearly so; inflorescence a terminal raceme; tepals usually white or blue, sometimes pink or brownish, very rarely pale yellow, free or fused; style 1, with 1 (often 3-lobed) stigma; ovary superior; fruit a capsule splitting along centre of ovary-cells.

19. ORNITHOGALUM L. - *Star-of-Bethlehems*
Flowers each with 1 bract; tepals free, white with green stripe(s) on outside; stamens inserted on receptacle, with flattened filaments.

1 Bracts longer than pedicels; filaments with 1 acute lobe at apex on either side of anther. Stems to 60cm. Intrd-natd; grassy places; scattered in C & S Br *Drooping Star-of-Bethlehem* - **O. nutans** L.
1 Bracts shorter than pedicels (at least on lower flowers); filaments without apical lobes 2
  2 Inflorescence corymbose; tepals >14mm. Stems to 30cm. Native; grassy places, rough ground and open woods; scattered throughout Br and CI, but perhaps native only in E En
Star-of-Bethlehem - **O. angustifolium** Boreau
  2 Inflorescence an elongated raceme; tepals <14mm. Stems to 80(100)cm. Native; woods and scrub; very local in SC En N to Hunts, rarely natd elsewhere in En
Spiked Star-of-Bethlehem - **O. pyrenaicum** L.

20. SCILLA L. - *Squills*
Flowers each with 0 or 1 bract; tepals free, blue, rarely white or pink; stamens inserted on base of perianth, with narrow or flattened filaments.

1 Bracts >4mm 2
1 Bracts 0 or <4mm 4
   2 Flowers usually >20; lower bracts ≥3cm. Stems to 50cm, with 20-100 suberect flowers. Intrd-natd; spreading where planted or neglected; very scattered in Br, Man and CI
Portuguese Squill - **S. peruviana** L.
   2 Flowers <15; bracts <3cm 3
3 Leaves 2-5mm wide; tepals 5-8mm. Stems to 15cm, with ≤12 suberect flowers; Native; dry short grassland near sea, especially cliff-tops; locally common on coasts of W Br from S Devon to Shetland, down E coast S to Cheviot, E coast of Ir *Spring Squill* - **S. verna** Huds.
3 Leaves 10-30mm wide; tepals 8-12mm. Stems to 40cm, with c.5-15 suberect flowers. Intrd-natd; open woodland; scattered from S Somerset and Berks to C Sc *Pyrenean Squill* - **S. liliohyacinthus** L.
   4 Flowers pendent; tepals 12-16mm. Stems to 20cm, with usually <5 ± pendent flowers. Intrd-natd; spreading where planted and neglected; scattered in Br, mainly SE En
Siberian Squill - **S. siberica** Haw.
   4 Flowers erect to patent; tepals 3-10mm 5
5 Flowering Jul-Sep without leaves; tepals 3-6mm. Stems to 25cm, with 4-20 suberect flowers. Native; short grassland usually near sea; local in CI, SW En scattered E to S Essex and Surrey
Autumn Squill - **S. autumnalis** L.
5 Flowering Feb-Apr with leaves; tepals 5-10mm 6
   6 Leaves (1)2(-3), sheathing flowering stem 1/4-1/2 way up from base. Stems to 20cm, with usually <10 ± erect flowers. Intrd-natd; churchyards and on banks; very scattered in S En, rare to N En
Alpine Squill - **S. bifolia** L.
   6 Leaves 3-7, sheathing flowering stem only at base 7
7 All pedicels <10mm; bracts c.1mm. Stems to 15cm, with 7-15(20) ± erect flowers. Intrd-natd; churchyards and open woods; N Somerset and W Suffolk *Greek Squill* - **S. messeniaca** Boiss.
7 Lower pedicels up to 2.5cm; bracts (1.5)2-3mm. Stems to 30cm, with (3)7-10(15) ± erect flowers. Intrd-natd; churchyards and open woods and on banks; very scattered in S En *Turkish Squill* - **S. bithynica** Boiss.

21. HYACINTHOIDES Heist. ex Fabr. - *Bluebells*
Flowers each with 2 bracts; tepals free, blue, sometimes white or pink; stamens inserted on or at base of perianth, with narrow filaments.

1 Tepals ± patent, 5-8mm; all stamens inserted at base of perianth. Stems to 40cm. Intrd-natd; in neglected old woodland; Ayrs, Dorset and S Essex *Italian Bluebell* - **H. italica** (L.) Rothm.
1 Tepals erect to erecto-patent, >10mm; at least 3 outer stamens fused

from base to >1/4 way up perianth 2
2 Racemes pendent at apex; perianth tubular for most of length; anthers cream. Stems to 50cm. Native; woods, hedgerows, shady banks, grassland in wetter regions; frequent to abundant
Bluebell - **H. non-scripta** (L.) Rothm.
2 Racemes erect; perianth bell-shaped; anthers same colour as tepals. Stems to 40cm. Intrd-natd; woods, copses, shady banks and field-borders; frequent in Br and CI
Spanish Bluebell - **H. hispanica** (Mill.) Rothm.

H. non-scripta x *H. hispanica* is more commonly grown in gardens than either parent; it often escapes and also arises naturally where natd *H. hispanica* meets native or natd *H. non-scripta*; it is fertile, forming a complete spectrum between the parents and often natd in absence of both.

22. HYACINTHUS L. - *Hyacinth*

Flowers with 1 minute bract; tepals fused into tube c. as long as lobes, blue, pink or white, very rarely pale yellow; stamens inserted on perianth-tube, the filaments almost wholly fused to it.

1 Stems to 30cm; racemes erect, with few to many, pendent to suberect flowers; perianth 10-35mm, the lobes very strongly recurved. Intrd-surv; long persistent where thrown out or neglected; scattered in S En *Hyacinth* - **H. orientalis** L.

23. CHIONODOXA Boiss. - *Glory-of-the-snows*

Flowers with 1 rudimentary or 0 bract; tepals fused into tube much shorter than lobes, blue, often with white central area; stamens inserted at apex of perianth-tube, fully exserted, with very flattened, white filaments.

1 Perianth wholly blue (but filaments white). Stems to 30cm, mostly with 6-16 flowers. Intrd-natd; occurrence as for *C. forbesii* but less common *Lesser Glory-of-the-snow* - **C. sardensis** Barr
1 Perianth blue with a white centre zone (and white filaments) 2
2 Stems mostly with 4-12 flowers; perianth bright blue distally. Stems to 30cm (often much less). Intrd-natd; spreading from seeds where neglected or thrown out; scattered in Br, mainly S En *Glory-of-the-snow* - **C. forbesii** Baker
2 Stems all or mostly with 1-2 flowers; perianth pale blue distally. Stems to 30cm. Intrd-natd; natd on grassy slope; Middlesex
Boissier's Glory-of the-snow - **C. luciliae** Boiss.

24. MUSCARI Mill. - *Grape-hyacinths*

Flowers with 1 minute or 0 bract, the apical group sterile, the lower ones fertile and often of different colour; tepals fused for most of length, blue to blackish-blue, with white lobes, or brownish; stamens inserted c.1/2 way up perianth-tube, included, with narrow filaments.

1 Fertile flowers brownish-buff, on pedicels mostly >5mm; apical sterile
flowers bright bluish-violet, some on pedicels >5mm. Stems to 60cm.
Intrd-natd; persistent weed of cultivated and rough ground,
dunes and open grassland; local in SW En, S Wa and CI, rare and
usually casual elsewhere *Tassel Hyacinth* - **M. comosum** (L.) Mill.
1 All flowers blue to blackish-blue, on pedicels <5mm 2
   2 Perianth of fertile flowers blackish-blue to dark violet-blue. Stems
to 30cm. Native; dry grassland, hedgebanks and field borders;
very local in E & W Suffolk and Cambs, rarely natd elsewhere
Grape-hyacinth - **M. neglectum** Ten.
   2 Perianth of fertile flowers bright blue. Stems to 30cm. Intrd-natd;
garden escape and throwout spreading vegetatively and by seed
on rough ground, banks and grassy places; scattered in Br,
mainly S but N to C Sc, and CI
Garden Grape-hyacinth - **M. armeniacum** Baker

SUBFAMILY 6 - ALLIOIDEAE (genera 25-29). Plant with a rhizome or a
bulb, mostly smelling of garlic or onion when fresh; leaves usually all basal,
sometimes on stem, linear to ± cylindrical, linear-oblong or elliptic;
inflorescence a terminal umbel with usually scarious spathe at base,
sometimes reduced to 1 flower or some or all flowers replaced by bulbils
but still with spathe; tepals various colours, free or fused proximally; style 1,
with capitate to 3-lobed stigma; ovary superior or semi-inferior; fruit a
capsule splitting along centre of ovary-cells.

25. ALLIUM L. - *Onions*

Plant with bulb(s), smelling of onion or garlic when fresh; leaves linear to ±
cylindric, or elliptic; flowers in umbel, some or all often replaced by bulbils;
tepals free or ± so, white to greenish, pink, purple or yellow; ovary superior;
ovules usually 2 per cell.

General key
1	Inflorescence consisting entirely of bulbils	*Key A*
1	Inflorescence with at least 1 flower	2
2	Inflorescence with bulbil(s) and flower(s)	*Key B*
2	Inflorescence with flowers only	*Key C*

Key A - Inflorescence consisting entirely of bulbils
1 Leaves circular to semi-circular or subcircular in section 2
1 Leaves obviously bifacial, flat to strongly keeled 4
   2 Stem hollow, inflated and bulging just below middle; leaves
usually >4mm wide. Stems to 1m, terete. Intrd-surv; garden
throwout or relic; scattered throughout Br *Onion* - **A. cepa** L.
   2 Stem solid or nearly so, not inflated; leaves <4mm wide 3
3 Spathe of 2 persistent valves each with apical attenuate part much
longer than basal part. Stems to 80cm, terete but slightly ridged.
Native; dry grassy places; scattered throughout En, very scattered
in Wa, Sc and (intrd) Ir *Field Garlic* - **A. oleraceum** L.

3 Spathe of 1 ± deciduous valve with apical attenuate part c. as long as basal part. Stems to 80cm, terete, finely ridged. Native; grassy places, rough ground, banks and waysides; common in S En, frequent to scattered in rest of BI *Wild Onion* - **A. vineale** L.
    4 Stems triangular in section; leaves all basal. Stems to 40cm, with acute angles. Intrd-natd; woods, grassy places, rough ground and waysides; scattered through much of En, E Ir and Sc *Few-flowered Garlic* - **A. paradoxum** (M. Bieb.) G. Don
    4 Stems ± circular in section; at least some leaves borne on stem 5
5 Leaves <4mm wide. Stems to 60cm, terete, faintly ridged. Intrd-natd; in rough ground, grassy places and waysides; scattered in Br and CE & NE Ir *Keeled Garlic* - **A. carinatum** L.
5 Leaves >5mm wide 6
    6 Leaves 2-5; main bulb single, with often numerous small bulblets outside its covering. Stems to 80cm, terete. Native; dry grassland and scrub; local in Br from Cheshire and S Lincs N to S Aberdeen, natd in SW Ir, rarely elsewhere *Sand Leek* - **A. scorodoprasum** L.
    6 Leaves 4-10; main bulb composed of several ± equal bulblets within common cover. Stems to 1m, terete, often coiled at first. Intrd-natd; casual where thrown out, rarely ± natd in coastal saline areas as in Caerns and W Lancs; very scattered in En and Wa *Garlic* - **A. sativum** L.

Key B - Inflorescence consisting of both flowers and bulbils
1 Leaves circular to semi-circular or subcircular in section 2
1 Leaves obviously bifacial, flat to strongly keeled 5
    2 Stem hollow, inflated and bulging just below middle; leaves usually >4mm wide (see Key A, couplet 2) **A. cepa**
    2 Stem solid or ± so, not inflated; leaves <4mm wide 3
3 Stamens shorter than tepals; filaments simple
(see Key A, couplet 3) **A. oleraceum**
3 Stamens longer than tepals; inner 3 filaments divided distally into 3 points, the middle one anther-bearing 4
    4 Spathe 1-valved; lateral points of inner 3 filaments >2x as long as central point (see Key A, couplet 3) **A. vineale**
    4 Spathe 2-valved; lateral points of inner 3 filaments <2x as long as central point. Stems to 80cm, terete, finely ridged. Native; on limestone rocks, W Gloucs, and on sandy waste ground by sea, Jersey, rarely natd elsewhere
Round-headed Leek - **A. sphaerocephalon** L.
5 Stems triangular in section (see Key A, couplet 4) **A. paradoxum**
5 Stems ± circular in section 6
    6 Tepals yellow. Stems to 45cm, terete. Intrd-natd; warm banks and hedgerows; scattered in S En, Jersey *Yellow Garlic* - **A. moly** L.
    6 Tepals pink to white, greenish or purplish 7
7 Filaments simple; leaves often <5mm wide 8
7 Inner 3 filaments divided distally into 3 points, the middle one anther-bearing; leaves ≥5mm wide 9

8 Stamens shorter than tepals; spathe shorter than pedicels. Stems to 75cm, terete. Intrd-natd; rough or cultivated ground, old dunes, hedgerows and waysides; frequent in SW En, S Wa and CI, scattered elsewhere **Rosy Garlic - A. roseum** L.

8 Stamens longer than tepals; spathe longer than pedicels
(see Key A, couplet 5) **A. carinatum**

9 Stamens longer than tepals 10
9 Stamens shorter than tepals 11

10 Bulb scarcely swollen at base, without bulblets; style shorter than tepals; spathe persistent at least until flowering. Stems to 1m, terete. Intrd-surv; casual where thrown out or a relic; very scattered in Br **Leek - A. porrum** L.

10 Bulb swollen at base, with bulblets around it within common cover; style longer than tepals; spathe usually deciduous before flowering. Stems to 2m, terete. Native; rocky or sandy places and rough ground near sea; very local in SW En, S & NW Wa, N & CW Ir and CI **Wild Leek - A. ampeloprasum** L.

11 Leaves 2-5; main bulb single, with often numerous small bulblets outside its cover; common part of inner 3 filaments 2-3x as long as central distal anther-bearing point
(see Key A, couplet 6) **A. scorodoprasum**

11 Leaves 4-10; main bulb composed of several ± equal bulblets within common cover; common part of inner 3 filaments c. as long as central distal anther-bearing point (see Key A, couplet 6) **A. sativum**

Key C - Inflorescence consisting entirely of flowers
1 Leaves circular to semi-circular or subcircular in section 2
1 Leaves obviously bifacial, flat to strongly keeled 5
    2 Stem hollow, inflated and bulging just below middle; leaves usually >4mm wide (see Key A, couplet 2) **A. cepa**
    2 Stem solid or nearly so, not inflated; leaves <4mm wide 3
3 Filaments simple; stamens shorter than tepals. Stems to 50cm, terete, hollow. Native; rocky ground, usually on limestone; local in SW & N En and S Wa, very scattered relic or throwout elsewhere
Chives - **A. schoenoprasum** L.
3 Inner 3 filaments divided distally into 3 points, the middle one anther-bearing; stamens at least as long as tepals 4
    4 Spathe 1-valved; lateral points of inner 3 filaments ≥2x as long as central point (see Key A, couplet 3) **A. vineale**
    4 Spathe 2-valved; lateral points of inner 3 filaments <2x as long as central point (see Key B, couplet 4) **A. sphaerocephalon**
5 Tepals yellow (see Key B, couplet 6) **A. moly**
5 Tepals white to pink, greenish or purplish 6
    6 Leaves with distinct petiole, the blade elliptic to narrowly so. Stems to 45cm, variously ± terete but ridged to triangular in section with obtuse angles. Native; woods and other damp shady places; frequent over most of BI *Ramsons* - **A. ursinum** L.
    6 Leaves without petiole, linear to filiform 7

7 Stem triangular in section 8
7 Stem ± circular in section 10
    8 Stigma simple; spathe 1-valved. Stems to 50cm, triangular in section with 2 edges much more acute than other. Intrd-natd; rough and cultivated ground, hedgebanks and waysides; frequent in SW En and CI, very scattered elsewhere in En and Man   *Neapolitan Garlic* - **A. neapolitanum** Cirillo
    8 Stigma 3-lobed; spathe 2-valved 9
9 Umbel 1-sided, with pendent flowers; tepals never opening >45°. Stems to 45cm, triangular in section with very acute angles. Intrd-natd; weed of rough, waste and cultivated ground, copses, hedgerows and waysides; common in SW En and CI, scattered elsewhere in En, Wa and Ir   *Three-cornered Garlic* - **A. triquetrum** L.
9 Umbel not 1-sided, with erect and pendent flowers; tepals opening >45° at first, less so later. Stems to 25cm, triangular in section with very acute angles. Intrd-surv; persistent in neglected estates, S Essex and Middlesex   *Italian Garlic* - **A. pendulinum** Ten.
    10 Inner 3 filaments divided distally into 3 points, the middle one anther-bearing 11
    10 Filaments simple 12
11 Bulb scarcely swollen at base, without bulblets; style shorter than tepals; spathe persistent at least until flowering
    (see Key B, couplet 10) **A. porrum**
11 Bulb swollen at base, with bulblets around it within common cover; style longer than tepals; spathe usually deciduous before flowering
    (see Key B, couplet 10) **A. ampeloprasum**
    12 Leaves conspicuously pubescent at edge. Stems to 45cm, terete. Intrd-surv; rough and cultivated ground, hedgebanks and waysides; very scattered in SW En and CI
    Hairy Garlic - **A. subhirsutum** L.
    12 Leaves glabrous 13
13 Stamens longer than tepals; leaves ≤3mm wide; spathe with valves much longer than pedicels   (see Key A, couplet 5) **A. carinatum**
13 Stamens shorter than tepals; leaves most or all >4mm wide; spathe with valves rarely as long as pedicels 14
    14 Leaves >2cm wide. Stems to 80cm, terete. Intrd-natd; rough ground; few places in S En   *Broad-leaved Leek* - **A. nigrum** L.
    14 Leaves <1.5cm wide 15
15 Covering of bulb minutely pitted; spathe with 1 primary valve, often deeply ≥2-lobed   (see Key B, couplet 8) **A. roseum**
15 Covering of bulb with undulating or net-like markings; spathe 2-valved. Stems to 40cm, terete. Intrd-surv; garden plant persistent in woods; W Kent   *American Garlic* - **A. unifolium** Kellogg

26. **NECTAROSCORDUM** Lindl. - *Honey Garlic*

Plant with bulb, smelling of garlic when fresh; leaves linear, strongly keeled; flowers in umbel, sweetly scented; tepals free, greenish-red; ovary semi-inferior; ovules numerous per cell.

1 Stems to 1.2m, terete; leaves 1-2cm wide; tepals 12-17mm; stamens shorter than tepals. Intrd-natd; rough ground; very scattered in En, mostly S (*Honey Garlic* - **N. siculum** (Ucria) Lindl.)   2
  2 Tepals greenish-red; a few fruiting pedicels conspicuously longer than all others   **N. siculum ssp. siculum**
  2 Tepals greenish-cream tinged with pink; most fruiting pedicels nearly as long as longest **N. siculum ssp. bulgaricum** (Janka) Stearn

27. NOTHOSCORDUM Kunth - *Honeybells*

Plant with bulb, not smelling of garlic or onion; leaves linear, scarcely or not keeled; flowers in umbel, sweetly scented; tepals fused at base, greenish-white with pink midrib; ovary superior; ovules numerous per cell.

1 Stems to 60cm, terete; leaves 4-15mm wide; tepals 8-14mm; stamens shorter than tepals. Intrd-natd; garden plant natd in rough and arable land and neglected estates; scattered in S & SW En, S Ir and CI   *Honeybells* - **N. borbonicum** Kunth

28. AGAPANTHUS L'Hér. - *African Lily*

Plant with short tuber-like rhizome, not smelling of garlic or onion; leaves oblong-linear, scarcely keeled; flowers in umbel, slightly zygomorphic; tepals fused in lower 1/2, bright blue, very rarely white; ovary superior; ovules numerous per cell.

1 Plant forming dense clump; stems to 1m, terete; leaves 20-70 x 1.5-5.5cm; perianth 26-50mm; stamens c. as long as perianth. Intrd-natd; on sandy soil by sea in Scillies, persistent in CI and Surrey
                                          African Lily - **A. praecox** Willd.

29. TRISTAGMA Poepp. - *Starflowers*

Plant with bulb, smelling of garlic when fresh; leaves linear, slightly keeled; flowers solitary, terminal, with spathe below, sweetly scented; tepals fused in lower 1/2, pale bluish-violet with dark midrib outside; ovary superior; ovules numerous per cell.

1 Stems to 35cm, terete; leaves 4-8mm wide; perianth 25-35mm, the tube 12-16mm, narrow, the lobes patent, acute, forming flower 30-45mm across. Intrd-natd; weed of cultivated and waste sandy ground; W Cornwall, Scillies and CI, rare casual or relic elsewhere
                                Spring Starflower - **T. uniflorum** (Lindl.) Traub

SUBFAMILY 7 - AMARYLLIDOIDEAE (genera 30-36). Plant with a bulb; leaves all basal, linear to narrowly elliptic or linear-oblong; inflorescence of 1 flower or a terminal umbel with usually scarious spathe at base; tepals white to yellow or orange, rarely pink, free or fused proximally, sometimes with funnel- or collar-shaped corona within the rows of tepals; style 1, with simple or slightly 3-lobed stigma; ovary inferior; fruit a capsule dehiscing irregularly or along centre of ovary cells, often slightly succulent.

30. AMARYLLIS L. - *Jersey Lily*
Flowers in umbel, erecto-patent, trumpet-shaped, slightly zygomorphic, without corona; tepals fused at base into short tube, pink, all ± similar; flowers appearing in late summer or autumn, before leaves.

1 Stems very stout, to 60cm; leaves 30-45 x 1.5-3cm, flat; perianth 5-10cm, with tube 1-1.5cm. Intrd-surv; relic in old fields, rough ground and sandy places; CI (especially Jersey), Scillies and W Cornwall *Jersey Lily* - **A. belladonna** L.

31. CRINUM L. - *Cape-lilies*
Flowers in umbel, erecto-patent, trumpet-shaped, actinomorphic, without corona; tepals fused at base into long tube, pink, all ± similar; flowers appearing in late summer or autumn, after leaves.

1 Stems to 80cm; leaves 60-120 x 4-10cm, channelled; perianth 12-20cm, incl. tube 6-10cm. Intrd-natd; garden escape on dunes; Alderney (*C. bulbispermum* (Burm. f.) Milne-Redh. & Schweick. x *C. moorei* Hook.) *Powell's Cape-lily* - **C. x powellii** Baker

32. STERNBERGIA Waldst. & Kit. - *Winter Daffodil*
Flowers solitary, erect, ± *Crocus*-like, actinomorphic, without a corona; tepals fused into narrow proximal tube, yellow, all ± similar; flowers appearing in autumn, ± with leaves.

Differs from *Crocus* in its ovary aerial at flowering, ± entire stigma, 6 stamens and leaves without central pale stripe.

1 Stems 2.5-10(20)cm; leaves 7-10 x 0.4-1.5cm; perianth with tube 0.5-2cm, the lobes 3-5.5cm. Intrd-natd; grassy slopes by sea; Jersey *Winter Daffodil* - **S. lutea** (L.) Spreng.

33. LEUCOJUM L. - *Snowflakes*
Flowers solitary or few in umbel, pendent, bell-shaped or bowl-shaped, actinomorphic, without a corona; tepals free, white with green or yellow patches, all ± similar; flowers appearing in late winter or spring, ± with leaves.

1 Stems to 40cm, with 1(-2) flowers; leaves up to 30 x 0.5-2.5cm; tepals 15-25mm, with green or yellow patch near apex. Possibly native; damp scrub and stream-banks; 2 sites in S Somerset and Dorset, rarely natd elsewhere *Spring Snowflake* - **L. vernum** L.
1 Stems to 60cm, with (1)2-5(7) flowers; leaves up to 50 x 0.5-1.5cm; tepals 10-22mm, each with green patch near apex (*Summer Snowflake* - **L. aestivum** L.) 2
    2 Stems with the 2 sharp edges remotely and often inconspicuously denticulate, at least in lower 1/2; flowers (2)3-5(7), 13-22mm. Native; wet meadows and willow scrub by rivers; very local in

S En N to Oxon and in S Ir, natd elsewhere in BI
L. aestivum ssp. aestivum
2 Stems with the 2 sharp edges entire throughout; flowers (1)2-4, 10-15mm. Intrd-natd; in damp places and rough ground; scattered through BI from CI to Easterness
L. aestivum ssp. pulchellum (Salisb.) Briq.

34. GALANTHUS L. - *Snowdrops*

Flowers solitary, pendent, actinomorphic, without a corona; tepals free, the inner white with green patch(es) and forming bell-shaped to bowl-shaped whorl, the outer white and spreading when in full flower; flowers appearing in late winter or spring, ± with leaves or before them, sometimes *flore pleno*.

1 Leaf upperside clear green, not glaucous. Stems to 25cm. Intrd-natd; grassy and marginal places; scattered in S En, Dunbarton (*G. ikariae* Baker)   *Green Snowdrop* - **G. latifolius** Rupr.
1 Leaf upperside glaucous, at least along central band   2
   2 Leaves with margins folded under at least along most of length, especially when young, glaucous on upper surface. Stems to 25cm. Intrd-natd; woods and damp grassland; local in S En (*Pleated Snowdrop* - **G. plicatus** M. Bieb.)   3
   2 Leaves flat, or with margins inrolled especially when young, or if weakly folded under then not wholly glaucous on upperside   4
3 Inner tepals with green patch at apex only   **G. plicatus ssp. plicatus**
3 Inner tepals with green patches at apex and base, sometimes partly joining up   **G. plicatus ssp. byzantinus** (Baker) D.A. Webb
   4 Leaves flat or folded under only at base as they unfold, ± linear, ≤1cm wide at flowering   5
   4 Leaves inrolled as they unfold, oblanceolate, at least one >1cm wide at flowering   6
5 Leaf upperside with glaucous central band and dull green lateral bands. Stems to 20cm; flowers in late autumn or early winter before leaves appear. Intrd-natd; grassy places; W Kent
    Queen Olga's Snowdrop - **G. reginae-olgae** Orph.
5 Leaf upperside wholly glaucous. Stems to 20cm. Possibly native in S En; woods, damp grassy places, banks and streamsides; scattered throughout Br and CI, rare in Ir   *Snowdrop* - **G. nivalis** L.
   6 Inner tepals with green patches at apex and base. Stems to 30cm. Intrd-natd; woods and damp grassland; local in S En, Man
    Greater Snowdrop - **G. elwesii** Hook. f.
   6 Inner tepals with green patch at apex only. Stems to 15cm. Intrd-natd; woods and damp grassland; very local in S En (*G. caucasicus* (Baker) Grossh.)   *Caucasian Snowdrop* - **G. alpinus** Sosn.

At least 3 combinations of hybrids occur in S En in mixed wild populations.

35. NARCISSUS L. - *Daffodils*

Flowers solitary or few in umbel, pendent to erecto-patent, actinomorphic, with a corona; tepals and corona fused to form hypanthial tube between base of tepals and apex of ovary; tepals white to yellow; corona white to yellow or orange; flowers appearing in spring, with or after leaves, sometimes *flore pleno*.

An extremely popular garden genus with numerous interspecific hybrids and thousands of cultivars, many with uncertain parentage. Many occur natd in fields, waysides, woods, rough ground, banks, etc., and are very difficult to classify. The descriptions mainly apply to the commonest variants of each taxon, but others (especially colour variants) occur.

1	Tepals reflexed back through 180° or almost so	2
1	Tepals patent to erecto-patent	3
	2 Hypanthial tube 10-20mm; corona c. as long as wide; upper 3 stamens and stigma exserted from corona; leaves subcylindrical, 1.5-3mm wide. Intrd-natd; well natd in Jethou (near Guernsey), less so in W Kent ***Angel's-tears* - N. triandrus** L.	
	2 Hypanthial tube 2-3mm; corona distinctly longer than wide; all stamens and stigma included in corona; leaves flat or grooved on upperside, 3-6mm wide. Intrd-natd; rarely natd; W Kent and Surrey ***Cyclamen-flowered Daffodil* - N. cyclamineus** DC.	
3	Hypanthial tube parallel-sided, sometimes abruptly expanded at apex; corona <10mm, wider than long; stamens of 2 distinct lengths	4
3	Hypanthial tube distinctly widening towards apex; corona usually >10mm, if <10mm then longer than tepals, c. as long as to much longer than wide; stamens all of same length or ± so	8
	4 Flower 1; corona yellow with sharply contrasting narrow red rim. Intrd-natd (***Pheasant's-eye Daffodil* - N. poeticus** L.)	5
	4 Flowers (1)2-8(20); corona white or yellow all over	6
5	Tepals usually 20-25mm, strongly overlapping; corona c.14mm across, ± discoid; lower 3 stamens included. Common, especially in S **N. poeticus ssp. poeticus**	
5	Tepals usually 22-30mm, scarcely overlapping; corona c.8-10mm across, shortly cylindrical; all 6 stamens partly exserted. Frequent relic in CI, Man **N. poeticus ssp. radiiflorus** (Salisb.) Baker	
	6 Flowers (1)2(-3); hypanthial tube ≥20mm; pollen sterile. Intrd-natd; rather frequent relic in most of BI, mainly S (*N. tazetta* x *N. poeticus*) ***Primrose-peerless* - N. x medioluteus** Mill.	
	6 Flowers (2)3-8(20); hypanthial tube ≤20mm; pollen fertile	7
7	Corona yellow; flowers ≤8(15). Intrd-surv; rather rare relic in SW En and CI ***Bunch-flowered Daffodil* - N. tazetta** L.	
7	Corona white; flowers ≤20. Intrd-surv; occurrence as for *N. tazetta* ***Paper-white Daffodil* - N. papyraceus** Ker Gawl.	
	8 Tepals linear to very narrowly triangular or lanceolate, ≤5mm wide. Intrd-natd; very distinctive sp. rarely natd; S En ***Hoop-petticoat Daffodil* - N. bulbocodium** L.	
	8 Tepals ovate or triangular-ovate to suborbicular, ≥1cm wide	9

9 Corona c. as long as tepals; hypanthial tube <2x as long as greatest width 10
9 Corona distinctly shorter than tepals; hypanthial tube >2x as long as greatest width 13
   10 Hypanthial tube 9-15mm. Intrd-natd; well natd in Man, less so in few places in S En *Lesser Daffodil* - **N. minor** L.
   10 Hypanthial tube 15-25mm (**N. pseudonarcissus** L.) 11
11 Tepals paler than corona. Stems to 50cm. Native; woods and grassland; local (but often abundant) in En, Wa and Jersey, cultivars commonly natd throughout BI
Daffodil - **N. pseudonarcissus ssp. pseudonarcissus**
11 Tepals and corona same colour 12
   12 Pedicels (between origin of spathe and base of ovary) mostly <15mm; leaves up to 30 x 1cm; tepals distinctly shorter than corona, not twisted. Intrd-natd; long natd in Pembs and Carms, Cards and probably elsewhere but overlooked *Tenby Daffodil* - **N. pseudonarcissus ssp. obvallaris** (Salisb.) A. Fern.
   12 Pedicels mostly >15mm; leaves up to 50 x 1.5cm; tepals as long as corona, twisted at base. Intrd-natd; common over most of BI
Spanish Daffodil - **N. pseudonarcissus ssp. major** (Curtis) Baker
13 Flower 1; corona usually conspicuously deeper in colour than tepals; leaves ± glaucous, ≥8mm wide; stem distinctly 2-edged. Intrd-natd; commonly natd in BI, especially S (*N. poeticus* x *N. pseudonarcissus*)
Nonesuch Daffodil - **N. x incomparabilis** Mill.
13 Flowers (1)2-4; tepals and corona the same colour; leaves green, ≤8mm wide; stem subterete. Intrd-natd; rarely natd in Scillies, Cornwall and CI (*N. jonquilla* x *N. pseudonarcissus*)
Campernelle Jonquil - **N. x odorus** L.

Other hybrids are cultivated and may become natd to varying degrees; perhaps the commonest is *N. pseudonarcissus* x *N. cyclamineus*, which is the parentage of many garden cultivars including the much-planted 'February Gold'.

36. PANCRATIUM L. - *Sea Daffodil*
Flowers several in umbel, erect to erecto-patent, actinomorphic, with a corona with toothed margin; tepals and corona white, fused to form a long slender greenish hypanthial tube above ovary; stamens fused to corona at base, exserted; flowers appearing in summer, after leaves.

1 Stems to 60cm; leaves up to 50 x 2cm, very glaucous; flowers very fragrant; hypanthial tube 6-8cm; tepals 3-5cm; corona 2-3cm. Intrd-surv; 5 plants on sand-dunes in S Devon *Sea Daffodil* - **P. maritimum** L.

SUBFAMILY 8 - ASPARAGOIDEAE (genera 37-38). Plant rhizomatous; leaves all on stems, reduced to small scales, replaced functionally by green stems (cladodes) arising from their axils; flowers dioecious or ± so; inflorescence an inconspicuous cluster of 1-few flowers borne in scale-leaf axil on main stem or on cladode; tepals greenish to yellowish-white, free or fused just at base, without corona; style 1 or ± 0, with capitate or 3-lobed stigma; ovary superior; fruit a spherical red berry.

37. ASPARAGUS L. - *Asparagus*
Stems herbaceous; cladodes borne in cluster of 4-10(more), cylindrical to slightly flattened; flowers 1-2(3) in axils of scale-leaves on main stems; tepals fused at base; berry 5-10mm.

1 (*Asparagus* - **A. officinalis** L.) 2
  2 Stems procumbent to decumbent, to 30(60)cm; cladodes on main lateral branches 4-10(15)mm, rigid, usually glaucous; pedicels 2-6(8)mm; seeds usually <5. Native; grassy sea-cliffs; very local in SW En and SW Wa, SE Ir and CI
      Wild Asparagus - **A. officinalis ssp. prostratus** (Dumort.) Corb.
  2 Stems erect, to 1.5(2)m; cladodes on main lateral branches (5)10-20(25)mm, flexible, usually green; pedicels 6-10(15)mm; seeds usually 5-6. Intrd-natd; dry sandy soils among sparse grass, often on maritime dunes; scattered throughout BI N to C Sc
      Garden Asparagus - **A. officinalis ssp. officinalis**

38. RUSCUS L. - *Butcher's-brooms*
Evergreen shrubs; cladodes borne singly, flattened and leaf-like; flowers 1-2 in axils of scale-leaves borne in centre of adaxial side of cladode; tepals free; berry 8-13mm.

1 Stems erect, to 75(100)cm, much branched, cladodes 1-3(4) x 0.4-1cm, with sharp spine at apex. Native; woods, hedgerows, rocky places on dry soils; rather local in CI and Br N to Norfolk and N Wa, natd elsewhere   *Butcher's-broom* - **R. aculeatus** L.
1 Stems oblique, to 40cm, simple; cladodes 3-10 x 1-3.3cm, spineless. Intrd-natd; shady places; W En
      Spineless Butcher's-broom - **R. hypoglossum** L.

SUBFAMILY 9 - ALSTROEMERIOIDEAE (genus 39). Plant with tuberous roots; leaves all on stems, lanceolate; inflorescence a terminal simple or compound umbel without a spathe; tepals orange with darker markings, free, without corona; flowers slightly zygomorphic, with curved stamens; style 1, with 3-lobed stigma; ovary inferior; fruit a capsule splitting along centre of ovary cells.

39. ALSTROEMERIA L. - *Peruvian Lily*

1 Stems erect, to 1m; leaves 7-10cm; tepals 4-6cm, orange with red spots and streaks; stamens shorter than tepals. Intrd-natd; grassy places, rough ground and old garden sites; scattered in Br N to Man and C Sc, mainly in N   *Peruvian Lily* - **A. aurea** Graham

163. IRIDACEAE - *Iris family*

Usually erect, mostly glabrous, herbaceous perennials, rhizomatous or with a corm or rarely a bulb or swollen roots; easily distinguished from Liliaceae by the 3 stamens (only the distinctive *Ruscus* in the latter family has 3 stamens).

1	Style-branches broad and petaloid; flowers *Iris*-like	2
1	Style-branches not petaloid, narrow; flowers not *Iris*-like	3
	2 Plant with rhizome or bulb; roots not tuberous; ovary 3-celled	**5. IRIS**
	2 Plant without rhizome or bulb; roots tuberous; ovary 1-celled	**4. HERMODACTYLUS**
3	Flowers 1-few, erect, arising direct from ground or on very short stems, *Crocus*-like	4
3	Flowers few-many, erect or laterally-directed, usually arising from aerial green stems in spikes or panicles, not *Crocus*-like	5
	4 Leaves subterete, without white line; perianth-tube <1cm, sheathed by green bract	**7. ROMULEA**
	4 Leaves flat, channelled and with central whitish line on upperside; perianth-tube >1.5cm, sheathed by white or brown bract	**8. CROCUS**
5	Perianth actinomorphic, with radially symmetrical lobes and straight tube; plant with or without a corm	6
5	Perianth zygomorphic, often with bilaterally symmetrical lobes but sometimes only so due to curved tube; plant with a corm	11
	6 Perianth-lobes fused proximally into tube >5mm	7
	6 Perianth-lobes completely free or fused proximally into tube <5mm	8
7	Perianth-tube >3cm, the lobes shorter; bracts <2cm, 3-toothed at apex, without dark streaks	**10. IXIA**
7	Perianth-tube <2cm, the lobes >2cm; bracts >2cm, deeply and jaggedly toothed at apex, with irregular dark longitudinal streaks	**11. SPARAXIS**
	8 Inner tepals c.2x as long as outer	**1. LIBERTIA**
	8 Inner and outer tepals same or nearly same length	9
9	Stem terete, arising from a corm; flowers sessile	**10. IXIA**
9	Stem flattened, narrowly winged, arising from rhizome or fibrous roots; flowers stalked	10
	10 Tepals twisting spirally after flowering; filaments free, arising from top of short perianth-tube	**3. ARISTEA**
	10 Tepals not twisting after flowering; filaments fused either just at base or for most of length, arising from base of perianth	**2. SISYRINCHIUM**
11	Style 3-branched, each branch bifid with the 6 ultimate branches longer than the 3 primary branches	12
11	Style unbranched with 3-lobed stigma, or style with 3 branches each unbranched or shortly bifid	13

12 Bracts >2cm; spike erect; seeds winged **6. WATSONIA**
12 Bracts <1.5cm; spike bent near base; seeds not winged **12. FREESIA**
13 Uppermost perianth-lobe >2x as long as rest **14. CHASMANTHE**
13 Uppermost perianth-lobe slightly longer to slightly shorter than rest 14
   14 Perianth-lobes strongly narrowed at base; style-branches
      widened distally **9. GLADIOLUS**
   14 Perianth-lobes not narrowed at base; style-branches filiform with
      minutely capitate to shortly bifid stigmas **13. CROCOSMIA**

1. LIBERTIA Spreng. - *Chilean-irises*

Plants with short rhizomes; leaves *Iris*-like; inflorescence a small terminal panicle; flowers actinomorphic; tepals white, free, the inner c.2x as long as outer but of similar shape; filaments slightly fused at base; style with 3 entire linear branches.

1 Inflorescence unbranched; flowers with pedicels shorter than bracts,
  c.25mm across; inner tepals 12-18mm. Stems erect, to 1.2m. Intrd-natd;
  rough ground, waysides and rocky lakeshore and coasts; Scillies,
  W Cornwall, Man, CW Sc, S Kerry   *Chilean-iris* - **L. formosa** Graham
1 Inflorescence branched; flowers with pedicels exceeding bracts by
  c.5-10mm, c.13mm across; inner tepals 6-9mm. Stems erect, to 1m.
  Intrd-surv; roadside ditches; Gigha (Kintyre)
                              Lesser Chilean-iris - **L. elegans** Poepp.

2. SISYRINCHIUM L. - *Blue-eyed-grasses*

Plants with fibrous roots and short or 0 rhizomes; leaves *Iris*-like; inflorescence a terminal cyme or a panicle of terminal and lateral cymes; flowers actinomorphic; tepals pale to bright yellow or blue, nearly free, ± equal; filaments slightly fused at base to fused for most of length; style with 3 entire linear branches.

1 Tepals blue   2
1 Tepals cream to yellow for most part   3
   2 At least some stems branched, each branch with 1 terminal
     inflorescence; perianth 15-20mm across, pale blue; pedicels arched
     to pendent in fruit. Stems to 50cm. Probably native; wet meadows
     and stony ground by lakes; very local in W Ir (W Cork to
     W Donegal)   *Blue-eyed-grass* - **S. bermudiana** L.
   2 Stem unbranched, with 1 terminal inflorescence; perianth 25-
     35mm across, violet-blue; pedicels erect in fruit. Stems to 50cm.
     Intrd-natd; grassy places, rough ground and waysides; scattered
     in Br N to Easterness, CI
                      American Blue-eyed-grass - **S. montanum** Greene
3 Stem unbranched, with terminal and several lateral cymes; leaves
  >1cm wide. Stems to 75cm. Intrd-natd; tips, waste ground, banks and
  waysides; scattered in S En from Surrey to Scillies
                              Pale Yellow-eyed-grass - **S. striatum** Sm.
3 Stem branched or unbranched, with 1 terminal cyme on each branch;

leaves ≤1cm wide — 4
4 Tepals bright yellow, stem unbranched. Stems to 60cm. Intrd-natd; damp grassy places near sea; Mons, Pembs, Co Wexford, W Galway *Yellow-eyed-grass* - **S. californicum** (Ker Gawl.) W.T. Aiton
4 Tepals cream to pale yellow; stem branched. Stems to 45cm. Intrd-surv; on gravelly paths; Jersey
Veined Yellow-eyed-grass - **S. laxum** Sims

3. ARISTEA Aiton - *Blue Corn-lily*
Plants with rhizomes; leaves *Iris*-like; inflorescence a loose terminal panicle of few-flowered clusters; flowers actinomorphic; tepals united proximally into tube <5mm, blue, ± equal; filaments free, arising from top of perianth-tube; style very slender, with 3-lobed stigma.

1 Stems to 60cm, flattened, bearing reduced leaves; leaves up to 60 x 1.2cm, linear; perianth 8-15mm. Intrd-natd; garden escape on rough ground; Tresco (Scillies) *Blue Corn-lily* - **A. ecklonii** Baker

4. HERMODACTYLUS Mill. - *Snake's-head Iris*
Plants with tuberous roots (no rhizomes, bulbs or corms); leaves subterete, very long, 4-angled; flowers solitary, terminal, actinomorphic, *Iris*-like; tepals united proximally into tube; tepals, stamens and styles ± like those of *Iris*.

1 Stems to 40cm; leaves longer than stems, up to 50cm x 3mm; outer tepals yellowish-green; inner tepals yellowish-green on claw, purplish-brown to blackish on blade. Intrd-natd; grassy places and hedgerows; rare in SW En *Snake's-head Iris* - **H. tuberosus** (L.) Mill.

5. IRIS L. - *Irises*
Plants with rhizomes or rarely bulbs; leaves *Iris*-like (vertical, flat, with 2 identical faces) or subterete or 4-angled; inflorescence terminal, rather simple, cymose; flowers actinomorphic; tepals united proximally into perianth-tube; outer tepals usually longer and wider than the inner, patent, recurved or reflexed, with a narrow proximal part (*claw*) and expanded distal part (*blade*); inner tepals usually erect, less differentiated into blade and claw; filaments free, borne at base of outer tepals; style with 3 long, broad, petaloid branches each with 2 lobes at apex beyond stigma, each covering a stamen.

1 Leaves subterete to slightly flattened, angled or channelled; plant with a bulb — 2
1 Leaves flat, not channelled or angled, vertical, with 2 identical faces; plant with rhizome — 4
2 Perianth-tube >10mm. Stems to 50cm. Intrd-surv; relic in old fields, rough ground and waste places; frequent in CI (*I. filifolia* Boiss. x *I. tingitana* Boiss. & Reut.) *Dutch Iris* - **I. x hollandica** hort.
2 Perianth-tube <10mm — 3

3 Leaves evergreen; claw of outer tepals ≤10mm wide, 1.5-2x as long as blade. Stems to 50cm. Intrd-surv; relic in old fields, rough ground and waste places; Scillies and CI   *Spanish Iris* - **I. xiphium** L.

3 Leaves dying down in winter; claw of outer tepals >20mm wide, no longer than blade. Stems to 50cm. Intrd-natd; in grassy places; Shetland and W Kent   *English Iris* - **I. latifolia** (Mill.) Voss

    4 Outer tepals bearded, i.e. with mass of stout multicellular hairs on inner face. Stems to 90cm. Intrd-natd; banks, rough and waste ground, waysides, old planted areas; frequent in C & S Br and CI   *Bearded Iris* - **I. germanica** L.

    4 Outer tepals not bearded, sometimes softly pubescent with unicellular hairs   **5**

5 Tepals predominantly yellow or yellow and white, without blue, purple, mauve or violet or only small spots or veining of it   **6**

5 Tepals predominantly of some shade of blue, purple, mauve or violet   **8**

    6 Leaves evergreen, dark green, with stinking smell when crushed; seeds bright orange; stems distinctly compressed. Stems to 80cm. Native; dry places in woods, hedges, banks and cliffs near sea, mostly on calcareous soils; locally frequent in CI and Br N to N Wa and Notts, natd elsewhere   *Stinking Iris* - **I. foetidissima** L.

    6 Leaves dying in winter, mid- to pale-green, not stinking; seeds brownish; stems subterete   **7**

7 Inner tepals white; outer tepals white with large yellow patch on blade; petaloid style-lobes subentire. Stems to 1.2m. Intrd-natd; fields, banks and limestone scrub; scattered in S En
   Turkish Iris - **I. orientalis** Mill.

7 Tepals yellow all over, the outer often with brownish or purple spots or veins; petaloid style-lobes deeply serrate. Stems to 1.5m. Native; wet meadows, fens and ditches, by lakes and rivers; common
   Yellow Iris - **I. pseudacorus** L.

    8 Flowering stems 0 or very short; perianth-tube 6-28cm; style-branches with yellow glands near margins. Intrd-surv; garden throwout by lane; S Somerset   *Algerian Iris* - **I. unguicularis** Poir.

    8 Flowering stems well developed; perianth-tube ≤2cm; style-branches without yellow glands   **9**

9 Leaves evergreen, dark green, with stinking smell when crushed; seeds bright orange   (see couplet 6) **I. foetidissima**

9 Leaves dying in winter, mid- to pale-green, not stinking; seeds brownish   **10**

    10 Stems hollow; perianth-tube 4-7mm; bracts brown and papery at flowering. Stems to 1.2m. Intrd-natd; rough, often wet or shaded, ground; scattered throughout Br N to Easterness
   Siberian Iris - **I. sibirica** L.

    10 Stems solid; perianth-tube 7-20mm; bracts at least partly green at flowering   **11**

11 Upper part of ovary sterile, narrower than ovary below and perianth-tube above, forming acuminate beak on capsule ≥5mm   **12**

11 Ovary without sterile apical part; capsule with 0 or short beak <5mm   **13**

5. IRIS

12 Capsule with beak 5-8mm, with 1 rib where 2 ovary-cells meet; leaves mostly <10mm wide; flowers mostly >8cm across. Stems to 90cm. Intrd-surv; persistent in swamp; W Kent
Japanese Iris - **I. ensata** Thunb.
12 Capsule with beak 8-16mm, with 2 ridges where 2 ovary-cells meet; leaves mostly >10mm wide; flowers mostly <8cm across. Stems to 90cm. Intrd-natd; wet places and by fen ditches; N Lincs, Dorset and N Somerset *Blue Iris* - **I. spuria** L.
13 Outer tepals glabrous on central patch; capsules setting many seeds. Stems to 1m. Intrd-natd; by lakes and rivers and in reed-swamps; scattered in En and N to C Sc *Purple Iris* - **I. versicolor** L.
13 Outer tepals pubescent on central patch; capsules setting 0-few seeds. Stems to 1m. Intrd-natd; in reed-swamp and rough pasture; by Lake Windermere (Westmorland) (*I. versicolor* x *I. virginica* L.)
Windermere Iris - **I. x robusta** E.S. Anderson

6. WATSONIA Mill. - *Bugle-lily*

Plants with corms; leaves *Iris*-like; inflorescence a spike; flowers slightly zygomorphic, with curved perianth-tube; tepals united into tube longer than the lobes, white, the lobes ± equal; filaments free, borne in perianth-tube; style very slender, with 3 deeply bifid branches.

1 Stems to 1m; leaves up to 60 x 4cm; perianth-tube 2.5-3.5cm; perianth-lobes 2.5-3.5cm. Intrd-natd; rough ground; Tresco (Scillies)
Bugle-lily - **W. borbonica** (Pourr.) Goldblatt

7. ROMULEA Maratti - *Sand Crocuses*

Plants with corms; leaves subterete, 4-grooved; flowers 1-several on very short stem, *Crocus*-like, actinomorphic; tepals united into tube much shorter than lobes, white or mauve, yellow inside at base, the lobes equal; filaments free, borne in perianth-tube; style very slender, with 3 bifid stigmas; ovary above ground at flowering.

1 Corm obliquely narrowed at base; leaves 5-10cm x 0.6-1mm, recurved; perianth 7-15mm incl. tube, usually mauve, sometimes white. Native; maritime sandy turf; very local near Dawlish (S Devon), common in all of CI *Sand Crocus* - **R. columnae** Sebast. & Mauri
1 Corm rounded at base; leaves 15-25cm x 1-2.5mm, erect or ± so; perianth 15-45mm incl. tube, white. Intrd-natd; at wall-base and in sparsely grassy area; Guernsey *Oniongrass* - **R. rosea** (L.) Eckl.

8. CROCUS L. - *Crocuses*

Plants with corms; leaves linear, ± flattened, with central whitish channel; flowers erect, 1-few on short underground pedicels that elongate at fruiting, actinomorphic; tepals united into long narrow tube, various colours, the lobes equal; filaments free, borne at apex of perianth-tube; style slender, with 3 (or more) branches near apex, each with variously divided stigmas; ovary subterranean at flowering.

Below the flower, and sheathing the ovary and part of the perianth-tube, are 1-2 bracts, borne immediately below the ovary, and 0-1 spathe, borne at the base of the pedicel. See *Colchicum* and *Sternbergia* (both Liliaceae) for differences.

1 Flowers appearing in autumn (Sep-Dec), often without leaves, never predominantly yellow   2
1 Flowers appearing in spring (Jan-Apr), usually with leaves, often predominantly yellow   7
    2 Anthers white to cream-coloured   3
    2 Anthers yellow   4
3 Throat of corolla uniformly deep yellow; filaments densely pubescent; style with many yellow or orange main branches. Intrd-natd; in churchyard; W Suffolk   **Hairy Crocus - C. pulchellus** Herb.
3 Throat of corolla whitish with yellow blotches; filaments glabrous to minutely pubescent; style with 3 cream to yellow main branches. Intrd-natd; in meadows and on grassy tracksides; Surrey and E & W Suffolk   **Kotschy's Crocus - C. kotschyanus** K. Koch
    4 Style with 3 main branches; throat of corolla uniformly yellow. Intrd-surv; persistent on grassy verge; Surrey
          Italian Crocus - C. longiflorus Raf.
    4 Style with many main branches; throat of corolla whitish to pale yellow   5
5 Corm with covering splitting into rings at base, not becoming fibrous; perianth conspicuously darker-veined outside. Intrd-natd; in churchyards and on waysides; Surrey, Berks and E Suffolk
        Bieberstein's Crocus - C. speciosus M. Bieb.
5 Corm with covering becoming fibrous, not splitting into rings at base; perianth usually not darker-veined outside   6
    6 Leaves developing well after flowering, 3-4 per shoot, 2-4mm wide; throat of corolla white to pale purple. Intrd-natd; fields, parks, grassy banks, the most thoroughly natd sp.; scattered in En and Wa, especially NW En, Kirkcudbrights
        Autumn Crocus - C. nudiflorus Sm.
    6 Leaves emerging during flowering, 5-7 per shoot, 0.5-2mm wide; throat of corolla white to pale yellow. Intrd-natd; grassland; Surrey   **Late Crocus - C. serotinus** Salisb.
7 Perianth predominantly pale to deep yellow, sometimes tinged or striped dark purple   8
7 Perianth predominantly white or pale mauve to dark purple, sometimes yellow on throat   10
    8 Corm with covering splitting horizontally into rings at base, scarcely vertically and not becoming fibrous or reticulated; ground colour of flowers creamy-white to yellow. Intrd-natd; in grassy places; scattered in S En
        Golden Crocus - C. chrysanthus (Herb.) Herb.
    8 Corm with covering splitting vertically, not horizontally, becoming fibrous or reticulated; ground colour of flowers yellow to deep

8. CROCUS

         yellow   9
9   Perianth-lobes uniformly bright yellow; leaves 0.5-1mm wide. Intrd-
     natd; in copse and on grassy verge; W Kent and Middlesex
                              Ankara Crocus - **C. ancyrensis** (Herb.) Maw
9   Perianth-lobes suffused or striped purplish-brown on outside, or if
     uniformly yellow then leaves 1-4mm wide. Intrd-natd; grassy places,
     meadows, churchyards, banks; scattered in En (*C. angustifolius*
     Weston x *C. flavus* Weston)   *Yellow Crocus* - **C. x stellaris** Haw.
       10 Throat of corolla yellow; spathe 0; bracts 2, white   11
       10 Throat of corolla white to mauve or purple; spathe 1, papery;
           bract 1, white   12
11   Corm with covering splitting horizontally into rings at base, scarcely
     vertically and not becoming fibrous or reticulated; corolla usually
     darker-veined or -striped on outside. Intrd-natd; on grassy verge;
     Middlesex   *Silvery Crocus* - **C. biflorus** Mill.
11   Corm with covering splitting vertically, not horizontally, becoming
     fibrous and reticulated; corolla not darker-veined or -striped on
     outside. Intrd-surv; persistent on common and in churchyard;
     Surrey   *Sieber's Crocus* - **C. sieberi** J. Gay
       12 Leaves mostly 2-3mm wide; flowers mauve to pale purple with
           white perianth-tube. Intrd-natd; habitats as for *C. vernus*; scattered
           in Br N to MW Yorks   *Early Crocus* - **C. tommasinianus** Herb.
       12 Leaves mostly 4-8mm wide; flowers white to deep purple, often
           with dark stripes outside; perianth-tube usually mauve to purple,
           white only if rest of perianth is white. Intrd-natd; the most
           commonly grown sp., grassy places, meadows, churchyards and
           banks; scattered in BI (*Spring Crocus* - **C. vernus** (L.) Hill)   13
13   Style usually equalling or exceeding stamens; tepal-lobes (2.5)3-5.5
     x 0.9-2cm, usually coloured, sometimes white   **C. vernus** ssp. **vernus**
13   Style usually distinctly shorter than stamens; tepal-lobes 1.5-3.5(5)
     x 0.4-1.2cm, very often white. Smaller, less handsome and rarer
     than ssp. *vernus*   **C. vernus** ssp. **albiflorus** (Schult.) Asch. & Graebn.

C. vernus x *C. tommasinianus* and *C. chrysanthus* x *C. biflorus* occur with the parents in SE En.

9. GLADIOLUS L. - *Gladioluses*

Plants with corms; leaves *Iris*-like; inflorescence a spike; flowers zygomorphic, with curved perianth-tube; tepals united into tube shorter than lobes, pinkish- to purplish-red, the lobes unequal, much narrowed at base; filaments free, borne on perianth-tube; style slender, with 3 short branches widened distally.

1   Stems to 50(90)cm, unbranched; flowers 3-8(10); perianth 3.5-5cm, the
     lobes 2.5-4 x 0.6-1.6cm. Native; among bracken in scrub; New Forest
     (S Hants)   *Wild Gladiolus* - **G. illyricus** W.D.J. Koch
1   Stems to 1m, often branched; flowers mostly 10-20; perianth 4-5.5cm,
     the lobes 3-4.5 x 1.5-2.5cm. Intrd-natd; old bulb-fields, field-margins,

roadsides and rough ground; scattered in extreme S En, frequent in CI and Scillies
Eastern Gladiolus - **G. communis** L.

10. IXIA L. - *Corn-lilies*

Plants with corms; leaves *Iris*-like; inflorescence a spike or a raceme of spikes; flowers actinomorphic; tepals united proximally into short or long tube, variously white or yellow to red, ± equal; filaments free, arising from top of or within perianth-tube; style very slender, with 3-lobed stigma.

1 Stems to 15(40)cm; perianth-tube 2-3mm; perianth-lobes mainly red but with white and/or yellow stripes. Intrd-natd; old bulb-fields or rough ground; Scillies *Red Corn-lily* - **I. campanulata** Houtt.
1 Stems to 1m; perianth-tube 3-7cm, very slender; perianth-lobes cream to pale yellow tinged with red. Intrd-surv; old bulb-fields; Scillies
Tubular Corn-lily - **I. paniculata** D. Delaroche

11. SPARAXIS Ker Gawl. - *Harlequinflowers*

Plants with corms; leaves *Iris*-like; inflorescence a spike; flowers ± actinomorphic, with brown, jaggedly-toothed, dark-streaked bracts; tepals united into straight tube shorter than lobes, mostly red and white, the lobes ± equal; filaments free, borne in perianth-tube, slightly asymmetrically arranged; style very slender, with 3 linear branches.

1 Stems to 45cm; perianth-tube 0.8-1.4cm; perianth-lobes 2-3cm, red or red-and-white striped, often yellow near base. Intrd-surv; persistent in and by old bulb-fields; Scillies
Plain Harlequinflower - **S. grandiflora** (D. Delaroche) Ker Gawl.

12. FREESIA Klatt - *Freesia*

Plants with corms; leaves *Iris*-like; inflorescence a spike; flowers slightly zygomorphic, with curved perianth-tube, sweetly scented; tepals united into tube longer than lobes, white, yellow, orange, pink, purple or mauve, the lobes ± equal; filaments free, borne in perianth-tube; style very slender, with 3 deeply bifid branches.

1 Stems to 40cm; perianth-tube 1.5-3cm; perianth-lobes 0.8-1.5cm. Intrd-surv; relic sometimes ± natd in and by old bulb-fields; Guernsey and Scillies *Freesia* - **F. x hybrida** L.H. Bailey

13. CROCOSMIA Planch. - *Montbretias*

Plants with corms that produce rhizomes; leaves *Iris*-like; inflorescence an often branched spike; flowers zygomorphic, with curved perianth-tube; tepals united into tube longer or shorter than lobes, orange to brick-red, the lobes rather unequal; filaments free, borne asymmetrically in perianth-tube; style slender, with 3 branches each entire to shortly bifid.

1 Leaves ribbed and pleated at least when young, at least some >3cm wide; perianth >4.5cm

13. CROCOSMIA

1 Leaves ribbed but not pleated, <3cm wide; perianth <4(5)cm 3
  2 Perianth-lobes c.1/2 as long as -tube or less, erecto-patent; stamens shorter than perianth. Stems to 1.2m. Intrd-natd; in marginal habitats, rough and waste ground; local in W Sc, W Ir, NW En, W Wa and S Br *Aunt-Eliza* - **C. paniculata** (Klatt) Goldblatt
  2 Perianth-lobes c. as long as -tube, widely spreading; stamens slightly exceeding perianth. Stems to 1.2m. Intrd-natd; in marginal habitats, rough and waste ground; E Cornwall and N Somerset *Giant Montbretia* - **C. masoniorum** (L. Bolus) N.E. Br.
3 Perianth-lobes c.1/2 as long as -tube, ± erect; perianth-tube very narrow at base, abruptly widened distally. Stems to 80cm. Intrd-natd; by roads, lakes and rivers; CW & SW Sc, Man, Caerns, E Cornwall and SW, MW & NW Ir *Potts' Montbretia* - **C. pottsii** (Baker) N.E. Br.
3 Perianth-lobes c. as long as -tube, ± patent; perianth-tube gradually expanded distally. Stems to 60cm. Intrd-natd; hedgerows, woods, by lakes and rivers, and on waste ground; scattered throughout BI, common in Ir, W Br and CI (*C. pottsii* x *C. aurea* (Hook.) Planch.) *Montbretia* - **C. x crocosmiiflora** (Lemoine) N.E. Br.

14. CHASMANTHE N.E. Br. - *Chasmanthe*

Plants with corms; leaves *Iris*-like; inflorescence a spike; flowers strongly zygomorphic, with curved perianth-tube; tepals united, variously red to orange, the lobes extremely unequal, the uppermost at least as long as tube and continuing its curvature, the 2 adjacent parallel to it but shorter, the 3 lower much shorter and slightly down-turned; filaments free, borne on perianth-tube; stamens conspicuously exserted; style long-exserted, very narrow, with 3 branches.

1 Stems to 1.3m; leaves to 80 x 3.5cm, with strong midrib; perianth-tube 3-3.5cm, yellow on lowerside, orange-red on upperside; uppermost perianth-lobe orange-red, 2-3.5cm. Intrd-natd; in damp shady places; Scillies, mainly Tresco *Chasmanthe* - **C. bicolor** (Ten.) N.E. Br.

164. AGAVACEAE - *Centuryplant family*

Perennials with thick, woody, sparsely branched stems; flower structure as in Liliaceae, but the huge rosettes of tough leaves, on the ground or at the ends of branches, separate Agavaceae at a glance.

1 Leaf-rosettes sessile on ground or ± so; perianth greenish- or brownish-yellow 2
1 Leaf-rosettes at ends of woody branches; perianth whitish 3
  2 Leaves with extremely strong spines at margins and apex **2. AGAVE**
  2 Leaves spineless **4. PHORMIUM**
3 Leaves spine-tipped, strongly recurved; perianth >4cm **1. YUCCA**
3 Leaves often sharply pointed but not spine-tipped, not recurved;

164. AGAVACEAE

perianth <1cm 3. CORDYLINE

1. YUCCA L. - *Spanish-daggers*

Stems usually branched, with leaf-rosettes at ends of branches; leaves with sharp spine at apex, entire or with few inconspicuous teeth; flowers hypogynous; perianth actinomorphic, bell-shaped, with lobes much longer than tube; fruit an indehiscent capsule.

1 Plant to 2(5)m; leaves up to 100 x 5cm, mostly strongly recurved; perianth creamy- or greenish-white, 5-8cm. Intrd-surv; on sand-dunes and in gravel-pits; Glam, S Devon and Worcs
Curved-leaved Spanish-dagger - **Y. recurvifolia** Salisb.

2. AGAVE L. - *Centuryplant*

Stems ± 0, the leaf-rosette sessile on ground; leaves with extremely sharp spine at apex and many more along margins; flowers epigynous; perianth ± actinomorphic, tubular, with tube much longer than lobes; fruit a dehiscent capsule.

1 Rosettes mostly 2-3m across, with massive succulent, tough, very spiny leaves 1-2m x 15-30cm; flowering stem rarely produced, to 7(12)m; perianth greenish-yellow, 7-10cm. Intrd-surv; very persistent where planted, sometimes surviving from suckers when main rosette dies after flowering; CI
Centuryplant - **A. americana** L.

3. CORDYLINE Juss. - *Cabbage-palm*

Stems well-developed, simple or branched, with leaf-rosettes at ends of branches; leaves entire, sharply pointed but without a spine; flowers hypogynous; perianth actinomorphic, with short tube and wide-spreading much longer lobes; fruit a berry, becoming dry with age.

1 Stems to 20m, becoming branched after first flowering; leaves up to 100 x 6cm; perianth white, 5-6mm, c.1cm across. Intrd-natd; much planted and very persistent by sea in W; producing seedlings in Man, CI and W Cornwall
Cabbage-palm - **C. australis** (G. Forst.) Endl.

4. PHORMIUM J.R. & G. Forst. - *New Zealand Flaxes*

Stems ± 0, the leaf-rosette sessile on ground; leaves entire, not spiny, folded proximally, nearly flat distally; flowers hypogynous; perianth slightly zygomorphic, with short tube and longer lobes, ± tubular; fruit a dehiscent capsule.

1 Rosettes 1-2m across, with suberect, extremely tough, fibrous leaves up to 3m x 12cm; flowering stem to 4m; perianth with outer lobes brownish-red, with inner lobes with not or slightly recurved tips. Intrd-natd; very persistent where planted on cliffs or rocky places by sea; W Cornwall, Scillies, W Cork, Man and CI, self-sown in Scillies
New Zealand Flax - **P. tenax** J.R. & G. Forst.

4. PHORMIUM

1 Smaller than *P. tenax*; leaves up to 2m x 7cm; flowering stem to 2m; perianth 2.5-4cm, with outer lobes greenish-yellow tinged with red, with ≥1 inner lobes strongly recurved at tip. Intrd-natd; planted and natd as for *P. tenax*; self-sown in Scillies
Lesser New Zealand Flax - **P. cookianum** Le Jol.

165. DIOSCOREACEAE - *Black Bryony family*

Glabrous, twining, herbaceous perennials with subterranean tuber; leaves alternate, simple, entire, petiolate; the only herbaceous twiner with dioecious inconspicuous flowers and red berries except *Bryonia* (Cucurbitaceae), which has pubescent, palmately lobed leaves and stem-tendrils.

1. TAMUS L. - *Black Bryony*

1 Stems to 5m; leaves c.5-15 x 4-11cm, broadly ovate, strongly cordate; perianth yellowish-green, 3-6mm across; berry 10-13mm across. Native; scrambling over hedges, shrubs and wood-margins; local in CI, common in Br N to Cumberland and Durham, rarely intrd elsewhere
Black Bryony - **T. communis** L.

166. ORCHIDACEAE - *Orchid family*

Erect, herbaceous perennials, sometimes ± chlorophyll-less saprophytes, with succulent roots, subterranean tubers or rhizomes; the distinctive flowers could be confused only by the very inexperienced with a few petaloid monocotyledons or dicotyledons (e.g. *Impatiens*, *Orobanche*, *Pinguicula*): the inferior ovary and 1(-2) stamens on a column will dispel the confusion.

The perianth has 6 free tepals in 2 whorls of 3, usually all petaloid (though sometimes greenish or brownish), the 3 outer ones ('sepals') similar, 2 of the inner ones ('petals') similar and often ± similar to 3 outer, the other (usually apparently the lowest but actually the uppermost due to twisting of the flower through 180°) usually strongly different and forming a lip (*labellum*) which is usually the largest and most conspicuous part of the flower, and often is extended from its base behind the flower into a hollow spur; stamens and stigmas borne on a special structure (*column*) in the centre of the flower; stamens 2 (*Cypripedium*) or 1 (others), each with sessile anther containing pollen as many single grains (*Cypripedium*) or as 2-4 masses (*pollinia*) (others): pollinia often stalked, often provided with a sticky pad (*viscidium*) (at base of stalk when present); ovary 1-celled; style 0; stigmas 3, either all receptive (*Cypripedium*) or 2 receptive and the third a sterile protrusion (*rostellum*) or ± 0 (others); fruit a capsule. Many of the spp. are prone to produce plants with unusual labellum-shapes or flower-colours (e.g. albinos), but these usually occur in populations of normal plants.

166. ORCHIDACEAE

Hybrids occur frequently within the tribe Orchideae, especially within *Dactylorhiza*, and should be looked for whenever ≥2 spp. of the tribe occur close together. Hybrids between 2 spp. with the same chromosome number are usually fertile, and even those between spp. with different chromosome numbers are not always completely sterile. 7 intergeneric hybrid combinations also occur, 6 of them between genera 14-18.

1	Plants saprophytic, without green leaves	2
1	Plants not saprophytic, with green leaves	4
2	Flowers not twisted upside down, hence labellum and spur directed ± upwards; spur >5mm; very rare	**4. EPIPOGIUM**
2	Flowers twisted upside down (as normal in orchids), hence labellum directed downwards; spur 0	3
3	Labellum 8-12mm, c.2x as long as other tepals, brown; flowers usually >20	**5. NEOTTIA**
3	Labellum ≤6mm, c. as long as other tepals, whitish-cream with reddish markings; flowers ≤12	**11. CORALLORRHIZA**
4	Spur present, sometimes very short	5
4	Spur 0	14
5	Labellum with 3 lobes, the central 3-6cm, linear, ribbon-like, in a loose spiral	**22. HIMANTOGLOSSUM**
5	Labellum without a central ribbon-like lobe ≥3cm	6
6	Spur <3mm	7
6	Spur >3mm	10
7	Flowers greenish-brown; labellum parallel-sided, with 3 short apical lobes	**17. COELOGLOSSUM**
7	Flowers white, cream or greenish-white, sometimes with red or pink markings; labellum not parallel-sided, with 1 or more laterally protruding lobes on either side	8
8	Central lobe of labellum entire; flowers never with pink or red tinge or markings	**15. PSEUDORCHIS**
8	Central lobe of labellum conspicuously 2-3-lobed at apex; flowers often with pink or red tinge or markings	9
9	Flowers white to pinkish; labellum <5mm, scarcely longer than outer tepals; Ir and Man only	**19. NEOTINEA**
9	Flowers white and dark purple, the latter predominating in unopened flowers hence at top of spike; labellum >5mm, c.2x as long as outer tepals; Br only	**20. ORCHIS**
10	Labellum linear, entire; flowers always white (or green-tinged)	**13. PLATANTHERA**
10	Labellum not linear, lobed (if scarcely so then very wide); flowers pure white only in rare albinos	11
11	Spur >(8)11mm, filiform, >6x as long as widest point	12
11	Spur ≤11mm, not filiform, <6x as long as widest point	13
12	Labellum plane, not raised into plates; spike ± cylindrical; flowers strongly scented; each pollinium becoming detached separately, each with a small stalk and basal sticky pad	**16. GYMNADENIA**
12	Labellum with 2 raised plates near its base; spike pyramidal;	

166. ORCHIDACEAE

  flowers not scented; each pollinium with its own stalk but later detached together on a common sticky pad **14. ANACAMPTIS**
13 Lower bracts herbaceous, green, often suffused purplish; labellum terminating in a single pointed to rounded tooth or lobe much smaller than or rarely nearly as large as the portions on either side
  18. DACTYLORHIZA
13 Bracts membranous, brown, often suffused purplish; labellum terminating in a forked lobe (often with a tooth in the notch) larger than lobes on either side, or in a small truncate or notched lobe smaller than portions on either side **20. ORCHIS**
  14 Labellum yellow, c.3cm, concavely bowl-shaped; other tepals maroon, 3-5cm; very rare **1. CYPRIPEDIUM**
  14 Labellum not concavely bowl-shaped; other tepals<2cm 15
15 Labellum velvety in texture, resembling insect's abdomen **23. OPHRYS**
15 Labellum not velvety, not resembling insect's abdomen 16
  16 At least some flowers not twisted upside down, hence labellum directed ± upwards 17
  16 Flowers all twisted upside down, hence labellum directed downwards 18
17 Labellum entire and flat at margin; leaves ≤2cm, with minute tubercles near apex **10. HAMMARBYA**
17 Labellum crenate or crisped at margin; leaves >2cm, without tubercles **9. LIPARIS**
  18 Labellum with 2 lobes at its apex exceeding all others 19
  18 Labellum with 1 lobe at its apex exceeding all others, sometimes this slightly notched 20
19 Leaves 2, on stem; labellum with 0 or 2 small lateral lobes much shorter than 2 apical lobes **6. LISTERA**
19 Leaves usually >2, the largest ones basal; labellum with 2 lateral lobes longer than 2 apical lobes **21. ACERAS**
  20 Flowers white to greenish- or yellowish-white, in 1-3 distinct spirals in spike, or if forming a strictly 1-sided spike then leaves all in basal rosette 21
  20 Flowers white or whitish only in rare albinos, not in distinct spirals, if forming a strictly 1-sided spike then leaves mainly on stems 22
21 Shortly rhizomatous; leaves conspicuously net-veined, in basal rosettes; labellum not frilly at edge **8. GOODYERA**
21 Tufted, with swollen roots; leaves not or inconspicuously net-veined, if in basal rosette then labellum frilly at edge **7. SPIRANTHES**
  22 Main leaves 2(-3), basal; labellum with apical lobe and 2 shorter laterals, not constricted **12. HERMINIUM**
  22 Main leaves (2)3-many, on stems (sometimes near base); labellum constricted at base of apical lobe, the proximal part often with 2 lateral lobes 23
23 Flowers erect to erecto-patent, sessile; proximal part of labellum partly wrapped round column **2. CEPHALANTHERA**
23 Flowers patent to pendent, stalked; proximal part of labellum

not wrapped round column **3. EPIPACTIS**

TRIBE 1 - CYPRIPEDIEAE (genus 1). Stamens 2 plus a large sterile projection (staminode); pollen dispersed as separate grains; receptive stigmas 3; labellum a deeply concave bowl; spur 0; flower single (very rarely 2), terminal.

1. CYPRIPEDIUM L. - *Lady's-slipper*

1 Stems to 30cm, rather pubescent. Native; north-facing grassy slope on limestone; 1 locality in MW Yorks, planted in a few sites in N En *Lady's-slipper* - **C. calceolus** L.

TRIBE 2 - NEOTTIEAE (genera 2-8). Fertile stamen 1; pollen dispersed as 2 often rather friable (or 2 each with 2 halves) pollinia; receptive stigmas 2; third stigma 0 or a sterile bulge (rostellum); stamen borne at back of column; pollinia sessile or with an apical stalk, with or without a sticky pad (pollinia with basal stalk and sticky pad in *Epipogium*); labellum of various shapes, often without well-marked lobes, often constricted near middle so delimiting proximal and distal parts; spur 0 (present in *Epipogium*).

2. CEPHALANTHERA Rich. - *Helleborines*

Shortly rhizomatous; leaves several, all on stem; flowers sessile or ± so, borne spirally, white or purplish-pink; spur 0; labellum constricted c.1/2 way into proximal and distal parts, neither markedly lobed, the proximal part partly wrapped round column; rostellum 0; pollinia without stalks.

1 Flowers purplish-pink; ovaries with glandular hairs; labellum acute. Stems to 60cm. Native; *Fagus* woods on chalk or limestone; very rare in N Hants, Bucks and E Gloucs   *Red Helleborine* - **C. rubra** (L.) Rich.
1 Flowers white with yellow or orange marks on labellum; ovaries glabrous; labellum obtuse 2
  2 Lower leaves ovate to rather narrowly so; bracts longer than ovaries; sepals obtuse. Stems to 60cm. Native; shady woods, on chalk and limestone; locally frequent in S En N to Northants and Herefs   *White Helleborine* - **C. damasonium** (Mill.) Druce
  2 Lower leaves lanceolate to narrowly elliptic-oblong; bracts shorter than ovaries; sepals acute. Stems to 60cm. Native; woods and shady places on calcareous soils; much less common but much more widespread than *C. damasonium*, scattered in most of Br and Ir   *Narrow-leaved Helleborine* - **C. longifolia** (L.) Fritsch

C. damasonium x *C. longifolia* = *C.* x *schulzei* E.G. Camus, Bergon & A. Camus occurred in 1974 and 1975 in S Hants with both parents.

3. EPIPACTIS Zinn - *Helleborines*

Rhizomatous, mostly shortly so; leaves several, all on stem; flowers distinctly pedicellate, borne spirally or ± on 1 side of stem, various dull

3. EPIPACTIS

colours; spur 0; labellum usually differentiated c.1/2 way into proximal and distal parts, neither markedly lobed, the proximal part ± cup-shaped and not wrapped round column; rostellum obvious and secreting a white sticky cap (*viscidium*), or minute and with 0 or vestigial viscidium; pollinia without stalks.

E. phyllanthes, *E. leptochila* and *E. youngiana* form a problematical complex of self-pollinated plants in which sp. limits are uncertain and disputed. They can be distinguished from the other spp. by having a rostellum which secretes little or no viscidium that usually withers soon after the flower opens. In the other (cross-pollinated) spp. the rostellum is obvious and the viscidium remains in the open flower until removed along with the pollinia by visiting insects. This can be effected with a match-stick or similar object; when the viscidium is touched and the object pulled away the pollinia are drawn out with it. In the self-pollinated spp. the pollinia crumble apart and cannot easily be pulled out whole.

1 Rhizome long; labellum strongly constricted separating proximal and distal portions, the proximal with erect triangular lobe on each side. Stems to 45(60)cm, pubescent. Native; fens, base-rich marshy fields, dune-slacks; locally frequent in Bl N to C Sc
Marsh Helleborine - E. palustris (L.) Crantz
1 Rhizome short or ± 0; labellum not or slightly constricted between proximal and distal portions, the proximal with 0 or obscure lateral lobes 2
   2 Inflorescence-axis glabrous or nearly so; flowers pendent as soon as they open; leaves often shorter than internodes. Stems to 40cm, glabrous or ± so. Native; woods on calcareous or sandy soils sometimes heavy-metal-polluted, and on dunes; scattered in Br N to Cheviot and Westmorland, very scattered in Ir
***Green-flowered Helleborine* - E. phyllanthes** G.E. Sm.
   2 Inflorescence-axis pubescent to densely so; at least younger flowers usually patent to erecto-patent; leaves usually longer than internodes 3
3 Ovary pubescent to densely so; perianth usually reddish-purple all over. Stems to 30(60)cm, densely whitish-pubescent. Native; limestone scrub, grassland, scree and rocky places; very locally frequent in CW Ir, N Sc, N Wa, N En S to Derbys
***Dark-red Helleborine* - E. atrorubens** (Hoffm.) Besser
3 Ovary glabrous to sparsely pubescent; perianth usually greenish often marked or tinged with pink, purple or violet, but not so coloured all over 4
   4 Upper leaves usually spirally arranged; rostellum secreting obvious, white, persistent viscidium; pollinia becoming detached as integral units 5
   4 Upper leaves usually obviously 2-ranked; rostellum without or with sparse, soon disappearing viscidium; pollinia crumbling apart 6
5 Leaves dark green, the lowest wider than long or almost so; distal

part of labellum wider than long, with 2 usually rough brownish bosses near base. Stems to 80(100)cm, pubescent above. Native; woods, scrub and hedgerows; frequent in Br and Ir except rare in N Sc ·*Broad-leaved Helleborine* - **E. helleborine** (L.) Crantz

5 Leaves greyish-green, often tinged violet, the lowest considerably longer than wide; distal part of labellum at least as long as wide, with 2 smoothly pleated pinkish bosses near base. Stems to 60(80)cm, often densely clumped, pubescent above. Native; woods on calcareous or sandy soil; frequent in SE & SC En N to Salop and Leics *Violet Helleborine* - **E. purpurata** Sm.

6 Rostellum >1/2 as long as anthers; stigma with 2 basal bosses, with the rostellum forming a 3-horned shape; ovary usually glabrous; petals pinkish; distal part of labellum wider than long. Stems to 60cm, pubescent above. Native; woodland on heavy, often heavy-metal-polluted, soils; S Northumb and Lanarks
Young's Helleborine - **E. youngiana** A.J. Richards & A.F. Porter

6 Rostellum ≤1/2 as long as anthers; stigma without marked basal bosses hence not 3-horned; ovary usually pubescent; petals pale green; distal part of labellum longer than wide or wider than long. Stems to 60cm, pubescent at least above. Native; woods mostly on calcareous or heavy-metal-polluted soils, river-gravels and dunes; locally scattered in Br N to Lanarks
Narrow-lipped Helleborine - **E. leptochila** (Godfery) Godfery

Hybrids of *E. helleborine* with *E. atrorubens* and *E. purpurata* occur rarely in Br where the parents meet.

4. EPIPOGIUM Borkh. - *Ghost Orchid*

Saprophytic, chlorophyll-less, with coral-like rhizome producing thin creeping rhizomes; leaves few, small and scale-like; flowers shortly pedicellate, not twisted upside down (hence spur and labellum point upwards), pale pink; spur present, c. as long as labellum; labellum with 2 short rounded lateral lobes at base; rostellum well developed; pollinia with basal stalk ending in viscidium.

1 Stems to 25cm, pinkish; flowers 1-2(4), c.15-20mm vertically across, patent to slightly pendent. Native; in deeply shaded woods on leaf-litter or rotten stumps; very rare in Herefs, Oxon and Bucks
Ghost Orchid - **E. aphyllum** Sw.

5. NEOTTIA Guett. - *Bird's-nest Orchid*

Saprophytic, chlorophyll-less, with very short rhizome wrapped with succulent roots; leaves few, small and scale-like; flowers shortly pedicellate, pale brown; spur 0; labellum divided apically into 2 lobes and with 2 small lateral teeth near base; other 5 tepals convergent to form loose hood; rostellum well developed; pollinia not stalked.

1 Stems to 50cm, brown; flowers numerous, crowded, c.15-20mm

vertically across, patent. Native; on leaf-litter in shady woods; scattered throughout most of Br and Ir
Bird's-nest Orchid - **N. nidus-avis** (L.) Rich.

6. LISTERA R. Br. - *Twayblades*
Shortly rhizomatous; leaves normally 2, in opposite pair on stem (on non-flowering stems the leaves are at the apex); flowers shortly pedicellate, yellowish-green to dull reddish; spur 0; labellum deeply divided apically into 2 lobes, sometimes with short tooth between them; rostellum well developed; pollinia not stalked.

1 Stems 20-60(75)cm; leaves ovate-elliptic, 5-20cm, with 3-5 prominent longitudinal veins; labellum yellowish-green, 7-15mm. Native; woods, hedgerows, grassy fields, dune-slacks, sometimes on *Calluna*-moors; frequent   *Common Twayblade* - **L. ovata** (L.) R. Br.
1 Stems to 10(25)cm; leaves triangular-ovate, 1-2.5cm, with prominent midrib; labellum dull reddish, 3.5-4.5mm. Native; upland woods and moors in usually wet, acid places, often among *Sphagnum* or under *Calluna*; frequent in Sc, scattered in Ir and in Br S to Derbys and N Devon   *Lesser Twayblade* - **L. cordata** (L.) R.Br.

7. SPIRANTHES Rich. - *Lady's-tresses*
With tuberous roots; leaves several, basal and on stem or ± all basal; flowers sessile, white with green markings, usually borne spirally in tight spike; spur 0; labellum ± unlobed, with slightly to markedly frilly distal edge, ± appressed to other tepals to form tubular or trumpet-shaped perianth; rostellum well developed; pollinia not stalked.

1 Leaves at flowering time obovate-elliptic, all in tight rosette adjacent to base of flowering stem which bears only reduced scale-leaves. Stems to 15(20)cm, with flowers in single spiral or 1-sided spike. Native; short permanent grassland and grassy dunes; locally frequent in BI N to NE Yorks, Man and Co Sligo
Autumn Lady's-tresses - **S. spiralis** (L.) Chevall.
1 Leaves at flowering time linear-lanceolate to -oblanceolate, around base of flowering stem and short way up it   2
  2 Flowers in 1 spiral row in spike, 6-8mm excl. ovary; bracts 6-9mm; leaves subacute to obtuse. Stems to 40cm. Native; marshy ground; Hants, Guernsey and Jersey, but extinct since 1959
Summer Lady's-tresses - **S. aestivalis** (Poir.) Rich.
  2 Flowers in 3 spiral rows in spike, 10-14mm excl. ovary; bracts 10-20(30)mm; leaves acute. Stems to 30cm. Native; marshy meadows near streams, rivers or lakes; extremely local in SW, W & NE Ir, CW & NW Sc, S Devon
Irish Lady's-tresses - **S. romanzoffiana** Cham.

8. GOODYERA R. Br. - *Creeping Lady's-tresses*
With rhizomes giving rise to sterile leaf-rosettes and flowering stems with

166. ORCHIDACEAE

basal leaf-rosette and reduced scale-like stem-leaves; flowers sessile, like those of *Spiranthes* but in weak spiral or 1-sided spike, and labellum with entire distal edge.

1 Stems to 20(25)cm; leaves ovate-elliptic, ± patent; flowers 3-5mm excl. ovary. Native; on barish ground under *Pinus* or *Betula* or rarely on moist dunes; local in N Br S to Cumberland, ?intrd in E & W Norfolk and E Suffolk *Creeping Lady's-tresses* - **G. repens** (L.) R.Br.

TRIBE 3 - EPIDENDREAE (genera 9-11). Fertile stamen 1; pollen dispersed as 4 (or 2, each with 2 halves) pollinia; receptive stigmas 2; third stigma a minute sterile bulge (rostellum); stamen borne at apex of column; pollinia sessile, with minute sticky pads; labellum rather small, simple or with 2 short lateral lobes; spur 0 or ± so.

9. LIPARIS Rich. - *Fen Orchid*
Leaves usually 2, on stem, green; stem with 2 basal tubers side-by-side; labellum directed upwards, downwards or any intermediate direction, frilly on margins, scarcely lobed.

1 Stems to 20cm; leaves 2.5-8cm, elliptic; flowers <20, yellowish-green, c.10mm vertically across. Native; wet peaty fens and dune-slacks; very local in N Devon, E & W Norfolk, Glam and Carms
Fen Orchid - **L. loeselii** (L.) Rich.

10. HAMMARBYA Kuntze - *Bog Orchid*
Leaves 2(-4), on stem, green; stem with 2 basal tubers, 1 above the other; labellum directed upwards, entire on margins, not lobed.

1 Stems to 8(12)cm; leaves 0.5-2cm, elliptic, with marginal fringe of tiny bulbils; flowers <20, yellowish-green, c.7mm vertically across. Native; on wet *Sphagnum* in bogs; sparse in CW & NW Sc, S Hants and C Wa, very rare elsewhere in Br and Ir *Bog Orchid* - **H. paludosa** (L.) Kuntze

11. CORALLORRHIZA Gagnebin - *Coralroot Orchid*
Yellowish-brown or yellowish-green, saprophytic; leaves all on stem, scale-like; stem with coral-like rhizome; labellum directed downwards, distinctly 3-lobed, entire on margins.

1 Stems to 20cm; leaves reduced to few sheaths on stem; flowers ≤12, yellowish-green tinged brown, c.6mm vertically across. Native; damp peaty or mossy ground under trees or shrubs in woods, scrub and dune-slacks; scattered in N Br S to MW Yorks
Coralroot Orchid - **C. trifida** Châtel.

TRIBE 4 - ORCHIDEAE (genera 12-23). Fertile stamen 1; pollen dispersed as 2 pollinia; receptive stigmas 2; third stigma a small to large sterile bulge (rostellum); stamen borne in front of column; pollinia on long (short in

Herminium) stalks each with a sticky pad or the 2 sharing 1 sticky pad; labellum often large and conspicuous, very variably lobed; spur 0 to very long.

12. HERMINIUM L. - *Musk Orchid*
Leaves 2(-4), near base of stem, plus usually 1 reduced leaf higher up; all tepals ± incurved; labellum narrow, with 2 short lateral lobes; the 2 petals rather similar but with shorter lobes; spur 0; plant with 1 ± globose underground tuber.

1 Stems to 15(25)cm; leaves elliptic-oblong, 2-7cm; flowers yellowish-green, c.6-8mm vertically across. Native; chalk and limestone grassland; local in S Br N to E Gloucs and Beds
Musk Orchid - **H. monorchis** (L.) R. Br.

13. PLATANTHERA Rich. - *Butterfly-orchids*
Leaves 2(-3), near base of stem, plus few reduced leaves higher up; flowers pure white to greenish-white; upper 3 tepals ± incurved; labellum linear-oblong, entire; 2 lateral sepals spreading; spur long and slender; plant with 2 ellipsoid underground tubers with tapering apices.

1 Flowers c.18-23mm transversely across; labellum 10-16mm; pollinia 3-4mm, divergent downwards; viscidia c.4mm apart; spur 19-28 x c.1mm. Stems to 60cm. Native; woods and (in N) in open grassland, usually on calcareous soils; locally frequent throughout Br and Ir, much commoner than *P. bifolia* in S
Greater Butterfly-orchid - **P. chlorantha** (Custer) Rchb.
1 Flowers c.11-18mm transversely across; labellum 6-10mm; pollinia c.2mm, parallel, the viscidia c. 1mm apart; spur 15-20 x c.1mm. Stems to 60cm. Native; similar habitats and distribution to *P. chlorantha*, commoner than it in N   *Lesser Butterfly-orchid* - **P. bifolia** (L.) Rich.

14. ANACAMPTIS Rich. - *Pyramidal Orchid*
Leaves several, decreasing in size up stem; upper 3 tepals ± incurved; labellum deeply and nearly equally 3-lobed, with 2 raised plates near its base; 2 lateral sepals spreading; spur long and slender; plant with 2 subglobose underground tubers.

See *Gymnadenia* for differences.

1 Stems to 60cm; leaves lanceolate, the lowest c.8-15cm; flowers in ± pyramidal dense spike, pinkish-purple (rarely white), c.10-12mm vertically across, with spur 12-14 x <1mm. Native; chalk and limestone grassland, calcareous dunes; locally frequent in Bl N to S Ebudes and Fife   *Pyramidal Orchid* - **A. pyramidalis** (L.) Rich.

14 x 16. ANACAMPTIS x GYMNADENIA = X GYMNANACAMPTIS Asch. & Graebn.
X *G. anacamptis* (F.H. Wilms) Asch. & Graebn. (*A. pyramidalis* x *G. conopsea*)

has been recorded from Hants, Gloucs and Co Durham; it has the labellum plates of *Anacamptis* and the scent and cylindrical spike of *Gymnadenia*.

15. PSEUDORCHIS Ség. - *Small-white Orchid*
Leaves several, decreasing in size up stem; upper 5 tepals ± incurved; labellum quite deeply and nearly equally 3-lobed, without raised plates at base; spur short, wide, rounded at apex; plant with cluster of tapering underground tubers.

1 Stems to 20(40)cm; leaves oblong-oblanceolate, the lowest 2.5-8cm; flowers in a cylindrical dense spike, creamy-white, c.2-4mm vertically across, with spur 2-3mm. Native; short grassland, usually base-rich and upland; frequent in C, W & N Sc, very scattered elsewhere in N Br, Ir and Wa   *Small-white Orchid* - **P. albida** (L.) Á. & D. Löve

15 x 16. PSEUDORCHIS x GYMNADENIA = X PSEUDADENIA P.F. Hunt

X P. schweinfurthii (A. Kern.) P.F. Hunt (*P. albida* x *G. conopsea*) has been recorded from several places in N Br and is still frequent in NW Sc with both parents; it is intermediate in size, perianth shape (especially spur) and colour (pale pink).

15 x 18. PSEUDORCHIS x DACTYLORHIZA = X PSEUDORHIZA P.F. Hunt

X P. bruniana (Brügger) P.F. Hunt (*P. albida* x *D. maculata*) was found in Orkney in 1977; it resembles *D. maculata* in stem and leaf characters, and *P. albida* in inflorescence shape, size and colour, but has intermediate floral characters.

16. GYMNADENIA R. Br. - *Fragrant Orchid*
Leaves several, decreasing in size up stem; upper 3 tepals ± incurved; labellum shallowly 3-lobed, without raised plates at base; 2 lateral sepals spreading; spur long and slender; plant with several divided tapering underground tubers.

Often grows with *Anacamptis*, but flowers earlier (little or no overlap); the shape of the tubers, labellum and spike, and the scented flowers, distinguish it.

1 Stems to 40(75)cm; leaves linear-lanceolate, the lowest c.6-15cm; flowers in ± cylindrical dense spike, sweetly scented, usually lilac-purple, c.8-12mm vertically across, with spur 8-17 x c.1mm. Native (*Fragrant Orchid* - **G. conopsea** (L.) R. Br.)   2
  2 Lateral sepals mostly 4-5 x c.2mm; labellum obscurely lobed, (3)3.5-4(5)mm wide. Flowers (7)8-10(12)mm horizontally across; spur (8)11-14(15)mm. Base-rich to -poor hilly grassland in Sc, W Wa and N & SW En, bogs in S Hants and E Sussex
                           G. conopsea ssp. **borealis** (Druce) F. Rose
  2 Lateral sepals mostly 5-7 x c.1mm; labellum conspicuously lobed,

(4.5)5.5-7(8)mm wide 3
3 Flowers (7)10-11(13)mm horizontally across; labellum scarcely wider than long, (4)5-6(6.5) x (4.5)5.5-6.5(7)mm; lateral sepals 5-6 x c.1mm; spur mostly 12-14mm. Dry chalk or limestone grassland; frequent in Br N to Co Durham, N Ir **G. conopsea ssp. conopsea**
3 Flowers (10)11-13(14.5)mm horizontally across; labellum much wider than long, (3)3.5-4(4.5) x (5.5)6.5-7(8)mm; lateral sepals 6-7 x c.1mm; spur mostly 14-16mm. Base-rich fens and usually N-facing chalk grassland; scattered and local in Br N to Westmorland, W Ross, scattered through Ir **G. conopsea ssp.densiflora**
(Wahlenb.) E.G. Camus, Bergon & A. Camus

16 x 17. GYMNADENIA x COELOGLOSSUM = X GYMNAGLOSSUM Rolfe

X G. jacksonii (Quirk) Rolfe (*G. conopsea* x *C. viride*) has been recorded sporadically throughout much of Br and Ir; the inflorescences resemble those of *Gymnadenia* but are tinged green and have a much shorter spur.

16 x 18. GYMNADENIA x DACTYLORHIZA = X DACTYLODENIA Garay & H.R. Sweet

Hybrids of this combination often resemble *Dactylorhiza* in general appearance but have usually faintly spotted leaves and scented flowers with a longer spur; the perianth is variously intermediate in details. Precise parentage is difficult to determine without knowledge of the sp. or spp. of *Dactylorhiza* present nearby, but hybrids involving *D. fuchsii*, *D. maculata*, *D. incarnata* and *D. praetermissa* have been reliably recorded in scattered localities throughout most of Br and Ir.

17. COELOGLOSSUM Hartm. - *Frog Orchid*

Leaves several, decreasing in size up stem; upper 5 tepals incurved; labellum oblong, shallowly 3-lobed near tip with the central lobe the shortest; spur very short, rounded; plant with 2 divided tapering underground tubers.

1 Stems to 20(35)cm; leaves elliptic-oblong, sometimes broadly so, the lowest c.1.5-5cm; flowers in ± cylindrical dense spike, yellowish-green tinged with reddish-brown, c.6-10mm vertically across; labellum 3.5-6mm; spur c.2mm. Native; grassland, especially on base-rich or calcareous soils; locally frequent throughout Br and Ir
Frog Orchid - **C. viride** (L.) Hartm.

17 x 18. COELOGLOSSUM X DACTYLORHIZA = X DACTYLOGLOSSUM P.F. Hunt & Summerh.

Hybrids of this combination usually have flowers of the colour of the *Dactylorhiza* parent variously tinged or overlaid with green; other characters of habit, leaves and perianth shape are variously intermediate. Precise parentage is difficult to determine without knowledge of the sp. or spp. of *Dactylorhiza* present nearby. Hybrids involving *D. fuchsii*, *D. maculata* and *D.*

purpurella have been reliably reported, others less certainly.

18. DACTYLORHIZA Nevski - *Marsh-orchids*

Leaves several, the lower sheathing stem, the upper transitional to bracts and not sheathing (though often clasping) stem; upper 3 tepals ± incurved; labellum usually shallowly 3-lobed, sometimes more deeply so or ± unlobed, nearly always as wide as or wider than long; 2 lateral sepals spreading, erect or bent down; spur down-pointed, usually <10mm, mostly rather wide; plant with divided tapering underground tubers.

A very difficult genus owing to ready hybridization between any of the spp., and the complex pattern of variation within most spp. whereby considerable differences between populations are often evident. Except with typical material it is often not possible to identify single specimens; before using the key the population should be surveyed and means of 5-10 non-extreme plants calculated.

1 Stem solid; leaves nearly always spotted; usually 2-6 reduced non-sheathing leaves present on stem transitional between main (sheathing) leaves and bracts; lateral sepals spreading horizontally or bent down; spur usually <2mm wide at midpoint   2
1 Stem usually hollow, at least below; leaves often not spotted; usually 0-2 reduced non-sheathing leaves present on stem transitional between main (sheathing) leaves and bracts; lateral sepals ± erect; spur usually >2mm wide at midpoint   3
   2 Labellum lobed c.1/2 way to base with central lobe usually exceeding the two laterals and ≥1/2 as wide as them; leaves mostly subacute to obtuse, with spots usually ± transversely elongated. Stems to 50(70)cm. Native; damp woods, banks and meadows, marshes and fens, usually on base-rich soil; ± common throughout BI   *Common Spotted-orchid* - **D. fuchsii** (Druce) Soó   687
   2 Labellum lobed much <1/2 way to base, with central lobe as long as or shorter than 2 laterals and much ≤1/2 as wide as them; leaves mostly narrowly acute to subacute, with usually ± circular spots. Stems to 40(50)cm. Native; damp peaty places in bogs, marshes and ditches; ± common throughout BI
                            Heath Spotted-orchid - **D. maculata** (L.) Soó   687
3 Leaves unspotted or with spots on both surfaces, yellowish-green, narrowly hooded at apex; labellum usually with markedly reflexed sides (if slightly reflexed then leaves with spots on both surfaces) hence appearing very narrow from front, usually with 2 distinct dark loop-shaped marks side by side. Stems to 40(80)cm. Native (*Early Marsh-orchid* - **D. incarnata** (L.) Soó)   4   687

FIG 687 - Labella of *Dactylorhiza*. 1, *D. incarnata* (4th in row, ssp. *ochroleuca*). 2, *D. praetermissa*. 3 (1st 3), *D. purpurella*; 3 (4th in row), *D. praetermissa* var. *junialis*. 4, *D. majalis* ssp. *occidentalis*. 5, *D. majalis* ssp. *cambrensis*. 6, *D. traunsteineri*. 7, *D. lapponica*. 8, *D. fuchsii*. 9, *D. maculata*. Drawings by R.H. Roberts.

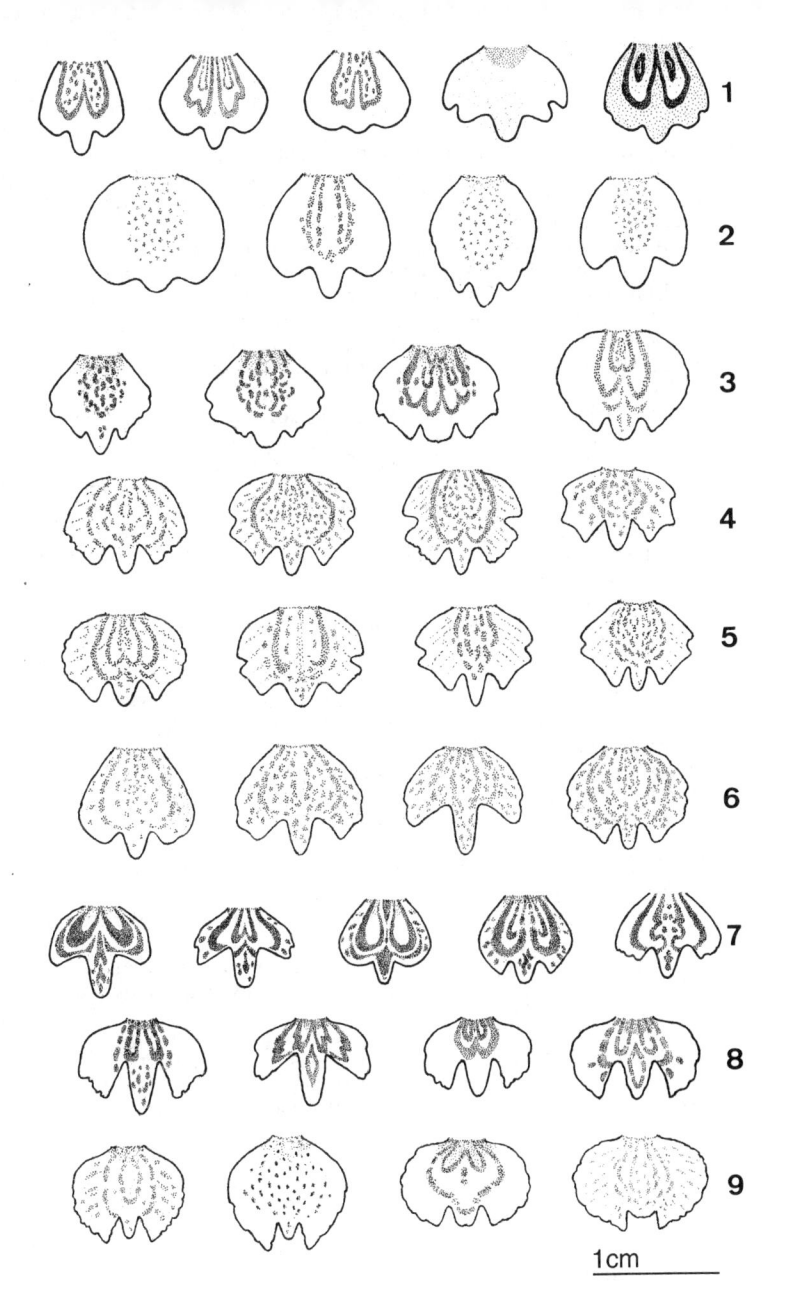

FIG 687 - see caption opposite

3   Leaves unspotted or with spots mostly on upperside, mid-, dark- or greyish-green, not or broadly hooded at apex; labellum usually without markedly reflexed sides, often nearly flat, usually without 2 distinct dark loops   9

   4   At least some plants with leaves with spots on both surfaces. Plants mostly 15-40cm; perianth ground-colour pinkish-mauve; labellum usually <7.5 x 8mm, usually with obvious central lobe. Marshes on limestone by lakes in WC Ir, neutral mountain flushes in W Ross   **D. incarnata ssp. cruenta** (O.F. Müll.) P.D. Sell

   4   Plants with unmarked leaves, rarely some with spots on upperside only   5

5   Perianth pale yellow or cream; labellum usually >6.5 x 8mm and with well-marked lobes, the lateral ones usually indented; lowest bract usually >30 x 20mm. Plants 20-50cm. Calcareous fens in E Anglia
D. incarnata ssp. ochroleuca (Boll) P.F. Hunt & Summerh.

5   Perianth variously pink to purple, rarely white or cream (if so then labellum usually <6.5 x 8mm and without indented lateral lobes and lowest bract usually <30 x 20mm)   6

   6   Ground-colour of perianth pink; bracts usually lacking anthocyanin   7

   6   Ground-colour of perianth red to purple; bracts usually strongly suffused with anthocyanin   8

7   Plants to 40cm with <6 leaves; labella usually <7 x 8.5mm, marked with lines; spurs usually <7.5mm. Wet meadows, fens and marshes on base-rich or neutral soils; locally frequent in En and Wa, extremely scattered in Sc and Ir   **D incarnata ssp. incarnata**

7   Plants to 50(80)cm with ≥6 leaves; labella usually >7 x 8.5mm, marked with dots; spurs usually >7.5mm. Base-rich fens and marshes; very local in E Norfolk and W Galway
D incarnata ssp. gemmana (Pugsley) P.D. Sell

   8   Plants mostly >20cm; perianth ground-colour reddish-purple. Bogs and other neutral to acid wet peaty places; scattered throughout Br and Ir   **D incarnata ssp. pulchella** (Druce) Soó

   8   Plants mostly <20cm; perianth ground-colour vivid ruby- or crimson-red. Damp base-rich sandy areas near sea, damp inland lake-shores in Ir; frequent throughout Ir, locally common by coast in W Br N to Shetland and in NE En
D. incarnata ssp. coccinea (Pugsley) Soó

9   Total number of leaves usually <5, the widest <1.5(2)cm wide; labellum usually distinctly 3-lobed, with central lobe distinctly exceeding 2 laterals and usually >1/2(1/3) as long as unlobed basal part   10

9   Total number of leaves usually ≥5 (except in some Scottish *D. majalis*), the widest >(1.5)2cm wide; labellum usually not strongly 3-lobed, if so then central lobe usually <1/3 as long as unlobed basal part (except *D. majalis*)   11

   10   Leaves with strong dark spots and rings on upperside; bracts predominantly green, suffused reddish-purple at margins and

apex, with dark spots and rings on upperside; transitional non-sheathing leaves (0-)2. Stems to 21cm. Native; slightly acidic to base-rich hillside flushes; very local in Westerness, Kintyre, N Ebudes and Outer Hebrides

Lapland Marsh-orchid - **D. lapponica** (Hartm.) Soó 687

10 Leaves with 0 or faint dark spots or rings on upperside; bracts strongly suffused reddish-purple all over, without dark spots or rings; transitional non-sheathing leaves 0-1(2). Stems to 30(40)cm. Native; calcareous fens and other damp base-rich grassy places; local in Ir, W Sc, NW & SW Wa, N En and E Anglia

Narrow-leaved Marsh-orchid - **D. traunsteineri** (Rchb.) Soó 687

11 Labellum with well-marked narrow central lobe usually 1/3-1/2 as long as unlobed basal part, with dark spots or lines mostly in central part but usually some extending almost to margins; leaves unmarked (parts of Ir only) or with strong spots or blotches mostly >2mm across. Native; marshes, fens, wet meadows and dune-slacks
(*Western Marsh-orchid* - **D. majalis** (Rchb.) P.F. Hunt & Summerh.) 12

11 Labellum usually obscurely lobed, with dark spots or lines usually confined to central part; leaves usually unmarked, sometimes with small spots <2mm across or with rings (rarely larger spots) 13

12 Stems mostly <20cm; largest leaves mostly <10cm x >2cm; spur usually <3.5mm wide at entrance. Scattered over Ir (mostly S & W), N Uist (Outer Hebrides)

D. majalis ssp. occidentalis (Pugsley) P.D. Sell 687

12 Stems mostly 20-30cm; largest leaves mostly >10cm x <2cm; spur usually >3.5mm wide at entrance. NW & MW Wa, SE Yorks, and NW & N Sc **D. majalis ssp. cambrensis** (R.H. Roberts) R.H. Roberts 687

13 Leaves usually broadly hooded at apex, their dark markings (if present) small spots <2mm across or rarely larger; labellum ± rhombic, usually reddish-purple, usually <7.5 x 9.5mm. Stems to 25(40)cm. Native; similar places to *D. praetermissa* (its northern counterpart); frequent in N Br S to SE Yorks, Derbys and Pembs, frequent in N Ir, scattered in S Ir

Northern Marsh-orchid - **D. purpurella** (T. & T.A. Stephenson) Soó 687

13 Leaves flat or slightly hooded at apex, their dark markings (if present) mainly as rings; labellum orbicular to transversely broadly elliptic, usually pinkish- or pale mauvish-purple, usually >7.5 x 9.5mm. Stems to 50(70)cm. Native; slightly acid to calcareous damp places in fens, marshes, bogs, meadows, gravel-pits and waste alkali-, colliery- and ash-tips; frequent in Br N to S Northumb and W Lancs, Cl *Southern Marsh-orchid* - **D. praetermissa** (Druce) Soó 687

Of the 8 spp., *D. fuchsii* and *D. incarnata* are diploid (2n=40) and the others are tetraploid (2n=80). Hybrids occur in almost any combination wherever mixed populations exist. Hybrids within a ploidy level are highly fertile, those between ploidy levels highly but not completely sterile; even in the case of triploid hybrids (2n=60) backcrossing and introgression can occur. Identification of hybrids involves careful examination of the characters of the putative parents in the sites concerned. The hybrids are intermediate in most characters, have low pollen fertility if triploid, and often show marked hybrid vigour. 18 hybrid combinations have been found. *D. fuchsii* x *D. praetermissa* = ***D. x grandis*** (Druce) P.F. Hunt occurs throughout the range of *D. praetermissa* in Br and is probably the commonest hybrid orchid in S Br. *D. maculata* x *D. purpurella* = ***D. x formosa*** (T. & T.A. Stephenson) Soó occurs throughout the range of *D. purpurella* in Br and Ir and is probably the commonest hybrid orchid in N Br and Ir.

19. NEOTINEA Rchb. f. - *Dense-flowered Orchid*

Leaves 2-3(4) near base of stem, with reduced leaves up stem; upper 5 tepals incurved; labellum with 2 large lateral lobes and larger terminal lobe usually shallowly subdivided at apex or with small tooth between the divisions; spur short, rounded at apex; plant with 2 ovoid underground tubers.

1 Stems to 30(40)cm; leaves oblong-elliptic, occasionally purple-spotted, 2-6cm; flowers in dense cylindrical spike, creamy-white or pink-tinged, c.5-6mm vertically across, with labellum 4-5mm, with spur c.2mm. Native; rocky and sandy grassy places and maritime dunes; very local in CW & SW Ir, very rare in W Donegal and Man
Dense-flowered Orchid - **N. maculata** (Desf.) Stearn

20. ORCHIS L. - *Orchids*

Leaves several, on stem, with few smaller ones above; upper 3 or upper 5 tepals incurved, the 2 lateral sepals incurved or erect to patent; labellum with 2 lateral lobes and a terminal lobe, the latter often larger than the laterals and usually 2-3-lobed at its apex; spur long or short, rounded to truncate or emarginate at apex; plant with 2 ovoid underground tubers.

1 Upper 3 tepals incurved to form a 'helmet'; 2 lateral sepals erect to patent 2
1 All 5 upper tepals incurved to form a 'helmet' 3
    2 Labellum with terminal lobe exceeded by 2 laterals, sometimes ± 0; spur shorter than ovary; bracts 3-veined or the lowest few 5-veined; leaves never spotted. Stems to 50(80)cm. Native; marshy meadows and by lakes; locally common in Jersey and Guernsey *Loose-flowered Orchid* - **O. laxiflora** Lam.
    2 Labellum with terminal lobe exceeding laterals; spur at least as long as ovary; bracts 1-veined or the lowest few 3-veined; leaves usually dark-spotted. Stems to 40(60)cm. Native; neutral or base-rich grassland, scrub and woods, usually in shade in S, in open in N; frequent to common *Early-purple Orchid* - **O. mascula** (L.) L.
3 Area of central lobe of labellum from smaller than to slightly larger than that of each lateral lobe 4
3 Area of central lobe of labellum at least 2x that of each lateral lobe 5
    4 Spur horizontal or directed upwards, ± as long as ovary; labellum mauvish-purple with paler, spotted, central area. Stems to

20(40)cm. Native; base-rich to neutral short undisturbed grassland; local over most of En, Wa, Ir and CI, Ayrs
Green-winged Orchid - **O. morio** L.

4 Spur directed downwards, <1/2 as long as ovary; labellum white with reddish-purple spots. Stems to 15(30)cm. Native; short grassland on chalk and limestone; extremely local over much of En, Glam **Burnt Orchid** - **O. ustulata** L.

5 Outside of all 3 sepals (forming 'helmet') dark reddish-purple, contrasting strongly with very pale labellum; 2 main sublobes of terminal lobe of labellum wider than long. Stems to 50(100)cm. Native; woods and scrub, rarely open grassland, on chalk; locally frequent on N Downs in E & W Kent, very scattered (mostly extinct) elsewhere in S En *Lady Orchid* - **O. purpurea** Huds.

5 Outside of sepals pale pinkish-purple, scarcely or not contrasting with labellum; 2 main sublobes of terminal lobe of labellum longer than wide 6

   6 Two main sublobes of terminal lobe of labellum oblong, >2x as wide as lateral lobes. Stems to 45(60)cm. Native; chalk grassland and old chalk-pit with invading trees and shrubs; 4 sites in Bucks, Oxon and W Suffolk *Military Orchid* - **O. militaris** L.

   6 Two main sublobes of terminal lobe of labellum linear, c. as wide as lateral lobes. Stems to 30(40)cm. Native; chalk grassland and open scrub; 4 sites in E Kent and Oxon
Monkey Orchid - **O. simia** Lam.

3 hybrids have been found very sporadically in mixed populations.

20 x 21. ORCHIS x ACERAS = X ORCHIACERAS E.G. Camus
Hybrids between *A. anthropophorum* and both *O. simia* and *O. purpurea* have been found in E Kent.

21. ACERAS R. Br. - *Man Orchid*
Leaves several, near stem-base, with few smaller ones above; upper 5 tepals incurved to form 'helmet'; labellum with 3 lobes, the lateral linear, the terminal larger and with 2 linear terminal sublobes; spur 0; plant with 2 ovoid underground tubers.

1 Stems to 40(50)cm; leaves narrowly elliptic-oblong, unspotted; flowers greenish-yellow, often tinged reddish-brown. Native; chalk and limestone grassland or scrub; local in SE En, scattered W to Dorset and N to S Lincs *Man Orchid* - **A. anthropophorum** (L.) W.T. Aiton

22. HIMANTOGLOSSUM W.D.J. Koch - *Lizard Orchid*
Leaves several, decreasing in size up stem; upper 5 tepals incurved to form 'helmet'; labellum very long and narrow, with 2 linear lateral lobes and a long terminal lobe spirally coiled at first and with 2 small sublobes at apex; spur short; plant with 2 ovoid underground tubers.

1 Stems to 70(90)cm; leaves elliptic-oblong, purple-mottled or not; flowers greyish-yellowish-green; labellum 4-7cm, the central lobe 3-6cm. Native; on calcareous soils in rough ground, dunes, scrub and marginal places usually among tall grass; scattered and sporadic in S & E En W to N Somerset and N to W Suffolk
Lizard Orchid - H. hircinum (L.) Spreng.

23. OPHRYS L. - *Orchids*
Leaves several, decreasing in size up stem; upper 5 tepals all patent, the 2 petals markedly different from the 3 sepals; labellum velvety in texture, subentire or with 2 small lateral lobes, the terminal lobe large and resembling an insect's abdomen; spur 0; plant with 2 ovoid underground tubers.

1 Labellum with distinct lateral lobes; the 2 petals filiform; labellum distinctly longer than wide. Stems to 60cm; sepals yellowish-green. Native; woods, scrub, grassland, spoil-heaps, fens and lakesides on calcareous soils; scattered throughout Br N to Westmorland and NE Yorks, C Ir   **Fly Orchid - O. insectifera** L.
1 Labellum with 0 or obscure lateral lobes; the 2 petals oblong to linear-oblong; labellum not or only just longer than wide   2
  2 Sepals yellowish- to brownish-green; the 2 petals yellowish-green, >1/2 as long as sepals. Stems to 20(35)cm. Native; grassland or spoil-heaps on chalk or limestone; very local from E Kent and W Suffolk to Dorset and W Gloucs
Early Spider-orchid - O. sphegodes Mill.
  2 Sepals pink or greenish-pink; the 2 petals pink or greenish-pink, ≤1/2 as long as sepals   3
3 Apex of labellum shortly bilobed, with the short simple projection between directed downwards and backwards (hence ± invisible from front of flower). Stems to 45(60)cm. Native; grassland, scrub, spoil-heaps and sand-dunes on calcareous or base-rich soils; locally frequent in Br N to Cumberland and Durham, CI, scattered in Ir
Bee Orchid - O. apifera Huds.
3 Apex of labellum shortly bilobed, with the short projection between directed prominently forwards and often 3-toothed. Stems to 35(55)cm. Native; short grassland on chalk; very local in E Kent
Late Spider-orchid - O. fuciflora (Crantz) Moench

3 hybrids have been found very sporadically in mixed populations.

GLOSSARY

For some special terms, used only in 1 or few families, direct reference to the relevant family (families) is made; in those cases the family description and the notes immediately following it should be consulted. Words in the definitions that are given in **bold** are themselves defined elsewhere in the glossary.

abaxial - of a lateral organ, the side away from the axis, normally the **lowerside**

achene - a dry, indehiscent, 1-seeded fruit, ± hard, with papery to leathery wall; **achene-pit**, see 139. Asteraceae

acicle - a slender prickle with scarcely widened base

actinomorphic - of a flower with radial (i.e. >1 plane of) symmetry

acuminate - gradually tapering to a point; Fig 699

acute - with point <90°; Fig 699

adaxial - of a lateral organ, the side towards the axis, normally the **upperside**

adherent - joined or fused

aerial - above-ground or above-water

agamospecies - a group of **apomictic** plants treated at the species level, but usually exhibiting a much narrower range of variation than a sexual species

alien - not **native**, introduced to a region deliberately or accidentally by man

alternate - lateral organs on an axis 1 per **node**, successive ones on opposite sides

anastomosing - dividing up and then joining again, usually applied to veins

androecium - the group of male parts of a flower; all the **stamens**; Fig 702

andromonoecious - having male and **bisexual** flowers on the same plant

annual - completing its life-cycle in ≤12 months (but often not within 1 calendar year)

anther - pollen-bearing part of a **stamen**, usually terminal on a stalk or **filament**; Fig 702

anthesis - flowering time; strictly pollen-shedding time

apiculus - a small, abruptly delimited point; **apiculate**, with an apiculus; Fig 699

apomictic - producing seed wholly female in origin, without fertilization

appendage - small extra protrusion or extension, such as on a **petal**, **sepal** or seed

appendix - see 151. Araceae

appressed - lying flat against another organ

aril - **succulent** covering around a seed, outside the **testa** (not the **pericarp**)
aristate - extended into a long bristle; Fig 699
ascending - sloping or curving upwards
auricle - basal extension of a leaf-blade, especially in Poaceae; Fig 572
awn - see 157. Poaceae
axil - angle between main and lateral axes; **axillary**, in the axil; see also **subtend**
axile - of a **placenta** formed by central axis of an **ovary** that is connected by **septa** to the wall; Fig 700

beak - a narrow, usually apical, projection
berry - a **succulent** fruit, the seeds usually >1 and without a stony coat
biennial - completing its life-cycle in >1 but <2 years, not flowering in the first year
bifacial - flattened, with 2 main surfaces, e.g. **lowerside** and **upperside**; cf. **unifacial**
bifid - divided into two, usually deeply, at apex
bifurcate - dividing into two branches
biotype - a genetically fixed variant of a **taxon** particularly adapted to some (usually environmental) condition
birdseed-alien - **alien** introduced as contaminant of birdseed
bisexual - of a plant or flower, bearing both sexes
blade - main part of a flat organ (e.g. **petal**, leaf); cf. **claw**, **petiole**; Figs 572, 699
bloom - delicate, waxy, easily removed covering to fruit, leaves etc.; see also **pruinose**
bract - modified, often scale-like, leaf **subtending** a flower, less often a branch; **bracteate**, with bract(s); Fig 701
bracteole - a supplementary or secondary **bract**, or a bract once removed; Fig 701
bud-scales - scales enclosing a bud before it expands
bulb - swollen underground organ consisting of condensed stem and **succulent** scale-leaves
bulbil - a small bulb or tuber, usually **axillary**, on an aerial part of the plant
bullate - with the surface raised into blister-like swellings

callus - see 157. Poaceae
calyx (plural **calyces**) - the outer **whorl(s)** of the **perianth**, if different from the inner; all the **sepals**; **calyx-tube**, **calyx-lobes**, the **proximal** fused and **distal** free parts of a calyx in which the **sepals** are partly fused; Fig 702
capillary - hair-like
capitate - head-like, such as a tight inflorescence on a stalk, a knob-like stigma on a style, or a stalked gland
capitulum - see 138. Dipsacaceae, 139. Asteraceae; Fig 701
capsule - a dry, many-seeded dehiscent fruit formed from >1 carpel
carpel - the basic female reproductive unit of Magnoliopsida, 1-many per flower, if >1 then **free** or fused; Fig 702

GLOSSARY

carpophore - a stalk-like sterile part of a flower between the **receptacle** and **carpels**, as in some Apiaceae and Caryophyllaceae; Fig 702

cartilaginous - cartilage-like in consistency, hard but easily cut with a knife, not green

caryopsis - see 157. Poaceae

casual - an **alien** plant not **naturalized**, persisting for only a short time

catkin - a condensed **spike** of reduced flowers on a long axis, often flexible and wind-pollinated

cell - of an **ovary**, the chambers into which it may be divided (often each one corresponding to a **carpel**); Fig 700

cladode - see 162. Liliaceae: Asparagoideae

clavate - club-shaped, slender and distally thickened

claw - **proximal**, narrow part of a flat organ such as a **petal**, bearing the **blade** distally

cleistogamous - of flowers, not opening, becoming self-pollinated in the bud stage

column - a stout stalk formed by fusion of various floral parts in, e.g., Orchidaceae, Geraniaceae, *Rosa*; **columnar**, column-like

commissure - see 111. Apiaceae

compound - not **simple**, of a leaf divided right to the **rhachis** into **leaflets**; Fig 700

compressed - flattened

cone - compact body composed of axis with lateral organs bearing spores or seeds, as in Lycopodiopsida, Equisetopsida, Pinopsida; **cone-scales**, the lateral organs of a cone

connective - part of **anther** connecting its 2 halves

contiguous - touching at the edges with no gap between

convergent - of ≥2 organs with apices closer together than their bases

cordate - of the base of a flat organ; see Fig 699

coriaceous - of leathery texture

corm - short, usually erect, swollen underground stem

corolla - the inner **whorls** of the **perianth**, if different from the outer; all the **petals**; **corolla-tube**, **corolla-lobes**, the **proximal** fused and **distal** free parts of a **corolla** in which the **petals** are partly fused; Fig 702

corona - a funnel- or collar-shaped petaloid outgrowth borne inside the tepals, as in *Narcissus*

corymb - a **raceme** in which the lower flowers have longer **pedicels**, producing a ± flat-topped **inflorescence**; **corymbose**, corymb-like though not necessarily strictly a corymb; Fig 701

cotyledon - the first leaves of a plant, 1 in Liliidae, usually 2 in Magnoliidae, 2-several in Pinopsida, usually quite different in appearance from all subsequent leaves

crenate - of the margin of a flat organ; see Fig 699

culm - see 157. Poaceae

cuneate - of the base of a flat organ; see Fig 699

cupule - see 41. Fagaceae

cuspidate - abruptly narrowed to a point; Fig 699

cyme - an **inflorescence** in which each flower terminates the growth of a branch, more **distal** flowers being produced by longer branches lateral to it; **cymose**, in the form of a cyme; Fig 701

decaploid - see **polyploid**
deciduous - not persistent, e.g. leaves falling in autumn or **petals** falling after **anthesis**
decumbent - **procumbent** but with the apex turning up to become **ascending** or **erect**
decurrent - of a lateral organ, having its base prolonged down the main axis
decussate - **opposite**, with successive pairs at right angles to each other
dehiscent - opening naturally
dentate - with a row of ± patent teeth; cf. serrate; Fig 699
denticulate - minutely or finely **dentate**
depressed-globose - similar to **globose** but wider than long
dichasium - cyme with 2 lateral branches at each **node**; Fig 701
dimorphic - occurring in 2 forms
dioecious - having the 2 sexes on different plants
diploid - having 2 matching sets of chromosomes, as in **sporophytic** tissue
disc - anything disc-shaped, e.g. top of *Rosa* or *Nuphar* fruit, nectar-secreting ring inside flower at base; **disc flowers**, see 139. Asteraceae
dissected - deeply divided up into segments
distal - at the end away from point of attachment; cf. **proximal**
divaricate - dividing into widely **divergent** branches
divergent - of ≥2 organs with apices further apart than their bases
dorsiventral - with distinct upperside and lowerside
drupe - a **succulent** or spongy fruit, the seeds usually 1 and with a stony coat
dry - not **succulent**

e- - without, e.g. **eglandular**, **ebracteate**
ellipsoid - a solid shape elliptic in side view
elliptic - a flat shape widest in middle and 1.2-3x as long as wide (if less **broadly** so; if more, **narrowly** so); Fig 699
emarginate - with a pronounced, angled notch at the apex; cf. **retuse**; Fig 699
embryo-sac - see **gametophyte**, **ovule**
endemic - confined to one particular area, i.e. (in this book) to BI
endosperm - In Magnoliopsida the nutritive tissue for the embryo in the developing seed; it might or might not remain as the food-store in the mature seed
entire - of the margin of a flat organ, not **toothed** or **lobed**; see Fig 699
epicalyx - organs on the outside of a flower, calyx-like but outside and additional to the **calyx**
epicormic - of new shoots borne direct from the trunk of a **tree**
epigynous - of a flower with an **inferior ovary**; Fig 702
erect - upright
erecto-patent - between **erect** and **patent**
escape - a plant growing outside a garden but having spread vegetatively or

GLOSSARY

by seed from one
evergreen - retaining leaves throughout the year
exceeding - longer than
exserted - protruding from

falcate - sickle- or scythe-shaped
false-fruit - an apparent **fruit** actually formed by tissue (e.g. **receptacle, bracts**) in addition to the real fruit
fascicle - a bunch or bundle, usually with short and indeterminate branching pattern; **fasciculate**, with or in the form of fascicles
fastigiate - a plant with upright branches forming a narrow outline
fibrous roots - a root system in which there is no main axis; cf. **tap-root**
filament - stalk part of a **stamen**; Fig 702
filiform - thread- or wire-like
flexuous - of a stem or hair, wavy
flore pleno - 'double' flower, with many more **petals** than normal, usually due to conversion of **stamens** to petals; in Asteraceae, a 'double' **capitulum**, with **disc flowers** all or many converted to **ray flowers**
floret - see 157. Poaceae; Fig 572
foliaceous - leaf-like, of an organ not normally thus
follicle - a dry, usually many-seeded fruit **dehiscent** along 1 side, formed from 1 **carpel**
free - separate, not fused to another organ or to one another except at point of origin
free-central - of a **placenta** formed by central axis of any **ovary** that is not connected by **septa** to the wall; Fig 702
fruit - the ripe, fertilized **ovary**, containing seeds

gametophyte - the **haploid** generation of a plant that bears the true sex-organs (that produce the gametes), in **pteridophytes** the **prothallus**, in **spermatophytes** the pollen grains (male) and **embryo-sac** (female)
glabrous - hair-less
gland - a secreting structure, usually round or ± so, on the surface of an organ, below the surface, or raised on a stalk
glandular - with the functions of, or bearing, glands
glaucous - bluish-white in colour (rather than green)
globose - spherical
glume - see 156. Cyperaceae, 157. Poaceae; Fig 572
grain-alien - alien introduced as contaminant of grain
granulose - with a fine sand-like surface texture
gynodioecious - having female and **bisexual** plants
gynoecium - the group of female parts of a flower; all the **carpels**; Fig 702

half-epigynous - of a flower with a **semi-inferior ovary**; Fig 702
haploid - having only 1 set of chromosomes, as in **gametophytic** tissue
hastate - of the base of a flat organ; Fig 699
heptaploid - see **polyploid**
herb - a plant dying down to ground-level each year

herbaceous - not woody, dying down each year; leaf-like as opposed to **woody**, horny, **scarious** or spongy

heterophyllous - having leaves of ≥2 distinct forms

heterosporous - having **spores** of 2 sorts (**megaspores**, female; and **microspores**, male), as in all **spermatophytes** and a few **pteridophytes**

heterostylous - having 2 forms (not sexes) of flower on different plants, the 2 sorts with different **styles** and/or **stigmas** (and pollen)

hexaploid - see **polyploid**

hilum - scar on a **seed** where it left its point of attachment

hispid - with harsh hairs or bristles

homosporous - having **spores** all of 1 sort, as in most **pteridophytes**

homostylous - not **heterostylous**

hypanthium - extension of **receptacle** above base of **ovary**, in **perigynous** and **epigynous** flowers; Fig 702

hyaline - thin and ± transparent

hypogynous - of a flower with a **superior ovary**, the **calyx**, **corolla** and **stamens** being inserted at the base of the ovary; Fig 702

imparipinnate - pinnate with an unpaired terminal **leaflet**; Fig 700

included - not exserted

incurved - curved inwards

indehiscent - not **dehiscent**

indusium - small flap or pocket of tissue covering groups of **sporangia** in many Pteropsida

inferior - of an **ovary** that is borne below the point of origin of the **sepals**, **petals** and **stamens** and is fused with the **receptacle** (**hypanthium**) surrounding it; Fig 702

inflated - of an organ that is dilated, leaving a gap between it and its contents

inflorescence - a group of flowers with their branching system and associated **bracts** and **bracteoles**; Fig 701

insertion - the position and form of the point of attachment of an organ

internode - the stem between adjacent **nodes**; Fig 572

introduced - a plant that owes its existence in this country to importation (deliberate or not) by man

introgression - the acquiring of characteristics by one species from another by hybridization followed by backcrossing

isodiametric - of any shape or organ, ± the same distance across in any plane

isophyllous - not **heterophyllous**

keel - a longitudinal ridge on an organ, like the keel of a boat; see also 79. Fabaceae

labellum - see 166. Orchidaceae

laciniate - irregularly and deeply toothed; Fig 699

laminar - in the form of a flat leaf

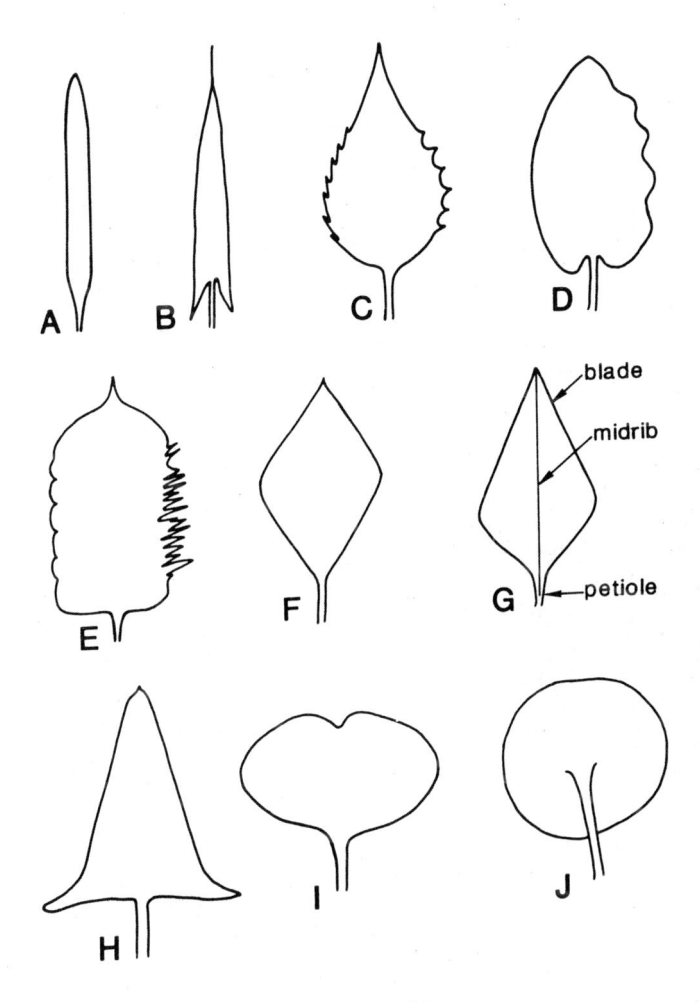

FIG 699 - Simple leaf-shapes.
A, **linear**, apex acute, base cuneate, margin entire.
B, **lanceolate**, apex aristate, base sagittate, margin entire.
C, **ovate**, apex acuminate, base rounded, margin serrate (left) and dentate (right). D, **elliptic**, apex obtuse, base cordate, margin entire (left) and sinuous (right). E, **oblong**, apex cuspidate, base truncate, margin crenate (left) and laciniate (right).
F, **rhombic**, apex mucronate, base cuneate, margin entire.
G, **trullate**, apex acute, base cuneate, margin entire.
H, **triangular**, apex apiculate, base hastate, margin entire.
I, **transversely elliptic**, apex emarginate, base rounded, margin entire.
J, **orbicular** and peltate, margin entire. Drawings by S. Ogden.

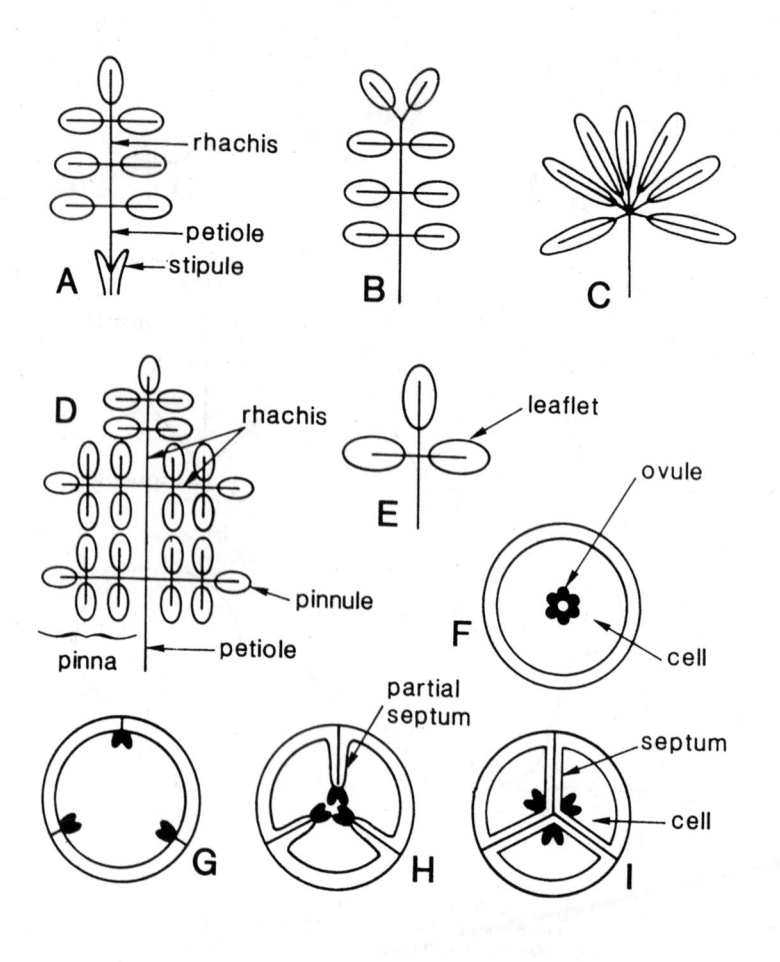

FIG 700 - A-E, Compound leaf-types. A, **pinnate** (imparipinnate, stipulate). B, **pinnate** (paripinnate). C, **palmate**. D, **2-pinnate**. E, **ternate**. F-I, ovaries in transverse section to show septa and placentation. F, 1-celled, **free-central** placentation. G, 1-celled, **parietal** placentation. H, 1-celled, **parietal** placentation. I, 3-celled, **axile** placentation. Drawings by S. Ogden.

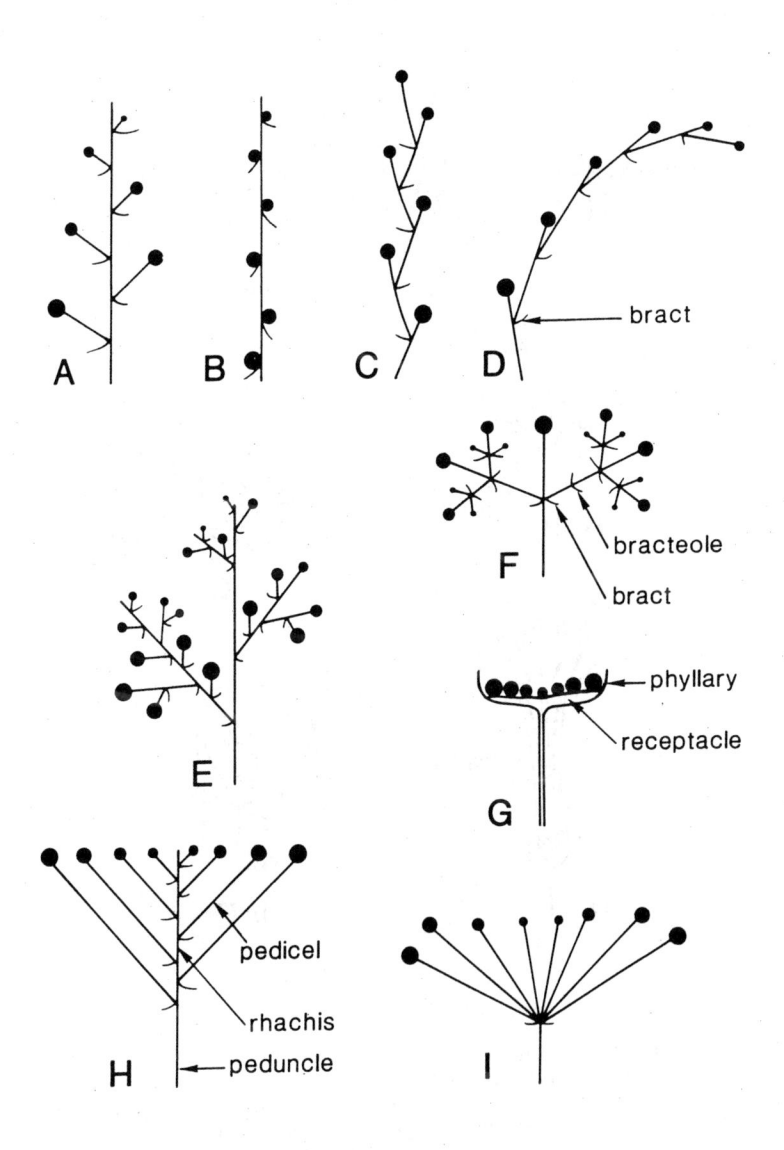

FIG 701 - Inflorescences. A, **raceme**. B, **spike**. C-D, **cymes (monochasial)**.
E, **panicle**. F, **cyme (dichasial)**. G, **capitulum**. H, **corymb**. I, **umbel**.
Drawings by S. Ogden.

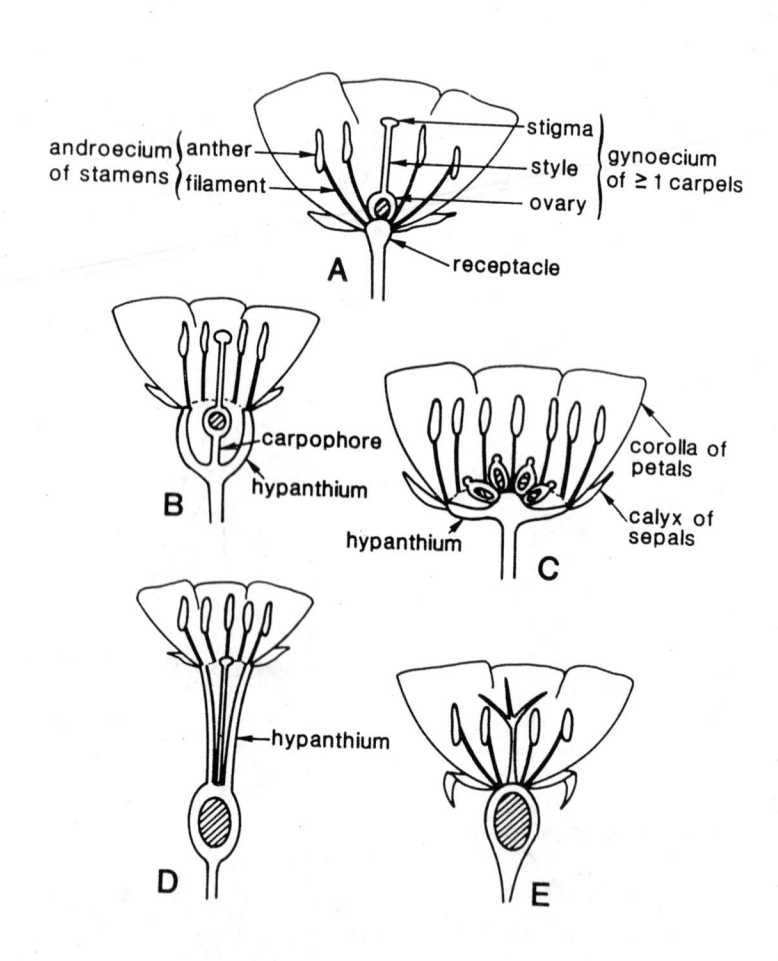

FIG 702 - Half-flowers to show ovary position and other parts.
A, **hypogynous flower**, ovary superior, carpels 1 or >1 and fused.
B, **perigynous flower** with cup-shaped hypanthium, ovary superior, with carpophore, carpels 1 or >1 and fused.
C, **perigynous flower** with flat hypanthium, ovary superior, carpels 4, free.
D, **epigynous flower** with tubular hypanthium, ovary inferior, carpels 1 or >1 and fused. E, **epigynous flower**, ovary inferior, carpels 3, fused.
Drawings by S. Ogden.

Lammas growth - extra, usually abnormal, growth put on in summer by some **trees** (around Lammas Day, 1st Aug)

lanceolate - very narrowly **ovate**, c.6x as long as wide; Fig 699

latex - milky juice

lax - loose or diffuse, not dense

leaflet - a division of a **compound** leaf; Fig 700

leaf-opposed - a lateral organ borne on the stem on opposite side from a leaf, not in a leaf-**axil** as usual

leaf-rosette - a radiating cluster of leaves, often at the base of a stem at soil level

legume - a usually dry, usually many-seeded **fruit dehiscent** along 2 sides, formed from 1 **carpel**

lemma - see 157. Poaceae; Fig 572

lenticel - small corky wart-like strucure on the surface of some stems and fruits

lenticular - lens-shaped; can vary from biconvex to biconcave

ligulate - see 139. Asteraceae

ligule - minute **membranous** flap at base of leaf of *Isoetes* or *Selaginella*; see also 139. Asteraceae, 157. Poaceae; Fig 572

limb - **distal** expanded part of a **calyx** or corolla, as distinct from the **tube** or **throat**

linear - long and narrow with ± parallel margins, i.e. extremely narrowly **oblong**; Fig 699

lip - part of the **distal** region of a **calyx** or corolla sharply differentiated from the rest due to fusion or close association of its parts

lobe - a substantial division of a leaf, **calyx** or corolla; cf. **tooth**

lodicule - see 157. Poaceae; Fig 572

long-shoot - stem of potentially unlimited growth, especially in **trees** or **shrubs**

lowerside - the under surface of a flat organ

lunate - crescent moon-shaped

mealy - with a floury texture

megasporangium - in a **heterosporous** plant, the **sporangia** bearing **megaspores**

megaspore - in a **heterosporous** plant, the female **spores** that give rise to female **gametophytes**

meiosis - special form of cell-division (in **sporangia**, **pollen-sacs** or **ovules**) in which the chromosome number is halved, producing **haploid spores**

membranous - like a membrane in consistency

mericarp - a 1-seeded portion formed by splitting up of a 2-many-seeded **fruit**, as in Geraniaceae, Apiaceae, Boraginaceae, Malvaceae, etc.; see also **schizocarp**

-merous - divided into or composed of a particular number of parts; hence **trimerous, tetramerous, pentamerous**

micron - a micrometre, i.e. one-thousandth of a millimetre or one-millionth of a metre

microsporangium - in a **heterosporous** plant, the **sporangia** bearing **microspores**

microspore - in a **heterosporous** plant, the male **spores** that give rise to male **gametophytes**

midrib - the central, main **vein**; Fig 699

monocarpic - living for >1 year, flowering and fruiting, and then dying

monochasium - **cyme** with 1 lateral branch at each **node**; Fig 701

monoecious - having separate male and female flowers on the same plant

monomorphic - not **polymorphic** (e.g. **dimorphic** or **trimorphic**), occurring in 1 form only

morph - one of several forms of a **polymorphic** taxon

mucronate - having a very short bristle-like tip; **mucro**, the tip itself; Fig 699

mycorrhiza - fungal cells that live within or intimately around the roots of vascular plants

naked - not enclosed

naturalized - an **alien** plant that has become established and self-perpetuating

native - opposite of alien, a plant that colonizied BI by natural means, often long ago, from other native areas

nec - nor, nor of

nectariferous - nectar-bearing

nectar-pit - a **nectariferous** pit

nectary - any **nectariferous** organ, usually a small knob or a modified **petal** or **stamen**

nodding - bent over and **pendent** at tip

node - the position on a stem where leaves, flowers or lateral stems arise; Fig 572

nonaploid - see **polyploid**

nothomorph - one of ≥2 variants of a particular hybrid

nut - a dry, indehiscent 1-seeded **fruit** with a hard **woody** wall, often large

nutlet - a small nut, or a **woody**-walled **mericarp**

ob- - the other way up from normal, usually flattened or widened at the **distal** rather than **proximal** end, e.g. **obovoid**, **obtrullate**

oblong - a flat shape with middle part ± parallel-sided, 1.2-3x as long as wide (if less, **broadly** so; if more, **narrowly** so); Fig 699

obtuse - with a point >90°; Fig 699

octoploid - see **polyploid**

oilseed-alien - **alien** introduced as contaminant of oilseed

opposite - of 2 organs arising laterally at 1 **node** on opposite sides of the stem

orbicular - a flat shape circular in outline; Fig 699

ovary - the basal part of the **gynoecium** containing the **ovules**; Figs 572, 700, 702

ovate - a flat shape widest nearer the base and 1.2-3x as long as wide (if less, **broadly** so; if more, **narrowly** so); Fig 699

ovoid - a solid shape **ovate** in side view

ovule - organ (inside the **ovary** in Magnoliopsida, **naked** in Pinopsida) that contains the **embryo-sac** (which in turn contains the egg) developing into the **seed** after fertilization; Fig 702

palea - see 157. Poaceae; Fig 572
palmate - a **compound** leaf, with >3 **leaflets** all arising at 1 point; Fig 700
panicle - a **compound** or much-branched **inflorescence**, either **racemose** or **cymose**; **paniculate**, in the form of a panicle; Fig 701
papilla - small nipple-like projection; **papillose** or **papillate**, covered with papillae
pappus - see 139. Asteraceae
parasite - plant that gets all or some of its nourishment by attachment (often but not always under the ground) to other plants
parietal - of a **placenta** formed by central axis of an **ovary** that is connected by **septa** to the wall; Fig 700
paripinnate - **pinnate** without an unpaired terminal **leaflet**; Fig 700
partial septum - a **septum** that is incomplete; Fig 700
patent - projecting ± at right-angles
pedicel - stalk of a flower; **pedicellate**, having a pedicel; Figs 572, 701
peduncle - stalk of a group of ≥2 flowers; **pedunculate**, having a peduncle; Fig 701
peltate - of a flat shape with its stalk arising from the plane surface, not the edge; Fig 699
pendent - hanging down
pentamerous - divided into or composed of five parts
pentaploid - see **polyploid**
perennial - living >2 years
perianth - the outer non-sexual covering layers of the flower (the **calyx** and **corolla** together), usually used when the calyx and corolla are not or little differentiated; **perianth-lobes**, **perianth-tube**, the **lobes** and **tube** of a partially fused perianth
pericarp - the wall of a **fruit**, originally the **ovary** wall
perigynous - of a flower with a **superior ovary** but with the **calyx, corolla** and **stamens** inserted above the base of the ovary on an extension of the **receptacle (hypanthium)** that is not fused with the ovary; Fig 702
persistent - remaining attached longer than normal
petal - one of the segments of the inner **whorl(s)** of the **perianth**; **petaloid**, petal-like; Fig 702
petiole - the stalk of a leaf; **petiolate**, with a petiole; Figs 699, 700
phyllary - see 139. Asteraceae; Fig 701
pinna - the primary division of a ≥2-**pinnate** leaf; Fig 700
pinnate - a **compound** leaf, with >3 **leaflets** arising in **opposite** pairs along the **rhachis**; 2-(etc)**pinnate**, pinnate with the **pinnae** pinnate again (etc.); see also **paripinnate, imparipinnate**; Fig 700
pinnule - the ultimate division of a ≥2-**pinnate** leaf, usually applied only in ferns; Fig 700
placenta - points of origin of **ovules** in an **ovary**; **placentation**, the arrangement of placentae; Fig 700

plastic - varying in form according to environmental conditions, not according to genetic characteristics

pollen-sac - the microsporangium of a **spermatophyte**; one of the chambers in an **anther** in which the pollen is formed

pollinium - see 166. Orchidaceae

polymorphic - occurring in 2 or more forms (not **monomorphic**)

polyploid - having >2 sets of chromosomes, e.g. 3(triploid), 4(tetraploid), 5(pentaploid), 6(hexaploid), 7(heptaploid), 8(octoploid), 9(nonaploid), 10(decaploid)

prickle - spiny outgrowth with a broadened base

pricklet - small, weak **prickle**

procumbent - trailing along the ground

proliferating - with **inflorescences** bearing plantlets instead of flowers or **fruits**

pro parte - partly; in part

prothallus - small **gametophyte** generation of a plant bearing the true sex-organs, mostly applied to the free-living gametophytes of **pteridophytes**

proximal - at the end near the point of attachment; cf. **distal**

pruinose - with a **bloom**

pteridophytes - ferns and fern allies, i.e. Lycopodiopsida, Equisetopsida, Pteropsida

pubescent - with hairs; **pubescence**, the hair covering

punctate - marked with dots or transparent spots

raceme - an **inflorescence** with the oldest flowers (or **spikelets** in Poaceae) the most **proximal** lateral ones and a potentially continuously growing apex; **racemose**, in the form of a raceme; Fig 701

radiate - see 139. Asteraceae

rank - a vertical file of lateral organs; **2-ranked**, etc., with 2 (etc.) ranks of lateral organs

ray - anything that radiates outwards, e.g. branches of an **umbel**, **stigma**-ridges in *Papaver* or *Nuphar*; **ray flowers**, see 139. Asteraceae

receptacle - the usually expanded, often cup-shaped or tubular, apical part of a **pedicel** on which the flower parts are inserted; in Asteraceae, the expanded apical part of the **capitulum**-stalk that bears all the flowers; **receptacular scales** or **bristles**, see 139. Asteraceae; Figs 701, 702

recurved - curved down or back

reflexed - bent down or back

reniform - kidney-shaped

resiniferous - producing resin; **resinous**, resin-like; **resin-duct**, microscopic canal producing resin

reticulate - forming or covered with a network

retuse - with a shallow blunt notch at the apex; cf. **emarginate**

revolute - rolled back or down

rhachilla - see 157. Poaceae; Fig 572

rhachis - the axis (not the stalk) of an **inflorescence** or **pinnate** leaf; Figs 700, 701

GLOSSARY

rhizome - an underground or ground-level, usually horizontal or down-growing stem, often ± swollen; **rhizomatous**, bearing or in the form of a rhizome; cf. **stolon**

rhombic - a flat shape, widest in the middle and ± angled (not rounded) there, 1.2-3x as long as wide (if less, **broadly** so; if more, **narrowly** so); Fig 699

rigid - stiff, not flexible

rostellum - see 166. Orchidaceae

rounded - without a point or angle; Fig 699

rugose - with a wrinkled surface

sagittate - of the base of a flat organ; see Fig 699

saprophyte - a plant deriving its nourishment from decaying organisms, usually leaf-mould

scabrid - rough to the touch, with minute **prickles** or bristly hairs

scale-leaf - a leaf reduced to a small scale

scape - a flowering stem of a plant in which all the leaves are basal, none on the scape

scarious - of thin, papery texture and not green

schizocarp - a **fruit** that breaks into 1-seeded portions or **mericarps**

sclerenchyma - **woody** tissue in a partly or mostly non-woody organ

scorpioid - a **monochasial cyme** that is coiled up like a scorpion's tail when young

scrambler - a plant sprawling over other plants, fences, etc.

seed - a fertilized **ovule**

self-compatible - self-fertile, able to self-fertilize

self-incompatible - self-sterile, not able to self-fertilize

semi-inferior - of an **ovary** of which the lower part is **inferior**, but the upper part is **free** and projects above the **sepals**, etc.

sensu lato - in the broad sense

sensu stricto - in the narrow sense

sepal - one of the segments of the outer **whorl(s)** of the **perianth**; **sepaloid**, sepal-like; Fig 702

septum - a wall or membrane dividing the **ovary** into **cells**; Fig 700

serrate - with a row of ± apically directed **teeth**; **serration**, that sort of toothing; cf. dentate; Fig 699

sessile - not stalked

sheath - see 157. Poaceae; Fig 572

short-shoot - a short stem of strictly limited growth, usually lateral on a **long-shoot**, especially on **trees** and **shrubs**

shrub - a **woody** plant that is not a **tree**

simple - not **compound**; Fig 699

sinuous - wavy, either of a hair or stalk, or of the margin of a leaf and then the sinuation in the same plane as the leaf surface; Fig 699

sinus - the space or indentation between 2 **lobes** or **teeth**; **basal sinus**, the sinus at the base of a leaf, either side of the **petiole** if present

solitary - borne singly

sorus - a group of **sporangia** in Pteropsida

soyabean-alien - **alien** introduced as contaminant of soyabeans

spadix - see 151. Araceae

spathe - an ensheathing **bract**, as in Lemnaceae, Araceae, Hydrocharitaceae

spathulate - paddle- or spoon-shaped

spermatophyte - a seed-plant, i.e. Pinopsida and Magnoliopsida

spike - a **racemose inflorescence** in which the flowers (or **spikelets** in Poaceae) have no stalks; **spikiform**, in the form of a spike; Fig 701

spikelet - see 50/1. Plumbaginaceae/*Limonium*, 156. Cyperaceae, 157. Poaceae; Fig 572

spine - a sharp, stiff, straight **woody** outgrowth, usually not greatly widened at base; **spinose**, spine-like; **spinulose**, diminutive of last; **spiny**, with spines

spiral - lateral organs on an axis 1 per **node**, successive ones not at 180° to each other

sporangium - a body producing **spores** in **pteridophytes**

spore - the **haploid** product of **meiotic** division, produced on the **sporophyte** and developing into the **gametophyte**

sporophyte - the **diploid** generation of a plant that bears the **sporangia** (**ovules** and **pollen-sacs** in Magnoliopsida); the main plant body of all vascular plants

spreading - growing out **divergently**, not straight or **erect**

spur - a protrusion or tubular or pouch-like outgrowth of any part of a flower

stamen - the basic male reproductive unit of Magnoliopsida, 1-many per flower, sometimes fused; Figs 572, 702

staminode - a sterile **stamen**, sometimes modified to perform some other function, e.g. that of a **petal** or **nectary**

standard - see 79. Fabaceae

stellate - star-shaped, with radiating arms

stem-leaves - leaves borne on the stem as opposed to basally

stigma - the apical part of a **gynoecium** that is receptive to pollen, simple to much branched; Figs 572, 702

stipule - one of a (usually) pair of appendages at the base of a leaf or its **petiole**, often but often not **foliaceous**; **stipulate**, with stipules; Fig 700

stolon - an **aerial** or **procumbent** stem, usually not swollen; **stoloniferous**, bearing stolons; cf. **rhizome**

style - the stalk on any **ovary** bearing the **stigma(s)**, sometimes absent; Fig 702

stylopodium - see 111. Apiaceae

sub- - almost, as in subacute, subglabrous, subglobose, subentire, subequal; sometimes under, as in subaquatic

subshrub - a **perennial** with a short **woody** surface stem producing **aerial herbaceous** stems

subtend - of a lateral organ, to have another organ in its **axil**

subulate - tapering ± constantly from a narrow base to a fine point

succulent - fleshy and juicy or pulpy

sucker - a new **aerial** shoot borne (often underground) on the roots of a **tree** or **shrub**

GLOSSARY

superior - of an **ovary** that is borne above the **calyx**, **corolla** and **stamens** or, if below or partly below them, then not fused laterally to the **receptacle**; Fig 702

survivor - an alien plant not naturalized, but long-persistent, usually a relic of planting

suture - a seam of a union, often splitting open in later development

tap-root - a main descending root bearing laterals; cf. **fibrous roots**

taxon - any taxonomic grouping, such as a genus or species

tendril - a spirally coiled thread-like outgrowth from a stem or leaf, used by the plant for support

tepal - one of the segments of the **perianth**, used when **sepals** and **petals** are not differentiated

terete - rounded in section

terminal - at the very apex

ternate - a **compound** leaf with 3 **leaflets**; 2-(etc.)**ternate**, ternate with the 3 divisions ternate again; Fig 700

testa - the outer coat of a seed

tetrad - a group of 4 **spores** or pollen grains formed by **meiotic** division

tetramerous - divided into or composed of four parts

tetraploid - see **polyploid**

throat - the opening where the **tube** joins the **limb** of a **corolla** or **calyx**

tiller - see 157. Poaceae; Fig 572

tomentose - a very dense often ± matted hair-covering

tooth - a shallow division of a leaf, **calyx** or **corolla**, or of the apex of a **capsule**; cf. **lobe**, **valve**

transverse - lying cross-ways; **transversely elliptic** (etc.), **elliptic** (etc.) but with the point of attachment at the side, not at one end; Fig 699

tree - a **woody** plant, usually ≥5m, with a single trunk

triangular - a flat shape widest at the base and 1.2-3x as long as wide (if less, **broadly** so; if more, **narrowly** so); Fig 699

trifid - divided into 3, usually deeply, at apex

trimerous - divided into or composed of three parts

trimorphic - occurring in 3 forms

tripartite - divided into 3 parts

triploid - see **polyploid**

trullate - a flat shape widest nearer the base and ± angled (not rounded) there, 1.2-3x as long as wide (if less, **broadly** so; if more, **narrowly** so); Fig 699

truncate - of the base or apex of a flat organ, straight or flat; Fig 699

tube - narrow, cylindrical, **proximal** part of a **calyx** or **corolla**, as distinct from the **limb**, **lobes** or **throat**

tuber - swollen roots or subterranean stems; **tuberous**, tuber-like

tubercle - a small ± spherical or **ellipsoid** swelling; **tuberculate**, with a surface texture covered in minute tubercles

tubular - in the form of a hollow cylinder; **tubular flowers**, see 139. Asteraceae

tufted - of elongated organs or stems that are clustered together

twig - ultimate branch of a woody stem

umbel - an **inflorescence** in which all the **pedicels** arise from one point; **compound umbel**, an umbel of umbels; **umbellate**, umbel-like; Fig 701

undulate - wavy at the edge in the plane at right-angles to the surface

unifacial - with only 1 surface, not with a **lowerside** and **upperside**; cf. **bifacial**

unisexual - of a plant or flower, bearing only 1 sex

upperside - the upper surface of a flat organ

valve - a deep division or **lobe** of a **capsule** apex; cf. **tooth**

vascular - pertaining to the **veins** or wood (i.e. conducting tissue) of an organ; **vascular bundle**, one anatomically discrete file of vascular tissue; **vascular plant**, a plant with vascular bundles, i.e. **pteridophytes** and **spermatophytes**

vegetative - not reproductive

vein - a strand of **vascular tissue** consisting of ≥1 **vascular bundles**; **venation**, the pattern of veins

verrucose - covered in small wart-like outgrowths

vicariant - a taxon that replaces a related one in a different area; **vicarious**, being a vicariant; **vicariance**, the state of being vicarious

viscidium - see 166. Orchidaceae

whorl - a group of lateral organs borne >2 at each **node**; **whorled**, in the form of a whorl

wing - any **membranous** or **foliaceous** extension of an organ, e.g. a stem, **seed** or **fruit**; **winged**, with a wing; see also 79. Fabaceae

woody - hard and wood-like, not quickly dying or withering

wool-alien - an **alien** introduced as a contaminant of raw wool imports

woolly - clothed with shaggy hairs

zygomorphic - of a flower with bilateral (i.e. 1 plane of) symmetry

INDEX

In the main only families and genera are indexed, without authorities. Latin family and generic names are given in bold-face capitals, and synonyms in non-bold lower case italics. English 'generic' names are entered in non-bold lower case, but in addition some English 'specific' names are added in order to differentiate cases where the English 'generic' name occurs in more than one Latin genus (e.g. Fescue).

If an English 'generic' name is not found it is worth searching under the English 'specific name', as the latter might in fact be part of the former (e.g. Hemp-nettle or Sowthistle); such cases are not cross-referenced. The page number provided refers to the first use under each genus. The keys to families and genera, and the illustrations, are not indexed.

ABIES, 27
Abraham-Isaac-Jacob, 375
ABUTILON, 144
ACACIA, 264
ACAENA, 231
Acaena, 231
ACANTHACEAE, 431
ACANTHUS, 431
ACER, 312
ACERACEAE, 312
ACERAS, 691
ACHILLEA, 498
Aconite, Winter, 64
ACONITUM, 64
ACORUS, 534
ACROPTILON, 468
ACTAEA, 65
Adder's-tongue, 11
ADIANTACEAE, 12
ADIANTUM, 12
ADONIS, 72
ADOXA, 448
ADOXACEAE, 448
AEGOPODIUM, 336
AEONIUM, 206
Aeonium, 206
AESCULUS, 311
AETHEORHIZA, 475

AETHUSA, 338
AGAPANTHUS, 659
AGAVACEAE, 673
AGAVE, 674
AGERATUM, 516
AGRIMONIA, 230
Agrimony, 230
  Bastard, 230
X **AGROPOGON**, 614
AGROSTEMMA, 119
AGROSTIS, 612
AILANTHUS, 313
AIRA, 610
AIZOACEAE, 91
Ajowan, 347
AJUGA, 388
Ake-ake, 494
Akiraho, 494
ALCEA, 143
ALCHEMILLA, 232
Alder, 89
Alexanders, 335
ALISMA, 519
ALISMATACEAE, 517
Alison, 175
  Hoary, 175
  Sweet, 175
Alkanet, 373

Alkanet (contd)
  False, 374
  Green, 374
ALLIARIA, 168
ALLIUM, 655
Allseed, 310
  Four-leaved, 117
Almond, 240
ALNUS, 89
ALOPECURUS, 617
Alpine-sedge, 555
ALSTROEMERIA, 664
Altar-lily, 535
ALTHAEA, 143
ALYSSUM, 175
Amaranth, 105
AMARANTHACEAE, 104
AMARANTHUS, 104
AMARYLLIS, 660
AMBROSIA, 509
AMELANCHIER, 248
American-spikenard, 325
AMMI, 348
AMMOPHILA, 615
Amomyrtus, 294
AMSINCKIA, 375
ANACAMPTIS, 683
ANACARDIACEAE, 313
ANAGALLIS, 201
ANAPHALIS, 486
ANCHUSA, 373
Anchusa, Garden, 374
ANDROMEDA, 192
ANEMONE, 65
Anemone, 65
ANETHUM, 344
Angel's-tears, 662
ANGELICA, 349
Angelica, 349
Angelica-tree, 325
Angels'-trumpets, 363
ANISANTHA, 622
ANOGRAMMA, 12
ANTENNARIA, 486
ANTHEMIS, 499
ANTHOXANTHUM, 610
ANTHRISCUS, 334
ANTHYLLIS, 270

ANTIRRHINUM, 410
APERA, 616
APHANES, 235
APIACEAE, 326
APIUM, 346
APOCYNACEAE, 357
APONOGETON, 522
APONOGETONACEAE, 522
Apple, 242
Apple-mint, 395
Apple-of-Peru, 358
APTENIA, 92
AQUIFOLIACEAE, 303
AQUILEGIA, 72
ARABIDOPSIS, 169
ARABIS, 173
Arabis, 174
ARACEAE, 534
ARALIA, 325
ARALIACEAE, 324
ARAUCARIA, 33
ARAUCARIACEAE, 33
ARBUTUS, 193
Archangel, Yellow, 384
ARCTIUM, 464
ARCTOSTAPHYLOS, 193
ARCTOTHECA, 484
AREMONIA, 230
ARENARIA, 110
ARGEMONE, 77
Argentine-pear, 359
ARISARUM, 536
ARISTEA, 667
ARISTOLOCHIA, 60
ARISTOLOCHIACEAE, 59
ARMERIA, 136
ARMORACIA, 172
ARNOSERIS, 471
ARONIA, 248
ARRHENATHERUM, 606
Arrowgrass, 523
Arrowhead, 518
ARTEMISIA, 496
Artichoke, Globe, 468
  Jerusalem, 512
ARUM, 535
Arum, Bog, 535
  Dragon, 536

INDEX

ARUNCUS, 220
Asarabacca, 59
ASARINA, 411
ASARUM, 59
Ash, 402
ASPARAGUS, 664
Asparagus, 664
Aspen, 151
ASPERUGO, 376
ASPERULA, 441
Asphodel, Bog, 647
  Scottish, 647
  White, 648
ASPHODELUS, 648
ASPLENIACEAE, 16
ASPLENIUM, 17
X **ASPLENOPHYLLITIS**, 18
ASTER, 490
Aster, China, 494
  Goldilocks, 491
  Mexican, 514
  Sea, 491
ASTERACEAE, 453
ASTILBE, 210
ASTRAGALUS, 269
ASTRANTIA, 332
Astrantia, 332
ATHYRIUM, 19
ATRIPLEX, 99
ATROPA, 359
Aubretia, 175
AUBRIETA, 175
AUCUBA, 301
Aunt-Eliza, 673
Auricula, 198
AVENA, 606
Avens, 229
  Mountain, 229
Awlwort, 182
Azalea, 191
  Trailing, 191
AZOLLA, 24
AZOLLACEAE, 24

Baby's-breath, 122
BACCHARIS, 495
BALDELLIA, 518
BALLOTA, 384

Balm, 390
  Bastard, 387
Balm-of-Gilead, 152
Balsam, 323
Balsam-poplar, 152
BALSAMINACEAE, 323
Bamboo, 587
  Arrow, 588
  Hairy, 588
  Narihira, 586
  Square-stemmed, 588
Baneberry, 65
BARBAREA, 170
Barberry, 73
Barley, 626
  Wood, 626
BARTSIA, 426
Bartsia, 426
  Alpine, 426
  Yellow, 426
Basil, Wild, 390
BASSIA, 99
Bastard-toadflax, 301
Bay, 59
Bayberry, 86
Beadplant, 440
Beak-sedge, 552
Bean, 267
  Broad, 274
Bear's-breech, 431
Bearberry, 193
Beard-grass, 616
BECKMANNIA, 618
Bedstraw, 441
  Tree, 440
Beech, 87
  Southern, 87
Beet, 101
Beetle-grass, 631
Beggarticks, 513
Bellflower, 435
  Ivy-leaved, 437
BELLIS, 495
Bent, 612
  Water, 617
BERBERIDACEAE, 73
BERBERIS, 73
BERGENIA, 210

Bermuda-buttercup, 314
Bermuda-grass, 634
BERTEROA, 175
BERULA, 336
BETA, 101
Betony, 383
BETULA, 88
BETULACEAE, 87
BIDENS, 512
Bilberry, 195
Bindweed, 364
    Field, 364
Birch, 88
Bird-in-a-bush, 78
Bird's-foot, 272
Bird's-foot-trefoil, 271
Bird's-nest, 197
Birthwort, 60
Bistort, 124
Bitter-cress, 172
Bitter-vetch, 276
    Wood, 274
Bittersweet, 361
Black-bindweed, 128
Black-eyed-Susan, 511
Black-grass, 618
Black-jack, 513
Black-poplar, 152
BLACKSTONIA, 355
Blackthorn, 241
Blackwood, Australian, 264
Bladder-fern, 19
Bladder-sedge, 565
Bladder-senna, 269
Bladdernut, 311
Bladderseed, 344
Bladderwort, 432
Blanketflower, 515
BLECHNACEAE, 23
BLECHNUM, 23
Bleeding-heart, 78
Blinks, 107
Blood-drop-emlets, 409
Blown-grass, 613
Blue-eyed-grass, 666
Blue-eyed-Mary, 378
Blue-gum, 294
Blue-sowthistle, 476

Bluebell, 653
Blueberry, 195
BLYSMUS, 551
Bog-laurel, 191
Bog-myrtle, 86
Bog-rosemary, 192
Bog-rush, 552
Bog-sedge, 568
Bogbean, 366
BOLBOSCHOENUS, 549
Borage, 374
BORAGINACEAE, 368
BORAGO, 374
Boston-ivy, 309
BOTRYCHIUM, 11
Box, 303
    Carpet, 304
BRACHIARIA, 637
BRACHYGLOTTIS, 506
BRACHYPODIUM, 624
Bracken, 15
Bramble, 222
BRASSICA, 183
BRASSICACEAE, 161
Bridal-spray, 219
Bridewort, 219
Bristle-grass, 639
BRIZA, 599
Broadleaf, New Zealand, 301
Brome, 619, 621, 622, 623
    False, 624
    Stiff, 624
BROMELIACEAE, 643
BROMOPSIS, 621
BROMUS, 619
Brooklime, 414
Brookweed, 202
Broom, 288
    Montpellier, 289
    Mount Etna, 289
    Spanish, 288
Broomrape, 429
BRUNNERA, 373
BRYONIA, 149
Bryony, Black, 675
    White, 149
Buck's-beard, 220
Buckler-fern, 21

INDEX

Buckthorn, 308
  Alder, 308
Buckwheat, 126
BUDDLEJA, 401
BUDDLEJACEAE, 401
Buffalo-bur, 361
Bugle, 388
Bugle-lily, 669
Bugloss, 373
Bugseed, 99
Bullwort, 348
Bulrush, 643
BUNIAS, 169
BUNIUM, 335
BUPLEURUM, 345
Bur-grass, 635
Bur-marigold, 512
Bur-reed, 642
Burdock, 464
Burnet, 230
Burnet-saxifrage, 336
Butcher's-broom, 664
BUTOMACEAE, 517
BUTOMUS, 517
Butterbur, 507
Buttercup, 67
Butterfly-bush, 401
Butterfly-orchid, 683
Butterwort, 431
Button-grass, 633
Buttonweed, 501
BUXACEAE, 303
BUXUS, 303

Cabbage, 183, 186, 187
Cabbage-palm, 674
CABOMBA, 61
CABOMBACEAE, 61
CAKILE, 186
CALAMAGROSTIS, 614
Calamint, 390
X CALAMMOPHILA, 615
CALCEOLARIA, 410
CALENDULA, 509
CALLA, 535
CALLISTEPHUS, 494
CALLITRICHACEAE, 397
CALLITRICHE, 397

CALLUNA, 193
CALOTIS, 489
CALTHA, 63
CALYSTEGIA, 364
CAMELINA, 178
CAMPANULA, 435
CAMPANULACEAE, 434
Campion, 119
  Rose, 119
Canary-grass, 611
Candytuft, 180
CANNABACEAE, 83
CANNABIS, 83
Canterbury-bells, 435
Cape-gooseberry, 360
Cape-lily, 660
Cape-pondweed, 522
CAPRIFOLIACEAE, 444
CAPSELLA, 179
CAPSICUM, 360
Caraway, 348
CARDAMINE, 172
Cardoon, 468
CARDUUS, 465
CAREX, 553
CARLINA, 464
CARPINUS, 90
CARPOBROTUS, 94
Carrot, 352
  Moon, 337
CARTHAMUS, 470
CARUM, 348
CARYOPHYLLACEAE, 108
CASTANEA, 87
Castor-oil-plant, 304
Cat-mint, 389
Cat's-ear, 472
Cat's-tail, 618
CATABROSA, 602
CATANANCHE, 471
CATAPODIUM, 602
Catchfly, 119, 120
  Berry, 121
Caterpillar-plant, 273
Caucasian-stonecrop, 207
Cedar, 29
CEDRUS, 29
Celandine, Greater, 77

Celandine (contd)
    Lesser, 67
CELASTRACEAE, 302
CELASTRUS, 303
Celery, 346
CENCHRUS, 641
CENTAUREA, 469
CENTAURIUM, 354
Centaury, 354
    Guernsey, 354
    Yellow, 353
CENTRANTHUS, 450
Centuryplant, 674
CEPHALANTHERA, 678
CEPHALARIA, 452
CERASTIUM, 113
CERATOCAPNOS, 79
CERATOCHLOA, 623
CERATOPHYLLACEAE, 61
CERATOPHYLLUM, 61
CETERACH, 18
CHAENOMELES, 242
CHAENORHINUM, 410
CHAEROPHYLLUM, 373
Chaffweed, 201
CHAMAECYPARIS, 32
CHAMAEMELUM, 498
CHAMERION, 297
Chamomile, 498, 499
Charlock, 185
CHASMANTHE, 673
Chasmanthe, 673
Checkerberry, 192
CHELIDONIUM, 77
CHENOPODIACEAE, 95
CHENOPODIUM, 95
Cherry, 239
Chervil, 333, 334
Chestnut, Sweet, 87
Chickweed, 112
    Jagged, 112
    Upright, 114
    Water, 114
Chickweed-wintergreen, 201
Chicory, 471
Chilean-iris, 666
CHIMONOBAMBUSA, 588
CHIONODOXA, 654

Chives, 657
CHLORIS, 634
CHOISYA, 313
Chokeberry, 248
CHRYSANTHEMUM, 499
CHRYSOCOMA, 492
CHRYSOSPLENIUM, 214
Cicely, Sweet, 334
CICENDIA, 353
CICER, 278
CICERBITA, 476
CICHORIUM, 471
CICUTA, 348
Cineraria, 505
Cinquefoil, 225
CIRCAEA, 300
CIRSIUM, 466
CISTACEAE, 145
CITRULLUS, 150
CLADIUM, 553
CLARKIA, 299
Clarkia, 299
Clary, 396
CLAYTONIA, 107
Cleavers, 442
CLEMATIS, 66
Clematis, 67
CLINOPODIUM, 390
Cloudberry, 222
Clover, 282
Club-rush, 550, 551
    Floating, 551
    Round-headed, 549
    Sea, 549
    Wood, 549
Clubmoss, 6
    Alpine, 6
    Fir, 5
    Lesser, 6
    Marsh, 5
CLUSIACEAE, 138
COCHLEARIA, 177
Cock's-eggs, 360
Cock's-foot, 602
Cocklebur, 510
Cockspur, 637
Cockspurthorn, 263
COELOGLOSSUM, 685

INDEX

COINCYA, 186
COLCHICUM, 649
Colt's-foot, 507
    Purple, 508
Columbine, 72
COLUTEA, 269
Comfrey, 371
COMMELINACEAE, 538
Coneflower, 511
CONIUM, 345
CONOPODIUM, 335
CONRINGIA, 183
CONSOLIDA, 65
CONVALLARIA, 651
CONVOLVULACEAE, 364
CONVOLVULUS, 364
CONYZA, 493
X **CONYZIGERON**, 493
COPROSMA, 440
Copse-bindweed, 128
Coral-necklace, 117
Coralbells, 214
Coralberry, 446
CORALLORRHIZA, 682
Coralroot, 173
Cord-grass, 635
CORDYLINE, 674
COREOPSIS, 513
Coriander, 335
CORIANDRUM, 335
CORISPERMUM, 99
Corn-lily, 672
    Blue, 667
CORNACEAE, 300
Corncockle, 119
Cornel, Dwarf, 300
Cornelian-cherry, 300
Cornflower, 470
Cornsalad, 449
CORNUS, 300
CORONILLA, 272
CORONOPUS, 182
CORREA, 314
CORRIGIOLA, 116
CORTADERIA, 630
CORYDALIS, 78, 79
    Climbing, 79
Corydalis, 78

CORYLUS, 90
CORYNEPHORUS, 609
COSMOS, 514
Costmary, 496
COTONEASTER, 249
Cotoneaster, 249
Cottongrass, 547
Cottonweed, 498
COTULA, 501
Couch, 624, 625
Cow-wheat, 419
Cowbane, 348
Cowberry, 195
Cowherb, 122
Cowslip, 198
Crab, 242
Crack-willow, 158
CRAMBE, 187
Cranberry, 195
Crane's-bill, 316
CRASSULA, 205
CRASSULACEAE, 205
+ **CRATAEGOMESPILUS**, 262
CRATAEGUS, 262
X **CRATAEMESPILUS**, 262
Creeping-Jenny, 200
CREPIS, 479
Cress, Garden, 181
    Hoary, 181
    Rosy, 174
    Shepherd's, 179
    Thale, 169
    Tower, 174
    Trefoil, 173
    Violet, 179
CRINUM, 660
CRITHMUM, 337
CROCOSMIA, 672
CROCUS, 669
Crocus, 669
    Sand, 669
Crosswort, 444
    Caucasian, 440
Crowberry, 189
Crowfoot, 70
CRUCIATA, 444
CRYPTOGRAMMA, 12
CRYPTOMERIA, 31

Cuckooflower, 173
CUCUBALUS, 121
Cucumber, 150
  Squirting, 150
CUCUMIS, 150
CUCURBITA, 150
CUCURBITACEAE, 149
Cudweed, 485, 486
Cumin, 346
CUMINUM, 346
Cup-grass, 638
Cupidone, 471
CUPRESSACEAE, 31
X **CUPRESSOCYPARIS**, 32
CUPRESSUS, 32
Currant, 204
CUSCUTA, 366
CUSCUTACEAE, 366
Cut-grass, 588
CYCLAMEN, 199
CYDONIA, 241
CYMBALARIA, 411
CYNARA, 468
CYNODON, 634
CYNOGLOSSUM, 378
CYNOGLOTTIS, 374
CYNOSURUS, 597
CYPERACEAE, 545
CYPERUS, 552
Cyphel, 111
Cypress, 32
  Leyland, 32
CYPRIPEDIUM, 678
CYRTOMIUM, 21
CYSTOPTERIS, 19
CYTISUS, 288

DABOECIA, 192
DACTYLIS, 602
DACTYLOCTENIUM, 633
X **DACTYLODENIA**, 685
X **DACTYLOGLOSSUM**, 685
DACTYLORHIZA, 686
Daffodil, 662
  Sea, 663
  Winter, 660
DAHLIA, 514
Dahlia, 514

Daisy, 495
  Bur, 489
  Crown, 499
  Oxeye, 500
  Seaside, 493
  Shasta, 500
Daisy-bush, 494
DAMASONIUM, 519
Dame's-violet, 170
Dandelion, 477
DANTHONIA, 629
DAPHNE, 293
DARMERA, 211
Darnel, 596
DATURA, 363
DAUCUS, 352
Day-lily, 648
Dead-nettle, 385
Deergrass, 547
DELAIREA, 505
DENNSTAEDTIACEAE, 15
Deodar, 29
DESCHAMPSIA, 608
DESCURAINIA, 168
DEUTZIA, 203
Deutzia, 203
Dewberry, 223
Dewplant, 92
  Deltoid-leaved, 93
  Pale, 93
  Purple, 93
  Shrubby, 92
DIANTHUS, 122
DIAPENSIA, 197
Diapensia, 197
DIAPENSIACEAE, 197
DICENTRA, 78
DICKSONIA, 15
DICKSONIACEAE, 15
DIGITALIS, 413
DIGITARIA, 640
Dill, 344
DIOSCOREACEAE, 675
DIPHASIASTRUM, 6
DIPLOTAXIS, 183
DIPSACACEAE, 450
DIPSACUS, 451
DISPHYMA, 93

INDEX

Dittander, 182
DITTRICHIA, 488
Dock, 129
Dodder, 366
Dog-rose, 237
Dog's-tail, 597
　Golden, 597
Dog's-tooth-violet, 649
Dog-violet, 147
Dogwood, 300
DORONICUM, 507
DOWNINGIA, 439
Downy-rose, 239
DRABA, 176
DRACUNCULUS, 536
Dragon's-teeth, 271
Dropseed, 633
Dropwort, 221
DROSANTHEMUM, 93
DROSERA, 145
DROSERACEAE, 145
DRYAS, 229
DRYOPTERIDACEAE, 20
DRYOPTERIS, 21
DUCHESNEA, 229
Duck-potato, 518
Duckweed, 537
　Greater, 537
　Rootless, 538
Dysentery-herb, 320

ECBALLIUM, 150
ECHINOCHLOA, 637
ECHINOPS, 463
ECHIUM, 370
Eelgrass, 533
EGERIA, 521
EHRHARTA, 589
ELAEAGNACEAE, 290
ELAEAGNUS, 290
ELATINACEAE, 137
ELATINE, 137
Elder, 445
Elecampane, 488
ELEOCHARIS, 548
ELEOGITON, 551
Elephant-ears, 210
ELEUSINE, 632

Elm, 81
ELODEA, 521
ELYMUS, 624
ELYTRIGIA, 625
X ELYTRORDEUM, 625
EMPETRACEAE, 189
EMPETRUM, 189
Enchanter's-nightshade, 300
EPILOBIUM, 295
EPIPACTIS, 678
EPIPOGIUM, 680
EQUISETACEAE, 8
EQUISETUM, 8
ERAGROSTIS, 631
ERANTHIS, 64
EREPSIA, 93
ERICA, 193
ERICACEAE, 189
ERIGERON, 492
ERINUS, 414
ERIOCAULACEAE, 538
ERIOCAULON, 538
ERIOCHLOA, 638
ERIOPHORUM, 547
ERODIUM, 321
EROPHILA, 176
ERUCA, 185
ERUCASTRUM, 185
ERYNGIUM, 333
Eryngo, 333
ERYSIMUM, 169
ERYTHRONIUM, 649
ESCALLONIA, 204
Escallonia, 204
ESCHSCHOLZIA, 77
EUCALYPTUS, 294
EUONYMUS, 302
EUPATORIUM, 515
EUPHORBIA, 305
EUPHORBIACEAE, 304
EUPHRASIA, 420
Evening-primrose, 298
Everlasting, Mountain, 486
　Pearly, 486
Everlasting-pea, 277
Everlastingflower, 487
EXACULUM, 354
Eyebright, 420

FABACEAE, 264
FAGACEAE, 86
FAGOPYRUM, 126
FAGUS, 87
FALCARIA, 348
FALLOPIA, 127
False-acacia, 267
False-buck's-beard, 210
FARGESIA, 587
FASCICULARIA, 643
Fat-hen, 99
FATSIA, 325
Fatsia, 325
Fen-sedge, 553
Fennel, 344
    False, 347
    Giant, 350
    Hog's, 350
Fenugreek, 280
Fern, Beech, 16
    Jersey, 12
    Kangaroo, 14
    Killarney, 13
    Lemon-scented, 16
    Limestone, 19
    Maidenhair, 12
    Marsh, 16
    Oak, 19
    Ostrich, 18
    Parsley, 12
    Ribbon, 13
    Royal, 12
    Sensitive, 18
    Water, 24
Fern-grass, 602
FERULA, 350
Fescue, 590, 596
FESTUCA, 590
X FESTULOLIUM, 595
X FESTULPIA, 595
Feverfew, 496
FICUS, 84
Fiddleneck, 375
Field-rose, 236
Field-speedwell, 417
Fig, 84
Figwort, 403
    Cape, 408

FILAGO, 485
Filbert, 91
FILIPENDULA, 221
Filmy-fern, 13
Finger-grass, 640
    Water, 638
Fir, 27
    Douglas, 28
Firethorn, 261
Flat-sedge, 551
Flax, 309
    New Zealand, 674
Fleabane, 487, 488, 492, 493
Fleawort, 505
Flixweed, 168
Flossflower, 516
Flowering-rush, 517
Fluellen, 412
FOENICULUM, 344
Fool's-water-cress, 346
Forget-me-not, 376
    Bur, 378
    Great, 373
    White, 375
FORSYTHIA, 402
Forsythia, 402
Fountain-bamboo, 586
Fox-and-cubs, 481
Fox-sedge, 557
Foxglove, 413
    Fairy, 414
Foxglove-tree, 408
Foxtail, 617
FRAGARIA, 228
FRANGULA, 308
FRANKENIA, 149
FRANKENIACEAE, 149
FRAXINUS, 402
FREESIA, 672
Freesia, 672
Fringecups, 214
FRITILLARIA, 650
Fritillary, 650
Frogbit, 520
FUCHSIA, 299
Fuchsia, 299
FUMARIA, 79
FUMARIACEAE, 78

INDEX

Fumitory, 79

GAGEA, 649
GAILLARDIA, 515
GALANTHUS, 661
GALEGA, 268
GALEOPSIS, 385
Galingale, 552
GALINSOGA, 512
GALIUM, 441
Gallant-soldier, 512
Garlic, 655
  Honey, 658
GASTRIDIUM, 615
GAUDINIA, 607
GAULTHERIA, 192
GAZANIA, 484
GENISTA, 289
Gentian, 355, 356
GENTIANA, 356
GENTIANACEAE, 353
GENTIANELLA, 355
GERANIACEAE, 316
GERANIUM, 316
Geranium, 322
German-ivy, 505
Germander, 388
GESNERIACEAE, 431
GEUM, 229
Giant-rhubarb, 292
GLADIOLUS, 671
Gladiolus, 671
Glasswort, 102
  Perennial, 102
GLAUCIUM, 77
GLAUX, 202
GLECHOMA, 389
Globe-thistle, 463
Globeflower, 63
Glory-of-the-snow, 654
GLYCERIA, 604
GLYCINE, 268
GNAPHALIUM, 486
Goat's-beard, 474
Goat's-rue, 268
Godetia, 299
Gold-of-pleasure, 178
Golden-samphire, 488

Golden-saxifrage, 214
Goldenrod, 490
Goldilocks, 492
Good-King-Henry, 96
GOODYERA, 681
Gooseberry, 204
Goosefoot, 95
Gorse, 289
  Spanish, 289
Grape-hyacinth, 654
Grape-vine, 308
Grass-of-Parnassus, 215
Grass-poly, 292
Greenweed, 289
GRINDELIA, 489
GRISELINIA, 301
GROENLANDIA, 532
Gromwell, 369
GROSSULARIACEAE, 203
Ground-elder, 336
Ground-ivy, 389
Ground-pine, 388
Groundsel, 503
  Tree, 495
Guelder-rose, 445
GUIZOTIA, 510
Gum, 294
Gumplant, 489
GUNNERA, 292
GUNNERACEAE, 292
GYMNADENIA, 684
X **GYMNAGLOSSUM**, 685
X **GYMNANACAMPTIS**, 683
GYMNOCARPIUM, 19
GYPSOPHILA, 122
Gypsywort, 392

HAINARDIA, 603
Hair-grass, 607, 608, 610
  Grey, 609
  Mediterranean, 608
Hairy-brome, 621
HALORAGACEAE, 291
HALORAGIS, 291
HAMMARBYA, 682
Hampshire-purslane, 298
Hard-fern, 23
Hard-grass, 603

Hard-grass (contd)
    One-glumed, 603
Hardhack, 220
Hare's-ear, 345
Hare's-tail, 616
Harebell, 436
Harlequinflower, 672
Hart's-tongue, 16
Hartwort, 351
Haw-medlar, 262
Hawk's-beard, 479
    Tuberous, 475
Hawkbit, 473
    Scaly, 472
Hawkweed, 482
Hawthorn, 262
Hazel, 90
Heath, 193
    Blue, 191
    Prickly, 192
    St Dabeoc's, 192
Heath-grass, 629
Heather, 193
    Bell, 194
HEBE, 418
Hebe, 418
HEDERA, 324
Hedge-parsley, 352
HEDYPNOIS, 472
HELENIUM, 515
HELIANTHEMUM, 145
HELIANTHUS, 511
HELICHRYSUM, 487
HELICTOTRICHON, 605
Heliotrope, Winter, 508
Hellebore, 63
Helleborine, 678
HELLEBORUS, 63
HEMEROCALLIS, 648
Hemlock, 345
Hemlock-spruce, 28
Hemp, 83
Hemp-agrimony, 515
Hemp-nettle, 385
Henbane, 359
HEPATICA, 66
HERACLEUM, 351
Herb-Paris, 652

Herb-Robert, 317
HERMINIUM, 683
HERMODACTYLUS, 667
HERNIARIA, 117
HESPERIS, 170
HEUCHERA, 214
HIBISCUS, 144
HIERACIUM, 482
HIEROCHLOE, 610
HIMANTOGLOSSUM, 691
HIPPOCASTANACEAE, 311
HIPPOCREPIS, 273
HIPPOPHAE, 290
HIPPURIDACEAE, 397
HIPPURIS, 397
HIRSCHFELDIA, 186
Hogweed, 351
HOHERIA, 141
HOLCUS, 609
Hollowroot, 78
Holly, 303
    New Zealand, 494
Holly-fern, 21
    House, 21
Hollyhock, 143
    Australian, 143
HOLODISCUS, 221
HOLOSTEUM, 112
Holy-grass, 610
HOMOGYNE, 508
HONCKENYA, 111
Honesty, 175
Honewort, 345
Honeybells, 659
Honeysuckle, 447
    Himalayan, 446
Hop, 83
HORDELYMUS, 626
HORDEUM, 626
Horehound, Black, 384
    White, 387
Hornbeam, 90
Horned-poppy, 77
HORNUNGIA, 179
Hornwort, 61
Horse-chestnut, 311
Horse-nettle, 361
Horse-radish, 172

INDEX

Horsetail, 8
Hottentot-fig, 94
HOTTONIA, 199
Hound's-tongue, 378
House-leek, 206
Huckleberry, 362
HUMULUS, 83
HUPERZIA, 5
Hutchinsia, 179
Hyacinth, 654
  Tassel, 655
HYACINTHOIDES, 653
HYACINTHUS, 654
HYDRANGEA, 203
Hydrangea, 203
HYDRANGEACEAE, 202
HYDRILLA, 521
HYDROCHARIS, 520
HYDROCHARITACEAE, 520
HYDROCOTYLE, 332
HYDROPHYLLACEAE, 368
HYMENOPHYLLACEAE, 13
HYMENOPHYLLUM, 13
HYOSCYAMUS, 359
HYPERICUM, 138
HYPOCHAERIS, 472
Hyssop, 391
HYSSOPUS, 391

IBERIS, 180
Iceplant, 92
Iceland-purslane, 126
ILEX, 303
ILLECEBRUM, 117
IMPATIENS, 323
Indian-rhubarb, 211
INULA, 487
IOCHROMA, 359
IPOMOEA, 365
IRIDACEAE, 665
IRIS, 667
Iris, 667
  Snake's-head, 667
ISATIS, 169
ISOETACEAE, 7
ISOETES, 7
ISOLEPIS, 551
IVA, 510

Ivy, 324
IXIA, 672

Jacob's-ladder, 367
Japanese-lantern, 360
JASIONE, 438
Jasmine, 402
JASMINUM, 402
Johnson-grass, 641
JONOPSIDIUM, 179
Jonquil, Campernelle, 663
JUGLANDACEAE, 85
JUGLANS, 86
JUNCACEAE, 538
JUNCAGINACEAE, 523
JUNCUS, 539
Juneberry, 248
Juniper, 33
JUNIPERUS, 33

KALMIA, 191
Kangaroo-apple, 361
Karo, 202
Kelch-grass, 629
KERRIA, 221
Kerria, 221
Ketmia, Bladder, 144
KICKXIA, 412
Knapweed, 469
  Russian, 468
KNAUTIA, 452
Knawel, 116
KNIPHOFIA, 648
Knotgrass, 126
Knotweed, 124, 127
KOBRESIA, 553
KOELERIA, 607
KOELREUTERIA, 311
KOENIGIA, 126
Kohuhu, 202
Koromiko, 418

Labrador-tea, 190
LABURNUM, 287
Laburnum, 287
LACTUCA, 475
Lady-fern, 19
Lady's-mantle, 232

Lady's-slipper, 678
Lady's-tresses, 681
   Creeping, 681
LAGAROSIPHON, 522
LAGURUS, 616
LAMARCKIA, 597
Lamb's-ear, 384
LAMIACEAE, 379
LAMIASTRUM, 384
LAMIUM, 385
LAMPRANTHUS, 92
LAPPULA, 378
LAPSANA, 471
Larch, 29
LARIX, 29
Larkspur, 65
LATHRAEA, 429
LATHYRUS, 276
LAURACEAE, 59
Laurel, 240
LAURUS, 59
Laurustinus, 446
LAVANDULA, 395
LAVATERA, 142
Lavender, 395
Lavender-cotton, 497
LEDUM, 190
Leek, 656
LEERSIA, 588
LEGOUSIA, 437
LEMNA, 537
LEMNACEAE, 536
LENS, 276
Lenten-rose, 64
LENTIBULARIACEAE, 431
Lentil, 276
LEONTODON, 473
LEONURUS, 384
Leopard's-bane, 507
Leopardplant, 506
LEPIDIUM, 180
Leptinella, 502
LEPTOCHLOA, 631
LEPTOSPERMUM, 293
Lettuce, 475
   Wall, 477
LEUCANTHEMELLA, 500
LEUCANTHEMUM, 500

LEUCOJUM, 660
LEVISTICUM, 350
LEYCESTERIA, 446
LEYMUS, 626
LIBERTIA, 666
LIGULARIA, 506
LIGUSTICUM, 349
LIGUSTRUM, 403
Lilac, 403
LILIACEAE, 644
LILIUM, 650
Lily, African, 659
   Kerry, 648
   Jersey, 660
   Martagon, 650
   May, 651
   Peruvian, 664
   Pyrenean, 650
   Snowdon, 649
Lily-of-the-valley, 651
   False, 652
Lime, 140
LIMNANTHACEAE, 323
LIMNANTHES, 323
LIMONIUM, 134
LIMOSELLA, 409
LINACEAE, 309
LINARIA, 412
LINNAEA, 446
LINUM, 309
LIPARIS, 682
Liquorice, Wild, 269
LISTERA, 681
LITHOSPERMUM, 369
Little-Robin, 317
LITTORELLA, 400
Liverleaf, 66
LLOYDIA, 649
LOBELIA, 438
Lobelia, 438
   Californian, 439
   Lawn, 439
LOBULARIA, 175
Loganberry, 223
LOISELEURIA, 191
LOLIUM, 595
London-rocket, 168
Londonpride, 212

INDEX

Longleaf, 348
LONICERA, 447
Loosestrife, 200
Lords-and-Ladies, 535
LOTUS, 271
Lousewort, 428
Lovage, 350
   Scots, 349
Love-grass, 631
Love-in-a-mist, 64
Love-lies-bleeding, 106
Lucerne, 281
LUDWIGIA, 298
LUMA, 294
LUNARIA, 175
Lungwort, 370
Lupin, 286
   False, 286
LUPINUS, 286
LURONIUM, 519
LUZULA, 543
LYCHNIS, 119
LYCIUM, 359
LYCOPERSICON, 360
LYCOPODIACEAE, 5
LYCOPODIELLA, 5
LYCOPODIUM, 6
LYCOPUS, 392
Lyme-grass, 626
LYSICHITON, 535
LYSIMACHIA, 200
LYTHRACEAE, 292
LYTHRUM, 292

MACLEAYA, 77
Madder, 444
   Field, 440
Madwort, 376
MAHONIA, 74
MAIANTHEMUM, 651
Maize, 641
MALCOLMIA, 170
Male-fern, 22
Mallow, 141
   Greek, 144
   New Zealand, 141
   Prairie, 144
   Prickly, 141
   Royal, 143
Maltese-Cross, 119
MALUS, 242
MALVA, 141
MALVACEAE, 141
Maple, 312
Mare's-tail, 397
Marigold, 509, 514
   Corn, 500
   Dwarf, 514
Marjoram, 391
Marram, 615
   Purple, 615
Marrow, 150
MARRUBIUM, 387
Marsh-bedstraw, 442
Marsh-elder, 510
Marsh-mallow, 143
Marsh-marigold, 63
Marsh-orchid, 686
Marshwort, 346
MARSILEACEAE, 13
Masterwort, 350
Mat-grass, 589
MATRICARIA, 500
MATTEUCCIA, 18
MATTHIOLA, 170
Mayweed, 500, 501
Meadow-foam, 323
Meadow-grass, 599
Meadow-rue, 73
Meadowsweet, 221
MECONOPSIS, 76
MEDICAGO, 280
Medick, 280
Medlar, 262
MELAMPYRUM, 419
MELICA, 604
Melick, 604
Melilot, 279
MELILOTUS, 279
MELISSA, 390
MELITTIS, 387
Melon, 150
   Water, 150
MENTHA, 392
MENYANTHACEAE, 366
MENYANTHES, 366

MERCURIALIS, 304
Mercury, 304
MERTENSIA, 375
MESPILUS, 262
MEUM, 344
Mexican-stonecrop, 207
Mexican-tea, 96
Mezereon, 293
MIBORA, 616
Michaelmas-daisy, 490
Mignonette, 188
MILIUM, 590
Milk-parsley, 350
    Cambridge, 349
Milk-vetch, 269
Milkwort, 310
Millet, 590, 636, 641
    Japanese, 637
    Shama, 637
    White, 637
MIMOSACEAE, 264
MIMULUS, 408
Mind-your-own-business, 85
Mint, 392
Mintweed, 396
MINUARTIA, 111
MISOPATES, 411
Mistletoe, 302
Mock-orange, 203
MOEHRINGIA, 110
MOENCHIA, 114
MOLINIA, 630
MONESES, 197
Moneywort, Cornish, 419
Monk's-hood, 64
Monk's-rhubarb, 130
Monkey-puzzle, 33
Monkeyflower, 408
MONOTROPA, 197
MONOTROPACEAE, 197
MONSONIA, 320
Montbretia, 672
MONTIA, 107
Moonwort, 11
Moor-grass, Blue, 603
    Purple, 630
MORACEAE, 84
Morning-glory, 365

MORUS, 84
Moschatel, 448
Motherwort, 384
Mountain-laurel, 191
Mountain-pine, 30
Mouse-ear, 113
Mouse-ear-hawkweed, 480
Mousetail, 72
Mousetailplant, 536
Mudwort, 409
MUEHLENBECKIA, 128
Mugwort, 496
Mulberry, 84
Mullein, 405
Mung-bean, 268
MUSCARI, 654
Musk, 409
Musk-mallow, 142
Mustard, Ball, 178
    Black, 185
    Chinese, 184
    Garlic, 168
    Hare's-ear, 183
    Hedge, 168
    Hoary, 186
    Russian, 167
    Tower, 174
    White, 185
MYCELIS, 477
MYOSOTIS, 376
MYOSOTON, 114
MYOSURUS, 72
MYRICA, 86
MYRICACEAE, 86
MYRIOPHYLLUM, 291
MYRRHIS, 334
MYRTACEAE, 293
Myrtle, Chilean, 294

Naiad, 533
NAJADACEAE, 532
NAJAS, 533
NARCISSUS, 662
NARDUS, 589
NARTHECIUM, 647
Nasturtium, 323
Navelwort, 205
NECTAROSCORDUM, 658

INDEX

Needle-grass, 589
NEOTINEA, 690
NEOTTIA, 680
NEPETA, 389
NERTERA, 440
NESLIA, 178
Nettle, 84
NICANDRA, 358
NICOTIANA, 363
NIGELLA, 64
Niger, 510
Nightshade, 361
    Deadly, 359
Ninebark, 218
Nipplewort, 471
Nit-grass, 615
NOTHOFAGUS, 87
NOTHOSCORDUM, 659
NUPHAR, 60
NYMPHAEA, 60
NYMPHAEACEAE, 60
NYMPHOIDES, 367

Oak, 87
Oat, 606
Oat-grass, 605
    False, 606
    French, 607
    Yellow, 607
Oceanspray, 221
OCHAGAVIA, 643
ODONTITES, 426
OEMLERIA, 241
OENANTHE, 337
OENOTHERA, 298
OLEACEAE, 401
OLEARIA, 494
Oleaster, 290
OMPHALODES, 378
ONAGRACEAE, 294
Onion, 655
Oniongrass, 669
ONOBRYCHIS, 270
ONOCLEA, 18
ONONIS, 278
ONOPORDUM, 467
OPHIOGLOSSACEAE, 11
OPHIOGLOSSUM, 11

OPHRYS, 692
Orache, 99
Orange, Mexican, 313
Orange-ball-tree, 401
X ORCHIACERAS, 691
Orchid, 690, 692
    Bird's-nest, 680
    Bog, 682
    Coralroot, 682
    Dense-flowered, 690
    Fen, 682
    Fragrant, 684
    Frog, 685
    Ghost, 680
    Lizard, 691
    Man, 691
    Musk, 683
    Pyramidal, 683
    Small-white, 684
ORCHIDACEAE, 675
ORCHIS, 690
Oregon-grape, 74
OREOPTERIS, 16
ORIGANUM, 391
ORNITHOGALUM, 652
ORNITHOPUS, 272
OROBANCHACEAE, 428
OROBANCHE, 429
Orpine, 207
ORTHILIA, 196
ORYZOPSIS, 590
OSCULARIA, 93
Osier, 157
OSMUNDA, 12
OSMUNDACEAE, 12
Osoberry, 241
OTANTHUS, 498
OXALIDACEAE, 314
OXALIS, 314
Oxeye, Autumn, 500
    Yellow, 489
Oxlip, 199
Oxtongue, 473
OXYRIA, 134
OXYTROPIS, 269
Oxytropis, 269
Oysterplant, 375

PACHYSANDRA, 304
PAEONIA, 137
PAEONIACEAE, 137
Pampas-grass, 630
PANCRATIUM, 663
PANICUM, 636
Pansy, 148
PAPAVER, 75
PAPAVERACEAE, 75
PARAPHOLIS, 603
PARENTUCELLIA, 426
PARIETARIA, 85
PARIS, 652
PARNASSIA, 215
Parrot's-feather, 291
Parsley, 347
    Cow, 334
    Fool's, 338
    Stone, 347
Parsley-piert, 235
Parsnip, 350
PARTHENOCISSUS, 309
PASPALUM, 638
Pasqueflower, 66
PASTINACA, 350
PAULOWNIA, 408
Pea, 276
    Chick, 278
    Garden, 278
    Scurfy, 268
Peach, 240
Pear, 242
Pearlwort, 115
PEDICULARIS, 428
PELARGONIUM, 322
Pellitory-of-the-wall, 85
Penny-cress, 180
Pennyroyal, 393
Pennywort, 332
PENTAGLOTTIS, 374
Peony, 137
Pepper, Sweet, 360
Pepper-saxifrage, 344
Peppermint, 394
Peppermint-gum, 294
Pepperwort, 180
PERICALLIS, 505
Periwinkle, 357

PERSICARIA, 124
Persicaria, 125
PETASITES, 507
PETRORHAGIA, 122
PETROSELINUM, 347
PETUNIA, 364
Petunia, 364
PEUCEDANUM, 350
PHACELIA, 368
Phacelia, 368
PHALARIS, 611
Phanerophlebia, 21
PHASEOLUS, 267
Pheasant's-eye, 72
PHEGOPTERIS, 16
PHILADELPHUS, 203
PHLEUM, 618
PHLOMIS, 386
PHLOX, 367
Phlox, 367
PHORMIUM, 674
PHOTINIA, 249
PHRAGMITES, 630
PHUOPSIS, 440
PHYGELIUS, 408
PHYLLITIS, 16
PHYLLODOCE, 191
Phymatodes, 14
PHYMATOSORUS, 14
PHYSALIS, 360
PHYSOCARPUS, 218
PHYSOSPERMUM, 344
PHYTEUMA, 437
PHYTOLACCA, 91
PHYTOLACCACEAE, 91
PICEA, 28
Pick-a-back-plant, 214
Pickerelweed, 644
PICRIS, 473
Pigmyweed, 205
Pignut, 335
    Great, 335
Pigweed, 104
Pillwort, 13
PILOSELLA, 480
PILULARIA, 13
Pimpernel, 201
    Yellow, 200

INDEX

PIMPINELLA, 336
PINACEAE, 26
Pine, 29
Pineappleweed, 501
PINGUICULA, 431
Pink, 122
Pink-sorrel, 315
Pinkweed, 125
PINUS, 29
Pipewort, 538
Pirri-pirri-bur, 231
PISUM, 278
Pitcherplant, 144
PITTOSPORACEAE, 202
PITTOSPORUM, 202
Pittosporum, 202
PLAGIOBOTHRYS, 375
Plane, 81
PLANTAGINACEAE, 399
PLANTAGO, 399
Plantain, 399
PLATANACEAE, 81
PLATANTHERA, 683
PLATANUS, 81
PLEIOBLASTUS, 587
Ploughman's-spikenard, 488
Plum, 241
PLUMBAGINACEAE, 134
Plume-poppy, 77
POA, 599
POACEAE, 571
Pokeweed, 91
POLEMONIACEAE, 367
POLEMONIUM, 367
POLYCARPON, 117
POLYGALA, 310
POLYGALACEAE, 310
POLYGONACEAE, 123
POLYGONATUM, 651
POLYGONUM, 126
POLYPODIACEAE, 14
POLYPODIUM, 14
Polypody, 14
POLYPOGON, 616
POLYSTICHUM, 20
Pond-sedge, 565
Pondweed, 523
    Horned, 533

    Opposite-leaved, 532
PONTEDERIA, 644
PONTEDERIACEAE, 644
Poplar, 151
Poppy, 75
    Californian, 77
    Mexican, 77
    Welsh, 76
POPULUS, 151
PORTULACA, 106
PORTULACACEAE, 106
POTAMOGETON, 523
POTAMOGETONACEAE, 523
Potato, 362
POTENTILLA, 225
PRATIA, 439
Pride-of-India, 311
Primrose, 198
Primrose-peerless, 662
PRIMULA, 198
PRIMULACEAE, 198
Prince's-feather, 106
Privet, 403
PRUNELLA, 389
PRUNUS, 239
X **PSEUDADENIA**, 684
PSEUDOFUMARIA, 79
PSEUDORCHIS, 684
X **PSEUDORHIZA**, 684
PSEUDOSASA, 588
PSEUDOTSUGA, 28
PSORALEA, 268
PTERIDACEAE, 13
PTERIDIUM, 15
PTERIS, 13
PTEROCARYA, 86
PUCCINELLIA, 598
PULICARIA, 488
PULMONARIA, 370
PULSATILLA, 66
Pumpkin, 150
Purple-loosestrife, 292
Purslane, 106, 107
PYRACANTHA, 261
Pyrenean-violet, 431
PYROLA, 196
PYROLACEAE, 196
PYRUS, 242

Quaking-grass, 599
Queensland-hemp, 141
QUERCUS, 87
Quillwort, 7
Quince, 241
    Chinese, 242
    Japanese, 242

RADIOLA, 310
Radish, 187
Ragged-Robin, 119
Ragweed, 509
Ragwort, 502, 506
    Chinese, 506
RAMONDA, 431
Ramping-fumitory, 79
Rampion, 437
Ramsons, 657
Rannoch-rush, 523
RANUNCULACEAE, 62
RANUNCULUS, 67
Rape, 184
RAPHANUS, 187
RAPISTRUM, 187
Raspberry, 222
Raspwort, Creeping, 291
Rauli, 87
Red-cedar, 31, 33
Red-hot-poker, 648
Red-knotgrass, 127
Redshank, 125
Redwood, 31
Reed, 630
REINECKEA, 652
Reineckea, 652
RESEDA, 188
RESEDACEAE, 188
Restharrow, 278
RHAMNACEAE, 308
RHAMNUS, 308
RHEUM, 128
RHINANTHUS, 427
Rhodes-grass, 634
RHODODENDRON, 190
Rhododendron, 190
Rhodostachys, 643
    Tresco, 643
Rhubarb, 128

RHUS, 313
RHYNCHOSPORA, 552
RIBES, 204
RICINUS, 304
RIDOLFIA, 347
Robin's-plantain, 493
ROBINIA, 267
Roblé, 87
Rock-cress, 173
Rock-rose, 145
    Spotted, 145
Rocket, 167
    Garden, 185
    Hairy, 185
    Sea, 186
RODGERSIA, 210
Rodgersia, 210
ROMULEA, 669
RORIPPA, 171
ROSA, 235
ROSACEAE, 215
Rose, 235
Rose-of-heaven, 120
Rose-of-Sharon, 138
Rosemary, 395
Roseroot, 206
ROSMARINUS, 395
ROSTRARIA, 608
Rowan, 243
RUBIA, 444
RUBIACEAE, 439
RUBUS, 222
RUDBECKIA, 511
RUMEX, 129
RUPPIA, 532
RUPPIACEAE, 532
Rupturewort, 117
RUSCHIA, 92
RUSCUS, 664
Rush, 539
Russian-vine, 128
Rustyback, 18
RUTACEAE, 313
Rye, 628
Rye-grass, 595
RYTIDOSPERMA, 629

Safflower, 470

INDEX

Saffron, Meadow, 649
Sage, 386
    Wood, 388
SAGINA, 115
SAGITTARIA, 518
Sainfoin, 270
SALICACEAE, 151
SALICORNIA, 102
SALIX, 153
Sally-my-handsome, 94
Salmonberry, 223
SALPICHROA, 360
Salsify, 474
SALSOLA, 104
Saltmarsh-grass, 598
Saltwort, 104
SALVIA, 396
SAMBUCUS, 445
SAMOLUS, 202
Samphire, Rock, 337
Sand-grass, 616
Sandbur, 641
Sandwort, 110, 111
    Sea, 111
    Three-nerved, 110
SANGUISORBA, 230
Sanicle, 332
SANICULA, 332
SANTALACEAE, 301
SANTOLINA, 497
SAPINDACEAE, 311
SAPONARIA, 121
SARCOCORNIA, 102
SARRACENIA, 144
SARRACENIACEAE, 144
SASA, 587
SASAELLA, 588
SATUREJA, 390
SAUSSUREA, 465
Savory, 390
Saw-wort, 468
    Alpine, 465
SAXIFRAGA, 211
SAXIFRAGACEAE, 209
Saxifrage, 211
SCABIOSA, 452
Scabious, 452
    Devil's-bit, 452
    Field, 452
    Giant, 452
SCANDIX, 334
SCHEUCHZERIA, 523
SCHEUCHZERIACEAE, 522
SCHISMUS, 629
SCHKUHRIA, 514
SCHOENOPLECTUS, 550
SCHOENUS, 552
SCILLA, 653
SCIRPOIDES, 549
SCIRPUS, 549
SCLERANTHUS, 116
SCOLYMUS, 470
Scorpion-vetch, 272
SCORPIURUS, 273
SCORZONERA, 474
SCROPHULARIA, 407
SCROPHULARIACEAE, 403
Scurvygrass, 177
SCUTELLARIA, 387
Sea-blite, 103
Sea-buckthorn, 290
Sea-fig, Angular, 94
    Lesser, 93
Sea-heath, 149
Sea-holly, 333
Sea-kale, 187
Sea-lavender, 134
Sea-milkwort, 202
Sea-purslane, 100
Sea-spurrey, 118
SECALE, 628
SECURIGERA, 273
Sedge, 553
    False, 553
SEDUM, 206
SELAGINELLA, 6
SELAGINELLACEAE, 6
Selfheal, 389
SELINUM, 349
SEMIARUNDINARIA, 586
SEMPERVIVUM, 206
SENECIO, 502
Senna, Scorpion, 273
SEQUOIA, 31
SEQUOIADENDRON, 31
SERIPHIDIUM, 496

Serradella, 272
SERRATULA, 468
Service-tree, 243
SESELI, 337
SESLERIA, 603
SETARIA, 639
Shaggy-soldier, 512
Shallon, 192
Sheep-laurel, 191
Sheep's-bit, 438
Sheep's-fescue, 594
Shepherd's-needle, 334
Shepherd's-purse, 179
SHERARDIA, 440
Shield-fern, 20
Shoreweed, 400
SIBBALDIA, 228
Sibbaldia, 228
SIBTHORPIA, 419
SIDA, 141
SIDALCEA, 144
SIGESBECKIA, 511
Signal-grass, 637, 638
SILAUM, 344
SILENE, 119
Silky-bent, 616
Silver-fir, 27
Silverweed, 226
SILYBUM, 468
SIMAROUBACEAE, 313
SIMETHIS, 648
SINACALIA, 506
SINAPIS, 185
Sinarundinaria, 587
SISON, 347
SISYMBRIUM, 167
SISYRINCHIUM, 666
SIUM, 336
Skullcap, 387
Skunk-cabbage, 535
Slipperwort, 410
Slough-grass, 618
Small-reed, 614
Smearwort, 60
Smilo-grass, 590
SMYRNIUM, 335
Snapdragon, 410
    Trailing, 411

Sneezeweed, 515
Sneezewort, 498
Snow-in-summer, 113
Snowberry, 446
Snowdrop, 661
Snowflake, 660
Soapwort, 121
Soft-brome, 620
Soft-grass, 609
Soft-rush, 541
SOLANACEAE, 357
SOLANUM, 361
SOLEIROLIA, 85
SOLIDAGO, 490
Solomon's-seal, 651
SONCHUS, 475
SORBARIA, 218
Sorbaria, 218
SORBUS, 243
SORGHUM, 641
Sorrel, 129
    Mountain, 134
Southernwood, 496
Sowthistle, 475
Sowbread, 199
Soyabean, 268
Spanish-dagger, 674
Spanish-needles, 513
SPARAXIS, 672
SPARGANIACEAE, 642
SPARGANIUM, 642
SPARTINA, 635
SPARTIUM, 288
Spatter-dock, 61
Spearwort, 68
Speedwell, 414
SPERGULA, 117
SPERGULARIA, 118
Spider-orchid, 692
Spiderwort, 538
Spignel, 344
Spike-rush, 548
Spinach, 99
    New Zealand, 94
    Tree, 98
SPINACIA, 99
Spindle, 302
SPIRAEA, 219

INDEX

Spiraea, 219
SPIRANTHES, 681
SPIRODELA, 537
Spleenwort, 17
SPOROBOLUS, 633
Spotted-laurel, 301
Spotted-orchid, 686
Spring-sedge, 559
Springbeauty, 107
Spruce, 28
Spurge, 305
Spurge-laurel, 293
Spurrey, 117
    Sand, 118
Squill, 653
Squinancywort, 441
St John's-wort, 138
St Patrick's-cabbage, 212
St Paul's-wort, 511
STACHYS, 382
Staff-vine, 303
STAPHYLEA, 311
STAPHYLEACEAE, 311
Star-of-Bethlehem, 649, 652
Star-thistle, 469
Starflower, 659
Starfruit, 519
Steeple-bush, 220
STELLARIA, 111
STERNBERGIA, 660
Stink-grass, 632
STIPA, 589
Stitchwort, 111
Stock, 170
    Virginia, 170
Stonecrop, 206
    Mossy, 205
Stork's-bill, 321
Stranvaesia, 249
Strapwort, 116
STRATIOTES, 520
Strawberry, 228
    Barren, 226
    Yellow-flowered, 229
Strawberry-blite, 96
Strawberry-tree, 193
SUAEDA, 103
SUBULARIA, 182

SUCCISA, 452
Succory, Lamb's, 471
Sumach, 313
Summer-cypress, 99
Sundew, 145
Sunflower, 511
Swede, 184
Sweet-briar, 238
Sweet-flag, 534
Sweet-grass, 604
Sweet-William, 123
Swine-cress, 182
Sword-fern, 21
Sycamore, 312
SYMPHORICARPOS, 446
SYMPHYTUM, 371
SYRINGA, 403

TAGETES, 514
TAMARICACEAE, 148
Tamarisk, 148
TAMARIX, 148
TAMUS, 675
TANACETUM, 495
Tansy, 495
Tapegrass, 522
TARAXACUM, 477
Tare, 275
Tarragon, 496
Tasmanian-fuchsia, 314
Tasselweed, 532
TAXACEAE, 34
TAXODIACEAE, 30
TAXUS, 34
Tea-tree, 293
Teaplant, 359
Tear-thumb, 125
Teasel, 451
TEESDALIA, 179
Teff, 632
TELEKIA, 489
TELLIMA, 214
TEPHROSERIS, 505
TETRAGONIA, 94
TETRAGONOLOBUS, 271
TEUCRIUM, 388
THALICTRUM, 73
Thamnocalamus, 587

THELYPTERIDACEAE, 15
THELYPTERIS, 16
THERMOPSIS, 286
THESIUM, 301
Thimbleberry, 222
Thistle, 465, 466
    Carline, 464
    Cotton, 467
    Golden, 470
    Milk, 468
THLASPI, 180
Thorn-apple, 363
Thorow-wax, 345
Thrift, 136
Throatwort, 437
THUJA, 33
Thyme, 391
    Basil, 391
THYMELAEACEAE, 293
THYMUS, 391
Tickseed, 513
TILIA, 140
TILIACEAE, 140
Timothy, 619
Toadflax, 410, 411, 412
Tobacco, 363
TOFIELDIA, 647
TOLMIEA, 214
Tomatillo, 360
Tomato, 360
Toothpick-plant, 348
Toothwort, 429
Tor-grass, 624
TORDYLIUM, 351
TORILIS, 352
Tormentil, 227
TRACHELIUM, 437
TRACHYSPERMUM, 347
TRACHYSTEMON, 375
TRADESCANTIA, 538
TRAGOPOGON, 474
TRAGUS, 635
Traveller's-joy, 66
Treacle-mustard, 169
Treasureflower, 484
    Plain, 484
Tree-fern, 15
Tree-mallow, 142

Tree-of-heaven, 313
Trefoil, 283
TRICHOMANES, 13
TRICHOPHORUM, 547
TRIENTALIS, 201
TRIFOLIUM, 282
TRIGLOCHIN, 523
TRIGONELLA, 280
TRINIA, 345
TRIPLEUROSPERMUM, 501
X **TRIPLEUROTHEMIS**, 499
TRISETUM, 607
TRISTAGMA, 659
Triticale, 628
X **TRITICOSECALE**, 628
TRITICUM, 628
TROLLIUS, 63
TROPAEOLACEAE, 323
TROPAEOLUM, 323
TSUGA, 28
TUBERARIA, 145
Tufted-sedge, 565
Tulip, 650
TULIPA, 650
Tunicflower, 122
Turnip, 184
Turnip-rape, 184
Tussac-grass, 600
TUSSILAGO, 507
Tussock-sedge, 557
Tutsan, 138
Twayblade, 681
Twinflower, 446
TYPHA, 643
TYPHACEAE, 643

ULEX, 289
ULMACEAE, 81
ULMUS, 81
UMBILICUS, 205
UROCHLOA, 638
URTICA, 84
URTICACEAE, 84
UTRICULARIA, 432

VACCARIA, 122
VACCINIUM, 195
Valerian, 450

VALERIANA, 450
VALERIANACEAE, 449
VALERIANELLA, 449
VALLISNERIA, 522
Velvetleaf, 144
Venus's-looking-glass, 437
VERBASCUM, 405
VERBENA, 379
VERBENACEAE, 379
Vernal-grass, 610
VERONICA, 414
Veronica, Hedge, 418
Vervain, 379
Vetch, 273
    Crown, 273
    Horseshoe, 273
    Kidney, 270
Vetchling, 276
VIBURNUM, 445
Viburnum, 445
VICIA, 273
VIGNA, 268
VINCA, 357
VIOLA, 146
VIOLACEAE, 146
Violet, 146
Violet-willow, 157
Viper's-bugloss, 370
Viper's-grass, 474
Virgin's-bower, 67
Virginia-creeper, 309
VISCACEAE, 302
VISCUM, 302
VITACEAE, 308
VITIS, 308
VULPIA, 596

WAHLENBERGIA, 437
Wall-rocket, 183
Wall-rue, 18
Wallaby-grass, 629
    Swamp, 605
Wallflower, 169
Walnut, 86
Wandering-jew, 538
Warty-cabbage, 169
Water-cress, 171
Water-crowfoot, 70

Water-dropwort, 337
Water-lily, Fringed, 367
    White, 60
    Yellow, 60
Water-milfoil, 291
Water-parsnip, Greater, 336
    Lesser, 336
Water-pepper, 125
Water-plantain, 519
    Floating, 519
    Lesser, 518
Water-purslane, 292
Water-shield, 61
Water-soldier, 520
Water-speedwell, 414
Water-starwort, 397
Water-violet, 199
Waterweed, 521
    Curly, 522
    Esthwaite, 521
    Large-flowered, 521
Waterwort, 137
WATSONIA, 669
Wayfaring-tree, 446
Weasel's-snout, 411
Weeping-grass, 589
WEIGELA, 447
Weigelia, 447
Weld, 188
Wellingtonia, 30
Wheat, 628
Whin, Petty, 289
White-elm, 82
Whitebeam, 243
Whitlowgrass, 176
Whorl-grass, 602
Wild-oat, 607
Willow, 153
Willowherb, 295
    Rosebay, 297
Windmill-grass, 634
Wineberry, 223
Wingnut, 86
Winter-cress, 170
Wintergreen, 196
    One-flowered, 197
    Serrated, 196
Wireplant, 128

Witch-grass, 636
Woad, 169
Wolf's-bane, 65
WOLFFIA, 538
Wood-rush, 543
Wood-sedge, 567
Wood-sorrel, 314
Woodruff, 442
    Blue, 441
    Pink, 441
WOODSIA, 20
Woodsia, 20
WOODSIACEAE, 18
Wormwood, 497
    Sea, 496
Woundwort, 382

XANTHIUM, 510

Yard-grass, 632
Yarrow, 498
Yellow-cress, 171
Yellow-eyed-grass, 666
Yellow-pine, 30
Yellow-rattle, 427
Yellow-sedge, 570
Yellow-sorrel, 314
Yellow-vetch, 274
Yellow-wort, 355
Yellow-woundwort, 383
Yew, 34
Yorkshire-fog, 609
YUCCA, 674
YUSHANIA, 586

ZANNICHELLIA, 533
ZANNICHELLIACEAE, 533
ZANTEDESCHIA, 535
ZEA, 641
ZOSTERA, 533
ZOSTERACEAE, 533

Vice-counties of the British Isles

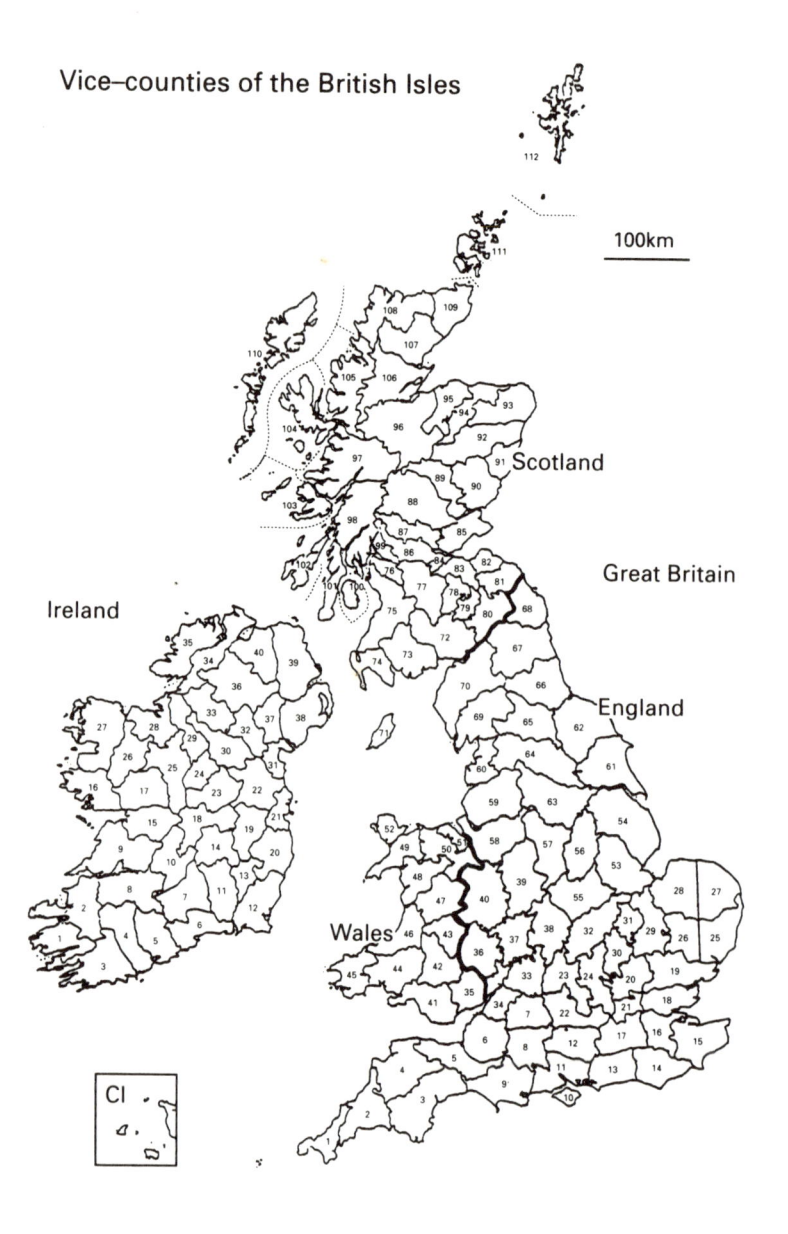